Automotive Technology for General Service Technicians

Automotive Technology for General Service Technicians

Ron Haefner
Columbus High School
Columbus, NE

Paul Leathers
Columbus High School
Columbus, NE

THOMSON

DELMAR LEARNING

Australia • Canada • Mexico • Singapore • Spain • United Kingdom • United States

Automotive Technology for General Service Technicians

Ron Haefner • Paul Leathers

Vice President, Technology and Trades ABU:
David Garza

Director of Learning Solutions:
Sandy Clark

Managing Editor:
Larry Main

Sr. Acquisitions Editor:
David Boelio

Product Manager:
Matthew Thouin

Marketing Director:
Deborah Yarnell

Marketing Manager:
Erin Coffin

Marketing Coordinator:
Patti Garrison

Director of Production:
Patty Stephan

Content Project Manager:
Cheri Plasse

Technology Project Manager:
Kevin Smith

Editorial Assistant:
Lauren Stone

Library of Congress Cataloging-in-Publication Data:
Haefner, Ron.
   Automotive technology for general service technicians / Ron Haefner, Paul Leathers.
       p. cm.
   ISBN 1-4180-1340-4
       1. Automobiles—maintenance and repair.   I. Leathers, Paul.   II. Title.

TL152.H1675 2006
629.28'72—dc22

2006505160

NOTICE TO THE READER

Publisher does not warrant or guarantee any of the products described herein or perform any independent analysis in connection with any of the product information contained herein. Publisher does not assume, and expressly disclaims, any obligation to obtain and include information other than that provided to it by the manufacturer.

The reader is expressly warned to consider and adopt all safety precautions that might be indicated by the activities herein and to avoid all potential hazards. By following the instructions contained herein, the reader willingly assumes all risks in connection with such instructions.

The publisher makes no representation or warranties of any kind, including but not limited to, the warranties of fitness for particular purpose or merchantability, nor are any such representations implied with respect to the material set forth herein, and the publisher takes no responsibility with respect to such material. The publisher shall not be liable for any special, consequential, or exemplary damages resulting, in whole or part, from the readers' use of, or reliance upon, this material.

Contents

CHAPTER 6

Basic Shop Procedures   77

CHAPTER 9

Using Service Information   123

CHAPTER 7

Measuring Systems, Measurements, and Fasteners   95

CHAPTER 10

Vehicle Inspection and Detailing   137

CHAPTER 8

Service Information   109

CHAPTER 11

Basic Electricity And Electrical Circuits   179

CHAPTER 18

Battery and Starting System Maintenance, Diagnosis, and Service 267

CHAPTER 19

Charging Systems 279

CHAPTER 20

Charging System Maintenance and Diagnosis 289

CHAPTER 21

Ignition Systems 299

CHAPTER 22

Ignition System Maintenance, Diagnosis, and Service 319

CHAPTER 23

Brake System Design and Operation 333

CHAPTER 24

Brake System Maintenance, Diagnosis, and Service 361

CHAPTER 25

Steering Columns, Steering Linkages, and Power Steering Pumps: Maintenance, Diagnosis, and Service 385

CHAPTER 26

Manual and Power Steering Gear Maintenance, Diagnosis, and Service 413

CHAPTER 27

Tires, Wheels, and Hubs 443

Preface

Thomson Delmar Learning is pleased to present *Automotive Technology for General Service Technicians*. This new text is designed for a student's first exposure to all areas of automotive technology and, authored by two experienced high-school teachers, is written with the needs and abilities of 9th- and 10th-graders in mind. It offers current, concise information on ASE and other subject areas, and combines classroom theory, diagnosis and repair into one easy-to-use volume. The book is purposefully designed to respond to the increasing demand for a general automotive service education, and is ideally suited for schools that have opted or intend to opt for certification under NATEF's General Service Technician Program Standards.

You will notice several differences from a traditional automotive textbook. Fifty-four short chapters divide complex material into chunks. The first ten chapters cover general topics for the General Service Technician and include an overview of the automotive business, "soft skills" such as developing professionalism and a good reputation, safety, tools and equipment usage, measurement systems, service information, and valeting. The bulk of the following chapters covers the more technical areas. The major topics here are electrical systems, brakes, suspension and steering, engines, fuel, emissions, computer controls, drivelines, HVAC and supplemental restraint systems. These chapters are grouped together with service chapters that follow the theory chapters. The size and organization of the chapters allows instructors to give reading assignments that students will find easier to digest.

While the text covers required topics found in the NATEF General Service Technician Program Standards, additional content areas have been added, owing to the importance the authors placed on them and the need for instructors to reference this material. A few of these areas include wheel alignment, emissions, and supplemental restraint systems.

The exact order of the chapters has been the topic of many discussions and in the end it was the authors' decision to present the chapters in this logical, but flexible way. The teaching order of the chapters will ultimately be determined by instructors' own experiences and program requirements. These shorter chapters can be taught in almost any sequence, allowing instructors to deliver the material in the order desired and in the manner that best fits the direction and pace of their individual program requirements.

Automotive Technology for General Service Technicians also features an art-intensive approach to its content to support today's visual learners. In other words, images drive the chapters. From line art to photos, you will find abundant figures that illustrate the systems, parts, and procedures under discussion. Also look for helpful text boxes that draw attention to key points in highlighted areas in the text. Building on the early chapters, learners will be ready to tackle the technical material in successive sections and test their knowledge with the ASE style practice questions at the end of each chapter.

We hope that you find *Automotive Technology for General Service Technicians* a valuable addition to your classroom. Good luck!

About the Authors

Ron Haefner teaches automotive technology at Columbus High School in Columbus, Nebraska. He earned two Bachelor of Science degrees, one in industrial technology and the other in industrial education, as well as a Master of Science degree in industrial education. As an automotive instructor, Haefner's record is exceptional—his students have been winners of manufacturer-sponsored state and national skills contests on nine separate occasions.

Paul Leathers teaches electronics to automotive and other students at Columbus High School in Columbus, NE. His students have won state SkillsUSA competitions in electronics in 2002, 2003, 2004 and 2005. Before embarking on his teaching career, Leathers spent ten years in the US Army as a combat engineer, and then owned and operated automotive and collision repair facilities. He is a Ford Master Technician, an ASE Master Technician in automotive and heavy truck, and he holds an ASE L1 Certification.

Acknowledgments

The authors and Thomson Delmar Learning extend special thanks to the teachers who reviewed the draft manuscript and whose invaluable feedback helped shape this text:

Roger E. Adams
Pittsburg State University
Pittsburg, KS

William DeKryger
Central Michigan University
Mount Pleasant, MI

Mike Elder
Pittsburg High School
Pittsburg, KS

Carl L. Hader
Grafton High School
Grafton, WI

David R. Haskins
South Shore Regional Vo-Tech
Westport, MA

Melvin Jensen
Dixie State College of Utah
St. George, UT

Terry Ireland
Murrieta Valley High School
Murrieta, CA

Rory Perrodin
Longview Community College
Kansas City, MO

Eddie Shumate
Forsyth Technical Community College
Winston-Salem, NC

Paul Strassner
Page County Technical Center
Luray, VA

Cecil Trumbo
Manatee Community College
Bradenton, FL

Kirk VanGelder
Clark County Skills Center
Vancouver, WA

Nick Wagoner
Central Community College
Columbus, NE

Stephen Walters
Springfield-Clark County Joint Vocational School
Springfield, OH

Richard Wittwer
Southern Utah University
Cedar City, UT

Jim Wrede
Traverse Bay Area Career Technology Center
Traverse City, MI

We would like to thank Gene Steffy Ford, O'Reilly Auto Parts, and Advance Auto Parts for their assistance and support. We would also like to thank Mikes Auto, AAA Auto Repair Network, AAA Nebraska, Snap-On, Ernst Auto Center GM/Toyota, and Phil Spady Chrysler Dodge.

Supplements

e.resource—An e.resource is available to instructors who want to make classroom lectures as efficient and engaging as possible. This CD-ROM includes an Instructor's Guide, PowerPoint™ presentations with images, an Image Library and a Computerized Test Bank with over 1,000 questions.

Student Workbook with NATEF Task Sheets—The accompanying Student Workbook provides chapter-by-chapter review activities such as crosswords, component labeling/identification and fill-in-the-blanks for excellent reinforcement of the chapter material. Part II includes nearly 100 Task Sheets that guide the student through common shop procedures for all automotive system areas. The Task Sheets reference all applicable NATEF and General Service Technician Standards.

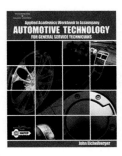

Applied Academics Workbook—An Applied Academics Workbook supplements the text with chapter activities in math, science and communications skills that are keyed to automotive examples. Task Sheets are provided for relevant chapters, where students can put this information to work in the shop.

CHAPTER

1 Automotive Business

Learning Objectives

After you have read, studied, and practiced the contents of this unit, you should be able to:

- List the types of automotive repair facilities.
- Describe the types of information that is used for a repair order.
- Define flat rate.
- Describe some of the employees obligations to the employer.
- List job responsibilities of the technician.

Key Terms

ASE
ATRA
Automotive technician
Customer service

Flat rate
iATN
Independent repair shops
NATEF

Quick-lube shops
Repair order
Specialty shops

INTRODUCTION

This chapter contains information regarding various types of automotive repair shops and the type of work done in each shop. As an automotive technology student, this information should assist you in deciding on the type of shop where you would like to work. This chapter discusses the following topics: job responsibilities for each shop position, completing and processing work orders, employer and employee obligations, job responsibilities for technicians, technician certification, certification of automotive training programs, job applications, resumes, and customer relations. It is very important for technicians to understand the job responsibilities of each shop employee and

how it relates to their position in the shop. Knowledge of ASE certification is needed to choose an area(s) of certification. Technicians know the value of ASE certification and should be familiar with the NATEF certification for automotive training programs. After you become a certified **automotive technician**, you will require frequent training to stay current with the fast-paced changes in automotive technology. When applying for a job, a complete, well thought-out application and resume will be beneficial. Technicians need to understand how to establish and maintain excellent customer relations. Information regarding several automotive trade organizations and the benefits to members is also discussed in this chapter. One of the best ways to receive the necessary up-to-date information is to join one or more trade organizations that provide current information to their members.

TYPES OF AUTOMOTIVE REPAIR SHOPS

It is important for students and technicians to understand the various types of automotive shops and automotive-related careers so you can decide which type of career path you would prefer to pursue. This knowledge should lead to greater job satisfaction.

Did You Know? *In 2004, the average salary for a technician with over 5 years experience was $47,948.00.*

Automotive Dealerships

Automotive dealerships are usually independently owned, but they have a contract to market and service the vehicles from one or more vehicle manufacturers. Small dealerships may sell and service only one type of vehicle, whereas larger dealerships usually sell and service several vehicle makes **(Figure 1-1)**. The dealership must conform to certain standards set by the vehicle manufacturer. These standards may include regulations regarding the size and layout of the facility, new

Figure 1-1 A dealership sells and services vehicles manufactured by one or more manufacturers.

vehicle sales policies, accounting and financing practices, prescribed parts inventories, and service procedures. The dealership usually has a general manager who is responsible for the entire operation. In some dealerships, the general manager may be the owner. The department managers are responsible to the general manager. Department managers usually include a business manager, sales manager, parts manager, and service manager. Business office staff may include accountants and a receptionist, and these personnel are responsible to the office manager. The sales personnel are responsible to the sales manager. In larger dealerships there may be separate managers for new vehicle sales and used vehicle sales. The parts manager is responsible for parts personnel, and the service manager is responsible for shop personnel including service writer(s), technicians, technician's assistants, and clean-up personnel. Dealership technicians are required to have a complete knowledge of the diagnostic and service procedures on the vehicles sold by the dealership, and the technician is required to perform warranty service on these vehicles.

Independent Repair Shops

Independent repair shops are the most common type of automotive repair facility **(Figure 1-2)**. Independent repair shops are privately owned and operated without being affiliated with a vehicle manufacturer, automotive parts manufacturer, or chain organization. Independent repair shops may

Figure 1-2 An independent automotive repair shop.

perform all types of repairs on a large variety of vehicles. In this type of operation, a wide variety of equipment is required, and the technicians must be knowledgeable regarding the diagnostic and repair procedures on many different vehicles. However, as automotive technology becomes more advanced, many of these shops are specializing in specific types of repairs such as under-vehicle work that includes suspension, steering, wheel alignment and balancing, and brakes. Other independent shops may choose to specialize in repairs on specific makes of cars and light-duty trucks. Independent repair shops usually stock some of the more popular automotive parts, and purchase the rest of the required parts from an automotive parts store or a dealership parts department.

Specialty Shops

Specialty shops specialize in one type of automotive repair work. Specialty shops include muffler shops, transmission shops, fuel injection and tune-up shops, brake shops, and tire shops. Specialty shops may be part of an automotive national or regional chain that has similar shops in other cities **(Figure 1-3)**. For example, many tire companies have a chain of tire stores across the United States. If specialty shops are part of an automotive chain, they are often an independently owned franchise. The owner must follow the policies of the

chain organization. Specialty tire shops sell and install tires and usually specialize in under-vehicle service such as wheel balancing and alignment, suspension repairs, and brake service. Muffler shops specialize in exhaust system repairs, but they may do other under-vehicle repair work such as suspension and brakes.

Quick-Lube Shops

Quick-lube shops specialize in lubrication work that consists of oil and filter changes and chassis lubrication **(Figure 1-4)**. Most quick-lube shops specialize in performing this work quickly while the owner or driver waits in the shop waiting room. Some quick-lube shops use pits rather than vehicle lifts because using a pit allows them to perform faster lubrication work. The vehicle is driven over a service pit, and one technician in the pit drains the oil, changes the filter, and lubricates the chassis. Another technician working under the hood checks the fluid levels and installs the new oil. Quick-lube shops usually stock different brands of oil so they can supply the type of oil requested by the customer. Quick-lube shops usually perform a series of checks while performing lubrication work. These checks may include fluid levels, light operation, belt condition, wiper blade condition, and tire condition. Some quick-lube shops perform minor service work such as replacing belts, lights, and wiper blades.

Figure 1-3 A specialty automotive repair shop may be part of a national or regional automotive chain.

Figure 1-4 A quick-lube shop.

Did You Know? *Customer service and satisfaction must be considered the top priority by all employees in any successful automotive business.*

SUCCESSFUL BUSINESS MANAGEMENT

The top priority in all departments of a successful business must be **customer service** and satisfaction. When the sales department sells a customer a new vehicle, the customer may be very impressed with the professionalism and expertise of the sales personnel. However, if the customer takes this new vehicle to the service department and encounters service personnel who are indifferent to the customer's concerns regarding the vehicle, where do you think the customer will purchase his or her next vehicle? All personnel must be proficient in public relations. A technician staff must be positive and polite at all times when dealing with customers, and must exhibit a concern for the customers and their vehicles. There must be excellent communication between all departments for the dealership to operate efficiently and profitably. Regular staff meetings are essential to voice concerns and provide communication between departments. Excellent communication between service personnel and customers is essential. Service personnel should always explain in basic terms the problem with a customer's vehicle, and also the cost of repairs. Some dealerships and shop owners have a policy of introducing their new customers to some of the service department personnel, including a technician. The customer appreciates personalized attention, and this provides the customer with a positive image of the business. It is easy to lose this personalized attention for the customer in large operations. Small touches like calling a customer by name, and showing courtesy improve public relations. Always placing fender, seat, and floor mat covers on customer vehicles while performing service work on the vehicle demonstrates this respect.

Service Manager

The service manager is responsible for the complete shop operation. The service manager is responsible for the general business operation, all shop personnel, and handling customer concerns. Large automotive shops may have a shop foreman who works under the supervision of the service manager. The service manager must have excellent public relations skills. The service manager must also have good organizational skills and extensive experience in the automotive service industry. The service manager is responsible for implementing the vehicle manufacturer's policies on warranty and service procedures. The service manager completes the necessary arrangements and scheduling for technician training. In most shops, the service manager is responsible for hiring shop personnel. The service manager is responsible for communication and cooperation with other departments and personnel in the business.

Service Writers

The service writer is extremely important in maintaining excellent customer relations because the person in this position is usually the first person to greet the customer. In a typical automotive repair shop, the service writer meets the customer and completes the **repair order**. Repair orders are one of the most important first steps in the chain of events in repairing a customer's vehicle accurately and on time. Generally, each facility will have its own repair work order, but all will need similar information and all will need to have some legal aspect authorizing repairs. In addition to detailing pertinent information regarding the repairs, the customer must authorize repairs. Most states require a signature; this in effect makes a work order a legal contract to provide services and a payment for them.

The service writer should always be neatly and cleanly dressed, and he or she must be polite and professional. Photo identification with the service writer's name clearly displayed helps the customer to know the service writer. The service writer should also call the customer by name. The service writer is responsible for completing the work order. The service writer must listen carefully to the customer's concerns. If the customer's problem is obvious from the operation of the vehicle with the engine running in the shop, the service writer does not need to question the vehicle owner any further regarding the problem. However, when the customer's complaint is not obvious from the operation of the vehicle while the engine is running in the shop, the service writer should politely question the customer regarding the problem. The service writer asks the customer about the exact symptoms and/or sounds related to the problem with the customer's vehicle; finds out from the customer if the problem occurs only at a certain vehicle speed or a specific temperature; and then writes an accurate description of the vehicle's problem and the necessary repairs on the work order to allow the service department technician to accurately diagnose and repair the problem and correct the customer's complaint. If the customer requests an estimate for the vehicle repairs, the service writer should obtain the estimate as quickly as possible. The service writer should be sure any other customer concerns are met. For example, the customer may require the service of the customer shuttle vehicle to get to his or her place of work. If the customer is going to wait for a quick repair, the service writer should be sure the customer knows where the customer waiting room is located.

Repair Orders

Repair orders may vary depending on the shop, but repair orders usually have this basic information:

1. Customer's name, address, and phone number(s)
2. Customer's signature
3. Vehicle make, model, year, and color
4. Vehicle identification number (VIN) (**Figure 1-5**)
5. Vehicle mileage
6. Engine displacement

Figure 1-5 The VIN is visible through the driver's side of the windshield.

7. Date and time
8. Service writer's name
9. Work order number
10. Labor rate
11. Estimate of repair costs
12. Accurate and concise description of the vehicle problem

In many shops, the repair orders are completed on a computer terminal, and the computer may automatically write the vehicle repair history on the repair order if the vehicle has a previous repair history in the shop's computer system.

Technicians and Repair Orders

The repair order informs the technician regarding the problem(s) with the vehicle. In many shops the technician has to enter a starting time on the repair order. This may be done on a computer terminal or by inserting the work order into the time clock. The technician may also have to enter a technician code number on the work order to indicate who worked on the vehicle. The technician's code number on the work order is also used to pay the technician for the repair job. The technician must diagnose and repair the problem(s) indicated on the repair order. When the technician obtains parts from the parts department to complete the repair, the technician must present the order number. The parts personnel enter the parts and the cost on the repair order. In some shops the technician is required to enter the completed repairs on the work order. For example, the description of the problem on the work order may be "A/C system inoperative." If the technician replaced the A/C compressor fuse to correct the problem, he or she may enter "Replaced A/C compressor fuse" on the order. Some shops require the service writer or shop foreman to sign the work order when the repair job is successfully completed. The work order is routed back to the cashier who calculates all the charges on the work order including the appropriate taxes. Some shops add a miscellaneous charge on the work order. A typical miscellaneous charge is 10 percent of the total charges on the work order. This miscellaneous charge is to cover the cost of small items such as bolts, cotter pins, grease, lubricants, sealers, and waste disposal costs that are not entered separately on the work order.

In many shops the technicians work on a flat-rate basis. In these shops the technician is paid a **flat rate** for each repair. In a dealership, the flat-rate time is set by the vehicle manufacturer. Independent shops use generic flat-rate manuals published by firms such as Mitchell Publications. If the flat-rate time is 2.0 hours for completing a specific vehicle repair, the customer is charged for 2.0 hours, and the technician is paid for 2.0 hours even though he or she completed the repair in 1.5 hours. Conversely, if the technician takes 2.5 hours to complete the job, the technician is paid for only 2.0 hours and the customer is charged for 2.0 hours. The flat-rate time is usually entered on the work order.

Employee to Employer Obligations

The ever-increasing electronics content on today's vehicles requires that technicians are familiar with the latest electronics technology. There are many different ways to obtain training on the latest automotive technology, but it is absolutely essential. Automotive training may be obtained by these methods:

1. Obtaining training information, service manuals, or bulletins from **OEM**s, independent parts and component manufacturers, or independent suppliers of service manuals and training books. After the information is obtained, it is essential that you read and study it.

> **Tech Tip** One of the best ways to keep up to date on automotive technology is to become a member of an automotive trade organization that provides information related to the area of automotive service in which you are working.

2. Join an organization dedicated to supplying information to automotive service technicians such as the ATRA.

3. Join an Internet organization such as the iATN, where you can communicate with other technicians and obtain the answers to service problems.

4. Download information available on the Internet from automotive equipment manufacturers. Many of these manufacturers provide operator's manuals and other information on their equipment for downloading purposes.

5. Attend satellite training seminars available from some independent automotive training organizations and OEMs.

6. Attend training seminars in your location sponsored by equipment and parts manufacturers or OEMs.

7. Attend training seminars sponsored by the automotive department at your local college.

The successful automotive technician must be committed to life-long training, but the technician's employer must also be committed to assisting technicians employed in his or her shop to obtain the necessary training. This assistance may be financial, providing time off work to attend training seminars, or arranging the necessary training programs. When you begin employment, you enter into a business agreement with your employer.

A business agreement involves an exchange of goods or services that have value. Although the automotive technician may not have a written agreement with his or her employer, the technician exchanges time, skills, and effort for money paid by the employer. Both the employee and the employer have obligations. The automotive technician's obligations include the following:

1. Productivity—As an automotive technician, you have a responsibility to your employer to make the best possible use of time on the job. Each job should be done in a reasonable length of time. Employees are paid for their skills, effort, and time.

2. Quality—Each repair job should be a quality job! Work should never be done in a careless manner. Nothing enhances customer relations like quality workmanship.

3. Teamwork—The shop staff are a team, and technicians and management personnel are team members. You should cooperate with and care about other team members. Each member of the

team should strive for harmonious relations with fellow workers. Cooperative teamwork helps improve shop efficiency, productivity, and customer relations. Customers may be "turned off" by bickering among shop personnel.

4. Honesty—Employers and customers expect and deserve honesty from automotive technicians. Honesty creates a feeling of trust among technicians, employers, and customers.

5. Loyalty—As an employee, you are obligated to act in the best interests of your employer, both on and off the job.

6. Attitude—Employees should maintain a positive attitude at all times. As in other professions, automotive technicians have days when it may be difficult to maintain a positive attitude. For example, there will be days when the technical problems on a certain vehicle are difficult to solve. However, a negative attitude certainly will not help the situation! A positive attitude has a positive effect on the job situation as well as on the customer and employer.

7. Responsibility—You are responsible for your conduct on the job and your work-related obligations. These obligations include always maintaining good workmanship and customer relations. Attention to details such as always placing fender, seat, and floor mat covers on customer vehicles prior to driving or working on the vehicle greatly improve customer relations.

8. Following directions—All of us like to do things "our way." Such action, however, may not be in the best interests of the shop, and as an employee you have an obligation to follow the supervisor's directions.

9. Punctuality and regular attendance—Employees have an obligation to be on time for work and to be regular in attendance on the job. It is very difficult for a business to operate successfully if it cannot count on its employees to be on the job at the appointed time.

10. Regulations: Automotive technicians should be familiar with all state and federal regulations pertaining to their job situation, such as the Occupational Safety and Health Act (OSHA) and hazardous waste disposal laws.

Employer to Employee Obligations

Employer to employee obligations include:

1. Wages—The employer has a responsibility to inform the employee regarding the exact amount of financial remuneration they will receive and when they will be paid.

2. Fringe benefits—A detailed description of all fringe benefits should be provided by the employer. These benefits may include holiday pay, sickness and accident insurance, and pension plans.

3. Working conditions—A clean, safe workplace must be provided by the employer. The shop must have adequate safety equipment and first-aid supplies. Employers must be certain that all shop personnel maintain the shop area and equipment to provide adequate safety and a healthy workplace atmosphere.

4. Employee instruction—Employers must provide employees with clear job descriptions, and be sure that each worker is aware of his or her obligations.

5. Employee supervision—Employers should inform their workers regarding the responsibilities of their immediate supervisors and other management personnel.

6. Employee training—Employers must make sure that each employee is familiar with the safe operation of all the equipment that they are required to use in their job situation. Since automotive technology is changing rapidly, employers should provide regular update training for their technicians. Under the right-to-know laws, employers are required to inform all employees about hazardous materials in the shop. Employees should be familiar with MSDS, which detail the labeling and handling of hazardous waste and the health problems if exposed to hazardous waste.

Job Responsibilities

An automotive technician has specific responsibilities regarding each job performed on a customer's vehicle. These job responsibilities include:

1. Do every job to the best of your ability. There is no place in the automotive service industry for careless workmanship! Automotive technicians and students must realize they have a very responsible job. During many repair jobs you, as a student or technician working on a customer's vehicle, actually have the customer's life and the safety of his or her vehicle in your hands. For example, if you are doing a brake job and leave the wheel nuts loose on one wheel, that wheel may fall off the vehicle at high speed. This could result in serious personal injury for the customer and others, plus extensive vehicle damage. If this type of disaster occurs, the individual who worked on the vehicle and the shop may be involved in an expensive legal action. As a student or technician working on customer vehicles, you are responsible for the safety of every vehicle that you work on! Even when careless work does not create a safety hazard, it leads to dissatisfied customers who often take their business to another shop. Nobody benefits when that happens.

2. Treat customers fairly and honestly on every repair job. Do not install parts that are unnecessary to complete the repair job.

3. Use published specifications; do not guess at adjustments.

4. Follow the service procedures in the service manual provided by the vehicle manufacturer or an independent manual publisher.

5. When the repair job is completed, always be sure the customer's complaint has been corrected. Retest the system that was repaired to verify for correct operation.

6. Do not be too concerned with work speed when you begin working as an automotive technician. Speed comes with experience.

THE JOB APPLICATION

Did You Know? Always be prepared for an interview when entering a business for a job application. People who are prepared will be hired before those who are not.

The first impression that you make on a prospective employer is very important! Your letter of application and resume are important when you are looking for employment. Some larger businesses have a standard job application form that must be filled out when applying for employment. This application form will request all the pertinent facts regarding your past employment experience, training, job preferences, special skills, and personal information. Always fill out the job application form completely and neatly. Even smaller businesses may request a resume and a letter of application from those seeking employment. Coming prepared with a brief portfolio of past accomplishments and work history is always a good idea.

The resume is a brief, one-page document that includes the following information:

1. Address and telephone number.

2. Work experience—List your previous work experience beginning with your most recent job. Most employers want a work history for the last 5 years. List the dates of employment, and include the names and addresses of previous employers.

3. References—List the names, addresses, and phone numbers of two or three previous employers or instructors who are familiar with your work and/or training. Always contact the people whom you would like to use for a reference before listing their names.

4. Education—Be sure to include the names and addresses of schools or colleges that you attended, and state the diplomas and/or degrees that you obtained. List any areas of specialization. Include industry classes that you have attended, and list any special awards that you have received.

5. Special skills—Include a description of any special skills that you have. For example, you may have specialized in electronic system diagnosis in a previous job, and as a result you have special skills in this area.

6. Hobbies and special interests—List any hobbies that you participate in. For example, you may be interested in building racing engines

and automotive racing. Although this may not be directly related to your work as a technician, this hobby indicates your extensive interest and participation in the automotive industry.

The letter of application is a brief, one-page document that contains these items:

1. The name and address of the person who is responsible for conducting the hiring interview. This is usually the service manager or personnel manager.
2. Your name, address, and phone number.
3. The job title of the position for which you are applying, with an explanation of how you found out about this job opportunity.
4. An explanation of why you would like this position.
5. Reasons why you believe that you would be successful in this position.

CUSTOMER RELATIONS

When dealing with customers, always remember the two Ps: positive and polite. Even on the days when we do not feel positive and polite! Perhaps there are two vehicles in the shop with time-consuming, difficult diagnostic problems, and the vehicle owners are requesting their cars. Even though we do not feel positive and polite, we must maintain these attitudes when talking to customers. These attitudes will certainly improve our customer relations and bring customers back to the shop the next time their vehicle needs repairs. Employees at all levels in an automotive shop should be courteous and maintain a positive attitude. Perhaps the customer's vehicle is older than average and not in very good mechanical condition. In some cases, the customer may not realize the poor condition of their vehicle. The customer's vehicle is important to him or her, and we should respect that fact. Rather than being negative about the condition of the customer's vehicle, the technician and all service personnel should see this customer and vehicle as an opportunity for a considerable amount of service work. Technicians and all service personnel should be interested in the customer's vehicle concerns, and these concerns must be corrected by the service work performed on the vehicle.

Automotive service personnel must be dedicated to quality workmanship. Repairs should not be done with the attitude that a repair is good enough to keep the vehicle running. Each repair should be the best possible repair to correct the customer's complaint. All service personnel must be committed to quality care of the customer's vehicle. This includes all personnel who will have any part in servicing the customer's vehicle. For example, in a large shop operation, quality care of customer vehicles applies to car jockeys who may drive these vehicles in and out of the shop. Do not smoke, eat, or drink in customer vehicles because the customer may be a non-smoker who does not appreciate the odor of cigarette smoke, or food in their vehicle. Always place seat covers and floor mat covers on the vehicle, and place covers on the fenders when working under the hood. All personnel in an automotive service shop must be committed to teamwork. Teamwork means working together for the improvement of the shop operation. This may mean helping a co-worker lift a heavy component. Perhaps a co-worker is having trouble diagnosing a difficult problem, and you have encountered and corrected this problem on a previous service job. Teamwork means sharing diagnostic knowledge! All automotive service personnel must be neat in appearance. Coveralls or lab coats should be changed as often as necessary to maintain a neat, clean appearance.

National Institute for Automotive Service Excellence (ASE)

The National Institute for Automotive Service Excellence (**ASE**) has provided voluntary testing and certification of automotive technicians on a national basis for many years. The ASE provides certification for technicians in various areas such as Automotive, Medium/Heavy Duty Truck, and Collision Repair. The image of the automotive service industry has been enhanced by the ASE certification program. More than 415,000 technicians now have current certifications and work in a wide variety of automotive service shops. ASE provides

Figure 1-6 ASE certification shoulder patches worn by automotive technicians and master technicians.

certification in eight areas of automotive repair: engine repair, automatic transmissions/transaxles, manual drive train and axles, suspension and steering, brakes, electrical systems, heating and air conditioning, and engine performance.

A technician may take the ASE test and become certified in any or all of the eight areas. When a technician passes an ASE test in one of the eight areas, an Automotive Technician's shoulder patch is issued by ASE. If a technician passes all eight tests, he or she receives a Master Technician's shoulder patch **(Figure 1-6)**. Retesting at 5-year intervals is required to remain certified. The certification test in each of the eight areas contains forty to eighty multiple-choice questions. The test questions are written by a panel of automotive service experts from various areas of automotive service including automotive instructors, service managers, automotive manufacturers' representatives, test equipment representatives, and certified technicians. The test questions are examined and checked for quality by a national sample of technicians.

On an ASE certification test, approximately 45 percent to 50 percent of the questions are Technician A and Technician B format, and the multiple-choice format is used in 40 percent to 45 percent of the questions. Fewer than 10 percent of ASE certification questions are an *except* format where the technician selects one incorrect answer out of four possible answers. ASE tests are designed to test the technician's understanding of automotive systems, testing and diagnostic knowledge, and ability to follow proper repair procedures.

ASE regulations demand that each technician must have 2 years of working experience in the automotive service industry prior to taking a certification test or tests. However, relevant formal training may be substituted for 1 year of working experience. Contact ASE for details regarding this substitution. ASE also provides certification tests in automotive specialty areas such as Advanced Engine Performance Specialist, Alternate Fuels Light Vehicle Compressed Natural Gas, Parts Specialist, Machinist-Cylinder Head Specialist, Machinist-Cylinder Block Specialist, and Machinist-Assembly Specialist. Shops that employ ASE-certified technicians display an official ASE blue seal of excellence. This blue seal increases the customer's awareness of the shop's commitment to quality service and the competency of certified technicians.

National Automotive Technicians Education Foundation (NATEF)

In 1978, the Industry Planning Council (IPC) and American Vocational Association (AVA) were concerned about the quality of automotive education across the United States. The IPC was composed of representatives from the automotive industry and vocational education. These two organizations directed a multi-year study that developed an automotive task list and an evaluation guide for automotive training programs. The ASE was selected to administer the evaluation of automotive training programs. **NATEF** was established as a separate foundation and an affiliate of ASE. NATEF evaluates and certifies automotive, autobody, and medium/heavy truck training programs. The evaluation of automotive training programs involves a comprehensive self-evaluation and an on-site evaluation.

During the self-evaluation program, instructors, administrators, and advisory committee members rate the automotive program using these standards:

1. Purpose
2. Administration
3. Learning resources
4. Finances
5. Student services
6. Instruction

7. Equipment

8. Facilities

9. Instructional staff

10. Cooperative work agreements

The self-evaluation helps to identify areas in the automotive training program that require improvement according to national standards. After the self-evaluation is completed, the automotive program submits the self-evaluation materials with an application to NATEF. An on-site evaluation is scheduled by NATEF if the training program appears to meet the required standards. An Evaluation Team Leader (ETL) and technicians in the training program conduct the on-site evaluation. The ETL must be an educator with ASE master-technician certification and NATEF program evaluation training. The ETL, in cooperation with the evaluation team, submits a report to NATEF. If the automotive training program meets the required standards, the program receives NATEF certification. After NATEF certification is obtained, the automotive training program may display the ASE-NATEF logo **(Figure 1-7)**. Automotive training programs must be recertified every 5 years. More than 1,000 automotive training programs have been certified by NATEF. Autobody and Medium/Heavy Duty Truck training programs are also certified by NATEF. In 1995, Ohio State University (OSU) conducted an extensive study regarding the effect of NATEF certification on student learning. OSU researchers concluded that NATEF certification has a significant positive effect on learning that takes place in automotive technician training programs.

Automatic Transmission Rebuilders Association (ATRA)

The Automatic Transmission Rebuilders Association (**ATRA**) was founded in 1954. ATRA provides technical information to transmission technicians and rebuilding shops. ATRA is the oldest and largest organization of independently owned transmission rebuilding firms with members in the United States, Canada, and twenty-two other countries. ATRA is dedicated to the welfare and improvement of the automatic transmission repair

Figure 1-7 The ASE-NATEF logo.

industry for the benefit of the motoring public. ATRA distributes technical and business information through bulletins, books, seminars, *GEARS* magazine, a yearly exposition, and Internet programs. Specific information may be downloaded from the ATRA website. This site may be accessed at http://www.atra-gears.com/.

The ATRA bookstore is exclusively for ATRA members and certified ATRA technicians. This bookstore supplies books related to business management including topics such as advertising, marketing, and taxes. Books, bulletins, technical manuals, and technical training videos are also available. The bookstore also markets ATRA member patches and chevrons. ATRA also provides customer bulletins on topics such as "What is a REMAN transmission?" ATRA also provides technician certification tests in these areas: Transmission Rebuilder, Transmission Remove and Replace Technician, and Transmission Diagnostician. These tests are provided at many locations throughout North America. The *GEARS* magazine published by ATRA provides information on technical shop and office management, safety and EPA concerns, new products and new industry developments, and current interests. ATRA provides technical seminars in many different locations. These seminars explain the popular transmissions used in vehicles built by the major manufacturers. These technical seminars provide transmission fixes, updates, and modifications to correct many transmission problems. ATRA's yearly Powertrain Industry Exposition provides technical update information, new product information, and demonstrations regarding new equipment or products, professional development, and networking opportunities.

Mobile Air Conditioning Society Worldwide (MACS Worldwide)

The Mobile Air Conditioning Society Worldwide (MACS Worldwide) was founded in 1981. MACS provides technical and business information through the *Automotive Cooling Journal*, *MACS Action*, and *MACS Service Reports*. MACS was founded to meet the automotive air conditioning (A/C) industry's need for comprehensive technical information, training, and communication. MACS Worldwide has over 1,600 members including A/C service shops, installers, distributors, component suppliers, and manufacturers. These members are located throughout the United States, Canada, and many other countries. The MACS website may be accessed at http://www.macsw.org. MACS provides members with up-to-date technical and business information through a number of publications. These publications include *Automotive Cooling Journal, MACS Action,* and *MACS Service Reports.* The *Automotive Cooling Journal* is published monthly and provides reports on people, industry trends, new industry developments, technical information, and industry events such as trade shows and expositions. *MACS Action* is published bimonthly and provides members with information on marketing, business management, and association news. *MACS Service Reports* are published monthly, and this technical publication provides A/C service bulletins, service tips, and specific technical information regarding A/C and cooling system service.

MACS provides training programs in these areas:

1. Automotive Air Conditioning, Diagnosis, and Service
2. Automotive Heating and Air Conditioning Test Preparation
3. Electronics and Electrical Control with an Introduction to Automatic Temperature Control
4. Guidelines for Automotive Air Conditioning Retrofit
5. Problem Servicing Mobile Air Conditioning Systems

MACS training programs are sold as a package that contains specific materials such as a student manual, instructor guide package, overhead transparencies, color slides, and a video.

Instructors who wish to purchase MACS training programs numbered 1 through 3 in the previous list must have a MACS Technician Training Institute (TTI) training certificate. If an instructor holds a current state-approved teaching certificate, MACS will issue a MACS TTI certificate to the instructor upon submission to MACS of the instructor's credentials. When an instructor does not have a current state-approved teaching certificate, the instructor must attend a seminar approved by MACS to obtain a MACS TTI certificate. The MACS Technician Training and Certification Program for Refrigerant Recovery and Recycling is approved by the EPA, and is offered as a self-study program conducted by mail or in a classroom environment with an instructor. Instructors must meet specific criteria to become a MACS Certification Trainer/Proctor. MACS sponsors an annual convention and trade show for members. This convention provides technical seminars, special speakers, and a display of new products and equipment at the trade show.

Automotive Service Association (ASA)

The Automotive Service Association (ASA) was founded in 1951 when a group of automotive shop owners recognized that the problems facing independent automotive repair businesses could be solved more efficiently through their association and cooperation. This recognition led to the formation of the Independent Garageman's Association of Texas that evolved through various organizations into the ASA. ASA promotes professionalism and excellence in the automotive repair industry through education, representation, and member services. ASA presently has 12,000 member shops and individuals representing 65,000 professionals in many different countries. The purpose of ASA is to advance the professionalism and excellence in the automotive repair industry through education, representation, and member services. ASA has a mechanical division and a collision division. The ASA website may be accessed at http://www.asashop.org. ASA publishes the monthly *AutoInc.* magazine. This magazine contains mechanical, collision, and management information. *AutoInc.* also includes a report from ASA's legislative office in Washington, D.C. ASA publishes a monthly bulletin entitled *Mechanical Division Dispatch and Collision Repair Report.* ASA sponsors the annual Congress of Automotive Repair and Service (CARS) for the mechanical division. This conference provides technical seminars, forums on current industry concerns such as inspection/maintenance (I/M) programs, and Automotive Management Institute (AMI) seminars. ASA also hosts an annual convention that provides division meetings, (AMI) seminars, and keynote speakers. ASA operates an automotive hotline called Identifix. This hotline service provides diagnostic information via telephone to solve specific automotive problems. A hotline service such as Identifix employs experts in various areas of automotive technology, and they have all the necessary service manuals and diagnostic computer software. Identifix offers other services such as faxing wiring diagrams upon request. Identifix publishes a bi-monthly *Identifix Update* bulletin containing many service tips.

Tech Tip *A hotline service can be very helpful when attempting to diagnose difficult automotive problems.*

International Automotive Technicians Network (iATN)

The International Automotive Technicians Network (**iATN**) is a group of over 43,000 automotive technicians in 138 countries. These members share technician knowledge and information with other members via the Internet. iATN is the largest network of automotive technicians in the world. The iATN mission is to promote the continued growth, success, and image of the professional automotive technician by providing a forum for the exchange of knowledge and the promotion of education, professionalism, and integrity. The iATN website may be accessed at http://www. iatn.net.

The iATN goals are as follows:

1. To provide programs and services that promote the professional growth and effectiveness of iATN members through the exchange of knowledge.
2. To provide an interactive, technology-rich learning environment.
3. To increase the public's understanding, appreciation, and respect for the Automotive Service Industry.
4. To promote public awareness of the iATN and its members.
5. To continually adapt to the changing needs of supporting iATN members.

In return iATN members agree to:

1. Uphold the highest standards of professionalism, competence, and integrity.
2. Pursue excellence through ongoing education and the exchange of knowledge and experience of fellow iATN members.
3. Maintain quality through the use of proper tools, equipment, parts, and procedures.
4. Promote public awareness of the importance of quality professional service and the advancing technology of the modern automobile.
5. Support and promote the mission and goals of the iATN.

The iATN is supported by a number of top companies, associations, and publications in the automotive industry. If an iATN member encounters an automotive problem, iATN suggests that the technician attempt to diagnose the problem using information from service manuals, data systems such as Mitchell-on-Demand, and technical hotlines. The technician can also access the iATN Fix database. If the technician still cannot solve the problem, a problem file may be sent to iATN outlining the necessary vehicle information and the details of the problem. This problem file is sent via the Internet to many iATN members. Any iATN member who has experienced and solved the same problem can post a fix file to iATN that outlines the answer to the automotive problem and the required fix procedure. An iATN member may configure his or her account to receive problem and fix files in only specific areas of automotive repair such as engine electronics and fuel systems. The iATN offers many other member benefits such as live conferencing and technical forums. These forums include technical discussion forums, technical theory forums, technical tip forums, and tool and equipment forums.

Summary

- Independent repair shops are privately owned, and they are not affiliated with vehicle manufacturers, automotive parts suppliers, or chain organizations.
- Specialty shops usually perform one type of automotive repair work.
- Quick-lube shops specialize in fast lubrication service.
- Automotive dealerships have a contract with one or more vehicle manufacturers to sell and service the manufacturer's vehicles.
- Technicians must understand the various job descriptions in a typical automotive repair shop.
- Technicians must be familiar with work orders and how they are processed in a typical shop.
- Technicians must have a knowledge of employer to employee obligations.
- Technicians must understand employee to employer obligations.
- Technicians must be familiar with their job responsibilities to perform efficiently in an automotive shop.

- Technicians must understand the ASE certification of automotive technicians.
- Technicians need to be familiar with NATEF certification of automotive training programs.
- Technicians must be able to complete job applications and resumes.
- Technicians must be able to establish and maintain good customer relations.
- The ATRA distributes technical and business information to automotive transmission technicians and rebuilding shops.
- MACS Worldwide provides technician information, training, and communication for the automotive A/C industry.
- The ASA provides membership service such as the monthly *AutoInc.* magazine and a yearly Congress of Automotive Repair and Service.
- The iATN provides forums for the exchange of information in the automotive industry.

Review Questions

1. In an automotive dealership, the vehicle manufacturer may set standards regarding all of these items *except*:

   A. The layout of the facility.

   B. New vehicle sales policies.

   C. Accounting and financing practices.

   D. Staff benefits policies.

2. When discussing dealership management, Technician A says customer service and satisfaction is extremely important.

Technician B says communication between dealership departments is not a high priority. Who is correct?

A. Technician A

B. Technician B

C. Both Technician A and Technician B

D. Neither Technician A nor Technician B

3. All of these statements about the ATRA are true *except*:

A. ATRA was founded in 1954.

B. ATRA has members in 24 countries.

C. ATRA distributes information on most automotive electronic systems.

D. ATRA publishes customer bulletins.

4. When discussing the iATN, Technician A says the iATN has members in 136 countries. Technician B says the iATN is helpful when diagnosing automotive problems. Who is correct?

A. Technician A

B. Technician B

C. Both Technician A and Technician B

D. Neither Technician A nor Technician B

5. Technician A says a wide variety of test equipment is required in an independent repair facility. Technician B says technicians working in an independent repair facility may be required to be knowledgeable regarding the diagnostic and service procedures on many different vehicles. Who is correct?

A. Technician A

B. Technician B

C. Both Technician A and Technician B

D. Neither Technician A nor Technician B

6. Technician A says an independent repair shop may specialize in specific types of repairs. Technician B says an independent repair shop may specialize in repairs on specific makes of vehicles. Who is correct?

A. Technician A

B. Technician B

C. Both Technician A and Technician B

D. Neither Technician A nor Technician B

7. Technician A says all dealership personnel must be proficient in public relations. Technician B says the top priority in all dealership departments must be customer service and satisfaction. Who is correct?

A. Technician A

B. Technician B

C. Both Technician A and Technician B

D. Neither Technician A nor Technician B

8. The ASA promotes professionalism and excellence in the automotive repair industry through all of these methods *except*:

A. Test equipment sales.

B. Education.

C. Representation.

D. Member services.

9. Technician A says the ASA operates an automotive hotline service called Fix-It-Right. Technician B says an automotive hotline service provides diagnostic information via telephone to solve specific automotive problems. Who is correct?

A. Technician A

B. Technician B

C. Both Technician A and Technician B

D. Neither Technician A nor Technician B

10. Quick-lube shops may also perform _____ service work.

11. The ATRA provides technician certification tests in these areas:

1. _____

2. _____

3. _____

12. ATRA publishes a magazine entitled _____.

13. Describe the advantages of ATRA membership for automotive technicians.

14. Describe the MACS publications and training programs for automotive technicians.

15. Explain the purposes of Identifix.

16. Describe the benefits of iATN membership for automotive technicians.

17. Technician A says the service writer is responsible for meeting the customers. Technician B says the service writer is responsible to the service manager. Who is correct?

    A. Technician A

    B. Technician B

    C. Both Technician A and Technician B

    D. Neither Technician A nor Technician B

18. Technician A says that 4 years of work experience in an automotive shop is required prior to ASE certification. Technician B says that ASE provides certification in specialty areas such as Cylinder Head Specialist. Who is correct?

    A. Technician A

    B. Technician B

    C. Both Technician A and Technician B

    D. Neither Technician A nor Technician B

19. Technician A says a NATEF-certified automotive training program must be recertified every 3 years. Technician B says prior to NATEF certification of an automotive training program, a self-evaluation must be completed by the program staff. Who is correct?

    A. Technician A

    B. Technician B

    C. Both Technician A and Technician B

    D. Neither Technician A nor Technician B

20. In a large automotive repair shop, Technician A says the service manager is responsible to the shop foreman. Technician B says the service manager is responsible to the cashier. Who is correct?

    A. Technician A

    B. Technician B

    C. Both Technician A and Technician B

    D. Neither Technician A nor Technician B

21. Service writers should follow all of these job performance guidelines *except*:

    A. The service writer should be neatly and cleanly dressed.

    B. The service writer should deal with the customer quickly without discussing the customer's concerns.

    C. The service writer should be professional and polite.

    D. The service writer should call the customer by name.

22. Technician A says the service writer should ask the customer about the exact symptoms and/or sounds related to the customer's complaint. Technician B says the service writer should find out if the customer's vehicle complaint occurs at a specific vehicle speed or engine temperature. Who is correct?

    A. Technician A

    B. Technician B

    C. Both Technician A and Technician B

    D. Neither Technician A nor Technician B

23. All of these statements regarding the service manager's responsibilities are true *except*:

    A. The service manager is responsible for implementing the vehicle manufacturer's warrantee policies and recommended service procedures.

    B. The service manager is responsible for hiring shop personnel.

    C. The service manager is responsible to the shop foreman.

    D. The service manager is responsible for communication with other departments in the business.

24. Technician A says the repair order should contain an accurate and precise description of the vehicle problem. Technician B says the repair

order should contain the customer's signature. Who is correct?

A. Technician A

B. Technician B

C. Both Technician A and Technician B

D. Neither Technician A nor Technician B

25. Technician A says a resume should contain personal medical records. Technician B says a resume should contain the job title for which you are applying. Who is correct?

A. Technician A

B. Technician B

C. Both Technician A and Technician B

D. Neither Technician A nor Technician B

26. In an automotive repair shop, the shop foreman supervises the _____.

27. In many automotive shops, technicians work on a _____ _____ basis.

28. Each ASE automotive certification test contains _____ to _____ multiple choice questions.

29. ASE automotive master technician status is obtained by passing _____ ASE certification tests.

30. Explain ten employee to employer obligations.

31. Describe six employer to employee obligations.

32. Describe a technician's job responsibilities when working in an automotive shop.

33. Describe the information that should be included on a resume.

2 Employee Skills for Career Success

Key Terms

Honesty

Professionalism

Reputation

Tech talk

INTRODUCTION

To be successful in any career, particularly auto repair, the single most important skill one can have is people skills. This book will focus on making you a skilled entry-level technician that will be able to apply diagnostic skills to repair modern automobiles. But when one really examines truly successful technicians they have two main reasons for their success. They are above average in their repair skills and they are good with customers. Often these technicians have a customer base that goes to them first whenever they need repairs or advice. This chapter will focus on the personal traits that make customers comfortable and at ease to become "regulars." In the business world this is referred to as "soft skills."

PERSONAL TRAITS OF SUCCESSFUL TECHNICIANS

Businesses spend countless amounts of money surveying customers to find out how well they did as a business, and why customers bring in repeat business to their prospective areas. Friendliness of personnel and how they made them feel at ease rate among the highest. In larger repair facilities or dealerships often a service writer or shop manager does most of the customer relations with the public, however, a large percentage of people like to talk directly to the person who will be fixing their auto and will ask to speak with the technician. Making the customer feel at ease and confident that they have made the right choice about who, and where their car is fixed is not something to be taken lightly, especially when one considers that to the average person their car is the second most expensive item they will ever own.

At this moment the technician needs to remember that the person they are talking to is not a technician and often doesn't understand the terminology and that **tech talk** immediately puts people on the defensive.

When a person brings in their car for repairs it is a similar experience psychologically to taking their children to the doctor; not only are some people that attached to their cars but they are in a vulnerable position emotionally. Having to trust an unknown person with repairs that can not only affect their safety, but often involve fairly large sums of money, and technology that they don't understand, puts people in a vulnerable position. The person standing in front of the technician needs to feel like the work will be done as carefully as if they were the technician's favorite grandparent.

Humor is sometimes a good tool to help ease the unease that customers feel; however, it can also be the worst one. More often than not, being very professional is the easiest way to ease the situation. Humor has an easy way of backfiring, particularly if the customer doesn't share your sense of humor and a seemingly harmless joke offends or turns off a potentially good customer. It is often best to not use humor until you get to personally know the customer and even then you must be careful in case someone listening might take offense at the remarks.

Reputation

Of all the tools and techniques that really good technicians have, the one they will most closely guard is their **reputation**. Although it takes time and experience to build a reputation, experienced technicians know that it is also the most fragile possession they have. Good shops will also have a reputation for excellence, along with the technicians who work there. Having a good reputation brings in customers when others have to advertise and hard sell their services. When you look at shops that are always busy with customers preferring to wait rather than go elsewhere, you will find that these places often have a reputation for excellence.

Rarely do these shops or technicians have the cheapest shop rates and more commonly they are the most expensive. These are the types of shops that will draw the best technicians and the best customers.

Integrity is about representing themselves and their shop in the best way possible. A technician with a high degree of integrity does not talk down about other technicians or brag up things that they have done, but, prefers to let their work speak for itself and their customers do it for them. For example, a new customer comes into the shop complaining about the poor service they received at another shop. The smart technician will maintain **professionalism** by not saying anything derogatory about the other technician or shop, for several reasons. The first is that this is possibly an opportunity to be professional and win a new customer, because, maybe they did receive poor service. And two, some people can be impossible to please and are just looking to justify that they really are an undesirable customer to deal with. The "bad treatment" could have been a verbal misunderstanding over repair specifics. Extra care should be taken when dealing with these situations.

Honesty is a way of doing business and dealing with people that isn't turned off at the end of the workday. When a person brings in a problem that is outside of their expertise they readily point out who is the best person for the job, knowing that they have done their customer a better service. If they make a mistake, they will fix the problem and adjust the bill if necessary so the customer isn't charged for their error. For example, a repair order was agreed to repair a faulty fuel injector, but in the process of the repair the technician discovers that it's only a bad connection. The honest technician will repair the connector and adjust the bill by not charging for an expensive injector, saving the customer money. The smart technician would also tell the customer about the change and why. This fosters customer loyalty because they understand that their best interests were taken into account and while they could have been easily overcharged, they weren't. With today's television channels often showing stories about people getting swindled by unscrupulous repair facilities it's no wonder that people are often apprehensive about getting their car fixed and wondering if they are going to be swindled next. Television has a powerful effect upon how we think; one such story as this and it takes a long time to erase that image from a persons mind, whether or not the story was true. For instance, a story was run on T.V. a few years ago about an epidemic of unscrupulous garages. To prove their point, the news crew would pull off the coil wire on their car and call a tow truck from a local garage. When the driver would arrive they would give a story about their "car troubles" and want him to fix it on the spot. When the driver would check to see that the car wouldn't start, they would get permission to tow it to their shop. Upon arriving at the shop the car would be diagnosed by a technician and the wire replaced. When they would get the bill the news crew would announce their proof that they had been swindled and tell their news audience how unscrupulous the particular shop was. In reality the shop had done nothing wrong. Purposely sabotaging your car and calling someone to tow and repair it will result in a legitimate repair bill. The shop owner had to pay for a tow truck, employ a driver for the truck, pay the technician to look at the car and reinstall the wire once it was found. These are honest expenses and to do them for nothing as the news crew expected, since it was just a wire, is unrealistic. If you go to a doctor's office for and complain of sickness, you will get billed whether you are sick or not.

Pride is not about bragging; that's boasting. Pride is about a sense of ownership in a job well done. The technician with a sense of pride about their work usually views each car as if it were their own. Treat each car like it was a vintage auto worth millions, no matter what its condition.

Personal touches are another way of showing pride; for instance, the technician who leaves a customer survey and their business cards after each job shows that they take pride in their work. Greeting customers and talking to them are signs of confidence that people want to see **(Figure 2-1)**. Taking the extra step ensures not only the customer's satisfaction, but your own as well. Technicians who take pride in their work show it. They will set higher standards for themselves and their work than is required to meet company or customer standards.

Figure 2-1 Greeting customers to discuss repairs is an excellent way to build trust.

One can easily see pride by the way a technician treats their own and the shop's tools.

Appearance is very often closely related to one's sense of pride. If you tour local repair facilities, both the best and worst, without having any work performed by the shop, the appearance of each will directly correspond to how people view them and their work performance **(Figure 2-2)**. Technicians are immediately judged by the public on their appearance; for instance, if two unknown technicians stand side by side and one has a dirty and unkempt appearance and the other has a neat appearance, almost no one would pick the unkempt one as the better technician. Even though automotive repair can sometimes be a dirty job, you must take the extra effort to remain presentable. This is why most shops will rarely pick uniforms with white in them and usually rely on darker colors **(Figure 2-3)**.

Manners are important. "Please" and "thank you" are important words to people. Using good manners is a way of showing that you respect the other person. There is an old cliché that says all you need to survive in the business world is a good introduction and handshake. While this is not entirely true it is good advice and is an important soft skill that is often underestimated **(Figure 2-4)**.

Figure 2-2 Which technician would you pick to work on your car?

Figure 2-3 Taking the extra step to keep a presentable appearance shows professionalism.

Figure 2-4 A good greeting and handshake shows confidence.

Employability

For the beginning entry-level technician, few things are more important than getting that first job. You've looked at all the shops and picked out the ones you have decided would be the best places to begin your career, but what will get your name on their employee list?

A good resume is considered by most to be the first step, but before you get to hand in the resume you will have been interviewed and often don't realize it. A person's first interview begins when they open the door of their prospective employer. How they present themselves goes a long way to determining if their resume is even considered. For instance, you can be a great technician but if you dress sloppy or act unprofessional to people, you won't be considered. For example, you walk into a shop to fill out a job application and the office manager is very busy. Rather than being patient, you make a snide comment. At this point you have just killed any chance of getting an interview. Remember, everyone has bad days but you are expected to deal with it and always be *polite* and *professional*. It would be better to either wait or ask for a time to come back. Also keep in mind that most repair shops are small businesses that are often owned and run by a family. The office manager could easily be the owner, their spouse, or a relative and might be the first or only stop in the hiring process.

It is a good idea to read a book and get advice when putting together your first resume. The basics

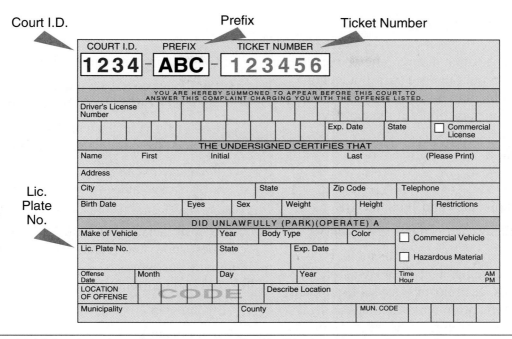

Figure 2-5 Vehicle violations demonstrate a lack of judgment and maturity.

of a good resume are more detailed than one chapter can give. Often for the first-time job seeker references will be harder to come by, and employers will look at work history. For example, a student who comes to class on time and prepared every day is considered to have a good work history. Teachers often make good references because they have a good understanding of their student's traits and work ethic.

Having a clean driving record is critical. This fact is often overlooked by most, except the employer and their insurance company. Part of your job as a technician will be to drive cars to test them and to verify that repairs are correct. The shop will have to carry insurance to cover any vehicles in their possession for repairs, and if you have a bad driving record the insurance company will either refuse to insure you or it will be very expensive. Good shops don't mind paying good wages as part of their way to make profits, but when insurance costs prohibit their profit because of a technician's bad habits, usually the technician with the poor judgment will either not be hired or will be fired. Most shops have a policy stating that a technician must have and *maintain* a clean driving record, so not only is having a good record important, keeping it clean is a must **(Figure 2-5)**. Offenses such as a DWI conviction often spell an immediate loss of employment for even the best technicians.

Summary

- Employing good people skills is a requirement for successful technicians.

- Avoid tech talk when discussing repairs with the customer.

- Avoid using humor when discussing repairs with customers.

- It can take years to establish a good reputation that attracts customers.

- Show customers that you are proud of your work.

- Customers like personal touches.

- It is very important to have and maintain a good driving record.

- Always be polite and professional.

Review Questions

1. Technician A says that maintaining an employer's good reputation is a part of a technician's job. Technician B says that a shop's reputation is the responsibility of the employer. Who is correct?

   A. Technician A

   B. Technician B

   C. Both Technician A and Technician B

   D. Neither Technician A nor Technician B

2. Technician A says that soft skills include explaining to a customer a repair procedure if they have questions. Technician B says a technician with a good reputation will often have a loyal customer base that returns whenever repairs or maintenance is needed. Who is correct?

   A. Technician A

   B. Technician B

   C. Both Technician A and Technician B

   D. Neither Technician A nor Technician B

3. Technician A says that technicians who have the best people skills are usually the most in demand. Technician B says you should try to be honest on most repairs. Who is correct?

   A. Technician A

   B. Technician B

   C. Both Technician A and Technician B

   D. Neither Technician A nor Technician B

4. All of these statements about your driving record are true *except*:

   A. Employers will look at your record for driving offenses.

   B. The more past offenses you have the more it will cost the employer to insure you.

   C. You may not be hired if there are too many driving violations.

   D. Having serious offenses after you are hired can be grounds for firing.

5. The technician who demonstrates pride in their work will (circle all that apply):

   A. Get the job done correctly the first time.

   B. Put fender covers on the most expensive vehicles to avoid scratching them.

   C. Look over vehicles while they are being serviced for possible extra repair work.

   D. Treat each vehicle as if it were an expensive vintage auto.

6. When discussing best ways to deal with nervous customers, Technician A says humor is the best approach. Technician B says that dealing with customers is only the service writer's job. Who is correct?

   A. Technician A

   B. Technician B

   C. Both Technician A and Technician B

   D. Neither Technician A nor Technician B

7. When discussing the repair estimate with a potential customer, all of the following are good things to remember *except*:

   A. A good joke will help.

   B. Try not to use language that is overly technical.

   C. Make sure you are presentable before you meet the customer.

   D. Use good manners; be professional.

8. All the following are good things to show at a job interview *except*:

   A. A good handshake when you greet the employer.

   B. Well groomed personal appearance.

   C. A well written resume.

   D. The latest fashions that are being worn at school.

9. Notifying a customer if a mistake has been made is a sign of _____.

10. Describe some of the ways a technician who takes pride in their work shows it.

11. Why could a DWI conviction result in immediate loss of employment?

CHAPTER

3 General Shop Safety

Learning Objectives

After you have read, studied, and practiced the contents of this unit, you should be able to:

- Understand the reasons for developing a personal shop safety plan.
- List potential shop safety hazards.
- Know basic shop safety rules.
- List steps to maintain air quality in the shop.
- Know where to find MSDS sheets and why they are important.

Key Terms

EPA

Hazardous waste materials

Liable

MSDS

OSHA

RCRA

Safe working habits

INTRODUCTION

Safety is extremely important in the automotive shop. The knowledge and practice of safety precautions prevent serious personal injury and expensive property damage. Automotive students and technicians must be familiar with shop hazards and shop safety rules. The first step in providing a safe shop is learning about shop hazards and safety rules. The second, and most important, step in this process is applying your knowledge of shop hazards and safety rules while working in the shop. In other words, you must actually develop **safe working habits** in the shop from your understanding of shop hazards and safety rules. When shop employees have a careless attitude toward safety, accidents are more likely to occur. All shop personnel must develop a serious attitude toward safety. The result of this serious attitude is that shop personnel will learn and adopt all shop safety rules.

Shop personnel must be familiar with their responsibilities and rights regarding hazardous waste disposal. These rights are explained in the right-to-know laws. Shop personnel must also be familiar with the types of hazardous materials in the automotive shop and the proper disposal methods for these materials according to state and federal regulations. Not only will proper disposal methods protect the technician. Failure to comply with the law can make the technician legally **liable** for damages.

OCCUPATIONAL SAFETY AND HEALTH ACT AND THE ENVIRONMENTAL PROTECTION AGENCY

The Occupational Safety and Health Act (**OSHA**) was passed by the United States government in 1970. The purposes of this legislation are:

1. To assist and encourage the citizens of the United States in their efforts to assure safe and healthful working conditions by providing research, information, education, and training in the field of occupational safety and health.

2. To assure safe and healthful working conditions for working men and women by authorizing enforcement of the standards developed under the Act. Since approximately 25 percent of workers are exposed to health and safety hazards on the job, the OSHA is necessary to monitor, control, and educate workers regarding health and safety in the workplace.

The Environmental Protection Agency (**EPA**) is the federal agency responsible for air and water quality in the United States. The EPA was established to compile and enforce regulations that apply to waste generated by service or manufacturing businesses. These regulations include waste disposal and landfills for household and industrial waste. The EPA is also responsible for vehicle emission standards and the enforcement of these standards as they apply to vehicle manufacturers and owners. In many states, vehicle owners must have a compulsory vehicle emission test at specific intervals.

SHOP HAZARDS

Service technicians and students encounter many hazards in an automotive shop. When these hazards are known, basic shop safety rules and procedures must be followed to avoid personal injury. Some of the hazards in an automotive shop include the following:

1. Flammable liquids such as gasoline and paint must be handled and stored properly in approved, closed containers.

2. Flammable materials such as oily rags must be stored properly in closed containers to avoid a fire hazard.

3. Batteries contain a corrosive sulfuric acid solution and produce explosive hydrogen gas while charging.

SAFETY TIP *The sulfuric acid solution in batteries is harmful to most types of clothing. This solution is a skin irritant and causes severe chemical burns if it contacts human eyes.*

4. Loose sewer and drain covers may cause foot or toe injuries.

5. Caustic liquids, such as those in hot cleaning tanks, are harmful to skin and eyes.

6. High-pressure air in the shop's compressed-air system can be very dangerous or fatal if it penetrates the skin and enters the bloodstream. High-pressure air released near the eyes causes severe eye injury.

7. Frayed cords on electrical equipment and lights may result in severe electrical shock.

8. **Hazardous waste material**, such as batteries, and caustic cleaning solutions must be handled with the adequate personal protection **(Figure 3-1)**.

9. Carbon monoxide from vehicle exhaust is poisonous.

Figure 3-1 Recommended safety clothing is required by law when handling hazardous materials.

Figure 3-2 Types of hearing protectors.

Figure 3-3 Shop safety equipment consists of safety goggles, respirator, welding shield, proper work clothes, hearing protection, welding gloves, work gloves, and safety shoes.

10. Loose clothing or long hair may become entangled in rotating parts on equipment or vehicles, resulting in serious injury.

11. Dust and vapors generated during some repair jobs are harmful. Asbestos dust, which may be released during brake lining service and clutch service, is a contributor to lung cancer.

12. High noise levels from shop equipment such as an air chisel may be harmful to the ears and adequate hearing protection must be worn **(Figure 3-2)**.

13. Oil, grease, water, or parts cleaning solutions on shop floors may cause someone to slip and fall, resulting in serious injury.

14. The incandescent bulbs used in some trouble lights may shatter when the light is dropped. This action may ignite flammable materials in the area and cause a fire. Most insurance companies now require the use of trouble lights with fluorescent or LED bulbs in the shop.

SHOP SAFETY RULES

The application of some basic shop rules helps prevent serious, expensive accidents. Failure to comply with shop rules may cause personal injury or expensive damage to vehicles and shop facilities. It is the responsibility of the employer and all shop employees to make sure that shop rules are understood and followed until these rules become automatic habits. The following basic shop rules should be followed:

1. Always wear safety glasses and other protective equipment that is required by a service procedure **(Figure 3-3)**. For example, a brake parts washer must be used to avoid breathing asbestos dust into the lungs. Asbestos dust is a known cause of lung cancer. Asbestos dust is encountered in manual transmission clutch facings and brake linings.

2. Tie long hair securely behind your head, and do not wear loose or torn clothing.

3. Do not wear rings, watches, or loose hanging jewelry. If jewelry such as a ring, metal watchband, or chain makes contact between an electrical terminal and ground, the jewelry becomes extremely hot, resulting in severe burns.

4. Do not work in the shop while under the influence of alcohol or drugs.

SAFETY TIP *Before starting a vehicle put your foot on the brake to stop the vehicle in case of a malfunction and the vehicle is started in gear. Never reach through a window to start a vehicle.*

5. Set the parking brake when working on a vehicle. If the vehicle has an automatic transmission, place the gear selector in park unless a service procedure requires another selector position. When the vehicle is equipped with a manual transmission, position the gear selector in neutral with the engine running, or in reverse with the engine stopped.

6. Always connect a shop exhaust hose to the vehicle tailpipe, and be sure the shop exhaust fan is running. If it is absolutely necessary to operate a vehicle without a shop exhaust pipe connected to the tailpipe, open the large shop door to provide adequate ventilation.

SAFETY TIP *Carbon monoxide in the vehicle exhaust may cause severe headaches and other medical problems. High concentrations of carbon monoxide may result in death!*

7. Keep hands, clothing, and wrenches away from rotating parts such as cooling fans. Remember that electric drive fans may start turning at any time, even with the ignition off.

8. Always leave the ignition switch off unless a service procedure requires another switch position.

9. Do not smoke in the shop. If the shop has designated smoking areas, smoke only in these areas.

10. Store oily rags and other discarded combustibles in covered metal containers designed for this purpose.

11. Always use the wrench or socket that fits properly on the bolt. Do not substitute metric for United States Customary (USC) wrenches, or vice versa.

12. Keep tools in good condition. For example, do not use a punch or chisel with a mushroomed end. When struck with a hammer, a piece of the mushroomed metal could break off, resulting in severe eye or other injury.

13. Do not leave power tools running and unattended.

14. Avoid contact with hot metal components, such as exhaust manifolds, other exhaust system components, radiators, and some air conditioning lines and hoses. Serious and painful burns can be avoided.

15. When a lubricant such as engine oil is drained, always wear heavy plastic gloves because the oil could be hot enough to cause burns.

16. Prior to getting under a vehicle, check to be sure the vehicle is placed securely on safety stands.

17. Operate all shop equipment, including lifts, according to the equipment manufacturer's recommended procedure. Do not operate equipment unless you are familiar with the correct operating procedure.

18. Do not run or engage in horseplay in the shop.

19. Obey all state and federal fire, safety, and environmental regulations.

20. Do not stand in front of or behind vehicles.

21. Always place fender, seat, and floor mat covers on a customer's vehicle before working on the car.

22. When one end of a vehicle is raised, place wheel chocks on both sides of the wheels remaining on the floor.

23. All shop employees must be familiar with the location of shop safety equipment.

24. Collect oil, fuel, brake fluid, and other liquids in the proper safety containers.

25. Use only approved cleaning fluids and equipment. Do not use gasoline to clean parts.

26. Be sure safety shields are in place on all rotating equipment.

27. All shop equipment must have regular scheduled maintenance and adjustment.

28. Some shops have safety lines around equipment. Always work within these lines when operating equipment.

29. Be sure the shop heating equipment is properly ventilated.

30. Post emergency numbers near the phone. These numbers should include a doctor, ambulance, poison control center, fire department, hospital, and police.

31. Do not leave hydraulic jack handles where someone may trip over them.

32. Keep work areas free of debris.

33. Inform the shop foreman of any safety dangers and suggestions for safety improvement.

34. Do not direct high-pressure air from an air gun against human skin or near the eyes.

SAFETY TIP *High-pressure air may penetrate the skin and enter the blood stream. Air in the blood stream may be fatal! High-pressure air discharged near the eyes may cause serious eye damage.*

35. All shop employees must wear proper footwear. Heavy-duty work boots or shoes with steel toes are the best footwear in an automotive shop.

SMOKING, ALCOHOL, AND DRUGS IN THE SHOP

Do not smoke when working in the shop. If the shop has designated smoking areas, smoke only in these areas. Do not smoke in customers' cars. A nonsmoker will not appreciate cigarette odor in the car. A spark from a cigarette or lighter may ignite flammable materials in the workplace. The use of drugs or alcohol must be avoided while working in the shop. Even a small amount of drugs or alcohol affects reaction time. This includes over-the-counter medications as well as prescription medications. When taking medications observe all warnings and cautions in the directions. Given the nature of the job where customer vehicles must be test driven to verify repairs, the consequences of driving impaired are very serious, most states treat any infractions the same as if the driver was impaired by alcohol. In an emergency situation, slow reaction time may cause personal injury. If a heavy object falls off the workbench and your reaction time is slowed by drugs or alcohol, you may not be able to move your foot out of the way in time to avoid a foot injury. When a fire starts in the workplace and you are a few seconds slower operating a fire extinguisher because of alcohol or drug use, it could make the difference between extinguishing a fire and having expensive fire damage.

AIR QUALITY

Vehicle exhaust contains small amounts of carbon monoxide, a colorless, odorless, poisonous gas. Weak concentrations of carbon monoxide in the shop air may cause drowsiness, nausea, and headaches. Strong concentrations of carbon monoxide can be fatal. All shop personnel are responsible for air quality in the shop. Shop management is responsible for an adequate exhaust system to remove exhaust fumes from the maximum number of vehicles that may be running in the shop at the same time. Technicians should never run a vehicle in the shop unless a shop exhaust hose is installed on the tailpipe of the vehicle. The exhaust fan must be switched on to remove exhaust fumes. If shop heaters or furnaces have restricted chimneys, they release carbon monoxide emissions into the shop air. Therefore, chimneys should be checked periodically for restriction and proper ventilation.

Monitors are available to measure the level of carbon monoxide in the shop. Some of these monitors read the amount of carbon monoxide present in the shop air; others provide an audible alarm if the concentration of carbon monoxide exceeds the danger level. Diesel exhaust contains some carbon monoxide, but particulate emissions are also present in the exhaust from these engines. Particulates are basically small carbon particles that can be harmful to the lungs.

The sulfuric acid solution in vehicle batteries is a corrosive, poisonous liquid. If a battery is charged with a fast charger at a high rate for a period of time, the battery becomes hot, and the sulfuric acid solution begins to boil. Under this condition, the battery may emit a strong sulfuric acid smell, and these fumes may be harmful to the lungs. When a battery is boiling, hydrogen gas is also emitted, presenting an extreme fire hazard; batteries have been known to explode under these conditions if a spark or flame ignites the gas. If this happens, the battery charger should be turned off or the charging rate should be reduced considerably. Some automotive clutch facings and brake linings contain asbestos. Never use compressed air to blow dirt from these components since this action disperses asbestos dust into the shop air where it may be inhaled by technicians and other people in the shop. A brake parts washer or a vacuum cleaner with a high efficiency particulate air (HEPA) filter must be used to clean the dust from these components.

Even though technicians take every precaution to maintain air quality in the shop, some undesirable gases may still reach the air. For example, exhaust manifolds may get oil on them during an engine overhaul. When the engine is started and these manifolds become hot, the oil burns off the manifolds and pollutes the shop air with oil smoke. Adequate shop ventilation must be provided to take care of this type of air contamination. Technicians may need to perform work that will generate dust, and while having a good shop ventilation system is a must, the technician will often need to wear protective equipment due to the proximity of the contaminates.

SHOP SAFETY EQUIPMENT

Technicians must understand shop safety equipment and the use of this equipment. Technicians must also know the location of safety equipment. When the technician understands safety equipment and knows the location of this equipment, accidents such as fires may be extinguished quickly. If the technician is not familiar with the use of safety equipment or does not know its location in the shop, he or she may require a longer time to put the fire extinguisher into operation. This delay may allow a fire to get out of control and become an expensive disaster.

Fire Prevention and Fire Extinguishers

An automotive shop is a dangerous place for a fire because a number of flammable liquids such as gasoline, engine oil, and transmission fluid are usually stored in the shop. The shop also contains a number of vehicles with gasoline or diesel fuel in the fuel tanks. Technicians must always practice fire prevention! For example, never turn on the ignition switch or crank the engine in a vehicle with the fuel line disconnected. If this action is taken, fuel will be discharged from the disconnected fuel line, and a spark from the ignition system may ignite this fuel. Oily rags must be stored in closed

containers. Oily rags may generate enough heat to self-ignite and start a fire. This process is called spontaneous combustion. When oily rags are stored in closed containers, they cannot receive enough oxygen to support a fire. Fire extinguishers are one of the most important pieces of safety equipment. All shop personnel must know the location of the fire extinguishers in the shop. If you have to waste time looking for an extinguisher after a fire starts, the fire could get out of control before you put the extinguisher into operation. Fire extinguishers should be located where they are easily accessible at all times. A decal on each fire extinguisher identifies the type of chemical in the extinguisher and provides operating information **(Figure 3-4)**. Shop personnel should be familiar with the following types of fires and fire extinguishers:

1. Class A fires are those involving ordinary combustible materials such as paper, wood, clothing, and textiles. Water, foam, and multipurpose dry chemical fire extinguishers are used on these types of fires.

2. Class B fires involve the burning of flammable liquids such as gasoline, oil, paint, solvents, and greases. These fires may also be extinguished with multipurpose dry chemical fire extinguishers. In addition, fire extinguishers containing halogen, or halon, may be used to extinguish class B fires. The chemicals in this type of extinguisher attach to the hydrogen,

Figure 3-4 Types of fire extinguishers.

hydroxide, and oxygen molecules to stop the combustion process almost instantly. However, the resultant gases from the use of halogen-type extinguishers are toxic and harmful to the operator of the extinguisher. A water-type fire extinguisher will cause the fire to spread even more.

3. Class C fires involve the burning of electrical equipment such as wires, motors, and switches. These fires are extinguished with multipurpose dry chemical fire extinguishers. Water or foam fire extinguishers will conduct electricity and cause electrical shock.

4. Class D fires involve the combustion of metal chips, turnings, and shavings. Special dry chemical fire extinguishers are the only type of extinguisher recommended for these fires.

Additional information regarding types of extinguishers for various types of fires is provided in **Figure 3-5**. Some multipurpose dry chemical fire extinguishers may be used on class A, class B, or class C fires.

SAFETY TIP *Do not look at the arc from an arc welder when someone else is welding, and always wear proper eye protection when operating an arc welder. Retina burns caused by unprotected welding are permanent.*

Causes of Eye Injuries

Eye injuries may occur in various ways in the automotive shop. Some of the more common eye accidents are:

	Class of fire	Typical fuel involved	Type of extinguisher
Class **A** Fires (green)	For ordinary combustibles Put out a class A fire by blowing its temperature or by coating the burning combustibles.	Wood Paper Cloth Rubber Plastics Rubbish Upholstery	Water* Foam* Multipurpose dry chemical
Class **B** Fires (red)	For flammable liquids Put out a class B fire by smothering it. Use an extinguisher that gives a blanketing, flame-interrupting effect; cover whole flaming liquid surface.	Gasoline Oil Grease Paint Lighter fluid	Foam* Carbon dioxide Halogenated agent Standard dry chemical Purple K dry chemical Multipurpose dry chemical
Class **C** Fires (blue)	For electrical equipment Put out a class C fire by shutting off power as quickly as possible and by always using a nonconducting extinguisher agent to prevent electric shock.	Motors Appliances Wiring Fuse boxes Switchboards	Carbon dioxide Halogenated agent Standard dry chemical Purple K dry chemical Multipurpose dry chemical
Class **D** Fires (yellow)	For combustible metals Put out a class D fire of metal chips, turnings, or shavings by smothering or coating with a specially designed extinguisher agent.	Aluminum Magnesium Potassium Sodium Titanium Zirconium	Dry powder extinguisher and agents only

*Cartridge-operated water, foam, and soda-acid types of extinguishers are no longer manufactured. These extinguishers should be removed from service when they become due for their next hydrostatic pressure test.

Figure 3-5 Fire extinguisher selection guide.

1. Thermal burns from excessive heat
2. Irradiation burns from excessive light, such as from an arc welder
3. Chemical burns from strong liquids or vapors such as battery electrolyte
4. Foreign material in the eye
5. Penetration of the eye by a sharp object
6. A blow from a blunt object

Wearing safety glasses with sideshields and observing shop safety rules will prevent most eye accidents.

Eyewash Fountains

If a chemical gets in your eyes, it must be washed out immediately to prevent a chemical burn. An eyewash fountain is the most effective way to wash the eyes. An eyewash fountain is similar to a drinking water fountain, but the eyewash fountain has water jets placed throughout the fountain top. Every shop should be equipped with some eyewash facility **(Figure 3-6)**. Be sure you know the location, and know how to use the eyewash fountain in the shop.

Safety Glasses and Face Shields

Wearing safety glasses or a face shield is one of the most important safety rules in an automotive shop. Face shields protect the face; safety glasses protect the eyes. When grinding, safety glasses must be worn, and a face shield can be worn. Shop insurance policies require the use of eye protection in the shop. Some automotive technicians have been blinded in one or both eyes because they did not bother to wear safety glasses. All safety glasses must be equipped with safety glass or plastic lenses, and they should provide some type of side protection **(Figure 3-7)**.

When selecting a pair of safety glasses, they should feel comfortable on your face. If they are uncomfortable, you may tend to take them off, leaving your eyes unprotected. A face shield should be worn when handling hazardous chemicals or when using an air or electric grinder or buffer **(Figure 3-8)**. For technicians who wear eyeglasses, special lenses and frames can be purchased that are impact resistant.

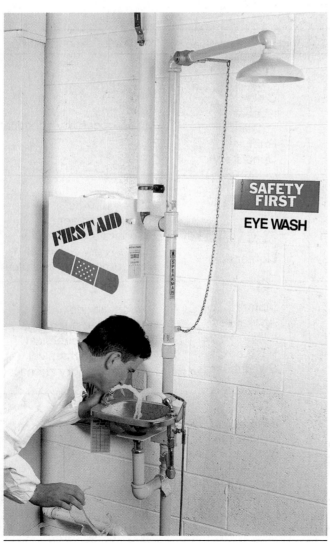

Figure 3-6 An eyewash fountain.

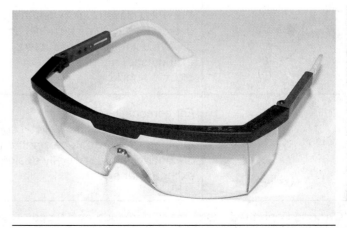

Figure 3-7 Safety glasses with side shields must be worn in the shop.

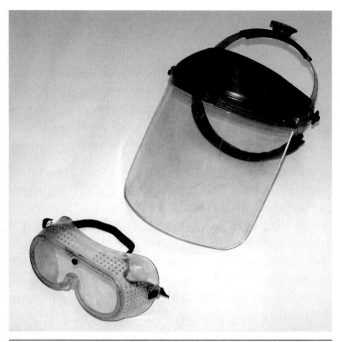

Figure 3-8 **Safety goggles and a face shield.**

First-Aid Kits

First-aid kits should be clearly identified and conveniently located **(Figure 3-9)**. These kits contain items such as bandages and ointment required for minor cuts. All shop personnel must be familiar with the location of first-aid kits. At least one of the shop personnel should have basic first-aid training. This person should be in charge of administering first aid and keeping first-aid kits filled.

SHOP LAYOUT

There are many different types of shops in the automotive service industry, including new car dealers, independent repair shops, specialty shops, service stations, and fleet shops.

The shop layout in any shop is important to maintain shop efficiency and contribute to safety. Every shop employee must be familiar with the location of all safety equipment in the shop. Shop layout includes bays for various types of repairs, space for equipment storage, and office locations. Most shops have specific bays for certain types of work, such as electrical repair, wheel alignment and tires, and machining **(Figure 3-10)**. Safety equipment such as fire extinguishers, first-aid kits, and eyewash fountains must be in easily accessible locations, and the location of each piece of safety equipment must be clearly marked. Areas such as the parts department and the parts cleaning area must be located so they are easily accessible from all areas of the shop. The service manager's office should also be centrally located. All shop personnel should familiarize themselves with the shop layout, especially the location of safety equipment. If you know the exact fire extinguisher locations, you may put an extinguisher into operation a few seconds faster. Those few seconds could make the difference between a fire that is quickly extinguished and one that becomes out of control, causing extensive damage and personal injury.

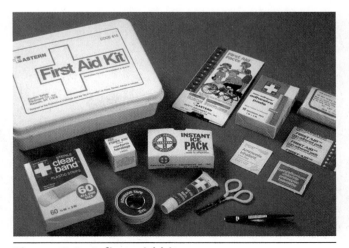

Figure 3-9 A first-aid kit.

Figure 3-10 An efficient shop layout.

The tools and equipment required for a certain type of work are stored in that specific bay. For example, the equipment for electrical and electronic service work is stored in the bay allotted to that type of repair. When certain bays are allotted to specific types of repair work, unnecessary equipment movement is eliminated. Each technician has his or her own tools on a portable roll cabinet that can be moved to the vehicle being repaired. Special tools are provided by the shop, and these tools may be located on tool boards attached to the wall. Other shops may have a tool room where special tools are located. Adequate workbench space must be provided in those bays where bench work is required.

HAZARDOUS WASTE DISPOSAL

Hazardous waste materials in automotive shops are chemicals or components that the shop no longer needs. These materials pose a danger to the environment and to people if they are disposed of in ordinary trashcans or sewers. However, it should be noted that no material is considered hazardous waste until the shop has finished using it and is ready to dispose of it. The EPA publishes a list of hazardous materials, which is included in the Code of Federal Regulations. Waste is considered hazardous if it is included on the EPA list of hazardous materials, or if it has one or more of these characteristics:

1. *Reactive*—Any material that reacts violently with water or other chemicals is considered hazardous. If a material releases cyanide gas, hydrogen sulphide gas, or similar gases when exposed to low-pH acid solutions, it is hazardous. A material that is reactive reacts with some other chemicals and gives off gas(es) during the reaction.

2. *Corrosive*—If a material burns the skin or dissolves metals and other materials, it is considered hazardous. A material that is corrosive causes another material to be gradually worn away by chemical action.

3. *Toxic*—Materials are hazardous if they leach one or more of eight heavy metals in concentrations greater than 100 times primary drinking water standard. A toxic substance is poisonous to animal or human life.

4. *Flammable*—A liquid is hazardous if it has a flash point below 140°F (60°C). A solid is hazardous if it ignites spontaneously. A substance that is flammable can be ignited spontaneously or by another source of heat or flame.

Federal and state laws control the disposal of hazardous waste materials. Every shop employee must be familiar with these laws. Hazardous waste disposal laws include the Resource Conservation and Recovery Act (**RCRA**). This law basically states that hazardous material users are responsible for hazardous materials from the time they become a waste until the proper waste disposal is completed.

Many automotive shops hire an independent hazardous waste hauler to dispose of hazardous waste material **(Figure 3-11)**. The shop owner or manager should have a written contract with the hazardous waste hauler. Rather than have hazardous waste material hauled to an approved hazardous waste disposal site, a shop may choose to recycle the material in the shop. An example of this would be a shop that has a machine to recycle used antifreeze, or a shop that is heated by using drained engine oil. Therefore, the user must store hazardous waste material properly and safely and be responsible for the transportation of this material until it arrives at an approved hazardous waste disposal site and is processed according to the law.

Figure 3-11 A hazardous waste hauler.

The RCRA controls these types of automotive waste:

1. Paint and body repair products waste
2. Solvents for parts and equipment cleaning
3. Batteries and battery acid
4. Mild acids used for metal cleaning and preparation
5. Waste oil, engine coolants, or antifreeze
6. Air-conditioning refrigerants
7. Engine oil filters

Never, under any circumstances, use these methods to dispose of hazardous waste material:

1. Pour hazardous wastes on weeds to kill them.
2. Pour hazardous wastes on gravel streets to prevent dust.
3. Throw hazardous wastes in a dumpster.
4. Dispose of hazardous wastes anywhere but an approved disposal site.
5. Pour hazardous wastes down sewers, toilets, sinks, or floor drains.

The right-to-know laws state that employees have a right to know when the materials they use at work are hazardous. The right-to-know laws started with the Hazard Communication Standard published by OSHA in 1983. This document was originally intended for chemical companies and manufacturers that required employees to handle hazardous materials in their work situation.

At the present time, most states have established their own right-to-know laws. Meanwhile, the federal courts have decided to apply these laws to all companies, including automotive service shops. Under the right-to-know laws, the employer has three responsibilities regarding the handling of hazardous materials by its employees. First, all employees must be trained about the types of hazardous materials they will encounter in the workplace.

Employees must be informed about their rights under legislation regarding the handling of hazardous materials. All hazardous materials must be properly labeled, and information about each hazardous material must be posted on material safety data sheets (**MSDS**), which are available from the manufacturer **(Figure 3-12)**. The employer has a responsibility to place MSDS where they are easily accessible by all employees. The MSDS provide extensive information about the hazardous material, such as:

1. Chemical name
2. Physical characteristics
3. Protective equipment required for handling
4. Explosion and fire hazards
5. Other incompatible materials
6. Health hazards such as signs and symptoms of exposure, medical conditions aggravated by exposure, and emergency and first-aid procedures
7. Safe handling precautions
8. Spill and leak procedures

Second, the employer has a responsibility to make sure that all hazardous materials are properly labeled. The label information must include health, fire, and reactivity hazards posed by the material, as well as the protective equipment necessary to handle the material. The manufacturer must supply all warning and precautionary information about hazardous materials, and this information must be read and understood by the employee before handling the material.

Third, employers are responsible for maintaining permanent files regarding hazardous materials. These files must include information on hazardous materials in the shop, proof of employee training programs, and information about accidents such as spills or leaks of hazardous materials. The employer's files must also include proof that employees' requests for hazardous material information such as MSDS have been met. A general right-to-know compliance procedure manual must be maintained by the employer.

MATERIAL SAFETY DATA SHEET

PRODUCT NAME: KLEAN-A-KARB (aerosol) #- HPMS 102068
PRODUCT: 5078, 5931, 6047T
(page 1 of 2)

1. Ingredients	CAS #	ACGIH TLV	OSHA PEL	OTHER LIMITS	%
Acetone	67-64-1	750ppm	750ppm		2-5
Xylene	1330-20-7	100ppm	100ppm		68-75
2-Butoxy Ethanol	111-76-2	25ppm	25ppm	(skin)	3-5
Methanol	67-56-1	200ppm	200ppm		3-5
Detergent	-	NA	NA		0-1
Propane	74-98-6	NA	1000ppm		10-20
Isobutane	75-28-5	NA	NA	1000ppm	10-20

2. PHYSICAL DATA : (without propellent)
Specific Gravity   : 0.865   Vapor Pressure   : ND
                                     % Volatile   : >99

Boiling Point   : 176°F Initial Evaporation Rate : Moderately Fast
Freezing Point   : ND   Vapor Density   : ND
Solubility: Partially soluble in water   pH   :NA
Appearance and Odor: A clear colorless liquid, aromatic odor

3. FIRE AND EXPLOSION DATA
Flashpoint   : −40°F   Method   : TCC
Flammable Limits propellent   LEL: 1.8   UEL: 9.5
Extinguishing Media: CO_2, dry chemical, foam
Unusual Hazards   : Aerosol cans may explode when heated above 120°F.

4. REACTIVITY AND STABILITY
Stability   : Stable
Hazardous decomposition products: CO_2, carbon monoxide (thermal)
Materials to avoid: Strong oxidizing agents and sources of ignition

5. PROTECTION INFORMATION
Ventilation : Use mechanical means to insure vapor concentration
              is below TLV.
Respiratory: Use self-contained breathing apparatus above TLV.

Gloves   : Solvent resistant   Eye and Face: Safety Glasses
Other Protective Equipment: Not normally required for aerosol product usage

Figure 3-12 Material safety data sheets (MSDS) inform employees about hazardous materials.

Summary

- The United States Occupational Safety and Health Act of 1970 assures safe and healthful working conditions and authorizes enforcement of safety standards.

- Many hazardous materials and conditions can exist in an automotive shop, including flammable liquids and materials, corrosive acid solutions, loose sewer covers, caustic liquids, high-pressure air, frayed electrical cords, hazardous waste materials, carbon monoxide, improper clothing, harmful vapors, high noise levels, and spills on shop floors.

- MSDS provide information regarding hazardous materials, labeling, and handling.

- The danger regarding hazardous conditions and materials may be avoided by eliminating shop hazards and applying the necessary shop rules and safety precautions.

- The automotive shop owner/management must supply the necessary shop safety equipment, and all shop personnel must be familiar with the location and operation of this equipment. Shop safety equipment includes gasoline safety cans, steel storage cabinets, combustible material containers, fire extinguishers, eyewash fountains, safety glasses and face shields, first-aid kits, and hazardous waste disposal containers.

Review Questions

1. While discussing shop hazards, Technician A says high-pressure air from an air gun may penetrate the skin. Technician B says air in the blood stream may be fatal. Who is correct?

   A. Technician A

   B. Technician B

   C. Both Technician A and Technician B

   D. Neither Technician A nor Technician B

2. While discussing hazardous waste disposal, Technician A says the right-to-know laws require employers to train employees regarding hazardous waste materials. Technician B says the right-to-know laws do not require employers to keep permanent records regarding hazardous waste disposal. Who is correct?

   A. Technician A

   B. Technician B

   C. Both Technician A and Technician B

   D. Neither Technician A nor Technician B

3. While discussing material safety data sheets (MSDS), Technician A says these sheets explain employers' and employees' responsibilities regarding hazardous waste handling and disposal. Technician B says these sheets contain specific information about hazardous materials. Who is correct?

   A. Technician A

   B. Technician B

   C. Both Technician A and Technician B

   D. Neither Technician A nor Technician B

4. While discussing hazardous material disposal, Technician A says certain types of hazardous waste material may be poured down a floor drain. Technician B says a shop is responsible for hazardous waste materials from the time they become waste until the proper waste disposal is completed. Who is correct?

   A. Technician A

   B. Technician B

   C. Both Technician A and Technician B

   D. Neither Technician A nor Technician B

5. According to health and safety inspection records, the percentage of workers who are exposed to health and safety hazards on the job is:

   A. 25 percent.

   B. 40 percent.

   C. 55 percent.

   D. 62 percent.

6. All of these statements about the Environmental Protection Agency (EPA) are true *except*:

   A. The EPA is responsible for air and water quality in the United States.

   B. The EPA was established to compile hazardous waste regulations.

   C. The EPA was established to compile vehicle manufacturing standards.

   D. The EPA was established to enforce hazardous waste regulations.

7. Technician A says loose clothing is dangerous because it may become entangled in rotating

components on vehicles or shop equipment. Technician B says asbestos dust may be generated in the shop air during brake service. Who is correct?

A. Technician A

B. Technician B

C. Both Technician A and Technician B

D. Neither Technician A nor Technician B

8. All of these statements about the danger of wearing finger rings or metal watch bands in an automotive shop are true *except*:

A. A ring or metal watchband may cause a chemical burn on the finger or arm.

B. A ring or metal watchband can make electrical contact between an electrical terminal and ground.

C. A ring or metal watchband may become very hot if high electric current flows through one of these components.

D. A ring or metal watchband may cause severe burns to a finger or arm.

9. Technician A says when a vehicle is parked in the shop, the ignition switch should be left on the OFF position unless a service procedure requires another switch position. Technician B says oily rags should be stored in containers without lids. Who is correct?

A. Technician A

B. Technician B

C. Both Technician A and Technician B

D. Neither Technician A nor Technician B

10. All of these statements about carbon monoxide in the shop air are true *except*:

A. Carbon monoxide causes headaches.

B. Carbon monoxide may enter the shop air from vehicle exhaust.

C. A restricted furnace chimney may cause carbon monoxide in the shop air.

D. Carbon monoxide contributes to lung cancer.

11. The poisonous gas in vehicle exhaust is

_____.

12. Never direct high-pressure shop air from an air gun against human _____ or

_____.

13. Wearing rings or other jewelry may cause severe _____.

14. Breathing asbestos dust may cause

_____ _____.

15. List four classes of fires and explain the type of fire extinguisher that should be used for each type of fire.

16. Explain two requirements related to location of safety equipment in the shop.

17. List five illegal methods of hazardous waste disposal.

18. List eight types of information found in MSDS related to hazardous materials.

4 Tool and Equipment Safety

Learning Objectives

After you have read, studied, and practiced the contents of this unit, you should be able to:

- Employ basic electrical safety in the shop.
- Understand the importance of proper handling of gasoline and other potentially hazardous materials.
- Understand fire prevention and the safe use of fire extinguishers.
- Develop a safety conscious attitude in the shop.
- Safely use tools, lifts, and jacks.

Key Terms

ASDM

Frame contact lift

Lifting points

Pneumatic tools

Spontaneous combustion

Torque

INTRODUCTION

Each person in an automotive shop must follow proper, safe procedures when handling flammable liquids and operating shop equipment. When all personnel in the automotive shop follow these procedures, personal injury, vehicle damage, and property damage may be prevented. All shop personnel must also be familiar with different types of fires that may occur in an automotive shop, and they must also understand the proper type of fire extinguisher to use on various fires. It is essential that shop personnel know how to operate fire extinguishers and to properly extinguish fires. Shop personnel must understand safe shop housekeeping procedures and safe vehicle operation in the shop.

ELECTRICAL SAFETY

In the automotive shop you will be using electric drills, shop lights, wheel balancers, wheel aligners, and battery chargers. Electrical safety precautions must be observed on this equipment, for example:

- Frayed cords on electrical equipment must be replaced or repaired immediately.

- All electrical cords from lights and electrical equipment must have a ground connection. The ground connector is the round terminal in a three-prong electrical plug. Do not use a two-prong adaptor to plug in a three-prong electrical cord. Three-prong electrical outlets are mandatory in all shops.

- Do not leave electrical equipment running and unattended.

GASOLINE SAFETY

Gasoline is a very explosive liquid! One exploding gallon of gasoline has a force equal to fourteen sticks of dynamite. It is the expanding vapors from gasoline that are extremely dangerous. These vapors are present even in cold temperatures. Vapors formed in gasoline tanks on cars are controlled, but vapors from a gasoline storage can may escape from the can, resulting in a hazardous situation. Therefore, gasoline storage containers must be placed in a well-ventilated space.

Approved gasoline storage cans have a flash-arresting screen at the outlet **(Figure 4-1)**. This screen prevents external ignition sources from igniting the gasoline within the can while the gasoline is being poured. Follow these safety precautions regarding gasoline containers:

1. Always use approved gasoline containers that are painted red for proper identification.

Figure 4-1 An approved gasoline container.

2. Do not fill gasoline containers completely full. Always leave the level of gasoline at least one inch from the top of the container. This allows for expansion of the gasoline at higher temperatures. If gasoline containers are completely full, the gasoline will expand when the temperature increases. This expansion forces gasoline from the can and creates a dangerous spill.

3. If gasoline containers must be stored, place them in a well-ventilated area such as a storage shed. Do not store gasoline containers in your home or in the trunk of a vehicle.

4. When a gasoline container must be transported, be sure it is secured against upsets.

5. Do not store a partially filled gasoline container for long periods of time because it may give off vapors and produce a potential danger.

6. Never leave gasoline containers open except while filling or pouring gasoline from the container.

7. Do not prime an engine with gasoline while cranking the engine.

8. Never use gasoline as a cleaning agent.

FIRE SAFETY

When fire safety rules are observed, personal injury and expensive fire damage to vehicles and property may be avoided. Follow these safety rules:

- Familiarize yourself with the location and operation of all shop fire extinguishers.

- If a fire extinguisher is used, report it to management so the extinguisher can be recharged.

- Do not use any type of open flame heater to heat the work area.

- Do not turn the ignition switch on or crank the engine with a gasoline line disconnected.

- Store all combustible materials such as gasoline, paint, and oily rags in approved safety containers.

- Clean up gasoline, oil, or grease spills immediately.

- Always wear clean shop clothes. Do not wear oil-soaked clothes.

- Do not allow sparks and flames near batteries.

- Be sure that welding tanks are securely fastened in an upright position.

- Do not block doors, stairways, or exits.

- Do not smoke when working on vehicles.

- Do not smoke, create sparks, or use an open flame near flammable materials or liquids.

- Store combustible shop supplies (such as solvents) in a closed steel cabinet.

- Store gasoline in approved safety containers.

- If a gasoline tank is removed from a vehicle, do not drag the tank on the shop floor.

- Know the approved fire escape route from your classroom or shop to the outside of the building.

- If a fire occurs, do not open doors or windows. This action creates extra draft, which makes the fire worse.

- Do not put water on a gasoline fire because the water will make the fire worse.

- Call the fire department as soon as a fire begins, and then attempt to extinguish the fire.

- If possible, stand 6 to 10 feet from the fire and aim the fire extinguisher nozzle at the base of the fire with a sweeping action.

- If a fire produces a great deal of smoke in the room, remain close to the floor to obtain oxygen and avoid breathing smoke.

- If the fire is too hot or the smoke makes breathing difficult, get out of the building.

- Do not re-enter a burning building.

- Keep solvent containers covered except when pouring from one container to another. When flammable liquids are transferred from bulk storage, the bulk container should be grounded to a permanent shop fixture such as a metal pipe. During this transfer process, the bulk container should be grounded to the portable container **(Figure 4-2)**. These ground wires prevent the buildup of a static electric charge, which could cause a spark and a disastrous explosion. Always discard or clean empty solvent containers because fumes in these containers are a fire hazard.

- Familiarize yourself with different types of fires and fire extinguishers, and know the type of extinguisher to use on each fire.

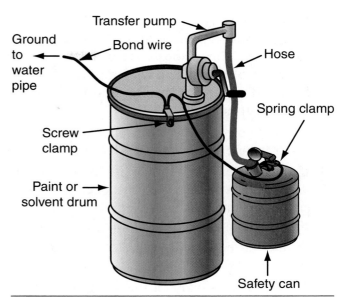

Figure 4-2 Safe procedures for flammable liquid transfer.

Using a Fire Extinguisher

Everyone working in the shop must know how to operate the fire extinguishers. There are several different types of fire extinguishers, but their operation usually involves the following steps:

1. Get as close as possible to the fire without jeopardizing your safety.

2. Grasp the extinguisher firmly and aim the extinguisher at the fire.

3. Pull the pin from the extinguisher handle.

4. Squeeze the handle to dispense the contents of the extinguisher.

5. Direct the fire extinguisher nozzle at the base of the fire, and dispense the contents of the extinguisher with a sweeping action back and forth across the fire. Most extinguishers discharge their contents in 8 to 25 seconds.

6. Always be sure the fire is extinguished.

7. Always keep an escape route open behind you so a quick exit is possible if the fire becomes out of control.

VEHICLE OPERATION

When driving a customer's vehicle, certain precautions must be observed to prevent accidents and maintain good customer relations:

1. Prior to starting and driving a vehicle, make sure the brakes are operational and fasten the safety belt. Pump the brake pedal to test its firmness, if a vehicle does not have adequate brakes it should be pushed into the shop to avoid an accident.

2. Check to be sure there is no person or object under the car before you start the engine.

3. If the vehicle is parked on a lift, be sure the lift is fully down and the lift arms, or components, are not in contact with the vehicle chassis.

4. Check to see if there are any objects directly in front of or behind the vehicle before driving away.

5. Always drive slowly in the shop, and watch carefully for personnel and other moving vehicles. Radios should be off and the windows rolled down.

6. Make sure the shop door is up high enough so there is plenty of clearance between the top of the vehicle and the door.

7. Watch the shop door to be certain that it is not coming down as you attempt to drive under the door.

8. If a road test is necessary, wear your seat belt, obey all traffic laws, and never drive in a reckless manner.

9. Do not squeal tires when accelerating or turning corners.

If the customer observes that service personnel take good care of his or her car by driving carefully and installing fender, seat, and floor mat covers, the service department image is greatly enhanced in the customer's eyes. These procedures impress upon the customer that shop personnel respect the car. Conversely, if grease spots are found on upholstery or fenders after service work is completed, the customer will probably think the shop is careless, not only in car care but also in service work quality.

HOUSEKEEPING SAFETY

Careful housekeeping habits prevent accidents and increase worker efficiency. Good housekeeping also helps impress upon the customer that

quality work is a priority in this shop. Follow these housekeeping rules:

- Keep aisles and walkways clear of tools, equipment, and other items.

- Be sure all sewer covers are securely in place.

- Keep floor surfaces free of oil, grease, water, and loose material.

- Sweep up under a vehicle before lowering the vehicle on the lift.

- Proper trash containers must be conveniently located, and these containers should be emptied regularly.

- Access to fire extinguishers must be unobstructed at all times, and fire extinguishers should be checked for proper charge at regular intervals.

- Tools must be kept clean and in good condition.

- When not in use, tools must be stored in their proper location.

- Oily rags must be stored in approved, covered containers **(Figure 4-3)**. A slow generation of heat occurs from oxidation of oil on these rags. Heat may continue to be generated until the ignition temperature is reached. The oil and the rags then begin to burn, causing a fire. This action is called **spontaneous combustion**. However, if the oily rags are in an airtight, approved container, the fire cannot receive enough oxygen to cause burning.

- Store paint, gasoline, and other flammable liquids in a closed steel cabinet **(Figure 4-4)**.

- Rotating components on equipment and machinery must have guards, and all shop

Figure 4-3 Store oily shop rags in an approved metal container.

Figure 4-4 An approved metal storage cabinet for combustible materials.

equipment should have regular service and adjustment schedules.

- Keep the workbenches clean. Do not leave heavy objects, such as used parts, on the bench after you are finished with them.

- Keep parts, tools, and materials in their proper location.

- When not in use, creepers must not be left on the shop floor. Creepers should be stored in a specific location.

- The shop should be well lighted, and all lights should be in working order.

- Frayed electrical cords on lights or equipment must be replaced.

- Walls and windows should be cleaned regularly.

- Stairs must be clean, well lighted, and free of loose material.

AIR BAG SAFETY

Technicians must be familiar with air bag safety rules. If air bag safety rules are not followed, expensive air bags may be accidentally deployed, and the technician and others may be injured. Follow these air bag safety rules:

1. When service is performed on any air bag system component, always disconnect the negative battery cable, isolate the cable end, and wait for the amount of time specified by the vehicle manufacturer before proceeding with the necessary diagnosis or service. The average waiting period is 2 minutes, but some vehicle manufacturers specify up to 10 minutes. Failure to observe this precaution may cause accidental air bag deployment and personal injury.

2. Replacement air bag system parts must have the same part number as the original part. Replacement parts of lesser or questionable quality must not be used. Improper or inferior components may result in inappropriate air bag deployment and injury to the vehicle occupants.

3. Do not strike or jar a sensor or an air bag system diagnostic monitor (**ASDM**). An ASDM is the computer that operates the air bag system. This may cause air bag deployment or make the sensor inoperative. Accidental air bag deployment may cause personal injury, and an inoperative sensor may result in air bag deployment failure, causing personal injury to vehicle occupants.

4. All sensors and mounting brackets must be properly torqued to ensure correct sensor operation before an air bag system is powered up. If sensor fasteners do not have the proper **torque**, improper air bag deployment may result in injury to vehicle occupants.

5. When working on the electrical system on an air bag-equipped vehicle, use only the vehicle manufacturer's recommended tools and service procedures. The use of improper tools or service procedures may cause accidental air bag deployment and personal injury. For example, do not use 12-volt or self-powered test lights when servicing the electrical system on an air bag-equipped vehicle.

LIFTING AND CARRYING

Many automotive service jobs require heavy lifting. Know your maximum weight lifting ability, and do not attempt to lift more than this weight. If a heavy part exceeds your weight lifting ability, have a co-worker help with the lifting job. Follow these steps when lifting or carrying an object:

1. If the object is going to be carried, be sure your path is free from loose parts or tools.

2. Position your feet close to the object; position your back reasonably straight for proper balance.

3. Your back and elbows should be kept as straight as possible. Continue to bend your knees until your hands reach the best lifting location on the object.

4. Be certain the container is in good condition. If a container falls apart during the lifting operation, parts may drop out of the container and result in foot injury or part damage.

5. Maintain a firm grip on the object; do not attempt to change your grip while lifting is in progress.

6. Straighten your legs to lift the object, and keep the object close to your body. Use leg muscles rather than back muscles **(Figure 4-5)**.

7. If you have to change direction of travel, turn your whole body. Do not twist.

8. Do not bend forward to place an object on a workbench or table. Position the object on the front surface of the workbench and slide it back. Do not pinch your fingers under the object while setting it on the front of the bench.

Figure 4-5 Use your legs muscles and keep your back straight when lifting heavy objects.

9. If the object must be placed on the floor or a low surface, bend your legs to lower the object. Do not bend your back forward because this movement strains back muscles.

10. When a heavy object must be placed on the floor, place suitable blocks under the object to prevent jamming your fingers.

SAFETY TIP *Improper lifting procedures may cause severe back injury.*

HAND TOOL SAFETY

Many shop accidents are caused by improper use and care of hand tools. Follow these safety steps when working with hand tools:

1. Maintain tools in good condition and keep them clean. Worn tools may slip and result in hand injury. If a hammer with a loose head is used, the head may fly off and cause personal injury or vehicle damage. If your hand slips off a greasy tool, it may cause some part of your body to hit the vehicle, causing injury.

2. Using the wrong tool for the job may damage the tool, fastener, or your hand if the tool slips. If you use a screwdriver as a chisel or pry bar, the blade may shatter causing serious personal injury.

3. Use sharp, pointed tools with caution. Always check your pockets before sitting on the vehicle seat. A screwdriver, punch, or chisel in the back pocket may put an expensive tear in the upholstery. Do not lean over fenders with sharp tools in your pockets.

4. Tools that are intended to be sharp should be kept sharp. A sharp chisel, for example, will do the job faster with less effort.

VEHICLE LIFT (HOIST) SAFETY

SAFETY TIP *When a vehicle is raised on a lift, the vehicle must be raised high enough to allow engagement of the lift locking mechanism. If the locking mechanism is not engaged, the lift may drop suddenly resulting in personal injury and/or vehicle damage.*

Special precautions and procedures must be followed when a vehicle is raised on a lift **(Figure 4-6)**. Follow these steps when operating a lift:

1. Always be sure the lift is completely lowered before driving a vehicle on or off the lift.

2. Do not hit or run over lift arms and adaptors when driving a vehicle on or off the lift. Have a co-worker guide you when driving a vehicle onto the lift. Do not stand in front of a lift with the car coming toward you.

3. Be sure the lift pads contact the car manufacturer's recommended **lifting points** shown in the service manual. If the proper lifting points are not used, components under the vehicle, such as brake lines, suspension components, or body parts, may be damaged. Failure to use the recommended lifting points may cause the vehicle to slip off the lift, resulting in severe vehicle damage and personal injury.

4. Before a vehicle is raised or lowered, close the doors, hood, and trunk lid.

5. When a vehicle has been lifted a short distance off the floor, stop the lift and check the contact between the hoist lift pads and the vehicle to be sure the lift pads are still on the recommended lifting points.

6. When a vehicle has been raised, be sure the safety mechanism is in place to prevent the lift from dropping accidentally.

7. Prior to lowering a vehicle, always make sure there are no objects, tools, or people under the vehicle. Ensure the doors are closed.

8. Do not rock a vehicle on a lift during a service job.

9. When a vehicle is raised, removal of some heavy components may cause vehicle imbalance. For example, since front-wheel-drive cars have the engine and transaxle at the front of the vehicle, these cars have most of their weight on the front end. Removing a heavy rear-end component on these cars may cause the back end of the car to rise off the lift. If this happens, the vehicle could fall off the lift!

SAFETY TIP *When doing heavy work on a vehicle on the lift use tall jack stands on the front and rear of the vehicle to prevent tipping.*

10. Do not raise a vehicle on a lift with people in the vehicle.

11. When raising pickup trucks or vans on a lift, remember these vehicles are higher than a passenger car. Be sure there is adequate clearance between the top of the vehicle and the shop ceiling, or components under the ceiling.

12. Do not raise a four-wheel-drive vehicle with a **frame contact lift** unless proper adaptors are used. Lifting a vehicle on a frame contact lift without the proper adaptors may damage axle joints.

13. Do not operate a front-wheel-drive vehicle that is raised on a frame contact lift. This may damage the front drive axles.

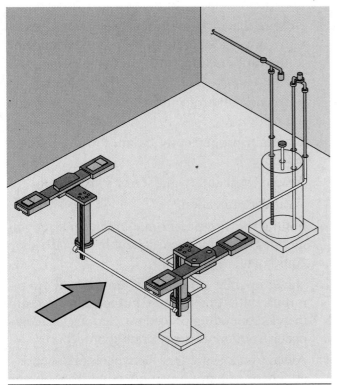

Figure 4-6 A two-post vehicle lift.

HYDRAULIC JACK AND SAFETY

Accidents involving the use of floor jacks and safety stands may be avoided if these safety precautions are followed:

1. Never work under a vehicle unless safety stands are placed securely under the vehicle chassis and the vehicle is resting on these stands **(Figure 4-7)**.

2. Prior to lifting a vehicle with a floor jack, be sure that the jack lift pad is positioned securely under a recommended lifting point on the vehicle. Lifting the front end of a vehicle with the jack placed under a radiator support may cause severe damage to the radiator and support.

3. Position the safety stands under a strong chassis member, such as the frame or axle housing. The safety stands must contact the vehicle manufacturer's recommended lifting points.

4. Since the floor jack is on wheels, the vehicle tends to move as it is lowered from a floor jack onto safety stands. Always be sure the safety stands remain under the chassis member during this operation, and be sure the safety stands do not tip. All the safety stand legs must remain in contact with the shop floor.

5. When the vehicle is lowered from the floor jack onto safety stands, remove the floor jack from under the vehicle. Never leave a jack handle sticking out from under a vehicle.

Figure 4-7 Safety stands.

Someone may trip over the handle and injure himself or herself.

POWER TOOL SAFETY

Power tools use electricity, shop air, or hydraulic pressure as a power source. Careless operation of power tools may cause personal injury or vehicle damage. Follow these steps for safe power tool operation:

1. Do not operate power tools with frayed electrical cords.

2. Be sure the power tool cord has a proper ground connection.

3. Do not stand on a wet floor while operating an electric power tool.

4. Always unplug an electric power tool before servicing the tool.

5. Do not leave a power tool running and unattended.

6. When using a power tool on small parts, do not hold the part in your hand. The part must be secured in a bench vise or with locking pliers.

7. Do not use a power tool on a job where the maximum capacity of the tool is exceeded.

8. Be sure that all power tools are in good condition; always operate these tools according to the tool manufacturer's recommended procedure.

9. Make sure all protective shields and guards are in position.

10. Maintain proper body balance while using a power tool.

11. Always wear safety glasses or a face shield.

12. Wear ear protection.

13. Follow the equipment manufacturer's recommended maintenance schedule for all shop equipment.

14. Never operate a power tool unless you are familiar with the tool manufacturer's recommended operating procedure. Serious accidents occur from improper operating procedures.

15. Always make sure that the wheels are securely attached and are in good condition on the bench grinder.

16. Keep fingers and clothing away from grinding and buffing wheels. When grinding or buffing a small part, hold the part with a pair of locking pliers.

17. Always make sure the sanding or buffing disk is securely attached to the sander pad.

18. Special heavy-duty sockets must be used on impact wrenches. If ordinary sockets are used on an impact wrench, they may break and cause serious personal injury.

COMPRESSED-AIR EQUIPMENT SAFETY

The shop air supply contains high-pressure air in the shop compressor and air supply lines. Serious injury or property damage may result from careless operation of compressed-air equipment. Follow these steps to improve safety:

1. Never operate an air chisel unless the tool is securely connected to the chisel with the proper retaining device.

2. Never direct a blast of air from an air gun against any part of your body. If air penetrates the skin and enters the bloodstream, it may cause very serious health problems and even death.

3. Safety glasses or a face shield should be worn for all shop tasks, including those tasks involving the use of compressed-air equipment.

4. Wear ear protection when using compressed-air equipment.

5. Always maintain air hoses and fittings in good condition. If an end suddenly blows off an air hose, the hose will whip around, possibly causing personal injury.

6. Use only air gun nozzles approved by OSHA.

7. Do not use an air gun to blow debris off clothing or hair.

8. Do not clean the workbench or floor with compressed air. This action may blow very small parts against your skin or into your eye. Small parts blown by compressed air may also cause vehicle damage. For example, if the car in the next stall has the air cleaner removed, a small part may find its way into the carburetor or throttle body. When the engine is started, this part will likely be pulled into the cylinder by engine vacuum, and the part will penetrate through the top of a piston.

9. Never spin bearings with compressed air because the bearing will rotate at extremely high speed. This may damage the bearing or cause it to disintegrate, causing personal injury.

10. All **pneumatic tools** must be operated according to the tool manufacturer's recommended operating procedure.

11. Follow the equipment manufacturer's recommended maintenance schedule for all compressed- air equipment.

CLEANING EQUIPMENT SAFETY AND ENVIRONMENTAL CONSIDERATIONS

All technicians are required to clean parts during their normal work routines. Face shields and protective gloves must be worn while operating cleaning equipment. In most states, environmental regulations require that the runoff from steam cleaning must be contained in the steam cleaning system. This runoff cannot be dumped into the sewer system. Since it is expensive to contain this runoff in the steam cleaning system, the popularity of steam cleaning has decreased. The solution in hot and cold cleaning tanks may be caustic, and contact between this solution and skin or eyes must be avoided. Parts cleaning often creates a slippery floor, and care must be taken when walking in the parts cleaning area. The floor in this area should be cleaned frequently. When the cleaning solution in hot or cold cleaning tanks is replaced, environmental regulations require that the old solution be handled as hazardous waste. Use caution when placing aluminum or aluminum alloy parts in a cleaning solution. Some cleaning solutions will damage these components. Always follow the cleaning equipment manufacturer's recommendations.

Parts Washers with Electromechanical Agitation

Some parts washers provide electromechanical agitation of the parts to provide improved cleaning action. These parts washers may be heated with gas or electricity. Various water-based hot tank cleaning solutions are available depending on the type of metals being cleaned. For example, Kleer-Flo Greasoff® number 1 powdered detergent is available for cleaning iron and steel. Non-heated electromechanical parts washers are also available, and these washers use cold cleaning solutions such as Kleer-Flo Degreasol® formulas.

Many cleaning solutions, such as Kleer-Flo Degreasol® 99R, contain no ingredients listed as hazardous by the EPA's RCRA Act. This cleaning solution is a blend of sulfur-free hydrocarbons, wetting agents, and detergents. Degreasol® 99R does not contain aromatic or chlorinated solvents, and it conforms to California's Rule 66 for clean air. Always use the cleaning solution recommended by the equipment manufacturer.

Cold Parts Washer with Agitation

Immersion Tank

Some parts washers have an agitator immersion chamber under the shelves that provides thorough parts cleaning. Folding work shelves provide a large upper cleaning area with a constant flow of solution from the dispensing hose. This cold parts washer operates on Degreasol® 99R cleaning solution.

Aqueous Parts Cleaning Tank

The aqueous parts cleaning tank system uses a water-based environmentally friendly cleaning solution rather than traditional solvents. The immersion tank is heated and agitated for effective parts cleaning (**Figure 4-8**). A sparger bar pumps a constant flow of cleaning solution across the surface to push floating oils away, and an integral skimmer removes these oils. This action prevents floating surface oils from re-depositing on cleaned parts.

HANDLING SHOP WASTES

The shop is responsible for hazardous waste until such waste is delivered to a hazardous waste site. Many shops contract a hazardous waste hauler to transport hazardous waste from the shop to government-approved recyclers or hazardous waste disposal sites. Always hire a properly licensed waste hauler, and be sure you know how the waste hauler is disposing of shop wastes. The shop owner is legally and financially responsible for the hazardous waste until it safely reaches the disposal site. Be sure to have a written contract with the hazardous waste hauler. The hazardous waste hauler fills out the necessary forms related to waste disposal and communicates with various state and federal agencies that are in charge of hazardous waste disposal regulations. Always keep all shipping bills from your hazardous waste hauler to prove you have recycled or disposed of hazardous waste material.

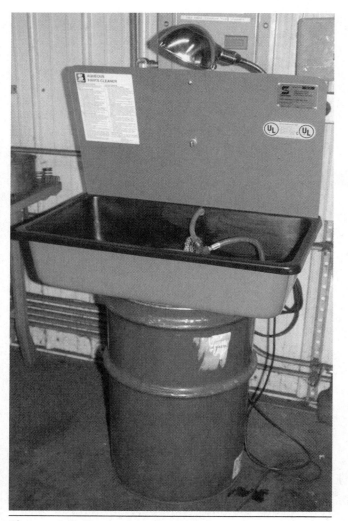

Figure 4-8 Parts cleaning tank.

Batteries

Batteries should always be recycled by shipping them to a designated recycler or back to the battery distributor. When defective batteries are stored on site, they should be kept in watertight, acid-resistant containers. Acid residue from batteries is hazardous because it is corrosive, and it may contain lead and other toxins. Inspect defective batteries for cracks and leaks. Spilled battery acid should be neutralized by covering it with baking soda, and then cleaned up in accordance with local regulations regarding the disposal of hazardous material.

Oil

Used oil is usually hauled to an oil recycling facility. Place oil drip pans under vehicles with oil leaks so oil does not drip onto the storage area **(Figure 4-9)**. In some states it is legal to burn used oil in a commercial space heater. State and local authorities must be contacted regarding regulations and permits.

Oil Filters

Used oil filters should be allowed to drain into an appropriate drip pan for 24 hours. After the draining process, oil filters should be squashed and recycled.

Figure 4-9 Used oil storage tank.

Solvents

Parts cleaning equipment that uses hazardous cleaning chemicals should be replaced with cleaning equipment that uses water-based degreasers. If hazardous chemicals are used in cleaning equipment, these chemicals must be recycled or disposed of as hazardous waste. Evaporation from spent cleaning chemicals is a contributor to ozone depletion and smog formation. Spent cleaning chemicals should be placed in closed, labeled containers and stored on drip pans or in diked areas. The storage area for waste materials should be covered to prevent rain from washing contaminants from stored materials into the ground water. This storage area may have to be fenced and locked if vandalism is a possibility.

Liquids

Engine coolant should be collected and recycled in an approved coolant recycling machine. Other liquids, such as brake fluid and transmission fluid, should be labeled and stored in the same area as solvents. Used brake fluid or transmission fluid should be recycled or disposed of as hazardous waste material.

Shop Towels

When dirty shop towels are stored on site, they should be placed in closed containers that are clearly marked, "contaminated shop towels only." Shop towels should be cleaned by a laundry service that has the capability to treat the wastewater generated by cleaning these towels.

Refrigerants

When servicing automotive air conditioning systems, it is illegal to vent refrigerants to the atmosphere. Certified equipment must be used to recover and recycle the refrigerant and recharge air conditioning systems. This service work must be performed by an EPA-certified technician.

> **Tech Tip** *All refrigerants should be clearly marked for type and other conditions such as Recycled, New, or Contaminated. This avoids costly mix-ups.*

INTERPRETING MATERIAL SAFETY DATA SHEETS (MSDS)

The product manufacturer's name and address is provided at the top of the MSDS **(Figure 4-10)**. The product manufacturer's phone number is also provided so they can be contacted in case of an emergency. The product name is provided with the chemical family name, plus any synonyms.

The ingredients in the hazardous material are listed in the ingredients section. The threshold limit value (TLV) and the permissible exposure limit (PEL) recommended by OSHA are listed in this section. The TLV and PEL values are the permissible concentrations of the hazardous material in the air to which a person may be exposed daily without known harmful effects. These values are usually expressed in parts per million (PPM). The percentage column indicates the percentage of the ingredient in relation to the total weight or volume of the hazardous product.

MATERIAL SAFETY DATA SHEET

PRODUCT NAME: KLEAN-A-KARB (aerosol) #- HPMS 102068
PRODUCT: 5078, 5931, 6047T
(page 1 of 2)

1. Ingredients	CAS #	ACGIH TLV	OSHA PEL	OTHER LIMITS	%
Acetone	67-64-1	750ppm	750ppm		2-5
Xylene	1330-20-7	100ppm	100ppm		68-75
2-Butoxy Ethanol	111-76-2	25ppm	25ppm	(skin)	3-5
Methanol	67-56-1	200ppm	200ppm		3-5
Detergent	-	NA	NA		0-1
Propane	74-98-6	NA	1000ppm		10-20
Isobutane	75-28-5	NA	NA	1000ppm	10-20

2. PHYSICAL DATA : (without propellent)

Specific Gravity : 0.865   Vapor Pressure : ND
                                 % Volatile : >99

Boiling Point : 176°F Initial   Evaporation Rate : Moderately Fast
Freezing Point : ND   Vapor Density : ND
Solubility: Partially soluble in water   pH :NA
Appearance and Odor: A clear colorless liquid, aromatic odor

3. FIRE AND EXPLOSION DATA

Flashpoint : −40°F   Method : TCC
Flammable Limits propellent LEL: 1.8   UEL: 9.5
Extinguishing Media: CO_2, dry chemical, foam
Unusual Hazards : Aerosol cans may explode when heated above 120°F.

4. REACTIVITY AND STABILITY

Stability : Stable
Hazardous decomposition products: CO_2, carbon monoxide (thermal)
Materials to avoid: Strong oxidizing agents and sources of ignition

5. PROTECTION INFORMATION

Ventilation : Use mechanical means to insure vapor concentration
is below TLV.
Respiratory: Use self-contained breathing apparatus above TLV.

Gloves : Solvent resistant   Eye and Face: Safety Glasses
Other Protective Equipment: Not normally required for aerosol product usage

---

Figure 4-10 Material safety data sheet (MSDS).

The *Physical Data section* of the MSDS provides information about the hazardous material such as the appearance and odor of the material. This information could help emergency personnel to recognize a hazardous material. This section also provides other information such as specific gravity, boiling point, and solubility.

The *Fire and Explosion Data section* may help you to prevent or fight a fire involving the hazardous material. The flash point listed in this section indicates the lowest temperature at which a liquid gives off enough vapor to ignite. The proper type of fire extinguisher that should be used to put out a fire involving this material is listed in the extinguishing media information. This section also provides information regarding any materials that should be kept away from a fire involving this hazardous material.

The *Reactivity and Stability* section provides information regarding the mixing of other material with the hazardous material. The stability classification indicates how the hazardous material resists chemical or physical change. This section also provides information about materials that may cause violent reactions when brought in contact with the hazardous material.

The *Protection Information* section supplies information regarding the necessary protective equipment required when you are handling the hazardous material. This section also provides information regarding the proper ventilation procedure when dealing with the hazardous material. Many MSDS also contain a section on leak and spill procedures. These procedures include the personal precautions to be observed if the hazardous material is spilled. This section also provides information about the proper procedure for disposal of the materials used in the clean up process.

Summary

- Electrical safety precautions include replacing frayed electrical cords, assuring that all electrical equipment has a ground connection, and not leaving electrical equipment running and unattended.

- Gasoline must be stored only in approved gasoline containers.

- Fire safety precautions include never turning on an ignition switch or cranking an engine with a fuel line disconnected, and storing all combustible materials in approved containers.

- Shop employees must be familiar with proper fire extinguisher operation.

- Safe vehicle operating procedures include always checking for adequate brake operation before driving the vehicle and making sure the shop door is up high enough before attempting to drive into or out of the shop.

- Safe housekeeping procedures include keeping floors free from oil, grease, water, and loose material, and providing unobstructed access to fire extinguishers and other shop safety equipment.

- Air bag safety precautions include always disconnecting the vehicle battery and waiting until the proper amount of time has elapsed before servicing an air bag system component. Use only the vehicle manufacturer's recommended tools for servicing these systems.

- Shop employees must follow proper lifting and carrying procedures to avoid back injury.

- Proper hand tool safety includes keeping tools in good condition and using the proper tool for the repair job being performed.

- Lift safety precautions include making sure the lift arms are connecting the vehicle manufacturer's specified vehicle lift points, being sure the lift safety mechanism is in place after the vehicle is raised on a lift.

- Hydraulic jack and safety stand safety precautions include always placing the jack lift pad on the vehicle manufacturer's specified lift point, and never exceeding the weight lifting capacity of the jack.

- Power tool safety procedures include always being sure the power tool has a ground connection. Do not operate power tools with frayed electrical cords.

- Compressed-air equipment safety precautions include never directing compressed air against or near human flesh or eyes. Always maintain air hoses and fittings in good condition.

- Cleaning equipment safety precautions include wearing face shields and protective gloves when

operating cleaning equipment, and keeping the floor clean and dry in the cleaning equipment area.

- Shop personnel must know the proper disposal procedures for all shop wastes.

- Shop personnel must be able to interpret MSDS.

Review Questions

1. While lifting heavy objects in the automotive shop:

   A. Bend your back to pick up the heavy object.

   B. Place your feet as far as possible from the object.

   C. Bend forward to place the object on the workbench.

   D. Straighten your legs to lift an object off the floor.

2. While discussing power tool safety, Technician A says an electric power tool cord does not require a ground. Technician B says frayed electric cords should be replaced. Who is correct?

   A. Technician A

   B. Technician B

   C. Both Technician A and Technician B

   D. Neither Technician A nor Technician B

3. While operating hydraulic equipment safely in the automotive shop, remember that:

   A. Safety stands have a maximum weight capacity.

   B. The driver's door should be open when raising a vehicle on a lift.

   C. A lift does not require a safety mechanism to prevent lift failure.

   D. Four-wheel-drive vehicles should be lifted on a frame contact lift.

4. All these shop rules are correct *except*:

   A. USC tools may be substituted for metric tools.

   B. Loose drain covers can cause severe foot injuries.

   C. Hands should be kept away from electric-drive cooling fans.

   D. Power tools should not be left running and unattended.

5. Technician A says the larger, flat connector is the ground connection in a three-prong electrical plug. Technician B says a two-prong adaptor may be used to plug in a three-prong end on an electrical cord. Who is correct?

   A. Technician A

   B. Technician B

   C. Both Technician A and Technician B

   D. Neither Technician A nor Technician B

6. All of these statements about gasoline containers are true *except*:

   A. Gasoline containers should be stored in a non-ventilated area.

   B. When transporting a filled gasoline container, it must be secured.

   C. Gasoline containers should be filled only to within 1 inch of the container top.

   D. Gasoline must be placed in approved gasoline containers.

7. When discussing shop safety and fires, Technician A says combustible shop supplies such as paint should be stored in a metal cabinet. Technician B says water should be used to extinguish a gasoline fire. Who is correct?

   A. Technician A

   B. Technician B

   C. Both Technician A and Technician B

   D. Neither Technician A nor Technician B

8. All of these statements about fire extinguishers and their use are true *except*:

   A. Direct the fire extinguisher nozzle at the base of a fire.

   B. To activate a fire extinguisher, pull the pin and rotate the handle.

   C. Many fire extinguishers discharge their contents in 8 to 25 seconds.

   D. When using a fire extinguisher to put out a fire, keep an escape route open behind you.

9. When discussing air bag system service and diagnosis, Technician A says the negative battery terminal should be disconnected and the vehicle manufacturer's specified time period elapsed before working on an air bag system. Technician B says all air bag sensor mounting bolts must be tightened to the specified torque to ensure proper sensor operation. Who is correct?

   A. Technician A

   B. Technician B

   C. Both Technician A and Technician B

   D. Neither Technician A nor Technician B

10. All of these statements about using a vehicle lift are true *except*:

    A. When raising a vehicle on a lift, the driver's door should be left open.

    B. The lift arms must contact the vehicle manufacturer's specified lift points.

    C. After the vehicle is raised, the lift safety mechanism must be in place.

    D. When lifted on a frame-contact lift, a four-wheel drive vehicle requires special adaptors.

11. The round terminal in a three-prong electrical plug is the _____ connection.

12. When servicing an air bag system, the technician must not _____ or _____ air bag sensors.

13. If battery acid is spilled, it may be neutralized by covering it with _____ _____.

14. Gasoline containers should be stored in a _____ _____ area.

15. Explain the proper safety precautions when filling gasoline containers.

16. Describe the proper fire extinguisher operating procedure.

17. Explain the necessary safety procedures when servicing air bag systems.

18. Describe the proper procedure for lifting heavy objects.

CHAPTER

5 Tools and Equipment

Learning Objectives

After you have read, studied, and practiced the contents of this unit, you should be able to:

- Explain the necessary precautions when operating a vehicle lift.

- Describe hydraulic jack and safety stand safety precautions.

- Explain the necessary safety precautions when using power tools.

- Describe the use of electrical and electronic test equipment.

Key Terms

Analog meters

Digital meters

Distributor ignition (DI)

Electronic ignition (EI)

High-impedance test light

Inductive pickup lead

International system (SI)

Lab scope

Multimeters

Press fit

Scan tool

Torque wrench

United States Customary System (USCS)

INTRODUCTION

Automotive technicians must use a variety of tools and be familiar with shop equipment, power tools, and special tools. This chapter provides information regarding the design and purpose of common hand tools. Information is also provided for basic shop equipment, power tools, and electronic equipment. Many precautions are provided that you need to know regarding the safe operation of shop equipment, power tools, and electronic equipment.

COMMON HAND TOOLS

Technicians must be familiar with different types of hand tools and the proper use of these tools. Using the proper tool for the job often saves time and allows a technician to perform their work more efficiently. Improper tool use may cause vehicle component damage and injury to the technician.

Wrenches

A set of box-end and open-end wrenches are included in a basic technician's tool set. A box-end

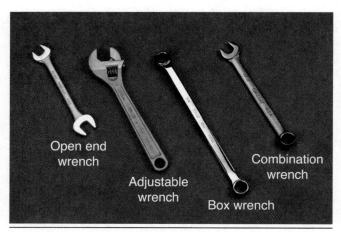

Figure 5-1 Common wrenches.

wrench **(Figure 5-1)** completely surrounds a nut or the head of a bolt and is less apt to slip and cause damage or injury. When there is very little space around a nut or bolt, it may be difficult to use a box-end wrench. An open-end wrench has an open, squared end that may be used where a box-end wrench will not fit. Open-end and box-end wrenches may have different sizes at either end. Box-end wrenches are available in either six points or twelve points.

Tech Tip *Other wrenches are available such as a ratcheting box end. These ratcheting box ends are available in box-end and combination wrenches.*

Twelve-point box-end wrenches allow you to work in smaller areas compared to six-point wrenches. An open-end wrench is often the best tool for rotating a nut or holding a bolt head. A technician's tool set should also include a set of combination wrenches that have an open-end wrench on one end and a box-end wrench on the opposite end. Both ends of these wrenches are the same size, and the open end and box end can be used alternately on the same bolt or nut. On most older domestic vehicles, the nuts and bolts require wrenches with United States Customary (USC) sizes. The **United States Customary System (USCS)** is a system of weights and measures used in the United States. These nut and bolts are usually manufactured in increments of 1/16 of an

inch. Most imported and newer domestic vehicles require metric-sized wrenches with increments of 1 millimeter (mm). The metric system of measurements may be referred to as the International System (SI).

Tech Tip *Metric and USC wrenches are not interchangeable. For example, a 9/16-inch wrench is 0.011 inches larger than a 14-millimeter nut. If the 9/16-inch wrench is used to turn or hold a 14-millimeter nut, the wrench may slip. This action may result in skinned knuckles for the technician and rounded-off shoulders on the nut. Only two sizes of metric and USC wrenches are nearly the same size. Eight millimeter is very close to the same size as 5/16 inch, and 19 millimeter is the almost the same size as 3/4 inch.*

The **International system (SI)** is a system of weights and measures in which every measurement is multiplied or divided by 10 to obtain larger or smaller units. Technicians must have complete sets of both USC and metric wrenches. Allen wrenches are used to tighten or loosen small bolts or setscrews with a machined hex-shaped recess in the head of the bolt or screw. The proper size Allen wrench fits snugly into the recess of the bolt head.

NOTE: An Allen wrench may be called a hex-head wrench.

To loosen or tighten line or tubing fittings, flare-nut wrenches should be used rather than open-end wrenches. These fittings are usually made from soft metal that distorts easily. Using an open-end wrench on line or tubing fittings tends to round off the shoulders on the fitting. Flare-nut wrenches are like a box-end wrench with an opening in the box end. This opening allows the flare-nut wrench to be slid over the line and placed on the line fitting. A flare-nut wrench surrounds most of the line fitting, and thus prevents the wrench from slipping on the fitting.

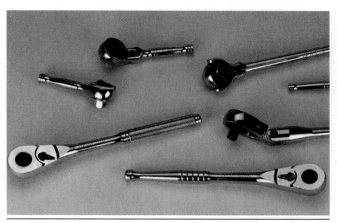

Figure 5-2 Ratchets come in various styles.

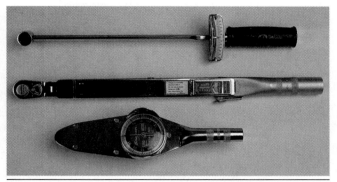

Figure 5-3 Three common types of torque wrenches are shown: a beam type, click type, and dial type.

Ratchets and Sockets

A typical technician's tool set contains USC and metric socket sets in 1/4-inch, 3/8-inch, and 1/2-inch drive. The drive size refers to the size of the square opening in the top of the socket. Each socket set will usually contain various extensions: a ratchet and a long breaker bar. The ratchet allows you to tighten or loosen a bolt without removing and resetting the wrench after you have turned it **(Figure 5-2)**. The long breaker bar is used to provide increased leverage when loosening tight bolts. Many sockets are designed with twelve points, but six-point and eight-point sockets are found in some socket sets. A six-point socket has stronger walls and improved grip compared to a twelve-point socket, but six-point sockets have half the positions of a twelve-point socket. Six-point sockets are mostly used on fasteners that are rusted or rounded. Eight-point sockets are available for use on square nuts or square-headed bolts. Universal joints are included in most socket sets. When loosening or tightening a fastener, a universal joint allows an angle between the socket and the ratchet and extension.

Torque Wrenches

A **torque wrench** measures the tightness of a bolt or nut. Vehicle manufacturers provide torque specifications in their service manuals for most fasteners on various components. The torque specifications are provided in foot-pounds (USC) or Newton-meters (metric). A foot-pound is the work

or pressure accomplished by a force of 1 pound applied through a distance of 1 foot. A Newton-meter is the work or pressure accomplished by a force of 1 kilogram applied through a distance of 1 meter. Torque wrench drive sizes are the same as socket drive sizes. Torque wrenches may be dial type, click type, beam type, or electronic. When using a dial-type torque wrench, the bolt torque is indicated on a dial as the bolt is tightened. A click-type torque wrench must be set to the specified torque. When the bolt torque reaches this setting, the wrench provides an audible click. On a beam-type torque wrench, the wrench bends as the bolt is tightened, and a pointer moves across a scale on the wrench to indicate the bolt torque. On an electronic torque wrench, the bolt torque is indicated on a digital display on the wrench as the bolt is tightened **(Figure 5-3)**.

> **Tech Tip** Torque wrenches may be used with a torque angle gauge. This allows the technician to torque a torque-to-yield fastener to a specified torque setting and then turn the fastener the specified number of degrees.

Screwdrivers

The flat-tip screwdriver is the most common type **(Figure 5-4)**. Blade-type screwdrivers are available in many different sizes depending on the size of fastener to be loosened or tightened. On this type of screwdriver, the blade tip fits into a slot in the fastener head. A Phillips screwdriver

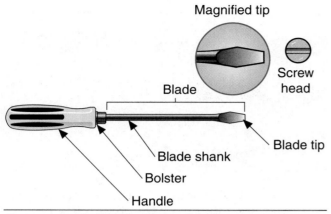

Figure 5-4 A flat-tip screwdriver.

has a cross point on the blade tip. This cross point has four surfaces that fit four matching recesses in the fastener head. The Phillips screwdriver provides more gripping power compared to a blade-type screwdriver **(Figure 5-5)**. A Torx-type screwdriver has a six-prong tip that fits snugly into a matching recess in the fastener head **(Figure 5-6)**. Torx-type screwdrivers are available in different sizes. All screwdrivers are available in different lengths from 2-inch stubby screwdrivers to 12 inches or longer. An offset screwdriver is a blade-type screwdriver with the blade tips positioned at a 90° angle in relation to the shank of the screwdriver, and the blades at opposite ends of the screwdriver

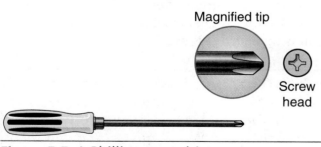

Figure 5-5 A Phillips screwdriver.

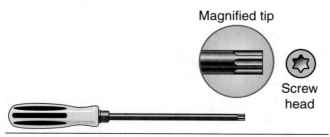

Figure 5-6 A Torx screwdriver.

are at right angles to each other. This offset screwdriver may be used to loosen or tighten fasteners that do not have direct access above the fastener head.

Pliers

Diagonal pliers have sharp cutting edges on the plier jaws. These pliers are used for cutting wire or removing cotter pins.

NOTE: Diagonal pliers may be called side cutters.

Combination pliers have a slip joint between the two jaws. The slip joint can be set for either of two jaw openings. Other types of pliers include channel lock, needle nose, snap ring, and vise grip **(Figure 5-7)**. Vise-grip pliers are sometimes called locking pliers. Rib-joint pliers contain several channels that provide different jaw openings. These pliers are used to grip large objects.

NOTE: Channel-lock pliers may be called water-pump pliers or slip-joint pliers.

Needle-nose pliers are used to grip objects in a small opening. Snap-ring pliers have special jaws that grip the ends of snap rings to remove and replace these components. Vise-grip pliers are designed so the jaws may be locked onto a component.

Hammers and Mallets

Ball-peen hammers are available in different weights. The average tool set contains at least two ball-peen hammers, one 8 ounce and possibly a 16 ounce. A soft-faced mallet is also required for tapping parts during removal or replacement. The soft-faced mallet may have plastic and lead or plastic and brass faces on the head of the mallet. When this type of mallet is used to tap components apart,

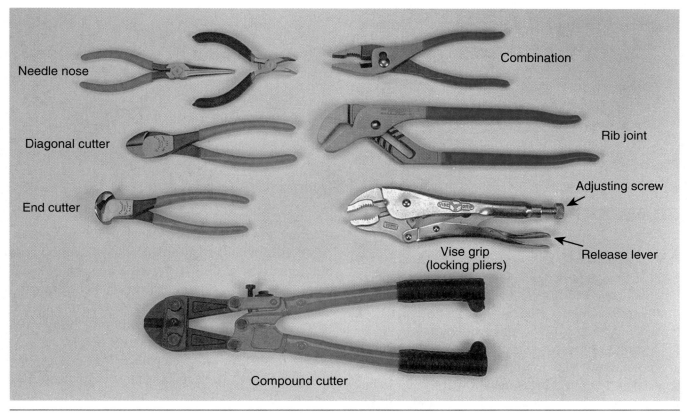

Figure 5-7 Common pliers.

it will not mark the component. If a ball-peen hammer is used to tap components apart, it will definitely mark and damage the components. Another type of hammer available is the brass hammer. It may also be used to tap apart components without marring the surface **(Figure 5-8)**. Dead-blow hammers are also available. These are usually a plastic hammer with steel shot on the inside of the head. This type of hammer does not bounce back as a ball-peen hammer might. The steel shot tends to transfer most of the force into the object being hammered, hence the name dead blow. These dead-blow hammers may also be purchased with a variety of hammer faces and are even available with steel ball-peen and steel hammer faces.

Punches and Chisels

A tool set contains a variety of punches and chisels. Drift punches are used to remove roll pins **(Figure 5-9)**. Brass-drift punches are available if component marking and damage is a concern. Tapered punches are used to align bolt holes. Center punches have a sharp tip that is used to indent a component in the proper location prior to drilling a hole in the component. This indent prevents drill wandering and component damage. Various types of chisels include: flat, cape, round-nose cape, and diamond point.

Brass hammer   Ball-peen hammer   Rubber mallet

Figure 5-8 Common types of hammers.

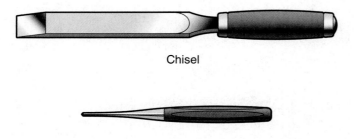

Figure 5-9 Common chisel and Punch.

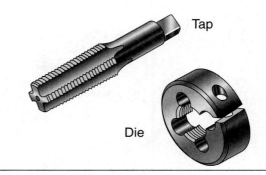

Figure 5-11 A tap and a die.

FILES, TAPS, AND DIES

Types of files in a tool set may include: flat, round, half-round, and triangular (**Figure 5-10**). Files may also be single cut or double cut. Single-cut files have the cutting grooves positioned diagonally across the face of the file. Double-cut files have the cutting grooves positioned diagonally in both directions across the face. Double-cut files are considered first cut or roughening files because they remove a great deal of metal. Single-cut files remove small amounts of metal, and they are considered finishing files. The most common threading tools are taps and dies (**Figure 5-11**). Taps and dies are available in USC and metric threads. USC taps and dies are also available in national fine (NF) or national coarse (NC). A tap is used to cut or restore internal threads in an

opening. Openings in metal castings may be classified as blind holes or through holes. A blind hole bottoms in the casting, whereas a through hole extends through the casting (**Figure 5-12**). A tapered tap is used in a through hole, and a bottoming tap is used to cut or restore threads in a blind hole (**Figure 5-13**). A pitch gauge may be placed over the bolt threads to determine the proper tap (**Figure 5-14**). The tap is rotated with a special handle that is designed to grasp the square top on the tap.

A die is a circular or hex-shaped tool that is designed to cut or restore external threads on a bolt, rod, or fastener. A die with the proper thread size must be selected. The die is mounted in a special turning handle. A screw extractor may be used to extract broken bolts or studs. Screw extractors are available in various sizes, and a specific drill size is specified for each extractor. The end of the broken bolt or stud is center punched and drilled with the specified drill. When the screw extractor is rotated, it is designed to grip the drilled opening in

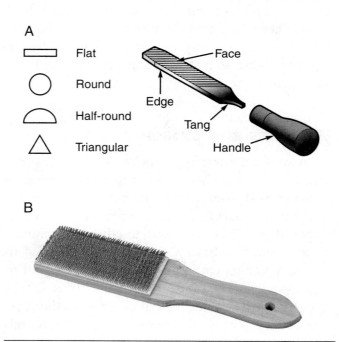

Figure 5-10 Various file types and a file card.

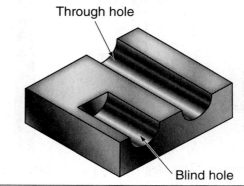

Figure 5-12 A through hole and a blind hole are shown for comparison.

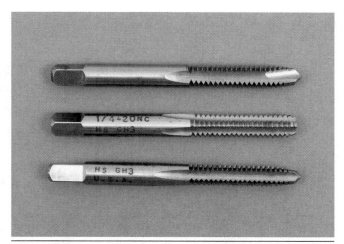

Figure 5-13 A bottoming tap and a taper tap.

Figure 5-14 A pitch gauge may be used to determine the proper tap or die selection.

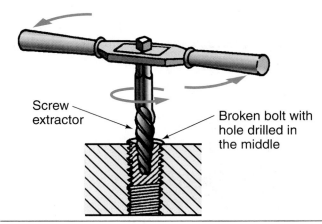

Screw extractor

Broken bolt with hole drilled in the middle

Figure 5-15 A screw extractor is used to remove broken bolts.

the bolt. The same turning handle that is used to rotate a tap may be used to rotate the screw extractor and remove the broken bolt **(Figure 5-15)**.

GEAR AND BEARING PULLERS

Many gears and bearings have a slight interference fit (press fit) when they are installed in a housing or on a shaft. For example, the inside diameter of a bore is 0.001 inch smaller than the outside diameter of the shaft that fits in the bore. When the shaft is installed in the bore, it must be pressed in to overcome the 0.001-inch interference fit. A **press fit** is present when a part is forced into an opening that is slightly smaller than the part itself to provide a tight fit. This press fit prevents the parts from moving on each other.

SAFETY TIP *Do not use a hammer to strike the hex-shaped end of a puller when the puller is tightened on a gear or bearing. This action may cause the puller to suddenly slip off the component, resulting in personal injury.*

The removal of gears and bearings must be done carefully to prevent damage to these components and the bore or shaft where they are mounted. Bearing pullers are designed to fit over the outer diameter of the bearing or through the center opening in the bearing to pull the bearing from its mounting location. The bearing puller must have the right jaws or adapters to fit the bearing or gear properly so the puller will not slip off the component being pulled. Various pullers are illustrated in **Figure 5-16**.

BUSHING AND SEAL PULLERS AND DRIVERS

Technicians have to use bushing and seal pullers and drivers when servicing components such as wheel hubs, transmissions, and differentials. Pullers are usually a threaded or slide-hammer type **(Figure 5-17)**. Bushings or seals may be damaged if the wrong tool is used for removal or installation. Seal drivers must fit the seal

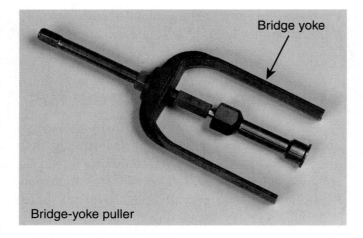

Bridge yoke

Bridge-yoke puller

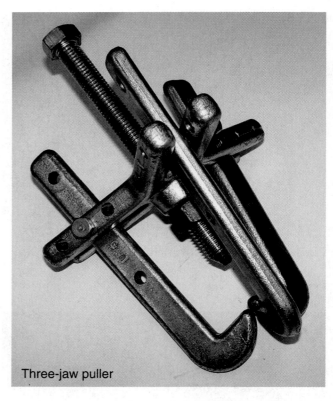

Three-jaw puller

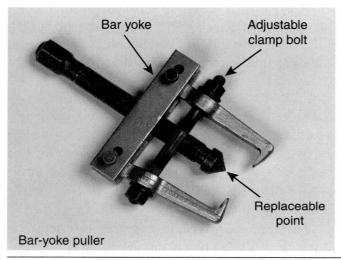

Bar yoke

Adjustable clamp bolt

Replaceable point

Bar-yoke puller

Figure 5-16 A bridge-yoke puller, bar-yoke puller, and a three-jaw puller are shown.

properly, and the seal must be started squarely into the housing **(Figure 5-18)**. Seal drivers are available in various diameters to fit squarely against the outside edge of different size seals. They also provide an internal recess that allows the puller to fit over a protruding shaft. The seal-driver handle is

tapped with a soft hammer to install the seal. Be sure the housing is clean and free from burrs before installing the seal. The outer diameter of some seals is coated with a sealer to prevent leaks between the seal case and the housing. If the seal case is not coated, a special sealer may be placed

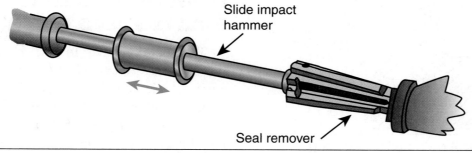

Slide impact hammer

Seal remover

Figure 5-17 Slide hammers may be used to remove seals.

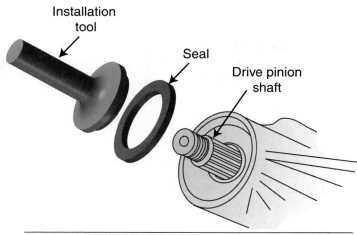

Installation tool

Seal

Drive pinion shaft

Figure 5-18 A seal driver is used to install a seal.

only on the outer diameter of the seal case. After the seal is installed, lubricate the seal lips with the vehicle manufacturer's specified lubricant.

POWER TOOLS AND SHOP EQUIPMENT

Power tools make a technician's job easier and increase production. Power tools operate faster and supply more torque compared to hand tools. Power tools also require increased safety measures. Power tools may be operated by air pressure (pneumatic), electricity, or hydraulic fluid. An air hose from the shop air supply must be connected to a pneumatic power tool.

SAFETY TIP *Improper use of power tools may cause serious personal injury. Never operate a power tool unless you are familiar with the manufacturer's recommended operating procedures.*

Pneumatic tools are commonly used by technicians and offer the following advantages:

- Deliver higher torque and higher RPM to help get the job done quickly
- Air tools are versatile
- Air tools have no shock hazard when operated in wet environments
- Weigh less and more compact than a comparable corded tool

- Require less maintenance
- Operate at lower temperatures due to the cooling from the air flow

However, electric power tools are usually less expensive than pneumatic tools. Power can be supplied from a conventional wall socket to a power tool. Electric power tools offer the following advantages:

- Less noise and vibration
- Widely used outside automotive repair facilities
- Wide array of tools for various applications
- No oily air exhaust
- Cord is relatively small

Another group of tools gaining popularity are cordless electric tools. Cordless impact drivers and wrenches are becoming handy and versatile power tools for driving a variety of fasteners. Improvements in recent years have made them compact, lightweight, and compared to a conventional cordless drill they can deliver and speed and power with good control. They also come in a wide variety allowing the technician to select the right tool for the application. The higher voltage rating allows an increase in power but also adds weight due to the larger batteries.

Impact Wrench

An impact wrench uses air pressure or electricity, both corded and cordless, to loosen or tighten a nut or bolt with a hammering action **(Figure 5-19)**.

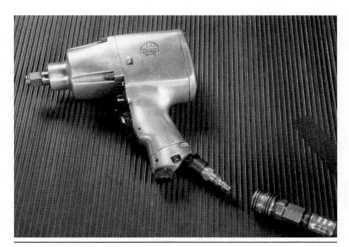

Figure 5-19 An air impact wrench.

Light-duty impact wrenches are available in drive sizes of 1/4 inch, 3/8 inch, and 1/2 inch. Heavy-duty impact wrenches are available in 3/4-inch and 1-inch drives.

> **Tech Tip** *Impact wrenches should not be used to tighten fasteners on components that may be damaged by the hammering force of the wrench.*

> **SAFETY TIP** *Impact wrenches require the use of thick-walled sockets to withstand the hammering force of the wrench. If conventional sockets are used on an impact wrench, they may shatter, and this action can result in personal injury.*

> **SAFETY TIP** *Hearing protection should be worn when operating noisy tools and equipment such as the air impact and the air hammer.*

Air ratchets are often used when loosening or tightening fasteners because they allow the technician to work faster than an ordinary ratchet. An air ratchet turns fasteners without a hammering force, and this tool can be used with conventional sockets. Air ratchets usually have a 3/8-inch drive. Air ratchets are not torque sensitive. Therefore, fasteners should be tightened snugly with an air ratchet and then tightened to the specified torque with a torque wrench.

> **Tech Tip** *Impact wrenches should not be used to tighten fasteners on components that may be damaged by the hammering force of the wrench.*

Blowgun

After parts are cleaned, they are blown off and dried with a blowgun. A blowgun snaps into the end of a shop air hose and directs airflow when a button or lever is pressed **(Figure 5-20)**. Always use OSHA-approved blowguns. Before using a

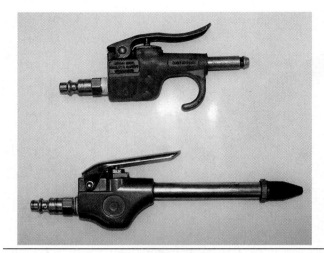

Figure 5-20 OSHA and non-OSHA approved air guns.

blowgun, be sure the air-bleed holes in the side of the gun are not plugged.

> **SAFETY TIP** *If airflow from a blowgun is directed near human flesh, the air may penetrate the skin and enter the blood stream. This action can result in serious medical problems or death.*

Bench Grinder

A bench grinder is usually bolted to the workbench **(Figure 5-21)**. The bench grinder must have shields and guards in place, and you must always wear face protection when using the grinder. Bench grinders are classified by wheel size. Wheel sizes of 6 to 10 inches are commonly used in automotive shops. Bench grinders may be equipped with three different types of wheels:

Figure 5-21 A bench grinder.

1. A grinding wheel is used for sharpening tools or deburring metal components.
2. A wire wheel brush is used for buffing or general cleaning, such as rust or paint removal.
3. A buffing wheel is used for polishing and buffing.

Utility Light

A utility light is used to supply adequate light in the immediate work area. A utility light may be powered from a conventional wall socket or it may be battery powered. Utility lights may be incandescent, fluorescent, or may be LED. In some shops, the utility lights are suspended from reels attached to the ceiling.

In many areas, insurance regulations demand the use of fluorescent trouble lights. Incandescent utility lights may shatter and burn if they are dropped, and this action can result in a fire. A cage must protect the bulb in an incandescent utility light. Some incandescent bulbs may be purchased with a tough rubber-like coating that will help reduce the formation of loose glass shards should the lightbulb break. Always keep the utility light cord away from rotating components. An incandescent utility light can burn carpet or upholstery. Be cautious when using this type of light inside a vehicle. Other names for a utility light include droplight and trouble light **(Figure 5-22)**.

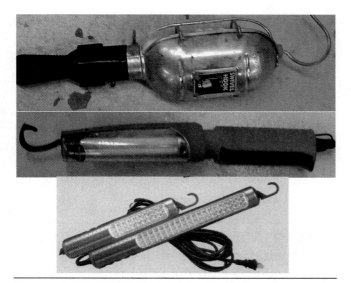

Figure 5-22 Utility lights, Incandescent, fluorescent, and LED types are shown.

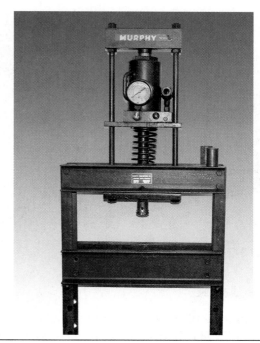

Figure 5-23 A hydraulic press.

Hydraulic Press

A hydraulic press is used to supply the necessary force to disassemble or reassemble press-fit components **(Figure 5-23)**. A hydraulic press uses a hydraulic cylinder and ram to remove and install precision-fit components from their mounting location. Although presses may be operated by hand, air pressure, or electricity, the hydraulic press is the most common. Most hydraulic presses are floor mounted, but smaller presses may be mounted on the bench or on a pedestal. A hydraulic cylinder and ram are mounted above the press bed. When the hydraulic pump is operated by hand, the ram extends against the work bed to exert pressure on the press-fit component. The component being pressed must be properly supported on the press bed to prevent component slipping. A shield must be in place around the component being pressed.

SAFETY TIP *When operating a hydraulic press, never operate the pump handle until the reading on the pressure gauge exceeds the maximum pressure rating of the press. If this pressure is exceeded, some part of the press may suddenly break, causing severe personal injury.*

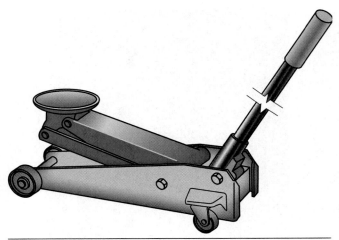

Figure 5-24 A hydraulic floor jack.

Floor Jack

A floor jack is used to raise a vehicle off the ground. The hydraulic floor jack is the most common jack, but air-operated jacks are available **(Figure 5-24)**. A floor jack uses hydraulic pressure supplied to a hydraulic cylinder, ram, and lift pad to lift one end or one corner of a vehicle. Floor jacks are rated according to the amount of weight they can lift safely. The jack handle is usually moved up and down to operate the hydraulic floor jack. The lift pad on the floor jack must be positioned under the manufacturer's specified lift point on the vehicle. A release lever on the jack handle is moved slowly to lower the floor jack. Do not leave a jack handle in the downward position where someone may trip over it.

SAFETY TIP *The maximum lifting capacity of a floor jack is usually written on the jack decal. Never lift a vehicle that is heavier than the weight rating of the floor jack. Lifting a vehicle that exceeds the rating of the jack may cause the jack to break or collapse, resulting in personal injury and/or vehicle damage.*

Safety Stands

After a vehicle is raised with a floor jack, it must be supported on safety stands before working under the vehicle **(Figure 5-25)**. The safety stand must be positioned under a strong structural part of the vehicle chassis. A service manual provides the location of vehicle lift and support points. Safety stands are

Figure 5-25 Typical safety stands (jack stands).

rated according to the amount of weight they will support. Never support a vehicle on a safety stand if the weight supported exceeds the rating of the safety stand. When a vehicle is lowered onto safety stands, be sure all four legs on each safety stand remain in contact with the floor. A general rule of thumb is to use safety stands whenever a jack is used.

SAFETY TIP *The maximum holding capacity of safety stands are usually written on the safety stand decal. Never support a vehicle that is heavier than the weight rating of the safety stand. Exceeding the capacity of the safety stands cause them to break or collapse, resulting in personal injury and/or vehicle damage.*

Vehicle Lift (Hoist)

A vehicle lift raises a vehicle so the technician can work under it. Automotive lifts come in a wide variety. They are available in two-post, four-post, mid-rise, low-rise, in-ground, portable, scissor, and several others.

Tech Tip *If the lift pad on a floor jack is not positioned under the vehicle manufacturer's specified lift point, undervehicle components may be damaged.*

Figure 5-26 A vehicle hoist.

The lift arms must be positioned under the vehicle manufacturer's recommended lift points before raising a vehicle **(Figure 5-26)**. Some lifts have an electric motor that drives a hydraulic pump to create fluid pressure and force the lift upward.

> **Tech Tip** *Always be sure the lift arms are securely positioned under the vehicle manufacturer's specified lift points before raising the vehicle. These lift points are illustrated in the service manual. If the lift arms are not positioned under the proper lift points, chassis components may be damaged.*

> **Tech Tip** *The vehicle doors, hood, and truck lid must be closed before raising a vehicle on a lift.*

Other lifts use shop air pressure to force the lift upward. If shop air pressure is used, it is applied to hydraulic fluid in the lift cylinder. A control lever is placed near the lift, which supplies shop air pressure to the lift cylinder. On an electric lift, a switch near the lift turns the lift motor on. After a vehicle is raised, always be sure the safety lock is engaged. When the safety lock is released, a lever is used to slowly lower the lift. Before lowering a lift, be sure there are no tools, equipment, or people under the lift.

Engine Lift

An engine lift or hoist, are common names for a hydraulic crane. It is used to remove an engine through the hood opening **(Figure 5-27)**. A sling is used to attach the hoist to the engine. The sling-attaching bolts must be strong enough to support the engine weight, and these bolts must be threaded in far enough to prevent the bolts from stripping out. Most service manuals provide the proper locations for sling attachment. The engine lift usually has an adjustable boom and legs. Adjust the legs out far enough to prevent the lift from tipping when the engine weight is supplied to the lift. Extending the boom lowers the lift capacity. After adjusting the boom and legs, be sure the lock pins in these components are properly installed and retained. When an engine is being lifted or lowered, do not place any part of your body under the lift. Once the engine is removed from the vehicle, immediately lower it onto the floor or install it on an engine stand.

NOTE: An engine lift may be called a cherry picker.

Figure 5-27 An engine hoist.

Figure 5-28 A typical tire changer (rim-clamp model).

Tire Changers

Tire changers are used to mount and demount tires **(Figure 5-28)**. These changers may be used on most common tire sizes. A wide variety of tire changers are available, and each one operates differently. Most tire changers are pneumatically powered with a few being powered by a combination of electricity and pneumatics. Most modern tire changers are of the rim-clamp style where the machine grips the rim of the tire rather than the center of the rim. Always follow the procedure in the equipment operator's manual and the directions provided by your instructor.

ELECTRICAL AND ELECTRONIC TEST EQUIPMENT

Technicians must be familiar with electrical and electronic test equipment. Using the proper electrical and electronic test equipment makes diagnosis faster and more accurate. When electrical or electronic equipment is used improperly, the circuit being diagnosed and the test equipment may be severely damaged.

Circuit Testers

Circuit testers are used to diagnose defects in electrical circuits. A large variety of circuit testers

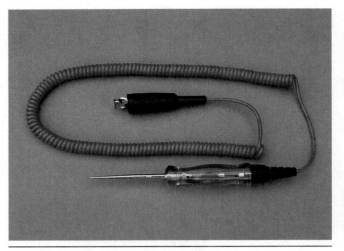

Figure 5-29 A 12-volt test light.

are available for automotive testing, but the most common circuit tester is the 12-volt test light. A sharp probe is molded into the handle on a 12-volt test light, and the upper part of the handle is transparent. A 12-volt bulb is mounted in the transparent handle **(Figure 5-29)**, and the probe is connected to the bulb terminal. A ground clip is connected to the wire extending from the handle, and this wire is attached to the ground side of the bulb. In most test situations, the ground clip is connected to ground on the vehicle, and the probe is connected to an electric circuit to determine if voltage is available. The bulb is illuminated if voltage is available at the probe. High-impedance test lights are available for diagnosing computer systems. A **high-impedance test light** contains a very small, high-resistance bulb. This type of test light is the only type used to test computer systems when the vehicle manufacturer's service procedure recommends the use of such equipment.

Tech Tip *Do not use a conventional 12-volt test light to diagnose computer system wires or components. The current draw of these test lights may damage computer system components.*

A self-powered test light is similar in appearance to a 12-volt test light, but this test light has an internal battery that supplies power to the circuit being tested **(Figure 5-30)**. These are sometimes called

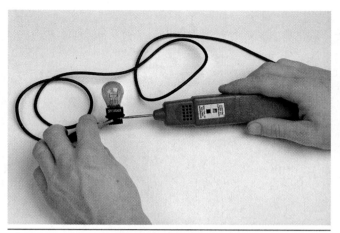

Figure 5-30 A self-powered test light.

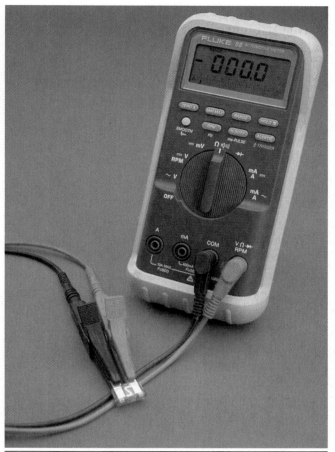

Figure 5-31 A digital multimeter.

continuity testers. In many automotive circuits, voltage is supplied to one end of the circuit, and the other end of the circuit is connected to ground. The ground side of the circuit may be called the negative side of the circuit, and the positive side of the circuit is connected to the battery positive terminal. When diagnosing an electrical circuit with a self-powered test light, the positive side of the circuit is disconnected, and the test light probe is connected to the circuit so the test light supplies voltage to the circuit in place of the battery. If the circuit is not open, the test light is illuminated because the other end of the circuit is connected to ground. When the circuit is open, the test light is not illuminated.

Tech Tip *If any type of circuit tester is used to diagnose an air bag system, accidental air bag deployment may occur. Use only the vehicle manufacturer's recommended equipment on these systems.*

Multimeter

Multimeters are small hand-held meters that provide the following readings in various scales: DC volts, AC volts, ohms, amperes, and milliamperes. Multimeters do not have heavy leads, so the maximum current flow reading on this type of meter is often 10 amperes. A control knob on the front of the meter is rotated to select the desired scale. Most multimeters are digital type, and some of these meters are auto-ranging, which means the

meter will automatically change to a higher scale if the reading goes above the highest value on the scale being used **(Figure 5-31)**. If the meter is not auto-ranging, the technician must manually select the next highest scale. If a multimeter is connected improperly, current flow through the meter may be excessive. Under this condition, an internal fuse will blow and protect the meter.

NOTE: A digital multimeter is referred to as a DMM.

Digital meters provide a digital reading, whereas **analog meters** have a movable pointer that moves across various scales. Digital multimeters have higher resistance than analog meters. Always use the type of meter recommended by the vehicle manufacturer. The meter leads are plugged into the appropriate terminals on the front of the meter, and the reading provided by the terminal position is indicated beside the terminal. The black test lead is usually plugged into the common (com) terminal. This terminal may be referred to as a ground terminal. Some multimeters have additional test capabilities such as diode condition, frequency or hertz (Hz), temperature, capacitance, inductive current clamps, engine rpm, and ignition dwell.

Tachometer

Analog tach-dwell meters are one of the most basic types of ignition test equipment. The colored tach-dwell meter lead is connected to the negative primary ignition coil terminal, and the black meter lead is connected to ground. A switch on the tach-dwell meter is used to select engine rpm or dwell. Ignition dwell is not adjustable on electronic-type distributor ignition (DI) or on electronic ignition (EI). Therefore, the dwell reading is not very useful for diagnostic purposes. The tachometer indicates engine rpm. A **distributor ignition (DI)** system uses a distributor to distribute spark from the coil secondary terminal to the spark plugs. An **electronic ignition (EI)** system does not have a distributor. This type of ignition system has a coil for each spark plug or pair of spark plugs.

> **NOTE:** Analog tachometers are seldom used today as digital tachometers are more versatile and popular.

Most tachometers are digital. This type of tachometer usually has an **inductive pickup lead** that is connected over the number 1 spark plug wire. Some may also be clamped over the primary wire of a COP ignition system. This type of tachometer is suitable for EI systems.

Timing Light

A timing light is used to check ignition timing with the engine running **(Figure 5-32)**. Voltage is supplied to the timing light from two leads connected to the battery terminals with the correct polarity. Most timing lights have a lead with an inductive clamp that is positioned over the number 1 spark plug wire. When the timing light trigger

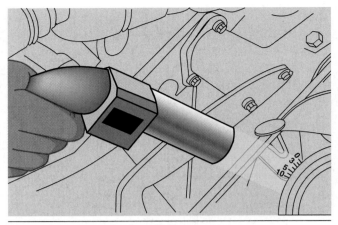

Figure 5-32 A timing light.

switch is pulled on with the engine idling, the timing light emits a beam of light each time the number 1 spark plug fires. The engine timing marks are usually positioned on the crankshaft pulley or on the flywheel. A stationary pointer or notch is located above the rotating timing marks. The timing marks are lines on the crankshaft pulley or flywheel that indicate various degrees of crankshaft rotation. One line represents top dead center (TDC), and other lines represent specific degrees of crankshaft rotation before top dead center (BTDC). Some engines have degree lines on the timing marks that indicate after top dead center (ATDC). On other timing marks, the degree lines are on the pointer, and the crankshaft has only one notch for a timing mark.

Always complete all of the vehicle manufacturer's recommended timing procedures before checking the ignition timing. For example, on some fuel-injected engines, an inline timing connector must be disconnected to prevent the computer from supplying spark advance while checking the base timing. With the engine idling at the specified rpm, the timing light beam is aimed at the timing marks. If necessary, rotate the distributor until the timing marks are located at the vehicle manufacturer's specified position. Be sure the distributor hold-down bolt is tightened to the specified torque after the timing is set. Timing adjustments are not possible on EI systems. Many timing lights have an advance knob that allows the technician to check the spark advance. With the engine running at the specified higher speed, rotate the timing advance knob on the timing light until the timing marks come back to TDC. The degree scale around the advance knob indicates the degrees of spark advance. Some timing lights have a digital readout in place of the degree scale around the advance knob.

Volt-Amp Tester

A volt-amp tester is used to test voltage and amperes in automotive circuits **(Figure 5-33)**. A volt-amp tester has the capability to perform a battery load test, a starter draw test, an alternator maximum output test, and an alternator voltage test. A typical volt-amp tester contains an ammeter, voltmeter, and a carbon pile load. Many volt-amp testers provide digital readings. A carbon pile load is a stack of carbon disks, and a control knob on the volt-amp tester tightens these disks together as the knob is rotated clockwise. When this knob is rotated counterclockwise to the off position, the carbon disks are not contacting each other. Two heavy leads are connected from the carbon pile load to the battery terminals with the correct polarity. The red, positive lead is connected to the positive battery terminal, and the black, negative lead is connected to the negative battery terminal. The ammeter usually has an inductive clamp that fits over the wire in which the amperes is being measured. This type of clamp reads the current flow from the magnetic strength surrounding the wire.

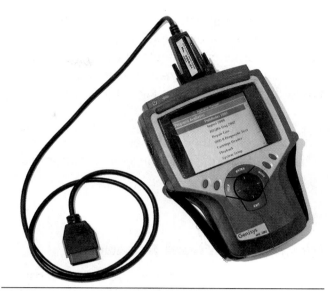

Figure 5-34 A scan tool.

Scan Tool

A **scan tool** is used for diagnosing automotive computer systems **(Figure 5-34)**. On many vehicles, the scan tool is connected to a data link connector (DLC). Vehicles manufactured since 1996 are equipped with on-board diagnostic II (OBD II) systems. On these systems, the DLC is mounted under the dash. On some older vehicles, the DLC is mounted under the hood. On many vehicles, the various computers are interconnected by data links, which are also connected to the DLC. These data links are referred to as Controller Area Networks (CAN). Therefore, when the scan tool is connected to the DLC, it can be used to diagnose several different computer systems on the vehicle such as the engine computer, transmission computer, body computer, antilock brake system (ABS) computer, suspension computer, and air-conditioning (A/C) computer. Most scan tools have removable modules, and the proper module for the vehicle and system being tested must be inserted in the scan tool before connecting the tool.

Lab Scope

A lab scope can be used to examine the electrical behavior of circuits. A **lab scope** provides waveforms or voltage traces representing the voltages in electronic circuits **(see Figure 22-18)**. However,

Figure 5-33 A volt-amp tester.

the lab scope is able to scan voltage signals much faster. For this reason, the lab scope waveform will indicate momentary defects in electronic components that many other types of test equipment may fail to display due to their slower sampling rate.

> *NOTE:* Graphing multimeters are also available that have many of the same functions as a lab scope.

Electronic Wheel Balancer

Electronic wheel balancers are used in most automotive shops **(Figure 5-35)**. Do not attempt to use this equipment until you have studied the operator's manual and your instructor has demonstrated the safe use of the balancer. The electronic wheel balancer is used to indicate the proper position and amount of wheel weight required to provide correct static and dynamic wheel balance. Many wheel balancers will have different balancing modes depending on where the technician needs to place the wheel weights. A few of the newest balancers can even measure the wheel with a simulated road force to measure the rolling resistance. Some problems that these measurements can help spot are tires with broken or defective belts.

SAFETY TIP *When using an electronic wheel balancer, always lower the safety shield over the tire and wheel before spinning the wheel. Failure to lower the safety shield may cause personal injury.*

Exhaust Analyzers

Exhaust analyzers are very valuable diagnostic tools. Five-gas analyzers read the levels of carbon monoxide (CO), hydrocarbons (HC), oxides of nitrogen (NOx), carbon dioxide (CO_2), and oxygen (O_2) **(Figure 5-36)**. Some shops are equipped with four-gas analyzers, which do not indicate NOx emissions. A pickup and hose assembly is connected from the vehicle tailpipe to the exhaust analyzer. When an exhaust analyzer is turned on, it performs an automatic warm-up and calibration procedure. A filter on the analyzer removes water and other particles from the exhaust before they enter the analyzer. If the filter or hose is restricted, a warning light is illuminated on the analyzer. The engine is usually warmed up to normal operating temperature before performing an exhaust emission analysis.

> *NOTE:* An exhaust analyzer may be called an infrared tester because some of these analyzers use infrared light to analyze the exhaust gases.

Figure 5-35 An electronic wheel balancer. This balancer is also equipped with a mechanism to measure the rolling resistance of the fire.

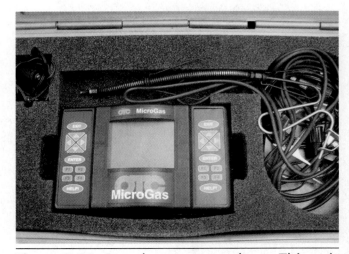

Figure 5-36 An exhaust gas analyzer. This unit is portable and may be taken with the vehicle during a test drive.

The levels of CO and HC in the exhaust are a direct indication of engine performance. For example, a high HC reading may indicate a misfiring cylinder, and a high CO reading may be caused by a rich air-fuel ratio. A high NOx reading may be caused by a malfunctioning exhaust gas recirculation (EGR) system. The levels of CO_2 and O_2 are affected very little by the catalytic converter on the vehicle. A high O_2 level indicates a lean air-fuel ratio, which may be caused by an intake manifold leak. Five-gas analyzers can be used to diagnose the following conditions: rich or lean air-fuel ratios, faulty injectors, catalytic converter malfunction, air pump malfunction, intake manifold leaks, improper evaporative (EVAP) system operation, improper EGR system operation, or defective engine conditions such as low cylinder compression or defective head gaskets. The five-gas exhaust analyzer is also very useful when diagnosing vehicles that failed a compulsory emission test.

Tech Tip *If the vehicle exhaust system is leaking, the exhaust analyzer readings will be inaccurate.*

Engine Analyzer

An engine analyzer contains all the necessary test equipment to perform a complete analysis of the engine and all the related engine systems **(Figure 5-37)**. A computer in the engine analyzer guides the technician through all the tests. Most engine analyzers contain the following test equipment and test capabilities: a cylinder output test, a pressure gauge, a vacuum gauge, a vacuum pump, a tachometer, a timing light/probe, a voltmeter, an ohmmeter, an ammeter, a carbon pile load, an oscilloscope, a scan tool, an exhaust/emissions analyzer, and a laboratory (lab) scope. An engine analyzer will test the battery, starting system, charging system, primary and secondary ignition systems, fuel system, electronic control systems, emission levels, and engine condition. The analyzer is connected to these

Figure 5-37 An engine analyzer.

systems by a variety of leads, inductive clamps, probes, and connectors. Several computers in the analyzer process the data received from these connections. Many engine analyzers have vehicle specifications on disk or CD. The technician must enter the necessary information regarding the vehicle being analyzed. A keyboard is used to submit commands or information into the analyzer. Most engine analyzers perform a complete series of tests and record the results automatically. A printer connected to the analyzer will print out all test results. The analyzer compares all the test results to specifications and identifies any test results that are not within specification. Many analyzers also provide diagnostic assistance for problems indicated by the out-of-specification readings. The technician may also select any test function or functions separately. Engine analyzer functions vary depending on the equipment manufacturer. The technician must be familiar with the engine analyzer in their shop.

Many of the functions of the engine analyzer have been replaced by the scan tool and OBDII. Cars made since 1996 have OBDII (onboard diagnostics), which will constantly monitor the various functions of the engine and emission systems. As a result the popularity of engine analyzers has declined in favor of the scan tool, lab scope, and exhaust gas analyzer.

Summary

- Wrenches include box-end, open-end, and combination type.

- Allen wrenches are hex-shaped tools that fit snugly into a matching recess in the fastener head. Flare-nut wrenches are like box-end wrenches with an opening cut through the box end.

- A ratchet allows the technician to tighten or loosen a fastener without removing and resetting the wrench after the fastener is rotated.

- Sockets may have twelve, eight, or six points.

- A universal joint allows an angle between the socket and ratchet or extension.

- A torque wrench measures the tightness of a bolt.

- Screwdrivers may be blade-type Phillips, Torx, or offset type.

- Types of pliers include combination, diagonal, and channel lock.

- A tap is used to thread openings in a casting (internal threads), and a die is used to cut or repair threads on a fastner (external threads).

- Air threads on a fastener (external threads).

- Gear and bearing pullers are used to remove gears or bearings that are mounted with a press fit.

- An impact wrench or air ratchet is used to remove and install fasteners.

- A blowgun is used to blow dry parts after the cleaning process.

- A bench grinder is used to grind or buff various components.

- A hydraulic press is used to remove and install press-fit components.

- A floor jack is used to raise one corner or one end of a vehicle.

- A vehicle lift is used to raise a vehicle so under-vehicle work may be performed.

- An engine lift is used to lift an engine during engine removal and replacement.

- Electrical test equipment includes circuit testers, multimeters, tachometers, timing lights, volt-amp testers, and scan tools.

- Electronic equipment includes wheel balancers, exhaust analyzers, and engine analyzers.

Review Questions

1. While discussing systems of weights and measures Technician A says the International system (SI) is called the metric system. Technician B says most late-model vehicles require metric-sized wrenches. Who is correct?

   A. Technician A

   B. Technician B

   C. Both Technician A and Technician B

   D. Neither Technician A nor Technician B

2. Power tools can be powered by

   A. pneumatic power.

   B. cordless or battery power.

   C. electricity.

   D. all of the above.

3. While discussing hand tools Technician A says flare-nut wrenches are used to loosen or tighten cylinder head bolts. Technician B says that flare-nut wrenches are used to loosen or tighten fuel and brake line fittings. Who is correct?

   A. Technician A

   B. Technician B

   C. Both Technician A and Technician B

   D. Neither Technician A nor Technician B

4. All of these statements about shop tools and equipment are true *except*:

   A. A tap is used to cut or restore the threads on a bolt.

   B. Single-cut files are considered finishing files.

C. A scan tool may be used to diagnose automotive computer systems.

D. A tire changer is usually pneumatically powered.

5. Technician A says Allen wrenches are hex-shaped. Technician B says USC sockets are available with a 5/8 drive. Who is correct?

A. Technician A

B. Technician B

C. Both Technician A and Technician B

D. Neither Technician A nor Technician B

6. All of these statements about sockets are true *except*:

A. Most sockets in a typical technician's tool set are designed with six and twelve points.

B. Compared to a twelve-point socket, a six-point socket has improved grip on a fastener.

C. Compared to a twelve-point socket, a six-point socket has stronger walls.

D. Twelve-point and six-point sockets have the same number of installation positions on a fastener.

7. Technician A says torque wrench drive sizes are the same as socket drive sizes. Technician B says torque wrenches may be dial type or torsion bar type. Who is correct?

A. Technician A

B. Technician B

C. Both Technician A and Technician B

D. Neither Technician A nor Technician B

8. Technician A says double-cut files are considered to be finishing files. Technician B says single-cut files have cutting grooves positioned straight across the face of the file. Who is correct?

A. Technician A

B. Technician B

C. Both Technician A and Technician B

D. Neither Technician A nor Technician B

9. All of these statements about impact wrenches are true *except*:

A. Conventional sockets should be used with a light duty impact wrench.

B. Impact wrenches may be operated by air pressure or electricity.

C. Impact wrenches loosen or tighten bolts or nuts with a hammering action.

D. A light-duty impact wrench may have a 1/2-inch drive.

10. Technician A says a conventional 12-volt test light is recommended to diagnose a computer circuit. Technician B says a conventional 12-volt test light is recommended to diagnose an air bag circuit. Who is correct?

A. Technician A

B. Technician B

C. Both Technician A and Technician B

D. Neither Technician A nor Technician B

11. When a fastener is tightened to the torque setting on a click-type torque wrench, the wrench provides an _____ _____.

12. A Torx driver has a _____ _____ tip.

13. A bottoming tap is used to cut or restore threads in a _____ hole.

14. Describe the design and purpose of an open-end wrench.

15. Explain how a self-powered test light may be used to test for an open circuit in a wire.

16. Explain the purpose of a scan tool, and describe the proper on-vehicle connection for this tool.

17. List the five gases indicated on a five-gas analyzer.

18. Explain the automotive systems and conditions that may be tested with an engine analyzer.

CHAPTER

6 Basic Shop Procedures

Learning Objectives

After you have read, studied, and practiced the contents of this unit, you should be able to:

- Demonstrate the proper use of an impact wrench.
- Use various types of torque wrenches.
- Use gear, bearing, and seal pullers.
- Remove broken studs and screws.
- Remove damaged nuts.
- Use an oxy-acetylene torch for heating.
- Jump-start a vehicle with a discharged battery.
- Demonstrate checking and changing respirator filters.
- Demonstrate proper battery charging.
- Use cleaning equipment properly.
- Use lifting equipment properly.

Key Terms

Aqueous

Center of gravity

Cross-threaded

Female-type quick disconnect coupling

Jump-start

Left-hand thread

Male-type quick disconnect coupling

Neutral flame

Memory saver

Quick disconnect coupling

Right-hand thread

Striker

INTRODUCTION

It is extremely important for automotive students and technicians to learn the proper use of shop equipment. Improper equipment use may cause personal injury, equipment damage, or vehicle damage. The first time students are required to use specific shop equipment, it is very important for them to learn the proper procedure for operating this equipment. When improper equipment operating procedures are used, it is more difficult to relearn the proper procedures.

USING SHOP TOOLS AND EQUIPMENT

Technicians must be familiar with the use of shop tools and equipment. When technicians lack this knowledge, personal injury may occur and expensive tools and equipment may be damaged. Knowledge of shop tools and equipment also allows technicians to work faster and more efficiently.

Connecting and Using an Impact Wrench

An impact wrench is a hand-held, reversible power tool for removing and tightening bolts and nuts. Impact wrenches may be powered by electricity or shop air pressure. Heavy-duty impact wrenches may deliver up to 450 foot-pounds (ft.-lb) (607.5 N·m) of torque. When an impact wrench is operating, the drive shaft on this tool may rotate at speeds from 2,000 to 14,000 rpm. When using an air operated impact wrench, the first step is to connect a shop air hose to the wrench. A **female-type quick disconnect coupling** on the end of the air hose is pushed onto the matching **male-type quick disconnect coupling** fitting on the impact wrench **(Figure 6-1)**. To release the **quick disconnect coupling**, set the impact tool on the workbench, grasp the air hose, and pull the outer, spring-loaded housing on the female part of the coupling toward the air hose.

When using an electric impact wrench, the electrical cord on the wrench must be plugged into a 110-volt electrical outlet. Always be sure the impact wrench electrical cord has a proper ground connection. Some electric impact wrenches are powered by a rechargeable battery **(Figure 6-2)**. Battery-powered impact wrenches are convenient

Figure 6-2 A cordless battery-operated impact wrench.

when working in areas where shop air pressure or electrical outlets are not available.

SAFETY TIP *Always inspect the cords for damage or frayed ends. Do not operate the tool until the damaged cord is replaced.*

If an electrical cord on an impact wrench does not have a proper ground connection, electrical shock and personal injury may occur when using the wrench. The most common size of an impact wrench drive shaft is 1/2 inch (12.7 millimeter). Heavy-duty impact wrench sockets must be used on an impact wrench **(Figure 6-3)**.

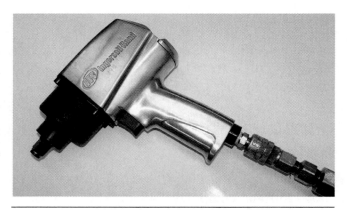

Figure 6-1 An air operated impact wrench and air hose coupling.

Figure 6-3 An impact socket used with an impact wrench.

If an ordinary socket is used on an impact wrench, the socket may shatter, resulting in personal injury. Using the wrong size socket on an impact wrench will damage the corners on the fastener being loosened or tightened.

Be sure the impact wrench switch is set for the desired rotation. When the proper size of socket is installed on the impact wrench drive, the socket is then placed over the fastener to be loosened. The trigger on the impact wrench is pulled to activate the wrench. When the trigger is pulled and the impact wrench drive shaft encounters the turning resistance of the fastener, a small spring-loaded hammer inside the impact wrench strikes an anvil attached to the impact wrench drive shaft. Each impact of the hammer against the anvil moves the socket and fastener in the desired direction to loosen and remove the fastener. When removing wheel and tire assemblies, always chalk-mark one of the wheel studs in relation to the wheel rim so the wheel may be re-installed in the original direction to maintain wheel and tire balance.

When installing and tightening a nut or bolt, always be sure the threads are in good condition. Damaged threads must be repaired with a tap or die. If damaged threads are not repaired, they may be stripped completely when the fastener is tightened, and this condition will not allow the fastener to be tightened to the specified torque. Many nuts are bi-directional, but other nuts must be installed in the proper direction. For example, many wheel nuts have a chamfer on the inner end that fits into tapered openings in the wheel rim to center the wheel **(Figure 6-4)**. These self-centering nuts must be

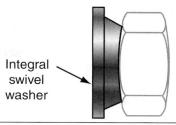

Figure 6-5 A wheel nut with an integral washer: this type of wheel nut is not self centering.

installed with the chamfer facing toward the wheel. Other non-self-centering wheel nuts have a flat integral swivel washer on the inner side of the nut. This swivel washer must be installed so it is facing toward the wheel **(Figure 6-5)**. In these applications, the wheel is centered by the hub design **(Figure 6-6)**. When installing a fastener, always determine if the fastener has a right-hand thread or a left-hand thread so you will know which direction to rotate the fastener. A fastener with a **right-hand thread** must be rotated clockwise when tightening the fastener. Most fasteners will be right hand. Left-hand wheel studs are usually stamped with an "L" in the center to designate the use of the left-hand thread. A fastener with a **left-hand thread** must be rotated counterclockwise to tighten the fastener.

Always start a fastener onto the threads by hand, and be sure the fastener rotates easily. If the fastener is not properly started onto the threads, the threads on the fastener will be cross-threaded and ruined when the fastener is tightened. A **cross-threaded** condition may also be defined as those threads that have been destroyed by starting a fastener onto its threads when the fastener is tipped slightly to one side and the threads on the fastener are not properly aligned with its matching threads. A cross-threaded condition may also be referred to as stripped threads.

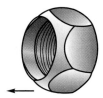

Figure 6-4 A wheel nut with a chamfered end. The chamfered end goes toward the wheel.

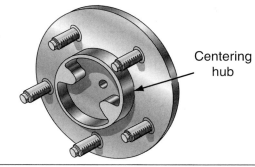

Figure 6-6 A hub is used to center the rim.

An impact wrench does not have torque-limiting capabilities. Therefore, fasteners can be over-tightened with an impact wrench, and this action may damage the components mounted by the fasteners. For example, wheel studs are mounted in the hub flange and extend through the brake rotor or drum and the wheel openings **(Figure 6-7)**. If the wheel nuts are over-tightened with an impact wrench, the brake rotor may be distorted, resulting in brake problems. To avoid this problem, tighten the fasteners snugly with the impact wrench, and then tighten the fasteners to the specified torque with a torque wrench. Some air-operated impact wrenches have an adjustable pressure that should be set to the low setting to avoid over tightening wheel nuts.

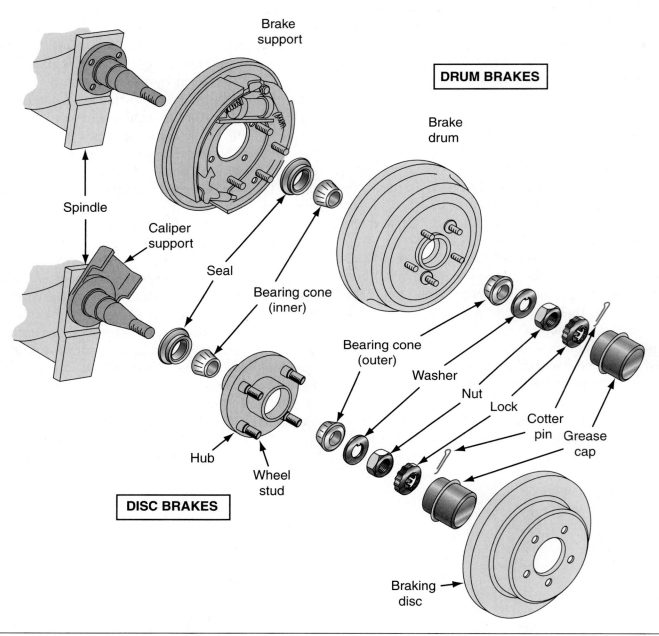

Figure 6-7 Various wheel studs for disc and drum brakes.

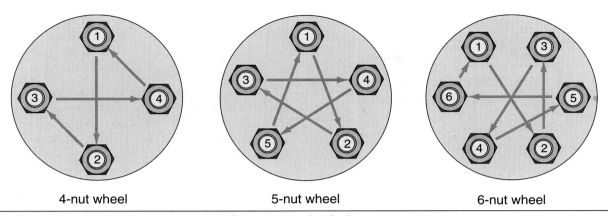

| 4-nut wheel | 5-nut wheel | 6-nut wheel |

Figure 6-8 Wheel torquing sequences for 4, 5, and 6 bolt patterns.

Some fasteners, such as wheel nuts, must be tightened in the proper sequence to avoid distorting the wheel rim **(Figure 6-8)**.

When air is compressed, moisture condenses in the air and tends to collect in the shop air lines. Develop a habit of oiling your air impact wrench daily. This action prevents internal rust, and extends the life of the impact wrench.

Using a Torque Wrench

A torque wrench allows the technician to tighten a fastener to the torque specified by the vehicle manufacturer. The specified torque provides the necessary clamping force between the components retained by the fasteners. Excessive torque may distort and warp components resulting in fluid leaks. Insufficient torque provides reduced clamping force. This condition may result in movement of the components retained by the fasteners, and this movement may cause damaged gaskets and leaks. Torque wrenches may be beam-type, click-type, dial-type, or digital-type **(Figure 6-9)**. Always apply steady pressure to the torque wrench until the specified torque is reached. If the torque wrench is jerked, the proper torque will not be achieved. Follow these steps when using a torque wrench:

1. Obtain the torque specifications from the vehicle manufacturer's service manual.

2. Familiarize yourself with any special torque procedures in the service manual, such as proper fastener tightening sequence.

3. Divide the torque specifications by three.

4. When using a click-type torque wrench, set the specified torque on the wrench handle.

5. Install the proper size of socket on the torque wrench drive.

6. Maintain the torque wrench at a 90-degree angle to the fastener being tightened.

7. Tighten the fastener to 1/3 of the specified torque.

8. Tighten the fasteners in proper sequence to 1/3 of the specified torque if applicable.

9. Tighten the fastener(s) to 2/3 of the specified torque.

10. Tighten the fastener(s) to within 10 ft.-lb (13 N·m) of the specified torque.

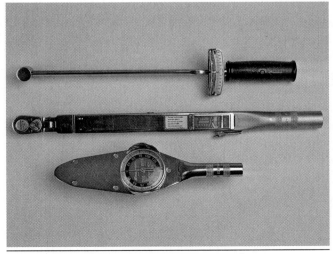

Figure 6-9 Three common types of torque wrenches are shown: a beam-type, click-type, and dial-type.

11. Tighten the fastener(s) to the specified torque.

12. Recheck the torque on the fastener(s).

> **Tech Tip** *Torque sticks may be used with an impact wrench when tightening wheel fasteners. The design of the torque stick limits the torque to the amount specified by the color of the torque stick. Use of a torque stick may not be approved by the vehicle manufacturer.*

USING GEARS, BEARINGS, AND SEALS

Pullers

Many gears and bearings are mounted on their shafts with a precision press-fit. This press-fit prevents any motion between the gear or bearing and the shaft on which it is mounted.

The absence of motion between these components prevents wear on the mating surfaces between the gear or bearing and the shaft. Always select the proper puller for the job. The puller jaws must make proper contact on the gear, bearing, or seal to be pulled **(Figure 6-10)**.

Some pullers have adjustable jaws to provide a proper jaw fit on various sizes of bearings, gears, or pulleys. Be sure the screw in the center of the puller does not damage the contacting surface on the shaft. For example, when pulling a crankshaft pulley, be sure the screw does not damage the threads in the center of the crankshaft. When

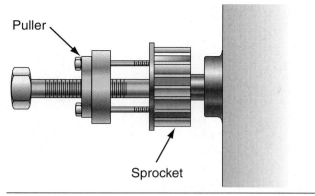

Figure 6-11 This puller has bolts extending through the puller into the gear being pulled.

pulling a gear or bearing, rotate the puller screw with steady, even pressure. Do not hammer or pry on the puller or the component being pulled. Some pullers have bolts that extend through the puller and thread into the gear being pulled **(Figure 6-11)**. Always be sure these bolts are threaded far enough into the gear so they do not pull out during the pulling operation. Some pullers have a slide hammer mounted on the puller shaft. The jaws of the puller are attached to the component to be pulled. In **Figure 6-12**, the jaws on a slide hammer puller are attached to pull a rear axle seal on a rear-wheel-drive vehicle. The technician pulls the slide hammer quickly against the outer end of the puller, and the resulting force pulls the component to which the jaws are attached.

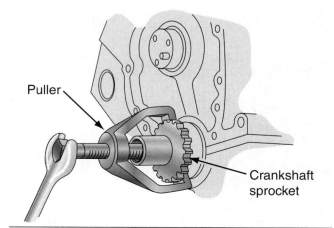

Figure 6-10 A puller is used to remove a crankshaft timing gear.

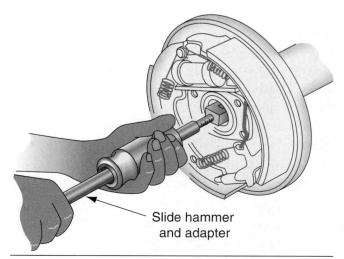

Figure 6-12 A slide hammer is used to remove a rear axle bearing.

When a puller is attached to a component, do not hammer on the puller or the component. This action may cause the puller to fly off the component resulting in personal injury.

Removing Broken Studs and Screws

Removing a broken stud or screw is usually a challenging and time consuming job for a technician. A great deal of time can be saved when the necessary precautions are taken to avoid breaking studs and screws. To avoid breaking fasteners, the first step is to spray all the fasteners with penetrating oil before attempting to remove the fasteners. This oil will work its way into the threads and dissolve rust and corrosion. The oil also lubricates the threads, which makes the fastener easier to remove. The second step to avoid breaking studs and screws is to always use the proper wrench. A box-end wrench or a socket grip the fastener more snugly than an open-end wrench. Third, be sure you are attempting to turn the fastener in the proper direction. Remember that a few automotive fasteners have left-hand threads. In some cases, the fastener may still break even though you followed the proper steps to avoid breakage.

When a fastener will not loosen with normal force, do not apply excessive force and break the fastener. If a fastener does not loosen with normal force, try tightening the fastener slightly, and then attempt to loosen the fastener. Sometimes this tightening action helps to loosen the fastener. When a fastener refuses to loosen, try heating the fastener with an oxy-acetylene torch. The heat will cause the nut to expand. When a cap screw will not loosen in a threaded block opening, heat the block casting in the area of the threaded opening. This heat may expand the block and allow the fastener to rotate. Use these methods to remove damaged or broken fasteners:

Do not spray penetrating oil on a fastener that has been heated and is still hot. Some of these oils are flammable, and this action could result in a sudden flame that ignites other combustible materials or causes personal injury.

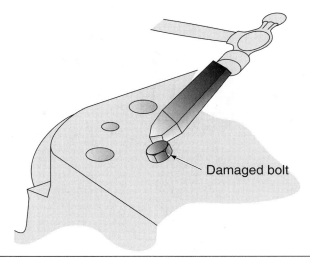

Figure 6-13 A hammer and chisel may be used to help loosen a fastener.

1. If the head on a fastener is damaged so a wrench slips on the head, a hammer and chisel may be used to tap on the side of the head and loosen the fastener **(Figure 6-13)**.

2. When a fastener does break off, if the fastener is sticking up above the casting it may be possible to install a stud remover over the fastener **(Figure 6-14)**. The stud remover is usually rotated with a 1/2-inch breaker bar.

3. Sometimes a nut may be arc welded on top of a broken fastener. Allow the nut to cool, and then rotate the nut to remove the fastener.

Figure 6-14 A stud remover.

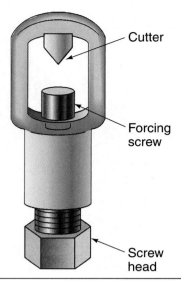

Cutter

Forcing screw

Screw head

Figure 6-15 A nut splitter is used to split a nut from a fastener.

4. When a nut is damaged, a nut splitter may be used to split the nut and remove it from the fastener **(Figure 6-15)**.

5. Broken fasteners may be removed with a left-hand twist drill and a reversible drill **(Figure 6-16)**.

6. Broken fasteners may be removed with a screw extractor. The first step in this process is to select the proper size drill bit for the size of extractor being used. Center punch the exact center of the broken fastener, and drill the

fastener. The drilled hole must be deep enough so the extractor does not bottom in the hole. Insert the extractor into the hole and rotate it counterclockwise with a turning handle to remove the fastener **(Figure 6-17)**.

7. If a broken fastener cannot be removed with a screw extractor, center punch it on the exact center of the broken fastener, and then drill all the way down through the fastener. Use progressively larger drill bits until all that remains of the fastener is a thin skin. Use a small metal pick to remove this thin skin from the threaded opening. Clean up the threaded opening with a bottoming tap.

Tech Tip *If a broken fastener will not loosen when attempting to rotate it with a screw extractor and turning handle, do not apply excessive force and break the extractor off in the fastener. Extractors are made from very hard, brittle steel, and the broken extractor will be very difficult to remove.*

Removing Damaged Nuts

Nuts and fasteners located under the vehicle become rusted and corroded after a period of time. Always spray penetrating oil on nuts before

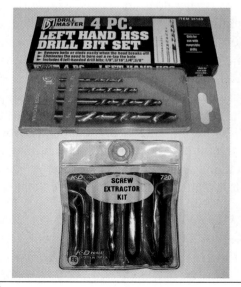

Figure 6-16 A screw extractor kit and left-hand twist drills.

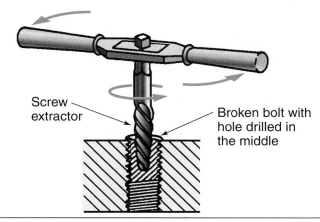

Figure 6-17 A screw extractor is used to remove broken bolts.

attempting to loosen them. Directions on the can of penetrating oil will indicate the length of time you should wait before attempting to loosen the nut after the oil application. If the nut still will not loosen after it is sprayed with penetrating oil, heat the nut with an oxy-acetylene torch.

Let the nut cool and then attempt to loosen it. If the corners of the nut are rounded off so a socket or box-end wrench slips on the nut, use a large pair of vise grips. Adjust the vise grips so they grip the nut tightly. Position your hand so that if the vise grips slip, your hand will not be injured. If the nut cannot be loosened using these methods, a nut splitter may be used to split the nut. The splitter is placed over the nut, and the screw on the splitter is rotated with a socket and ratchet until the nut is split. Do not rotate the screw on the splitter until the splitter damages the fastener threads. When installing nuts or cap screws in locations subjected to excessive heat and/or road splash, coat the threads with anti-seize compound before installation to prevent rust and corrosion on the fastener threads. The next technician who has to remove these fasteners will find they are easy to loosen.

USING AN OXY-ACETYLENE TORCH FOR HEATING

Technicians are often required to use an oxy-acetylene torch for heating, cutting, or welding. The first basic step in using this equipment is to learn how to light and adjust the torch for heating. Oxy-acetylene torch kits have two pressurized tanks. One tank contains oxygen and the other tank contains acetylene. In this explanation, we are assuming that the gauges are properly mounted on the tanks and all the hoses and the torch body are installed correctly on the end of the hoses. Most oxy-acetylene welders have a green hose and a red hose. The green hose is connected to the oxygen gauge and the red hose is attached to the oxy-acetylene gauge. The torch body has an oxygen control valve that adjusts the flow of oxygen from the green hose into the torch and an acetylene control valve that adjusts the flow of acetylene from the red hose into the torch **(Figure 6-18)**.

SAFETY TIP *The flame from an oxy-acetylene torch produces dangerous ultraviolet light rays and is extremely hot. Always wear approved welding goggles and gloves when using this equipment. Never allow the flame near any part of your body, clothes, or any ignitable material. Always use a striker to light the torch. Do not use matches or a cigarette lighter to light an oxy-acetylene torch. This action places your hand near the flame, which could result in severe burns to your hand.*

These valves are usually marked oxygen (OX) and acetylene (AC). Follow these steps to light the oxy-acetylene torch:

1. Be sure the valves on the oxygen and acetylene cylinders are fully closed by turning them clockwise until they are bottomed.

2. Observe the pressure gauges on the acetylene and oxygen cylinders. The high-pressure gauge in each set indicates the gas pressure in the cylinder and the low-pressure gauge indicates the pressure supplied to the torch. Be sure all the gauges indicate zero pressure.

3. If the gauges do not read zero, be sure the cylinder valves are closed and open the oxygen valve and the acetylene valve on the torch handle. Rotate the control knobs counterclockwise on

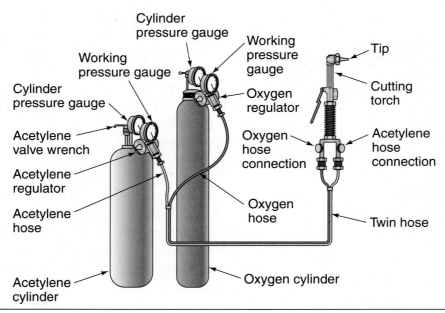

Figure 6-18 Major components of an oxy-acetylene torch.

both gauge sets until all the gauges read zero. Then close the oxygen and acetylene valves on the torch handle.

4. Install the proper end on the torch handle for the heating job you have to perform. Typically there are several heating tips with an oxy-acetylene torch. The smallest tip has a very small orifice designed for heating small components. The largest tip has a large orifice that is intended to produce a wider flame for heating larger components. The tips are threaded onto the end of the torch handle. Install the required tip snugly on the torch handle.

5. Close both valves on the torch handle and fully open the valve on the oxygen cylinder to its back-seated position by rotating it counterclockwise. Turn the control knob on the oxygen gauge set clockwise until the low-pressure gauge indicates 10 psi supplied to the torch.

6. Open the valve on the acetylene cylinder by rotating it 1/4 turn counterclockwise, and turn the control knob on the acetylene gauge set clockwise until the low-pressure gauge indicates 5 psi applied to the torch.

7. Put on a pair of welding gloves and a pair of welding goggles.

8. Aim the torch handle and tip away from the oxy-acetylene welder components, any part of your body, and any combustible materials.

9. Open the torch handle acetylene valve slightly, and operate the striker to create a spark and light the torch. A **striker** has a spring-type striker bar with a flint attached under the tip of the bar. When the striker is squeezed, the flint moves across a rough surface to create a spark.

10. Turn the torch acetylene valve on slowly until the black, sooty flame turns to a yellow flame.

11. Turn the torch oxygen valve slowly until the flame changes to a blue color. Continue to slowly open the torch oxygen valve until you can see a small, distinct inner cone right at the torch orifice in the center of the large flame. This type of flame is desirable for heating and it may be referred to as a **neutral flame.**

12. To shut the torch off, close the torch handle acetylene valve (red hose) first to extinguish the flame; then turn the torch oxygen valve off (green hose).

13. Turn valves on the oxygen and acetylene cylinders clockwise until they are fully closed.

14. Open the acetylene and oxygen valves on the torch and rotate the control knobs counterclockwise on both the acetylene and oxygen gauge sets until these knobs feel loose. Both gauges should read 0 psi. This will bleed the hoses of any pressurized gasses.

15. Close the acetylene and oxygen valves on the torch.

> **SAFETY TIP** *Acetylene gas is explosive! To avoid personal injury and possible explosions and fire, always use an oxyacetylene welder under the supervision of your instructor until you are familiar with the use of this equipment.*

> **SAFETY TIP** *The acetylene (fuel) hose is red and the oxygen hose is green. Oxygen can greatly promote combustion. Never use the oxygen or acetylene to blow dust away. Never service or operate the torch with oily or greasy hands or clothing.*

CHECKING AND CHANGING RESPIRATOR FILTERS

A respirator is worn in the shop to protect the technician from hazardous airborne particles such as asbestos dust, paint spray, and solvent vapors. The respirator is held on the face by straps attached to the respirator that fit behind the technician's head. As you breathe, air is taken into the respirator through two circular filters on each side of the respirator **(Figure 6-19)**. The filters keep hazardous particles

from entering your lungs. Air is expelled through an outlet in the lower center of the respirator.

The filter retainers on each side of the respirator are usually rotated counterclockwise to remove the retainers and filters. If there is any buildup of particles, dirt, or paint on the filters, they should be replaced with the proper filters for the hazardous material to which you are exposed. Filters may also have a limited time of use once they are put into service. Always consult the manufacturer's recommendations for filter usage and limits so the new filters fit properly with no air gaps between the filters and their seats on the respirator. The straps must hold the respirator snugly on your head. Be sure the respirator provides an adequate seal on your face.

> **SAFETY TIP** *Be sure the respirator contains the proper filter for the hazardous material to which you are subjected. The respirator may not provide adequate protection if it has the wrong filter.*

JUMP-STARTING A VEHICLE WITH A DISCHARGED BATTERY

A technician is sometimes required to **jump-start** a vehicle with a discharged battery. This seems like a basic operation, but if it is not done properly, very expensive computer damage may occur on the boost vehicle or the vehicle with the discharged battery. Jump-starting a vehicle with a discharged battery may be called battery boosting. Always wear eye protection when connecting a booster battery to a discharged battery. Be sure the discharged battery is not frozen.

> **Tech Tip** *If the battery has removable caps, then observe the electrolyte to see if it is frozen. If the battery is sealed, then use an infrared temperature gun to verify that the battery case is above 30° F before charging or jump-starting.*

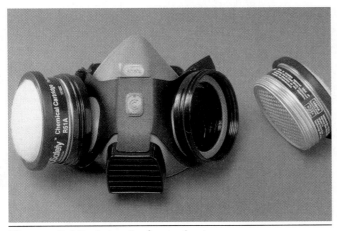

Figure 6-19 A typical respirator.

Boosting a frozen battery may cause a battery explosion. Do not lean over the battery when connecting the booster cables. Do not allow any bumper or body contact between the boost vehicle and the vehicle being boosted. Never jump-start a vehicle that has a battery in which the electrolyte has frozen. Follow these steps to connect the booster cables from the battery in the boost vehicle to the discharged battery:

1. Set the parking brake on both vehicles. Place the automatic transmission in park or a manual transmission in neutral.
2. Turn off all the electrical accessories on both vehicles.
3. Connect one end of the positive booster cable to the positive terminal of the discharged battery.
4. Connect the other end of the positive booster cable to the positive terminal of the battery in the boost vehicle.
5. Connect one end of the negative booster cable to the negative battery terminal in the boost vehicle.
6. Connect the other end of the negative booster cable to a good ground on the engine of the vehicle being boosted **(Figure 6-20)**.
7. Start the vehicle with the discharged battery.

8. Disconnect the negative booster cable from the engine in the vehicle with the discharged battery, and then disconnect the other end of this booster cable from the negative battery terminal in the boost vehicle.
9. Disconnect the end of the positive booster cable from the positive battery terminal in the boost vehicle, and then disconnect the other end of this cable from the positive battery terminal in the vehicle being boosted.

SAFETY TIP *Connecting booster cables with wrong polarity may damage electronic equipment on the boost vehicle or the vehicle being boosted. Loose booster cable connections may create a spark causing the battery in either vehicle to explode, resulting in serious personal injury and paint damage on the vehicle.*

Tech Tip *Booster packs are now widely available to jump-start a vehicle.*

BATTERY CHARGING

The vehicle battery serves two functions. It is principally used to start the engine and provides power to the electrical system when the charging

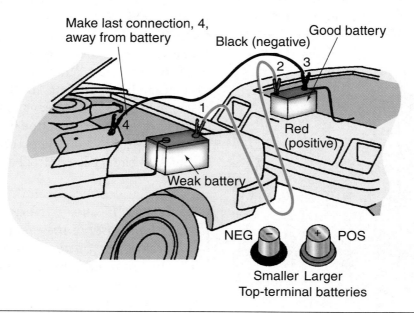

Figure 6-20 A jump-starting procedure.

system is not operating. Secondly it is used to filter or stabilize power and to provide extra power for the vehicles electrical system when electrical loads exceed the capability of the charging system.

There will be times when it will be necessary to use a battery charger to recharge a battery. These times are usually when the battery has lost a significant portion of its charge due to a failed charging system, lights or accessories inadvertently left on, extended testing of the vehicle during service work, or tests performed on the battery itself. The charging system is not meant to recharge a battery in these instances. A battery charger can recharge the battery at lower charge rates for extended periods of time **(Figure 6-21)**.

All battery chargers operate on the same principle. An electric current is produced to reverse the chemical reaction within the battery cells. It is recommended that the technician unhook the battery ground cable before recharging the battery. Though not absolutely necessary, this will help to minimize the possibility of damage to the charging system or other electronic components in the vehicle due to reversed polarity.

Tech Tip *Remove the battery ground cable before charging the battery.*

Some battery chargers are available with polarity protection to prevent reverse charging of the battery. If the battery ground is disconnected, then

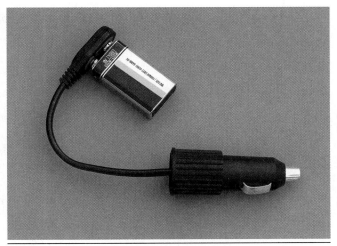

Figure 6-22 Various memory savers.

the use of a **memory saver** may be necessary. It is a simple tool to use; it saves computer memory on clocks, radio settings, and fuel injection systems. They usually plug into a cigarette lighter and are powered by either a small 9V battery, a wall-powered transformer, or are attached to a spare battery **(Figure 6-22)**.

Tech Tip *Always make sure the battery charger is turned off and unplugged before attaching or removing the battery charger cables to the battery terminals.*

General charging procedures are as follows:

1. Clean the battery of any dirt, dust, and corrosion. Clean the terminals if necessary.
2. Visually inspect the battery for damage. If the battery is serviceable, then inspect the electrolyte level and add distilled water if necessary.
3. Disconnect the battery ground if the battery is to be charged while still in the vehicle and use a memory saver if needed. If this is not practical, make sure to observe the correct polarity or use a battery charger with polarity protection.
4. Determine the correct charging current and time for the battery. Some chargers may have built-in devices to determine charge current and time **(Figure 6-23)**.
5. Make sure the charger main switch and timer are in the OFF position.

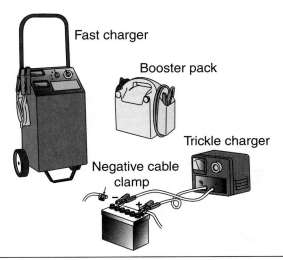

Fast charger

Booster pack

Trickle charger

Negative cable clamp

Figure 6-21 Charging a battery with a battery charger. Also a booster pack may be used to start a car with a weak battery.

Typical Charging Rates for Fully Discharged Batteries

Reserve Capacity Rating	20-Hour Rating	5 Amperes	10 Amperes	20 Amperes	30 Amperes	40 Amperes
75 Minutes or less	50 Ampere-Hours or less	10 Hours	5 Hours	2½ Hours	2 Hours	
Above 75 to 115 Minutes	Above 50 to 75 Ampere-Hours	15 Hours	7½ Hours	3½ Hours	2½ Hours	2 Hours
Above 115 to 160 Minutes	Above 75 to 100 Ampere-Hours	20 Hours	10 Hours	5 Hours	3 Hours	2½ Hours
Above 160 to 245 Minutes	Above 100 to 150 Ampere-Hours	30 Hours	15 Hours	7½ Hours	5 Hours	3½ Hours

Figure 6-23 Battery-charging rate chart.

6. Connect the leads to the proper battery terminals. The positive (+) cable of the charger should be attached to the positive (+) battery terminal. The negative (−) cable of the charger should be attached to the negative (−) battery terminal.

7. Plug the charger into an electrical outlet.

8. Turn the main power switch ON.

9. Set the proper charging current or charging rate.

10. Turn the timer to the desired charge rate.

SAFETY TIP *Always charge batteries in a well ventilated area.*

Slow Charging

Slow charging rates are preferred for battery charging. The slower rate allows for a more complete reversal of the chemical reaction and is easiest on the battery. Slow charging is preferable if time allows. Follow the general procedures for battery charging, along with the following:

1. Determine the charge rate. The maximum charge rate should not be more than 10 percent of the battery's capacity in Amp hour (Ahr) or not more than 10 amps. The charger may have a slow or low position.

2. Check the battery voltage occasionally. Turn the charger off and measure the battery voltage. If the voltage is 12.6 or higher, then the battery is considered to be fully charged.

Fast Charging

Fast charging rates may be necessary when time does not permit a slow charge. Fast charging may shorten the life of a battery as it charges the battery with a higher charging current for a shorter period of time. Some low-maintenance batteries cannot be fast charged. Follow the general procedures for battery charging, along with the following:

1. Determine the charge rate. The charger may have a fast or high position.

2. Set the timer for the desired time.

3. After the timer turns off, check the battery voltage. If the voltage is 12.6 or higher, then the battery is considered to be fully charged.

CLEANING EQUIPMENT OPERATION

Cleaning equipment comes in several categories. Pressure washers, parts cleaners, and bead blasters are but a few of those available for

Figure 6-24 A brake parts washer.

use in the shop. Any cleaning equipment must comply with state and federal wastewater laws where applicable. The most common cleaning equipment found in shops is the brake parts washer **(Figure 6-24)** and a conventional or aqueous parts washer **(Figure 6-25)**. Many parts washers are using an **aqueous** or water-based biodegradable cleaning solution rather than a solvent-based solution. Some conventional parts washers may be converted to an aqueous solution, but unfortunately most cannot. Contact the equipment manufacturer for information regarding the proper cleaning solution for use in the parts washer. You must read all safety and opera-

Figure 6-25 A conventional parts washer and an aqueous parts washer. *(Courtesy of Steffy Ford, Columbus NE)*

tion instructions for the particular equipment to be used. No matter what the particular style of equipment is to be used, there are some key operating guidelines:

- Read and follow all of the manufacturer's recommendations.
- Ensure that the shop manual permits cleaning using parts washers.
- Pre-clean parts if necessary. Remove any unnecessary grease, oil, gasket materials, etc.
- Maintain equipment filters as recommended by the manufacturer. Keep filters and oil skimmers operational with routine maintenance.
- Maintain solutions at recommended concentrations.
- Test solution in the parts washer regularly and add detergent when necessary.
- Dispose of washing solution in compliance with state and federal laws.

LIFTING EQUIPMENT OPERATION

A wide variety of vehicle lifts are used throughout the automotive service industry. Safe vehicle lifting requires an ongoing and consistent safety and health program, regular maintenance, and periodic worker training. These are the responsibility of the employer and it is the employer's duty to see that no one operates or maintains a lift without proper training.

For the standards pertaining to automotive and hoist safety, consult the ANSI Standard B153.1-Safety Requirements for the Construction, Care and Use of Automotive Lifts. The ANSI Standard (ANSI/ALI B153.1), since its inception in 1974, has been the benchmark for the construction of automotive lifts.

The following is some general information for the safe operation of automotive lifts:

1. Read and follow all safety instructions and operating procedures for the lift you are operating.

2. Do not block open or override safety devices.

3. Inspect your lift daily. Never operate it with broken or damaged parts such as lift arms that are

cracked, bent, or will not lock into place. Inspect chains and cables that are slack, deformed, corroded, cut, bent, or excessively worn.

4. Do not modify the lift with components not approved by the manufacturer. Never overload the lift. The manufacturer's rated capacity is shown on a nameplate fastened to the lift. If the nameplate is missing or the information is not readable due to wear, check immediately with the manufacturer's representative before using.

5. Before lifting any vehicle, know how to find its **center of gravity**. The center of gravity is the point between the front and the rear of the vehicle where the weight is distributed equally.

6. Always check to make sure the vehicle is not loaded with items or materials that might cause it to tip when raised.

7. Always keep the lift area free of obstructions, cords, hoses, grease, oil, trash, and other debris. Always inspect the area around you before carefully driving on or off the lift. Always perform a walk-around inspection.

8. Before driving the vehicle over the lift, position arms and supports to provide unobstructed clearance. Do not run over lift arms, adapters, or axle supports.

9. Load the vehicle on the lift carefully. Properly position lift supports to contact at the lifting points recommended by the vehicle manufacturer. Check the condition of the vehicle's lifting surfaces. Inspect the vehicle's lifting points for damage, rust, oil, dirt, undercoating, or other contaminants.

10. Raise the lift until supports contact the vehicle and check supports for secure contact with the vehicle. Raise the vehicle approximately one foot off the ground and shake it by pushing gently on the front or rear bumper to make sure it is stable. Raise the lift to the desired working height and engage the locking device.

11. Removal or installation of vehicle components may cause a critical shift in the center of gravity resulting in vehicle instability.

12. Always use under-hoist safety stands under the vehicle's frame or suspension for support and to help stabilize the vehicle **(Figure 6-26)**.

SAFETY TIP *When a vehicle is raised on a lift, the vehicle must be raised high enough to allow engagement of the lift locking mechanism. If the locking mechanism is not engaged, the lift may drop suddenly resulting in personal injury and/or vehicle damage.*

Figure 6-26 A vehicle on a lift with safety stands in use.

Summary

- An impact wrench is a reversible power tool for removing and tightening bolts and nuts.

- Impact wrenches may be powered by electricity or shop air pressure.

- A torque wrench is used to tighten fasteners to the specified torque.

- Torque wrenches may be beam-type, click-type, or dial-type.

- A puller is used to remove press-fit gears or bearings from their mounting location.

- Broken fasteners may be removed with a left-hand drill bit and a reversible drill or a screw extractor.

■ Damaged nuts may be removed by splitting them with a nut splitter.

■ When using an oxy-acetylene torch for heating, the oxygen pressure regulator should be set to 10 psi and the oxy-acetylene pressure regulator should be adjusted to 5 psi.

■ Always wear welding gloves and goggles when using an oxy-acetylene torch.

■ A respirator prevents hazardous materials in the shop air from entering your lungs.

■ When jump-starting a vehicle with a discharged battery, the booster cables must be connected with the proper polarity to the boost vehicle and the vehicle being boosted.

■ When jump-starting a vehicle with a discharged battery, the negative booster cable must be connected to a good ground on the engine in the vehicle being boosted.

Review Questions

1. Technician A says an impact wrench has torque-limiting capabilities. Technician B says the drive shaft on an impact wrench may turn at speeds between 2,000 and 14,000 rpm. Who is correct?

   A. Technician A

   B. Technician B

   C. Both Technician A and Technician B

   D. Neither Technician A nor Technician B

2. Technician A says fasteners may be over-tightened and damaged with an impact wrench. Technician B says a fastener should be started onto its threads with an impact wrench. Who is correct?

   A. Technician A

   B. Technician B

   C. Both Technician A and Technician B

   D. Neither Technician A nor Technician B

3. Technician A says broken fasteners may be removed with a left-hand drill bit and a reversible drill. Technician B says broken fasteners may be removed with a screw extractor. Who is correct?

   A. Technician A

   B. Technician B

   C. Both Technician A and Technician B

   D. Neither Technician A nor Technician B

4. When jump-starting a vehicle with a discharged battery, Technician A says the negative booster cable should be connected to an engine ground in the vehicle being boosted. Technician B says when jump-starting a vehicle with a discharged battery, the bumpers should be touching on the boost vehicle and the vehicle being boosted. Who is correct?

   A. Technician A

   B. Technician B

   C. Both Technician A and Technician B

   D. Neither Technician A nor Technician B

5. All of these statements about electric and air-operated impact wrenches are true *except*:

   A. A female-type quick disconnect coupling is threaded into an air-operated impact wrench.

   B. Some electric-operated impact wrenches are powered by a rechargeable battery.

   C. A technician may receive an electric shock from a 110-volt electric impact wrench with a defective ground connection.

   D. Electric and air-operated impact wrenches are reversible.

6. Technician A says a nut may become cross-threaded if it is not started by hand on a fastener. Technician B says improper tightening of the wheel nuts with an impact wrench may cause the rotor to warp. Who is correct?

A. Technician A

B. Technician B

C. Both Technician A and Technician B

D. Neither Technician A nor Technician B

7. When removing tight or seized fasteners, all of these precautionary steps should be taken to avoid breaking the fastener *except*:

A. Spray the fasteners with penetrating oil.

B. Be sure you are turning the fastener in the proper direction.

C. Be sure you are using the proper wrench on the fastener.

D. Apply excessive force to the wrench to loosen the fastener.

8. Technician A says broken fasteners may be removed by welding a nut on top of the fastener. Technician B says a damaged nut may be removed by splitting the nut with a nut splitter. Who is correct?

A. Technician A

B. Technician B

C. Both Technician A and Technician B

D. Neither Technician A nor Technician B

9. All of these statements about the use of oxy-acetylene welding equipment are true *except*:

A. The red hose is connected to the oxy-acetylene regulator.

B. Oxy-acetylene gas is poisonous.

C. Typically the pressure on the low-pressure oxygen gauge should be set to 10 psi.

D. Typically the pressure on the low-pressure oxy-acetylene gauge should be set to 5 psi.

10. The most common size of impact wrench drive shaft used in an automotive shop is:

A. 1/4 inch.

B. 1/2 inch.

C. 3/4 inch.

D. 7/8 inch.

11. A torque wrench may be _____ type, _____ type, _____ type, or _____ type.

12. A puller is used to remove components that have a _____.

13. When using an oxy-acetylene torch, the green hose is connected to the _____ regulator.

14. After using an oxy-acetylene torch, the torch valves should be opened and the control knobs on the oxygen and oxy-acetylene gauges rotated _____ until the gauges read _____ psi.

15. Explain the precautions that must be observed when installing and tightening a fastener with an impact wrench.

16. Explain the results of insufficient and excessive fastener torque.

17. Describe the first step that should be taken when removing a fastener with seized, rusted threads.

18. When using an electric impact wrench, explain two precautions that must be observed to avoid electrical shock.

19. Describe how a memory saver works.

20. Explain the difference between fast and slow charging.

CHAPTER

7

Measuring Systems, Measurements, and Fasteners

Learning Objectives

After you have read, studied, and practiced the contents of this unit, you should be able to:

- Understand USC and metric measurement systems.

- Perform measurements with a feeler gauge.

- Accurately measure components with an outside micrometer.

- Perform accurate measurements with a dial indicator.

- List four different types of fastener threads.

- Explain bolt diameter, pitch, length, thread depth, and grade marks.

- Describe the advantage of torque-to-yield bolts.

- Describe the proper tightening procedure for torque-to-yield bolts.

Key Terms

Bolt diameter

Bolt length

Bolt grade

Outside micrometer

Pitch

SI

Thread depth

Torque-to-yield bolt

Unified National Series

USCS

INTRODUCTION

Measuring systems, precision measurements, and fasteners are explained in this chapter. Automotive technicians must have an understanding of measuring systems. The fasteners used on modern cars have metric-sized heads. However, many of the vehicle manufacturer's specifications are provided in USCS measurements. Therefore, technicians must be familiar with both the USCS and metric systems of measurement. Technicians also

need to be familiar with the procedure for using measuring tools to perform precision measurements. Technicians must be able to identify bolt and nut strength markings so they always use an equivalent replacement fastener. An understanding of different types of fasteners is also a necessity for automotive technicians. For example, technicians must know that a number of automotive bolts require replacement each time they are removed. Many cylinder head bolts are an example of this type of fastener.

MEASURING SYSTEMS

Two systems of weights and measures are commonly used in the United States. One system of weights and measures is the US Customary System (**USCS**). Common measurements are the inch, foot, yard, and mile. In this system, the quart and gallon are common measurements for volume, and ounce, pound, and ton are measurements for weight. A second system of weights and measures is referred to as the metric system. The basic USCS linear measurement is the inch, while corresponding linear measurements in the metric system are the millimeter and the centimeter (**Figure 7-1**). Each unit of measurement in the metric system is related to the other metric units by a factor of ten. Thus, every metric unit can be multiplied or divided by ten to obtain larger units (multiples) or smaller units (submultiples). For example, the meter can be divided by one hundred to obtain centimeters (1/100 meter) or millimeters (1/1000 meter). The U.S. government passed the Metric Conversion Act in 1975 in an attempt to move American industry and the general public to accept and adopt the metric system.

NOTE: The metric system may be called the **SI** (System International).

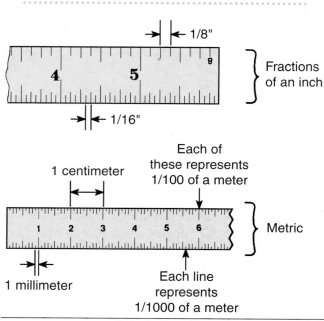

Figure 7-1 The USCS linear inch and the metric millimeter and centimeter are compared.

The automotive industry has adopted the metric system, and in recent years, most bolts, nuts, and fittings on vehicles have been changed to metric. During the early 1980s, some vehicles had a mix of USC and metric bolts. Import vehicles have used the metric system for many years. Although the automotive industry has changed to the metric system, the general public in the United States has been slow to convert from the USCS to the metric system. One of the factors involved in this change is cost. What would it cost to change every highway distance and speed sign in the United States to read kilometers? The answer to that question is probably hundreds of millions or billions of dollars.

NOTE: A metric conversion calculator provides fast, accurate metric-to-USCS conversions or vice versa.

Service technicians must be able to work with both the USCS and metric systems. Some common equivalents between the metric and USCS are listed in **Figure 7-2**. In the USCS, phrases such as 1/8 inch are used for measurements. The metric system uses a set of prefixes for this purpose. For example, the prefix kilo indicates 1,000; this prefix indicates there are 1,000 meters in a kilometer. Common prefixes in the metric system are listed in **Figure 7-3**.

Measurement of Mass

In the metric system, mass is measured in grams, kilograms, or tonnes. One thousand grams equals 1 kilogram. In the USCS, mass is measured in ounces, pounds, or tons. When converting pounds to kilograms, 1 pound equals 0.453 kilogram.

Measurement of Length

In the metric system, length is measured in millimeters, centimeters, meters, or kilometers. For example, ten millimeters (mm) equals 1 centimeter (cm). In the USCS, length is measured in inches, feet, yards, or miles. Some of the conversion factors for distance are shown in **Figure 7-4**.

1 meter (m) = 39.37 inches

1 centimeter (cm) = 0.3937 inch

1 millimeter (mm) = 0.03937 inch

1 inch = 2.54 cm

1 inch = 25.4 mm

1 ft. lb. = 1.35 newton meters (Nm)

1 in. lb. = 0.112 Nm

1 in. hg. = 3.38 kilopascals (kPa)

1 psi = 6.89 kPa

1 mile = 1.6 kilometer (m)

1 hp = 0.746 kilowatt (kW)

degrees F − 32 divided by 1.8 = degrees Celsius (C);

example, 212°F − 32 = 180 divided by 1.8 = 100°C

Figure 7-2 Metric and USCS equivalents.

NAME	SYMBOL	MEANING
mega	M	one million
kilo	k	one thousand
hecto	h	one hundred
deca	da	ten
deci	d	one tenth of
centi	c	one hundredth of
milli	m	one thousandth of
micro	μ	one millionth of

Figure 7-3 Prefixes commonly used in the metric system.

1 inch = 25.4 millimeters

1 foot = 30.48 centimeters

1 yard = 0.91 meter

1 mile = 1.60 kilometers

Figure 7-4 Measurements of length or distance.

Measurement of Volume

In the metric system, volume is measured in milliliters, cubic centimeters, and liters. One cubic centimeter equals 1 milliliter. If a cube has a length, depth, and height of 10 centimeters (cm), the volume of the cube is 10 centimeter × 10 centimeter × 10 centimeter = 1,000 cm$_3$ = 1 liter. When volume conversions are made between the two systems, 1 cubic inch equals 16.38 cubic centimeters. If an engine has a displacement of 350 cubic inches, 350 × 16.38 = 5733 cubic centimeters, and 5733/1,000 = 5.7 liters.

PRECISION MEASUREMENTS

Precision measurements are required in many areas of automotive service, such as engine repair, brake service, and steering diagnosis. If precision measurements are inaccurate, component operation is adversely affected and premature component failure may occur. Therefore, technicians must be familiar with these measurements.

Measuring Space with a Feeler Gauge and a Machinist's Rule

Feeler gauges are one of the most common automotive measuring tools. Many feeler gauge sets are thin strips of metal of varying precision thickness **(Figure 7-5)**. Wiretype feeler gauge sets contain round wires rather than metal strips **(Figure 7-6)**. A USCS set of round wire feeler gauges is marked in thousands of an inch, and this marking indicates

Figure 7-5 Various strip-type feeler gauge sets.

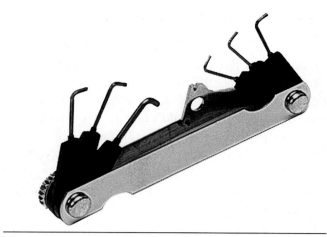

Figure 7-6 A round wire feeler gauge set.

the thickness of the gauge. Wire feeler gauges are commonly used to measure the spark plug gap between the center and ground electrode (**Figure 7-7**). The spark from the ignition system jumps across this gap to ignite the air-fuel mixture in the cylinder. The vehicle manufacturer provides spark plug gap specifications. When installing new spark plugs or servicing used spark plugs, the gap must be measured and adjusted if necessary. When learning to use feeler gauges, it is helpful to practice measuring and adjusting the gaps on some used spark plugs. Select a typical feeler gauge size for measuring spark plug gaps, such as 0.045 inch. Attempt to insert the feeler gauge between the spark plug electrodes. Bend the ground electrode as required until the selected feeler gauge slides between

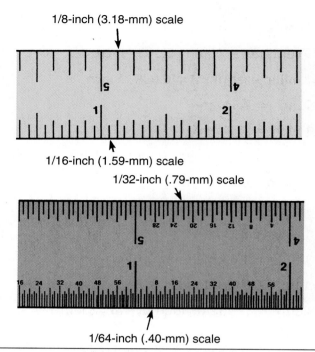

Figure 7-8 Graduations on a machinist's rule.

the spark plug electrodes with a light push fit. Be careful not to damage the center electrode or spark plug insulator when bending the ground electrode. A machinist's rule is commonly used for automotive measurements that do not require a close tolerance. A machinist's rule has four scales, two on each side of the rule. USCS scales include increments 1/8, 1/16, 1/32, and 1/64 inch (**Figure 7-8**).

> **Tech Tip** *There are 25.4 mm in 1 in. 1mm equals approximately .039 in. and 10 mm equals .394 in. Millimeters can be converted to inches by multiplying by .03937.*

> **Tech Tip** *Some spark plug feeler gauge sets have a bending tool that fits over the ground electrode when bending this component.*

Using an Outside Micrometer

When learning to use an outside micrometer, you can practice measurements on a used engine valve. When using an outside micrometer, insert two fingers through the micrometer frame and rotate

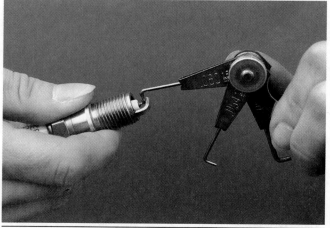

Figure 7-7 Measuring a spark plug gap with a round feeler gauge.

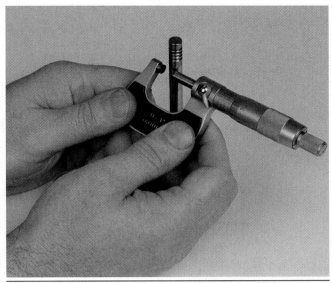

Figure 7-9 Proper hand position for holding a micrometer.

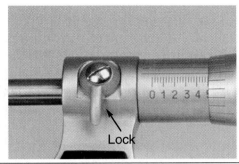

Figure 7-11 A micrometer with a lock lever.

the spindle with the thumb and forefinger on the same hand **(Figure 7-9)**. An **outside micrometer** is designed to measure the outside diameter of various components. Always use a clean shop towel to wipe any dust or dirt from the measuring surfaces on the micrometer spindle and anvil. Any dirt on these surfaces causes inaccurate readings. Place the spindle and anvil over the valve stem and slowly rotate the thimble until the spindle and anvil contact the valve stem **(Figure 7-10)**. The spindle and anvil must be positioned at a right angle to the valve stem, and these components must be centered on the diameter of this stem. Move the micrometer with a slight rocking action as you turn the thimble the last few thousandths of an inch. Rotate the thimble with a light rotating force until the spindle and anvil contact the valve stem lightly and squarely. You will know by feel when the spindle and anvil are centered on the valve stem diameter. Some micrometers have a lock that prevents the thimble from moving and changing the reading after the micrometer is positioned properly. Flip the lock lever over and lock the micrometer reading **(Figure 7-11)**. With the reading locked on the micrometer, the spindle and anvil should slide over the valve stem with a slight drag. Until you become used to the feel of a micrometer when it is properly positioned for component measurement, the thimble ratchet may be used to rotate the thimble **(Figure 7-12)**. When the anvil and spindle are properly positioned on a component with the correct force, the ratchet clicks and the spindle stops turning. This action prevents overtightening of the spindle on the component being measured.

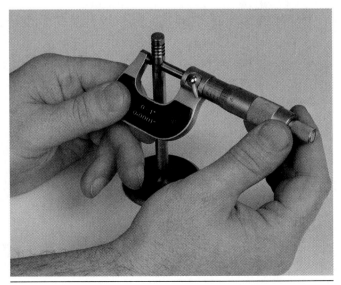

Figure 7-10 Positioning a micrometer to measure valve stem diameter.

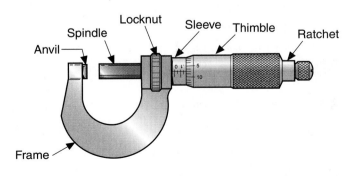

Figure 7-12 Major parts of an outside micrometer.

Tech Tip *Micrometers are available in different sizes. A 0- to 1-inch micrometer is designed to measure components with an outside diameter up to 1 inch. A 1- to 2-inch micrometer is designed to measure components with an outside diameter between 1 and 2 inches. Micrometers are precision instruments that must be kept clean and dry and treated with care. Always store the micrometer in the proper storage box. Do not turn a micrometer spindle downward onto a component with any force. This action will cause an inaccurate reading and it may also damage the micrometer.*

Tech Tip *Micrometers are available in metric or SI units as well.*

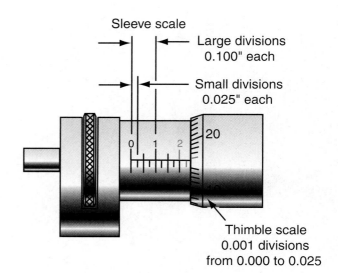

Sleeve scale
Large divisions 0.100" each
Small divisions 0.025" each
Thimble scale 0.001 divisions from 0.000 to 0.025

4 steps to read, add together:	Example above:
1. Select frame size	0 -1
2. Large barrel divisions	X 0.100 = 0.200"
3. Small barrel divisions	X 0.025 = 0.025"
4. Thimble divisions	X 0.001 = 0.016"
Reading	0.241"

Figure 7-13 A micrometer reading is shown.

Interpreting a Micrometer Reading

After the reading from a component is locked on the micrometer, the next step is to interpret the reading. Each graduation on the sleeve represents 0.025 inch, and each number on the sleeve represents 0.100 inch. In **Figure 7-13**, the 2 is exposed on the sleeve indicating 0.200 inch. Next count the number of exposed small divisions. Each division indicated 0.025 inch. Multiply the number of divisions by 0.025. The number of divisions from the thimble reading in Figure 7-13 is 1. 1 × 0.025 = 0.025.

Next, add the reading on the thimble to the reading on the sleeve. Each sleeve graduation indicates 0.001 inch. The thimble reading in Figure 7-13 is 0.013 inch.

Therefore, the actual micrometer reading is 0.200 + 0.025 + 0.016 = 0.241 inch.

Using a Dial Indicator

Dial indicators are usually sold in sets containing the dial indicator, a magnetic base, and other attaching arms **(Figure 7-14)**. Many automotive measurements, such as brake rotor runout and ball joint wear, require the use of a dial indicator.

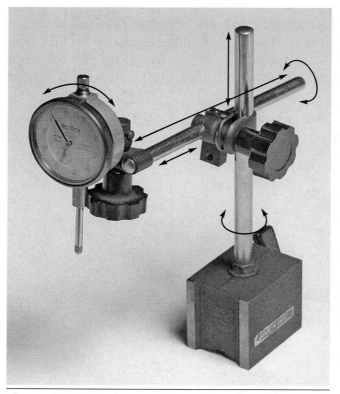

Figure 7-14 Various movements of a dial indicator with a magnetic base.

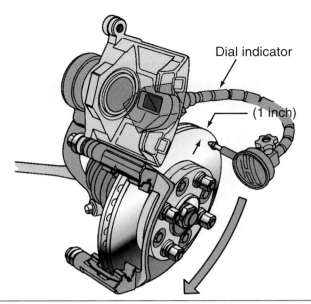

Figure 7-15 A dial indicator installed to measure brake rotor runout.

Figure 7-16 Zeroing the dial indicator.

The magnetic base is used to hold the dial indicator onto a ferrous metal component. The dial indicator may also be attached with special vise grips or C-clamps. When mounting a dial indicator, the support arm length should be kept to a minimum. Long support arms allow a slight dial indicator movement that causes inaccurate readings. The dial indicator should be mounted so it is at a 90° angle to the component being measured. Be sure the dial indicator mounting is secure and does not allow any indicator movement. In **Figure 7-15**, the dial indictor is set up to measure brake rotor runout. Always be sure the brake rotor and the dial indicator stem and plunger are clean. Dirt on these surfaces causes inaccurate readings. In this figure, a dial indicator with a magnetic base is attached to the lower end of the front strut. Be sure the dial indicator is at a 90° angle to the brake rotor. Position the dial indicator plunger against the rotor surface so approximately one-half of the stem is sticking out of the indicator. After the plunger is properly positioned, rotate the lock to prevent dial indicator movement. Gently grasp the movable dial indicator face and rotate this face until the zero position on the face is aligned with the dial indicator pointer **(Figure 7-16)**. Each dial indicator division indicates 0.001 inch. The pointer may swing to the right or left of zero when taking a dial indicator reading. Slowly rotate the brake rotor and observe the dial indicator. The pointer may move

twenty divisions on the right side of zero during part of a brake rotor rotation **(Figure 7-17)**. When the brake rotor is rotated for the remainder of a revolution, the dial indicator pointer may move to ten divisions to the left of zero. In this example, the total brake rotor runout is 0.020 + 0.010 = 0.030 inch. A small, separate pointer in the dial indicator face indicates the number of times the large gauge pointer makes a complete revolution.

FASTENERS

There are many different types of fasteners used throughout the vehicle. The most common are threaded fasteners, including bolts, studs, screws, and nuts. When servicing automotive components,

Figure 7-17 A dial indicator reading.

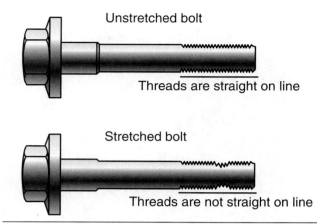

Figure 7-18 Check all bolts for stretch and other damage before reusing them.

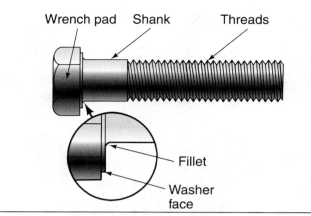

Figure 7-19 A typical bolt.

these fasteners must be inspected for thread damage, fillet damage, and stretch before they can be reused (**Figure 7-18**). In addition, many threaded fasteners are not designed for reuse and the service manual should be referenced for the manufacturer's recommendations. If a threaded fastener requires replacement, select a fastener the same diameter, thread pitch, strength, and length as the original. All bolts in the same connection must be of the same grade. Use nut grades that match their respective bolts, and use the correct washers and pins as originally equipped. Torque the fasteners to the specified value, and use torque-to-yield bolts where specified. Threaded fasteners used in automotive applications are classified by the **Unified National Series** using four basic categories: Unified National Coarse (UNC or NC), Unified National Fine (UNF or NF), Unified National Extrafine (UNEF or NEF), and Unified National Pipe Thread (UNPT or NPT). In recent years, the automotive industry has switched to the use of metric fasteners.

> **NOTE:** Unified National Fine-Thread or Unified National Course-Thread bolts may be referred to as fine-thread or coarse-thread bolts.

Metric threads are classified as course or fine, as denoted by an SI or ISO lettering. The most common type of threaded fastener used on the

vehicle is the bolt (**Figure 7-19**). To understand proper selection of a fastener, terminology must be defined (**Figure 7-20**). The head of the bolt is used to torque the fastener. Several head designs are used, including hex, Torx, slot, and spline. **Bolt diameter** is the measure across the threaded area or shank. The **pitch** in the USCS is the number of threads per inch. In the metric system, thread pitch is a measure of the distance (in millimeters) between two adjacent threads. **Bolt length** is the distance from the bottom of the head to the end of the bolt. The **bolt grade** denotes its strength and is used to designate the amount of stress the bolt can withstand. The grade of the bolt depends upon the material it is constructed from, bolt diameter, and thread depth. Grade marks are placed on the top of the head in the USCS to identify the bolt's strength (**Figure 7-21**). In the metric system, the strength of the bolt is identified by a property class number on the head (**Figure 7-22**). The larger the property class number, the greater the tensile strength of the fastener. **Thread depth** is the height of the thread from its base to the top of its peak. Like bolts, nuts are graded according to their tensile strength (**Figure 7-23**). As discussed earlier, the nut grade must be matched to the bolt grade. The strength of the connection is only as strong as the lowest grade used. For example, if a grade 8 bolt is used with a grade 5 nut, the connection is only a grade 5. Proper torque of the fastener is important to prevent thread damage and to provide the correct clamping forces. The service manual provides the manufacturer's recommended torque value and

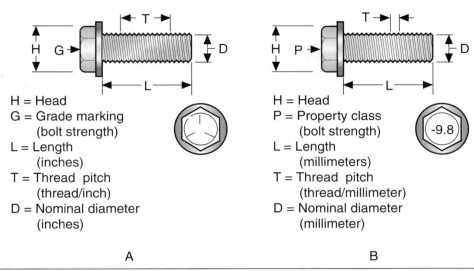

H = Head
G = Grade marking
     (bolt strength)
L = Length
     (inches)
T = Thread pitch
     (thread/inch)
D = Nominal diameter
     (inches)

H = Head
P = Property class
     (bolt strength)
L = Length
     (millimeters)
T = Thread pitch
     (thread/millimeter)
D = Nominal diameter
     (millimeter)

A   B

Figure 7-20 (A) USCS and (B) metric bolt terminology.

SAE grade markings					
Definition	No lines: unmarked indeterminate quality SAE grades 0-1-2	3 lines: common commercial quality Automotive and AN bolts SAE grade 5	4 lines: medium commercial quality Automotive and AN bolts SAE grade 6	5 lines: rarely used SAE grade 7	6 lines: best commercial quality NAS and aircraft screws SAE grade 8
Material	Low carbon steel	Med. carbon steel tempered	Med. carbon steel quenched and tempered	Med. carbon alloy steel	Med. carbon alloy steel quenched and tempered
Tensile strength	65,000 psi	120,000 psi	140,000 psi	140,000 psi	150,000 psi

Figure 7-21 Bolt grade identification marks.

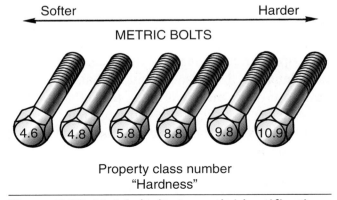

Softer   Harder

METRIC BOLTS

4.6   4.8   5.8   8.8   9.8   10.9

Property class number
"Hardness"

Figure 7-22 Metric bolt strength identification.

tightening sequence for most fasteners used in the engine or other components. The amount of torque a fastener can withstand is based on its tensile strength **(Figure 7-24)**. In order to obtain proper torque, the fastener's threads must be cleaned and may require light lubrication.

Torque-to-Yield Bolts

Automotive components, such as engines and transaxles, are designed with very close tolerances. These tolerances require an equal amount of

Inch system		Metric system	
Grade	Identification	Class	Identification
Hex nut grade 5	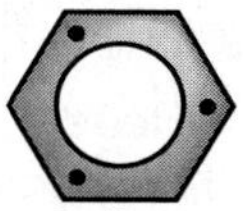 3 dots	Hex nut property grade 9	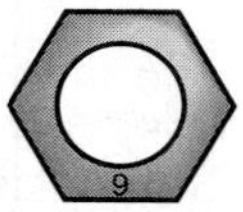 Arabic 9
Hex nut grade 8	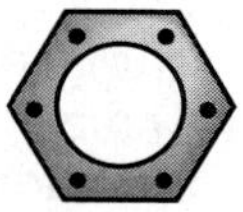 6 dots	Hex nut property grade 10	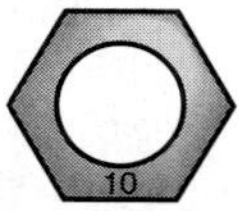 Arabic 10
Increasing dots represent increasing strength.		Can also have blue finish or paint dab on hex flat. Increasing numbers represent increasing strength.	

Figure 7-23 Nut grade markings.

STANDARD BOLT AND NUT TORQUE SPECIFICATIONS

Size Nut or Bolt	Torque (foot-pounds)	Size Nut or Bolt	Torque (foot-pounds)	Size Nut or Bolt	Torque (foot-pounds)
1/4–20	7–9	7/16–20	57–61	3/4–10	240–250
1/4–28	8–10	1/2–13	71–75	3/4–16	290–300
5/16–18	13–17	1/2–20	83–93	7/8–9	410–420
5/16–24	15–19	9/16–12	90–100	7/8–14	475–485
3/8–16	30–35	9/16–18	107–117	1–8	580–590
3/8–24	35–39	5/8–11	137–147	1–14	685–695
7/16–14	46–50	5/8–18	168–178		

Figure 7-24 Standard nut and bolt torque specifications.

clamping forces at mating surfaces. Normal head bolt torque values have a calculated 25-percent safety factor, that is, they are torqued to only 75 percent of the bolt's maximum proof load (**Figure 7-25**). Using the chart, it can be seen that a small difference between torque values at the bolt head can result in a large difference in clamping forces. Since torque is actually force used to turn a fastener against friction, the actual clamping forces can vary even at the same torque value. Up to about 25 ft.-lb (35 N·m) of torque, the clamping force is pretty constant. However, above this point, variation of actual clamping forces at the same torque value can be as high as 200 percent. This is due to variations in thread conditions or dirt and oil in some threads. Up to 90 percent of the torque is used up by friction, leaving 10 percent for the actual clamping. The result could be that some bolts have to provide more clamping force than others. This uneven clamping force can cause bolt, component, and gasket failure. To compensate and correct for these factors, many manufacturers use torque-to-yield bolts. A **torque-to-yield bolt** is a bolt that has been tightened to a specified yield or stretch point. The yield point of identical bolts does not vary much. A bolt that has been torqued to its yield

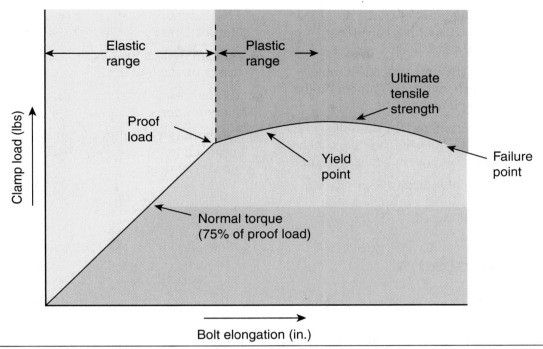

Figure 7-25 Relationship between proper clamp load and bolt failure.

point can be rotated an additional amount without any increase in clamping force. When a set of torque-to-yield fasteners is used, the torque is actually set to a point above the yield point of the bolt. This assures the set of fasteners will have an even clamping force. Manufacturers vary on specifications and procedures for securing torque-to-yield bolts. Always refer to the service manual for exact procedures. In most instances, a torque wrench is first used to tighten the bolts to their yield point. Next, the bolt is turned an additional number of degrees as specified in the service manual. In the

graph shown in Figure 7-25, notice that a bolt can be elongated considerably at its yield point before it reaches its failure point. Also notice that the clamp load is consistent between the proof load and the failure point of the bolt. Bolts that are torqued to their yield points have been stretched beyond their elastic limit and require replacement whenever they are removed or loosened.

Tech Tip *Torque-to-yield bolts must be replaced each time they are removed.*

Summary

- Every unit in the metric system can be divided or multiplied by ten to obtain smaller units or larger units.

- In the metric system, length is measured in millimeters, centimeters, meters, or kilometers.

- In the USC system, length is measured in inches, feet, yards, or miles.

- In the metric system, volume is measured in milliliters, cubic centimeters, and liters.

- In the USC system, volume is measured in cubic inches, quarts, and gallons.

- Spark plug gaps are measured with a round wire feeler gauge.

- A 1- to 2-inch outside micrometer is designed to measure components with an outside diameter between 1 and 2 inches.

- Each rotation of a USC micrometer thimble represents 0.025 inch, and each number on the sleeve indicates 0.100 inch.

- When reading a micrometer, the thimble reading is added to the sleeve reading to obtain the diameter of the component being measured.

- When mounting a dial indicator, the indicator stem should be at a 90-degree angle to the component being measured.

- The dial indicator must be securely mounted to eliminate movement of the indicator.

- Bolt strength in the USCS is indicated by grade marks on the bolt head.

- Torque-to-yield bolts provide a more uniform clamping force.

- Torque-to-yield bolts are tightened to the specified torque and then rotated a specific number of degrees.

Review Questions

1. Technician A says that in the metric system 1,000 meters equals 1 kilometer. Technician B says 1 cubic inch is equal to 16.38 cubic centimeters. Who is correct?

   A. Technician A

   B. Technician B

   C. Both Technician A and Technician B

   D. Neither Technician A nor Technician B

2. Technician A says that in the metric system mass is measured in kilograms. Technician B says 1 kilogram equals 0.453 pounds. Who is correct?

   A. Technician A

   B. Technician B

   C. Both Technician A and Technician B

   D. Neither Technician A nor Technician B

3. Technician A says to grind the center electrode when adjusting spark plug gaps. Technician B says if a spark plug gap is adjusted properly, the specified feeler gauge should fit between the electrodes with a light drag. Who is correct?

   A. Technician A

   B. Technician B

   C. Both Technician A and Technician B

   D. Neither Technician A nor Technician B

4. Technician A says a 1- to 2-inch outside micrometer will measure the outside diameter of a shaft with a 0.500-inch diameter. Technician B says when reading an outside micrometer, the thimble reading must be subtracted from the sleeve reading. Who is correct?

   A. Technician A

   B. Technician B

   C. Both Technician A and Technician B

   D. Neither Technician A nor Technician B

5. All of these statements about the metric system of measurements are true *except*:

   A. The basic linear measurement is the meter.

   B. Each unit of measurement is related to the other units by a factor of 20.

   C. The prefix kilo indicates 1,000.

   D. The liter is a measurement for volume.

6. Technician A says a machinist's rule is used for automotive measurements that require close tolerances of .001 in. Technician B says a feeler gauge may be used to measure the gap or clearance between two components. Who is correct?

   A. Technician A

   B. Technician B

   C. Both Technician A and Technician B

   D. Neither Technician A nor Technician B

7. All of these statements about using an outside micrometer are true *except*:

   A. The surfaces on the micrometer spindle and anvil must be clean.

   B. The spindle and anvil must be positioned at a right angle to the component being measured.

   C. The thimble should be turned with a light rotating force.

   D. A 1- to 2-inch micrometer will measure a component with a diameter of 2.25 inches.

8. When measuring the diameter of a shaft with a USCS micrometer, the number 4 is exposed on the sleeve but no other lines on the sleeve are visible after the 4, and the thimble reading is 0.020. The shaft diameter is:

   A. 0.380 inch.

   B. 0.400 inch.

   C. 0.420 inch.

   D. 0.440 inch.

9. While measuring brake rotor runout with a dial indicator, Technician A says the full length of the dial indicator stem should be sticking out of the indicator. Technician B says the dial indicator stem should be positioned at a 90-degree angle to the brake rotor. Who is correct?

   A. Technician A

   B. Technician B

   C. Both Technician A and Technician B

   D. Neither Technician A nor Technician B

10. All of these statements about fastener replacement are true *except*:

    A. Torque-to-yield bolts may be reused if they are removed.

    B. All bolts in the same connection must have the same grade.

    C. Nut grades must match their respective bolts.

    D. Fasteners must be tightened to the specified torque in the proper sequence.

11. When an outside micrometer is adjusted to measure the diameter of a round component, the spindle and anvil should slide over the component with a _____.

12. The ratchet on a micrometer may be used to rotate the _____.

13. If the sleeve reading on a micrometer is 0.150 inch and the thimble reading is 0.011 inch, the micrometer reading is _____ inch.

14. A dial indicator may be mounted on a metal component with a _____ base.

15. How may a dial indicator be mounted on a non-ferrous metal component?

16. Describe the proper procedure for mounting a dial indicator to measure brake rotor runout.

17. After the dial indicator is properly mounted, describe the proper procedure for measuring the brake rotor runout.

18. Explain the advantage of torque-to-yield bolts, and give one example of where these bolts are used in a vehicle.

19. Describe the proper procedure for tightening torque-to-yield bolts.

CHAPTER

8

Service Information

Learning Objectives

After you have read, studied, and practiced the contents of this unit, you should be able to:

- Explain the importance of service information as it relates to vehicle repair.

- Explain some of the information found in an owner's manual.

- Describe the general layout of a vehicle manufacturer's service manual.

- Explain the usual differences between a typical vehicle manufacturer's service manual and a generic service manual.

- Describe how electronic service information is stored and accessed.

- Explain the purpose of computer software.

- Explain the meaning of flash programming.

- Describe the purposes of service bulletins.

- Explain how a labor estimating guide is used when preparing a vehicle repair estimate.

Key Terms

Compact disks (CDs)

Digital versatile disks (DVDs)

Diagnostic link connector (DLC)

Diagnostic procedure charts

Flash programming

Technical service bulletins (TSB)

Vehicle Emission Control Information (VECI)

Vehicle Identification Number (VIN)

INTRODUCTION

Access to service information is absolutely essential when repairing modern vehicles. Without the necessary service information, technicians may waste a great deal of time diagnosing a problem that is explained in a service bulletin. In many cases it is impossible to diagnose electrical/electronic problems without a wiring diagram for the circuit being diagnosed. The technician must also have access to specifications for the vehicle being repaired. For example, without the proper torque specifications and procedures, expensive components in engines or transaxles may be warped and ruined.

TYPES OF SERVICE INFORMATION

Vehicle manufacturers supply an owner's manual with each vehicle. This manual contains basic service information such as using the jack supplied with the vehicle and changing a tire and wheel. The owner's manual includes information related to checking fluid levels and tire pressure. Information regarding the type of lubricants required for the vehicle and necessary fluid capacities are also included in the owner's manual.

Service manuals are one of the most common types of service information. Vehicle manufacturers or aftermarket suppliers of automotive information and components provide service manuals. With the increasing vehicle complexity in recent years, service manuals have become extensive. Some vehicle manufacturer's service manuals for specific vehicles now contain as many as five volumes with several hundred pages in each volume. Because of the large increase in the number of pages in service manuals, many vehicle manufacturers are now providing their service manuals on **compact disks (CDs)** or on **digital versatile disks (DVDs).** Placing service manuals on CDs or DVDs greatly reduces the storage space requirements for these manuals.

Vehicle manufacturers and some independent automotive service groups publish service bulletins. These bulletins contain service information related to the diagnosis and repair of a specific vehicle problem. Some service bulletins contain information regarding improved, modified parts that are required to eliminate a specific vehicle defect or complaint. An increasing amount of service information is available electronically on CD/DVDs, or on a computer network. Most vehicle manufacturers make extensive use of computer networks to provide service information to their dealerships. The dealership computers can access and/or download service information from the vehicle manufacturer's computer network.

OWNER'S MANUAL

The owner's manual provides valuable information for the vehicle owner. For example, this

Figure 8-1 The owner's manual can give valuable information about the common service specifications such as vehicle instrumentation, capacities, trailer towing, etc.

manual provides information regarding the proper use of vehicle restraints, such as seat belts, air bags, and child restraints. An index is included in the owner's manual so the desired information can be accessed quickly. Customer assistance and warranty information is also provided in the owner's manual **(Figure 8-1)**.

The amount of information in the owner's manual may vary depending on the vehicle make and model year. Vehicles have become more complex in recent years, and owner's manuals have been expanded to include basic information regarding the proper operation of vehicle systems and components.

The owner's manual includes information regarding vehicle instrumentation and warning systems. The use of the vehicle controls such as switches and levers are explained in the owner's manual. The manual also provides many cautions so the driver/owner understands the results of improper system or vehicle operation. Maintenance schedules, fluid capacities, and the recommended lubricants and fluids are provided in the owner's manual. The owner's manual provides information on checking all the fluid levels and adding the proper fluids.

The owner's manual provides information regarding the towing capacity of the vehicle and includes towing precautions. Basic service information is included such as changing tires, wiper blades, or lightbulbs. The owner's manual also provides information on proper cleaning and maintenance of the vehicle interior and exterior. Valuable information regarding the location, identification, and replacement of electrical fuses is also included.

MANUFACTURER'S SERVICE MANUAL

The service manual is one of the most important tools for today's technician. It provides information concerning component identification, service procedures, specifications, and diagnostic information. In addition, the service manual provides information concerning wiring harness connections and routing, wiring diagrams, component location, and fluid capacities. The service manual provides an explanation of the **Vehicle Identification Number (VIN) (Figure 8-2)**. The VIN is a -series of letters and numbers that identifies specific vehicle parameters such as the vehicle make, model, model year, and size of the engine. The VIN information is essential when ordering parts.

Most service manuals published by vehicle manufacturers now have a standard format **(Figure 8-3)**. The service manual usually provides illustrations to guide the technician through the

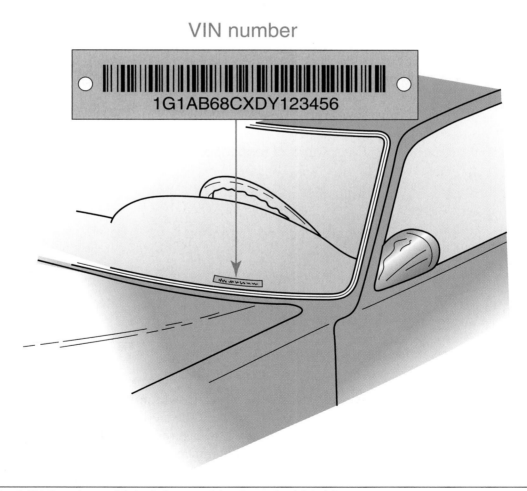

VIN number

1G1AB68CXDY123456

Figure 8-2 The VIN for the vehicle is located in the windshield area.

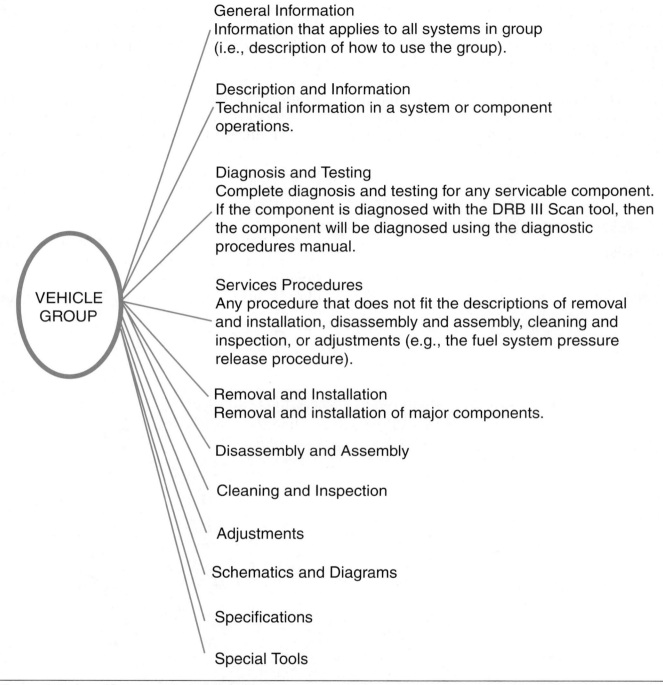

General Information
Information that applies to all systems in group
(i.e., description of how to use the group).

Description and Information
Technical information in a system or component
operations.

Diagnosis and Testing
Complete diagnosis and testing for any servicable component.
If the component is diagnosed with the DRB III Scan tool, then
the component will be diagnosed using the diagnostic
procedures manual.

Services Procedures
Any procedure that does not fit the descriptions of removal
and installation, disassembly and assembly, cleaning and
inspection, or adjustments (e.g., the fuel system pressure
release procedure).

Removal and Installation
Removal and installation of major components.

Disassembly and Assembly

Cleaning and Inspection

Adjustments

Schematics and Diagrams

Specifications

Special Tools

VEHICLE GROUP

Figure 8-3 A uniform service manual layout.

service operation **(Figure 8-4)**. Always use the correct manual for the vehicle and system being serviced. Follow each step in the service procedure. It is important to follow these steps as the manufacturer has outlined these as the proper order in which to complete repairs. Do not skip any of the steps!

Measurements such as torque, endplay, and clearance specifications are located in or near the service manual text or procedural information.

Ball Joint Boot Replacement

NOTE: The upper control arm ball joint, lower control arm ball joint, and knuckle upper ball joint are attached with the boot retainer to improve the sealing efficiency of the boot.

1. Remove the set ring and boot.

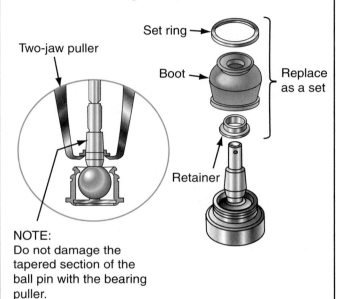

NOTE:
Do not damage the tapered section of the ball pin with the bearing puller.

2. Remove the retainer.

NOTE: The knuckle lower ball joint does not have a retainer.

3. Pack the interior of the boot and lip with grease.

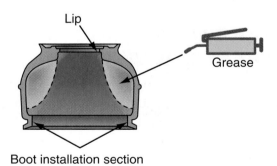

CAUTION: Do not contaminate the boot installation section with grease.

4. Wipe the grease off the sliding surface of the ball pin and pack with fresh grease.

5. Insert the new retainer lightly into the ball joint pin.

NOTE: When installing the ball joint, press the retainer into the ball joint pin.

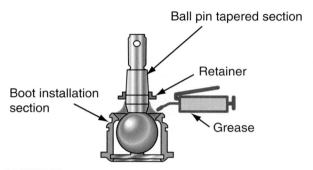

CAUTION:

Keep grease off the boot installation section and the tapered section of the ball pin.

Do not allow dust, dirt, or other foreign materials to enter the boot.

6. Install the boot in the groove of the boot installation section securely, then bleed air.

7. Adjust the special tool with the adjusting bolt until the end of the tool aligns with the groove on the boot.

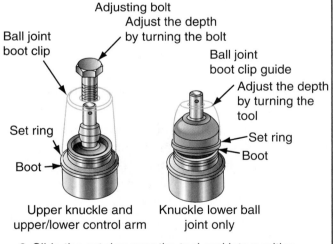

8. Slide the set ring over the tool and into position.

CAUTION: After installing the boot, check the ball pin tapered section and threads for grease contamination and wipe them if necessary.

Figure 8-4 Illustrations in a service manual guide the technician through service procedures.

TORQUE SPECIFICATIONS

Fan Assembly Motor-to-Fan..3.3 Nm (29 lbs. in.)
Fan-to-Radiator Support Bolt ....9 Nm (80lbs. in.)
Hose Clamps
    Heater Hose ....1.7 Nm (15 lbs. in.)
    Radiator Hose.. 3.4 Nm (30 lbs. in.)
Lower Air Deflector to Impact Bar.. 2 Nm (18 lbs. ft.)
Throttle Body Inlet Pipe Bolt 3.1L (VIN T) .... 25 Nm (18 lbs. ft.)
Transmission Oil Cooler Fittings at Radiator .... 27 Nm. (20 lbs. ft.)
Trans. Oil Cooler Pipe Connections (Alum. Radiator).............. 20 Nm (15 lbs. ft.)
Radiator Outlet Pipe to Block 2.01 (VIN K)........................ .. 27 Nm (20 lbs. ft.)
Radiator to Radiator Support Bolts..10 Nm (90 lbs. in.)
Thermostat Housing Bolts
    2.0L (VIN K) & 3.1L (VIN T)................................. 27 Nm (20 lbs. ft.)
Coolant Pump-to-Block Bolts
    2.0L (VIN K).. 25 Nm (19 lbs. ft.)
    3.1L (VIN T).. 24 Nm (18 lbs. ft.)
Coolant Pump-to-Front Cover Bolts
    3.1L (VIN T) ....10 Nm (90 lbs. in.)
Coolant Pump Pulley-to-Pump Bolts
    2.0L (VIN K) .... 24 Nm (17 lbs. ft.)
    3.1L (VIN T) ....21 Nm (15 lbs. ft.)
Surge Tank Bolts ....4 Nm (15 lbs. ft.)
Surge Tank Pipe to Block Bolt 3.1L (VIN T)............................... 8 Nm (70 lbs. in.)
Temperature Sending or Gauge Unit ....27 Nm (20 lbs. ft.)

Figure 8-5 This picture shows a shop manual page that gives vehicle torque specifications.

Specification tables are usually provided at the end of the procedural information or component area **(Figure 8-5)**. Because the service manual is divided into a number of main component and system areas, a table of contents is provided at the front of the manual to provide quick access to the desired information. Each component area or system is covered in a section of the service manual **(Figure 8-6)**. At the beginning of each section in the manual, a smaller table of contents guides you to the information regarding the specific system or component being serviced. The service manual may be divided into several volumes because of the extensive amount of information required to service today's vehicles.

TROUBLESHOOTING AND DIAGNOSTIC TABLES

Diagnostic information in each section of the service manual is usually provided in diagnostic procedure charts **(Figure 8-7)**. **Diagnostic procedure charts** provide the necessary diagnostic steps in the proper order to diagnose specific problems in vehicle systems. The test results obtained in a specific diagnostic step guide the technician to the next appropriate step. When the technician follows the appropriate diagnostic procedure chart, unnecessary diagnostic steps are avoided. Following these steps helps the technician to arrive at the correct repair in the most efficient time.

TABLE OF CONTENTS	SECTION NUMBER
GENERAL INFO. AND LUBE	
General Information	0A
Maintenance and Lubrication	0B
HEATING AND AIR COND.	
Heating and Vent. (nonA/C)	1A
Air Conditioning System	1B
V-5 A/C Compressor Overhaul	1D3
BUMPERS AND FRONT BODY PANELS	
Bumpers (See 10-4)	
Fr. End Body Panels (See 10-5)	
STEERING, SUSPENSION, TIRES, AND WHEELS	
Diagnosis	3
Wheel Alignment	3A
Power Steering Gear & Pump	3B1
Front Suspension	3C
Rear Suspension	3D
Tires and Wheels	3E
Steering Col. On-Vehicle Service	3F
Steering Col. - Std. Unit Repair	3F1
Steering Col. - Tilt, Unit Repair	3F2
DRIVE AXLES	
Drive Axles	4D
BRAKES	
General Info. - Diagnosis and On-Car Service	5
Compact Master Cylinder	5A1
Disc Brake Caliper	5B2
Drum Brake - Anchor Plate	5C2
Power Brake Booster Assembly	5D2
ENGINES	
General Information	6
2.0 Liter I-4 Engine	6A1
3.1 Liter V6 Engine	6A3
Cooling System	6B
Fuel System	6C
Engine Electrical - General	6D
Battery	6D1
Cranking System	6D2
Charging System	6D3
Ignition System	6D4
Engine Wiring	6D5
Driveability & Emissions - Gen.	6E
Driveability & Emissions - TBI	6E2
Driveability & Emissions - PFI	6E3
Exhaust System	6F

TABLE OF CONTENTS	SECTION NUMBER
TRANSAXLE	
Auto.Transaxel On-Car Serv.	7A
Auto. Trans. - Hydraulic Diagnosis	3T40-HD
Auto. Trans. - Unit Repair	3T40
Man. Trans. On-Car Service	7B
5-Sp. 5TM40 Man. Trans. Unit Repair	7B1
5-Sp. Isuzu Man. Trans. Unit Repair	7B2
Clutch	7C
CHASSIS ELECTRICAL, INSTRUMENT PANEL, & WASHER WIPER	
Electrical Diagnosis	8A
Lighting and Horns	8B
Instrument Panel and Console	8C
Windshield Wiper/Washer	8E5
ACCESSORIES	
Audio System	9A
Cruise Control	9B
Engine Block Heater	9C
BODY SERVICE	
General Body Service	10-1
Stationary Glass	10-2
Underbody	10-3
Bumpers	10-4
Body Front End	10-5
Doors	10-6
Rear Quarters	10-7
Body Rear End	10-8
Roof & Convertible Top	10-9
Seats	10-10
Safety Belts	10-11
Body Wiring	10-12
Unibody Collision Repair	11-1
Welded Panel Replacement	11-2
INDEX	
Alphabetical Index	

Figure 8-6 The table of contents directs you to the major systems and component areas in the service manual.

		YES	**NO**
A	1 Connect a scan tool to the data link connector. 2. Turn the key "ON", with the engine "OFF." Program the scan tool for the vehicle Does the scan tool display data?	Go to step C	Go to step B
B	Select "Diagnostic Circuit Check." Class 2 Message Monitor. Does the scan tool show other computers as active?	Go to NO Scan Tool Data	Go to Body and Accessories
C	Does the engine start and continue to run?	Go to step D	Go to Engine Cranks Does Not Run
D	Does the engine continue to run after the vehicle has exceeded 3.2 km/h (2 mph)?	Go to step E	Go to Body and Accessories

E	Are DTC's stores in memory?
	
	Priority for diagnosis
	
	    Fuel enable DTCs (P1626, P1631)
	    PCM malfunction DTCs (P0601, P0602)
	    System voltage DTC (P0560, P1635, P1639)
	    Component level DTCs (switches, sensors, ODMs)
	    System level DTCs (fuel trim, misfire, EGR flow,
	            TWC system, EVAP system,
	            idle control system, and HO2S
	            response/transition)
		Go to
		applicable
		DTC
			Go to step F

F	Observe the DTC information. If DTC status "Last Test Failed," Test Failed, Test Failed This Ignition, MIL Request, or History DTCs are set, save DTC Freeze Frame and/or Failure Records information Were any DTCs displayed?	Go to the applicable DTC table	Go to step G
G	Compare scan data with the values shown in the "Engine Scan Tool Data list." Are the values normal or within typical ranges?	Go to Symptoms	Go to Diagnostic Aids and Test Descriptions

Figure 8-7 This picture shows a shop manual DLC scan tool diagnostic chart.

Tech Tip *When using diagnostic proce-dure charts in a service manual, do not skip steps in the procedure unless you are instructed to do so. Skipping steps in a diag-nostic procedure may lead to inaccurate diagnosis and wasted time.*

GENERIC SERVICE MANUALS

Independent automotive service organizations or aftermarket automotive component suppliers usually publish generic service manuals. Generic service manuals are usually abbreviated com-pared to vehicle manufacturer's service manuals

Figure 8-8 A good assortment of service manuals is an essential shop resource.

Figure 8-9 Computers are replacing printed service manuals in many shops.

(Figure 8-8). A generic service manual may include coverage of several years of a specific vehicle make. Some generic service manuals may nclude coverage of several vehicle models supplied by the same manufacturer. Later generic service manuals may include updated steps or tips to follow during service procedures.

ELECTRONIC SERVICE INFORMATION

Service and parts information can also be provided through computer services **(Figure 8-9)**. Computerized service information may be provided on CD/DVDs. Service information on CD/DVDs can be stored and accessed more easily than information in service manuals. Computers may also be connected to a network to obtain service information. Using the computer keyboard, light pen, mouse, or touch-sensitive screen, the technician chooses from a series of menus on the computer monitor. When the desired information is accessed, it may be printed out for detailed study or taken to the vehicle for reference.

COMPUTER SOFTWARE

Computer software is the program in a computer that is responsible for proper operation of the system to which the computer is connected. A higher percentage of new-vehicle cost each year is allocated for the rapid increase in the complexity of computer software. The computer software contains extensive information regarding the operation of the computer system under all possible operating conditions. When many computers are replaced, the necessary software must be downloaded from the appropriate scan tool or computer system **(Figure 8-10)**. This process is usually referred to as **flash programming**. Flash programming may also be used to reprogram onboard computers to correct specific computer system operational problems.

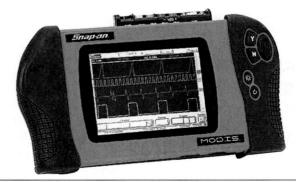

Figure 8-10 A scan tool is used to communicate with the vehicles onboard computers.

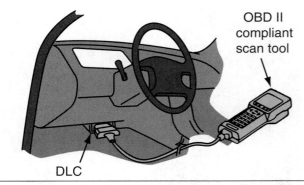

Figure 8-11 The DLC is the connection for the scan tool, used to help determine any trouble codes held in the memory of the vehicles onboard computers.

Tech Tip *As a precaution, never connect or disconnect the scan tool cable connection to the DLC with the ignition switch on. This action may cause computer damage or scan tool damage. The ignition switch must be turned off when connecting or disconnecting the scan tool.*

After a new computer is installed in a vehicle system, the scan tool is connected to a **diagnostic link connector (DLC)** that is usually located under the vehicle instrument panel **(Figure 8-11)**. Vehicles manufactured since 1996 have OBD II systems that have a 16-terminal DLC positioned under the driver's side of the instrument panel. Data links that interconnect each computer on the vehicle are connected to the DLC to allow access to any vehicle computer. When the scan tool is connected to the DLC, the appropriate buttons on this tool are pressed to download the necessary software into the appropriate computer.

TECHNICAL SERVICE BULLETINS

The vehicle manufacturer may improve or revise service procedures at any time. Many **technical service bulletins (TSB)** provide information regarding modified components and service procedures to correct specific problems on certain vehicles. When diagnosing automotive problems,

technicians must have up-to-date TSB information to provide fast, accurate diagnosis. The vehicle manufacturer's TSBs are made available to dealerships in printed form and/or on CD/DVD. Some vehicle manufacturers make their TSBs available for sale to independent automotive shops. Some TSBs provide up-to-date corrections for service manuals. If a significant number of corrections are required, a second or revised edition of the manual may be published. When service information is provided on CD/DVD, the CD/DVDs are updated frequently to include the latest TSB information. Some independent publishers of automotive service information produce CD/DVDs that contain TSB information for many different vehicle makes and model years.

LABOR ESTIMATING GUIDES

Many automotive repair shops provide the customer with an estimate of vehicle repair costs. Parts and labor costs are the two main expenses in any vehicle repair. The labor is the cost of the time required for the technician to repair the customer's vehicle. An estimating guide is used to provide an estimate of the labor costs required to repair a vehicle. An estimating guide lists the time required to perform all the possible repair jobs on different vehicles. The service person greeting the customer finds the number of hours to complete the necessary repair on the vehicle, and this number of flat-rate hours is then multiplied by the shop labor rate. For example, if the estimating guide indicates that 2.5 hours are required to repair the vehicle and the shop labor rate is $80.00 per hour, the estimated labor cost is 2.5 × 80 for a total labor cost of $200.00. The cost of automotive parts changes frequently. Therefore, when providing a parts cost estimate, it is more accurate to phone a parts store and obtain the exact price of the components required to repair the vehicle. The technician may also access parts price and availability over the parts store computer network, usually via the Internet.

NOTE: An estimating guide may be called a flat rate manual.

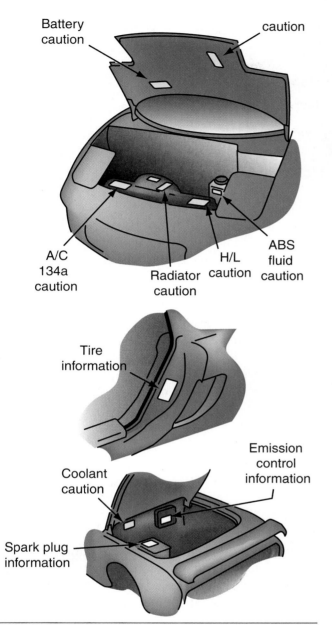

Figure 8-12 Vehicle warning labels.

Figure 8-13 A typical Vehicle Emission Control Information label.

Labor and parts costs plus all applicable taxes and miscellaneous charges are then added together to compile the parts and labor cost for repairing the customer's vehicle.

VEHICLE SERVICE AND WARNING PLACARDS

Each vehicle has a number of service and warning placards. Many of these placards are decals and labels adhered to the body structure in the engine compartment (**Figure 8-12**) as well as other key areas of the vehicle. The warning labels provide warnings that must be observed when servicing various vehicle systems. The emission control label is also mounted on the engine or in the engine compartment. This label provides valuable information regarding the engine size and the emission equipment on the engine. The **Vehicle Emission Control Information (VECI)** label also lists some engine specifications, such as spark plug type and gap, idle speed, and valve lash (**Figure 8-13**). The vacuum hose routing label is also mounted in the underhood area. This label illustrates the vacuum hose connections and routing for the vehicle (**Figure 8-14**). The vacuum hose routing diagram is very helpful when improper vacuum hose connections are suspected.

Estimates for the customer can be handwritten but are usually computerized. This allows the service writer or the technician to generate an estimate quickly and accurately. These estimates can be generated either with the customer or from a vehicle inspection. Computerized service information often has a provision for generating estimates.

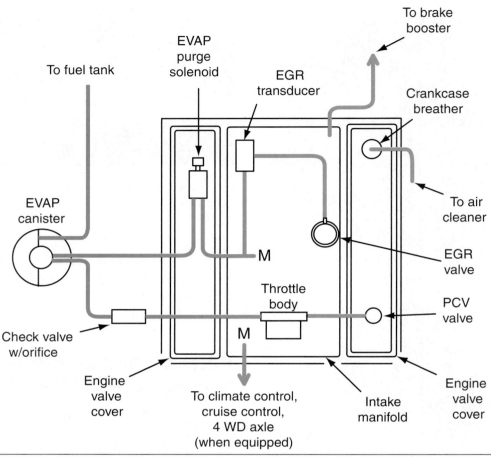

Figure 8-14 This drawing shows the vacuum hose routing for a vehicle.

Summary

- Sources of automotive service information include owner's manuals, service manuals, service bulletins, electronic media such as CD/DVDs and computer disks, computer networks, and service bulletins.

- Compared to printed service manuals, CD/DVDs provide greatly increased storage capacity.

- The owner's manual provides basic service and maintenance information.

- Service manuals contain information such as component identification, service procedures, specifications, and diagnostic information.

- Vehicle manufacturers or independent automotive service organizations may publish service manuals.

- Generic service manuals may cover several model years of a certain vehicle.

- Data for programming or reprogramming computers is stored in computer software and is downloaded from a scan tool or PC into an on-board automotive computer.

- Service bulletins contain valuable information regarding modified components and/or service procedures to correct specific vehicle problems.

- Labor-estimating guides list the labor time required to perform specific repairs on various vehicles.

- Vehicle service decals provide essential information regarding emission systems and vacuum hose routing.

- Vehicle warning labels make owners and technicians aware of precautions that must be observed when operating or servicing a vehicle.

Review Questions

1. Technician A says a vehicle owner's manual provides information about transaxle overhaul. Technician B says a vehicle owner's manual contains information about the type of oil that must be used in the engine. Who is correct?

   A. Technician A

   B. Technician B

   C. Both Technician A and Technician B

   D. Neither Technician A nor Technician B

2. Information about cleaning and maintenance of the vehicle interior and exterior is found in the:

   A. service manual.

   B. generic service manual.

   C. troubleshooting table.

   D. owner's manual.

3. Technician A says a CD/DVD has more information storage capacity than a printed service manual. Technician B says service bulletin information may be supplied on CD/DVDs. Who is correct?

   A. Technician A

   B. Technician B

   C. Both Technician A and Technician B

   D. Neither Technician A nor Technician B

4. Technician A says a labor-estimating guide provides labor costs for specific vehicle repairs. Technician B says a labor-estimating guide lists the cost of parts that are required for each vehicle repair. Who is correct?

   A. Technician A

   B. Technician B

   C. Both Technician A and Technician B

   D. Neither Technician A nor Technician B

5. All of these statements about owner's manuals are true *except*:

   A. The owner's manual includes information regarding tire changing.

   B. The owner's manual includes information regarding cooling system capacity.

   C. The owner's manual includes information regarding fuel injector service.

   D. The owner's manual includes information regarding the type of engine oil.

6. Technician A says a service bulletin may provide information about improved parts that are designed to correct a specific vehicle problem. Technician B says service bulletins may provide specific diagnostic and service information to correct a certain vehicle problem. Who is correct?

   A. Technician A

   B. Technician B

   C. Both Technician A and Technician B

   D. Neither Technician A nor Technician B

7. Technician A says the owner's manual contains vehicle warranty information. Technician B says the owner's manual contains information regarding vehicle warning systems. Who is correct?

   A. Technician A

   B. Technician B

   C. Both Technician A and Technician B

   D. Neither Technician A nor Technician B

8. Vehicle manufacturer's service manuals provide information on:

   A. proper cleaning and maintenance of the vehicle interior.

   B. diagnostic and service procedures.

   C. maintenance schedules.

   D. lightbulb replacement.

9. Technician A says service manuals for a specific vehicle are limited to one volume. Technician B says some service manuals are available on CD/DVD. Who is correct?

   A. Technician A

   B. Technician B

   C. Both Technician A and Technician B

   D. Neither Technician A nor Technician B

10. All of these statements about generic service manuals are true *except*:

   A. Generic service manuals are usually abbreviated compared to other service manuals.

   B. Generic service manuals may include coverage of several years of a specific vehicle make.

   C. Generic service manuals may include coverage of several vehicle models supplied by the same manufacturer.

   D. Vehicle manufacturers usually publish generic service manuals.

11. The emission control label provides information about the engine size and the _____ equipment on the vehicle.

12. Many emission control labels provide some engine _____.

13. The vacuum hose routing label is mounted in the _____ area.

14. Specification tables are provided in the _____ manual.

15. List ten information topics that are discussed in an owner's manual.

16. Describe how specific service information is accessed in a vehicle manufacturer's service manual.

17. Explain the meaning of flash programming as it relates to automotive computers.

18. Describe the purpose of a labor-estimating guide.

CHAPTER

9 Using Service Information

Learning Objectives

After you have read, studied, and practiced the contents of this unit, you should be able to:

- Describe the information on a typical emission control label.

- Interpret VINs.

- Locate the specified fluid capacities in a service manual.

- Find the vehicle specifications in a service manual.

- Describe two different types of maintenance schedules.

- List six different automotive service topics provided electronically on CDs or DVDs.

Key Terms

American Petroleum Institute (API) rating

Hybrid organic additive technology (HOAT)

Inorganic additive technology (IAT)

Organic additive technology (OAT)

Society of Automotive Engineers (SAE) viscosity rating

INTRODUCTION

Once you are familiar with the various sources of service information, you must understand how to access and use this information. When you are not familiar with the proper use of service information, you can waste valuable time searching for the required information. You must also understand where to locate the best and quickest source for the required information. Using brief or inadequate service information may lead to improper diagnostic and service procedures that do not locate and correct the cause of the customer's complaint.

VEHICLE IDENTIFICATION NUMBER (VIN) INTERPRETATION

When completing a work order, ordering parts, or servicing a vehicle, certain facts must be known about the vehicle. The vehicle manufacturer, make, and model year must be known. The vehicle manufacturer and make are often on the nameplate attached to the exterior of the body. Examples of vehicle manufacturer and make are Ford Mustang, Chrysler Concorde, and Honda Acura. The customer can also supply the necessary information regarding

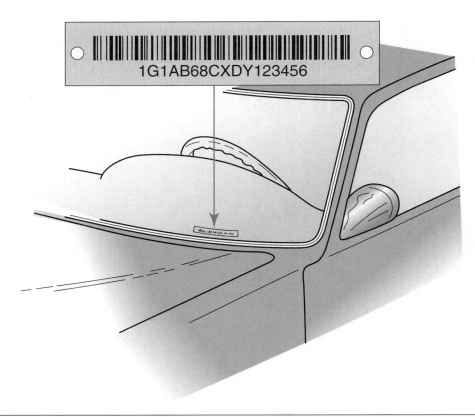

Figure 9-1 Vehicle Identification Number (VIN) location.

the vehicle manufacturer, make, and model year. This information is also on the vehicle registration.

Much like Social Security numbers, every vehicle has a different VIN. VIN plates are located on the dashboard and can be viewed through the windshield. The VIN is a series of letters and numbers that identify specific vehicle parameters such as the vehicle make, model, model year, and size of engine. On modern vehicles, the VIN number is mounted on the top of the dash, and this number is visible through the lower-left side of the windshield **(Figure 9-1)**. A close-up of a VIN plate is shown in **Figure 9-2**.

The VIN mounting location makes it more difficult for criminals to change the VIN. An explanation of the VIN is provided in the vehicle

Figure 9-2 Vehicle Identification Number (VIN) plate close-up.

manufacturer's service manual usually in the introduction or general information section. The introduction or general information section is the first section in a service manual. If the manual has more than one volume, the introduction or general information section is in volume one.

A VIN can be interpreted to show the information it contains. The following are the general explanations of the VIN and how to read them:

> **Tech Tip** *In North America, the right or left side of a vehicle is always determined from the driver's seat.*

1st character—Identifies the country in which the vehicle was manufactured. For example: U.S.A.(1or 4), Canada(2), Mexico(3), Japan(J), Korea(K), England(S), Germany(W), Italy(Z).

2nd character—Identifies the manufacturer. For example, Audi(A), BMW(B), Buick(4), Cadillac(6), Chevrolet(1), Chrysler(C), Dodge(B), Ford(F), GM Canada(7), General Motors(G), Honda(H), Jaguar(A), Lincoln(L), Mercedes Benz(D), Mercury(M), Nissan(N), Oldsmobile(3), Pontiac(2or5), Plymouth(P), Saturn(8), Toyota(T), VW(V), Volvo(V).

3rd character—Identifies vehicle type or manufacturing division.

4th to 8th characters—VDS (Vehicle Descriptor Section). These 5 characters occupy positions 4 through 8 of the VIN and may be used by the manufacturer to identify attributes of the vehicle. Identifies vehicle features such as body style, engine type, model, and series.

8th character—VDS (Engine). This character occupies position 8 of the VIN and is part of the VDS described earlier. It is used by the manufacturer to identify the engine used in the vehicle.

9th character—The check digit "character or digit 9" in the sequence of a VIN built beginning with model year 1981 (when the 17-character digit format was established) can best be described as identifying the VIN accuracy.

A check digit shall be part of each VIN (since 1981) and shall appear in position nine (9) of the VIN on the vehicle and on any transfer documents containing the VIN prepared by the manufacturer to be given to the first owner for purposes other than resale. Thus, the VINs of any two vehicles manufactured within a 30-year period shall not be identical. The check digit means a single number or letter "x" used to verify the accuracy of the transcription of the vehicle identification number.

After all other characters in the VIN have been determined by the manufacturer, the check digit is calculated by carrying out a mathematical computation specified by the government. This is based on VIN position, sample VIN, assigned value code, weight factor, and then multiply assigned value times weight factors. The values are added and the total is divided by 11. The remainder is the check digit number. The correct numeric remainder—zero through nine (0–9)—will appear. However, if the remainder is 10, the letter "X" is used to designate the check digit value/number.

10th character—Identifies the model year. For example: 1988(J), 1989(K), 1990(L), 1991(M), 1992(N), 1993(P), 1994(R), 1995(S), 1996(T), 1997(V), 1998(W), 1999(X), 2000(Y)——— 2001(1), 2002(2), 2003(3), 2004(4), 2005(5), 2006(6), etc.

11th character—Identifies the assembly plant for the vehicle.

12th to 17th characters—VIS. The last 8 characters of the VIN are used for the identification of a specific vehicle. The last 4 characters shall always be numeric. Identifies the sequence of the vehicle for production as it rolled off the manufacturer's assembly line.

Some imported vehicles have a slightly different VIN number interpretation **(Figure 9-3)**. On

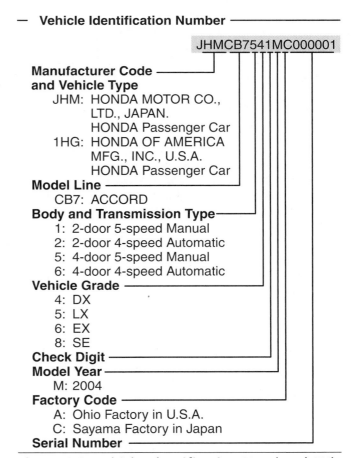

Figure 9-3 Vehicle Identification Number (VIN) interpretation.

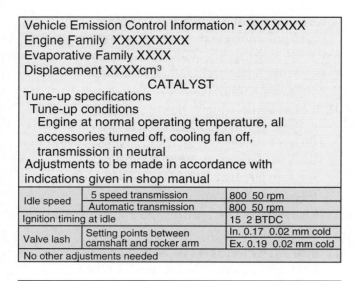

Figure 9-4 A vehicle emission control label.

ENGINE FAMILY:

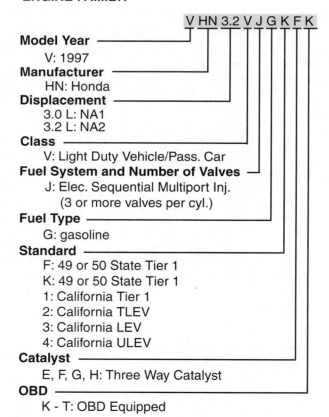

Figure 9-5 An engine family number interpretation.

EVAPORATIVE FAMILY:

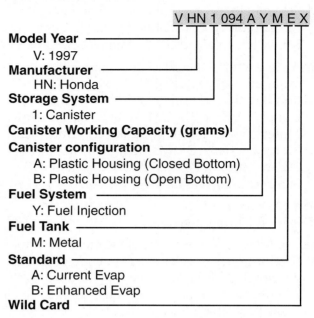

Figure 9-6 An evaporative family number interpretation.

these vehicles, some of the information related to engine family and evaporative family is interpreted from numbers on the emission label. A series of letters and numbers for the engine family and the evaporative family are displayed on the emission label **(Figure 9-4)**.

Interpretations of the engine family and evaporative family numbers are provided in the service manual **(Figure 9-5 and Figure 9-6)**. In the engine family number, the standard digit indicates the emission standards the vehicle is designed to meet. If the vehicle has a forty-nine-state emission rating, the vehicle will meet emission standards in the forty-nine states other than California. A fifty-state emission designation indicates the vehicle will meet emission standards in all fifty states. Any of the California emission designations indicate the vehicle is designed to meet that specific California emission standard. The following are some of the abbreviation explanations regarding California emission standards: TLEV—Transitional low emission vehicle; LEV—Low emission vehicle; ULEV—Ultra-low emission vehicle; SULEV—Super low emission vehicle.

TABLE OF CONTENTS	SECTION NUMBER
GENERAL INFOR. AND LUBE	
General Information	0A
Maintenance and Lubrication	0B
HEATING AND AIR CONDITIONING	
Heating and Ventilation (Non-A/C)	1A
Air Conditioning System	1B
V-5 A/C Compressor Overhaul	1D3
STEERING, SUSPENSION, TIRES, AND WHEELS	
Diagnosis	3
Wheel Alignment	3A
Power Steering Gear & Pump	3B1
Front Suspension	3C
Rear Suspension	3D
Tires and Wheels	3E
Steering Column, On-Vehicle Service	3F
Steering Column—Std., Unit Repair	3F1
Steering Column—Tilt, Unit Repair	3F2
DRIVE AXLES	
Drive Axles	4D
BRAKES	
General Information—Diagnosis and On-Car Service	5
Compact Master Cylinder	5A1
Disk Brake Caliper	5B2
Drum Brake—Anchor Plate	5C2
Power Brake Booster Assembly	5D2
ENGINES	
General Information	6
2.0 Liter L-4 Engine	6A1
3.1 Liter V6 Engine	6A3
Cooling System	6B
Fuel System	6C
Engine Electrical—General	6D
Battery	6D1
Cranking System	6D2
Charging System	6D3
Ignition System	6D4
Engine Wiring	6D5
Driveability and Emissions—General	6E
Driveability and Emissions—TBI	6E2
Driveability and Emissions—PFI	6E3
Exhaust System	6F
TRANSAXLE	
Auto. Transaxle On-Car Service	7A
Auto. Trans.—Hydraulic Diagnosis	3T40-HD
Auto. Trans.—Unit Repair	3T40
Man. Trans.— On-Car Service	7B
5-Sp. 5TM40 Man. Trans. Unit Repair	7B1
5-Sp. Isuzu Man. Trans. Unit Repair	7B2
Clutch	7C
CHASSIS ELECTRICAL, INSTRUMENT PANEL & WIPER/WASHER	
Electrical Diagnosis	8A
Lighting & Horns	8B
Instrument Panel & Console	8C
Windshield Wiper/Washer	8E5

Figure 9-7 A service manual general index.

LOCATING CAPACITIES AND FLUID REQUIREMENTS IN A SERVICE MANUAL

A general index is located at the front of a service manual. This index directs you to the main sections in the manual (**Figure 9-7**). The fluid capacities and requirements are often located in the maintenance or lubrication and maintenance section of the vehicle manufacturer's service manual. In some service manuals, the fluid capacities are listed with other specifications at the end of each section. Fluid capacities are also provided in the owner's manual. Technicians must know the specified fluid capacities so they do not overfill or underfill various components.

Typical capacities for a 2002 PT Cruiser are shown in **Table 9-1**.

Fluid requirements are usually listed in the lubrication and maintenance section of the service manual. In some service manuals, the fluid

Table 9-1	
Description	**Specification**
Fuel tank	57L (15 gal.)
Engine oil	1.6L, 4.5L (4.8 qts.) with filter change
Engine oil	2.0L, 2.2L, 4.3L (4.5 qts.) with filter change
Engine oil	2.4L, 4.8L (5.0 qts.) with filter change
Cooling system	7.0L (7.4 qts.) includes recovery bottle and heater
Automatic transaxle	3.8L (4.0 qts.) service fill
Automatic transaxle	8.1L (8.6 qts.) overhaul fill with converter empty
Manual transaxle, NV T350	2.4–2.7L (2.5–2.8 qts.)

1 - Engine
2 - Transmission
3 - Brake Fluid
4 - Powersteering Fluid
5 - Manual Steering Fluid
6 - Tierods
7 - CV Joints
8 - Steering Joints

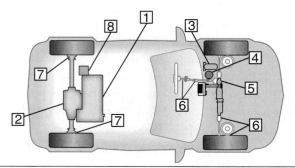

Figure 9-8 Typical vehicle lubrication points.

requirements are listed with other specifications at the end of each section. These requirements may be provided in text form or in a chart. A typical lubrication requirement chart is provided in **Figure 9-8**.

The lubrication chart lists the engine oil grade and viscosity ratings at various temperatures. **Figure 9-9** illustrates different labels and seals on oil containers. Engine oil with the proper American Petroleum Institute (API) rating and Society of Automotive Engineers (SAE) viscosity rating must be used. The **American Petroleum Institute (API) rating** is a universal engine oil rating that classifies oils according to the type of service for which the oil is intended. The **Society of Automotive Engineers (SAE) viscosity rating** is a universal rating for engine oil viscosity in relation to the atmospheric temperature in which the oil will be operating.

> **Tech Tip** *Using engine oil with an improper grade or viscosity rating may cause engine damage and premature engine failure.*

Technicians must be familiar with engine coolant requirements. Older coolant formulations were based on ethylene glycol using an **inorganic additive technology (IAT)**. The IAT coolant is the familiar green coolant that many of us have seen.

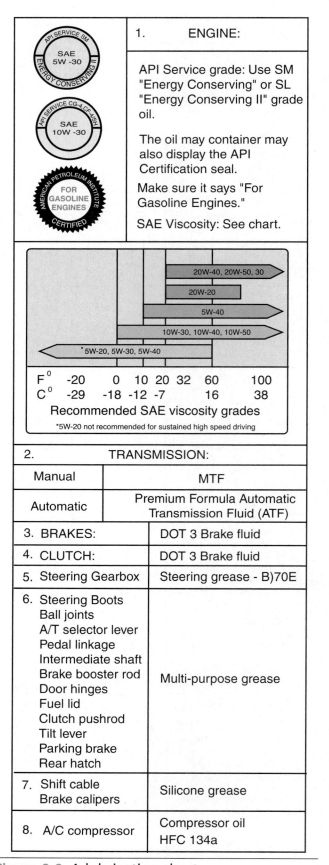

Figure 9-9 A lubrication chart.

In recent years, most engines have aluminum cylinder heads and some engines also have aluminum cylinder blocks. Some radiators on these engines also have aluminum cores. In these cooling systems, most vehicle manufacturers now recommend an ethylene glycol-based coolant with **organic additive technology (OAT)**. This type of coolant is usually orange in color and contains special additives to minimize corrosion in the cooling system. European manufacturers implemented long-life coolants with either hybrid (BMW, Mercedes, Volvo) or OAT technologies (Opel, Ford Europe, VW). The major Asian manufacturers have adopted OAT types in the last year as extended drain fluids.

Organic acid technology (OAT) is based primarily on carbon-based molecules, typically organic acids, to protect cooling system metal. These fluids do not wear out in service, but may not protect quickly in fast corrosion conditions like boiling. OAT is not backward compatible with older green coolant engines.

Tech Tip *Mixing orange- or red-colored antifreeze with green antifreeze containing a HOAT additive reduces the corrosion protection of the coolant and may cause premature cooling system corrosion plus water pump seal failure. If different types of antifreeze are mixed, the cooling system should be completely flushed and filled with the proper mixture of the vehicle manufacturer's specified coolant.*

Thus was developed a new type of coolant called **hybrid organic additive technology (HOAT)**. Hybrid coolant uses both inorganic and organic, carbon-based additives for long-life protection. The HOAT coolant comes in a variety of colors depending on the manufacturer's request. Daimler Chrysler uses orange and Ford uses yellow. The idea with hybrids is to provide excellent all around corrosion protection and extended drain intervals. Hybrids generally can replace green coolant in older vehicles.

A cooling system with a OAT or HOAT additive usually has a 5-year/100,000-mile warranty. In most climates where below-freezing temperatures are encountered, the coolant is a mixture of 50 percent ethylene glycol and 50 percent water. This mixture provides a coolant freeze point of $-35°F$ ($-37°C$). The coolant specified in the vehicle manufacturer's service manual must be used. It is important not to mix different types of engine coolants as each has a different formulation and additive package.

Tech Tip *Coolant recycling machines are available to clean and condition coolant. Coolant must be disposed of according to hazardous material disposal regulations.*

LOCATING VEHICLE SPECIFICATIONS AND MAINTENANCE SCHEDULES IN A SERVICE MANUAL

Technicians must use proper specifications for the vehicle being serviced. For example, if the proper torque specifications are not used, expensive aluminum castings may be warped, resulting in fluid leaks. In many service manuals, the specifications are located at the end of each section. The index for each section lists the page number where the specifications are located.

The specifications provide all the necessary measurements, adjustments, and torque values for the components being serviced **(Figure 9-10)**. Some service manuals provide specifications at the beginning of each subheading within a section. Other service manuals insert the specifications at the appropriate place in each service procedure. Technicians and service personnel must use maintenance schedules when servicing vehicles. Often customers are not aware that certain maintenance should be performed on their vehicle because of the mileage on the vehicle or the time since the last service interval. The service writer should check the vehicle mileage and past service record against the maintenance schedule to determine if routine maintenance is required on the vehicle.

Performing the maintenance listed in the maintenance schedule will maintain safe and trouble-free operation of the vehicle. The maintenance schedules are usually provided in the lubrication and maintenance section of the service manual. These maintenance schedules may be listed in charts **(Figure 9-11)**. In some

GENERAL MOTORS ENGINES
3.0L, 3.8L, and 3.8L "3800" V6 (Continued)
Engine Specifications (Continued)

Crankshaft Main and Connecting Rod Bearings

Engine	Main Bearings				Connecting Rod Bearings		
	Journal Diam. In. (mm)	Clearance In. (mm)	Thrust Bearing	Crankshaft End Play In. (mm)	Journal Diam. In. (mm)	Clearance In. (mm)	Side Play In. (mm)
All Models	[1]2.4988-2.4998 (63.469-63.494)	.0003-.0018 (.008-.005)	2	.003-.009 (.08-.23)	[1]2.2457-2.2499 (57.117-57.147)	.0003-.0028 (.008-.071)	.003-.015 (.076-.38)

[1]Maximum taper is .0003" (.008mm).

Pistons, Pins, and Rings

Engine	Pistons	Pins		Rings		
	Clearance In. (mm)	Piston Fit In. (mm)	Rod Fit In. (mm)	Ring No.	End Gap In. (mm)	Side Clearance In. (mm)
All Models	[1].0013-.0035 (.033-.089)	.0004-.0007 (.010-.018)	.00075-.00125 (.019-.032)	1	.010-.020 (.254-.508)	.003-.005 (.08-.13)
				2	.010-.020 (.254-.508)	.003-.005 (.08-.13)
				3	.015-.035 (.381-.889)	.0035 (.09)

[1]Measured at bottom of piston skirt. Clearance for 3.8L turbo is .001-.003" (.03-.08 mm).

Valve Springs

Engine	Free Length In. (mm)	Pressure Lbs. @ In. (Kg @ mm)	
		Valve Closed	Valve Open
3.8L (VIN C)	2.03 51.6	100-110@1.73 (45-49@44)	214-136@1.30 (97-61@33)
3.0L & 3.8L (VIN 3)	2.03 51.6	85-95@1.73 (39-42@44)	175-195@1.34 (79.1-88.2@34.04)
3.8L (VIN 7)	2.03 (51.6)	74-82@173 (33-37@44)	175-195@1.34 (79.1-88.2@34.04)

Camshaft

Engine	Journal Diam. In. (mm)	Clearance In. (mm)	Lobe Lift In. (mm)
3.0L	1.785-1.786 (45.34-45.36)	.0005-.0025 (.013-.064)	Int. .210 (5.334) Exh. .240 (6.096)
3.8L (VIN C)	1.785-1.786 (45.34-45.36)	.0005-.0025 (.013-.064)	[1].272 (6.909)
3.8L (VIN 3)	1.785-1.786 (45.34-45.36)	.0005-.0025 (.013-.064)	[1].245 (6.223)
3.8L (VIN 7)	1.785-1.786 (45.34-45.36)	.0005-.0025 (.013-.064)	....

[1] Specification applies to both intake and exhaust

Caution: Following specifications apply only to 3.0L (VIN L), 3.8L (VIN 3), 3.8L (VIN 7), and 3.8 "3800" (VIN C) engines.

TIGHTENING SPECIFICATIONS

Application	Ft. Lbs. (N.m)
Camshaft Sprocket Bolts	20 (27)
Balance Shaft Retainer Bolts	27 (37)
Balance Shaft Gear Bolt	45 (61)
Connecting Rod Bolts	45 (61)
Cylinder Head Bolts	[1]60 (81)
Exhaust Manifold Bolts	37 (50)
Flywheel-to-Crankshaft Bolts	60 (81)
Front Engine Cover Bolts	22 (30)
Harmonic Balancer Bolt	219 (298)
Intake Manifold Bolts	[2]
Main Bearing Cap Bolts	100 (136)
Oil Pan Bolts	14 (19)
Outlet Exhaust Elbow-to-Turbo Housing	13 (17)
Pulley-to-Harmonic Balancer Bolts	20 (27)
Outlet Exhaust Right Side Exhaust Manifold-to-Turbo Housing	20 (27)
Rocker Arm Pedestal Bolts	37 (51)
Timing Chain Damper Bolt	14 (19)
Water Pump Bolts	13 (18)

[1]Maximum torque is given. Follow specified procedure and sequence.
[2]Tighten bolts to 80 INCH lbs. (9 N.m.)

Figure 9-10 This page shows manufacturer's specifications for fit, torque, etc.

Service at the indicated mileage or time whichever comes first.	Miles x 1,000	15	30	45	60	75	90	105	120	Note
	km x 1,000	24	48	72	96	120	144	168	192	
	months	12	24	36	48	60	72	84	96	
Replace engine oil	Replace every 7,500 miles or 12 months									Capacity 5.0L.
Replace engine oil filter		●	●	●	●	●	●	●	●	
Check engine oil and coolant	Check oil and coolant at each fuel stop									Check for leaks.
Replace air cleaner element			●		●		●		●	
Inspect valve clearance							●			Intake 0.15-0.19mm Exhaust 0.17-0.21mm Measure when cold
Replace spark plugs								●		Gap: 1.0-1.1mm
Replace timing belt/inspect water pump								●		
Inspect and adjust drive belts			●		●		●		●	Check for cracks. Check for belt deflection at center of belt. Alternator 11.0-13.5mm A/C Compressor 10-12mm
Replace fuel filter					●				●	
Inspect idle speed								●		800 +/-rpm (MT neutral) 780 +/-rpm (AT neutral)
Replace engine coolant				●		●		●		Capacity 12.0L
Replace transmission fluid						●				Manual trans. MTF 2.65L Automatic trans. ATF 2.9L
Inspect front and rear brakes		●	●	●	●	●	●	●	●	Check brake pad thickness and movement. Check caliper for leakage.
Replace brake fluid (including ABS)				●			●			DOT3 brake fluid. Check that fluid is between upper and lower marks in reservoir.
Check parking brake adjustment		●	●	●	●	●	●	●	●	Engaged 10 to 14 notches
Rotate tires (Check inflation and condition at least once per month)	Rotate every 7,500 miles									Rotation method shown in owner's manual.

Service at the indicated mileage or time whichever comes first.	Miles x 1,000	15	30	45	60	75	90	105	120	Note
	km x 1,000	24	48	72	96	120	144	168	192	
	months	12	24	36	48	60	72	84	96	
Visually inspect the following items										
Tie-rod ends, steering gear box, and boots										Check for steering linkage looseness. Check condition of boots and fluid leakage.
Suspension components										Check for bolt tightness. Check ball joint boots.
Driveshaft boots										Check condition of boots.
Brake hoses and lines (including ABS)										Check for leakage.
All fluid levels and condition of fluid		●	●	●	●	●	●	●	●	Check for levels, condition of fluid, and leakage.
Cooling system hoses and connections										Check all hoses for leakage and damage. Check fan operation.
Exhaust system										Check catalytic converter heat shield, exhaust pipe, and muffler for damage and leaks.
Fuel lines and connections										Check for leaks.
Inspect air bag system	10 years after production									

Figure 9-11 A vehicle maintenance schedule.

service manuals, the maintenance schedules are in text format. The maintenance schedules are also in the owner's manual. Some vehicle manufacturers provide maintenance schedules for normal vehicle operation and separate maintenance schedules for severe vehicle operating conditions.

Some vehicle manufacturers define severe operating conditions as follows:

1. Driving less than 5 miles (8 km) per trip, or in freezing temperatures, driving less than 10 miles (16 km) per trip.
2. Driving in extremely hot temperatures above 90°F (32°C).
3. Extensive idling or long periods of stop-and-go driving.
4. Trailer towing, driving with a car-top carrier, or driving continually in mountainous conditions.
5. Driving on muddy, dusty, or icy roads on which salt has been applied.

Maintenance schedules for severe operating conditions require more frequent vehicle servicing such as changing the oil, oil filter, and air filter at lower mileage intervals.

USING GENERIC SERVICE MANUALS

Many generic service manuals have the same format as the vehicle manufacturer's service manuals. This format was explained previously in this chapter. Generic service manuals usually cover several models and model years. For example, one generic service manual covers 1985 through 1992 Volkswagen GTI, Golf, and Jetta.

Compared to vehicle manufacturer's service manuals, generic manuals tend to be abbreviated. Some generic service manual publishers produce a series of service manuals to cover specific vehicle classifications such as domestic cars, domestic light trucks and vans, and import cars, trucks, and vans **(Figure 9-12)**. The series of manuals based on domestic cars has separate manuals covering each of these topics: Engine

MECHANICAL MANUALS	Page
Mech. Parts and Labor Estimating Manuals	19
Service and Repair Manuals:	
Domestic Cars	20
Domestic Light Trucks, and Vans	22
Import Cars, Light Trucks, and Vans	23
Medium and Heavy Duty Truck Manuals	25
Transmission Manuals	26
Specialty Manuals	27
Training Products	27
ASE Test Preparation Manuals	28

Figure 9-12 A sample of categories from a generic service manual.

Performance, Electrical, Chassis, Engine, Heating and Air Conditioning, and Electrical Component Locator. These generic service manuals cover several model years of domestic vehicles. Some publishers produce CDs or DVDs containing automotive service information. This information may also be available via the Internet.

The automotive service information subjects available electronically from one publisher has the following features:

- Repair Package
- Estimator Package
- Communications Package
- Shop Manager
- Transmission Package
- Truck Package
- Vintage Service Information

When one of these electronic service information systems is installed and opened in a computer, it begins with a simple home page from which the desired information is selected. The special features of OnDemand5 Repair are as follows:

- Powerful search engine.
- Categories within Repair and Estimator are streamlined like never before! Now users can go from Repair to Estimator without having to reselect the category.

ONDEMAND repair gives you more coverage.	
• Accessories	• Engine Performance
• Air Bags	• Maintenance
• Air Conditioning & Heating	• Recalls
• Brakes	• Steering
• Clutches	• Suspension
• Diagnostics	• Technical Service Bulletins
• Drive Axles	• Transmission Servicing & Electronic Diagnosis
• Electrical	
• Emission Controls	• Wheel Alignment
• Engine Cooling	• Wiring Diagrams

Figure 9-13 Mitchell OnDemand5 Repair information sections.

- New fluid capacities.
- Single click access to technical service bulletins.
- Single click access to maintenance.

The service information sections included in this category are listed in **Figure 9-13**.

Summary

- The VIN is a series of letters and numbers that identifies vehicle parameters such as the make, model, model year, and size of engine. The vehicle emission label provides information about the emission equipment on the vehicle and the emission standards that the vehicle is designed to meet.

- Fluid capacities and fluid requirements are usually provided in the service manual's lubrication and maintenance section.

- Using an engine oil or coolant other than the ones specified in the service manual may cause premature engine wear and cooling system corrosion and failure.

- In engines with aluminum cylinder heads and cylinder blocks, the cooling system requires a coolant with hybrid organic additive technology (HOAT). Most service manuals have a general index at the front to the manual that lists the sections in the manual.

- A sub-index at the beginning of each section identifies specific information within the section.

- In many service manuals, the specifications are located at the end of each section.

- Maintenance schedules are usually located in the lubrication and maintenance section of a service manual.

- Separate maintenance schedules may be provided for normal and severe operating conditions.

- Some generic service manuals have the same format as vehicle manufacturer's service manuals.

- Some generic manual publishers provide automotive service information on CDs, DVDs, or via the Internet.

Review Questions

1. Technician A says the VIN number is located on the top of the dash on the left side. Technician B says the tenth digit in the VIN identifies the model year of the vehicle. Who is correct?

   A. Technician A

   B. Technician B

   C. Both Technician A and Technician B

   D. Neither Technician A nor Technician B

2. Technician A says the eighth digit in the VIN identifies the engine size. Technician B says the Check Digit is the ninth digit in the VIN. Who is correct?

   A. Technician A

   B. Technician B

   C. Both Technician A and Technician B

   D. Neither Technician A nor Technician B

3. Technician A says if a vehicle emission label indicates a forty-nine-state emission designation, the vehicle will meet California emission standards. Technician B says the TLEV abbreviation stands for top-level emission venue. Who is correct?

   A. Technician A

   B. Technician B

   C. Both Technician A and Technician B

   D. Neither Technician A nor Technician B

4. Technician A says the fluid capacities are usually provided in the lubrication and maintenance section of a service manual. Technician B says technicians must be familiar with fluid capacities and fluid requirements for the vehicle they are servicing. Who is correct?

   A. Technician A

   B. Technician B

   C. Both Technician A and Technician B

   D. Neither Technician A nor Technician B

5. When discussing VIN interpretation, Technician A says the first digit in the VIN indicates the vehicle country of origin. Technician B says the fourth digit in the VIN indicates vehicle type. Who is correct?

   A. Technician A

   B. Technician B

   C. Both Technician A and Technician B

   D. Neither Technician A nor Technician B

6. All of these statements about vehicle fluid capacities are true *except*:

   A. Fluid capacities are usually located in the service manual at the beginning of the engine section.

   B. Fluid capacities are provided in the owner's manual.

   C. Technicians must know the proper fluid capacities to avoid overfilling various components and systems.

   D. In some service manuals, the fluid capacities are located in the maintenance and lubrication section.

7. All of these statements about engine oil ratings are true *except*:

   A. The API oil grade rating indicates the type of service for which the oil is intended.

   B. The SAE viscosity oil rating indicates the atmospheric temperature for which the oil is suitable.

   C. Engine oils with different API or SAE viscosity ratings may be mixed.

   D. An engine oil chart in the service manual or owner's manual indicates the proper oil grade and viscosity rating at different temperatures.

8. An engine coolant solution containing 50 percent ethylene glycol and 50 percent water provides protection against freezing to:

   A. 0°F.

   B. −12°F.

   C. −20°F.

   D. −35°F.

9. Technician A says a green-colored antifreeze contains hybrid organic additive technology. Technician B says an antifreeze containing hybrid organic additive technology has special additives to prevent cooling system leaks. Who is correct?

   A. Technician A

   B. Technician B

   C. Both Technician A and Technician B

   D. Neither Technician A nor Technician B

10. According to vehicle manufacturers, severe vehicle operating conditions that require different maintenance schedules include all of these operating conditions *except*:

    A. trailer towing or driving continuously in mountainous road conditions.

    B. extensive idling or long periods of stop-and-go driving.

    C. driving continuously at low elevations of sea level to 1,000 feet.

    D. driving in extremely hot temperatures above 90°F.

11. The last six digits in a VIN are assigned to the vehicle by the _____.

12. The ninth digit in the VIN is not made available to the _____.

13. A coolant with HOAT is specified for engines with _____ cylinder heads and blocks.

14. A coolant with HOAT additive is _____ in color.

15. Explain where the specifications are usually located in a service manual.

16. Define severe operating conditions as they apply to maintenance schedules.

17. Explain the difference between normal maintenance schedules and severe operation maintenance schedules.

18. Describe the basic differences between generic service manuals and vehicle manufacturer's service manuals.

CHAPTER

10 Vehicle Inspection and Detailing

Learning Objectives

After you have read, studied, and practiced the contents of this unit, you should be able to:

- Check and adjust engine oil level and condition.
- Check and adjust transmission oil level and condition.
- Inspect and manually adjust engine accessory drive belts.
- Check and adjust brake and power steering fluid levels.
- Check and adjust coolant levels and test coolant quality in a vehicle with a recovery reservoir.
- Inspect all under-body components.
- Check and adjust tire pressures to correct levels.
- Check tire for correct wear pattern and ensure there are no embedded foreign objects in the tread.
- Inspect and refill a windshield washer fluid reservoir.

- Check the condition of windshield wiper blades and arms and change a windshield wiper blade.
- Check the lighting systems.
- Check an automotive security system on a vehicle.
- Carry out a preliminary visual inspection.
- Lubricate door hinges and locks and adjust jambs.
- Check interior trim.
- Check the condition and security of automotive seatbelts.
- Check the driver's seat for security and condition.
- Wash a vehicle using the appropriate detergent and methods.
- Wax a vehicle using the appropriate methods.

Key Terms

Belt glazing

Dynamic imbalance

Electrolysis

Glycol-based brake fluid

Hydrometer

Hygroscopic

Mineral-based brake fluid

Serpentine-belt

Silicon-based brake fluid

Static imbalance

V-belt

INTRODUCTION

The car should have a thorough inspection by a qualified technician. Inspecting and detailing the vehicle is an important aspect of customer satisfaction. These are frequently the only area of the car a customer will see. These services are quick and easy to perform. Any problems detected in this type of service ultimately can save the customer time, money and inconvenience in the long run.

UNDER-HOOD INSPECTION

The under-hood area of a vehicle contains many varied and complex systems. This section investigates some simple checks that you will undertake while performing an under-hood check. The major areas to check under the hood are engine lubrication, engine drive belts, hoses, brake fluid, and coolant **(Figure 10-1)**.

Engine Lubrication

One of the most important systems requiring regular maintenance is the engine lubrication system. This section investigates the concepts of friction and viscosity and demonstrates the simple procedures to use when servicing this system. The fundamentals of the lubrication system include the following:

- Friction
- Functions of oil

Figure 10-1 There are many systems to inspect under the hood of a vehicle.

- Viscosity
- Additives
- Oil indicators

Friction

Friction resists the movement of one surface over another. Friction may be desirable but many times it is not. Rough surface features that lock together cause friction. These features can be microscopically small, which is why even surfaces that seem to be smooth can experience friction. Friction can be reduced but never eliminated. Friction is always measured for two surfaces, using what is called the coefficient of friction. A low coefficient of friction for two surfaces means they can move easily over each other. A high coefficient of friction for two surfaces means they cannot move easily over each other.

NOTE: The main functions of oil are to lubricate, cool engine parts, clean, seal the piston rings, and cushion the force of the power stroke on the engine bearings.

Checking Engine Oil

It will be necessary to check and adjust engine oil level and check oil condition.

The following are safety recommendations to use when checking oil level:

- If the engine has been running, be careful not to burn yourself on the exhaust manifold or any other hot part of the engine when reaching for the dipstick. Remember, the dipstick and the oil on it will also be hot.
- Do not allow oil from the dipstick to drip onto engine components. Dripping oil from the dipstick will smoke or burn if it falls on any hot engine areas.
- Make sure that the hood is properly secured.
- Always make sure that you wear the appropriate personal protection equipment before starting the job.
- Always make sure that your work area/ environment is as safe as you can make it. Do not use damaged, broken, or worn out workshop equipment.

■ Always follow any manufacturer's personal safety instructions to prevent damage to the vehicle you are working on.

■ Make sure that you understand and observe all regulations and personal safety procedures when carrying out these tasks.

The following are points to note when checking oil level:

■ Make sure the vehicle is on a level surface and the engine is turned off for at least three minutes before taking a reading. If you don't, you'll get inaccurate readings. The oil will collect in the sump when the engine is off.

■ Typically, the amount of oil needed to raise the oil level from the low mark on the dipstick to the high mark is about a quart. This varies, so always check the shop manual to determine the correct quantity. Never fill the engine with oil above the full indicator of the dipstick! This could cause oil aeration and severe engine damage.

■ Although fresh oil is translucent, and oil that needs to be replaced looks black and dirty, it is often difficult to assess the condition of engine oil simply by its color. Oil loses its clean, fresh look very quickly and yet may still be serviceable. The best guide to changing oil is determining the vehicle's mileage and period of time since the last oil change.

■ If the oil on the dipstick is not blackish in color but looks milky gray, this could indicate that there is some water (or coolant) being mixed into the oil. There may be a serious problem somewhere in the engine, such as a leaking head gasket, and you should report this to your supervisor immediately.

■ Engine operating conditions can also influence the oil's condition. For instance, continuously stopping and starting the engine with very small operating cycles can cause condensation inside the engine. An extreme case of this will cause very rapid oil deterioration, and will require frequent oil changes.

■ Don't forget to replace the filler cap after topping up the oil.

The following is a general procedure to follow when checking oil level:

1. Locate dipstick. The dipstick is located on the side of the engine block and is usually very

Figure 10-2 Locate the engine dipstick.

easy to find, with a distinctively shaped or brightly colored handle. Many dipsticks will have yellow writing or have an oil can icon **(Figure 10-2)**.

2. Remove dipstick and wipe clean. Remove the dipstick, catching any drops of oil on a rag, and wipe it clean. There are markings on the lower end of the stick to indicate whether the oil level needs to be topped.

3. Take the oil level reading. Replace the dipstick and push it back down into the sump as far as it will go. Remove it again, and the level of oil in the oil pan will be clearly visible on the stick. If the level is below the full or topmost mark, then you should top up the engine to that level with fresh oil **(Figure 10-3)**.

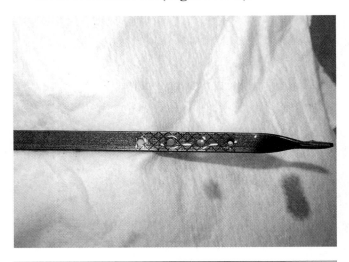

Figure 10-3 The oil should be between the add and the full marks.

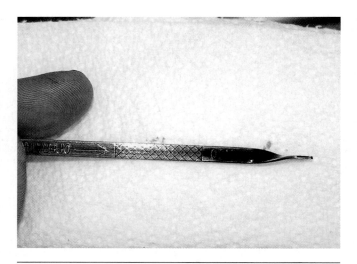

Figure 10-4 If the condition of the oil is poor then an oil change will be necessary.

4. Check condition of oil. If the oil appears very black and dirty, it may have lost some of its protective and lubricating qualities, and may need to be completely changed. Check the service record or ask the customer when the oil was last changed **(Figure 10-4)**.

5. Adjust level if necessary. If additional oil is needed, estimate the amount by checking the service manual guide to the dipstick markings. Unscrew the filler cap at the top of the engine, and using a funnel to avoid spillage, gently pour the oil into the engine.

6. Recheck the dipstick level. Replace the oil filler cap, and check the dipstick again to make sure the level of oil in the engine is now correct.

Power Steering Fluid

The following are safety recommendations to use when checking and adjusting power steering fluid:

- Turn engine off before checking fluid level.

- Make sure that you understand and observe all regulations and personal safety procedures when carrying out these tasks. If you are unsure of what they are, ask your supervisor.

- The power steering fluid should be at operating temperature to ensure an accurate reading of the fluid level.

- Always refer to the shop manual for the recommended type of power steering fluid.

- Oils and fluids tend to expand in volume when they are heated.

- The manufacturer has specified an optimum level of the power steering fluid and it is assumed that when the fluid is hot, it will be at this level. If the fluid is cold, it may be at a lower level.

- Don't forget to replace the filler cap after topping up the oil.

The following is a general procedure to follow when checking and adjusting power steering fluid:

1. Prepare the steering system for a fluid check. With the engine at the normal idle speed, you should turn the steering wheel from lock to lock a number of times. This will ensure the fluid is hot and the level is more accurate. Now, turn the engine off.

2. Check power steering fluid. Locate the power steering pump in your vehicle by checking the owner's manual. Unscrew the cap on top of the pump and check if the fluid reaches the full mark on the dipstick **(Figure 10-5)**.

3. Top-up power steering fluid. If the level is low, check your owner's manual or shop manual to see what kind of fluid your power steering pump requires, and fill it to the proper level. If the vehicle has a plastic reservoir, check the markers to see if any fluid needs to be added. Remember that using the wrong type of fluid can result in damage to the various seals in the system.

Figure 10-5 The fluid should reach the full mark on the dipstick.

Engine Drive Belt Inspection

Engine drive belts can drive the camshaft pulleys, power steering pumps, air conditioning compressors, alternators, and fans. With constant use, these belts can become worn and require replacing. When replacing a drive belt you must follow the manufacturer's guidelines, to make sure that the belt is correctly fitted and will not damage the engine or its components. An important part of this process is inspecting the belt for any signs of deterioration.

Inspecting and Adjusting an Engine Drive Belt

The following are safety recommendations to use when inspecting and adjusting an engine drive belt:

- Never try to inspect belts with the engine running.

- Always make sure that you wear the appropriate personal protection equipment before starting the job. It is very easy to hurt yourself even when the most exhaustive protection measures are taken.

- Always make sure that your work area/ environment is as safe as you can make it. Do not use damaged, broken, or worn out workshop equipment.

- Always follow the manufacturer's personal safety instructions to prevent damage to the vehicle you are working on.

- Make sure that you understand and observe all regulations and personal safety procedures when carrying out these tasks. If you are unsure of what they are, ask your supervisor.

The two types of drive belts are the **V-belt** and the **serpentine** or multi-grooved belt **(Figure 10-6)**. A V-belt sits inside a deep v-shaped groove in the pulley wheel. The sides of the V-belt contact the sides of the groove. Serpentine-belts have a flat profile with a number of grooves running lengthwise along the belt. These grooves are the exact reverse of the grooves in the outer edge of the pulley wheels; they increase the contact surface area, as well as prevent the belt from slipping off the wheel as it rotates.

There are a number of conditions to look for on a drive belt. These are cracks, oil contamination,

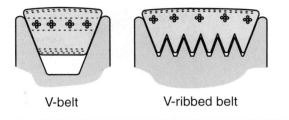

V-belt   V-ribbed belt

Figure 10-6 A cross section of a V-belt and a serpentine-belt.

glazing, torn, and bottoming-out. Cracks in a belt indicate that it is getting ready to fail and should be replaced. A belt that has been soaked in oil will not grip properly on the pulleys and will slip **(Figure 10-7)**. If the oil contamination is severe enough for this to happen, replace the belt **(Figure 10-8)**. **Belt glazing** is shininess on the surface of the belt, which comes in contact with the pulley. If the belt is very worn, the glazing could be caused by the belt "bottoming-out" and it

Figure 10-7 Cracks in a serpentine-belt.

Figure 10-8 An oil soaked serpentine-belt.

Figure 10-9 A V-belt showing signs of glazing.

Figure 10-10 A serpentine-belt showing signs of glazing.

Figure 10-11 A serpentine-belt showing tearing.

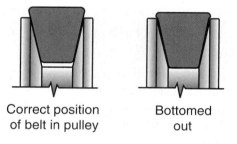

Correct position   Bottomed
of belt in pulley   out

Figure 10-12 A worn V-belt will contact the bottom of the pulley.

should be replaced **(Figure 10-9 and Figure 10-10)**. If it is not old and worn, glazing could simply indicate that the belt is not tight enough. Tightening the belt may be all that is necessary, depending on how severe the glazing is. Torn or split belts are unserviceable and should be replaced immediately **(Figure 10-11)**. When a V-belt becomes very worn, the bottom of the V-shape may contact the bottom of the groove in the pulley, preventing the sides of the belt from making good contact with the sides of the pulley groove **(Figure 10-12)**. This reduced friction causes the belt to slip and heat up, which in turn will cause the belt to glaze. A belt worn enough to bottom-out should be replaced.

Many vehicles require the technician to manually adjust the tension on the belt. Other vehicles have an automatic spring tensioning system. Depending on the system used on the particular vehicle, you should always follow the manufacturer's service instructions. There are a number of different types of tension gauges **(Figure 10-13)**. Follow the operating instructions on the tool. A tension gauge allows the technician to accurately set belt tension. If you don't have a tension gauge, you can estimate the tension by pushing the belt inwards with your hand. If it's correctly tensioned, you should be able to deflect the belt about 1/2 in. (1.25 cm) for each 12 in. (30 cm) of belt.

The following is a general procedure to follow when checking belts:

1. Inspect and check belt condition. Twist the belt so that you can see the underside of the V-shape

Figure 10-13 A belt tension gauge.

or the ribs on a serpentine-belt. Look for signs of wear and damage. You may need a flashlight to see these clearly. A cracked, glazed, or torn belt will need to be replaced.

2. Check tension. Check the belt tension by attaching the tension gauge to the longest belt span, and pulling it to measure the tension. Compare your reading to the specifications in the vehicle workshop manual.

3. Choose the correct tools. Select the correct wrench to loosen the tension adjustment fastener. This is usually on the alternator mounting or on a separate idler pulley wheel. You will also need a pry bar, which is a metal bar you can use as a lever to apply tension on the belt.

4. Adjust belt tension. Loosen the adjustment fastener. Then wedge the pry bar between the alternator and a strong part of the engine and pull in the direction that will apply tension to the belt. Tighten the adjustment fastener.

5. Check tension again and readjust if necessary. Check the tension again with the gauge, and if necessary loosen the fastener and adjust the belt again until it is at the correct tension for the vehicle.

6. Start the engine. Start the engine, and observe the belt to make sure it is properly seated and operating correctly.

7. Stop the engine and recheck the belt tension.

Brake Fluid Inspection

Brake fluid transmits pressure from the master cylinder to the brake actuators at the wheels. When brakes operate, they produce a large amount of heat. A disc brake, in a medium-sized vehicle, will produce enough heat to boil a liter of water, when stopping from around 40 mph (60 km/h). Brake fluid has a high boiling point, so it can operate in this extreme environment without breaking down or deteriorating. Brake fluid is **hygroscopic**; it absorbs moisture. Moisture will lower the boiling point, reducing the ability of the fluid to work at high temperatures. This is why brake fluid must be changed regularly, typically between 12 and 24-month maintenance intervals. Most brake fluids are **glycol-based** (DOT3, DOT4, DOT5.1); however, **silicon-based** (DOT5) fluids are also available. Some other vehicles may use brake fluids with a **mineral-based fluid**. Always confirm the proper brake fluid type before adding fluid to the brake system **(Figure 10-14)**.

While silicon fluids are not hygroscopic, they are usually not suitable replacements for brake systems designed to use other brake fluids. Do not use silicon-based brake fluid in vehicles fitted with ABS unless specifically authorized or recommended by the vehicle manufacturer. Conventional brake fluid should not be mixed with mineral oil fluid. Serious brake system damage will occur.

Figure 10-14 Add the proper type of brake fluid to the brake system.

Brake fluid can damage the paintwork by softening the base paint. If you spill any brake fluid on the paintwork, you must dilute it with fresh water and not wipe the paint until it hardens again.

> **Tech Tip** *DOT 3, DOT 4, and DOT 5.1 are amber in color. DOT 5 is usually dyed purple.*

> **Tech Tip** *Always use the manufacturer's recommended brake fluid. Using a different type of brake fluid may cause component damage and brake system failure.*

Several factors can influence vehicle braking. These factors include road surface, road conditions, the weight of the vehicle, the load on the wheels during stopping, different maneuvers, and the tires on the vehicle. An effective braking system takes all of these factors into account. A basic hydraulic braking system has two main sections—the brake assemblies at the wheels, and the hydraulic system that applies them. There is a brake for when the vehicle is in motion, usually a foot brake. And a park brake for when it's stationary, usually operated by hand. Some systems have all drum brakes. Some have disc brakes on the front wheels, and drum brakes on the rear. Others have all disc brakes. A basic braking system has a brake pedal, a master cylinder to provide hydraulic pressure, brake lines and hoses to connect the master cylinder to the brake assemblies, fluid to transmit force from the master cylinder to the wheel cylinders of the brake assemblies, and the brake assemblies, drum or disc, that stop the wheels.

The driver pushes the brake pedal; it applies force to the piston in the master cylinder. The piston applies pressure to the fluid in the cylinder, the lines transfer the pressure to the wheel cylinders, and the wheel cylinders at the wheel assemblies apply the brakes. Force is transmitted through the fluid. For cylinders the same size, the force transmitted from one is the same value as the force applied to the other. By using cylinders of different sizes, forces can be increased or reduced. In an actual braking system, the master cylinder is smaller than the wheel cylinders, so the force at all of the wheel cylinders is increased, but the distance the pistons move is decreased. When brakes are applied to a moving vehicle, they absorb the vehicle's kinetic energy. Friction between the braking surfaces converts this energy into heat. In drum brakes, the wheel cylinders force brake linings against the inside of the brake drum. In disc brakes, pads are forced against a brake disc. In both systems, heat spreads into other parts and the atmosphere, so brake linings and drums, pads and discs, must withstand high temperatures and high pressures. On modern vehicles this basic system has some refinements, such as a power booster. This assists the driver during brake application.

The brake fluid transmits hydraulic pressure from the master cylinder to the wheels. It is a special fluid with special properties. Most are a mixture of glycerin and alcohol, called glycol, with additives to give it the characteristics that are needed. It must have the correct viscosity for hot and cold conditions. Its boiling point must be higher than the temperature reached by the system. It must not damage seals, gaskets, or hoses or cause corrosion. Glycol-based fluids meet most requirements, although they do damage paint and they absorb moisture. This is important because as moisture is absorbed, it lowers the boiling point of the fluid. Brake fluids should not be mixed with mineral-based oils or solvents. If contamination is suspected, the braking system must be drained and flushed with a suitable solvent, and rubber components replaced.

> **Tech Tip** *If DOT 3, DOT 4, and DOT 5.1 brake fluid are dark amber in color, then the fluid is likely contaminated.*

The following are safety recommendations to use when checking and adjusting brake fluid:

- Never use any petroleum or mineral-based products, such as gasoline or kerosene to clean a braking system or its components. They are not compatible, and will result in a failure of the braking system and its components. This may result in injury to the passengers or damage to the vehicle.

- Make sure that you understand and observe all regulations and personal safety procedures when carrying out these tasks. If you are unsure of what they are, ask your supervisor.

- Brake fluid can damage the paintwork by softening the base paint.
- Brake fluid has a hygroscopic nature; it will absorb moisture rapidly. Always replace the cover or lid as soon as possible on the master cylinder and the brake fluid container.
- The higher the DOT number of the brake fluid, the higher its boiling point.
- Do not mix any DOT 5 silicone-based brake fluid with a DOT 3 or 4 glycol-based fluid because they are incompatible.
- As glycol-based brake fluid absorbs moisture, its boiling point is lowered. This can cause the brakes to be less effective or may cause them to fail to brake properly.
- Brake fluid is stored in the master cylinder. If you are unsure of its location, consult the vehicle's shop manual or the owner's manual.
- If the vehicle has antilock brakes, consult the owner's manual before filling the cylinder.
- If brake fluid splashes into your eyes, rinse it out with tap water immediately.
- Do not swallow brake fluid. It is toxic.
- Brake fluid reservoirs will indicate the maximum and minimum levels with a marker on the side, or level bars inside the container.

The following is a general procedure to follow when checking and adjusting brake fluid:

1. Check brake fluid. Wipe around the master cylinder top cover to prevent any dirt from entering the system. Open the top of the master cylinder by removing the plastic lid, or by prying the retaining clamp from a metal cover. Look inside the master cylinder. The fluid should be up to the FULL line on the side of the cylinder or within half an inch of the top of each chamber **(Figure 10-15)**.

2. Adjust brake fluid. Add the manufacturer's recommended brake fluid only if needed. Replace the master cylinder cover and check that it is fully seated.

3. Final inspection. Check for any leaks around the master cylinder with a flashlight. Dilute any brake fluid that may have been spilled onto the painted bodywork with fresh clean water. Do not rub the fluid with a cloth as the paint may be softened and easily damaged.

Figure 10-15 The brake fluid should be between the add and the full mark.

Coolant Inspection

One of the most important systems requiring regular maintenance is the engine cooling system. This section investigates the simple concepts and procedures to use when servicing the cooling system.

Cooling Principles

Modern vehicle cooling systems are made up of a number of different components that work in unison to provide an efficient method of cooling the engine at an appropriate operating temperature. Combustion of the air-fuel mixture in the cylinders generates heat, which produces high pressure, to force the piston down in the power stroke. Not all of this heat can be converted into useful work on the piston, and it must be removed to prevent seizure of moving parts. The cooling system is also responsible for maintaining proper operating temperature of the engine. Most engines are liquid-cooled. A liquid-cooled system uses coolant—a fluid that contains special chemicals mixed with water. Coolant flows through passages in the engine and through a radiator. The radiator accepts hot coolant from the engine, and lowers its temperature. Air flowing around and through the radiator takes heat from the coolant. The lower-temperature coolant is returned to the engine through a coolant or water pump.

Coolant

Engines once used plain water, which is cheap and usually easily available. That's where the names "water pump" and "water-jacket" came from. Water absorbs more heat for its volume than any other liquid, which is another good reason for using water **(Figure 10-16)**. But by itself, water causes problems. Ordinary impurities in tap water are harmful to engines. And even when water is pure, it still contains extra oxygen, which reacts with metals to cause corrosion and rust. Keeping unwanted oxygen out of a cooling system is one reason why modern cooling systems are sealed from air.

Water also allows **electrolysis**, an electrical and chemical process that corrodes metals. In modern cooling systems, chemicals called inhibitors are added to water to limit or inhibit corrosion **(Figure 10–17)**. Other additives are

Figure 10-17 A thermostat housing shows signs of corrosion.

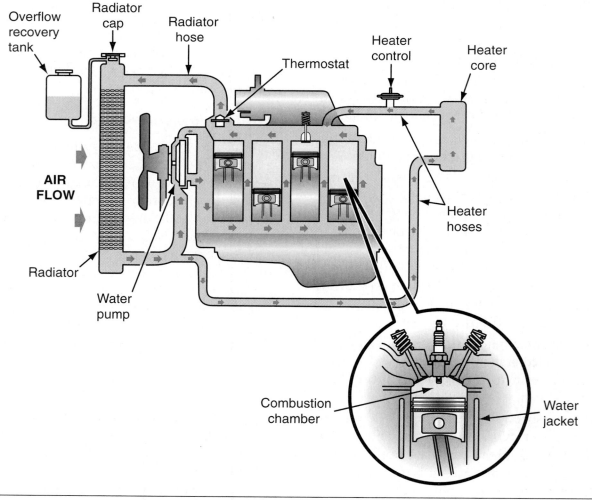

Figure 10-16 The cooling system removes excess heat from the engine.

used to make it harder for the water in the coolant to boil. Another harmful property of water is that it expands as it freezes. This is a problem for liquid-cooled engines, when temperatures drop below freezing. In a cold engine that is stopped, water in the cooling system can freeze and expand, with enough force to crack engine blocks and radiators.

An additive called antifreeze lowers the freezing point of water, to try to keep it below the outside temperature. The mixture of water and antifreeze is called coolant. The proper mixture of water and antifreeze can lower the freezing point to –34°F and raise the boiling point to 225°F.

Radiator Pressure Caps

If a coolant boils, it can be as serious for an engine as having it freeze. Boiling coolant in the water jackets becomes a vapor. No liquid is left in contact with the cylinder walls or head. Heat transfer by conduction stops. Heat builds up, and that can cause serious damage. One way to prevent this is with a radiator-pressure cap that uses pressure to change the temperature at which water boils. As coolant temperature rises, the coolant expands, and pressure in the radiator rises up above atmospheric pressure, and that lifts the boiling point of the water. Engine temperature keeps rising, and the coolant expands further. Pressure builds against a spring-loaded valve in the radiator cap, until at a preset pressure, the valve opens. In a recovery system, the hot coolant flows out into an overflow container. As the engine cools, coolant contracts and pressure in the radiator drops. Atmospheric pressure in the overflow container then opens a second valve, a vacuum vent valve, and overflow coolant flows back into the radiator. And that stops atmospheric pressure from collapsing the radiator hoses.

Recovery Systems

A recovery system maintains the proper coolant level in the system at all times. As engine temperature rises, coolant expands **(Figure 10-18)**. Pressure builds against a valve in the radiator cap until, at a preset pressure, the valve opens. Hot

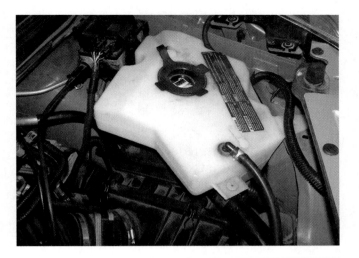

Figure 10-18 A coolant recovery tank recovers coolant as it expands and returns it to the cooling system when the engine cools.

coolant flows out into an overflow container. As the engine cools, coolant contracts and pressure in the radiator drops. Atmospheric pressure in the overflow container opens a second valve, and overflow coolant flows back into the radiator. No coolant is lost and excess air is kept out of the system. Like water, air contains oxygen, which reacts with metals to form corrosion.

The following are safety recommendations to use when checking and adjusting coolant:

- Always be very careful when opening a radiator cap, because the cap keeps the coolant under pressure to raise its boiling point. Sometimes, even the pressure in a warm engine can force the coolant to spurt out when the cap is released.

- Always cover the cooling system pressure cap with a rag to catch any hot spray.

- Always wear eye protection.

- Never remove a radiator cap on an overheated engine; wait for it to cool down first.

- Always make sure that you wear the appropriate personal protection equipment before starting the job. It is very easy to hurt yourself even when the most exhaustive protection measures are taken.

- Always make sure that your work area/ environment is as safe as you can make it. Do not use damaged, broken, or worn out workshop equipment.

- Always follow any manufacturer's personal safety instructions, to prevent damage to the vehicle you are servicing.

- Make sure that you understand and observe all regulations and personal safety procedures when carrying out these tasks. If you are unsure of what they are, ask your supervisor.

- There are two correct level marks on the reservoir, because the coolant in the system expands and contracts in volume, depending on how hot it is. The coolant level should be at the lower mark, when the vehicle is cold. The coolant level should be at the upper mark, when the coolant is hot.

- If the reservoir is completely empty, add coolant until the level is up to the appropriate mark for the engine temperature. Then run the engine until it is at its normal operating temperature, and check the level again. You will probably need to adjust the level again.

- For each 10-kPa increase in the radiator cap operating pressure, it will increase the boiling point of the coolant by 2.6°C (1 PSI for each 3.25°F).

SAFETY TIP *Never remove a radiator cap on an engine that has been running. Always allow the engine to cool before removing the radiator cap. Severe personal injury could result if the radiator cap is removed from a hot cooling system.*

The following is a general procedure to follow when checking coolant:

1. Check fluid level. Most modern vehicles have a coolant system that uses a transparent recovery tank as a coolant reservoir. Check the level of coolant in this reservoir. If the engine is hot, the level should be visible at the upper mark **(Figure 10-19)**. If the engine is cold, it should be at the lower mark **(Figure 10-20)**.

2. Check protection level with a **hydrometer**. Before adding new coolant, check the specific gravity of the coolant in the system with a coolant hydrometer. Draw some coolant up into the hydrometer and read the mark on the

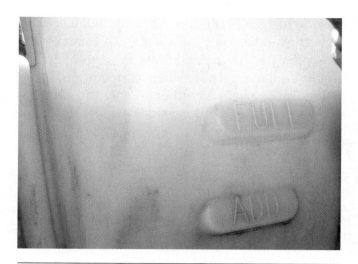

Figure 10-19 The level should be visible at the upper mark when the engine is hot.

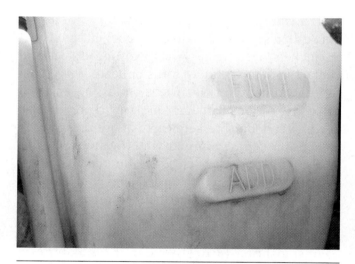

Figure 10-20 The level should be visible at the lower mark when the engine is cold.

float at the level of the fluid in the chamber. This will indicate the freezing point of the coolant mixture in the system, so you can tell if it has the right proportions of antifreeze and water.

3. Adjust fluid level. Check the service manual for the recommended type and mixture of coolant that will produce an appropriate level of protection for the conditions where the vehicle will be used. Use a funnel to add enough coolant to bring the level up to the appropriate mark. Replace the coolant reservoir cap **(Figure 10-21)**.

Figure 10-21 Be sure to properly install the pressure cap when cooling system service is completed.

Figure 10-22 Raise the vehicle to inspect under body components, brake components, and wheels and tires.

same safety considerations are applied to vehicle tire servicing.

Tech Tip *Coolant test strips are available that can test freezing point, boiling point, and acidity.*

Under-Hood Visual Inspection

Other items to visually inspect under the hood are the ignition, wiring, fuel, and emission systems. The ignition and wiring systems should be clean and free from any oil or other fluid contamination. They should be properly routed and be in place with the proper retainers and clips. The fuel system should have no external leaks and there should be no raw fuel odor. The emission system should appear intact and hoses should be free from oil or other contamination and should be free from cracks or other visual defects. Any problems in these systems should be brought to the attention of your supervisor.

UNDER-VEHICLE INSPECTION

Items to inspect under the vehicle are under-body components, brake components, and wheels and tires **(Figure 10-22)**.

The steering system is a complex network of linkages, gears, and support structures that are subject to intense road shock. It is vital that the system works properly for safety reasons. The

Under-Body Components

The following are points to note when inspecting under-body components:

- This is a systematic visual inspection of all major vehicle systems. Be prepared to note down any faults to discuss later with your supervisor.

- The steering and suspension inspection includes tie-rods, tire and wheel assembly, suspension bushings, and shock absorbers.

- You will be checking the brake friction material, lines, and hoses.

- In the transmission area, you will be looking for fluid leaks, tightening mounting bolts, and inspecting the clutch mechanism or shift linkage.

- Clamps and bolts may need tightening on the exhaust system and manifold pipe. You will also be looking for signs of exhaust leaks, corrosion, or deterioration.

- You will be checking for any excess movement in driveline shafts.

- Look for leaks around the differential and check the rear shock absorbers or leaf springs.

- The fuel tank must be secure and fuel lines inspected for damage or abrasion.

The following is a general procedure to follow when inspecting under-body components:

1. Begin under-vehicle visual inspection. Safely raise the vehicle to a comfortable working height. Be ready to record any faults you find, and begin your inspection at either end of the vehicle. Whichever end you choose, work systematically in one direction. Note any problems you find and discuss them with your supervisor. Pay particular attention to any fluid leaks, which will probably be the easiest problems to spot.

2. Check steering area. Locate the tie-rods and move them through their operating arc. The action should be smooth without binding. The tire and wheel assembly should also move in a forward and backward direction to detect lateral movement in the tie-rod end. Look carefully for missing or torn rubber boots around tie-rod ends or the steering shaft. At the same time, check the security of the steering mountings. Inspect any rubber suspension bushings for swelling or damage, and check shock absorbers for signs of damage or leaks. Inspect wiring harness that is accessible for any obvious damage **(Figure 10-23)**.

3. Check transmission area. Check the transmission mounting bolts for tightness. Trace and

Figure 10-24 Check the transmission for any leaks.

record the source of fluid leaks, if you find any. With a manual transmission, check the clutch operating mechanism for looseness or binding. For an automatic transmission, check the shift linkage for smoothness of operation. If the transmission is electronically controlled, check any wiring for obvious damage **(Figure 10-24)**.

4. Check the exhaust system. Check the tightness of the flange bolts on the engine manifold pipe. It is also important to make sure all the clamps on the exhaust system are tight. If there is an exhaust leak, it is usually identified by a blackish soot deposit at the source of the leak. Examine the catalytic converter, muffler, and resonator for any signs of corrosion or deterioration. Check the tail pipe for any corrosion, and looseness in the mounting brackets or hangers **(Figure 10-25)**.

5. Check brake components. Remove the wheels and inspect the brake friction material (brake linings), brake lines, and brake hoses. Note the thickness of the brake friction material. Check the brake hoses, looking for signs of cracking or abrasions.

6. Check parking brake cables. Inspect the parking brake cable to make sure it is not frayed, damaged, or binding **(Figure 10-26)**.

Figure 10-23 Inspect tie rod ends for excess movement or torn boots.

Figure 10-25 Inspect the muffler as well as the exhaust pipe.

Figure 10-26 Check the parking brake cables.

7. Check driveline shafts. On any rear wheel drive vehicles, including pick-ups and SUVs, inspect the drive shaft universal joints for signs of excess movement or rust. Rusty powder marks near the front, and rear universal joints could indicate a rusted and/or seized universal joint. To check for wear, rotate the shaft and flange in opposite directions. There should be no movement in the joint. On four-wheel drive vehicles, repeat this procedure on the front drive shaft universal joints.

8. Check differential and rear suspension area. On rear-wheel drive vehicles the rear axle housing supports the differential unit. On front-wheel drives, the differential is usually located in the transaxle housing. Inspect the pinion shaft oil seal for any obvious signs of leakage. Next, check the rear shock absorbers for any signs of physical damage or fluid leaks. Tighten all the suspension mounting bolts to the proper torque, noting any bolt that is loose. Inspect the suspension mounting bushes for any signs of deterioration or damage. This will include any control arms or struts. If the vehicle is fitted with leaf springs, inspect the springs for any cracks or misalignment. On a front-wheel drive vehicle, inspect the rear strut assembly for any physical damage or signs of fluid leaks. Inspect the brake hoses for any obvious signs of cracking or abrasion **(Figure 10-27)**.

9. Check fuel tank area. Inspect fuel tank mounting bolts or retaining clamp bolts, noting any bolt that is loose. Carefully check all the fuel lines for any signs of damage or abrasions that may cause a leak.

10. Discuss problems with your supervisor. After completing the inspection, discuss your list of problems with your supervisor to see what action can be taken to fix the problems.

Wheel and Tire Inspection

Tires are designed to maintain friction between the vehicle and the road. Tires must be appropriate for the vehicle, the conditions of the road, and they must be inflated properly. In this section you will

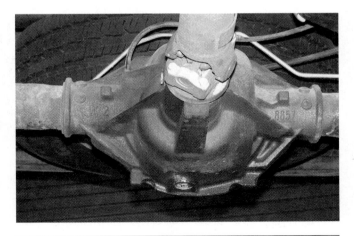

Figure 10-27 Inspect differentials for any leaks.

Figure 10-28 Tires are an important component of the vehicle.

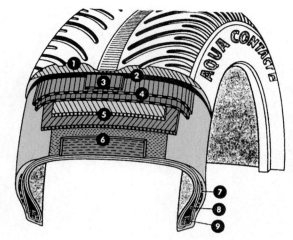

1. CAP tread	5. Two steel-cord plies
2. Tread base	6. Two rayon-carcass plies
3. Nylon wound aquachannel breaker	7. Double nylon bead reinforcement
4. Two-ply nylon wound breaker	8. Bead filler
	9. Bead core

Figure 10-30 A typical radial tire components.

learn about the different types of tires you will service in the workshop, and how to maintain them properly **(Figure 10-28)**.

Tire Overview

The tire provides a cushion between the vehicle and the road to reduce the transmission of road shocks. It also provides friction to allow the vehicle to perform its normal operations. Modern tires are manufactured from a range of materials. The rubber is mainly synthetic. Three types of tire construction are bias, belted bias, and radial **(Figure 10-29)**. Most passenger cars now use radial tires **(Figure 10-30)**.

And radials are replacing bias tires on four-wheel-drives and heavy vehicles. Tube tires require an inner tube to seal the air inside the tire. Tubeless tires eliminate the inner tube by making the complete wheel and tire assembly airtight. A special, airtight valve assembly is needed. This can be a tight fit into the rim, or it can be held with a nut and sealing washers.

Tires can be identified by markings on the sidewalls. This typically includes the manufacturer's

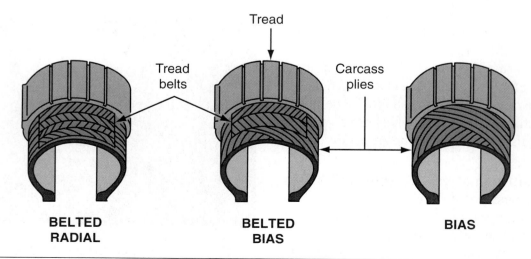

Figure 10-29 Three types of tire construction: radial, belted bias, and bias.

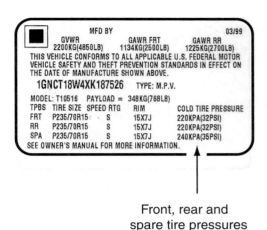

Front, rear and
spare tire pressures

Figure 10-31 A typical tire placard.

Figure 10-33 A tire pressure gauge is used to determine proper tire inflation.

name, the rim size, the type of tire construction, aspect ratio, maximum load and speed, and, in some cases, intended use. Regulations cover the allowable dimensions for wheels and tires on a particular vehicle. These specifications are usually set out on the tire placard attached to the vehicle. Incorrectly selected wheels and tires can overload wheel bearings and change steering characteristics. The tire placard lists the wheel and tire sizes and tire inflation pressures approved by the manufacturer for the vehicle **(Figure 10-31)**.Using other wheels and tires may be illegal.

Tire Principles

The sidewalls of radial-ply tires bulge where the tire meets the road, making it difficult to estimate inflation pressure visually **(Figure 10-32)**. It needs to be checked with an accurate tire gauge.

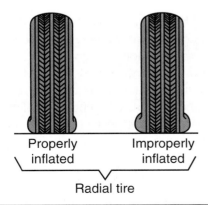

Figure 10-32 It is difficult to visually distinguish between a properly and improperly inflated radial tire.

Using correct inflation pressures extends tire life, and is vital for safety **(Figure 10-33)**. Sidewalls of an under-inflated tire flex too far, which pushes the center section of the tread up and away from the road surface. This causes wear at the shoulders of the tire. In an over-inflated tire, the sidewalls are straightened, which pulls the edges of the tread away from the road, and causes wear at the center of the tread.

Tech Tip *Properly inflated tires will last longer and give better fuel economy for the driver.*

Two methods are used to retain air inside the wheel assembly. These are the tube-type method and the tubeless method. A tube-type tire uses an inner tube, which provides an airtight container inside the tire. A tubeless tire is lined with a soft rubber layer to form an airtight seal. This inner liner also seals against small penetrations, letting air escape only relatively slowly. When a tubeless tire is fitted, an airtight valve assembly is used. It can be a tight fit into the rim, or be held with a nut and sealing washers.

A tire and wheel assembly must be balanced. As the wheel rotates, centrifugal force acts outwards. Any part heavier than the rest will vibrate vertically, with the heavy area slapping the road surface with each turn of the wheel. This is called **static** or single plane imbalance.

***

NOTE: The rapid movement of the wheel assembly vertically or up and down is called *tramp.*

***

Dynamic imbalance causes the wheel assembly to turn inwards, and then outwards, with each half revolution. As speed rises, rapid side movement of the front wheels causes a sideways vibration, or wheel wobble effect, at the front of the vehicle. This situation would occur if the wheel had extra weight on the outside and extra weight on the opposite side on the inside of the wheel. These conditions must be corrected to prevent cupping or dishing of the tread, and reduced tread life. Wheels should always be dynamically balanced using the proper equipment and procedures **(Figure 10-34)**.

***

NOTE: The rapid movement of the wheel assembly turn inwards and outwards is called *shimmy.*

***

Tread life can also be reduced by incorrect wheel alignment. The feathered edge of a tire indicates an incorrect toe setting. And wear on the one shoulder of the tire could be due to incorrect

Figure 10-34 A dynamic tire balancer.

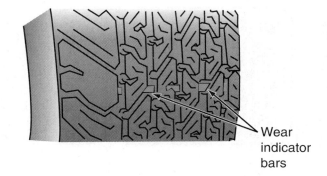

Wear indicator bars

Figure 10-35 Tread wear bars are molded into the tire tread. When the wear bar is flush with the tread surface it indicates the minimum tread depth has been reached, which is 2/32 in.

camber setting. Most passenger car tires have tread-wear indicators molded into the tread pattern. They generally provide an indication when the depth of a tire groove falls to 2/32 of an inch or 1.5 millimeters **(Figure 10-35)**.

Control of a vehicle in any weather conditions depends on frictional forces generated between the tires and road surface. On a dry road, a smooth rubber surface can provide a high coefficient of friction, sufficient to maintain a degree of control during braking, accelerating, and cornering. In wet conditions, the coefficient of friction between a smooth tire and the road surface falls to an extremely low value. Grooves in the tread pattern clear water away from the contact patch area. This allows a relatively "dry area" to be formed, and for road adhesion to be maintained.

Wheel and Tire Service

The following are safety recommendations to use when checking and adjusting tire pressure:

- If you check the tire pressures after the vehicle has been driven a long distance and the tires are hot, do not release the excess pressure. If you bleed the tire down to the manufacturer's recommendation, it will be under-inflated when the tire is cold or at normal operating temperature. This could cause the tire sidewall to fail.

- Never inflate a tire above the manufacturer's recommended maximum pressure. Make sure that you understand and observe all regulations and personal safety procedures when carrying out these tasks. If you are unsure of what they are, ask your supervisor.

Figure 10-36 The technician should have room and a good air supply.

- Tires get hotter as they are being driven. Driving on under-inflated tires can cause serious damage to the tires. The sidewalls are subjected to excessive flexing that can generate a great deal of internal heat.
- Check the owner's manual or the tire placard usually located on the driver's side door jamb for correct pressures.
- The tire pressures should be checked when the tires are cold. On average, the pressure in a tire

will increase by about 12.5 kPa for each 2°C (1 pound per square inch (psi) for each 10°F) the tire is above its normal operating temperature.

> **Tech Tip** *The tire placard is a decal detailing the proper inflation pressures for the vehicle's tires and is usually located on the driver's door post.*

The following is a general procedure to follow when checking and adjusting tire pressure:

1. Prepare the vehicle and equipment. Park the vehicle so you can reach all four tires with the air hose. Check the recommended pressures in the owner's manual or on the tire placard usually located on the driver's side door or surrounding location **(Figure 10-36)**.

2. Check the tire sidewall markings **(Figure 10-37)**. Check the tire specifications and maximum load carrying capacity. These details can normally be found on the drivers' side of the vehicle on a decal near the door pillar. If the tire markings do not meet the specifications of the vehicle, tell your supervisor and ask for their direction. This can have a serious effect on the performance of the tires and the vehicle.

Figure 10-37 Typical tire sidewall information. (*Courtesy of Goodyear Tire Company*)

3. Check and adjust pressure. Check the pressure when the tires are cold. Remove the cap from the tire valve on the first tire. Use a reliable tire gauge to check the air pressure in the tire. A quality pocket-type pencil gauge is ideal for this purpose. If there is a gauge attached to the air filler, this is more convenient. If you need to add air, use short bursts with the air hose, so you do not over inflate the tire. Re-check the tire pressure after filling it, and replace the cap on the tire valve. Repeat the process for the other tires.

> **Tech Tip** *Air pressure gauges should be checked for accurate pressure readings on a regular basis. The air filler with an attached gauge should be also checked for accurate pressure readings on a regular basis.*

Checking Tire Wear

The following are safety recommendations to use when checking the tire wear pattern:

- Make sure that you understand and observe all regulations and personal safety procedures when carrying out these tasks. If you are unsure of what they are, ask your supervisor.

- Some manufacturers supply a temporary spare rim assembly, instead of a full-size spare. It is not intended for long-term use or high speed, but it must have adequate tread.

The following is a general procedure to follow when checking tire wear:

1. Check for foreign objects and pressure. Inspect the tires for embedded objects in treads and remove them, and look for signs of wear on all wheels, including the spare. Check the pressure in the tires.

2. Check tread wear depth. Most tires have wear indicator bars incorporated into the tread pattern. Inspect the wear indicator bars. Tires should have at least 1/16 of an inch, or two millimeters of tread remaining. The wear indicator bars are normally set at this depth. If the tread is worn down to that level or below, they are unserviceable and must be replaced.

3. Check tread wear pattern. Check the wear patterns with the vehicle's shop manual to indicate the types of wear that have occurred. Causes of uneven wear can include faulty shock absorbers, incorrect front alignment angles, worn steering or suspension parts, and wheels out of balance. Uneven tread and bald spots can indicate over- or under-inflated tires and poor alignment.

4. Check tire for damage. Inspect the sidewalls of the tires for signs of cracking from impacts with blunt objects. Carefully examine the tread area for separation. This is usually identified as bubbles under the tread area. Spin the wheel and see if it is running true. If it is wobbling as it rotates, report it to your supervisor.

PERIPHERAL SYSTEM INSPECTION

Peripheral systems may not be essential to the operation of the vehicle, but they are necessary for legal and safety reasons. This section investigates the simple procedures you will undertake while inspecting peripheral systems. Peripheral systems include heating elements, vehicle doors, ignition switches, windshield service, lighting service, and security systems. It is important to be aware of the peripheral elements of a vehicle and inspect them regularly as part of a regular service.

Heating Elements

Many vehicles contain a heated rear window. The heating is performed by an element that is almost invisible when viewed from the driver's seat. Close inspection reveals that there are many small resistance wires embedded within the glass and crossing from one side to the other **(Figure 10-38)**. When the rear defroster switch is turned on, electrical current flows through the wire, causing it to heat. This gradually clears a fogged glass, allowing the driver to see through it. Heating elements can also be fitted to exterior mirrors to keep them free of frosting and fogging. More recent vehicle designs use a similar wire embedded within fixed glass as a radio antenna. The wire is usually separate to the resistance wire

Figure 10-38 A window heating grid element.

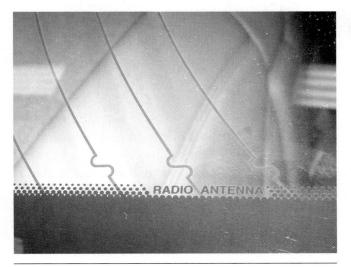

Figure 10-39 An antenna built into the window.

Figure 10-40 A rear hatch opening.

and can be located in any fixed window, including the windscreen. Some manufacturers fit the embedded antenna in more than one window **(Figure 10-39)**.

Vehicle Doors

A vehicle body contains many openings besides the vehicle doors. They include engine compartment hoods, hatch and tailgate openings, fuel doors, and battery access covers. All of these openings have to be secured and may require a remote switch or lever to be activated. In some cases the activation is by cable; others may use electric or vacuum solenoids. Some rear hatch openings have a hinged window incorporated, so the owner can have easy access without opening the entire door **(Figure 10-40)**. Engine compartment hoods may use single or double acting

hinges. Their release may be located inside the passenger compartment, under the dash, in the glove compartment, or on a doorjamb **(Figure 10-41)**. Some manufacturers are no longer fitting an engine compartment release to their vehicles. The hoods on these vehicles can be released only by the manufacturer's scan tool or by a service key fitted into a secluded opening on the vehicle body (usually behind a manufacturer's badge). Before opening the engine hood on these vehicles, check the owner's manual, or the manufacturer's manual, as unauthorized access may void any vehicle warranty.

Figure 10-41 A typical hood release inside a vehicle.

Ignition Switches

The ignition switch has more functions than simply starting the vehicle. The common points on an ignition switch usually include:

- Accessories
- Lock
- Off (not on all vehicles)
- On (run)
- Start

The key can be removed only from the "Lock" position. When this occurs, all non-essential electrical circuits are disabled, and the steering column lock is enabled. If fitted, the theft deterrent system is normally activated at this time. Many modern vehicles also include an "Off" ignition point. Turning the key from "Lock" to the "Off" position unlocks the steering column, but it does not enable any electrical systems or disable the theft deterrent system **(Figure 10-42)**.

The "Accessories" position allows power to be supplied to the vehicle entertainment system, blower fan, and in some vehicles the electric windows and sunroof. This is mainly for passenger convenience but prolonged use of any of these without the motor running will drain the battery **(Figure 10-43)**.

When the switch is turned to the "On" position all warning lamps on the instrument panel should

Figure 10-43 An ignition switch in the "ACC" position.

illuminate. This is to test the operation of the lamps **(Figure 10-44)**. On vehicles that are not fitted with a theft deterrent system, this position also activates the ignition and other electrical systems required for engine starting. Vehicles that are fitted with a theft deterrent system may not normally activate these systems until the key is turned to the "Start" position.

NOTE: Theft deterrent systems are sometimes referred to as an *Immobilizer.*

Figure 10-42 An ignition switch in the "LOCK" position.

Figure 10-44 An ignition switch in the "ON" position.

The "Start" position activates the starter motor solenoid, which enables the engine to crank and start the engine. Vehicles without a security system will be able to start immediately, as the required electrical systems will have already been activated when the key was turned to the "On" position. In vehicles fitted with a security system, a number of the essential electrical systems are not normally enabled until the key is turned to the "Start" position. When this occurs communication between various electronic modules determines whether to allow the engine to start. The security system is deactivated and the ignition, fuel, and charging systems are enabled. There may be a slight delay while this communication is occurring. In many automatic transmission vehicles, the ignition switch also has a transmission shift interlock device connected to it. In these vehicles the gear selector must be moved into the PARK location before the key can be removed from the lock in the OFF position. In the same way the transmission shift lever cannot be moved into gear until the engine has started.

Windshield Service

A clean windshield is important for safe driving. Vehicle maintenance should include this important servicing procedure. The windshield washer reservoir should be filled with the correct cleaning fluid and wiper blades should be inspected.

Windshield Washer Fluid

The following are some points to note when checking windshield washer fluid:
- Use only an approved windshield washer fluid.
- Never use laundry detergents in systems. Chemicals in the detergent may damage paintwork.

The following is a general procedure to follow when checking and adjusting windshield washer fluid:

1. Locate washer fluid reservoir. Locate the windshield wiper fluid container. This is normally found under the hood and is usually a plastic reservoir **(Figure 10-45)**.

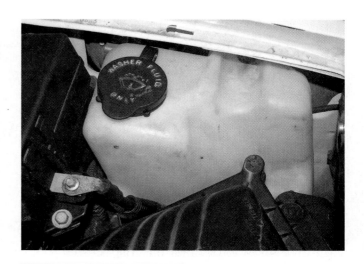

Figure 10-45 A windshield washer reservoir.

2. Check the fluid level. Some vehicles have electrical sensors incorporated into the reservoir. An indicator lamp on the inside of the vehicle will show the driver when the system needs to be filled. In most vehicles the fluid level is visible and a marker on the reservoir indicates the correct fluid level.

3. Fill reservoir. Choose an appropriate washer fluid. Some fluids are concentrated and need to be mixed with water. Never use laundry detergent to top up the reservoir, as the chemicals in the detergent can be detrimental to the vehicle paintwork. Take care not to spill the fluid, and fill the reservoir to between the minimum and maximum fill marks.

Wiper Blades

You should also check the condition of windshield wiper blades and arms and change the windshield wiper blades when necessary.

The following are some points to note when checking windshield wiper blades and arms:
- Never operate the wipers when they are dry because this may damage the blades or scratch the surface of the windshield.
- Never bend the arms to make better contact with the windshield. The arms are pre-tensioned by the manufacturer and damage could result. If the arms seem to have lost their spring tension, obtain a suitable replacement.

Figure 10-46 A wiper blade is inspected.

The following is a general procedure to follow when checking windshield wiper blades and arms:

1. Check windshield wiper blades. Lift the wiper arm away from the windshield and inspect the condition of the blades (**Figure 10-46**). Look for damage or loss of resilience in the material. Wet the windshield with a spray, or with the washers and switch the wipers on. If the windshield is being wiped clean, do not replace the blades. If the wiper blades are not wiping the glass evenly, or are smearing the windshield, you will need to replace the blades.

2. Remove blade assembly. Remove the blade assembly. Depending on the vehicle, you may need to remove the wiper arm from its mounting, or you may be able to just undo a spring clip and remove the blade insert.

3. Obtain replacement blade. Check the shop manual and obtain the correct size replacement. New developments in blade design means that the profile of the new blade may not be identical to the original.

4. Install inserts. Feed the new inserts into the wiper arm, and make sure the clips fit snugly and engage properly.

5. Test blades. Wet the windshield again, and operate the wipers to check their performance, making sure they remove the water evenly.

Tech Tip *Most wiper blades have an operational life of 6 to 12 months.*

LIGHTING SERVICE

The lighting system is an important and often overlooked part of the vehicle. The importance of the lighting system working efficiently cannot be overemphasized.

Lighting System Components

Key areas of the lighting system to inspect are:

- Headlights
- Park lights and taillights
- Stoplights
- Turn signal lights
- Reverse lights

NOTE: The terms *light* and *lamp* are interchangeable.

Headlights

In headlights, two filaments are necessary to provide for high and low beam function. These must be positioned correctly in relation to the highly polished reflector. This is called focusing and is carried out during manufacture. The high beam filament is positioned at the focal point of the reflector, to project the maximum amount of light forward and parallel to the reflector axis. This light is then shaped by the lens, which is made up of a pattern of many small glass lenses and prisms. These lenses and prisms bend the light horizontally and vertically to achieve the desired pattern for road illumination. The low beam is placed above and slightly to one side of the high beam filament. Mounting the low beam filament in this position produces a beam of light that is projected downwards and towards the curb. With this arrangement the high beam filament produces the best possible light output, while the low beam filament gives a downward and wider beam, which should not blind oncoming drivers (**Figure 10-47**).

A semi-sealed beam headlight uses a replaceable bulb with a collar. The collar locates the

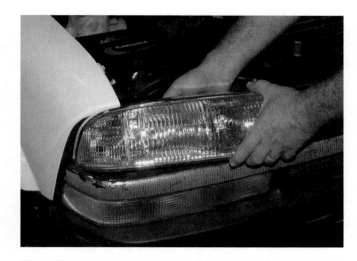

Figure 10-47 A headlight assembly is inspected.

bulb in the headlight and correctly positions the filaments. A sealed beam headlight has a highly polished aluminized glass reflector, which is then fused to the optic lens. This forms a completely sealed unit, which has the filaments accurately positioned in relation to the reflector. When a filament fails in a sealed beam light, the whole unit must be replaced. Some headlight bulbs have a partial shield below the low beam filament. This shield stops light from the filament striking the lower part of the reflector. The shield provides the primary shape of the low beam. The final shaping of the beam is carried out by the lens and provides a low beam that is asymmetrical.

> **Tech Tip** *Sealed beam lights should be checked for proper aim when replaced.*

Park Lights and Taillights

For motor vehicles and trailers, two red tail lamps operate when the headlight switch is in the park position and the headlight position. The two lights are located close to the widest points of the vehicle, so that other drivers can see vehicle width. The bulbs are connected in parallel to each other, so that the failure of one filament will not cause total circuit failure. A license plate illumination lamp is usually connected in parallel to the taillights and operates whenever the taillights are on. Taillights are usually incorporated in a cluster assembly at the rear of the vehicle.

Government regulations control the height of the lamps and their brightness. The park lights, sometimes called clearance lamps, are located at the front end of the vehicle, and are used at nighttime when the vehicle is parked on the side of the road. They use low wattage bulbs and may have a lens or diffuser that makes the emitted light wide spread **(Figure 10-48)**. In some cases, the park lights are incorporated in the headlight assembly. The park lights operate when the light switch is moved to the first position. For safety reasons the park and taillights continue to operate when the light switch is moved to the headlight position. The bulbs are connected in parallel with each other. The circuit for the park and taillights includes the battery, fusible links and fuses, park light switch, the lights at each corner of the vehicle and the license plate light, the wiring to connect the components together, and the ground circuit to complete the circuit to the battery through the vehicle frame. When the park light switch is closed, current flows from the battery, through the fusible link to the park light switch, where it is fed through the fuse to the front park lights and to the rear tail and license plate lights. After passing through the filaments, the current path is completed through the frame of the vehicle to the negative battery terminal.

Figure 10-48 A typical bulb used in the exterior lighting system of a vehicle.

> **Tech Tip** *Many newer vehicles may use electronic modules to control many of the lighting functions.*

> **Tech Tip** *Always use the correct replacement lamps for the vehicle.*

Stoplights

Stoplights are red lights fitted to the rear of the vehicle. They are usually incorporated in the taillight cluster, although many vehicles have a higher additional stop light mounted on top of the trunk lid or on the rear window called a high-level stop lamp. The stoplights are activated whenever the driver operates the foot brake to slow down, or to stop the vehicle. The circuit consists of the battery, fusible links and fuses, a stoplight switch, stoplight bulbs, wiring to connect the components, and the ground circuit to return current from the filaments to the battery.

When the operator of the vehicle depresses the brake pedal, a switch mounted on the pedal support is closed. This allows electrical current to flow from the battery through the fuse, through the switch, to the brake light filaments, and to return to the battery by the ground system. When the driver releases the pedal, it returns to the rest position and open circuits the stoplight circuit. The flow of electrical current stops and the brake lamps are extinguished.

> **Tech Tip** *Some late model cars are using LED lighting for some stoplamps.*

Turn Signal Lights

Turn signals are amber lights located on the extreme corners of the vehicle. A column-mounted switch, operated by the driver, directs a pulsing current to the turn signal lights on one side of the vehicle or the other. These pulsing lights warn other road users of the vehicles intended change of direction. Once activated they continue to operate until the switch is cancelled either by the operator or by a canceling mechanism in the switch. The canceling mechanism operates after the steering wheel is returned to the straight-ahead position from a turning position.

The circuit consists of the battery, fusible links and fuses, the ignition switch, the flasher unit, a three-position switch used as the turn signal switch, the lights at the front and rear of the vehicle, pilot lights mounted in the instrument cluster to indicate to the driver which way the switch has been operated, wiring to connect all of the components, and the ground circuit to return the electrical current to the battery. If the turn signal switch is turned to indicate a right-hand turn, current from the battery flows through the fusible link to the ignition switch, where it is directed through a fuse to the flasher unit. The flasher unit uses a timing circuit to pulse the current flowing out of the flasher unit 60 to 120 times per minute. This pulsing current is directed through the turn signal switch to the right-hand turn signal lights at the front and rear of the vehicle, causing the lamps to flash on and off. A pilot light on the instrument cluster also pulsates. The operation of the flasher unit also produces a clicking sound to warn the driver that the turn signals are in operation.

When the turn signal switch is returned to the "off" or central position, no current flows through the flasher unit so the timer circuit is switched off. When the turn signal switch is turned in the opposite direction, it directs the pulsing current to the left-hand lights at the front, and rear of the vehicle as well as the left-hand pilot light on the instrument cluster. Vehicles are equipped with hazard warning lights. This circuit is similar to the turn signal lights except that it simultaneously causes a pulsing in all exterior turn signal lights, and both pilot lights on the instrument panel. These can warn other road users that a hazardous condition exists, or that the vehicle is standing or parked in a dangerous position on the side of the road.

Reverse Light

The reversing lights are white lights fitted to the rear of a vehicle. They provide the driver with additional lighting behind the vehicle during low light conditions, and also alert other drivers to the fact that the vehicle is to be reversed. The circuit consists of the battery, fuses and fusible links, the ignition switch, the reversing light switch on the transmission, reversing light filaments, wiring to connect the components, and the ground circuit to allow current to return to the battery through the vehicle frame.

When the ignition switch is on and the vehicle is placed in reverse, gear current flows from the battery, through the ignition switch, to the closed reversing light switch on the transmission. Electrical current flows across the closed switch to the reversing lights, and then returns to the battery by the earth return system.

Checking Lighting Systems

The following are some points to note when checking the lighting system:

- The lamps (lightbulbs) will become extremely hot after a few seconds of operation. Use caution as these lamps can cause serious burns.

- Make sure that you understand and observe all regulations and personal safety procedures when carrying out these tasks. If you are unsure of what they are, ask your supervisor.

- Be sure to work in a systematic manner, or you could miss a faulty bulb or another component.

- A vehicle may have warning lights that will activate only if that circuit is in use. You may need to turn that circuit on to see the warning light.

The following is a general procedure to follow when checking the lighting system:

1. Check instrumentation. In a darkened area, turn on the ignition. The dash warning lights should be displayed. Start the engine. If any warning light stays on when the engine is started, it could indicate a problem in one of the car's safety or mechanical systems. If you are unsure about what any of the warning lights mean, consult the manufacturer's manual.

2. Check the car horn. Check the horn during the lighting check. Make sure the car horn is working. If the horn is not working, locate it under the hood with the help of the manufacturer's

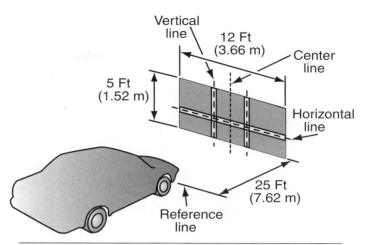

Figure 10-49 Headlights may be checked for alignment on an aiming screen.

manual. Check the wiring to make sure there is a good contact. If necessary, use a DVM to isolate the fault.

3. Check rear lights. Have someone stand behind the vehicle to report any problems, then turn the ignition on. Switch on the park lights and taillights. Do the same for left and right turn signal lights. Depress the brake pedal to make sure the brake lights work.

4. Check front lights. With somebody in front of the vehicle, make sure the high and low headlight beams, the park lights, and the turn signals are all working properly. Headlights may also be checked on a headlight alignment screen **(Figure 10-49)**.

5. Check interior lights. With the interior light switch in the correct position, open the driver's side door to make sure the interior lights work. If any of these lights do not operate, you may need to replace a bulb or fuse. Check the fuse first, using a DVM to check continuity. If the fuse is at fault, you should report this to your supervisor, as there could be a more serious fault in the vehicle's wiring system.

CHECKING A SECURITY SYSTEM

Make sure that you understand and observe all regulations and personal safety procedures when

Figure 10-50 A remote for a security system.

checking the security system. If you are unsure of what they are, seek assistance. Many of these systems use a remote control **(Figure 10-50)**. Some systems are movement sensitive, so it may be difficult to try to start the car from the inside. Your movement may set off the alarm. Some vehicle security transponders may not operate in all environments. The remote controls may not operate, properly if there is strong electromagnetic interference in the area. If a remote does not seem to operate, verify that it belongs to the vehicle you are attempting to unlock. It may be that someone mislabeled the key and remote from another vehicle. Most remotes are designed to cope with normal operating environments. Use caution and do not drop them into liquid or subject them to extreme temperatures or force.

The following is a general procedure to follow when checking a security system:

1. Set the security system. This section deals with vehicle security and alarms. When the system is armed, an LED type light usually blinks on the dash. In some cases you will also hear an audible tone. Sit inside the vehicle with the doors shut. Lock the doors and activate the security system. This is usually done with a transponder. The transponder that turns the security system on and off is usually a small battery-powered device attached to the key ring, or built in to the ignition key itself.

2. Test the security system. Start the engine. If the security system immobilizes the engine,

then this action should not be possible. If the security system is an audible alarm, it should now go off. If the security system works properly, move the vehicle to a different location—for instance, outside the workshop—and test it again. A transponder can suffer from electronic interference and may work better in some locations than others.

3. Check transponder operation. If the security test fails, check whether the transponder has power and is operating correctly. If it operates the door locks as well as the security system, locking and unlocking the vehicle will confirm that the transponder is working.

4. Replace transponder battery. If the transponder is not working properly, you might need to replace its battery. To do this, you will need to refer to the specific manufacturer's specifications for changing the battery to re-test the unit.

5. Action on failure. If either the security system or the transponder is not working, you should report it to your supervisor.

GENERAL INSPECTION

A general inspection of the vehicle allows the technician to check the cosmetic appearance, which may also reveal other areas of concern in the vehicle. These inspections may become important to the customer as they may help to track trends in the vehicle's overall condition. Some of these items are an important safety issue such as seat belt mechanisms. The areas covered during a general inspection are a visual inspection, checking door hinges, checking interior trim, checking seat belts, and checking vehicle seats. Any thorough inspection includes a vehicle's interior. It is not just for cosmetics but also for safety, as a defect or concern may have a hidden underlying cause.

Visual Inspection

The following is a general procedure to follow when performing a visual inspection:

1. Getting started. Once a month or prior to any long trip, a vehicle should be checked for overall

roadworthiness. The following simple measures will ensure that the vehicle is able to undertake the trip, and provide a warning of potential problems that may need further attention.

2. Areas to be checked. Begin your inspection by walking around the vehicle and observing any obvious items that need attention. Check the body condition to make sure that all the body components are secure. Open and close doors to check that they are operating correctly. Inspect the bumpers or fenders and ensure that they are secure. Inspect the external mirrors to ensure that they are secure and not broken.

Checking Door Hinges

The following are some points to note when checking door hinges and locks and adjusting door jamb/jambs:

- Different manufacturers use different forms of latches and locking mechanisms.

- Always make sure that you have the correct service manual for the job you are working on.

- On vehicles with heavy doors, such as some older vehicles and convertibles, the door latch contains a locating wedge or pin. This locates the door when closed, and takes some of the strain off the hinges.

The following is a general procedure to follow when checking door hinges and locks and adjusting door jamb/jambs:

1. Check door for hinge wear. Open the door to approximately 45° to the car body. Wear gloves when performing this test, in case there are any sharp edges underneath the door. Put your hand under the base of the door, and lift it gently against the hinges to see if there is excess wear in the pins. Some limited amount of wear is always present, and you will usually be able to move the door up and down a little at its outer edge. An excessive amount of movement means the door is worn and has "dropped." Report this to your supervisor, as the hinges or door jambs may need to be adjusted **(Figure 10-51)**.

2. Check door closure. Gently push the door closed. If the door shuts smoothly with a

Figure 10-51 Checking for looseness in door hinges.

distinct "click," then the door is catching the locking or latch mechanism as it is designed to do. Push the door in further, and you should hear a second "click." This indicates that the door is adjusted correctly **(Figure 10-52)**.

3. Lubricate hinges. Fully open each door. Using a manufacturer-approved lubricant, apply it to the tops of each hinge. Generally, lower hinges have a door check roller, cam and spring incorporated into the assembly. You should also lubricate the roller part of the assembly. Now swing the door gently back and forth to allow the lubricant to penetrate into the hinges. Wipe

Figure 10-52 Check the doors for proper operation.

any excess lubricant from the external parts of the hinges.

4. Lubricate lock and latch mechanism. Using a manufacturer's approved lubricant, apply it to each locking or latch mechanism. Generally the locking and latch mechanism is in the door itself, so the lock needs to be worked to allow the lubricant to penetrate into the mechanism. Operate the latch by opening and closing the door, and from both the outside and the inside of the vehicle. Wipe any excess lubricant from the external parts of the locking mechanism.

Checking Interior Trim

The following are some points to note when checking, removing, and replacing interior trim components:

- Different vehicles have different methods of trim fixture, although they all follow similar methods of removal. The vehicle manufacturer's manual will provide tips on the methods of removal.

- Many cars have electric window winder mechanisms with activation switches on the trim panels. Always make sure that you use the recommended procedure, so that you do not trap wires between the door inner panel and the trim panel. This could cause an electrical short and damage the vehicle.

The following is a general procedure to follow when checking, removing, and replacing interior trim components:

1. Check the attachment method. There are a number of different ways trim panels can be mounted in place, so refer to the shop manual for details of the panel attachment methods used in the vehicle **(Figure 10-53)**.

2. Remove fittings. Remove any external fixtures on the panel, such as armrests or window winders, and place them in a safe clean place. These components may be held in place by screws or clips. Try not to damage them during removal.

3. Obtain replacement clips. Many panels are held in place with clips in what look like keyholes or are tree shaped. These clips are normally made out of plastic and they can easily be broken

Figure 10-53 Carefully remove the trim panel following the manufacturers procedure.

during removal, so always make sure you have some replacement clips for the re-assembly process. It is better to sacrifice a few replaceable clips than to risk damaging the panel itself.

4. Remove panel. Most manufacturers have a special tool for removing the panel clips. Gently slide the tool under the panel and work your way along it until you locate a clip. Remove the tool and then replace it over the clip. Press down and the clip should pop out of its location. Repeat the process around the panel until it is completely unfastened. Lower the window, place your hands on the side of the panel, and wiggle it upwards. This should lift the panel from its seating, allowing you to remove it.

5. Replace panel. To replace the panel, make sure that all the panel clips are in place. Now lift the panel back in, and hook it over the inner door panel. Line up the clips with their location points and firmly push them into place. You will here a distinctive "click" as they lock in.

6. Replace fittings. Replace the armrests and any other components, such as window winder mechanisms. Clean off any finger marks with an approved solvent and a cloth that will not damage the panel material.

Checking a Seat Belt

Checking the seat belts is an important safety consideration. While these devices seem relatively

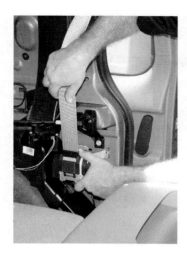

Figure 10-54 Inspect the seat belt for any defects.

simple, their proper functioning is important for the safety and comfort of the vehicle driver and passengers. Many vehicles have additional motorized mechanisms to help the seat belts operate. Always ensure that manufacturer's instructions are followed when lubricating and servicing these types of units. Careful inspection should reveal any readily observable flaws and defects. Any seat belt that has been involved in a severe impact may have been weakened and should be replaced **(Figure 10-54)**. The following is a general procedure to follow when checking seat belts:

1. Check anchor points of seat belt. There are three anchor points on "lap and sash" type seat belts and two on "lap only" type seat belts. First, check the side anchor bolt that holds the buckle end in place. Make sure it is sound and secure. Next, check the security of the bolt on the retractor housing. This is usually located near the bottom of the door panel next to the seat. Above the retractor is the upper anchor bolt. These bolts are sometimes adjustable in a slider unit. Check that this is also secure.

2. Inspect seat belt for fraying or wearing. Examine the seat belt material to see if there is any fraying or excessive wear. If the belt is damaged, then report this to your supervisor, as the belt should be replaced.

3. Check seat belt retraction. Pull the seat belt out slowly to check for smooth action. There

should not be any sticking, which might indicate a faulty retractor. Do this three times to make sure that it is consistently smooth in operation.

4. Check seat belt locking. Pull the seat belt out quickly to see that it locks and releases. The belt should lock when pulled suddenly, which is vital during emergency braking or collisions.

5. Check retractor operation. If the belt does not retract smoothly, remove the retractor housing and test it again. If the belt now retracts smoothly, then the retractor housing may need to be replaced.

Checking Vehicle Seats

An inspection of the vehicle seats should reveal their security and condition. A damaged or broken seat could cause a safety risk for the driver and passengers. If the existing seat cover is damaged, it is sometimes possible to make small repairs with some fabric or vinyl glue, depending on the original material. This may stop damaged areas or small tears from getting any worse. If you do this, follow the instructions on how to apply the glue **(Figure 10-55)**. If the seat fabric is badly ripped, or you can see the springs or seat frame, you should take the seat to a specialized upholstery repair shop. In many places, it is a legal requirement that the seat be in a "safe and secure condition," even if a seat cover has been placed over the damaged seat. Poor seat condition can be the subject of

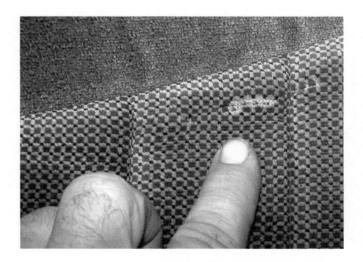

Figure 10-55 Inspect the seats for any nicks, tears, or other damage.

a safety defect, making the vehicle unsafe for the road. Always check your local laws.

The following is a general procedure to follow when checking vehicle seats:

1. Inspect visually. If a removable seat cover has been fitted, remove it so that you can see the original fabric. Inspect the seat for overall condition. The important thing to note is if the springs or frame inside the seat are visible through the fabric. If there are nicks or small tears, then note them down so that they can be reported to the customer.

2. Check the seat security. Now, look at the seat runners and their fixtures. Check to see that they are mounted according to the manufacturer's recommendations, and that the mounting points are free from corrosion or rust.

3. Check the seat operation. Check that the seat adjustment features operate correctly, as shown in the driver's operation manual. If any of them fails to function properly, make a note on your work order and report the fault to your supervisor.

VEHICLE DETAILING

At times it may be necessary to detail a customer vehicle. Detailing a customer's vehicle may be done at the request of the customer or may be provided by the repair facility either as an included service or as a courtesy. Some businesses specialize in detailing vehicles and this makes up the bulk of their business. It is important to understand the correct procedure to wash and wax a vehicle. Potentially hazardous cleaners may be used, so it is vital that the correct safety procedures are carried out.

Personal and Environmental Hazards

Cleaning products, especially solvents, can be toxic or flammable, particularly when used in high concentrations. It is important to comply with your federal, state, and local regulations concerning the safe handling and disposal of substances that can be hazardous to health **(Figure 10-56)**. Wherever possible, use the least amount of materials and in

Figure 10-56 Follow federal, state, and local laws regarding permissible disposal of substances.

the smallest quantities practical for any given task. Pour only small amounts of fluid onto a pad or cloth and keep the lids of any product containers closed when you are not actually pouring from them **(Figure 10-57)**.

Always mix any detergents according to manufacturer specifications. Cleaning chemicals and detergents is a potential environmental hazard. Make sure that the area used for washing and detailing a vehicle meets all local environmental regulations, particularly in regard to preventing contaminants from flowing into storm water drainage systems

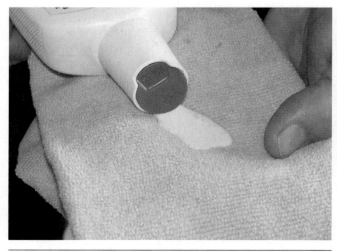

Figure 10-57 Use the minimal amount of chemicals necessary.

Figure 10-58 A trap has been installed beneath this drain to meet legal requirements.

(Figure 10-58). In the example in Figure 10-58, at a vehicle dealership, the waste run-off is collected in underground grease traps (below the metal cover plates) and treated before disposal.

Keep the work area ventilated. Using toxic substances in enclosed areas is especially hazardous. Keep your inside work area well ventilated, and when working inside vehicles leave all the doors and the sunroof, if equipped, wide open. When ventilation is poor, it may be necessary to use flexible ducting to remove fumes and vapors from inside vehicles you are working on.

Some solvents used in cleaning a vehicle are highly volatile and can explode. Remove all sources of ignition from the work area. It is good practice to disconnect the battery **(Figure 10-59)**. Wear

Figure 10-59 Unhook the battery when detailing a vehicle.

appropriate protective clothing including eyewear, and impervious gloves to protect hands and forearms. If you splash solvent onto items of clothing, remove them and hang them in a safe open place to dry out. If you are not sure how hazardous a particular chemical product is, refer to the relevant MSDS.

Washing a Vehicle

1. Rinse with plain water **(Figure 10-60)**. Pressure clean the outside of the vehicle thoroughly with plain water to remove all loose dust and street dirt **(Figure 10-61)**. Wash under the wheel arches and inside the engine compartment.

Figure 10-60 Start with a plain water washing.

Figure 10-61 Wash under the wheel arches.

2. Degrease the paintwork. Some degreasing products are suitable only for new or near new paintwork, and can increase the risk of damage to older paint surfaces. If the paintwork is faded or damaged, consult your chemical supplier for the most suitable degreasing product. Mix the degreaser concentrate as recommended by the manufacturer **(Figure 10-62)**. Spray the diluted degreaser all over the outside of the vehicle, including the road wheels and under the wheel arches. Let it work on the paintwork for a minute or so, to soften hardened or baked-on dirt and grease **(Figure 10-63)**. Rinse the degreaser off with water, using a pressure cleaner.

3. Inspect the surface of the vehicle. If there are any signs of contaminants still on the paintwork, re-apply the degreaser, then wash it off again **(Figure 10-64)**. You may need to carry out these steps a number of times until the contaminants soften and are rinsed away. Ensure that all the degreaser solution has been thoroughly rinsed off. Do not scratch the contaminant off with a scraper as this could easily damage the painted surface.

4. Hand wash with soap solution designed for automotive paint. Ensure that the concentrate has been mixed according to the manufacturer's specifications **(Figure 10-65)**. Start at the top of the vehicle and work down. Do not rub the surface hard, and use your other hand to

Figure 10-62 Use a degreaser under the wheel arches if necessary.

Figure 10-64 Use a degreaser if necessary.

Figure 10-63 Wash the vehicle with the proper detergent.

Figure 10-65 Hand wash using the proper detergent.

Figure 10-66 Observe the surface while hand washing.

Figure 10-67 Rinse the vehicle with clean water.

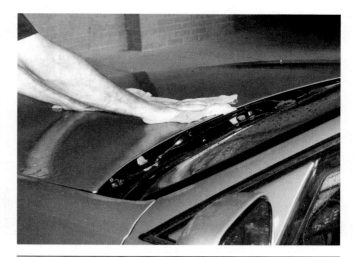

Figure 10-68 Dry the vehicle with a chamois.

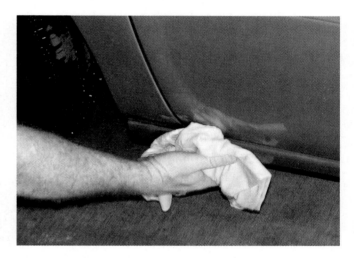

Figure 10-69 Dry any water that may have dripped out from behind body moldings.

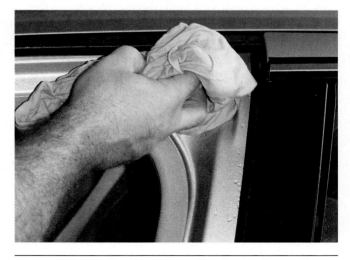

Figure 10-70 Dry the door jambs and weather stripping.

feel for the smoothness of the paint surface **(Figure 10-66)**. Allow the detergent to work on the paintwork for a minute or so before rinsing **(Figure 10-67)**. Rinse off the detergent suds with fresh water. Normal water pressure from a standard hose is adequate for this task.

5. Dry the vehicle **(Figure 10-68)**. Use two chamois cloths for drying: one for the paintwork, and the other for the road wheels and wheel arches.

6. After drying off the water, check around the vehicle and clean off any water that may have come away from vehicle gutters and crevices **(Figure 10-69)**. Dry inside all the door jamb/jambs, in the trunk, and any other location where water could seep **(Figure 10-70)**. If you

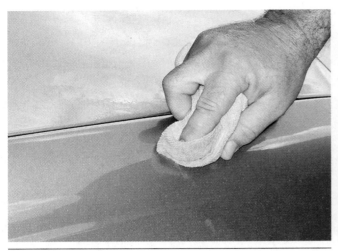

Figure 10-71 Use a cleaner and gently remove any stubborn deposits.

Figure 10-72 Hand polish the vehicle with minimal chemical usage.

find any spots of dirt still attached to the paint surface, gently remove them with a clean rag and a little kerosene **(Figure 10-71)**.

7. Wash the chamois cloths using detergent and the pressure cleaner every day, to ensure that they stay free of grease and dirt. Do not allow the paintwork chamois ever to touch the ground, as it could pick up dirt particles and other contaminants that will scratch the paintwork.

Tech Tip *Always use a detergent or soap designed for washing vehicles. Many detergents and soaps are very abrasive. Dishwashing soap is very abrasive and may damage vehicle paint.*

Waxing a Vehicle

1. Hand wax the vehicle **(Figure 10-72)**. Use a piece of new, thin soft foam rubber to apply the wax. Even clean cloth rags can have hard cotton stitching that may scratch the paint surface. Apply the wax using a very light circular rubbing action. Do not rub hard because this can score the paint surface **(Figure 10-73)**. Avoid applying wax to any rubber or plastic surfaces, such as the window seals or windshield washer nozzles. This can leave behind a

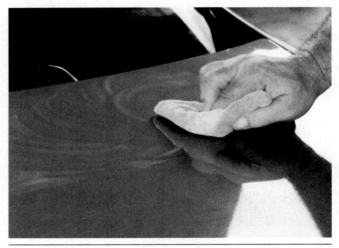

Figure 10-73 Use a light circular motion when waxing.

powdery residue; to remove it, you may need to wash and dry the vehicle again.

2. Remove the wax. Soft cotton flannel is a good material for wax removal, but check the material carefully before you use it, to be sure that there is nothing in or on the cloth that could damage the paint surfaces. It is a good idea to have a similar cloth in each hand. In this way, while you wax with one hand, you can lean on the vehicle to support yourself with the other hand without leaving greasy handprints. You can swap waxing hands quickly if one gets tired **(Figure 10-74)**. Check along the vehicle to ensure that you have removed all the wax

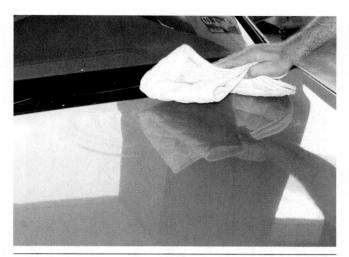

Figure 10-74 Use clean cloths when removing the wax.

Figure 10-76 Clean the glass surfaces of any unwanted stickers or decals.

Figure 10-75 Inspect the finish for imperfections.

Figure 10-77 Use disposable towels for glass cleaning.

and that no imperfections show up in the paint or the panels **(Figure 10-75)**.

3. Clean glass surfaces **(Figure 10-76)**. If there are any stickers on the glass area that should not remain there, remove them carefully with a razor blade, dampening them first if necessary. Use soft paper to clean the glass surfaces **(Figure 10-77)**. Using glass cleaner spray, carefully clean all the glass surfaces on the vehicle. Remember, you may need to lower the side windows slightly to effectively clean the whole glass surface **(Figure 10-78)**.

4. Finish the job. Clean and brush the tires with a suitable solution that will protect the

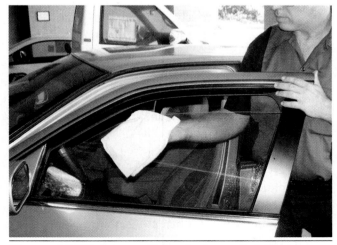

Figure 10-78 Lower the window slightly to clean the entire glass surface.

rubber and enhance its appearance. Wipe off any excess solution from the paintwork or wheels **(Figure 10-79)**.

5. Vacuum the interior of the vehicle. The vehicle is now ready for delivery to the customer.

Figure 10-79 Finish the detailing by cleaning the wheels.

Summary

The inspections discussed here are not meant to inspect every component of the vehicle but rather to inspect as many systems as possible in a given amount of time. They are also meant to cover those items that many customers did themselves in the past but may not do today. Vehicles have gotten much better in recent years and this has led to a "hands off" approach by many customers as far as basic inspection and maintenance is concerned. General inspections are a way for the shop to alert the customer to a potential problem before it becomes serious or costly. It is also a way for the shop to show an interest in the customer's vehicle as well as the potential to generate additional work for the technician.

Some inspections may be more comprehensive and intensive than those covered in this chapter. Many dealerships will have comprehensive vehicle inspections for used vehicles, particularly those vehicles being sold as a "certified" vehicle. The certification process may include as many as 150 different inspection points and will cover mechanical and appearance items. These inspection items are often set by the manufacturer to cover their particular vehicles.

Review Questions

1. Technician A says automotive engines rely on a high coefficient of friction for effective performance. Technician B says if the oil appears very black and dirty, it may have lost some of its protective and lubricating qualities. Who is correct?

A. Technician A

B. Technician B

C. Both Technician A and Technician B

D. Neither Technician A nor Technician B

2. Technician A says some of the functions of oil are to lubricate, cool engine parts, and clean. Technician B says some of the functions of oil are to seal the piston rings, and cushion the force of the power stroke on the engine bearings. Who is correct?

   A. Technician A

   B. Technician B

   C. Both Technician A and Technician B

   D. Neither Technician A nor Technician B

3. Technician A says lower viscosity fluids flow more easily than higher viscosity liquids. Technician B says higher viscosity fluids flow more easily than lower viscosity liquids. Who is correct?

   A. Technician A

   B. Technician B

   C. Both Technician A and Technician B

   D. Neither Technician A nor Technician B

4. Overfilling an engine with lubricating oil can:

   A. create a higher than normal oil pressure.

   B. cause the oil to churn or aerate, thus increasing oil pressure.

   C. cause the oil seals to leak due to the higher than normal oil pressure.

   D. cause the crankshaft to whip it into a foam and flood the seals.

5. Several factors can influence vehicle braking, including:

   A. road surface, road conditions, the un-sprung mass of the vehicle, load on the vehicle during stopping, and the brand of tires.

   B. road surface, road conditions, the total weight of the vehicle, load on the vehicle during stopping, and the condition of the tires.

   C. road surface, road conditions, the weight of the vehicle, load on the vehicle during stopping, and the brand of tires.

6. Technician A says to ensure that only clean brake fluid is used. Technician B says to ensure that only clean brake fluid of the correct grade or DOT type is used. Who is correct?

   A. Technician A

   B. Technician B

   C. Both Technician A and Technician B

   D. Neither Technician A nor Technician B

7. Technician A says the two marks on the reservoir of the master cylinder represent the minimum and maximum fluid levels. Technician B says that the lower mark indicates the level for cold fluid and the upper mark indicates the level for hot fluid. Who is correct?

   A. Technician A

   B. Technician B

   C. Both Technician A and Technician B

   D. Neither Technician A nor Technician B

8. The following are roles of the cooling system *except*:

   A. remove heat from the engine.

   B. keep the engine warm.

   C. lubricate internal components.

   D. heat the passenger compartment.

9. Technician A says that coolant is a chemical used in place of water. Technician B says that coolant is a fluid containing chemicals mixed with water. Who is correct?

   A. Technician A

   B. Technician B

   C. Both Technician A and Technician B

   D. Neither Technician A nor Technician B

10. The function of the radiator is to accept:

   A. hot coolant and raise its temperature.

   B. hot coolant and lower its temperature.

   C. cold coolant and raise its temperature.

   D. cold coolant and lower its temperature.

11. Technician A says that one of the functions of a cooling system pressure cap is to increase the boiling point of the coolant in proportion to the spring tension. Technician B says that one of the functions of the cooling system pressure cap is to protect the chemical additives of some coolants from breaking down. Who is correct?

   A. Technician A

   B. Technician B

   C. Both Technician A and Technician B

   D. Neither Technician A nor Technician B

12. What is the main function of a separate recovery tank?

   A. to maintain proper coolant level in the system at all times.

   B. to stabilize the coolant temperature.

13. The size of a tire must satisfy some basic conditions. These include all of the following *except*:

   A. having a suitable load carrying capacity for the vehicle.

   B. being compatible with the manufacturer's brand recommendation.

   C. having a load carrying capacity.

   D. being compatible with the manufacturer's desired size.

14. Technician A says the tire provides a cushion between the vehicle and the road to reduce the transmission of road shocks and friction to allow the vehicle to perform its normal operations. Technician B says the tire provides a cushion between the vehicle and the road to reduce the transmission of road shocks and reduction in friction to allow the driver to corner faster and harder if required. Who is correct?

   A. Technician A

   B. Technician B

   C. Both Technician A and Technician B

   D. Neither Technician A nor Technician B

15. When considering the vehicle's tire and its function, the weight of the vehicle is supported by which component of the tire?

   A. the sidewalls.

   B. the tread area.

   C. the tire bead.

   D. the tire air pressure.

16. There are three basic types of tire construction used on most passenger vehicles. They are:

   A. low profile, high profile, and standard profile.

   B. off-road, on-road, and hybrid.

   C. Firestone, Perelli, and Goodyear.

   D. bias, belted bias, and radial.

17. What kind of tire is the most commonly used?

   A. bias.

   B. belted bias.

   C. radial.

18. When checking tire pressures, you must ensure that the gauge is accurate. If the gauge is reading high, it could lead to the tire being:

   A. under-inflated.

   B. over-inflated.

   C. correctly inflated by adjusting the amount of air pressure put into the tire.

19. The minimum tread for a tire is:

   A. approximately one-sixteenth of an inch or 1.5 mm and when the wear bars are visible.

   B. approximately one-sixteenth of an inch or 1.5 mm and/or when the wear bars are flush with the remaining tread.

   C. approximately one-eighth of an inch or 3 mm and when the wear bars are visible.

20. When is the security system usually activated?

   A. lock.

   B. accessories.

   C. on.

   D. start.

21. At which position is the battery most likely to be drained through prolonged use?

    A. lock.

    B. accessories.

    C. on.

    D. start.

22. The park and tail lamp bulbs are connected in:

    A. series.

    B. series–parallel.

    C. parallel.

23. The park and tail lamp bulbs are located:

    A. as close as practical to the center of the vehicle.

    B. as far apart as possible.

    C. as close as practical to the maximum width of the vehicle.

24. The reversing lamp bulbs are connected in:

    A. series.

    B. series–parallel.

    C. parallel.

25. The reversing lamp bulbs are activated by:

    A. the driver.

    B. a transmission-mounted switch.

    C. the vehicle detecting a reverse direction of motion.

    D. the automatic transmission.

26. Indicators or turn lights are used to let other road users know that:

    A. The driver is about to make a right-hand turn.

    B. The driver is about to make a left-hand turn.

    C. The driver is intending to make either a right- or left-hand turn depending on which side the lights are activated on.

    D. The driver is about to change the direction of travel depending on which side the lights are activated on.

27. Why should you disconnect the battery prior to cleaning a vehicle with solvents?

    A. Because some solvents are highly volatile and can explode.

    B. Because solvents in contact with the battery can cause it to short out the electrical wiring of the vehicle.

    C. Because solvents can corrode the battery terminals.

Basic Electricity and Electrical Circuits

Learning Objectives

After you have read, studied, and practiced the contents of this unit, you should be able to:

- Describe the parts of an atom.
- Explain the importance of the number of electrons on the valance ring of an atom.
- Describe how electrons flow through a conductor.
- Define conductor, semiconductor, and insulator.

- Explain voltage, amperes, and ohms in an electric circuit.
- Explain how a voltage drop test works.
- Describe the main features of a series circuit.
- Explain the factors that determine the strength of an electromagnet.

Key Terms

Alternating current (AC)	Element	Ohms
Amperes	Electrons	Periodic Table
Atom	Electromagnetic induction	Protons
Compound	Insulator	Semiconductor
Conductor	Molecule	Valence ring
Direct current (DC)	Neutrons	Voltage

INTRODUCTION

An understanding of basic electricity is absolutely essential before a study of more complex electrical/electronic systems. It is very difficult to understand electrical/electronic systems if you do not have a clear understanding of basic electricity. If you are familiar with the principles of basic electricity and how they apply to the operation of electrical/electronic systems, then the understanding of these systems becomes much easier.

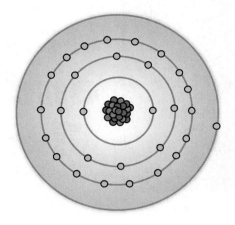

Figure 11-2 Structure of a copper atom.

ATOMIC STRUCTURE

An **atom** may be defined as the smallest particle of an element in which all the chemical characteristics of the element are present. Atoms are very small particles that cannot be seen even with a powerful electron microscope that magnifies millions of times. Atoms contain even smaller particles called **protons**, **neutrons**, and **electrons**. Protons and neutrons are located at the center or nucleus of each atom. Protons contain a positive electrical charge and neutrons add weight to the atom. Heavier elements contain more neutrons.

> *Did You Know?* **All electricity is really the measurement of the flow of electrons from one atom to the next.**

Electrons circle around the nucleus or center of the atom in various orbits **(Figure 11-1)**. For example, a hydrogen atom contains one proton in the nucleus and one electron orbiting around the

proton. The hydrogen atom does not have any neutrons. A copper atom has four orbits or rings with two, eight, eighteen, and one electron on these orbits **(Figure 11-2)**. Atoms may have from one to seven orbits with different numbers of electrons on each orbit. Electrons have a negative electrical charge. Elements are listed in the **Periodic Table** according to their number of protons and electrons. For example, hydrogen is number one on this scale and copper is number twenty-nine. The nucleus does not always have the same number of protons and neutrons. A copper atom has twenty-nine protons and thirty-five neutrons.

The outer orbit on an atom is called the **valence ring**. The number of electrons on the valence ring determines the electrical characteristics of the element. If the valence ring has one, two, or three electrons, the element is classified as a good **conductor**, because the electron(s) on the valence ring move easily from one atom to another. When an atom has four valence electrons, it is classified as a **semiconductor**. Semiconductors have special electrical characteristics. The characteristic that makes semiconductors special is the ability to be a conductor at certain times and an insulator at others. This is generally determined by the amount of energy applied to it. This energy can be in the form of heat or voltage. These materials are used to manufacture diodes, transistors, and carbon fiber spark plug wires. If an element has five or more

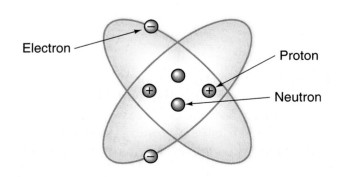

Figure 11-1 Parts of an atom.

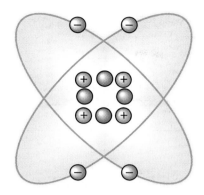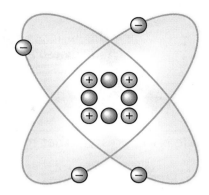

Figure 11-3 If an atom has the same number of electrons and protons, it is in balance.

valence electrons, these electrons will not move easily and the element is classified as an **insulator**. If an atom has the same number of protons and electrons, the atom is in balance **(Figure 11-3)**. Atoms always try to maintain their proper balance with the same number of electrons and protons. If an atom loses an electron, it will try to attract an electron from another atom. When an atom has more protons than electrons, it is unbalanced and the positively charged protons will immediately attract other electrons from another atom.

ELEMENTS, COMPOUNDS, AND MOLECULES

As stated previously, an **element** is a material with only one type of atom. For example, pure copper contains only copper atoms. A **compound** is a material containing two or more types of atoms. For example, water (H_2O) contains hydrogen and oxygen. A **molecule** is the smallest particle of a compound that retains all the characteristics of the compound.

ELECTRIC CURRENT FLOW

A massing of electrons must occur at one point in an electric circuit and a lack of electrons must be present at another point in the circuit before electrons will move through the circuit. The massing of electrons may be referred to as electrical pressure. The electrical circuit must also be complete between the massing of electrons and the lack of electrons. When these requirements are present in a circuit, the electrons begin moving from the massing of electrons into the atoms in the conductor **(Figure 11-4)**. If an atom has more electrons than protons, it immediately repels an electron to another atom. When an atom lacks an electron, it attracts an electron from another atom. Current flow may be defined as the mass movement of valence electrons from atom to atom in a conductor.

ELECTRIC CIRCUIT MEASUREMENTS
Voltage

Voltage is the electrical pressure caused by a massing of electrons at one point in an electrical

Conductor

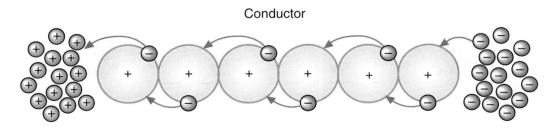

Figure 11-4 Electrons flow from atom to atom in a conductor.

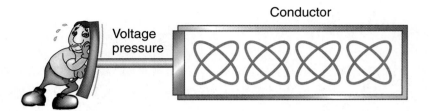

Figure 11-5 Voltage is the pressure that moves electrons.

circuit. Voltage causes the electrons to move through a circuit (Figure 11-5). Voltage is a measurement for electrical pressure difference. Voltage is the energy that is applied to an atom to force it to move its valence electrons; without this pressure these electrons would not move to any degree necessary to be useful.

If a voltmeter is connected across the terminals of an automotive battery, the voltmeter may indicate 12.6 volts if the battery is fully charged. This reading indicates there are 12.6 volts at one battery terminal in relation to 0 volts at the opposite battery terminal.

Amperes

Amperes is a measurement for the rate of electron flow or the amount of current flowing through a circuit (Figure 11-6). Amperes will continue to flow through a circuit as long as the massing of electrons and lack of electrons are maintained in the circuit. There are two theories regarding the direction of electron movement or current flow. The electron theory says that electrons are negatively charged and electrons move from a massing of electrons (negative charges) to a lack of electrons (positive charges). The

conventional theory says that for illustration purposes in the automotive industry we assume that current flows from positive to negative through a circuit. Excessive amounts of electrons (amps) flowing through a conductor will cause heat buildup, this in turn will raise the resistance in a conductor and drop the voltage in a conductor. This action will force the circuit to increase the electrons flowing to maintain the circuit, thereby increasing the heat. This cycle can happen very quickly with devastating results to wiring and electrical components if the right size and type of conductor is not used.

> **Tech Tip** If 1 ampere of current is flowing through a circuit, 6.25 billion billion electrons are passing a given point in the circuit in 1 second.

Direct current (DC) flows in only one direction. Most automotive circuits operate on direct current. **Alternating current (AC)** flows alternately in one direction and then in the opposite direction. The windings in the alternator stator have AC flowing in them, but this AC is rectified

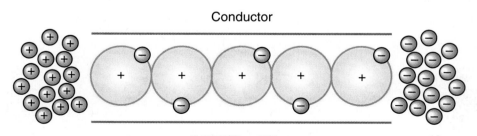

6.25 Billion billion
electrons per second = one ampere

Figure 11-6 Amperes is a measurement of electron movement through a circuit.

to DC by the diodes in the alternator. Therefore, DC is delivered from the alternator to the battery and electrical components on the vehicle.

Ohms

Electrical resistance may be considered the opposition to electron movement in a circuit and this resistance is measured in **ohms.** The size, length, type, and temperature of a conductor determines its resistance. For example, a smaller diameter wire has a higher resistance to electron movement compared to a large diameter wire. Material is another significant contributor to resistance both positively and adversely. Copper is an excellent conductive material with low resistance, however oxygen is a poor conductor with high resistance.

> **Did You Know?** *Ohm's Law states that it takes 1 volt, to move 1 amp, with 1 ohm of resistance.*

In order for oxygen to conduct electricity the voltage required to force it to conduct must be very high, in the thousands of volts, to conduct across a spark plug gap to the millions for a bolt of lightning to strike the earth. When metals combine with oxygen we get oxides like iron oxide (rust) to aluminum oxide (corrosion); these combined materials are also poor conductors and will add resistance to any circuit they are in, reducing efficiency or stopping the flow of electrons if enough is present.

Watts

Watts is an electrical measurement that is calculated by multiplying the amperes by the volts. Although watts are not measured when servicing vehicles, it plays an important role in electrical system design. For example, if the total amperage load of all the electrical accessories on a vehicle is 250 amperes in a 12-volt electrical system, the total watts of energy required are $12 \times 250 = 3,000$ watts. Therefore, the charging circuit on the vehicle must be capable of supplying over 3,000 watts.

OHM'S LAW

Ohm's Law states that the current flow in a circuit is directly proportional to the voltage and inversely

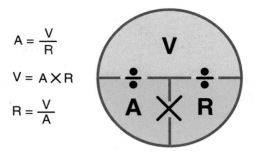

$$A = \frac{V}{R}$$

$$V = A \times R$$

$$R = \frac{V}{A}$$

Figure 11-7 Ohm's Law formula.

proportional to the resistance. This indicates that an increase in voltage causes a corresponding increase in current flow, whereas a decrease in voltage reduces current flow. Conversely, an increase in resistance decreases current flow and a decrease in resistance increases current flow. Ohm's Law may be used to calculate a value in a circuit if the other two values are known. In this formula, voltage is indicated by V, amperes is represented by A, and R indicates resistance **(Figure 11-7)**. If 12 volts are supplied to a circuit and the resistance in the circuit is 2 ohms, the current flow is $12 \div 2 = 6$ amperes. When a circuit has 4 amperes and 3 ohms resistance, the voltage is $4 \times 3 = 12$ volts. If a circuit has 12 volts and a current flow of 1.5 amperes, the resistance is $12 \div 1.5 = 8$ ohms.

VOLTAGE DROP

Voltage drop is the difference in voltage across a resistance when current flows through the resistance. There is always some voltage drop when current flows through a resistance. For example, less than 0.05-volt drop across most switch contacts is normal. However, higher-than-normal resistance causes high voltage drop in a circuit. High resistance and high voltage drop result in reduced current flow in a circuit. When a set of switch contacts becomes pitted and corroded, they may have a 2-volt drop, which reduces current flow in the circuit. This voltage loss is caused by the high resistance in the corrosion. The high resistance allows the contacts to act like another load in the circuit, the same as if you added additional components to a circuit to the point it ceases to function correctly.

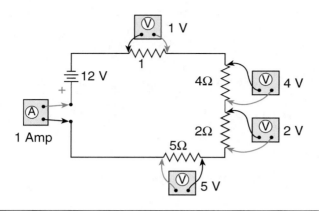

Figure 11-8 A series circuit.

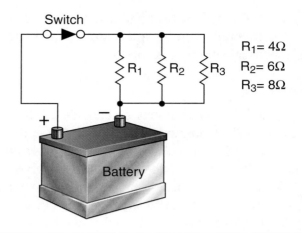

Figure 11-9 A parallel circuit.

SERIES CIRCUIT

In a series circuit, the same current flows through all the resistances in the circuit **(Figure 11-8)**. A series circuit has these features:

- The same current flows through all the loads in the circuit because there is only one path for current flow.

- The total resistance is the sum of all resistances in the circuit.

- The voltage drop across each load depends on the ohm value of that resistance.

- The sum of the voltage drops across each load equals the source voltage.

The heater blower circuit is an example of a series circuit. The resistors in the blower circuit that control blower speed are switched in series with the blower motor. By adding resistance the motor will slow down because it receives less voltage as the resistance is increased.

PARALLEL CIRCUIT

In a parallel circuit, each resistance is a separate path for current flow **(Figure 11-9)**. These facts may be summarized regarding parallel circuits:

- Each load has a separate path for current flow.

- The amount of current flow through each load depends on the amount of resistance in that part of the circuit.

- Equal full source voltage is supplied to each load and equal full source voltage is dropped across each load.

- The total resistance is always less than the ohm value of the lowest resistor in the circuit.

The light circuit on a vehicle is an example of a parallel circuit. Each of the lights is connected in parallel to the battery so each gets the full battery voltage.

SERIES-PARALLEL CIRCUIT

In a series-parallel circuit, a component is connected in series with the parallel circuit. Many automotive electrical circuits are series-parallel circuits. For example, the instrument panel lights are connected parallel to the battery, but the variable resistor that dims the instrument panel lights is connected in series with these lights.

ELECTROMAGNETS

A permanent magnet has an invisible magnetic field surrounding the magnet. These lines of force move from the north pole of the magnet to the south pole. One of the basic magnetic principles is that like magnetic poles repel each other and unlike magnetic poles have an attracting force **(Figure 11-10)**. This principle is used in a starting

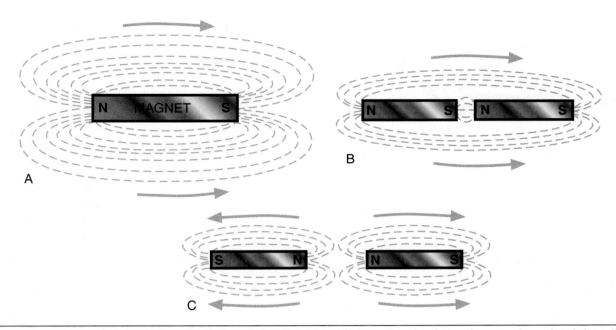

Figure 11-10 Basic Magnetic principles: (A) All magnets have a north and a south pole, (B) unlike poles attract, and (C) like poles repel each other.

motor. An electromagnet is manufactured by winding a coil of wire around a metal core. When current flows through a straight wire, a magnetic field surrounds the wire. This magnetic field is concentric to the wire and the same strength at any point along the wire. In the coil of wire on an electromagnet, the magnetic strength of each loop of wire adds to the strength of the other loops and the magnetic field surrounds the entire coil. An iron core placed in the center of the coil helps to concentrate the magnetic field because iron is a better conductor for magnetic lines of force compared to air **(Figure 11-11)**. The strength of an electromagnet is determined by the number of turns on the coil and the amount of current flow through the coil. Increasing either of these factors provides a stronger magnetic field around the electromagnet **(Figure 11-12)**. The magnetic strength is calculated by multiplying the amperes by the number of turns. For example, 1 ampere × 1,000 turns = 1,000 ampere turns. Electromagnets are used in relays and solenoids. If the current flow through the coil is reversed, the polarity of the coil is also reversed.

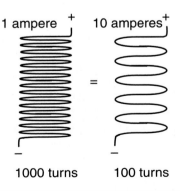

Figure 11-12 The strength of an electromagnet is determined by the number of coils, the distance between the coils, and the amount of amps flowing through the coils.

Figure 11-11 Putting a metal core in the center of an electromagnet concentrates the lines of force.

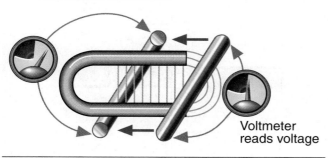
Conductor movement

Voltmeter reads voltage

Figure 11-13 A voltage is induced when either a conductor is moved through a magnetic field or a field is moved over a conductor.

ELECTROMAGNETIC INDUCTION

When a conductor is moved through a magnetic field, a voltage is induced in the conductor **(Figure 11-13)**. This process is called **electromagnetic induction**. If the conductor is connected to a complete circuit, current flows through the circuit.

> ***Did You Know?*** *During electromagnetic induction, it does not matter if the conductor is moved through a magnetic field or the magnetic field is moved across the conductor; voltage is still induced in the conductor.*

During electromagnetic induction, the amount of voltage induced in a conductor is determined by four factors:

- The number of conductors.
- The strength of the magnetic field.
- The angle at which the field crosses the conductor, a 90-degree angle is optimum.
- The speed of motion.

In an alternator, the revolving magnetic field of the rotor cuts across the stator windings and induces voltage in these windings. This voltage supplies current to the battery and electrical accessories on the vehicle. In an ignition coil, the windings are stationary and the magnetic field moves across these windings to induce voltage in the windings. This voltage is used to fire the spark plugs. Induction is generally wanted, however, in certain cases it can be harmful. For instance, if high voltage spark plug wires are allowed too close when one wire is fired current can be induced in another that is to close. This can cause a poor running engine at best and damage at its worst. To prevent this, manufacturers always space wires a certain distance in retainers or clips and will also turn some at right angles if one wire must cross another.

> ***Did You Know?*** *After magnetic lines or force pass through a piece of hard steel, the steel retains magnetism. After magnetic lines of force pass through a piece of soft iron, it becomes demagnetized immediately.*

Summary

- Protons and electrons are located at the center or nucleus of an atom. The protons have a positive electrical charge and the neutrons do not have any electrical charge, but they add weight to the atom.

- Electrons have a negative electrical charge and they move in orbits around the nucleus of an atom.

- An element is a liquid, solid, or a gas with only one type of atom.

- A compound contains two or more different types of atoms.

- The outer orbit on an atom is called the valence ring and the number of electrons on this ring determines the electrical characteristics of the element.

- A good conductor is a material with one, two, or three valence electrons.

- A semiconductor has four valence electrons.

- An insulator has five or more valence electrons.

- When current flows through a conductor, the valence electrons move from atom to atom in the conductor.

- Voltage is a measurement for electrical pressure difference.

- Amperes is a measurement for the amount of current flow in a circuit.

- Ohms are a measurement for the opposition to current flow in a circuit.

- In a series circuit, the same current flows through all the loads in the circuit.

- In a parallel circuit, each part of the circuit is a separate path for current flow.

- In a series-parallel circuit, a component is connected in series with the parallel components in the circuit.

- The number of coils and the current flow through the winding determines the strength of an electromagnet.

- During electromagnetic induction, the amount of voltage inducted in the conductor is determined by the strength of the magnetic field, the speed of motion, the angle at which the field cuts the conductor the angle at which the field crosses the conductor, and the number of conductors.

Review Questions

1. Technician A says protons have a positive electrical charge. Technician B says electrons are located in the nucleus of an atom. Who is correct?

   A. Technician A

   B. Technician B

   C. Both Technician A and Technician B

   D. Neither Technician A nor Technician B

2. Technician A says if the atoms in a material have four valence electrons, the material is a good conductor. Technician B says if the atoms in a material have five valence electrons, the material is a semiconductor. Who is correct?

   A. Technician A

   B. Technician B

   C. Both Technician A and Technician B

   D. Neither Technician A nor Technician B

3. Technician A says if an atom has the same number of protons and electrons, it is considered to be in balance. Technician B says if an atom loses an electron, it will attract an electron from another atom. Who is correct?

   A. Technician A

   B. Technician B

   C. Both Technician A and Technician B

   D. Neither Technician A nor Technician B

4. Technician A says an element has two or more types of atoms. Technician B says a molecule is the smallest particle of a compound. Who is correct?

   A. Technician A

   B. Technician B

   C. Both Technician A and Technician B

   D. Neither Technician A nor Technician B

5. All of these statements about atoms are true *except*:

   A. Protons are located in the nucleus of an atom.

   B. Neutrons have a positive electrical charge.

   C. Electrons move in orbits around the nucleus.

   D. Electrons have a negative electrical charge.

6. Technician A says the outer ring on an atom is called the valence ring. Technician B says the number of electrons on the valence ring of each atom determines the electrical characteristics of an element. Who is correct?

   A. Technician A

   B. Technician B

   C. Both Technician A and Technician B

   D. Neither Technician A nor Technician B

7. All of these statements about insulators, conductors, and semiconductors are true *except*:

   A. A good conductor has one valence electron on each atom.

   B. An element with four valence electrons is a semiconductor.

   C. Semiconductors are used to manufacture transistors.

   D. An element with three valence electrons is an insulator.

8. Technician A says voltage is a measurement for the amount of energy in an electric circuit. Technician B says amperes are a measurement for resistance in an electrical circuit. Who is correct?

   A. Technician A

   B. Technician B

   C. Both Technician A and Technician B

   D. Neither Technician A nor Technician B

9. All of these statements about electrical circuits and Ohm's Law are true *except*:

   A. A voltage increase causes an increase in current flow.

   B. An increase in resistance causes a decrease in current flow.

   C. If a circuit has 5 amperes of current and 2.5 ohms of resistance, the voltage is 12.5.

   D. If a circuit has 14 volts and 7 ohms of resistance, the current flow is 2.5 amperes.

10. In an electric circuit with three unequal resistances connected in series:

    A. The total resistance is the sum of all three resistances.

    B. The voltage drop across each resistance is the same.

    C. The current flow varies in each resistance.

    D. The sum of the voltage drops across each resistance is higher than the source voltage.

11. Voltage is a measurement for _____.

12. The rate of electron flow is measured in _____.

13. The opposition to current flow in a circuit is measured in _____.

14. When current flows in one direction only, it is called _____ current.

15. Explain the result of increasing the voltage in an electrical circuit.

16. Describe the result of decreasing the resistance in an electrical circuit.

17. List the factors that determine the strength of an electromagnet.

18. List the factors that determine the amount of voltage induced during electromagnetic induction.

12 Light Circuits

Learning Objectives

After you have read, studied, and practiced the contents of this unit, you should be able to:

- Describe the design of single- and dual-filament lightbulbs.
- Describe sealed beam design.
- Describe the operation of halogen and high intensity discharge headlights.
- Describe the headlight switch operation in the PARK position.
- Explain the headlight switch operation in the HEADLIGHT position.
- Describe the current flow through the headlights during high beam operation.
- Explain the operation of a circuit breaker.
- Describe the operation of the signal light switch and related circuit during a turn.
- Explain the operation of the interior light circuit.

Key Terms

Circuit breaker

Flash-to-pass feature

Fuse

Halogen

High intensity discharge (HID)

Incandescense

INTRODUCTION

Technicians must understand basic light circuits to be able to maintain, diagnose, and service these systems. When the operation of these circuits is understood, diagnosing becomes much easier and faster.

LAMPS

An automotive lightbulb usually contains one or two filaments. In a single filament bulb, the terminal is connected to one side of the filament and the opposite end of the filament is usually connected to the bulb case **(Figure 12-1)**. Voltage is supplied to the bulb terminal and current flows through the filament to the bulb case. The circuit is completed from the bulb case through the vehicle ground back to the battery. The indexing pins on the sides of the case retain the bulb in the socket. Many automotive bulbs have two filaments and two terminals that supply voltage to the filaments. These dual-filament bulbs serve two purposes such as stop and tail lights. The indexing pins position the bulb terminals properly in the socket. A

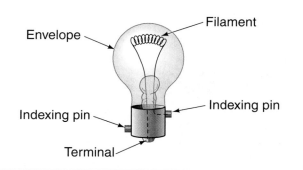

Figure 12-1 A single filament bulb.

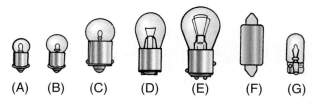

(A) (B) (C) (D) (E) (F) (G)

A,B Miniature bayonet for indicator and
instrument lights

C — Single contact bayonet for license
and courtesy lights

D — Double contact bayonet for trunk
and underhood lights

E — Double contact bayonet with
staggered indexing lugs
for stop, turn signals, and
brake lights

F — Cartridge type for dome lights

G — Wedge base for instrument lights

Figure 12-2 Various types of automotive bulbs.

variety of different bulbs are used in a typical vehicle
(Figure 12-2). When current flows through a bulb
filament, it becomes very hot. The electrical energy
in the filament is changed to heat energy and this
action is so intense that the filament glows and
gives off light. This process of changing electrical
energy to heat energy that produces light is called
incandescence. The filament is surrounded by a
vacuum that prevents overheating and destruction
of the filament. When a bulb is manufactured, a
vacuum is sealed inside the glass envelope surrounding the bulb.

SEALED BEAM HEADLIGHTS

Sealed beam headlights may be round or rectangular shaped. Sealed beam headlights have a parabolic

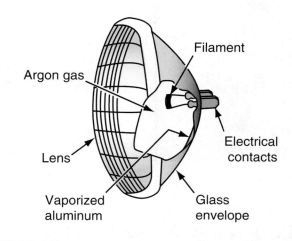

Figure 12-3 Sealed beam headlight design.

reflector sprayed with vaporized aluminum in the
rear of the sealed beam. This reflector is fused to
a glass lens in the manufacturing process **(Figure
12-3)**. All the oxygen is removed from the sealed
beam and then it is filled with argon gas. If oxygen
were allowed to remain in the sealed beam, the filament would become oxidized and burn out quickly.
Sealed beams may contain one or two filaments. If
the sealed beam operates on both high and low
beam, it has two filaments and three terminals.
Some sealed beams that operate only on high beam
contain a single filament and two terminals.

The light from the filament in a sealed beam is
reflected from the reflector through concave
prisms in the lens **(Figure 12-4)**. The prisms in
the lens direct the light beam downward in a flat,

Figure 12-4 The prisms in a sealed beam
headlight.

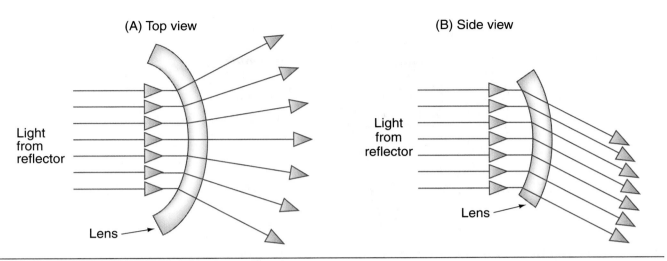

Figure 12-5 The prism directs the light beam into (A) a flat, horizontal pattern, and (B) downward.

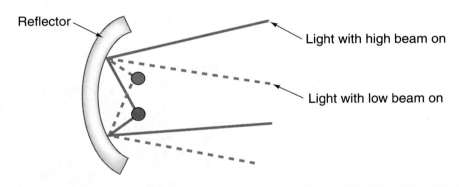

Figure 12-6 The location of the filament in a headlight controls the light beam projection.

horizontal pattern **(Figure 12-5)**. The filaments are precisely located in the reflector to properly direct the light. If a sealed beam has two filaments, the lower filament is for high beam and the upper filament is for low beam **(Figure 12-6)**.

> **Tech Tip** *Because the bulb in a halogen headlight is self-contained, a cracked lens does not prevent headlight operation. However, a cracked lens should be replaced because it results in poor light quality.*

HALOGEN HEADLIGHTS

Many newer vehicles have halogen headlights. This type of headlight contains a small bulb filled with iodine vapor. The bulb has a glass or plastic envelope surrounding a tungsten filament. The

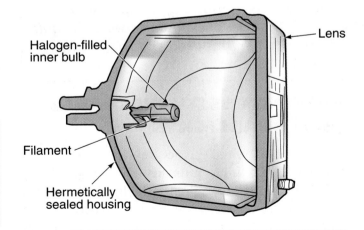

Figure 12-7 A halogen headlight.

bulb is installed in a sealed glass housing **(Figure 12-7)**. **Halogen** is a term for a group of chemically related nonmetallic elements including chlorine,

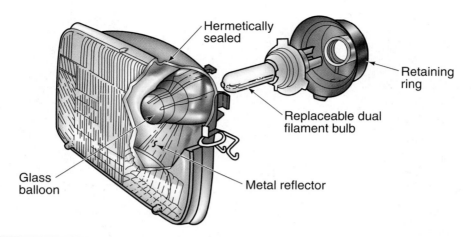

Figure 12-8 A composite headlight with a replaceable halogen bulb.

fluorine, and iodine. The tungsten filament can withstand higher temperatures and burn brighter because of the halogen added to the bulb. Halogen headlights produce approximately 25 percent more light compared to sealed beam headlights. Many vehicles are presently equipped with composite headlights and replaceable halogen bulbs **(Figure 12-8)**. The composite headlights allow the vehicle manufacturers to design the headlights in various shapes to conform to more aerodynamic body styling. For example, some composite headlights wrap around the front corner of the vehicle.

> **Tech Tip** *When replacing a halogen or HID bulb avoid touching the glass with your finger. The oil will eventually cause the bulb to crack when it is used and shorten its service life.*

HIGH INTENSITY DISCHARGE (HID) HEADLIGHTS

In recent years, some vehicles have been equipped with **high intensity discharge (HID)** headlights. In HID headlights, the light is produced from high voltage arcs across an air gap between two electrodes. An inert gas in the headlight amplifies the light provided by the high voltage arcing. Approximately 15,000 to 25,000 volts are required to initially force current across the air gap between the electrodes. Once this gap is bridged with a stream of electrons, about 80 volts are required to maintain the current flow across the gap. A voltage booster and controller are required to provide this

Figure 12-9 A high intensity discharge (HID) headlight assembly. Courtesy of Steffy Ford, Columbus NE.

higher voltage. Compared to halogen headlights, the HID lights provide approximately three times more light, draw less current, and last twice as long. The improved light output from HID lights allows these lights to be smaller and this makes it possible for the vehicle manufacturers to be more flexible in front-end body styling **(Figure 12-9)**.

HEADLIGHT AND DIMMER SWITCHES

The headlight switch is usually mounted on the front of the instrument panel **(Figure 12-10)**. Some headlight switches are pulled outward to turn on the lights, whereas other switches require a rotary or push-button action to turn on the lights. Most headlight switches have two positions. In the first

Figure 12-10 A push-pull, dash-mounted headlight switch.

(PARK) position, the switch turns on the park, tail, side-marker, and instrument panel lights **(Figure 12-11)** and in the second position the headlights are also turned on **(Figure 12-12)**. Many push-pull-type headlight switches contain a variable resistor that is connected in the instrument panel light circuit. The headlight switch knob is rotated to operate the variable resistor and vary the voltage supplied to the instrument panel lights. The action of the variable resistor allows the driver to dim or brighten the instrument panel lights as desired. If the vehicle has a rotary-type headlight switch, the variable resistor

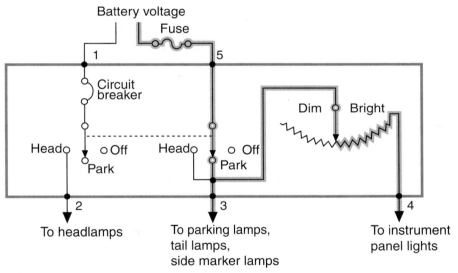

Figure 12-11 Current flow through a headlight switch in the PARK position.

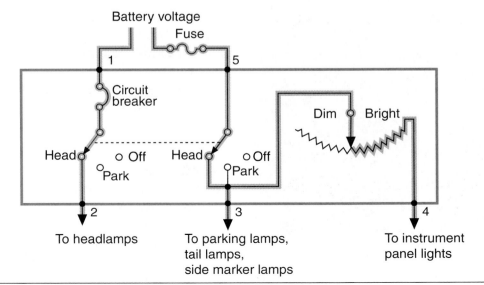

Figure 12-12 Current flow through a headlight switch in the HEADLIGHT position.

is mounted separately beside the headlight switch. A thumb wheel in the instrument panel operates this type of variable resistor.

In older vehicles, the dimmer switch is mounted on the floor pan and the driver's left foot was used to operate the dimmer switch. On most of these vehicles, the dimmer switch was mounted in a compartment so it was not exposed to road splash. However, in this location the dimmer switch was subject to moisture and corrosion. In recent years, the dimmer switch has been mounted in the steering column and operated by pulling the signal light lever upward toward the driver **(Figure 12-13)**. In late-model vehicles the turn signal lever can have most of the switch functions that were commonly mounted as individual switches in the dash in the past. Everything from windshield wiper functions to the cruise control is all contained within the one integrated switch.

NOTE: Many dimmer switches mounted in the steering column are combined with other switches, such as the signal light switch and wipe/wash switch. This combined switch may be called an *integrated switch* or a *multi-function switch*.

Figure 12-13 A multi-function switch.

This allows the manufacturer to save on multiple switches across the line. Often one switch will fit several makes and models, not only saving on inventory costs but with fewer different light and dash control systems technicians have fewer systems to be familiar with saving repair time.

HEADLIGHT AND PARK LIGHT CIRCUITS

Two terminals on the headlight switch are supplied with voltage directly from the battery. The circuit that supplies voltage to the headlight

> **Tech Tip** *When diagnosing a light circuit, you must have the proper light wiring diagram for the vehicle being serviced.*

circuit is not fused externally from the switch. However, a circuit breaker inside the headlight switch is connected in the headlight circuit. An external fuse is connected in the circuit to the headlight switch A **circuit breaker** is a protection device that opens an electric circuit if excessive current flows through the circuit. This action prevents damage to wiring and circuit components. After a circuit breaker cools, it will reset itself and close the electrical circuit. A **fuse** is a protection device containing a fusible strip that burns out and opens an electric circuit if excessive current flows through the circuit. This action protects circuit wiring and components. Burned-out fuses must be replaced because they have no reset action.

Fuses, relays, and flashers can be located in both an under-hood fuse and relay center and under

> **Tech Tip** *Some vehicles have more than one fuse and relay center. The location of these centers is shown in the service manual. Some vehicles have the flasher(s) mounted separately from the fuse and relay center.*

the dash panel **(Figure 12-14)**. Each component in this center is identified on the center cover. On older vehicles the fuse and relay center is commonly located under the instrument panel near the driver's side.

When the headlight switch is moved to the PARK position, voltage is supplied through the fuse and the headlight switch contacts to the park, tail, and side-marker lights. Voltage is also supplied through the variable resistor to the instrument panel

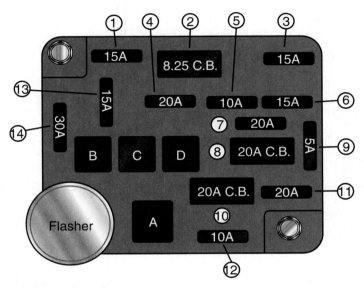

Fuses and Circuit Breakers
1. Stoplights, emergency warning system, speed control module, cornering light relays, and trailer tow relays
2. Windshield wiper/washer
3. Tail, park, license, coach, cluster illumination, side marker lights, and trailer tow relay
4. Trunk lid release, cornering lights, speed control, chime, heated backlite, and control A/C clutch
5. Electric heated mirror
6. Courtesy lights, clock feed, trunk light, miles-to-empty, ignition key warning chime, garage door opener, autolamp module, keyless entry, illuminated entry system, visor mirror light, and electric mirrors
7. Radio, power antenna, CB radio
8. Power seats, power door locks, keyless entry system, and door cigar lighters
9. Instrument panel lights and illuminated outside mirror
10. Power windows, sun roof relay, and power window relay
11. Horns and front and instrument panel cigar lighters
12. Warning lights, seat belt chimes, throttle solenoid positioner, autolamps system, and low fuel module
13. Turn signal lights, back up lights, trailer tow relays, keyless entry module, illuminated entry module, and cornering lights
14. ATC blower motor

Relays
A. Engine main relay
B. MFI main relay
C. Fan relay
D. Horn relay

Figure 12-14 A fuse and relay center.

lights. Each instrument panel bulb is grounded to the instrument panel. The variable resistor controls the brilliance of the instrument panel lights by varying the voltage supplied to these lights. When the headlight switch is moved to the second position, voltage is also supplied through the headlight switch contacts to the headlights. This voltage is supplied from the headlight switch to the dimmer switch. If the dimmer switch is on the low beam position, current flows through the low beam contacts in the dimmer switch to the low beams in both headlights to ground **(Figure 12-15)**. When the dimmer switch is in the high beam position, current flows through the high beam contacts in the dimmer switch to the high beam filaments in both headlights **(Figure 12-16)**. Current also flows from the high beam circuit to the high beam indicator bulb in the instrument panel. The illumination of this bulb reminds the driver that the highlights are on high beam.

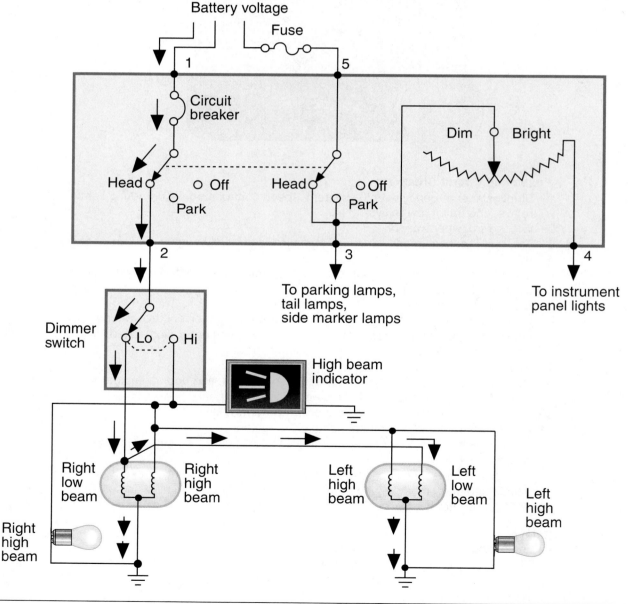

Figure 12-15 A low beam headlight circuit.

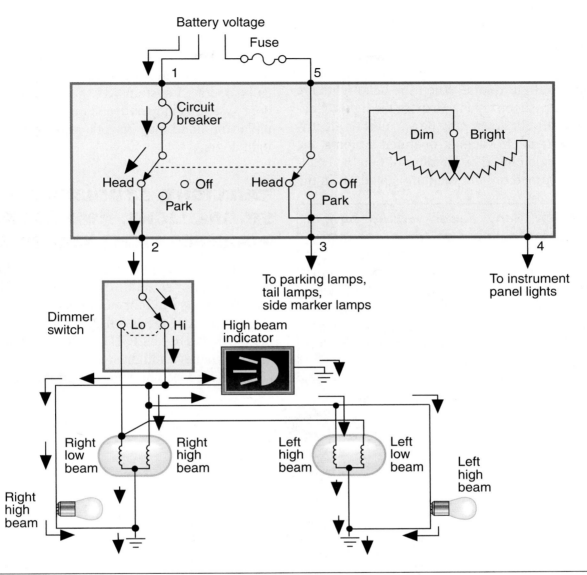

Figure 12-16 A high beam headlight circuit.

Many headlight systems have a **flash-to-pass feature**. During daylight hours when a driver wants to pass a vehicle in front, this feature may be used to turn the headlights on and signal the driver in front regarding the intended pass. To operate the flash-to-pass feature, the signal light lever is pulled upward, but not so far that it clicks. Regardless of the dimmer switch position, the headlights are on high beam when activated by the flash-to-pass feature, and the high beam indicator light in the instrument panel is illuminated. The headlights remain on as long as the signal light lever is held upward. Release the signal light lever to turn the headlights off.

Most vehicles have a light buzzer or beeper. If the lights are on, the ignition switch is off, and one of the front doors is opened, the buzzer sounds to remind the driver to shut off the lights. Some buzzers are activated if the lights are on and the ignition switch is turned off. The buzzer is often mounted on or near the fuse panel.

Some vehicles have automatic headlight dimmers. These vehicles have a light sensor mounted behind the grill. When the lights of an approaching vehicle supply a specific amount of light to the light sensor, the headlights dim automatically.

CONCEALED HEADLIGHT SYSTEMS

Some vehicles have the headlights concealed behind headlight doors. When the headlights are turned on, the doors open to expose the lights. Concealed headlight doors may be vacuum or electrically operated. On vacuum-operated systems, the doors are held closed by vacuum. When the headlights are turned on, the vacuum is bled off in the

> **Tech Tip** *Many newer vehicles have a light sensor mounted on top of the instrument panel. This sensor turns on the front and rear lights automatically when ambient light decreases to a specific level. This is very useful when driving through tunnels because the lights come on automatically without the driver trying to suddenly turn on the light switch.*

concealed door system. If there is a vacuum leak in these systems, one or both headlight doors may slowly open with the headlights shut off. An electric motor operates the headlight doors in electrically operated systems. If the electric headlight door system is inoperative, a manual override knob under the hood may be rotated to open the headlight doors.

TAILLIGHT, STOPLIGHT, SIGNAL LIGHT, AND HAZARD WARNING LIGHT CIRCUITS

Many vehicles have dual stop and taillight filaments in all or some of the rear lights. The same bulb filaments used for stoplights are also used for signal lights. If the headlight switch is moved to the PARK or HEADLIGHT position, voltage is supplied from the headlight switch to the taillight bulbs **(Figure 12-17)**. The signal light switch contains

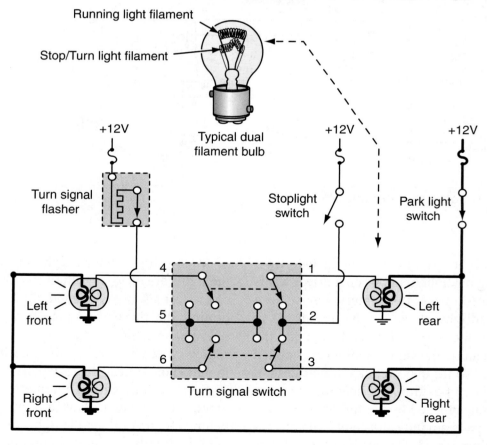

Figure 12-17 A signal light, stoplight, and turn signal circuit.

three triangular-shaped contacts. If the signal light switch is in the center position so the signal lights are not operating, the center contact in the signal light switch completes the circuit between the brake light switch input terminal and the rear stoplight terminals. If the driver depresses the brake pedal, current flows through these signal light switch contacts to the rear stoplights and the center high-mounted stoplight (CHMSL) bulbs.

When the signal light switch is moved to the right turn position, the contacts in the signal light switch are moved to the left. This contact movement shifts the right switch contact so it completes the circuit between the signal light flasher and the terminal to right rear signal lights. The right switch contact also completes the circuit between the flasher and the right dash turn indicator light and right front signal light **(Figure 12-18)**. When the circuit is completed through the flasher, it begins opening and closing the circuit to turn the right signal lights on and off. Many flashers contain a set of contacts mounted on a bimetallic strip that bends when heated. A heating coil is mounted on the bimetallic strip. Current flow through the heating

coil heats the bimetallic strip and opens the contacts. This strip immediately cools and closes the contacts. The heating and cooling of this bimetallic strip causes the flasher contacts to open and close at a specific rate. Some vehicles have a solid-state electronic flasher. This type of flasher provides light flashes at the same speed regardless of the current flow through the signal lights.

In the right turn position, the center switch contact maintains the circuit between the brake light switch and the left rear stoplight. Therefore, if the driver depresses the brake pedal, current flows through this circuit to the left rear stoplight and the CHMSL. The brake light switch is a mechanical switch operated by movement of the brake pedal. In the right turn position, the left switch contact is not completing any circuit.

When the signal light switch is moved to the left turn position, the contacts in the signal light switch are moved to the right. This contact movement shifts the left switch contact so it completes the circuit between the signal light flasher and the terminal to left rear signal lights. The right switch contact also completes the circuit between the

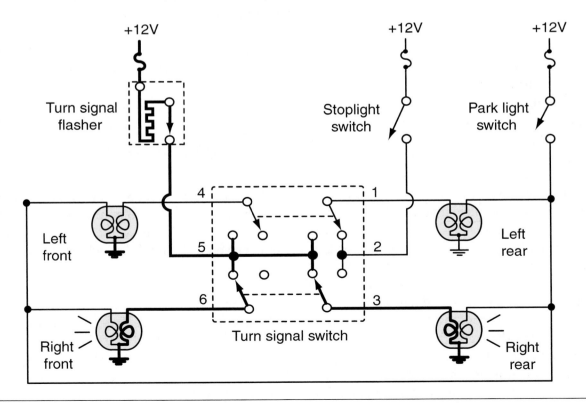

Figure 12-18 The position of the signal light switch contacts during a right turn.

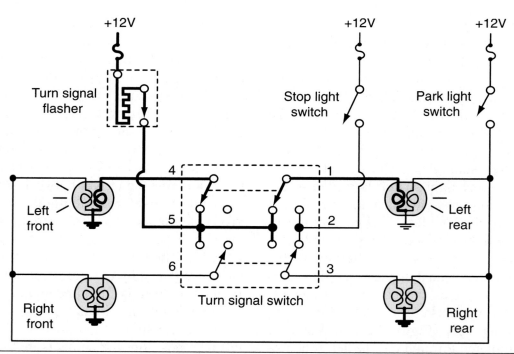

Figure 12-19 The position of the signal light switch contacts during a left turn.

flasher and the dash turn indicator light and left front signal light **(Figure 12-19)**. When the circuit is completed through the flasher, it begins opening and closing the circuit to turn the left signal lights on and off. In the left turn position, the center switch contact maintains the circuit between the brake light switch and the right rear stoplight. Therefore, if the driver depresses the brake pedal, current flows through this circuit to the right rear stoplight and the CHMSL.

When the steering wheel is rotated back to the straight ahead position after a turn, a cam mounted on the steering shaft returns the signal light switch to the off position to cancel the signal light operation. Signal light switches have a special feature that allows the driver to move the signal light switch lever enough to operate the signal lights on either side without latching the signal light switch into position.

When this special feature is used, the driver moves the signal light switch lever a small amount to flash the signal lights several times on the desired side to indicate an impending lane change. When the driver releases the signal light switch lever, the switch returns to the off position without the operation of the canceling mechanism. During a lane change, the steering wheel is not turned enough to operate the canceling mechanism.

The hazard-warning switch is usually mounted on the steering column. This switch is connected into the signal light circuit **(Figure 12-20)**. When the hazard-warning switch is pressed, all front and rear signal lights flash together. The hazard-warning switch may be used when a vehicle is parked in an emergency situation. The signal lights continue flashing as long as the hazard-warning switch is pressed.

INTERIOR LIGHTS

Interior lights usually include under-dash lights, door lights, map lights, and a dome light. Doorjamb switches are connected in the interior light circuit. Some vehicles have doorjamb switches on the front doors, whereas other vehicles

Tech Tip *Some vehicles have separate signal light and hazard warning flashers. Some vehicles have separate stop signal lights on the rear of the vehicle and some vehicles have amber signal lights on the rear.*

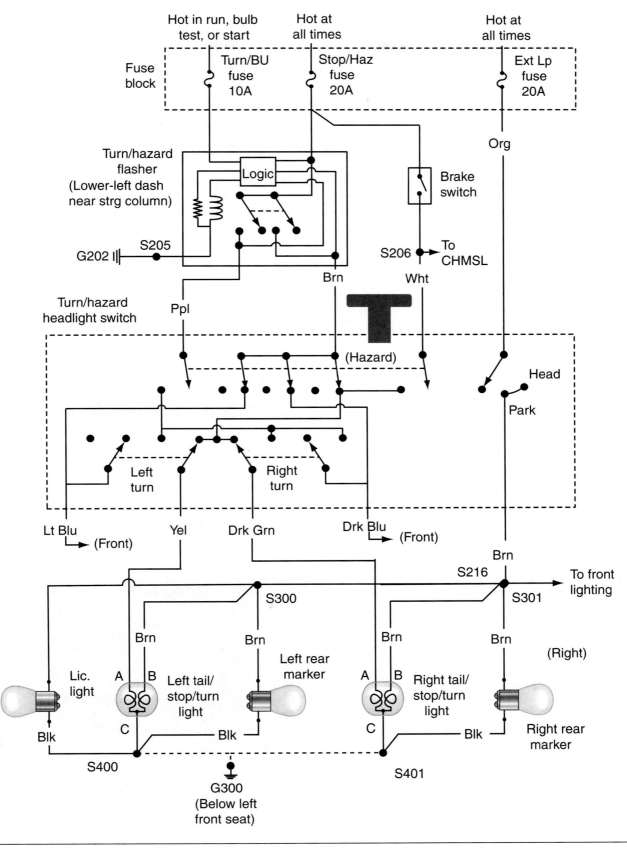

Figure 12-20 A hazard warning switch circuit.

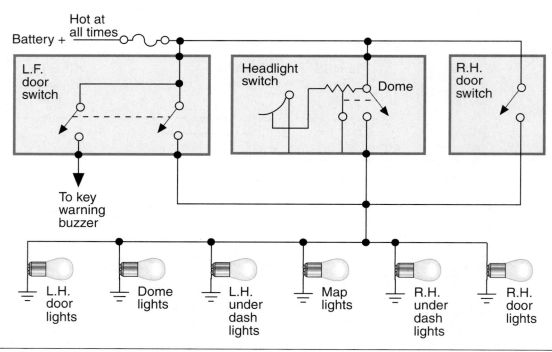

Figure 12-21 An interior light circuit.

Tech Tip *On some vehicles, the interior lightbulbs are insulated and the ground side of these bulbs is connected to the doorjamb switches. When a door is opened, the doorjamb switch completes the circuit to ground. In these circuits, the doorjamb switches have a single wire and the interior lights have two wires.*

SAFETY TIP *Interior lights may be called courtesy lights.*

have these switches on the front and rear doors. The doorjamb switches are normally open. When a door is opened, the doorjamb switch closes and supplies voltage to the interior lights. These lights are grounded to the vehicle chassis so they are illuminated when voltage is supplied to them **(Figure 12-21)**. On some vehicles the interior lights are turned on when the headlight switch knob is rotated fully counterclockwise. Other vehicles have a separate switch mounted near the dome light to turn on the dome light and interior lights.

DAYTIME RUNNING LIGHTS

On some vehicles, daytime running lights (DRLs) are separate lights mounted on the front of the vehicle. These lights have replaceable bulbs that provide more light than a stoplight but less than a headlight **(Figure 12-22)**. The DRLs are on when driving during the day. When the headlights are turned on the daytime running

Figure 12-22 Daytime running lights.

lights are turned off. DRLs are intended to warn drivers that a vehicle is approaching. On some vehicles, the DRLs are not illuminated until the transmission selector is placed in drive or reverse. This allows service personnel to work on the vehicle in the shop with the DRLs off. Other daytime running lights are illuminated when the engine is started during the day, but these lights are turned off if the parking brake is pressed. On certain makes, the high beam headlights are used as DRLs. When the high beam headlights are illuminated as DRLs, a resistor in the circuit reduces current flow and light brilliance to about one-half the light brilliance on high beam.

BACKUP LIGHTS

Backup lights have clear lenses and these lights are mounted on the rear of the vehicle. When the ignition switch is on and the transmission selector is placed in reverse, the backup lights are turned on **(Figure 12-23)**. The backup light switch is mounted on the transmission selector linkage or in the transmission. The backup lights provide some illumination behind the vehicle when backing up at night. These lights also warn anyone behind the vehicle about the backup action.

> **Tech Tip** *On some vehicles, the high beam headlights are used as DRLs. When the high beam headlights are illuminated as DRLs, a resistor in the circuit reduces current flow and light brilliance to about one-half the light brilliance on high beam.*

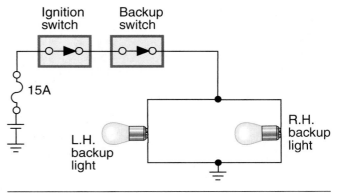

Figure 12–23 A backup light circuit.

Summary

- Automotive bulbs may have single or dual filaments.
- Automotive bulbs contain a vacuum to prevent overheating the filament.
- During the manufacturing process, oxygen is removed from sealed beams and they are filled with argon gas.
- Sealed beams may contain only high beam filaments or both high and low beam filaments.
- Many halogen headlights have a replaceable bulb.
- Halogen headlights provide more light than conventional sealed beams.
- HID headlights produce light when high voltage arcs across an air gap. These lights contain an inert gas that amplifies the light.

- HID headlights provide more light, draw less current, and last longer than other types of headlights.
- A driver-operated variable resistor controls instrument panel light brilliance by changing the voltage supplied to these lights.
- The dimmer switch directs voltage to the high or low beam headlights depending on the dimmer switch position.
- Many headlight switches contain a circuit breaker that is connected in the headlight circuit.
- Doors for concealed headlights may be vacuum or electrically operated.
- The signal light switch connects the flasher to the signal lights on one side of the vehicle depending on the signal light switch position.

- Most flashers contain a bimetallic strip and a heating coil.

- Some vehicles have separate signal light and hazard warning flashers.

- When the hazard warning switch is pressed, all signal lights begin flashing.

- The interior lights are switched on and off by the doorjamb switches.

Review Questions

1. Technician A says in a dual-filament bulb, one side of both light filaments is connected to the bulb case. Technician B says the dual filaments in a rear bulb may be used for stop and tail-lights. Who is correct?

   A. Technician A

   B. Technician B

   C. Both Technician A and Technician B

   D. Neither Technician A nor Technician B

2. Technician A says a taillight bulb contains a vacuum. Technician B says if the glass envelope leaks on a bulb, the filament will burn out quickly. Who is correct?

   A. Technician A

   B. Technician B

   C. Both Technician A and Technician B

   D. Neither Technician A nor Technician B

3. Technician A says composite headlights have replaceable halogen bulbs. Technician B says a halogen headlight provides reduced light compared to a conventional sealed beam. Who is correct?

   A. Technician A

   B. Technician B

   C. Both Technician A and Technician B

   D. Neither Technician A nor Technician B

4. Technician A says an HID headlight has a halogen bulb. Technician B says an HID headlight uses more current than a conventional sealed beam. Who is correct?

   A. Technician A

   B. Technician B

   C. Both Technician A and Technician B

   D. Neither Technician A nor Technician B

5. All of these statements about sealed beam headlights are true *except*:

   A. Sealed beam headlights are filled with argon gas.

   B. Sealed beam headlights may contain one or two filaments.

   C. A dual-filament sealed beam headlight has two terminals.

   D. In the dual-filament sealed beam headlight, the upper filament is for the low beam.

6. All of these statements about halogen headlights are true *except*:

   A. The tungsten filament in a halogen headlight can withstand higher temperatures than a sealed beam filament.

   B. Halogen headlights have a replaceable bulb.

   C. A cracked headlight lens results in poor light quality.

   D. A cracked headlight lens causes rapid headlight failure.

7. While discussing HID headlights, Technician A says HID headlights have a long-life replaceable bulb. Technician B says an HID light contains an inert gas that magnifies light. Who is correct?

   A. Technician A

   B. Technician B

   C. Both Technician A and Technician B

   D. Neither Technician A nor Technician B

8. When a push-pull-type headlight switch is pulled outward to the first position:

   A. The tail, park, and instrument panel lights are turned on.

   B. The high beam headlights are turned on.

C. The low beam headlights are turned on.

D. Voltage is supplied to the dimmer switch.

9. In a headlight circuit with a circuit breaker, the wire connected between the dimmer switch and one of the high-beam headlights is shorted to ground. Technician A says this condition will cause the circuit breaker to open the headlight circuit. Technician B says this problem will burn out the high-beam headlights. Who is correct?

A. Technician A

B. Technician B

C. Both Technician A and Technician B

D. Neither Technician A nor Technician B

10. All of these statements about concealed headlights are true *except*:

A. Concealed headlight doors may be operated electrically or by vacuum.

B. Vacuum-operated headlight doors are held open by vacuum.

C. A vacuum leak in a headlight door system may cause the doors to open.

D. A manual override knob may be rotated to force electrically operated headlight doors open.

11. To supply the necessary voltage to HID headlights, a voltage _____ and _____ are required.

12. On modern vehicles, the dimmer switch is usually operated by the _____.

13. If the headlight switch is in the PARK position, voltage is supplied through this switch to the _____, _____, _____, and _____ lights.

14. Many headlight switches have an internal _____ _____ connected in the headlight circuit.

15. Describe the current flow through the headlight circuit when the headlight switch is on and the dimmer switch is on high beam.

16. Describe the operation of the signal light switch and related circuit during a left turn.

17. Explain the operation of the interior lights when a front door is opened.

18. Explain the operation of the hazard warning lights.

Light Circuit Maintenance, Diagnosis, and Service

Learning Objectives

After you have read, studied, and practiced the contents of this unit, you should be able to:

- Explain voltage drop in a headlight circuit.
- Describe the precautions to be observed when servicing halogen bulbs.
- Describe general headlight aiming procedures.
- Explain how to test bulbs and fuses.

Key Terms

Aiming pads   Ohmmeter   Voltage drop

INTRODUCTION

It is extremely important for all the lights on a vehicle to be operating properly. For example, if the taillights on a vehicle are not working, a driver approaching the vehicle from the rear after dark may not see the vehicle soon enough to avoid an accident. Inoperative brake lights may cause a vehicle to be hit from the rear because the driver of the vehicle approaching from the rear did not realize the vehicle in front was going to stop. If one headlight is not working or the headlights are aimed too low, the driver has reduced visibility at night and this may result in a collision. Vehicle owners may not be aware that some

of the lights on their vehicle are not working. Therefore, the lights should be checked when the vehicle is brought into the shop for service and any deficiencies should be brought to the owners attention.

HEADLIGHT MAINTENANCE, DIAGNOSIS, AND SERVICE

Headlight maintenance involves checking the headlights each time a vehicle is brought to the shop for service. Be sure to check the headlights on high, low beam, and any other headlight features such as cornering lamps.

Headlight Diagnosis

Be sure the battery is fully charged and the battery terminals are clean and tight before any light diagnosis. When the engine is running, charging circuit voltage is supplied to all the lights that are turned on. Therefore, improper charging circuit operation affects the light circuits. For example, if the headlights become considerably brighter when the engine rpm is increased, this could indicate that charging voltage may be excessive or that there are more amps being used by the electrical system than the alternator can provide at low rpm's. Normal charging voltage is between 14.2 volts and 14.8 volts with the engine at normal operating temperature. Excessively high charging circuit voltage may also cause repeated failure of headlights and other lights. Excessively high charging system voltage also results in too much gassing of water from the battery. High resistance in the headlight circuit reduces current flow and causes dim headlights. If the high resistance is in the ground of one headlight, only that light is dim. A voltage drop test with the headlights on measures high resistance in the headlight circuit. The **voltage drop** is the measurement of the difference in voltage between two points in a circuit. Being able to accurately perform a good voltage drop test is a very valuable skill that is used on all areas of the vehicle that have electrical systems and is worth the extra time put in practicing.

Select a low scale on the voltmeter and connect the voltmeter across the part of the circuit in which high resistance is suspected **(Figure 13-1)**. For

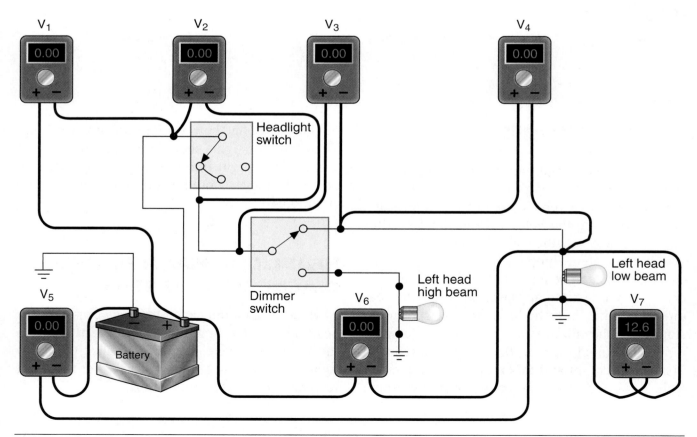

Figure 13-1 Measuring voltage drop in a headlight circuit.

example, to measure the voltage drop across the headlight ground, connect the voltmeter positive lead to the headlight ground and the negative lead to the negative battery terminal. The small measurement that is read is the amount lost by the connections in the ground circuit. If this exceeds specifications, a more thorough inspection is necessary to find the source of the loss. In order to narrow down a more specific source multiple voltage drop tests will need to be performed on smaller segments of the circuit. In most parts of the headlight circuit, except the headlight itself, voltage drop over 0.2 volts is excessive. Excessive voltage drop through a connector is the most common fault in the light system. Pay careful attention to connectors that are exposed to harsh environments such as fog light or headlight connectors that are repeatedly subjected to water and road chemicals. Headlight diagnosis is provided in **Figure 13-2**.

Symptom	Possible Cause	Remedy
Headlights do not light.	Loose wiring connections	Check and secure connections at headlight switch and dash panel connector.
	Open circuit in wiring	Check power to and from headlight switch. Repair as necessary.
	Worn or damaged headlight switch	Verify condition. Replace headlight switch if necessary.
One headlight does not work.	Loose wiring connections	Secure connections to headlight and ground.
	Sealed beam bulb burned out	Replace bulb.
	Corroded socket	Repair or replace, as required.
All headlights out; park and taillights are okay.	Loose wiring connections	Check and secure connections at dimmer switch and headlight switch.
	Worn or damaged dimmer switch	Check dimmer switch operation. Inspect for corroded connector. Replace, if required.
	Worn or damaged headlight switch	Verify condition. Replace headlight switch as necessary.
	Open circuit in writing or poor ground	Repair if required.
Both low beam or both high beam headlights do not work.	Loose wiring connections	Check and secure connection at dimmer switch and headlight switch.
		Check dimmer switch operation. Inspect for corroded connector. Replace if required.

Figure 13-2 Headlight diagnosis.

Headlight Service

The most common headlight service is changing and aiming headlights. The headlight replacement procedure varies depending on the vehicle and the type of headlights. Be sure the headlights are turned off before a headlight replacement procedure. When these lights are turned off, there is no voltage at the lights; therefore, the battery disconnecting is not required. The following is a typical sealed beam headlight replacement procedure:

1. Place fender covers over the vehicle in the work area.

2. Remove the bezel retaining screws and remove the bezel surrounding the headlight.

3. Remove the four sealed beam retaining ring screws but do not turn the headlight adjusting screws **(Figure 13-3)**.

4. Remove the sealed beam retaining ring and the sealed beam.

> **Tech Tip** *The retaining ring must be rotated about 1/8 of a turn counterclockwise to remove the retaining ring and bulb on some composite headlights.*

5. Disconnect the wiring connector from the back of the sealed beam.

6. Inspect the terminals in the wiring connector for corrosion and clean these terminals as required.

7. Place dielectric grease on the connector terminals and on the new sealed beam terminals to prevent corrosion.

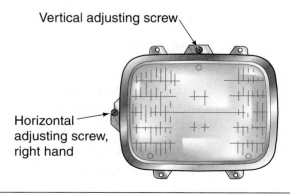

Figure 13-3 Headlight adjusting screws.

8. Install the wiring connector on the sealed beam terminals and be sure this connector is fully seated.

9. Install the new sealed beam with the embossed number at the top.

10. Install the sealed beam retaining ring and screws.

11. Turn the headlight switch on and be sure the sealed beam operates on both low and high beam.

12. Install the headlight bezel and retaining screws.

13. Check and readjust the headlight aim as necessary.

SAFETY TIP *When halogen headlight bulbs are illuminated, they become very hot. Do not touch halogen bulbs or attaching hardware immediately after the headlights have been turned off. Do not attempt to clean or replace halogen bulbs when the lights are on.*

Halogen Bulb Replacement in Composite Headlights

On some composite headlights, the bulb wiring terminals are not accessible behind the headlights. On most of these lights, two steel pins on top of the headlight assembly may be removed to allow the complete assembly to come forward out of the chassis. This action allows access to the wiring connectors and bulbs.

> ***Did You Know?*** *Do not touch the glass envelope on a halogen bulb with your fingers. When oil from your skin is deposited on this envelope, bulb life is shortened.*

The following is a typical replacement procedure for a halogen bulb in a composite headlight:

1. Be sure the headlight switch is off.

2. Place fender covers over the vehicle body in the work area.

3. Remove the cap over the rear of the bulb **(Figure 13-4)**.

4. Remove the wiring connector from the bulb and remove the retaining spring.

5. Remove the bulb from the composite headlight.

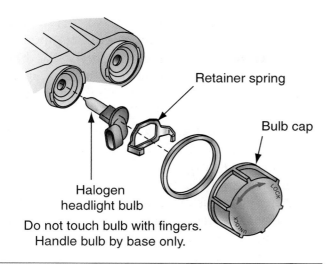

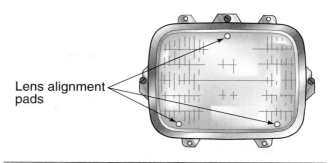

Figure 13-5 Headlight aiming pads.

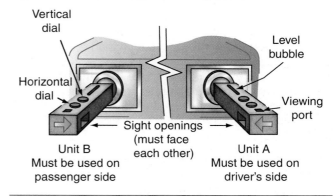

Do not touch bulb with fingers.
Handle bulb by base only.

Figure 13-4 Removing a halogen bulb.

6. Inspect the wiring connector for corrosion and clean as required.

7. Install the new bulb in the headlight and install the retaining ring.

8. Install the wiring connector and the bulb cap.

9. Turn the headlight switch on and be sure the headlight operates on low and high beam.

Headlight Aiming

Headlight aim is very important! A headlight that is misaimed by one degree downward reduces the driver's vision distance by 156 feet. If the headlight aim is too high, the headlights tend to reduce the vision of oncoming drivers creating a possible driving hazard. The following is a typical procedure for aligning sealed beam headlights:

1. Park the vehicle on a level floor. Be sure the vehicle has a normal load and the specified tire inflation. As weight is added to the rear of a vehicle the headlights will start to point higher.

2. Install fender covers over the vehicle body in the work area.

3. Select the proper headlight aimer adapter for the headlights being aligned. These adapters must fit over the aiming pads on the headlights **(Figure 13-5)**. **Aiming pads** are small projections on the front of a headlight lens.

4. Be sure the sealed beam outer surface is clean. Attach the headlight aimers to the sealed beams using the suction cups on the aimers. Be

Figure 13-6 Headlight aimers installed on the headlights.

sure the aimer sight openings face each other **(Figure 13-6)**.

5. Zero the horizontal adjustment dial on one of the aimers.

6. Be sure the split image target lines are visible in the aimer viewport.

7. Turn the headlight horizontal adjustment screw **(Figure 13-7)** until the target lines in the view port are aligned **(Figure 13-8).**

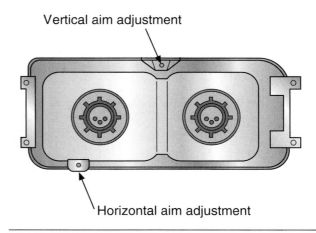

Figure 13-7 Headlight horizontal and vertical adjustment screws.

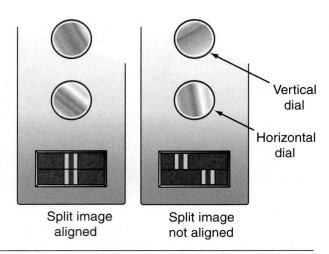

Split image   Split image
aligned   not aligned

Figure 13-8 Target lines in the headlight aimer indicate horizontal headlight alignment.

8. Repeat steps 5 through 7 on the opposite headlight.

9. Set the vertical adjustment dial to zero on one of the aimers and then turn the vertical aim adjustment screw until the bubble is centered in the aimer.

10. Repeat step 9 on the opposite headlight.

Many composite headlights have curved outer surfaces and special adapters are required to mount the headlight aimers properly on these headlights. Each composite headlight assembly has numbers molded into the headlight beside each aiming pad **(Figure 13-9)**. Each adjustment rod setting on the aimer adapter must match the number on the composite headlight aiming pad on which the adjustment rod will be placed. The adjustment rod setting must be locked in this position. After the adjustment rod setting is completed, the aimers are attached to the composite headlights and the headlight adjustment is performed using the same procedure for sealed beams.

Some vehicles do not have aiming pads on the headlights. These headlights may be aligned with the vehicle parked on a level floor 25 feet back from a blank wall. The vehicle must have a normal load and the specified tire inflation. The centerline of the vehicle and the vertical and horizontal centerlines of each headlight must be marked on the wall and the vehicle must be perpendicular to the wall **(Figure 13-10)**. With the headlights on

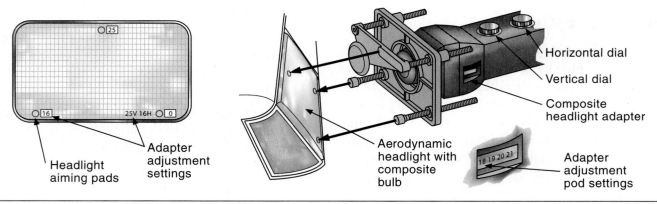

Figure 13-9 Composite headlight aiming pads and related numbers.

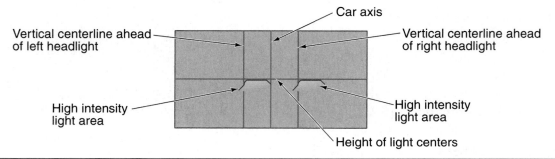

Figure 13-10 Headlight alignment using light beam projection.

high beam, the center of each light beam should be at the specified location on the wall. The amount of headlight beam drop may be specified by state or federal regulations. Rotate the headlight aiming screws as necessary to place the light beams at the specified location.

Some late-model vehicles have levels built into the headlight assemblies. This spirit level allows the driver to adjust the headlights in relation to the load in the vehicle. If a heavy load is placed in the trunk, the headlight beams become too high. Under this condition the driver may adjust the vertical headlight aiming screws to level the bubbles in each spirit level. After the load is removed from the trunk, the headlight beams will be too low and the driver may turn the vertical aiming screws to re-center the bubbles in the spirit levels.

TAILLIGHT, STOPLIGHT, AND PARK LIGHT CIRCUIT MAINTENANCE, DIAGNOSIS, AND SERVICE

The maintenance on any light circuit is basically the same. This maintenance includes regular inspections to determine if all the lights are working properly. Light circuit inspections should also include checking the wiring harness for frayed insulation, broken harness retainers, and improper contact between the harness and other components.

Diagnosis of tail, stop, and park lights is provided in **Figure 13-11**.

Follow these steps to remove a park light or front turn signal bulb:

1. Be sure the lights are turned off.
2. Remove the screw that retains the front signal and park light assembly to the chassis **(Figure 13-12)**.
3. Pull the front signal and park light assembly forward as far as the wiring harness will allow.
4. Turn the bulb socket 45° counterclockwise to remove the socket from the housing **(Figure 13-13)**.

5. Pull the bulb from the socket.
6. Inspect the light socket for corrosion on the socket and terminals. Clean the socket and terminals as necessary; if the socket was filled with dielectric grease, replace the grease before reinstalling the bulb.
7. Install the new bulb in the socket, re-install the socket in the light assembly, and then re-install the light assembly in the chassis.

> **Tech Tip** *Some bulbs may be pulled straight forward out of their socket. Bulbs with indexing pins must be rotated counterclockwise to remove them from the socket.*

8. Install and tighten the light assembly retaining screw.
9. Be sure all the lights operate properly after the light assembly is installed.

> **Tech Tip** *Replacement bulbs must have the same part number as the original bulb.*

SIGNAL LIGHT AND HAZARD WARNING LIGHT CIRCUIT MAINTENANCE, DIAGNOSIS, AND SERVICE

Maintenance of signal lights and hazard warning lights is similar to the maintenance of other light circuits. When inspecting signal and hazard warning lights, always be sure they flash at the proper

> **Tech Tip** *Some flashers are designed to flash at the same speed regardless of the current flow through the flasher. These flashers do not flash the signal lights at a different speed if a bulb is burned out on one side of the vehicle.*

Symptom	Possible Cause	Remedy
One taillight out.	Bulb burned out	Replace bulb.
	Open wiring or poor ground	Repair as necessary.
	Corroded bulb socket	Repair or replace socket.
All taillights and maker lamps out; headlights okay.	Loose wiring connections	Secure wiring connections where accessible.
	Open wiring or poor ground	Check operation of front park and marker lamps. Repair as necessary.
	Blown fuse	Replace fuse.
	Damaged headlight switch	Verify condition. Replace headlight switch if necessary.
Stoplights do not work.	Fuse or circuit breaker burned out	Replace fuse or circuit breaker. If fuse or circuit breaker blows again, check for short circuit.
	Worn or damaged turn signal circuit	Check turn signal operation. Repair as necessary.
	Loose wiring connections	Secure connection at stoplight switch.
	Worn or damaged stoplight switch	Replace switch.
	Open circuit in wiring	Repair as required.
Stoplights stay on continuously.	Damaged stoplight switch	Disconnect wiring connector from switch. If lamp goes out, replace switch.
	Switch out of adjustment	Adjust switch.
	Internal short circuit in wiring	If lamp stays on, check for internal short circuit. Repair as necessary.
One parking lamp out.	Bulb burned out	Replace bulb.
	Open wiring or poor ground	Repair as necessary.
	Corroded bulb socket	Repair or replace socket.
All parking lamps out.	Loose wiring connections	Secure wiring connections.
	Open wiring or poor ground	Repair as necessary.
	Bad switch	Replace switch.

Figure 13-11 Tail, stop, and park light diagnosis.

speed. If the signal lights flash normally on one side of the vehicle but they flash faster or slower on the opposite side, a problem is indicated in the circuit. The cause of the different flashing rates on the two sides of the vehicle may be a burned out signal light-bulb or resistance in the circuit.

Signal light diagnosis is provided in **Figure 13-14** and hazard warning light diagnosis is listed in

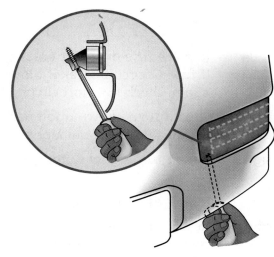

Figure 13-12 Removing the front park and turn signal light assembly.

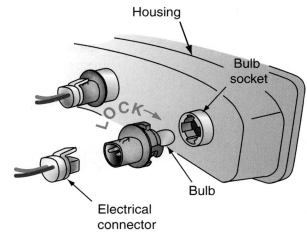

Figure 13-13 Removing the signal and park light sockets and bulbs.

Symptom	Possible Cause	Remedy
Turn signal lamps do not light.	Fuse or circuit breaker burned out	Replace fuse or circuit breaker. If fuse or circuit breaker blows again, check for short circuit.
	Worn or damaged turn signal flasher	Substitute a known good flasher. Replace if required.
	Loose wiring connections	Secure connections where accessible.
	Open circuit in wiring or poor ground	Repair as required.
	Damaged turn signal switch	Check continuity of switch assembly. Replace turn signal switch and wiring assembly as necessary.
Turn signal lamps light but do not flash	Worn or damaged turn signal flasher	Substitute a known good flasher. Replace if required.
	Poor ground	Repair ground.
Front turn signal lamps do not light	Loose wiring connector or open circuit	Repair wiring as required.
Rear turn signal lamps do not light.	Loose wiring connector or open circuit	Repair wiring as required.
One turn signal lamp does not light.	Bulb burned out	Replace bulb.
	Open circuit in wiring or poor ground	Repair as required.

Figure 13-14 Signal light diagnosis.

Symptom	Possible Cause	Remedy
Hazard flasher lamps do not flash.	Fuse or circuit breaker burned out	Replace fuse or circuit breaker. If fuse or circuit breaker blows again, check for short circuit.
	Worn or damaged hazard flasher	Substitute a known good flasher. Replace flasher if damaged.
	Worn or damaged turn signal operation	Repair turn signal system.
	Open circuit in wiring	Repair as required.
	Worn or damaged hazard flasher switch	Repair or replace turn signal switch and wiring assembly which includes hazard flasher.

Figure 13-15 Hazard warning light diagnosis.

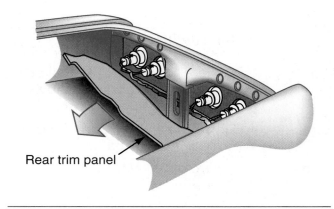

Rear trim panel

Figure 13-16

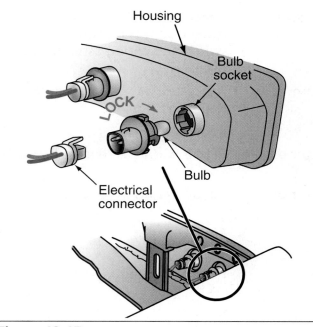

Housing

Bulb socket

LOCK

Bulb

Electrical connector

Figure 13-17

Figure 13-15. A typical procedure for changing a rear signal lightbulb is the following:

1. Open the trunk lid and remove the rear trim panel **(Figure 13-16)**.
2. Turn the light socket 45° counterclockwise on the bulb to be replaced **(Figure 13-17)**.
3. Remove the bulb from the socket.
4. Inspect the light socket and terminals for corrosion and clean as necessary.
5. Install the new bulb in the socket and install the socket and trim panel.

TESTING BULBS AND FUSES

Bulbs and fuses may be tested with an **ohmmeter**; however, a quick visual check often points out faulty bulbs easily. When visually inspecting a bulb look for signs of a burned filament often the inside of the bulb turns black or the filament is missing. When testing a dual-filament bulb with the digital ohmmeter, connect the leads from one of the bulb terminals to the case and then connect the meter leads from the other bulb terminal to the case. An infinite reading with either meter connection indicates a burned out bulb filament.

When testing a fuse, remove the fuse from the fuse panel and connect the ohmmeter leads across the fuse terminals. A zero ohm reading indicates a satisfactory fuse and a high or infinite reading indicates an open fuse. Fuses and circuit breakers may be tested using the same procedure and the test results are also the same. Many technicians find that using a test light is a very efficient way to test fuses quickly. By checking for power on both sides of the fuse they know that the fuse conducts electricity and is good.

SAFETY TIP *Never connect an ohmmeter to a circuit with voltage supplied to the circuit. This action may damage the ohmmeter.*

Summary

- High charging voltage may cause the headlights to become excessively bright when engine rpm is increased.

- High resistance in a headlight circuit reduces current flow and causes dim lights.

- If only one headlight is dim, the ground circuit on that light likely has high resistance.

- Measure the voltage drop in the headlight circuit to check the circuit resistance.

- Some moisture formation in non-vented composite headlights is normal.

- The glass envelope on halogen lightbulbs should not be touched with your hands.

- Many light sockets are removed by rotating them 45° counterclockwise.

- Headlights should be properly aimed vertically and horizontally.

- Digital ohmmeters do not require calibration.

- An ohmmeter may be damaged if it is connected to a circuit with voltage supplied to the circuit.

- When an ohmmeter is connected to fuse terminals, a low reading indicates a satisfactory fuse.

- When an ohmmeter is connected across a filament in a bulb, an infinite ohmmeter reading indicates an open filament.

Review Questions

1. Technician A says if all the headlights are dim, the alternator may be inoperative. Technician B says if all the headlights are dim, there may be high resistance in one headlight ground connection. Who is correct?

    A. Technician A

    B. Technician B

    C. Both Technician A and Technician B

    D. Neither Technician A nor Technician B

2. Technician A says when installing halogen bulbs, hold the glass envelope with your fingers. Technician B says some moisture formation in composite headlights is normal. Who is correct?

    A. Technician A

    B. Technician B

    C. Both Technician A and Technician B

    D. Neither Technician A nor Technician B

3. Technician A says when aiming sealed beam headlights, the aimer adapters must fit over the aiming pads on the sealed beams. Technician B says suction cups hold the aimers onto the sealed beams. Who is correct?

    A. Technician A

    B. Technician B

    C. Both Technician A and Technician B

    D. Neither Technician A nor Technician B

4. Technician A says headlights may be aimed with the headlight beams projected against a wall and the vehicle parked 50 feet back from the wall. Technician B says vehicle load and tire inflation should be normal before aiming the headlights. Who is correct?

    A. Technician A

    B. Technician B

    C. Both Technician A and Technician B

    D. Neither Technician A nor Technician B

5. A vehicle with two headlights has one headlight that is dim on low and high beam and one headlight that has normal brilliance. The most likely cause of this problem is:

   A. low charging system voltage.

   B. high resistance in the battery ground cable.

   C. high resistance in the dimmer switch.

   D. high resistance in the dim headlight ground circuit.

6. When measuring the voltage drop across the ground circuit from a headlight to the battery negative terminal, an acceptable voltage drop reading is:

   A. 0.2 volt.

   B. 0.8 volt.

   C. 1.0 volts.

   D. 1.2 volts.

7. All of these statements about servicing halogen headlightbulbs are true *except*:

   A. When illuminated, halogen bulbs become very hot.

   B. If a halogen bulb is burned out, the complete headlight assembly must be replaced.

   C. Do not attempt to clean or replace a halogen bulb with the headlights on.

   D. Do not touch the glass envelope in a halogen bulb with your fingers.

8. The signal lights on the left side of a vehicle flash faster than the signal lights on the right side of the vehicle.

   The most likely cause of this problem is:

   A. a burned out signal lightbulb on the left side of the vehicle.

   B. a defective signal light flasher.

   C. an open circuit in the signal light fuse.

   D. high resistance between the signal light fuse and the fuse holder terminals.

9. When using an ohmmeter to test the fuse in the stop light circuit, Technician A says the brake lights should be on. Technician B says an open fuse provides a very low reading on the ohmmeter. Who is correct?

   A. Technician A

   B. Technician B

   C. Both Technician A and Technician B

   D. Neither Technician A nor Technician B

10. After the headlights are turned on high beam, in a headlight circuit with a circuit breaker, both headlights keep going off and on. The most likely cause of this problem is:

    A. a defective circuit breaker.

    B. a grounded wire between the dimmer switch high beam terminal and the headlights.

    C. an open ground wire on one of the headlights.

    D. a loose high beam wiring connection on one headlight.

11. When the ohmmeter leads are connected to the fuse terminals, a very low ohmmeter reading indicates the fuse is _____ .

12. When testing bulb filaments, the ohmmeter leads are connected from one of the bulb terminals to the _____ .

13. When an ohmmeter is connected across a bulb filament, an infinite reading indicates the filament is _____ .

14. If the ohmmeter leads are connected across the terminals on a circuit breaker, a high reading indicates the breaker is _____ .

15. Explain the importance of proper headlight aim.

16. Explain the cause of excessive headlight brilliance when the engine rpm is increased.

17. Describe the proper test procedure for resistance in a headlight circuit.

18. Describe the ohmmeter test procedure for a circuit breaker.

CHAPTER

14 Indicator Lights and Gauges

Learning Objectives

After you have read, studied, and practiced the contents of this unit, you should be able to:

- Explain the operation of oil pressure indicator and charge indicator lights.

- Describe the operation of a bimetallic gauge.

- Explain the purpose and operation of an instrument voltage limiter.

- Describe the proper ammeter and voltmeter connections in an electrical circuit.

Key Terms

Bimetallic strip

Malfunction indicator lamp (MIL)

Normally open contacts

Proving circuit

Thermistor

INTRODUCTION

When you understand normal indicator light and gauge operation, you are able to diagnose problems in these circuits. Technicians must be able to quickly and accurately diagnose whether a defect is in the indicator light or gauge itself or in the system that the indicator light or gauge is monitoring.

OIL PRESSURE INDICATOR LIGHTS

The sending unit for the oil pressure indicator light is an on/off switch with a set of normally closed contacts. This sending unit is usually threaded into an opening in the main oil gallery of the engine block **(Figure 14-1)**. Normally closed

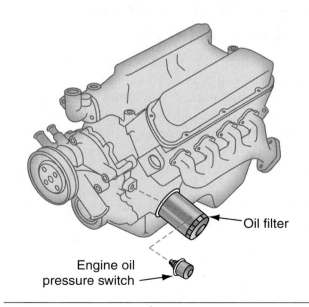

Figure 14-1 An oil pressure switch.

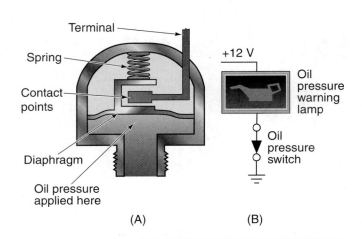

Figure 14-2 Oil pressure switch internal design.

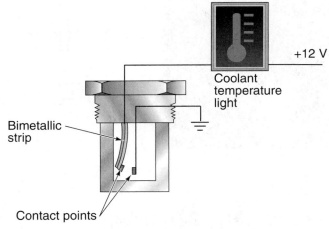

Figure 14-3 A temperature sending switch and related circuit.

contacts are closed with no pressure supplied to the unit and opened when pressure is supplied to the unit. Full pressure from the lubrication system is supplied to this sending unit. If less than 3-psi (20.6-kPa) oil pressure is supplied to the oil sending unit, the contacts in this unit are closed. Under this condition, current flows through the oil pressure indicator light and the sending unit contacts to ground and the light is on. When the engine is started, oil pressure is supplied to the diaphragm in the oil sending unit. If the oil pressure exceeds 3 psi (20.6 kPa), the sending unit contacts are forced open and the oil indicator light goes out **(Figure 14-2)**.

ENGINE TEMPERATURE WARNING LIGHTS

The sending unit for the engine temperature warning light contains a bimetallic strip and a set of normally open contacts. **Normally open contacts** remain in the open position until they are acted upon by temperature or pressure. When the engine temperature is below a specific temperature, the contacts in the temperature sending unit remain open and the temperature warning light remains off **(Figure 14-3)**. If the coolant temperature increases to a specific overheated condition, the bimetallic strip bends and closes the contacts in the sending unit. As long as the overheated condition is present, the sending unit contacts remain closed and the temperature warning light is on. Many temperature warning lights have a **proving circuit** in the ignition

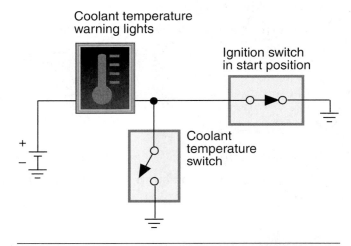

Figure 14-4 A coolant temperature switch proving circuit.

switch. When the ignition switch is in the start position, the proving circuit contacts in the ignition switch ground the temperature warning light **(Figure 14-4)**. This illuminates the temperature warning light while cranking the engine and proves that the temperature warning lightbulb is operating.

CHARGE INDICATOR LIGHTS

Some charge indicator lights have a resistor connected in parallel with the bulb. When the ignition switch is turned on, current flows through the charge indicator bulb and parallel resistor to one of the alternator terminals. This current flows through

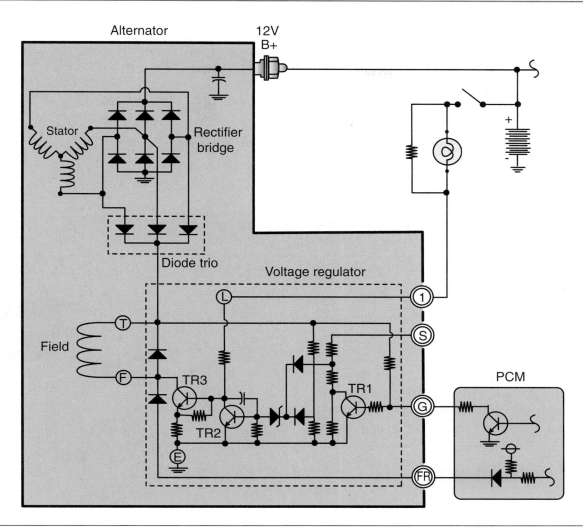

Figure 14-5 A charge indicator wiring circuit.

the alternator number 1 terminal to the alternator field coil and electronic voltage regulator to ground and the charge indicator light is on **(Figure 14-5)**. Once the engine starts, approximately 14.2 volts are supplied from the alternator battery terminal to the battery and electrical system. This same voltage is also supplied from the alternator stator windings through the diode trio to the field coil and to the number 1 alternator terminal. Because equal voltage is supplied to both sides of the charge indicator bulb, this light remains off.

Brake Warning Light

The red brake warning light in the instrument panel is connected to a switch in the combination brake valve. The combination brake valve also contains two hydraulic valves, the metering valve and the proportioning valve. These hydraulic valves are discussed in Chapter 44. All vehicles manufactured since 1967 have dual master cylinders. Pressure from each master cylinder piston is supplied to opposite ends of the combination valve piston that operates the brake warning light. If the master cylinder fluid level is satisfactory and both master cylinder pistons supply the same pressure, the piston in the brake warning light circuit remains centered. Under this condition, the brake warning switch is open and the warning light is off. If a fluid leak occurs and pressure from one master cylinder piston is low, the brake warning light piston moves toward the low pressure side of the piston. This action grounds the brake warning lightbulb through the switch and

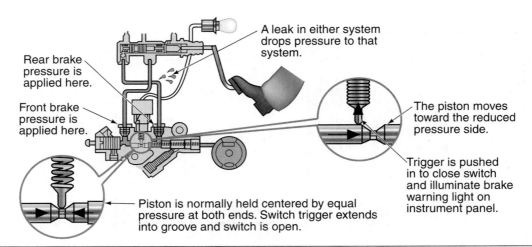

Figure 14-6 A brake warning light circuit.

the bulb is illuminated **(Figure 14-6)**. This circuit also has a prove-out circuit for bulb check when cranking.

> **Tech Tip** *On many vehicles, the red brake warning light is also illuminated if the parking brake is applied, the brake pad sensors detect a worn lining, or the brake fluid level is low.*

TYPES OF GAUGES

Various types of gauges are used in instrument panels. Each type of gauge has different operating principles and technicians must understand gauge operation to accurately diagnose and service each type of gauge and its related circuit.

Bimetallic Gauges

Most vehicles built prior to 1980 are equipped with bimetallic gauges. In this type of gauge, the needle is linked to a bimetallic strip. A **bimetallic strip** contains two different metals fused together. As a bimetallic strip is heated, these metals expand at different rates and cause the strip to bend. When this strip is heated, it pushes the needle across the gauge scale. A heating coil surrounds the bimetallic strip and the amount of heat supplied to this strip depends on the current flow through the heating coil **(Figure 14-7)**. The gauge sending unit contains a variable resistor that controls the current flow through the heating coil. Bimetallic gauges must have voltage limiters in their circuits to operate

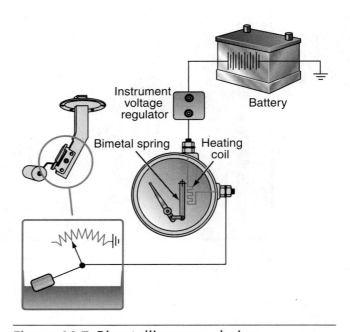

Figure 14-7 Bimetallic gauge design.

properly; almost all of this type of gauge operates on 5 volts. Bimetallic gauges most commonly were used as fuel and temperature gauges.

Balancing Coil Gauge

Some gauges contain a low coil and a high coil. The gauge needle is pivoted between the two coils and a permanent magnet is mounted on the needle. When the ignition switch is turned on, voltage is supplied between the two coils. The low coil is grounded and the high coil is connected to the sending unit **(Figure 14-8)**. If the sending unit has high resistance, more of the current flows through the

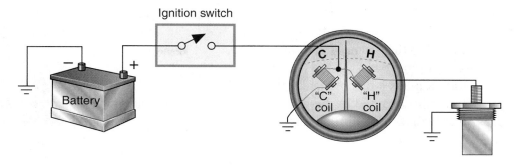

Figure 14-8 A balancing coil temperature gauge.

low coil to ground because this coil has lower resistance than the high coil and the sending unit. The magnetic field of the low coil attracts the magnet on the needle and moves the needle to the low position. When the sending unit has low resistance, most of the current flows through the high coil and the sending unit. Under this condition, the magnetic field of the high coil attracts the pointer magnet so the pointer moves to the high position. In some balancing coil gauges, voltage is supplied to the low or empty coil and the sending unit is connected between the empty coil and the full coil. The full coil is grounded **(Figure 14-9)**. For instance, when the fuel level is low in the tank, the sending unit has low resistance. Under this condition, current flows through the empty coil and the sending unit to ground and the magnetic field of the empty coil attracts the needle near the empty position. If the fuel level in the tank is high, the sending unit has high resistance. Under this condition, the current flows through the empty coil and the full coil to ground. Since the full coil has more coils of wire, it develops the stronger magnetic field and attracts the needle near the full position. A common test for this type of gauge was to remove the power lead from the sending unit, this would send the gauge to either high or low and then ground the wire sending the gauge to the opposite indicator. This test would confirm that the gauge and its wiring were okay and then the tech would test the sending unit.

Oil Pressure Gauges

The sender for an oil pressure gauge contains a diaphragm and a variable resistor. Engine oil pressure is supplied to the sending unit diaphragm. As the engine oil pressure increases, the sending unit diaphragm moves upward and the contact arm slides along the resistor **(Figure 14-10)**. An increase in oil pressure reduces sending unit resistance and a

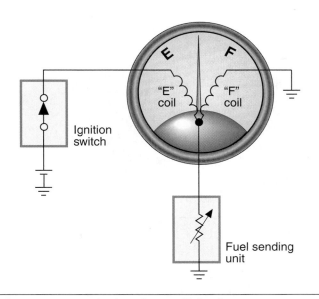

Figure 14-9 A balancing coil fuel gauge.

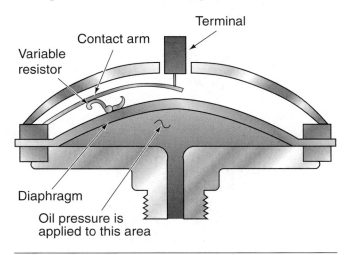

Figure 14-10 An oil pressure sending unit with a variable resistor.

decrease in oil pressure increases sending unit resistance. The change in resistance would affect the amount of voltage returned to the gauge in the same way that the fuel gauge system works.

A second type of oil gauge is a mechanical system. In this type of gauge oil is piped to the gauge itself and the pressure itself works similar to any other air or fluid pressure gauge.

Engine Temperature Gauge

The temperature sending unit is threaded into an opening in the cooling system (**Figure 14-11**). This sending unit is often mounted in the top of the intake manifold or in the cylinder head. The lower end of the sending unit is in contact with engine coolant. The sending unit for most temperature gauges contains a resistor disc called a thermistor (**Figure 14-12**). A **thermistor** is a special resistor that changes resistance in relation to temperature. At low temperatures, the thermistor has high resistance and as the coolant temperature increases, the resistance decreases. This would cause the balancing coil gauge to read hot or cool depending on the resistance at the sender.

Fuel Level Gauge

The fuel level gauge contains a float mounted on an arm attached to the gauge. The float moves up and down with the fuel level in the tank. As the

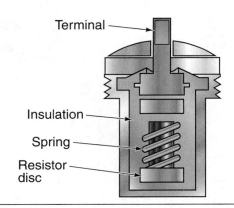

Figure 14-12 Coolant temperature sending unit design.

float moves up and down, it moves a sliding contact on a variable resistor in the sending unit. High fuel level in the tank results in low sending unit resistance and a low fuel level increases sending unit resistance (**Figure 14-13**).

Tech Tip *The fuel sending unit that is connected between the empty and full coils on a balancing coil fuel gauge operates the opposite way compared to the sending unit for a bimetallic gauge. The sending unit for this type of balancing coil gauge has high resistance when the fuel level in the tank is high.*

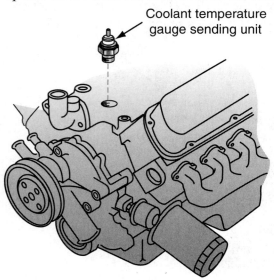

Figure 14-11 Coolant temperature gauge sending unit location.

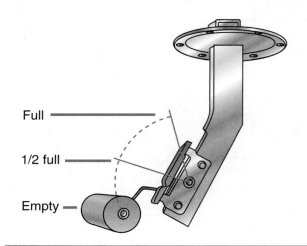

Float position	F	1/2	E
Resistance Ω	2 - 5	25.5 - 39.5	105 - 110

Figure 14-13 A fuel gauge sending unit.

Instrument Voltage Limiters

An instrument voltage limiter may be connected to bimetallic gauges. This limiter contains a set of contacts mounted on a bimetallic strip. The voltage supply to the gauges is connected through the limiter contacts. A heating coil surrounds the bimetallic strip and this heating coil is connected to ground on the limiter.

> **Tech Tip** *The instrument voltage limiter must be grounded on the instrument panel and this panel must have a satisfactory ground connection to the battery. If the instrument voltage limiter does not have a satisfactory ground, the limiter contacts remain closed and this action supplies 12 volts to the gauges. This voltage will damage the gauges very quickly.*

When the ignition switch is turned on, voltage is supplied through the limiter contacts to the gauges. Current also flows through the heating coil to ground. The heating coil heats the bimetallic strip very quickly and the limiter contacts open the circuit to the gauges and also to the heating coil. The bimetallic strip cools quickly and the contacts close. The voltage limiter supplies a pulsating 5 volts to the gauges regardless of the input voltage and this provides more stable gauge operation **(Figure 14-14)**.

Speedometers and Odometers

A mechanical speedometer has a cable drive from the transmission output shaft. A gear on the speedometer cable drive is meshed with a gear on the transmission output shaft **(Figure 14-15)**. Therefore, output shaft rotation turns the speedometer cable. This cable is connected to a permanent magnet surrounded by a metal drum in the speedometer.

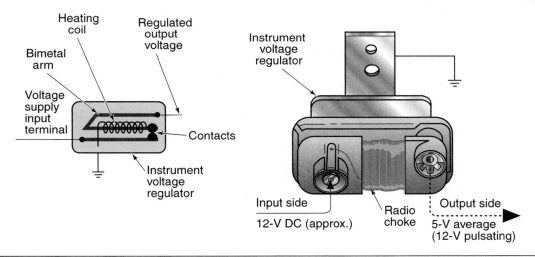

Figure 14-14 An instrument voltage limiter.

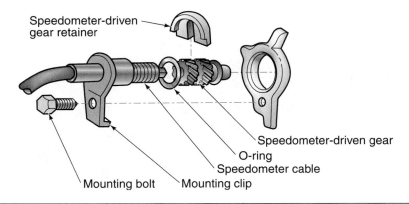

Figure 14-15 A speedometer cable drive.

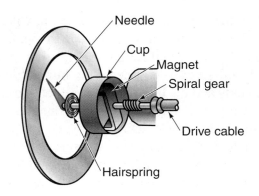

Figure 14-16 Speedometer design.

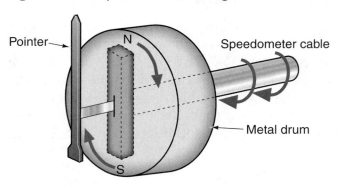

Figure 14-17 Speedometer operation.

The speedometer needle is attached to the drum, but there is no direct mechanical connection between the cable and the speedometer needle **(Figure 14-16)**. When the speedometer cable rotates the permanent magnet, a rotating magnetic field is created around the drum. This magnetic field pulls the drum and speedometer needle in a rotary motion to provide an accurate speedometer reading **(Figure 14-17)**. An odometer is driven by a worm gear drive from the speedometer cable **(Figure 14-18)**. Some odometers have six wheels and numbers from zero to nine are stamped on the outer surface around the wheels.

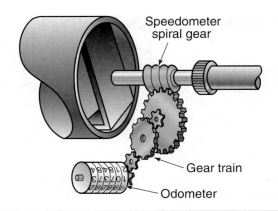

Figure 14-18 An odometer and drive mechanism.

These wheels are designed so when the right wheel makes one revolution, the wheel on the left moves one position. Odometers with seven wheels indicate mileage over 100,000.

> **Tech Tip** *Many modern vehicles are equipped with electronic speedometers. These speedometers may be analog or digital and they do not require a cable. The vehicle speed sensor (VSS) mounted in the transmission sends a voltage signal to the computer and the computer controls the speedometer reading.*

Tachometers

Some vehicles have tachometers that read engine rpm. The tachometer usually receives voltage input signals from the primary ignition system **(Figure 14-19)**. These voltage input signals are in direct proportion to engine speed and the tachometer changes these signals to an rpm reading.

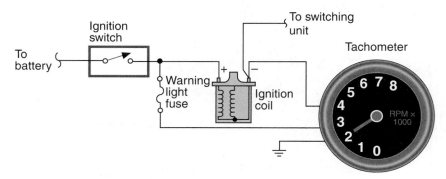

Figure 14-19 A tachometer circuit.

Voltmeters and Ammeters

Some vehicles have a voltmeter to indicate the charging system condition. The voltmeter has high internal resistance and it must be connected in parallel. Because a voltmeter has high internal resistance, it draws a very low current. The voltmeter leads are connected from the battery positive circuit to ground. On some vehicles, the voltmeter is connected to the battery positive circuit at the ignition switch. On these systems, there is no voltage supplied to the voltmeter when the ignition switch is turned off **(Figure 14-20)**. With the engine running, the voltmeter should indicate 13.2 volts to 15.2 volts. If the voltage is below 13.2 volts, the alternator voltage is too low and the battery will become discharged. When the voltage is above 15.2

volts, the alternator voltage is too high and the battery is being overcharged. Excessive charging voltage may also burn out electrical accessories on the vehicle. Voltmeters are only a general indicator of battery charge condition, and do not indicate the ability of an alternator to charge properly, just the amount of voltage in the battery.

Some vehicles have an ammeter to indicate charging circuit and electrical system operation. The ammeter is connected in series in the circuit between the alternator battery terminal and the positive battery terminal. The ammeter reads the amount of current flow from the alternator into the battery with the engine running. The alternator side of the ammeter is also connected to a junction block and this block connects the ammeter to the ignition switch and the vehicle accessories **(Figure 14-21)**. Therefore, the

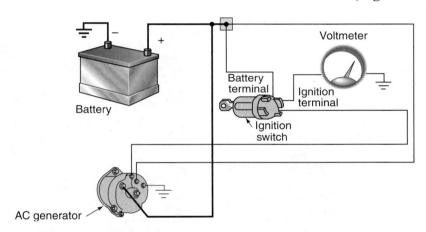

Figure 14-20 Voltmeter connections.

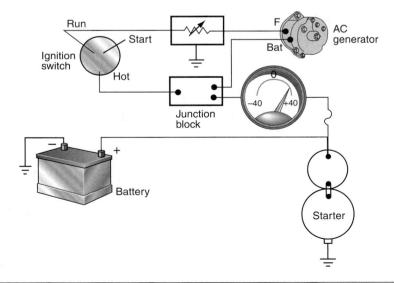

Figure 14-21 Ammeter connections.

ammeter will also indicate the current flow out of the battery and through the accessories with the engine stopped or with the engine running and the alternator inoperative.

Digital Gauges

Late-model vehicles are almost exclusively equipped with digital gauges **(Figure 14-22)**. These gauges are generally fed their inputs from the vehicle's onboard computer. As vehicles have become more computerized, so too have the gauge systems. For instance, the speedometer is no longer a mechanical gauge that is subject to the wear and inaccuracies of a cable system; instead the transmission either has a sensor for vehicle speed that is fed to the computer or the ABS wheel speed sensors are used. The computer then decides where to point the needle or give a numeric display. Since most vehicles are now equipped with ABS, the transmission speed sensor is unnecessary. Wheel speed sensors that used to control only the ABS system

now have their data used to indicate speed as well.

Since a computerized engine was monitoring engine temperature already, additional warning systems were eliminated because the computer could use its sensor data to indicate where the gauge should read to correspond to current conditions.

Digital gauges and their inputs are usually tested in the same manner as other automotive computer systems. Often car manufacturers build test modes into their diagnostic software for each model. In addition, when there is a malfunction the **malfunction indicator lamp (MIL)** **(Figure 14-23)** is often illuminated since the input is often needed to monitor or adjust engine performance.

Figure 14-23 A malfunction indicator lamp (MIL).

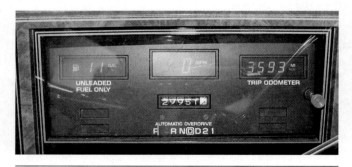

Figure 14-22 A digital gauge set up.

Summary

■ The oil pressure switch opens and turns the oil pressure indicator light off when the oil pressure reaches 3 psi (20.6 kPa).

■ When the engine is running, equal voltage on each side of the charge indicator light keeps the light off.

■ If one-half of the master cylinder is low on brake fluid, the brake switch in the combination valve closes and this action turns on the red brake warning light.

■ Bimetallic gauges contain a bimetallic strip wrapped with a heating coil and are linked to the gauge needle.

■ A balancing coil gauge contains two coils and the magnetism of these coils determines the gauge needle position by attracting a permanent magnet on the gauge needle.

■ A fuel gauge sending unit contains a float linked to a variable resistor. As the float moves up and down with the fuel level in the tank, the resistance changes in the variable resistor.

■ An instrument voltage limiter limits voltage to the gauges to 5 volts regardless of the input voltage.

■ An instrument voltage limiter contains a bimetallic strip surrounded by a heating coil and a set of normally closed contacts.

■ A mechanical speedometer has a cable driven by a set of gears on the transmission output shaft.

■ The odometer in a mechanical speedometer is driven through a gear set from the speedometer cable.

■ Many tachometers use a voltage input from the primary ignition circuit.

■ A voltmeter contains high internal resistance and it is connected in parallel to the circuit.

■ An ammeter contains low internal resistance and it is connected in series in the circuit.

■ Digital gauges use various computer inputs to indicate proper gauges readings.

Review Questions

1. Technician A says an oil pressure switch contains a set of normally open contacts. Technician B says the oil pressure warning light should be on if the engine is running and the oil pressure is 8 psi (55 kPa). Who is correct?

   A. Technician A

   B. Technician B

   C. Both Technician A and Technician B

   D. Neither Technician A nor Technician B

2. Technician A says the charge indicator light is grounded through the alternator field circuit if the ignition switch is on and the engine is not running. Technician B says when the engine starts, equal voltage on both sides of the charge indicator bulb turns this light off. Who is correct?

   A. Technician A

   B. Technician B

   C. Both Technician A and Technician B

   D. Neither Technician A nor Technician B

3. Technician A says the red brake warning light is on if the parking brake is applied. Technician B says the red brake warning light is on if both halves of the master cylinder are overfilled with brake fluid. Who is correct?

   A. Technician A

   B. Technician B

   C. Both Technician A and Technician B

   D. Neither Technician A nor Technician B

4. Technician A says in a bimetallic-type temperature gauge, current through the gauge and sending unit increases as the temperature decreases. Technician B says in a bimetallic-type temperature gauge, the resistance of the sending unit decreases as the coolant temperature decreases. Who is correct?

   A. Technician A

   B. Technician B

   C. Both Technician A and Technician B

   D. Neither Technician A nor Technician B

5. In a typical oil pressure indicator light circuit with the engine idling at normal operating temperature, the oil pressure indicator light is on if the oil pressure is below:

   A. 18 psi.

   B. 15 psi.

   C. 9 psi.

   D. 3 psi.

6. All of these statements about engine temperature warning light circuits are true *except*:

   A. A proving circuit illuminates the warning light when cranking the engine.

   B. The sending unit contains a set of normally closed contacts.

   C. The sending unit contains a bimetallic strip.

   D. The warning light is illuminated when the sending unit contacts close.

7. When discussing charge indicator light circuits, Technician A says with the engine running, the charge indicator light may be turned off by unequal voltage on each side of the light. Technician B says when the engine starts, 12.6 volts are supplied to one side of the light and 14.2 volts are supplied to the other side of the light. Who is correct?

   A. Technician A

   B. Technician B

   C. Both Technician A and Technician B

   D. Neither Technician A nor Technician B

8. All of these statements about red brake warning lights are true *except*:

   A. The red brake warning light is on if the parking brake is applied.

   B. The red brake warning light is on if there is no fluid in one-half of the master cylinder.

   C. The switch in the combination valve is closed each time the brakes are applied.

   D. The red brake warning light is connected to the switch in the combination valve.

9. A vehicle has a bimetallic-type fuel gauge that indicates FULL regardless of the amount of fuel in the tank. The most likely cause of this problem is:

   A. an open wire between the fuel gauge and the fuel tank.

   B. a grounded wire between the ignition switch and the fuel gauge.

   C. an open circuit in the gauge's fuse.

   D. a grounded wire between the fuel gauge and the fuel tank.

10. A vehicle has a bimetallic-type oil pressure gauge and an oil pressure sending unit containing a diaphragm and a variable resistor. Technician A says the resistance of the variable resistor increases as the oil pressure decreases. Technician B says the gauge pointer moves to the HIGH position as the resistance of the variable resistor decreases. Who is correct?

    A. Technician A

    B. Technician B

    C. Both Technician A and Technician B

    D. Neither Technician A nor Technician B

11. In a fuel gauge sending unit used with a bimetallic gauge, the resistance in the sending unit _____ as the tank is filled with fuel.

12. A temperature sending unit used with a bimetallic gauge contains a _____.

13. A fuel level sending unit contains a _____ linked to a _____.

14. When the contacts open in an instrument voltage limiter, the current flow through the heating coil on the bimetallic strip is _____.

15. Explain the basic operation of a mechanical speedometer.

16. Explain the proper voltmeter connection in an electrical circuit.

17. Explain the operation of an ammeter in the instrument panel.

CHAPTER

15

Indicator Light Maintenance, Diagnosis, and Service

Learning Objectives

After you have read, studied, and practiced the contents of this unit, you should be able to:

- Diagnose oil pressure indicator light problems.
- Diagnose charge indicator light problems.
- Test thermistor-type gauge sending units.

Key Terms

Combination valve   **Printed circuit board**

INTRODUCTION

Technicians must understand normal indicator light and gauge operation to detect and diagnose improper operation of these components. On vehicle models whose systems are unfamiliar, you should start by reading the owner's manual to understand proper system operation. When you understand normal indicator light and gauge operation, you can often detect problems in the circuits monitored by these indicator lights and gauges. Detecting these problems before they become serious usually saves the customer money and inconvenience, and demonstrates that you have the customer's interests at heart.

INDICATOR LIGHTS

Indicator lights should be inspected for proper operation when a vehicle is brought into the shop for minor service. Most shops have a policy of inspecting indicator light operation each time a vehicle is brought into the shop for lubrication service. A customer may not be aware that an indicator light is not working. For example, if the oil pressure indicator lightbulb is burned out, the oil pressure warning light never comes on even with the ignition switch on and the engine not running. If a customer does not notice this condition, the engine could be ruined from low oil pressure and the customer would not be aware of the problem until after the engine is severely damaged. Therefore, during routine lubrication service it is a good shop policy to inspect all the warning lights for proper operation. When one or more indicator lights is not working, the first step in diagnosing the problem is to visually inspect the indicator lights system for obvious damage such as broken wires, disconnected sending units, and contact with other components. Inspect the indicator light electrical connections for corrosion and looseness and checking for system power are all good starts to proper diagnosis.

The following technical diagnosis is general in nature and intentionally not vehicle specific. Since most modern vehicles have similar operating characteristics general diagnostic skills will allow you to successfully repair most systems on the majority of makes and models. Always remember to start with the proper manual specific to each model and year.

OIL INDICATOR LIGHTS
Oil Indicator Light Diagnosis

If the oil pressure indicator light does not come on with the ignition switch on and the engine not running, remove the wire from the oil sender switch and use a jumper wire to connect this wire to ground. If the indicator light does not come on under this condition, check the bulb, the voltage input to the bulb, and the wire from the light to the oil sender switch. When the oil indicator light is on with the oil sender wire grounded and the ignition switch on, the bulb and connecting wires are satisfactory. Therefore the problem must be in the oil sender switch. With the oil sender switch wire disconnected and the engine not running, connect an ohmmeter from this switch terminal to an engine ground near the terminal. A high or infinite reading indicates a defective oil sender switch. If the oil pressure indicator light is on with the engine running, the engine oil pressure could be very low. Under this condition, shut the engine off and check the oil level. Inspect the oil for contamination with gasoline or coolant. If the oil level is very low but the oil does not appear to be contaminated, the proper grade and viscosity of oil should be added to the crankcase until the oil level is at the FULL mark on the dipstick. If the oil indicator light now goes out with the engine running and there is no evidence of engine oil leaks, the vehicle may be driven. However, the engine should be checked as soon as possible for the cause of the excessive oil consumption and an oil pressure test performed to ensure no damage resulted from running the engine on low oil pressure.

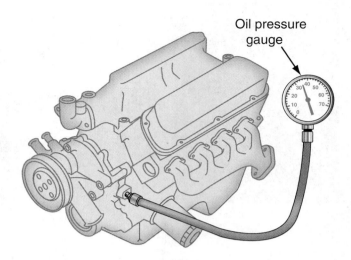

Figure 15-1 Oil pressure testing.

When the oil indicator light is on with the engine running and the oil level and condition are satisfactory, the vehicle should be towed to a service facility. Under this condition, the engine oil pressure may be extremely low or zero and severe engine damage may result from running the engine. When this condition is present, the oil sending switch should be removed and a test gauge installed in place of this switch **(Figure 15-1)**. If the engine oil pressure meets or exceeds the vehicle manufacturer's specifications, replace the oil sending switch. If the engine oil pressure is below the vehicle manufacturer's specifications, the oil pump and pump pickup, crankshaft, and camshaft bearings should be inspected and measured.

Tech Tip *On some engines, the oil pump is bolted to the outside of the engine block and this pump may be removed without removing the engine oil pan.*

Oil Indicator Light Service

If it is necessary to change the oil sending switch, special oil sender sockets are available to fit the serrations on the sender. Use a ratchet or ratchet and extension to turn the oil sender socket. When installing the new sending switch, coat the sender threads with pipe thread lubricant and always torque this switch to the specified torque.

NOTE: Pipe thread lubricant and sealer may be called *pipe dope.*

If the bulb in the oil pressure indicator light is burned out, it may be necessary to remove the retaining screws and pull the instrument panel partially out of the dash to gain access to the charge indicator bulb. Always disconnect the battery negative cable and wait the specified time period (2 to 10 minutes) before removing the instrument panel. Many lightbulbs are retained in the back of the instrument panel with a plastic retainer **(Figure 15-2)**. Rotate the retainer counterclockwise to release the bulb. Install the new bulb and turn the retainer clockwise to hold the bulb in place. Since it is often a time-consuming job to remove the instrument panel, it is always a good idea to test all bulbs in the dash while it is apart. Most instrument panels have a printed circuit board. A **printed circuit board** is made from a thin insulating material and electrical solder or metal tracks are imbedded into the insulating material. The tracks in the printed circuit board connect the bulbs and gauges to the external circuits. A wiring connector plugged into the printed circuit board connects the external circuits to this board.

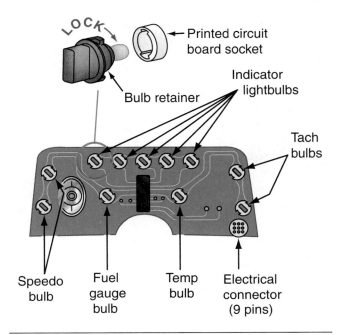

Figure 15-2 Indicator bulbs in the instrument panel.

Tech Tip *Disconnecting the battery prevents accidental grounding of any circuit during instrument panel removal. This grounding could damage expensive electronic components. Waiting for the specified time period allows the backup power supply to power down in the air bag system. This power supply provides voltage to deploy the air bags if the battery is disconnected in a collision. If this procedure is not followed, the air bags could be deployed accidentally while servicing the instrument panel.*

Tech Tip *Disconnecting the battery erases the station programming in the vehicle radio. Before disconnecting the battery, it is a good idea to write down the station programming selections on AM and FM. After the battery is reconnected, the original radio station programming may be restored. Disconnecting the battery also erases the engine computer memory. After the battery is reconnected, the engine operation may be somewhat erratic until the engine computer relearns the input sensor data. These problems with erasing computer memories when a battery is disconnected may be eliminated by connecting an appropriate power supply to the vehicle's cigarette lighter before disconnecting the battery.*

Tech Tip *Do not reconnect the negative battery cable and turn the ignition switch on to test the indicator bulb with the instrument panel pulled forward. If the vehicle has an instrument voltage limiter, this limiter is not grounded under this condition. When the battery is reconnected and the ignition switch is turned on, the instrument voltage limiter supplies 12 volts to the gauges, resulting in gauge damage. Always reinstall the instrument panel and tighten all the retaining screws before reconnecting the battery and turning the ignition switch on.*

CHARGE INDICATOR LIGHTS

The charge indicator light should be illuminated when the ignition switch is on and the engine is not running. When the engine starts, the charge indicator light should go out and it should remain off as long as the engine is running. Each time minor service is performed on a vehicle, the charge indicator light operation should be checked. The customer may not notice the charge indicator light is on with the engine running. If this condition is present and the alternator is not working, the battery may become discharged and fail to start the vehicle. This could result in an expensive service call for the customer. During a routine check of the charge indicator light operation, inspect the alternator wiring for worn insulation and loose or corroded terminals.

Charge Indicator Light Diagnosis

If the charge indicator light does not come on with the ignition switch on and the engine not running, consult a vehicle repair manual to locate the proper method to test the system. This is often done by removing the connector at the alternator and grounding one of the wires **(Figure 15-3)**. If the indicator light does not come on under this condition, check the bulb, the voltage input to the bulb, and the wire from the light to the alternator terminal. When the charge indicator light is on with the terminal grounded and the ignition switch on, the bulb and connecting wires are satisfactory. If the indicator light comes on after grounding but not when it is connected to the alternator, the fault is with either the connection at the alternator or the alternator itself.

When the charge indicator light is on with the engine running, stop the engine and use a belt tension gauge to measure the belt tension. If the belt tension is satisfactory, test the alternator output to prove the alternator is operating properly. If the alternator output is less than specified, the alternator must be repaired or replaced. Alternator diagnosis is explained in a later chapter.

If the charge indicator light is illuminated dimly with the engine running, remember this light is kept off by equal voltage on each side of the light with the engine running. High resistance between the alternator battery terminal and the junction block or high resistance between the alternator terminal and the charge indicator light causes low voltage on one side of the charge indicator bulb.

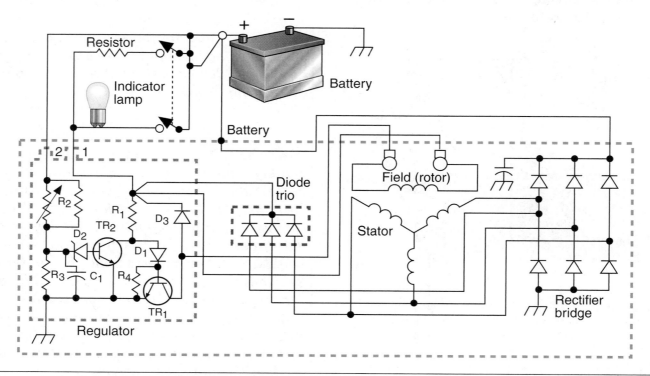

Figure 15-3 Charge indicator light and related circuit.

Unequal voltage across the charge indicator light causes a low current through this light. Measure the voltage drop across the circuits mentioned above to locate the high resistance.

Charge Indicator Light Service

If the charge indicator bulb must be replaced, follow the same procedure and precautions that were explained previously regarding oil pressure indicator light service.

BRAKE WARNING LIGHTS

The red brake warning light should be checked for proper operation when performing minor service on a vehicle. With the parking brake applied, the red brake warning light should be on. If this light is not on, remove the wiring connector from the parking brake switch, turn the ignition switch on, and ground the wire from the red brake warning light. Also, check the bulb, the voltage supply to the bulb, and the wire from the bulb to the parking brake switch.

If the red brake warning light is on with the engine running, remove the wire from the brake switch in the combination valve. If the brake warning light goes out under this condition, there is unequal pressure between the two halves of the master cylinder. Check the fluid levels in both halves of the master cylinder. Fill both halves of the master cylinder to the proper level with the brake fluid specified by the vehicle manufacturer.

The **combination valve** contains a proportioning valve, a metering valve, and a switch that grounds the red brake warning light if there is unequal pressure in the two halves of the master cylinder. The combination valve is usually mounted near the master cylinder. Follow the general service guidelines from the chapter on servicing brake systems.

Tech Tip *If one half of the master cylinder is low on brake fluid, it may be an indicator that the disk brakes pads are worn down and should be inspected.*

Did You Know? *If the red brake warning light is still on, the brake pad sensor may have detected a worn brake lining if equipped with pad sensors.*

Tech Tip *Many ABSs have an amber and a red brake warning light. The amber ABS warning light is illuminated if the ABS computer senses a defect in the ABS electronic system. Specific ABS problems cause the ABS computer to illuminate both the amber and red brake warning lights. The red brake warning light is also illuminated by unequal pressure between the two halves of the master cylinder or a parking brake application.*

If the brake warning lightbulb must be replaced, follow the same procedure and precautions that were explained previously regarding oil pressure indicator light service.

GAUGES AND RELATED CIRCUITS

Gauges should be inspected for proper operation when a vehicle is in the shop for minor service. If all the gauges are inoperative or erratic, the instrument voltage limiter and anything the gauge system shares in common, such as power feed wires or grounds, should be tested. When only one gauge is inoperative or erratic, concentrate the diagnosis on that gauge and its related circuit. Prior to gauge diagnosis, always inspect the wiring harness and electrical connections on the gauge-sending units. Repair any damaged wiring or corroded, loose connections.

Gauge Diagnosis

When all the gauges are inoperative or erratic, use a voltmeter to test the instrument voltage limiter. Connect a voltmeter from the battery side of the limiter to ground. With the ignition switch on, the input voltage to the limiter should be 12 volts or more. If this voltage is zero, test the gauge fuse. When the input voltage to the limiter is satisfactory,

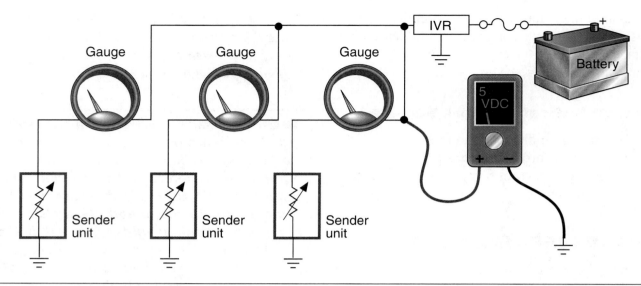

Figure 15-4 Testing the instrument voltage limiter.

connect the voltmeter from the gauge side of the limiter to ground (**Figure 15-4**). With the ignition switch on, the voltage supplied by the limiter should be a pulsating 5 volts. Replace the limiter if it does not supply the specified voltage. Test for a proper voltage feed to ensure that the gauges are receiving the correct amount. Perform a voltage drop on all grounds commonly shared.

To test individual gauges, disconnect the wire from the sending unit and connect the gauge tester from the sending unit wire to ground (**Figure 15-5**). Always connect and use the gauge tester as specified by the equipment manufacturer. With the gauge tester connected from the sending unit wire to ground, turn the ignition switch on and rotate the specified control knob on the tester while observing the gauge operation. The gauge tester contains a variable resistor(s) operated by the control knobs. As the control knob is rotated, the resistance in the tester is varied and the gauge should move from low to high. If the gauge does not move from low to high, test the wire from the gauge to the sending unit. When this wire is satisfactory, replace the gauge.

Sending Unit Diagnosis

If a sending unit contains a variable resistance, such as a fuel gauge sending unit, connect an ohmmeter from the sending unit terminal to ground on the sending unit case. When the float arm on the

Figure 15-5 A gauge tester.

fuel gauge sending unit is moved from the low fuel to the high fuel position, the ohmmeter should read the specified ohms resistance. If the sending unit does not provide the specified resistance, replace the sending unit. Thermistor-type sending units may be tested with an ohmmeter. With the

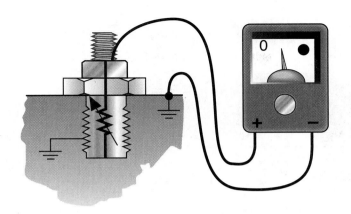

Figure 15-6 Testing a thermistor-type sending unit.

ohmmeter leads connected from the sending unit terminal to the case, the ohmmeter should indicate the specified resistance in relation to the sending unit temperature **(Figure 15-6)**.

> **Tech Tip** *Most fuel tank sending units have a wire connected from the sending unit case to a ground on the chassis. If this ground wire is damaged, corroded, or has an open circuit, the fuel gauge may be inoperative or erratic. With the ignition switch off, use an ohmmeter to test this wire. It should have very low resistance.*

If the temperature gauge sending unit must be replaced, always drain some of the coolant from the cooling system before loosening this unit. If the

> **Tech Tip** *A grounded or open wire between the fuel gauge and the sending unit causes different gauge readings depending on the type of fuel gauge. With a bimetallic-type fuel gauge, a grounded sending unit wire provides a continually high gauge reading and an open circuit in this wire provides a continually low gauge reading. Some balancing coil fuel gauges provide the same reaction to a grounded or open sending unit wire. Other balancing coil fuel gauges provide a high reading with an open sending unit wire and a low reading with a grounded sending unit wire.*

engine is at normal operating temperature, use extreme caution because the cooling system is pressurized. Test to see if the system is pressurized by squeezing one of the radiator hoses; if it is hard, do not open the system. A system can sometimes be cooled off and the pressure dropped by spraying cold water on the radiator. If a gauge must be replaced, follow the same procedure and precautions that were explained previously regarding oil pressure indicator light service.

Voltmeter and Ammeter Maintenance, Diagnosis, and Service

Voltmeters and ammeters usually do not require any maintenance or service. However, you should inspect the ammeter or voltmeter operation to detect problems in other circuits. For example, if you have just tested the battery and found it to be fully charged but the ammeter indicates a very high charging rate, the alternator voltage may be higher than specified. This condition results in excessive gassing of water from the battery and possibly burned out lights and electrical accessories. This problem should be verified by testing the charging system output voltage. If this voltage is higher than specified, the customer should be advised before proceeding with the repairs. When this problem is detected before the battery and electrical accessories are damaged, the customer can save some money and possible inconvenience.

Speedometer and Odometer Maintenance, Diagnosis, and Service

Speedometers and odometers require a minimum amount of maintenance. The speedometer cable may require periodic lubrication. A dry speedometer cable provides a rasping or chirping sound, especially in cold weather. A dry cable may also cause an erratic speedometer reading. Disconnect the cable from the back of the speedometer and remove the inner cable. When the inner cable cannot be removed by pulling it from the upper end of the cable, it may be necessary to remove the cable retainer from the transmission or transaxle. Inspect the inner cable for kinks and damage. Lubricate the cable with an approved lubricant and reinstall it in the outer casing. If the

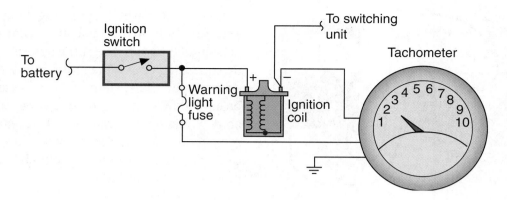

Figure 15-7 A tachometer circuit.

speedometer and odometer are inoperative, the inner cable or the drive gear is likely broken. If the speedometer is working properly but the odometer is inoperative, it must be replaced.

Did You Know? Federal and state laws prevent tampering with odometer readings. If an odometer must be replaced, it must be set to the same mileage as the defective odometer or a door sticker must be installed indicating the reading on the defective odometer.

Tachometer Maintenance, Diagnosis, and Service

If a tachometer is inoperative, inaccurate, or erratic, test the fuse and all the wires connected to the tachometer (**Figure 15-7**). Be sure all these wires are in satisfactory condition. There is no service on a tachometer if the fuse and all the wires are satisfactory; replace the tachometer.

NOTE: A tachometer may be called a tach.

Summary

- If the oil pressure indicator light is on with the engine running, the oil level and oil pressure may be excessively low or the oil may be contaminated. The oil may also have the wrong viscosity or the oil pressure switch may be defective.

- If the oil pressure indicator light does not come on with the ignition switch on and the engine not running, the bulb may be burned out, the oil pressure switch may be defective, the electrical circuit may be open, or there is no voltage supply to the light.

- A test gauge may be installed in place of the oil pressure switch to test oil pressure.

- Before attempting to remove an instrument panel, always disconnect the battery negative cable and wait for the time period specified by the vehicle manufacturer.

- If the charge indicator light is on with the engine running, the belt tension and alternator output should be tested.

- If the charge indicator bulb is not on with the ignition on and the engine not running, the bulb may be burned out, the alternator field circuit or external wiring to the bulb may have an open circuit, or the voltage supply to bulb may be zero. With the engine running, a glowing charge indicator light may be caused by high resistance in the charging circuit.

- If the red brake warning light is on with the engine running, there may be unequal pressure between the halves of the master cylinder, the parking brake may be on, or the ABS computer may have illuminated this light. Gauges may be tested with a gauge tester containing a variable resistor.

■ Sending units containing a variable resistor or a thermistor may be tested with an ohmmeter.

■ A dry speedometer cable causes a rasping noise, especially when cold.

Review Questions

1. The oil indicator light is on with the engine running. Technician A says the oil pressure switch may be defective. Technician B says the oil may be contaminated with gasoline. Who is correct?

   A. Technician A

   B. Technician B

   C. Both Technician A and Technician B

   D. Neither Technician A nor Technician B

2. Technician A says a test gauge may be installed in place of the oil pressure switch to test engine oil pressure. Technician B says before removing an instrument panel, always disconnect the battery positive cable. Who is correct?

   A. Technician A

   B. Technician B

   C. Both Technician A and Technician B

   D. Neither Technician A nor Technician B

3. The charge indicator light is not on with the ignition switch on and the engine not running. Technician A says the alternator field circuit may be open. Technician B says the fuse link in the alternator battery wire may have an open circuit. Who is correct?

   A. Technician A

   B. Technician B

   C. Both Technician A and Technician B

   D. Neither Technician A nor Technician B

4. All of the gauges read lower than normal on a vehicle with an instrument voltage limiter. Technician A says the voltage limiter may not be grounded on the instrument panel. Technician B says the wire to the fuel gauge sending unit may be grounded. Who is correct?

   A. Technician A

   B. Technician B

   C. Both Technician A and Technician B

   D. Neither Technician A nor Technician B

5. An oil indicator light does not come on when the ignition switch is turned on and the engine is not running. When the wire connected to the oil sending unit is grounded with the ignition switch on, the oil indicator light is illuminated. The most likely cause of this problem is:

   A. an open circuit in the oil sending unit.

   B. an open circuit in the wire between the oil indicator light and the oil sending unit.

   C. an open circuit between the ignition switch and the oil indicator light.

   D. an open circuit in the indicator light fuse.

6. All of these defects may be a cause of very low engine oil pressure *except*:

   A. worn main bearings.

   B. worn camshaft bearings.

   C. an air leak in the oil pump pickup.

   D. a leaking rear main bearing seal.

7. When diagnosing an instrument panel containing an instrument voltage limiter, all the gauges are tested and found to be defective. Technician A says the instrument voltage limiter may be defective. Technician B says the ignition switch may have been turned on with the instrument panel not grounded. Who is correct?

   A. Technician A

   B. Technician B

   C. Both Technician A and Technician B

   D. Neither Technician A nor Technician B

8. A charge indicator light is illuminated dimly with the engine running. The most likely cause of the problem is:

   A. excessive resistance in the charge indicator bulb.

   B. excessive resistance between the ignition switch and the charge indicator bulb.

   C. high resistance between the alternator and the battery ground.

   D. a defective voltage regulator, resulting in low alternator voltage.

9. With the ignition switch on, a typical voltage output from an instrument voltage limiter is a pulsating:

   A. 1 volt to 2 volts.

   B. 3 volts to 3.5 volts.

   C. 5 volts to 7 volts.

   D. 9 volts to 12 volts.

10. When a fuel gauge sending unit is removed from the fuel tank, it should be tested with:

    A. a voltmeter.

    B. an ohmmeter.

    C. an ammeter.

    D. a graphing voltmeter.

11. On a vehicle with ABS, the red brake warning light is illuminated with the engine running. List three causes of this problem.

    A. _____

    B. _____

    C. _____

12. A gauge tester contains a _____

    _____ .

13. A fuel gauge sending unit may be tested with a(n) _____ .

14. Federal and state laws prohibit_____ with odometer readings.

15. Explain the procedure to diagnose an oil pressure indicator light that is on with the engine running.

16. Describe the cause of a charge indicator light glowing with the engine running.

17. Explain the results of a grounded fuel gauge sending unit wire on a vehicle with a bimetallic-type fuel gauge.

18. Describe the results of excessively high charging system voltage.

CHAPTER

16 Basic Wiring Repair

Learning Objectives

After you have read, studied, and practiced the contents of this unit, you should be able to:

- Identify and test for short and open circuits.
- Identify the cause of the wiring failure.
- List steps to prevent future wiring damage.
- Explain advantages of soldering.
- Explain advantages of using solderless connectors.

Key Terms

Excessive resistance	Open circuit	Short circuit
Intermittent faults	Rosin core solder	Splicing

INTRODUCTION

Diagnosis and repair of a vehicle electrical system is a key component of a technician's job. As vehicle systems evolve and become more complex, part of this changing technology is that mechanical systems are being replaced by electrical components. These electrical components are far more efficient and use less energy than their mechanical counterpart. The ability to repair the electrical wiring and its components becomes more necessary than ever.

TYPES OF ELECTRICAL FAULTS

There are three main classifications of electrical system failures. The first are called **open circuits**.

An open circuit is a disconnection in the circuit; a broken or cut wire is an example of an open circuit. Current cannot flow through an open circuit. If an ohms test is performed upon the open circuit, it will read unlimited resistance. When a voltage drop is performed on the circuit with an open circuit, it will read full circuit voltage even though the system does not work.

Short circuits are the second main electrical wiring fault. A short circuit is a condition where the power side or the ground side bypasses the load or controls. A short is usually indicated when a fuse blows immediately when replaced or a device is turned on, or when a device will not turn off. Switches that will not turn off a device are shorted. In the case where there is not a fuse to protect the circuit or the short bypasses it, melted

Figure 16-1 A short circuit can cause wire to become dangerously hot and melt.

or burnt wiring often results **(Figure 16-1)**. Short circuits on the power side of the electrical system are often the most serious of electrical faults due to the possibility of a fire.

Excessive resistance is the last group of faults. A system with excessive resistance will run slower or bulbs will hardly light and be very dim. This is a condition where there is something in the circuit that is using part of the current and voltage that should go to the rest of the devices. Poor connectors, corrosion, and faulty parts all can add to much resistance into a circuit. A visual inspection and a voltage drop test are the best ways to find excessive resistance.

FINDING WIRING FAULTS

Damage to the wiring is most often related to causes from an external source; more than one type of damage can occur at a time, so when the problem is found, a good inspection of the cause of damage is important to prevent future problems. When repairing a vehicle that has electrical system faults, a good visual inspection in the particular area of concern often shows the problem. It is not uncommon to find electrical faults or their general area just by questioning the vehicle owner. Questions such as "When did the trouble start?" or "Was any other work performed at the time the trouble began?" can be valuable time-saving clues.

When a basic visual inspection does not reveal the source of the electrical problem, you will need to consult an automotive diagnostic manual. This will help pinpoint the cause of the concern, such as a component or its wiring. If the wiring is the cause of the problem, you will need to use a set of wiring schematics that match the vehicle in question to help trace the wire in the vehicle **(Figure 16-2)**. Wiring schematics contain diagrams that show the routing and location of the wiring systems. The wiring color-coding **(Figure 16-3)**, connectors, and ground locations are shown to save time and ease diagnosis **(Figure 16-4)**.

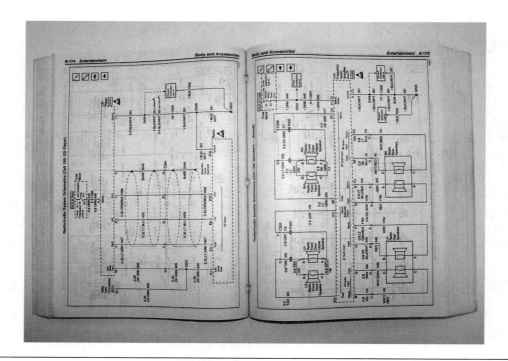

Figure 16-2 A quality electrical diagnostic manual.

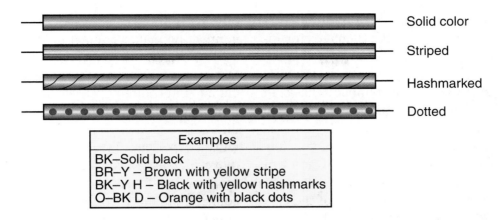

Examples
BK–Solid black
BR–Y – Brown with yellow stripe
BK–Y H – Black with yellow hashmarks
O–BK D – Orange with black dots

Figure 16-3 Manufacturers use various color schemes to mark wiring circuits.

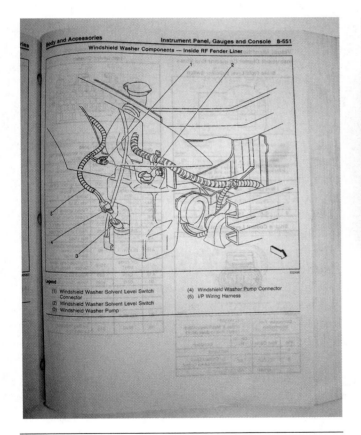

Figure 16-4 Using an electrical locator manual to find connectors.

Figure 16-5 Wiring can be damaged when it is allowed to move.

CAUSES OF WIRING DAMAGE

Friction Damage

Rubbing or friction damage causes the wiring harness to wear through. This can be caused by the wiring coming into contact with moving parts, such as engine belts or by long-term vibration in places that the harness is not adequately secured. This can result in either open or short circuits. Occasionally both conditions can exist in the same wiring harness at different times; for instance a wire rubs through causing it to open, but since it is not secured properly it moves around and occasionally it touches something else shorting out and blowing a fuse in the process. Once the source of the friction damage is found the wiring must be adequately secured or moved to a safe location to prevent future damage **(Figure 16-5)**.

Corrosion

Corrosion damage in the wiring circuit will cause high resistance or an open circuit. Wires or connections that are corroded often have a scaly discoloration. The color of the discoloration gives an indication of the cause of the damage; for instance, white corrosion if usually caused by battery acid. Green corrosion is often caused by the wire being exposed to moisture on a continuous basis. In most cases the initial cause is often an opening in the insulation that protects the wire. When corroded wires or connectors are the problem, they will need to be cut out and replaced. Corrosion will travel inside the insulation so you will need to cut the wire back far enough to remove all of the corroded section. Technicians can cause future problems by not properly resealing the insulation when performing electrical tests that require puncturing the harness or not sealing wiring repairs.

Flexing

Certain portions of a vehicle's wiring system are bent or flexed back and forth on a regular basis. The wiring that spans from the vehicle's body to the doors is one example. These sections, over time, can break and cause an open circuit. Particularly the driver's door is subject to a lot of wear by flexing and bending **(Figure 16-6)**. When repairing the broken sections that were damaged by flexing it is important to splice in additional wire to avoid shortening the existing wiring harness.

Loose Connections

Connections that become loose tend to make poor connections at best and often cause open

Figure 16-6 Wiring that is flexed repeatedly can break over time.

circuits in the wiring system. Occasionally if the connection makes contact intermittently, a device will work only on those occasions when the contact is made. Problems that appear occasionally are called "intermittents" describing when they function. **Intermittent faults** can often be the most difficult to locate. To locate an intermittent problem an experienced technician will flex and lightly move the suspected wire and connectors to help pinpoint the loose or broken one. Once a loose connection is located the internal connections are either gently bent to tighten them or it is replaced. Ground wires are often bolted directly to the vehicle body and over time will occasionally vibrate loose; before retightening it is best to lightly sand them to assure good metal-to-metal contact.

Heat Damage

Heat damage is often caused by two major problems: overloaded wiring and heat from an outside source. When copper wire is overheated the most obvious indicator is melted insulation. However, insulation doesn't always melt, and if brittle off-colored wire is encountered during repairs, suspect that the wiring has been overheated. Once the insulation is melted the circuit can then short out if it touches something of the opposite polarity. Putting the wrong size fuse in can result in overloading a circuit and ruining the wiring. Wire that has become brittle or discolored from being overheated must be replaced, since it will no longer be able to carry its rated current load safely.

Exhaust manifolds are an example of an outside heat source that will ruin wire. This is often the result of poor routing or not being held in place properly.

METHODS OF REPAIR
Soldering

Repairing a wiring harness by hot soldering and covering the repair with shrink tubing is the most preferred method of repair. This type of repair is the strongest and with the heat shrink tubing properly installed is completely sealed to moisture and air, preventing future corrosion problems. Soldering is a skill that takes a little practice to become proficient at **(Figures 16-7** through **16-13)**.

Figure 16-7 Cutting out the damaged wire.

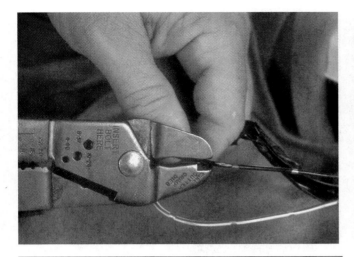

Figure 16-10 Soldering the wires together.

Figure 16-8 Stripping back enough wire to solder.

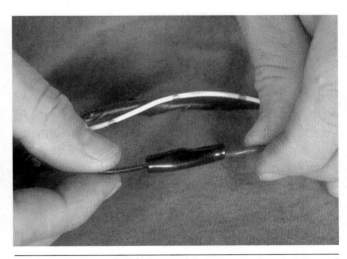

Figure 16-11 Slide the tubing over the soldered joint.

Figure 16-9 Slide the heat shrink tubing over the wire to be repaired.

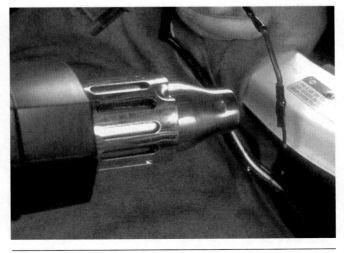

Figure 16-12 Use a heat gun to shrink the tubing.

Figure 16-13 Complete the repair by taping the loom back and securing it.

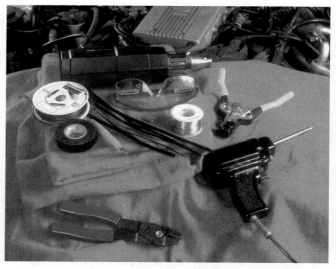

Figure 16-15 General repair tools to solder wire include: soldering iron, heat gun, wire strippers, safety glasses, solder, and electrical tape.

For repair of electrical wiring systems only **rosin core solder** can be used **(Figure 16-14)**. All core type solders have a hollow center in the wire filled with a flux that cleans and helps the solder bond with the material being soldered. Once the wire to be soldered has been stripped, the heat shrink tubing is slipped over one end, the wires are twisted together, and then soldered. After the wire cools the heat shrink tubing is slid over the joint and a heat gun is used to heat and shrink the tubing onto the wire **(Figure 16-15)**.

Splicing is repairing the wire by adding additional wire into the electrical system, often to replace damaged wire that has been cut out **(Figure 16-16)**. The basic repair is the same as a normal solder repair; however, care must be taken

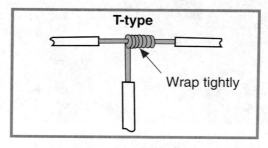

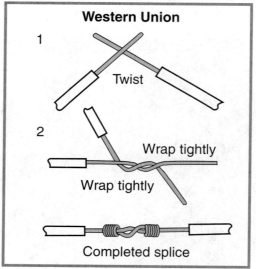

Figure 16-16 Splicing wire by twisting it before soldering.

Figure 16-14 A roll of rosin core solder.

to ensure that enough wire has been added and that the wire is of the proper gauge to handle the rated current load for the repaired section. If possible try to use wire of the same color, for instance if the repaired wire is red try to use a red wire for repairs. Some wire colors are not commonly available, such as ones with stripes or slashes; in these cases it is a good idea to use some tape to label the original color in case of future trouble to help avoid confusion.

Solderless Repair

The best method of repair is always to solder a wiring system; however, it is not always possible or safe to do so. For example, if you are repairing the wiring under the dash on a vehicle and there is insufficient room to safely maneuver a hot soldering iron, then solderless connectors are the best option possible **(Figure 16-17)**.

Crimp on style connectors are good for those areas that will not be exposed to moisture that would corrode the wire repair, and when crimped properly are equally as strong **(Figure 16-18)**.

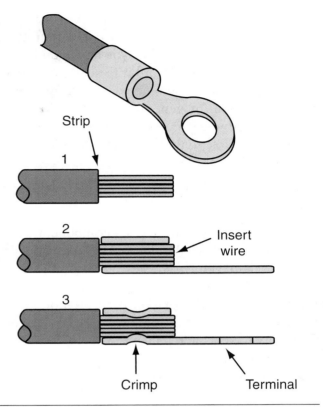

Figure 16-18 Steps to using solderless connectors.

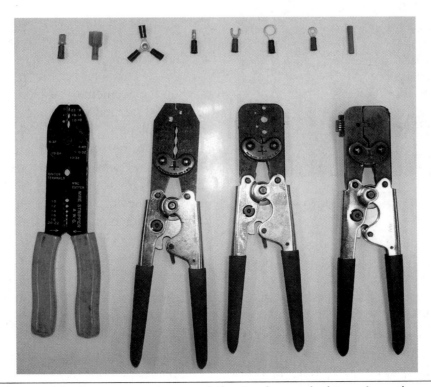

Figure 16-17 Various solderless connectors and the tools needed to crimp them.

When the soldering iron cannot be used and the wire will be exposed to moisture, certain types of crimp on connectors allow you to slip on a piece of shrink tubing first in order to seal the connection. These types of connectors were originally cost prohibitive for general use but as they have become more widespread in use, the cost has become comparable to the extra labor to solder.

Connector Repair

Electrical connectors become damaged from heat, mishandling, and age. If a component shorts out and over heats the electrical connector can become damaged from heat, often melting it **(Figure 16-19)**. Both the major vehicle manufacturers and some aftermarket parts suppliers have replacement connectors and their plastic retainers. Generally the plastic retainer will be pre-wired and the technician will cut the old one off and splice the new one on **(Figure 16-20)**. Mishandling often

Figure 16-20 A new wiring connector.

Figure 16-19 A wiring connector damaged by excessive current flow.

results in the retaining clip being broken off. If the part is not replaced, it has the risk of working loose later and causing problems.

Age often plays a factor in the connectors needing replacement or repair. Repeated exposure to heat, moisture, and vibration over time can cause the small connectors to become loose and the plastic retainers to become brittle to the touch. When the connector pins become loose they can often be "tightened" by using a small pick and slightly bending the connector to make it tight. If they cannot be tightened, then replacement is the only option.

Summary

- The three main types of electrical faults are short circuits, open circuits, and excessive resistance.

- A voltage drop test and an ohms test are the most common electrical tests.

- Technicians should question the vehicle owner to help pinpoint the trouble faster.

- A good electrical circuit manual should be used to diagnose electrical system faults.

- Wiring damage can be caused by friction, corrosion, flexing, and excessive heat.

- Soldering and using solderless connectors are the main methods of repair to a wiring system.

Review Questions

1. The left high beam headlight does not work. Technician A says the headlight may need to be replaced. Technician B says there may be an open circuit in the wire between the high beam and the light. Who is correct?

   A. Technician A

   B. Technician B

   C. Both Technician A and Technician B

   D. Neither Technician A nor Technician B

2. Technician A says a short circuit can cause a fuse to blow. Technician B says that an open circuit will cause a fuse to blow. Who is correct?

   A. Technician A

   B. Technician B

   C. Both Technician A and Technician B

   D. Neither Technician A nor Technician B

3. One headlight is always dimmer than the other and the cause is determined to be high resistance from a corroded wire. Technician A says the corrosion is possibly from moisture that is often sprayed on the wires when it's raining. Technician B says all the vehicles wiring will need to be replaced. Who is correct?

   A. Technician A

   B. Technician B

   C. Both Technician A and Technician B

   D. Neither Technician A nor Technician B

4. While replacing a broken connector by splicing on a new one, Technician A says the heat shrink tubing should be put on before the wires are soldered. Technician B says that solderless connectors can be used since the heat shrink tubing will seal out moisture. Who is correct?

   A. Technician A

   B. Technician B

   C. Both Technician A and Technician B

   D. Neither Technician A nor Technician B

5. An electrical fault that causes a circuit's fuse to blow is traced to a section of wire that is missing the insulation. Technician A says that friction from allowing the wires to rub wore off the insulation. Technician B says that heat caused by an open circuit melted the insulation off. Who is right?

   A. Technician A

   B. Technician B

   C. Both Technician A and Technician B

   D. Neither Technician A nor Technician B

6. When testing a circuit for faults with an ohm meter a very high reading is noted. Technician A says that a very high resistance reading will indicate a short circuit. Technician B says that high resistance readings are normal in wiring. Who is right?

   A. Technician A

   B. Technician B

   C. Both Technician A and Technician B

   D. Neither Technician A nor Technician B

7. While repairing a section of wire, Technician A says that it is important to keep the wire the same length after the repair. Technician B says that one should try to use wire of the same size and color. Who is correct?

   A. Technician A

   B. Technician B

   C. Both Technician A and Technician B

   D. Neither Technician A nor Technician B

8. A section of wire will need to be replaced because it has melted. What is the most likely cause of the failure?

   A. Excessive movement.

   B. Repeated exposure to moisture.

   C. Allowing it to rub against a moving belt.

   D. A defective part that shorted.

9. If a technician is unable to use wire with the same color insulation, it is always a good idea to _____.

10. A section of wire has become corroded. List three causes of this problem.

   A. _____

   B. _____

   C. _____

11. Why is it important to use wire of the same gauge in a repair?

12. A component that sometimes works and at other times does not, is often called a(n) _____ problem.

13. Repeatedly bending a wire will cause what type of fault?

17 Battery and Starting Systems

Learning Objectives

After you have read, studied, and practiced the contents of this unit, you should be able to:

- Describe battery design, including battery plate and cell groups.
- Explain battery operation during the charge and discharge cycles.
- Explain two common battery ratings.
- Describe basic starting motor armature and field coil design.
- Describe the operation of a starter solenoid.
- Explain the operation of an overrunning clutch starter drive while engaging and disengaging.

Key Terms

Battery plates

Cell group

Cold cranking ampere (CCA)

Electrolyte

Solenoid

Starter drive

INTRODUCTION

You must understand battery operation to comprehend how the battery operates in the charging circuit. For example, if the battery is gassing excessively, the charging voltage must be too high. You also need to understand that water is vented from the battery during the normal charging process. Therefore, only water should be added to the battery cells. You also need to understand starting motor design and operation so you can accurately diagnose starting motor and system defects.

BATTERY DESIGN

The battery case holds and protects the battery components and **electrolyte (Figure 17-1)**. Separating walls in the case form a separate reservoir for each cell. Battery cases have sediment spaces in the bottom of the case, which provides spaces for material shed off the plates. If this space were not provided, the material that is shed would make contact between the plates and short them together.

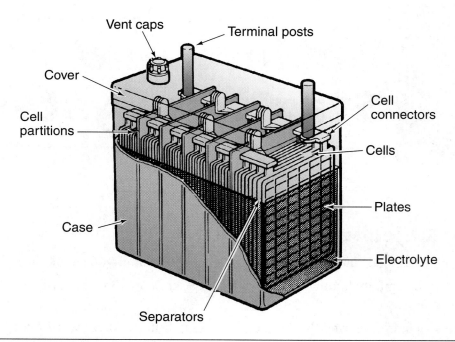

Figure 17-1 Battery internal design.

Cover

The battery cover is permanently sealed to the top of the case. The cover has openings for the terminal posts and vent holes for the venting of gases. Some batteries have the terminals extending through the side of the case and other batteries have terminals in the sides and top. Serviceable batteries have removable vent caps that may be removed to add water or test the electrolyte. These vent caps may be individual, strip type, or box type.

Plates

Battery plates contain a grid or coarse screen. In some batteries, the plate grids are made from lead mixed with antimony, which stiffens the lead. **Battery plates** are a coarse grid on which the active plate material is pasted. Many batteries manufactured today have plate grids made from lead and calcium. In some plate grids, the horizontal and vertical grids are positioned at right angles to each other. Other plate grids have vertical support bars mounted at an angle to the other vertical grids. The active plate materials are pasted on the grids. The material on the positive grids is lead peroxide and the negative plate material is sponge lead. A tab on the top of each plate grid allows the plate to be connected to the cell connector.

Cell Group

A **cell group** contains a number of alternately spaced negative and positive plates. A negative plate is positioned on the outside of each cell group; thus, each cell group has an odd number of plates. For example, a cell group may have five positive plates and six negative plates. Thin porous separators are positioned between the negative and positive plates. Separators are positioned between each pair of battery plates to keep the plates from touching each other. Separators are made from fiberglass sheets or plastic envelopes that fit between or over the plates. The separators prevent the plates from touching each other and shorting together electrically. Cell connectors are lead burned to the positive and negative plate tabs in each cell group. These cell connectors also extend through the partitions between the cells and the negative plates and are connected to the positive plates in the next cell. The negative plates in one end cell and the positive plates in the opposite end cell are connected to terminal posts that extend through the top or sides of the battery.

Terminal Posts

On top-terminal batteries, large round terminal posts extend through the top of the battery. The positive post is larger than the negative post. These terminal posts are sealed into the battery cover to prevent electrolyte leakage around the posts. POS or NEG is stamped on the battery cover beside the proper terminal post. The battery cable ends are clamped to the terminal posts. On side-terminal batteries, the battery terminals are threaded and the cables are bolted to the terminals.

Electrolyte

The electrolyte is a mixture of approximately 64 percent water (H_2O) and 36 percent sulfuric acid (H_2SO_4). The electrolyte reacts with the plate materials to produce voltage. The electrolyte must contain enough sulfuric acid so it does not freeze in extremely cold temperatures. If the battery contained a higher percentage of sulfuric acid, this acid would attach and deteriorate the plate grids too quickly.

SAFETY TIP *Sulfuric acid is a very strong corrosive liquid. It is damaging to human eyes and skin. Always wear face and eye protection and protective gloves and clothing when handling batteries or electrolyte. If electrolyte contacts your skin or eyes, immediately flush with clean water and obtain medical attention.*

SAFETY TIP *Battery electrolyte is very damaging to vehicle paint surfaces, upholstery, and clothing. Never allow electrolyte to contact any of these items.*

BATTERY OPERATION

If one negative plate and one positive plate are placed in an electrolyte solution, a chemical reaction takes place that causes about 2.1 volts difference between the plates. Because a 12-volt battery contains six cells, the total voltage of a fully

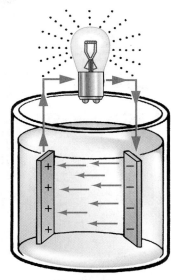

Discharging

Figure 17-2 Current flows from one battery plate to the other when a conductor with some resistance is connected between the plates.

charged battery is 12.6 volts. If an electrical load, such as a lightbulb, is connected to the two battery plates, the higher voltage on one plate forces current through the bulb to the opposite plate with the lower voltage at a controlled rate **(Figure 17-2)**. This action is called *battery discharging*. A battery cell with two plates will not supply much current flow and so plates are connected in parallel within the cell case to increase total current capability.

If a voltage source, such as an alternator, is connected to the battery plates, the alternator forces current flow into one plate and out of the other battery plate **(Figure 17-3)**. This action is called *battery charging*. The voltage of the alternator must be higher than the battery voltage to allow the alternator to force current through the battery. In a fully charged battery, the positive plate material is lead peroxide (PbO_2) and the negative plate material is lead (Pb) **(Figure 17-4)**. When an electrical load is connected across the battery terminals, the voltage difference between the plate materials forces current through the electrical load. As the battery discharges, the H_2SO_4 breaks up in the electrolyte and the sulfate (SO_4) goes to

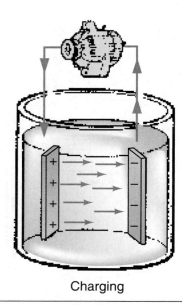

Charging

Figure 17-3 When the engine is running, current flows from the alternator through the battery plates.

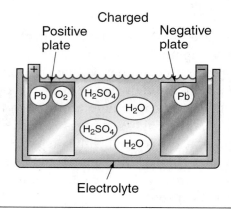

Charged

Positive plate   Negative plate

Electrolyte

Figure 17-4 Plate materials and electrolyte content in a fully charged battery.

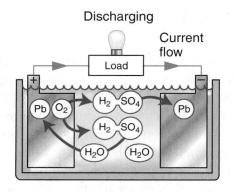

Discharging

Current flow

Load

Figure 17-5 Chemical action in a battery while discharging.

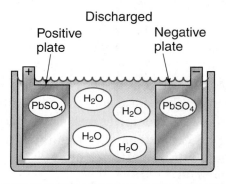

Discharged

Positive plate   Negative plate

Figure 17-6 Plate materials and electrolyte content in a discharged battery.

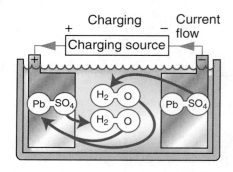

Charging   Current flow

Charging source

Figure 17-7 Chemical action in a battery while charging.

both plates, where it joins with the Pb to form lead sulfate ($PbSO_4$) on both plates. The O_2 on the positive plates joins with the hydrogen (H) in the electrolyte to form (H_2O) **(Figure 17-5)**. As the battery discharges, the percentage of water increases in the electrolyte until it is fully discharged. In a fully discharged battery, both sets of plates are coated with $PbSO_4$ and the electrolyte is mostly of H_2O with very little acid **(Figure 17-6)**.

When a charging source, such as the alternator, or a battery charger is charging the battery, the SO_4 comes off the plates and joins with the H in the

electrolyte to form H_2SO_4 **(Figure 17-7)**. The H_2O breaks up and the O goes to the positive plates, where it joins with the Pb to form lead peroxide (PbO_2).

When a battery is charging, hydrogen gas escapes at the negative plates and oxygen gas is given off at the positive plates. When combined, these two gases form water (H_2O). Because of the possibility

Tech Tip *Do not let a battery over-heat during charging. The plates will become damaged and shorten the battery's useful life. If the battery case is becoming hot, turn off the charger and let it cool down, then continue charging at a lower rate.*

that not all the hydrogen gas will have recombined with the oxygen to form water, caution should be taken to prevent sparks near a charging battery. The SO_4 is always on the plates or in the electrolyte. Because H_2O is the only liquid gassed from a battery, this is the only liquid that should be added to a battery.

Low-Maintenance Batteries

Low-maintenance batteries are designed to minimize heat and water loss. The filler caps may be removed to add water to the cells. In many batteries, the electrolyte level should be at a split ring round indicator below the filler cap in each cell. The water level must be above the battery plates. The average interval for adding water to a battery is approximately 15,000 miles. If the battery requires frequent addition of water, the charging system voltage may be too high.

Maintenance-Free Batteries

Maintenance-free batteries are designed to reduce internal heat and water loss. The cells in these batteries are vented but the filler caps cannot be removed and it is not possible to add water to the battery. Most maintenance free batteries have a built-in hydrometer in the top of the battery to indicate battery state of charge. If this hydrometer indicates a green dot, the battery is sufficiently charged for test purposes. When the hydrometer appears dark, the battery charge is below 65 percent, and a clear hydrometer indicates a low electrolyte level **(Figure 17-8)**.

BATTERY RATINGS

The **cold cranking ampere (CCA)** rating indicates the amperes that a battery will deliver at 0°F

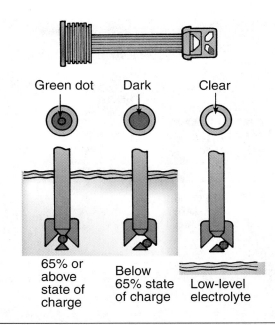

Green dot   Dark   Clear

65% or above state of charge — Below 65% state of charge — Low-level electrolyte

Figure 17-8 A hydrometer in the cover of a maintenance free battery.

(−18°C) for 30 seconds while maintaining a voltage above 1.2 volts per cell or 7.2 volts for the complete battery. The CCA rating for most automotive batteries is between 350 and 800 amperes **(Figure 17-9)**. CCA is the most common battery rating used.

Reserve Capacity Rating

The reserve capacity rating is the length of time in minutes that a fully charged battery at 80°F (27°C) will deliver 25 amperes with the voltage remaining above 1.75 volts per cell or 10.5 volts for the complete battery. Many reserve capacity ratings are between 55 and 125 minutes.

STARTING MOTOR ELECTROMAGNETIC PRINCIPLES

Starting motors are designed to have very low resistance and draw a very high current flow, which produces a high torque for short time periods. The field coils in a starting motor are made of heavy copper wire and are wound on steel pole shoes that are bolted to the starter case. One end of the field windings is connected to the

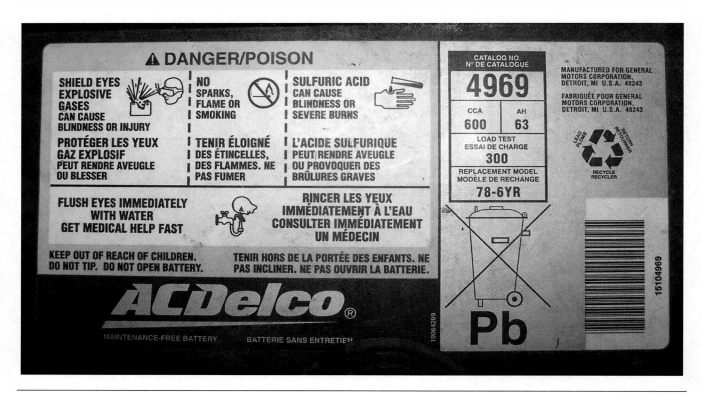

Figure 17-9 A battery rating label.

positive battery terminal and the other end of these windings is connected to an insulated brush in the starting motor. A copper commutator bar is soldered to each end of the armature winding. This winding is positioned between the pole shoes. The commutator bars are insulated from each other and the insulated brush contacts one commutator bar. A ground brush completes the circuit from the other commutator bar to ground. Current in the starting circuit flows from the positive battery terminal through the field coils and armature winding and returns through the ground brush to the negative battery terminal. Because the field coils are connected in series with each other and with the armature winding, the same amount of current must flow through the field coils and the armature winding. A high current flows through the field coils and armature winding because these heavy windings have very low resistance. The high current flow through the armature winding creates strong magnetic fields around the sides of this winding. Since the current flows in opposite directions in the two sides of the armature winding, the magnetic fields surrounding

the sides of this winding flow in opposite directions. The high current flow through the field coils creates a strong magnetic field between the pole shoes and the armature winding is positioned in this magnetic field **(Figure 17-10)**. The magnetic field around the upper side of the armature winding is moving in a counterclockwise direction.

On the top side of this winding, the magnetic field around the armature winding is moving in the same direction as the field between the pole shoes. Therefore, these magnetic fields join together and a very strong magnetic force is created above this side of the armature winding **(Figure 17-11)**. Below the upper side of the armature winding, the magnetic field around the armature winding is moving in the opposite direction to that of the magnetic field between the pole shoes. Magnetic fields moving in opposite directions in the same space cancel each other. Therefore, a very weak magnetic force is created below the upper side of the armature winding.

The magnetic field around the lower side of the armature winding is moving in a clockwise

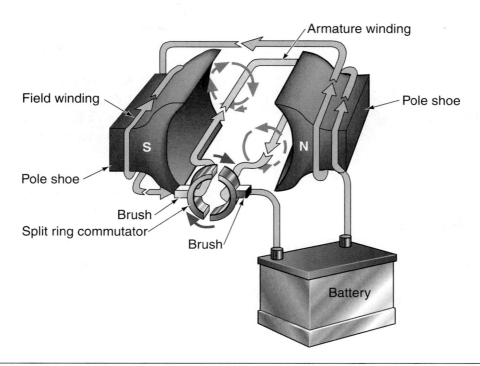

Figure 17-10 Starting motor field coils and armature winding.

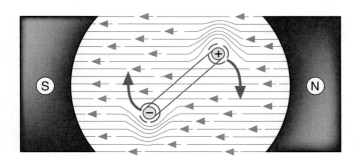

Figure 17-11 Reaction between the magnetic field between the pole shoes and the magnetic fields around the armature windings causes armature rotation.

Starter Armature Design

An armature has many windings mounted in a laminated metal core that is pressed onto the shaft **(Figure 17-12)**. Each winding is insulated from the other windings and from the mounting slots in the metal core. The ends of the windings are soldered to the commutator bars that are insulated from each other and from the armature shaft. Some armature windings are always in the magnetic field between the pole shoes to provide continuous armature rotation.

Did You Know? *A laminated armature core magnetizes and de-magnetizes faster than a core manufactured from a single piece of iron.*

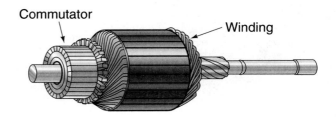

Figure 17-12 A starter armature.

direction. Because the magnetic field between the pole shoes is moving in the same direction as the magnetic field around the armature winding below this winding, a very strong magnetic force is created under the winding. The magnetic force above the lower side of the winding is cancelled because the magnetic lines of force in this area are moving in opposite directions. The strong magnetic forces above and below the sides of the armature winding cause the armature winding to rotate because this winding is mounted on a steel shaft and is supported on bushings.

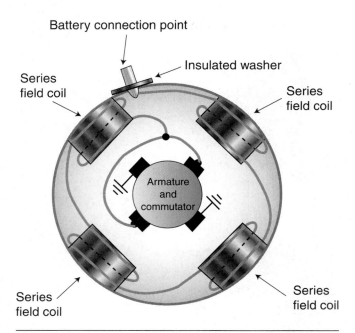

Figure 17-13 Series field coils.

Starter Field Coil Design

Some automotive starting motors have four field coils connected in series **(Figure 17-13)**. All four of these windings contain heavy copper wire with very low resistance. Other starting motors have three series field coils and one parallel or shunt field coil **(Figure 17-14)**. The shunt field coil has many turns of fine wire. While cranking the engine, the

shunt coil has the same magnetic strength as the series field coils. The shunt coil creates a strong magnetic field from a low current flow through many turns of wire, whereas the series field coils develop their strong magnetic fields from a high current flow through a few turns of wire. When the engine starts, the starter turns freely for a few seconds until the starter drive is pulled out of mesh. This is called an overrunning starter condition. Under this condition, the armature could rotate at a very high speed. When the starter overruns, the battery voltage increases because the high starter current load is removed from the battery. When this action occurs, the current flow through the shunt coil increases and strengthens the magnetic strength around this coil. The strong magnetic field of the shunt coil creates an induced opposing voltage in the armature windings that opposes the battery voltage and reduces the current flow through the series field coils and armature. This action reduces starter armature speed and protects the armature. If the armature rotates at excessive speed, the heavy armature windings may be thrown from their armature slots, resulting in severe armature damage. Some starting motors have permanent magnets in place of wound field coils **(Figure 17-15)**.

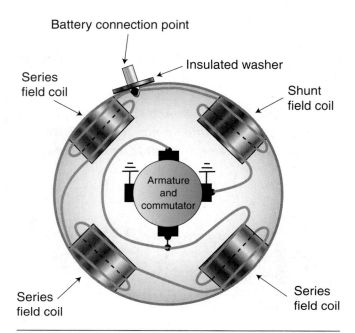

Figure 17-14 Series-shunt field coils.

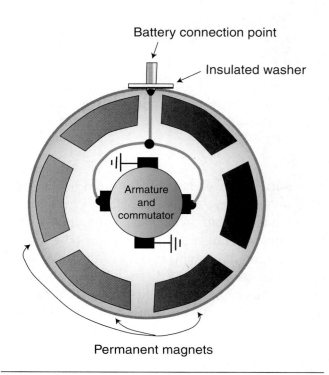

Figure 17-15 Permanent magnet field coils.

In this type of starting motor, electrical defects in the wound field coils are eliminated. A permanent magnet starting motor is also more compact compared to other starting motors. The permanent magnets are manufactured from an alloy of boron, neodymium, and iron. These permanent magnets have much greater strength than ordinary steel magnets. The permanent magnets are attached to the starting motor housing. These starters usually have more permanent magnets compared to the number of field coils in a conventional starter because no space is required for the windings. A permanent magnet motor operates in the same way as a conventional starting motor.

SOLENOIDS

Many starting motors have a solenoid that completes the electric circuit between the battery and starting motor. The solenoid also pulls the starter drive into mesh with the flywheel ring gear. The **solenoid** is an electro-mechanical device that moves the starter drive into mesh with the flywheel ring gear and opens and closes the electrical circuit

between the battery and the starting motor. The solenoid contains a heavy pull-in coil connected from the solenoid terminal to the main starting motor terminal. The finer hold-in coil is connected from the solenoid terminal to ground on the solenoid case. The solenoid terminal is connected to the start terminal on the ignition switch.

When the ignition switch is turned to the start position, current flows from this switch to the terminal on the starter solenoid. From this terminal, the current flows through a heavy pull winding to the main starting motor terminal. This current flows through the series field coils and the armature windings to ground, but there is not enough current to cause armature rotation. A small amount of current also flows through the hold-in winding to ground on the solenoid case. The combined magnetic fields of the two solenoid windings attracts the solenoid plunger and moves this plunger ahead (**Figure 17-16**). Forward plunger movement pulls the starter drive into mesh with the flywheel ring gear and further plunger movement pushes the solenoid disk against the two large solenoid terminals. When this contact is completed, a very high

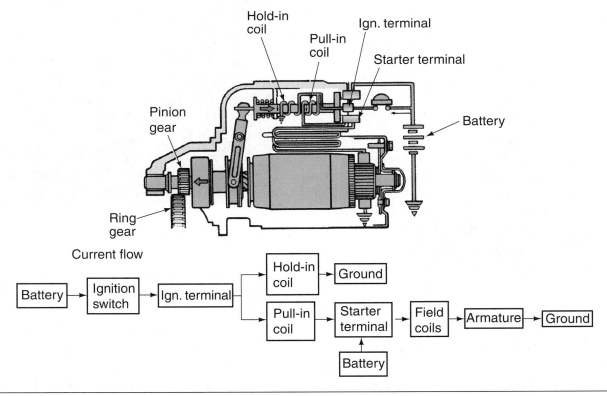

Figure 17-16 Solenoid current flow while engaging.

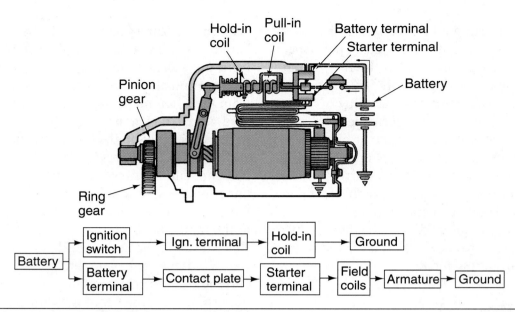

Figure 17-17 Solenoid current flow with starting motor engaged.

current flows from the battery positive terminal through the battery cable and solenoid disk and terminals into the starting motor. Under this condition, the starting motor begins cranking the engine. Once the solenoid plunger is pulled forward, current no longer flows through the pull-in winding because equal voltage is supplied through the solenoid disk to the other end of the pull-in winding (**Figure 17-17**). Under this condition, the hold-in coil magnetic field is strong enough to hold the plunger in the engaged position.

When the engine starts, the driver releases the ignition switch to the on position. Since current no longer flows through the ignition switch to the solenoid terminal there is no magnetic field around the hold-in coil the plunger return spring pushes the disk away from the solenoid terminals and current flow to the starting motor is stopped. The plunger return spring pulls the drive out of mesh with the flywheel ring gear and returns the plunger to its disengaged position.

STARTER RELAYS

Some starting motor systems have a starter relay that completes the electrical circuit between the battery and the starting motor (**Figure 17-18**). Many of these starting motors have a moveable

Figure 17-18 A starter relay.

pole shoe that pulls the starter drive into mesh with the flywheel ring gear (**Figure 17-19**). The magnetic switch contains a high-resistance winding. When the ignition switch is turned to the start position, current flows through the ignition switch to the starter relay terminal. Current then flows from this terminal through the magnetic switch winding and the neutral safety switch to

Figure 17-19 A movable pole shoe starter.

ground **(Figure 17-20)**. This current flow creates a magnetic field around the starter relay winding, which attracts the relay plunger. Plunger movement forces the disk against the relay terminals connected to the positive battery cable and the starting motor. A very high current now flows through the magnetic switch disk to the starting motor.

A shunt field coil and a holding coil surround the moveable pole shoe. The shunt field coil contains many turns of fine, high-resistance wire. The holding coil is grounded through a set of contacts

on top of the starting motor and this coil is also connected in series with one of the other field coils.

When the starter relay closes, most of the current flows through this winding and the contacts to ground because the holding coil is the path of least resistance. This high current flow creates a strong magnetic field and pulls the moveable pole shoe downward. A fork on the back of this pole shoe moves the starter drive into mesh with the flywheel ring gear. When the pole shoe is fully downward, it pushes the contacts open. Under this condition, current cannot flow through the holding coil and the contacts to ground. Current now flows through the series field coils and armature and the starter begins to crank the engine. The magnetic field around the shunt coil is strong enough to hold the moveable pole shoe downward. When the driver releases the ignition switch from the start to the on position, current flow no longer flows through the starter relay winding. A spring pushes the disk away from the relay terminals and a spring on the moveable pole shoe pushes the pole shoe upward and pulls the starter drive out of mesh with the flywheel ring gear. The neutral safety switch is mounted on the transmission linkage or in the transmission. The neutral safety switch is closed when the gear

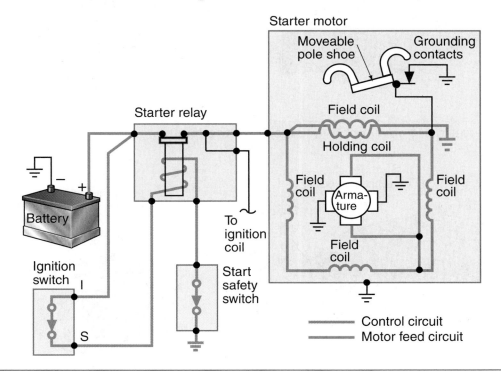

Figure 17-20 A starter relay circuit.

selector is in PARK or NEUTRAL, but this switch is open in other gear selector positions. Therefore, the gear selector must be in PARK or NEUTRAL to provide a ground on the starter relay winding and allow the starting motor to operate.

> **Tech Tip** *Some neutral safety switches are mounted on top of the steering column under the dash. Vehicles with a manual transmission usually have a clutch switch in place of the neutral safety switch.*

STARTER DRIVES

The overrunning clutch is the most common type of starter drive. The starter drive housing is mounted on armature shaft splines. The **starter drive** connects and disconnects the armature shaft and the flywheel ring gear. The starter drive housing contains four spring-loaded rollers in tapered grooves. The drive gear has a machined shoulder on the rear part of the gear. This machined shoulder is in the center of the rollers. When the drive is engaged with the flywheel ring gear and the armature begins to rotate, the rollers are turned so they move to the narrow end of the tapered grooves. Under this condition, rollers are jammed between the tapered grooves and the machined shoulder on the gear so the drive housing and the gear must rotate as a unit **(Figure 17-21)**. Therefore, the armature shaft rotates the solenoid housing and gear to crank the engine.

When the engine starts, the flywheel ring gear momentarily rotates the drive gear faster than the drive housing. This action rotates the rollers so they move into the wide end of the tapered grooves. Under this condition, the drive gear free wheels so it does not turn the armature at high speed and damage the armature **(Figure 17-22)**.

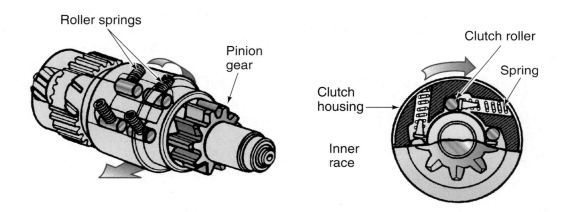

Figure 17-21 Starter drive operation while cranking the engine.

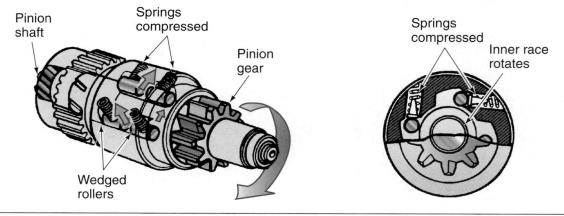

Figure 17-22 Starter drive operation while overrunning.

SAFETY TIP *The ratio between the starter drive gear and the flywheel ring gear is usually between 15:1 and 20:1. If this ratio is 20:1 and the drive is allowed to turn the armature momentarily when the engine starts, an engine speed of 1,000 rpm would rotate the armature at 20,000 rpm. This speed would throw the armature windings from their slots and destroy the starting motor.*

Tech Tip *Some engines have a smaller diameter flywheel. To compensate for this design, a gear reduction is provided between the starter armature and the drive.*

Summary

- The battery case contains a separate compartment for each battery cell.
- Battery plate grids are manufactured from lead and antimony or lead and calcium.
- The active material on positive battery plates is lead peroxide (PbO_2).
- The active material on negative battery plates is sponge lead (Pb).
- A battery cell always contains an odd number of plates.
- Separators made from fiberglass are positioned between each pair of battery plates.
- The battery cells are connected in series so the voltage of the cells adds together.
- Battery electrolyte contains about 64 percent water and 36 percent sulfuric acid.
- When a battery is charging, hydrogen is given off at the negative plates and oxygen is given off at the positive plates.
- The most common battery ratings are the cold cranking ampere (CCA) rating and the reserve capacity rating.

- A starting motor develops torque from the interaction of the magnetic field between the pole shoes and the magnetic fields around the armature windings.
- A starting motor has very low resistance and draws a very high current flow.
- A starting motor is designed to develop a very high torque for a short time period.
- Some starting motors have permanent magnet fields in place of field coil windings.
- A starter solenoid moves the starter drive into mesh with the flywheel ring gear and completes the electrical circuit between the battery and the starting motor.
- A starter relay is used in some starting circuits to complete the circuit between the battery and the starting motor.
- A starter drive connects and disconnects the armature with the flywheel ring gear.

Review Questions

1. Technician A says in a fully charged battery, the material on the positive plates is lead peroxide. Technician B says in a discharged battery, the material on the negative plates is sponge lead. Who is correct?

   A. Technician A

   B. Technician B

   C. Both Technician A and Technician B

   D. Neither Technician A nor Technician B

2. Technician A says the positive plates in one cell are connected to the positive plates in the next cell. Technician B says in a discharged battery the electrolyte contains 36 percent sulfuric acid. Who is correct?

   A. Technician A

   B. Technician B

   C. Both Technician A and Technician B

   D. Neither Technician A nor Technician B

3. Technician A says during battery discharge the oxygen (O_2) comes off the positive plates and joins with the hydrogen (H) in the electrolyte to form water (H_2O). Technician B says during battery discharge, both negative and positive plate materials are changed to lead sulfate ($PbSO_4$). Who is correct?

   A. Technician A

   B. Technician B

   C. Both Technician A and Technician B

   D. Neither Technician A nor Technician B

4. Technician A says the need to add water frequently to a battery is caused by low charging system voltage. Technician B says water may be added to a maintenance-free battery. Who is correct?

   A. Technician A

   B. Technician B

   C. Both Technician A and Technician B

   D. Neither Technician A nor Technician B

5. All of these statements about a lead-acid battery are true *except*:

   A. The active material on the negative plates is sponge lead.

   B. The active material on the positive plates is lead peroxide.

   C. Each battery cell group has an even number of plates.

   D. The electrolyte solution contains 64 percent water.

6. Separators in a battery cell:

   A. keep the plates from touching the bottom of the case.

   B. keep the plates from touching each other.

   C. prevent the electrolyte from touching the plates.

   D. prevent the cell connectors from touching each other.

7. The voltage of a fully charged lead-acid battery cell is:

   A. 1.5 volts.

   B. 1.8 volts.

   C. 2.0 volts.

   D. 2.1 volts.

8. While discussing lead-acid battery operation, Technician A says that when a battery is discharged, both plates are coated with lead sulfate. Technician B says that when a battery is discharged, the electrolyte has a high sulfuric acid content. Who is correct?

   A. Technician A

   B. Technician B

   C. Both Technician A and Technician B

   D. Neither Technician A nor Technician B

9. A lead-acid battery may be explosive when exposed to sparks of a flame because:

   A. petroleum products are used in the manufacture of the battery case.

   B. sulfuric acid is vented from the battery during the discharge process.

   C. hydrogen gas is vented from the battery during the charging process.

   D. lead peroxide gas slowly escapes from the battery when the engine is not running.

10. All of these statements about starting motor design are true *except*:

    A. The armature windings are insulated from the commutator bars.

    B. The armature windings are insulated from the armature core.

    C. The commutator bars are insulated from each other.

    D. Typically, four brushes contact the commutator bars.

11. If a built-in battery hydrometer contains a green dot, the battery is_____.

12. The CCA rating indicates the amperes that a battery will deliver at 0°F for 30 seconds with the battery voltage remaining above _____ volts.

13. In a starting motor, a _____ _____ is soldered to each end of the armature windings.

14. The shunt coil in a starting motor limits armature speed when the starter is _____.

15. Explain the operation of a starter solenoid while engaging and after engagement.

16. Describe the operation of a starter relay used with a moveable pole shoe starting motor.

17. Explain the operation of a starter drive while the starter is engaging and disengaging.

18. Explain battery operation while charging.

 18 Battery and Starting System Maintenance, Diagnosis, and Service

Learning Objectives

After you have read, studied, and practiced the contents of this unit, you should be able to:

- Perform basic battery maintenance and tests.
- Perform battery charge procedures.
- Diagnose and test starting motors.
- Perform starting motor on-vehicle electrical tests.

Key Terms

Hydrometer

Specific gravity

Voltage drop

Open circuit voltage

INTRODUCTION

The battery is the heart of the vehicle's electrical system and the condition of the battery affects the operation of the entire electrical system. When the engine is not running, the battery supplies voltage and current to operate any electrical accessories that are turned on. The battery supplies the high current required by the starting motor to crank the engine, and the power to maintain on-board computer memories when the engine is not running. Therefore, battery and cable condition are very important to provide proper electrical system operation.

Starting motor maintenance, diagnosis, and service are also extremely important. If the starting motor is defective, the engine may be prevented from starting, resulting in an expensive service call and/or tow bill.

BATTERY MAINTENANCE

The battery should be kept clean and dry and the battery terminals and cable ends should be clean and tight. Moisture on top of a battery and corrosion on the battery cable ends causes a battery to slowly discharge because a very low current leaks through the corrosion and moisture from one battery terminal to the other. The battery should be removed and cleaned with a baking soda and water solution. The baking soda helps to neutralize the sulfuric acid.

Figure 18-1 Cleaning battery terminals and cable ends.

When servicing a battery, always wear a face shield and protective gloves. Sulfuric acid is very damaging to human skin and eyes. The sulfuric acid solution in a battery is also damaging to clothing and automotive paint and upholstery.

When disconnecting the battery cables, always disconnect the negative cable first. If the wrench slips and connects with the vehicle's body, a short will not occur, possibly resulting in the battery damaging the car's finish or exploding.

Always reconnect the positive cable first, followed by the negative cable. The battery terminals and cable ends should be cleaned with a battery terminal cleaner **(Figure 18-1)**. Inspect the battery case for damage and leaks. If these conditions are present, battery replacement is necessary. Be sure the battery hold down is in satisfactory condition **(Figure 18-2)**. A badly eroded and

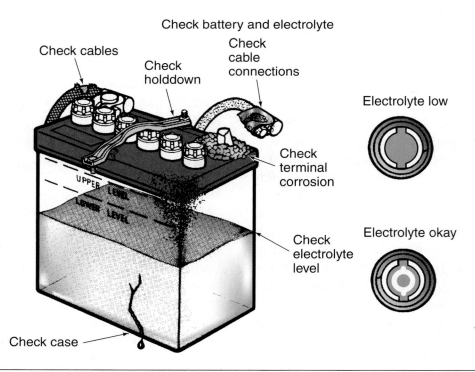

Figure 18-2 Battery inspection and maintenance.

SAFETY TIP *Never disconnect the positive cable first. If the wrench slips and makes contact between the positive cable and ground, a very high current flows through the wrench. This action may heat the wrench and burn your hand and possibly cause the battery to explode.*

SAFETY TIP *Never create sparks near a battery. Hydrogen gas given off while charging a battery is explosive. Even though the battery has not been charged for a period of time, some hydrogen gas may still be present in the battery cell vents.*

corroded battery tray should be replaced. If the battery hold down does not hold the battery securely, the battery may bounce out of place while driving. This action may allow contact between the battery and the cooling fan, resulting in damage to the battery case. Always use a battery carrying strap to lift and carry the battery **(Figure 18-3)**. Check the electrolyte level by removing the filler caps or observing the level through the translucent case. If the filler caps are removable, add water to the cells as necessary. When the electrolyte level is below the top of the plates in a maintenance-free battery, replace the battery. If the water loss from a battery is excessive and the case is not leaking, check the charging circuit for high voltage.

BATTERY DIAGNOSIS AND SERVICE

Battery diagnosis and service is extremely important to supply adequate battery life and avoid time-consuming and frustrating experiences. For example, if a battery is being undercharged by the charging system and this problem is not diagnosed and corrected, a no-crank condition will likely occur.

Hydrometer Test

If the battery has a built-in hydrometer, observe the indicator color **(Figure 18-4)**. A green indicator means the battery is over 65 percent charged, and may be tested in its present condition. If the hydrometer indicator is dark, the battery is less than 65 percent charged. Under this condition, the battery should be charged and load tested. A clear hydrometer indicator means the battery is low on electrolyte. If this condition is present on a battery with removable filler caps, add water to the cell(s) as required. When a clear hydrometer is present in a maintenance-free battery, replace the battery. Keep in mind that on batteries with a built-in hydrometer it is only an indicator of one cell not all six, so although the built-in hydrometer may indicate a good battery, it is not 100 percent accurate. A portable hydrometer may be used to test the state of charge in a battery with removable filler caps

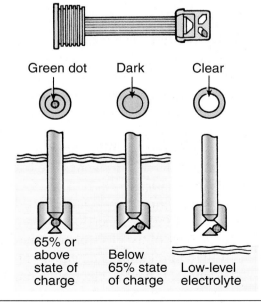

Green dot   Dark   Clear

65% or above state of charge   Below 65% state of charge   Low-level electrolyte

Figure 18-4 A built-in battery hydrometer.

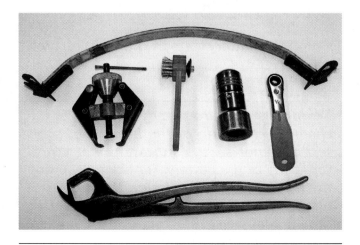

Figure 18-3 Battery service tools.

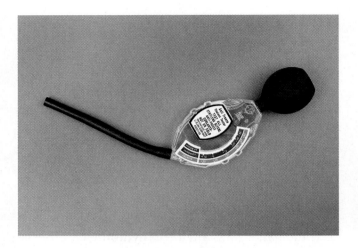

Figure 18-5 Portable hydrometers.

Figure 18-6 Specific gravity scale on hydrometer float.

(Figure 18-5). A **hydrometer** is designed to test the specific gravity of a liquid.

Place the hydrometer pickup into the electrolyte in a battery cell. Squeeze the hydrometer bulb and draw enough electrolyte into the hydrometer until the float moves upward and floats freely.

> **Tech Tip** *Variations of more than .050 specific gravity points indicates a dead cell. Battery replacement is necessary.*

Do not allow the top of the float to contact the top of the hydrometer because this causes an inaccurate reading. Read the specific gravity of the electrolyte at the level of the electrolyte on the float. A specific gravity scale is provided on the upper float extension **(Figure 18-6)**. Repeat the hydrometer test on all the cells. If the specific gravity is 1.265, the battery is fully charged, whereas a specific gravity of 1.190 indicates the battery is 50 percent charged **(Figure 18-7)**. The variation in specific gravity readings between the battery cells should not exceed 0.050 specific gravity points. The **specific gravity** of the electrolyte is the weight of the electrolyte in relation to an equal volume of water. For example, if battery electrolyte has 1.265 specific gravity, a quart of electrolyte is 1.265 times heavier than a quart of water.

STATE OF CHARGE	SPECIFIC GRAVITY
100%	1.265
75%	1.225
50%	1.190
25%	1.155
DEAD	1.120

Figure 18-7 Battery state of charge in relation to specific gravity.

> **Tech Tip** *A hydrometer test indicates the chemical condition of the battery, but this test does not always indicate the electrical capability of the battery. For example, if one of the cell connectors inside the battery is not making electrical contact, the specific gravity may be satisfactory but the battery will not deliver any current.*

Some hydrometers have a built-in thermometer and temperature correction scale. If the battery is very cold, the electrolyte is heavier and denser than at room temperature. Therefore, the electrolyte moves the float higher in the hydrometer. As indicated on the temperature correction scale, for every 10 degrees of electrolyte temperature below 80°F, 4 specific gravity points must be subtracted from the hydrometer reading. Conversely, if the battery is hot, the electrolyte is thin and less dense. Therefore, the electrolyte does not move the electrolyte as high in the hydrometer. If the electrolyte is hot, for every 10 degrees of electrolyte temperature above 80°F, add 4 specific gravity points to the hydrometer reading.

Open Circuit Voltage Test

If the battery has recently been charged, perform a load test to stabilize the battery voltage before completing an open circuit voltage test. **Open circuit voltage** is the voltage available when a circuit is open and there is no current flow in the circuit. Connect the voltmeter leads to the battery terminals with the correct polarity and read the open circuit voltage on the voltmeter. Compare the battery voltage reading to specifications. If the battery is fully charged, the open circuit voltage should be 12.6 volts. When the battery is partially charged, the voltage reading should be as specified in **Figure 18-8**.

Typical Open-Circuit Voltage and Specific Values		
Change level	Specific gravity	Voltage (12)
100%	1.265	12.7
75%	1.225	12.4
50%	1.190	12.2
25%	1.155	12.0
Discharged	1.120	11.9

Figure 18-8 Battery voltage output gives a general state of charge and condition.

Load Test

The load test indicates the capability of the battery to deliver a specific current flow with the battery voltage remaining above a certain value. Therefore, the load test indicates the electrical condition of the battery.

> **Tech Tip** *To obtain accurate load test results, a hydrometer test should indicate that the battery is 65 percent charged and the battery temperature must be above 0°F (−18°C). Do not load test maintenance-free batteries below this temperature.*

Follow these general steps to perform a load test:

1. Connect the load tester cables to the battery terminals with the proper polarity.
2. Connect the inductive ammeter clamp over the negative load tester cable **(Figure 18-9)**.
3. Rotate the load tester control knob clockwise until the ammeter reading is one-half the CCA rating of the battery.
4. Maintain this load on the battery for 15 seconds.
5. At the end of the 15 seconds, read the voltmeter reading and then rotate the tester control knob fully counterclockwise to the off position.
6. Disconnect the load tester cables and compare the voltage reading obtained in step 5 to specifications. A satisfactory battery with the electrolyte temperature at 70°F (21°C) has a load test voltage of 9.6 volts or above. If the battery is fully charged and fails the load test, replace the battery.
7. Watch the speed at which the voltage level climbs back to a normal range after the load is removed. A good battery will quickly recover near normal voltage levels. It is best to observe this on a couple of known good batteries before observing on a questionable one to get an idea of the rate of return.

Battery Drain Test

If the customer complains about the battery becoming discharged after the engine has not been

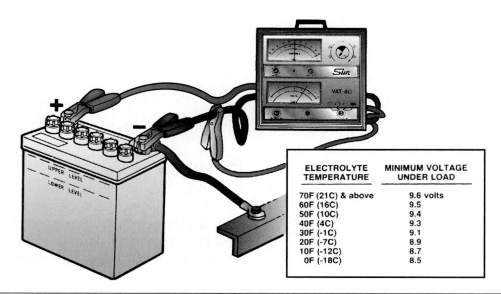

Figure 18-9 Battery load test connections and specifications.

ELECTROLYTE TEMPERATURE	MINIMUM VOLTAGE UNDER LOAD
70F (21C) & above	9.6 volts
60F (16C)	9.5
50F (10C)	9.4
40F (4C)	9.3
30F (-1C)	9.1
20F (-7C)	8.9
10F (-12C)	8.7
0F (-18C)	8.5

started for a period of several days, there may be an electrical drain on the battery through one of the vehicle electrical systems. The battery drain test allows the technician to locate the cause of battery electrical drain.

Tech Tip *Nearly all computers have a very low electrical drain to maintain the computer memory. Modern vehicles have a significant number of on-board computers and the electrical drain through these computers adds up. The technician must know how to measure battery drain and determine if this drain is satisfactory or excessive.*

Tech Tip *On vehicles without on-board computers, some technicians used to connect a test light in series between the negative battery terminal and the negative battery cable to test battery drain. This method is not accurate on computer-equipped vehicles because the test light does not indicate the exact drain in milliamperes.*

Follow this procedure to perform a battery drain test with a DVOM:

1. Be sure all the electrical accessories are turned off and the vehicle's doors are closed. If the

Tech Tip *An inductive amp clamp will do this test much quicker. It will need to be capable of reading milliamps. Follow the instructions that pertain to your particular clamp.*

vehicle has an under-hood light, remove the bulb from this light.

2. Disconnect the negative battery cable and connect the battery drain test switch to the negative battery terminal.

3. Connect the negative battery cable to the outer end of the test switch **(Figure 18-10)**.

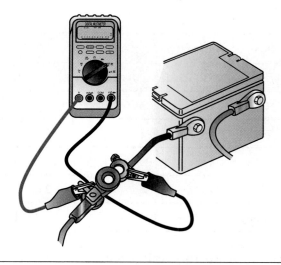

Figure 18-10 Battery drain test connections.

4. Connect a digital multimeter to the test switch terminals and select the highest ammeter scale on the multimeter.

5. Be sure the test switch is closed and start the engine. With the test switch closed, the starter current flows through this switch.

6. Operate the vehicle until the engine is at normal operating temperature and turn on all the electrical accessories.

7. Turn off all the electrical accessories and close the vehicle doors with the driver's window down. Turn the ignition switch off and remove the ignition key.

8. Turn the test switch to the open position. Current now flows from the battery through the ammeter to the electrical system.

9. Note the ammeter reading. Many computers gradually decrease their current draw as they enter a SLEEP mode. Therefore, the ammeter reading may gradually decrease.

....

NOTE: A slow battery drain may be called a *parasitic drain.* Always follow the vehicle manufacturer's recommended procedure for testing battery drain. On some vehicles, it is necessary to connect a scan tool to the DLC under the dash and perform a power down procedure on the body computer module (BCM) during this procedure.

....

10. Wait 20 minutes to allow the computers to enter the SLEEP mode. If a door is opened during this test, you will have to wait for the system to time out again. Switch the ammeter to a lower scale and read the battery drain in milliamps. Compare the battery drain to specifications. If specifications are not available, 50 milliamps drain or less is the average acceptable drain. However, it is not uncommon for luxury cars to have normal draws close to 100 milliamps. These cars generally have all the options available and it is not uncommon for them to drain the battery after several days of not being used.

11. Disconnect all the fuses and circuit breakers one at a time and observe the ammeter. When a fuse is disconnected and the milliamp drain decreases, the circuit connected to that fuse is causing the battery drain. Look in a wiring schematic manual to see what components are on that circuit before proceeding.

12. Close the test switch and disconnect the multimeter leads. Disconnect the test switch and reconnect the negative battery cable. You will have to reset all devices that have a memory, such as the radio, before returning the car to the customer.

Battery Charging

Slowly charging a battery has two advantages. This method of battery charging brings a battery to a fully charged condition and it reduces the possibility of overcharging and/or overheating the battery. A slow charger usually supplies a charging rate of 3 amperes to 10 amperes. Several batteries may be connected in series on a shop slow charger. When batteries are connected in series, the negative terminal on one battery is connected to the positive terminal on the next battery. The slow charging process should be continued until the battery specific gravity reaches 1.265.

Tech Tip *Disconnecting a battery cable(s) erases the memories in on-board computers, including the radio station programming. If the memory in the PCM (engine computer) is erased, the engine may have a rough idle problem and other adverse operating conditions until the computer relearns the sensor data. These problems can be avoided by connecting a 9 volts to 12 volts power supply to the cigarette lighter before disconnecting the battery cables.*

Tech Tip *When fast charging a battery in a vehicle, disconnect the battery cables and connect the charger to the battery terminals. This action prevents the charger voltage from being applied to the on-board computers.*

SAFETY TIP *Always charge batteries in a well-ventilated area.*

An average fast charger has a maximum output of 50 to 100 amperes. Fast charging a battery is much faster than slow charging, but fast charging has some disadvantages. Fast charging may overheat and damage a battery and a battery should not be fully charged at a high rate on a fast charger. When the specific gravity reaches 1.225, reduce the charging rate to less than 10 amperes to avoid gassing excessive water from the battery. When fast charging a battery, never allow the electrolyte temperature to exceed 125°F (52°C). Damage to battery plate grids and active plate materials occurs above this temperature.

Tech Tip *It is better to recharge a dead battery first before starting a vehicle. Using the vehicle's alternator to recharge a dead battery can overwork the alternator.*

STARTING MOTOR MAINTENANCE

The most important starting motor maintenance involves making sure the battery and starting motor cable connections are clean and tight. Inspect starting motor cables for worn insulation caused by improper contact with other components. Replace and reroute these cables as required. Be sure the starter mounting bolts are tight. Some starting motors have a ground cable from one of the starting motor mounting bolts to a bolt on the engine block. Be sure the bolts securing this cable are tight.

Starting Motor Diagnosis

Each time a vehicle is in the shop for lubrication or minor service, always listen to the starting motor as it starts the engine. If the starting motor cranks the engine slowly, the battery may be partly discharged, the cables may have high resistance, or the starting motor may be defective.

NOTE: A starting motor that cranks the engine slowly may be called a *dragging starter*.

Test the battery as explained previously and test the starter circuit and starting motor as described in the following section. If the starting motor provides a grinding noise when cranking the engine, the drive gear and/or flywheel ring gear may be worn. Worn starting motor bushings also cause a noisy starting motor. When the starter armature spins but the engine fails to crank, the starter drive may be slipping or sticking on the armature shaft. If a clicking noise is heard when the ignition switch is turned to the START position, the battery cables may not be making proper electrical contact on the battery terminals. A defective solenoid disk, terminals, and a low charge on a battery can also cause a clicking noise when the ignition switch is turned to the START position.

Starting Motor and Solenoid Diagnosis and Testing

When the starting motor slowly cranks the engine, first test the battery and ensure it is in satisfactory condition and fully charged. If the battery is fully charged, the problem may be caused by high resistance in the battery and/or starting motor cables or terminals. To test these cables, measure the voltage drop across each cable. Before measuring the voltage drop, disconnect the voltage supply to the primary ignition system to disable the ignition system and prevent the engine from starting. If the engine is fuel injected, disconnect the wiring connector from the fuel pump relay or remove the fuel pump fuse if equipped to prevent the injectors from discharging fuel into the intake manifold while cranking the engine.

Tech Tip *Many ignition and fuel pump systems can be disabled by removing the corresponding fuse.*

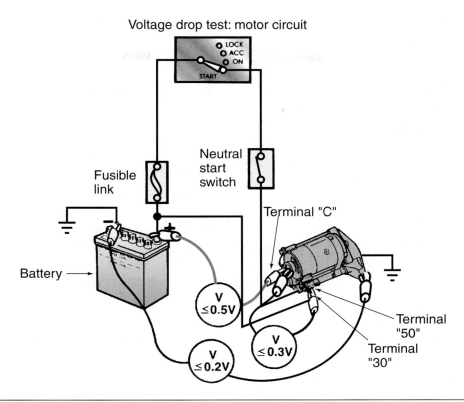

Voltage drop test: motor circuit

Figure 18-11 Starter circuit voltage drop tests.

Starting Circuit Resistance Tests

Connect the voltmeter leads across the cable to be tested and crank the engine. For example, to test resistance and voltage drop across the cable from the positive battery terminal to the solenoid terminal, connect the voltmeter leads across to the ends of this cable. Remember, a **voltage drop** is the voltage difference between two points in an electrical circuit. While cranking the engine, the voltage drop across this cable should not exceed 0.5 volt **(Figure 18-11)**. If the voltage drop across the cable exceeds specifications, the cable should be replaced. Voltage drop from the starting motor ground to the battery ground terminal should not exceed 0.2 volt. If the battery cables are satisfactory and the starter cranks slowly, measure the starter current draw.

Starter Current Draw Test

To measure the starter current draw, connect the battery starter tester cables to the battery terminals with the correct polarity. Place the inductive ammeter clamp from the battery load tester over the

negative battery cable **(Figure 18-12)**. Place the test selector switch in the STARTING TEST position. Disable the ignition and fuel system and crank the engine for 10 to 15 seconds and observe the ammeter and voltmeter reading. If the starting motor is satisfactory, the voltage remains above 9.6 volts and the starter current draw is within the

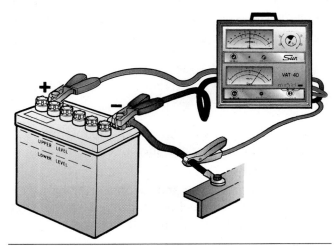

Figure 18-12 Starter current draw test connections.

vehicle manufacturer's specifications. An average V8 engine has a starter draw of 200 amperes, whereas the normal starter draw on a V6 engine is 150 amperes, and a 4-cylinder engine may have a starter draw of 125 amperes. If the starter current draw is excessive, remove and replace or test and repair the starter. Keep in mind that these are averages and some engines use more; also, the amp draw will be significantly higher when the system is cold.

> **Tech Tip** *Some starting motors have a shim(s) between the starter housing and the flywheel housing to provide proper starter alignment. When removing a starting motor, always check to see if these shims are present and note the position of the shims because they must be installed in their original position.*

Starting Motor Service and Inspection

If the starting motor has excessive current draw or any of the other problems described previously, the common practice in the automotive service industry is to replace the starting motor rather than to perform disassembly, testing, and repair procedures. Always disconnect the battery negative cable before removing a starting motor and reconnect this battery cable after the starting motor is reinstalled. When removing a starting motor, check for shims between the starting motor and the flywheel housing. If shims are present, the same

number of shims must be reinstalled with the starting motor. Always be sure the starting motor mounting surface on the flywheel housing is clean.

Solenoid Tests

Inspect the solenoid disk and terminals for a burned condition. If the solenoid disk and terminals are burned, replace the solenoid. To test the hold-in winding, connect a pair of ohmmeter leads from the solenoid S terminal to ground on the solenoid case **(Figure 18-13)**. A low ohmmeter reading indicates the hold-in winding is not open and an infinite ohmmeter reading indicates an open hold-in winding. Connect a pair of ohmmeter leads from the solenoid S terminal to the M terminal. A low ohmmeter reading indicates the pull-in winding is not open and the pull-in winding is open if the ohmmeter reading is infinite.

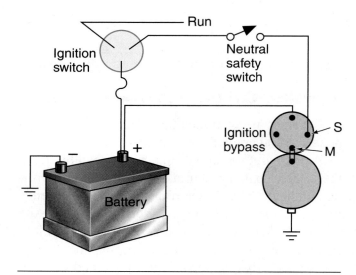

Figure 18-13 Solenoid terminal identification.

Summary

- One important aspect of battery maintenance is keeping the battery and the battery terminals clean and dry.

- When disconnecting and reconnecting battery terminals, disconnect the negative terminal first and reconnect this terminal last.

- A hydrometer test indicates the chemical condition of a battery.

- The specific gravity of a fully charged battery should be 1.265.

- When a battery is fully charged, it should have an open circuit voltage of 12.6 volts.

- A battery load test indicates the capability of the battery to deliver high current flow.

- A battery drain test measures the milliamperes of current flow out of a battery with the ignition switch and all electrical accessories turned off.

- When fast charging a battery, the battery temperature should not exceed 125°F (52°C).

- If a starting motor cranks the engine slowly, the battery may be discharged or defective, the battery and starting motor cables may have high resistance, or the starting motor may be defective.

- When the armature spins but the engine fails to crank, the starter drive is slipping or sticking on the armature shaft.

- Starting circuit resistance may be measured by testing the voltage drop across the cables and components in the circuit.

Review Questions

1. Technician A says when disconnecting battery cables, always disconnect the negative battery cable first. Technician B says when reconnecting battery cables, always connect the negative battery cable first. Who is correct?

   A. Technician A

   B. Technician B

   C. Both Technician A and Technician B

   D. Neither Technician A nor Technician B

2. Technician A says the specific gravity of the electrolyte in a fully charged battery is 1.225. Technician B says if the specific gravity is above 1.225, a battery may be charged at a high ampere rate on a fast charger. Who is correct?

   A. Technician A

   B. Technician B

   C. Both Technician A and Technician B

   D. Neither Technician A nor Technician B

3. Technician A says an open circuit voltage of 12.6 volts indicates a fully charged battery. Technician B says a load test voltage of 9 volts on a fully charged battery at 70°F (21°C) indicates a satisfactory battery. Who is correct?

   A. Technician A

   B. Technician B

   C. Both Technician A and Technician B

   D. Neither Technician A nor Technician B

4. Technician A says during a load test, the battery should be discharged at one-half the CCA rating. Technician B says during a load test, the load should be applied to the battery for 15 seconds. Who is correct?

   A. Technician A

   B. Technician B

   C. Both Technician A and Technician B

   D. Neither Technician A nor Technician B

5. All of these statements about battery service are true *except*:

   A. Sulfuric acid in battery electrolyte is damaging to human skin and eyes.

   B. Sulfuric acid in battery electrolyte is damaging to paint and upholstery.

   C. Moisture on a battery top has no effect on battery operation.

   D. The battery top may be washed with a baking soda and water solution.

6. If the battery case is not leaking, the most likely cause of excessive electrolyte loss from the battery is:

   A. a slipping alternator belt.

   B. excessively high charging system voltage.

   C. high resistance in the battery terminals.

   D. excessive resistance at the alternator battery terminal.

7. A battery has just been disconnected from a fast charger. The specific gravity of the electrolyte is 1.220 and the battery temperature is 120°F (48.8°C). The actual battery specific gravity is:

   A. 1.236.

   B. 1.242.

   C. 1.248.

   D. 1.250.

8. The maximum variation in specific gravity readings between the cells in a battery is:

   A. 0.050 points.

   B. 0.075 points.

   C. 0.100 points.

   D. 0.120 points.

9. The specific gravity of the electrolyte in a battery is 1.170 on all the cells with the battery temperature at 70°F. When the battery is load tested, the battery voltage is 8.6 volts at the end of the load test. Technician A says the battery should be charged and retested. Technician B says the battery is defective and should be replaced. Who is correct?

   A. Technician A

   B. Technician B

   C. Both Technician A and Technician B

   D. Neither Technician A nor Technician B

10. When a drain test is performed on a battery, an acceptable battery drain is:

    A. 450 milliamperes.

    B. 375 milliamperes.

    C. 250 milliamperes.

    D. 50 milliamperes.

11. Excessive battery drain should be tested using a(n) _____.

12. Immediately after the test switch is opened during a battery drain test, the current drain _____.

13. Several batteries may be connected in _____ on a shop slow charger.

14. Starting circuit resistance may be tested by measuring the _____ _____ in the circuit.

15. Explain the resulting starting motor operation if the starter drive is slipping.

16. Explain the ohmmeter connection for testing a solenoid pull-in winding for an open circuit.

17. Describe the procedure for testing voltage drop across the starting motor ground circuit.

18. Describe the ohmmeter connections for testing a solenoid hold-in coil.

CHAPTER

19 Charging Systems

Learning Objectives

After you have read, studied, and practiced the contents of this unit, you should be able to:

- List the basic alternator circuit parts and their function.
- Explain the function of a diode.
- Describe the function of a voltage regulator.
- Describe how A/C current is converted to D/C in an alternator.

Key Terms

Diode trio

Forward bias

Rectify

Reverse bias

Thermistor

INTRODUCTION

The purpose of the alternator is to supply current to recharge the battery and power the electrical accessories on the vehicle. Alternator operation is extremely important to provide proper operation of the electrical accessories and maintain the battery in a fully charged condition. If the alternator voltage is too high, it forces excessive current flow through the electrical accessories and the battery, resulting in damaged electrical components and excessive battery gassing. When the alternator voltage is too low, it does not supply enough current to the electrical accessories and the battery. This may lead to improper operation of some electrical accessories and a discharged battery.

ALTERNATOR DESIGN

The alternator contains a rotor and stator mounted between two end housings. Bearings in the end housings support the rotor shaft and allow rotor rotation **(Figure 19-1)**. The drive end of the rotor shaft extends through the drive end frame and a pulley is bolted or pressed onto this end of the rotor shaft. A cooling fan attached to the belt pulley forces cool air into the alternator to cool it. Cooling fans are often built into the case to aid in dissipating heat. The cooling fan rotates with the pulley to move air through the alternator and cool the alternator diodes **(Figure 19-2)**. A drive belt surrounding the alternator pulley and the crankshaft pulley rotate the rotor. The drive belt may

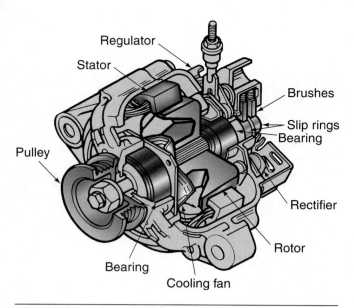

Figure 19-1 Alternator components.

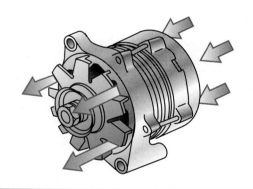

Figure 19-2 The cooling fan moves air through the alternator to cool the diodes.

also drive other components such as a power steering pump or air-conditioning compressor. The stator is mounted in a frame and is bolted between the two end frames so that it is stationary. The stator contains three insulated windings mounted in stator frame slots. There is a very small clearance between the rotor poles and the inside diameter of the stator frame. A rectifier plate is mounted in the slip ring end frame.

Rotor

An insulated field winding is mounted on a spool that is pressed onto the rotor shaft. Two metal poles are pressed onto the rotor shaft on each side of the winding and these poles have interlacing fingers positioned above the winding **(Figure 19-3)**.

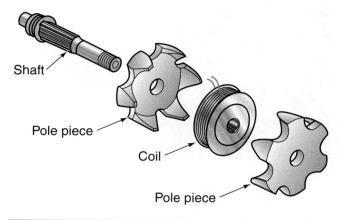

Figure 19-3 Rotor design.

Two insulated copper slip rings are mounted on the end of the rotor shaft and the ends of the field winding are connected to these slip rings. In some alternators, both brushes are insulated and connected to the two alternator field terminals. In other alternators, one brush is grounded and the other brush is connected to a field terminal.

Stator

The stator assembly contains three insulated windings. These windings are mounted in insulated stator frame slots. Some stators have wye-connected windings and in other stators the windings are delta connected. Wye-connected stator windings have three ends of these windings connected together in a "Y" shaped connection **(Figure 19-4)** and the other three ends of these windings are connected to the diodes. In a delta-connected stator, the end of one stator winding is connected to the next stator winding **(Figure 19-5)**. In this type of stator winding, the junction where each pair of stator windings is connected is attached to the diodes.

Diodes

A diode is made of two semiconductor materials joined together. A semiconductor material is usually manufactured from silicon. In the manufacturing process, the silicon is mixed with other elements such as boron and phosphorus. The unique mixture of semiconductor materials allows current to flow in one direction only. This condition may be referred to as **forward bias**. Since alternators produce alternating current it must be converted to

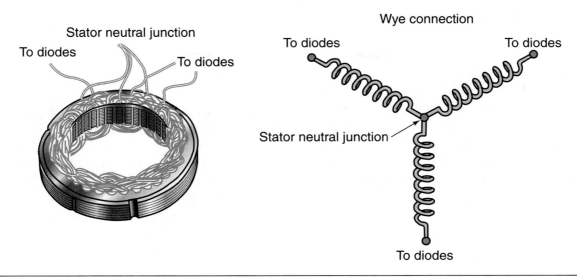

Figure 19-4 A wye-connected stator.

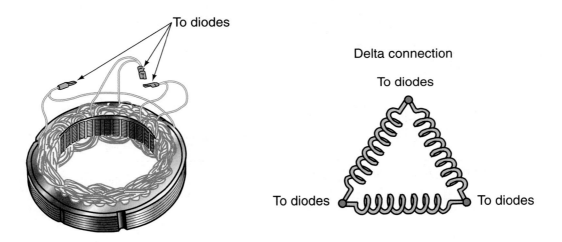

Figure 19-5 A delta-connected stator.

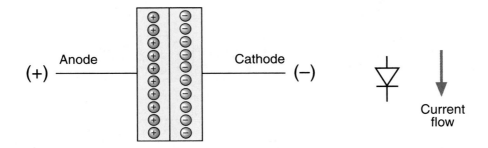

Figure 19-6 A diode and its symbol.

D/C before leaving the alternator and put into the cars electrical system. Groups of diodes are used to **rectify** the current so all electrons are moving in the same direction **(Figure 19-6)**. Generally there are six diodes used to rectify current in an alternator, three are put in forward bias to allow current to pass when the polarity is positive. Three diodes are put in **reverse bias** so that when the alternating current has switched polarities, they do not allow the charge to change direction, effectively converting A/C to D/C by changing the direction of current flow to one direction only.

A rectifier assembly containing six diodes is mounted on a heat sink in the slip ring end of the alternator. Three of these diodes are mounted on the insulated plate that is connected to the alternator battery terminal. The other three diodes are mounted on the side that is grounded to the end frame. The ends of the stator windings are connected to the diodes **(Figure 19-7)**.

NOTE: A diode may be called a *PN junction*.

ALTERNATOR OPERATION

The alternator battery terminal is connected through a 12-gauge wire to the positive battery terminal **(Figure 19-8)**. Therefore, battery voltage is available at the alternator battery terminal with the ignition switch off. A fusible link is usually

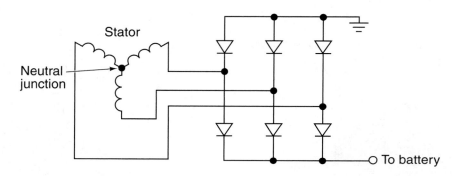

Figure 19-7 Stator windings connected to the diodes.

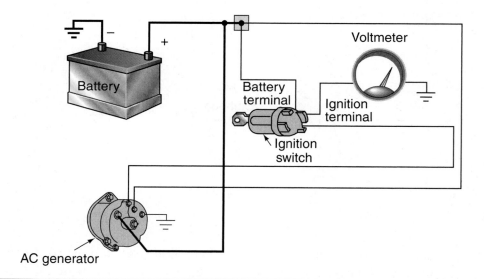

Figure 19-8 The alternator battery terminal is connected to the positive battery terminal in the charging circuit.

connected in the wire attached to the alternator battery terminal. This fusible link protects the battery wire and wiring harness if this wire is accidentally shorted to ground **(Figure 19-9)**.

In alternator circuits, voltage is supplied from the ignition switch to the alternator field terminal when the ignition switch is turned on **(Figure 19-10)**. Current then flows through the insulated brush, slip ring, field winding, and the other slip ring and brush to ground. This current flow through the field winding creates a magnetic field around the rotor. The interlacing fingers on one side of the rotor become north poles and the interlacing fingers on the opposite side of the rotor become south poles. The magnetic lines of force travel from the

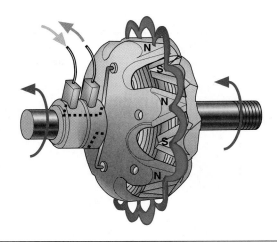

Figure 19-11 When current flows through the field coil, a magnetic field is created between the rotor poles.

north to the south poles on the interlacing fingers **(Figure 19-11)**.

NOTE: An alternator may be called an *AC generator*.

When the engine starts, the rotor revolves inside the stator and the rotor magnetic field cuts across the stator windings. Notice that the rotor poles around the circumference of the rotor are alternately north and south poles. Therefore, each stator winding is influenced by a north and a south pole followed by a south and a north pole. When the alternate magnetic poles on the rotor cut across a stator winding, an A/C is produced in the winding **(Figure 19-12)**. Since the

Figure 19-9 A fusible link.

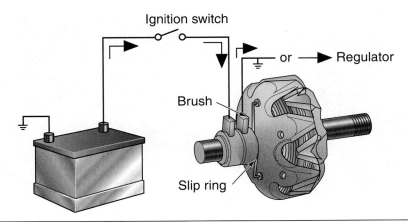

Figure 19-10 When the ignition switch is turned on, current flows through the field circuit.

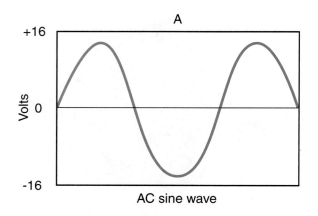

Figure 19-12 AC is produced in the stator windings.

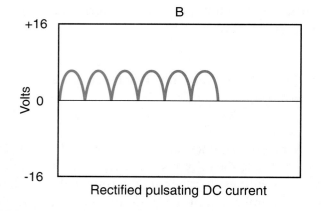

Figure 19-13 The AC in the stator windings is changed to DC by the diodes.

battery and electrical accessories on the vehicle must be supplied with DC, the AC in the stator windings must be rectified to DC **(Figure 19-13)**. This rectification is accomplished by the alternator diodes. The diodes change the AC current in the stator windings to a flow of DC current through the battery and electrical accessories.

VOLTAGE REGULATOR OPERATION

Modern vehicles are equipped with electronic voltage regulators. These regulators are often mounted inside the alternator, on the outside of the alternator end frame, or in the vehicle's PCM. If the voltage regulator is mounted in the vehicle's PCM, the operation is generally the same as to how the charging is controlled in internal voltage regulated alternators. Electronic regulators are non-adjustable. Some older vehicles have electronic regulators mounted externally from the alternator **(Figure 19-14)** that could be adjusted; however, when they need repair, they are generally replaced with newer non-adjustable regulators.

Tech Tip *Daimler-Chrysler vehicles manufactured since 1985 have the voltage regulator designed into the PCM. If the voltage regulator is defective, the PCM must be replaced.*

Figure 19-14 Externally mounted voltage regulators.

NOTE: The term *voltage regulator* is somewhat misleading. If something is regulated, it is prevented from going too high or too low. The voltage regulator only limits the alternator voltage and prevents it from becoming too high.

Electronic voltage regulator circuits vary depending on the vehicle. Always use the charging circuit wiring diagram for the vehicle you are working on. For most vehicles the regulator circuit works similarly. When the ignition switch is turned on, current flows through the ignition switch, charge indicator bulb, and parallel resistor to the corresponding alternator terminal. From this location, current flows through the integral electronic voltage regulator, slip rings, brushes, and field coil and the transistor in the regulator **(Figure 19-15)**. Current flows through this transistor to ground. This current flow creates a magnetic field around the rotor. When the engine starts, the rotor magnetic field induces voltage in the stator windings and current begins to flow from the stator windings through the rectifier to the battery positive terminal. Current also flows from the stator windings through the diode trio in the alternator. Under this condition, equal voltage is supplied to both sides of the charge indicator bulb and this bulb goes off. Current also flows through the diode trio and the slip rings, brushes, and field coil to TR1. Since TR1 is turned on, current flows through this transistor to ground **(Figure 19-16)**. Under this condition, field current and rotor magnetic strength are higher and the voltage in the stator windings increases. The **diode trio** is a small assembly containing three diodes. These diodes are connected from the stator terminals to the number 1 terminal and prevent alternating current from passing into the regulatory circuit.

The alternator number 2 terminal is connected to the positive battery terminal. Therefore, alternator voltage is sensed at this terminal when the engine is running. When the alternator voltage reaches a predetermined value, TR2 in the voltage regulator is turned on and TR1 is turned off **(Figure 19-17)**. The regulator typically limits the alternator voltage to between 13.8 volts to 14.8 volts.

Did You Know? *The resistance of the field coil limits the alternator field current. Because this coil typically has 3 to 4 ohms of resistance, the maximum field current is 3 to 4 amperes.*

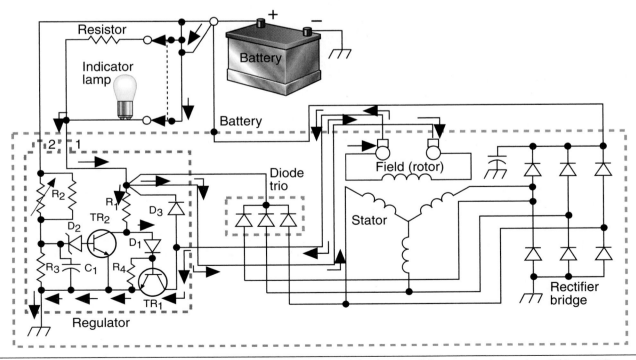

Figure 19-15 Current flow through the field circuit with the ignition switch in the ON position.

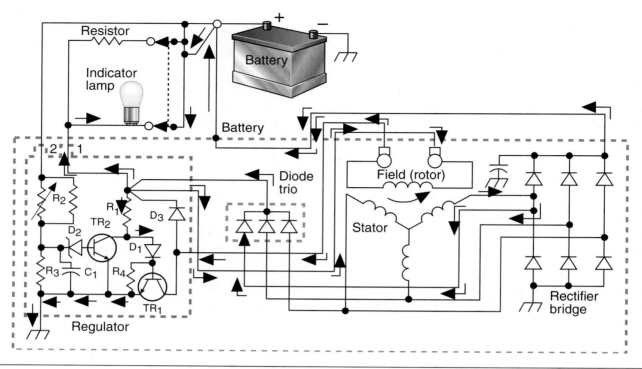

Figure 19-16 Current flow through the field coil and voltage regulator with the engine running and the field current turned on by the regulator.

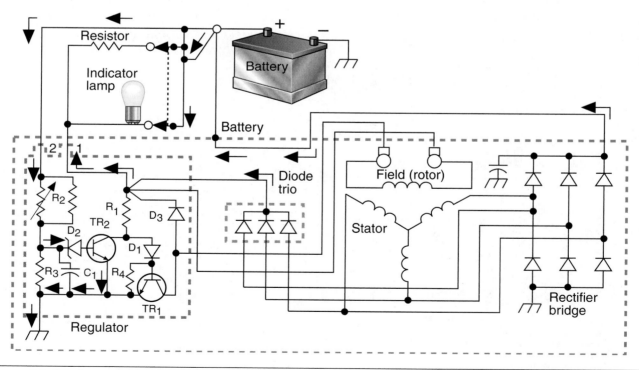

Figure 19-17 Current through the regulator when the regulator turns the current flow OFF through the field coil.

Resistor R2 in the voltage regulator is a thermistor with a parallel resistor. The thermistor allows the regulator to provide a higher alternator voltage when the atmospheric temperature is cold. This action compensates for additional resistance in the battery when it is cold and maintains the charging rate from the alternator through a cold battery. A **thermistor** is a special resistor that changes resistance in relation to temperature. When the thermistor is cold, its resistance increases.

Summary

- The cooling fan moves air through the alternator to cool the diodes.

- The ends of the field winding are connected to the slip rings.

- The field winding and the slip rings are insulated from the rotor shaft and poles.

- Three positive diodes in the rectifier assembly are mounted in an insulated heat sink and are connected to the alternator battery terminal.

- Three negative diodes in the rectifier assembly are grounded to the alternator end frame.

- The revolving rotor magnetic field induces an AC in the stator windings.

- The diodes change the AC in the stator windings to a DC for battery and electrical accessories.

- The electronic voltage regulator cycles the field current on and off to control field current and magnetic strength around the rotor. This action limits the voltage induced in the stator windings.

- The electronic voltage regulator provides a higher alternator voltage in cold weather to compensate for additional resistance in the battery under this condition.

Review Questions

1. Technician A says a typical alternator stator contains four windings. Technician B says in a wye-connected stator, one end of each stator winding is connected to the diodes. Who is correct?

   A. Technician A

   B. Technician B

   C. Both Technician A and Technician B

   D. Neither Technician A nor Technician B

2. Technician A says a diode conducts current when it is connected to a voltage source with forward bias. Technician B says a forward bias connection occurs when the positive side of a voltage source is connected to the positive side of a diode and the negative side of the voltage source is connected to the negative side of the diode. Who is correct?

   A. Technician A

   B. Technician B

   C. Both Technician A and Technician B

   D. Neither Technician A nor Technician B

3. Technician A says a diode contains three semiconductor materials mounted together. Technician B says if a voltage source is connected to a diode with reverse bias, the diode will be damaged. Who is correct?

   A. Technician A

   B. Technician B

   C. Both Technician A and Technician B

   D. Neither Technician A nor Technician B

4. Technician A says the wire from the alternator battery terminal to the positive battery terminal is protected by a fuse link. Technician B says if the ignition switch is turned off, the voltage at the alternator battery terminal is zero. Who is correct?

   A. Technician A

   B. Technician B

   C. Both Technician A and Technician B

   D. Neither Technician A nor Technician B

5. All these statements about alternator rotor and stator design are true *except*:

   A. The ends of the field winding are connected to the rotor slip rings.

   B. The two rotor slip rings are insulated from each other.

   C. The wye connection in a stator is connected to the stator frame.

   D. In a wye-connected stator, three ends of the stator windings are connected to the diodes.

6. All of these statements about alternator diodes are true *except*:

   A. The rectifier assembly in an alternator usually contains four positive diodes and two negative diodes.

   B. The diodes in an alternator change the AC voltage and current in the stator windings to DC voltage and current.

   C. A diode contains two semiconductor materials joined together.

   D. When phosphorus is melted into a silicon wafer, a negative material is created with an excess of electrons.

7. The magnetic field surrounding the alternator rotor:

   A. is created by current flow through the stator windings.

   B. induces a DC voltage in the stator windings.

   C. is present when the ignition switch is in the off position.

   D. travels from the north to the south poles on the interlacing rotor fingers.

8. When discussing electronic alternator voltage regulators, Technician A says most electronic voltage regulators are adjustable. Technician B says the electronic voltage regulator is connected in the circuit from the alternator battery terminal to the positive battery terminal. Who is correct?

   A. Technician A

   B. Technician B

   C. Both Technician A and Technician B

   D. Neither Technician A nor Technician B

9. An electronic alternator voltage regulator increases the alternator voltage by:

   A. weakening the magnetic field surrounding the rotor.

   B. increasing the field current.

   C. increasing the resistance in the regulator ground circuit.

   D. decreasing the resistance in the stator circuit.

10. A vehicle frequently experiences burned out lightbulbs. Technician A says to check the alternator voltage regulator setting. Technician B says to measure the resistance between the alternator battery terminal and the positive battery terminal. Who is correct?

    A. Technician A

    B. Technician B

    C. Both Technician A and Technician B

    D. Neither Technician A nor Technician B

11. The two ends of the field coil in the rotor are connected to the _____ .

12. In a wye-connected stator, three ends of the stator windings are connected to the diodes and the other three ends of these windings are connected _____ .

13. The diodes change _____ current in the stator windings to _____ current for the battery and accessories.

14. The voltage regulator limits the alternator voltage by controlling the rotor _____ .

15. Describe the design of a delta-connected stator.

16. Describe the design of an alternator rotor.

17. Explain how an electronic regulator limits alternator voltage.

18. Explain the purpose of a thermistor in an electronic regulator.

CHAPTER

20 Charging System Maintenance and Diagnosis

Learning Objectives

After you have read, studied, and practiced the contents of this unit, you should be able to:

- Inspect alternator belts and adjust belt tension.
- Diagnose charging system problems.

Key Terms

Alternator output test

Full fielding

Ribbed serpentine belt

Diagnostic trouble code (DTC)

INTRODUCTION

You must understand charging system design and operation to perform accurate diagnosis of these systems. Understanding proper charging system diagnostic procedures provides accurate diagnosis of these systems without damaging the charging system components and other electrical/electronic systems. Charging system maintenance, diagnosis, and service are explained in this chapter, including the newer systems in which the PCM controls the alternator voltage. The diagnosis of these late-model charging systems using DTCs is also explained.

ALTERNATOR AND VOLTAGE REGULATOR MAINTENANCE

Alternator belt condition and tension are extremely important for satisfactory alternator operation. A loose belt causes low alternator output, which may result in a discharged battery. A loose,

dry, or worn belt may cause squealing and chirping noises, especially during engine acceleration and cold start up.

Checking Alternator Belt Condition and Tension

The alternator belt should be inspected for cracking, oil, worn or glazed edges, tears, and splits **(Figure 20-1)**. If any of these conditions is present, the belt should be replaced.

Since the friction surfaces are on the sides of a V-belt, wear occurs in this area. If the belt edges are worn, the belt may be rubbing on the bottom of the pulley. This condition causes belt slipping and belt replacement is necessary to correct this problem.

> **Tech Tip** *Whenever the belts must be removed always check all pulleys and tensioners on the engine. Some problems that may not be noticed otherwise can be repaired before serious damage happens.*

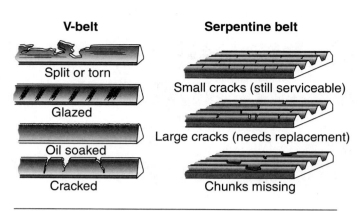

V-belt

Split or torn

Glazed

Oil soaked

Cracked

Serpentine belt

Small cracks (still serviceable)

Large cracks (needs replacement)

Chunks missing

Figure 20-1 Defective alternator belt conditions.

With the engine stopped, belt deflection may be measured to test belt tension. The maximum deflection should be 0.5 inch per foot of free span. The belt tension may be measured with a belt tension gauge placed over the center of the longest belt span **(Figure 20-2)**. The tension indicated on the gauge should equal the vehicle manufacturer's specifications.

If the belt requires tightening, follow this procedure:

1. With the engine stopped, loosen the alternator tension adjusting bolt in the alternator bracket.
2. Loosen the alternator mounting bolts.
3. Check the alternator bracket and mounting bolt for excessive wear. If these bolts or bolt openings are worn, replacement is necessary.

4. Pry against the alternator housing with a pry bar to tighten the alternator belt.
5. Hold the alternator in the position obtained in step 4 and tighten the tension-adjusting bolt in the alternator bracket.
6. Retest the belt tension with the tension gauge. If the belt does not have the specified tension, repeat step 1 through step 5.
7. Tighten the tension-adjusting bolt and the alternator mounting bolt to the specified torque.

Some alternators have a **ribbed serpentine belt**. Most serpentine belts have an automatic tensioning pulley; therefore, a belt tension adjustment is not required. The belt should be inspected to be sure it is properly installed on each pulley in the belt drive system **(Figure 20-3)**. The tension of a serpentine belt may be measured with a belt tension gauge in the same way as the tension of a V-belt. As the belt wears or stretches, the spring moves the tensioner pulley to maintain the belt tension. Some of these tensioners have a belt length scale that indicates new belt range and used belt range **(Figure 20-4)**. If the indicator is out of the used belt length range, belt replacement is required. Many belt tensioners have a 1/2-inch drive opening in which a ratchet or flex handle may be installed to move the tensioner pulley off the belt during belt replacement **(Figure 20-5)**. Belt pulleys must be properly aligned to minimize belt wear. The edges

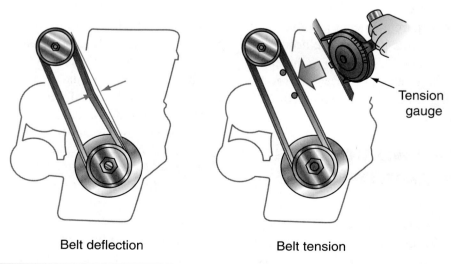

Belt deflection

Belt tension

Tension gauge

Figure 20-2 Methods of testing belt tension.

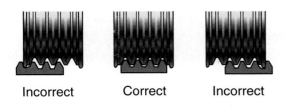

Figure 20-3 Proper and improper serpentine belt installation.

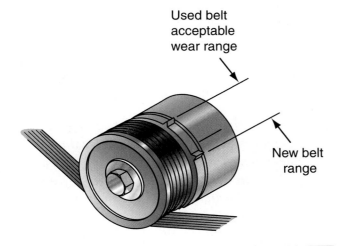

Figure 20-4 A belt tension scale.

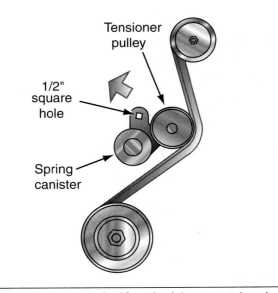

Figure 20-5 A one-half inch drive opening in the tensioner pulley.

of the pulleys must be in line when a straightedge is placed on the pulleys **(Figure 20-6)**. A misaligned alternator pulley may cause repeated belt failure. It is not uncommon for serpentine belts to

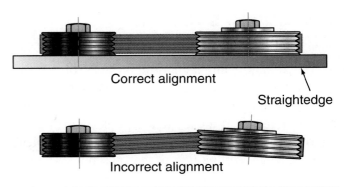

Figure 20-6 Checking pulley alignment.

quickly develop small cracks. All other replacement criteria are the same as for V-belts. The most common indicators of the need to replace are pieces of missing belt ribbing and small stones imbedded in the belt. During replacement carefully check the small grooves in the pulleys for debris; small pieces of the old belt can stick to the pulleys and cause the new belt to slip and squeal if not removed.

General Charging System Maintenance

When a vehicle is in the shop for minor service, the operation of the charge indicator light should be checked. If the charge indicator light is on with the engine running, further alternator diagnosis is required. Inspect the alternator wiring for worn insulation and loose or corroded connections. Be sure the insulating boot is in place over the alternator battery terminal. The electronic voltage regulator is usually mounted integrally in the alternator or on the back of the alternator housing. On cars built after the late 1960s there is no voltage regulator adjustment. If the regulator is mounted on the alternator end frame, be sure the regulator mounting bolts are tight and check any wiring terminals on the regulator for looseness and corrosion. With the engine running, listen for any unusual alternator noises. A growling noise may be caused by defective alternator bearings. Inspect the battery electrolyte level. If this level is excessively low, the alternator voltage may be too high. When the engine is accelerated, observe the brilliance of the headlights. If the lights become brighter or dimmer with engine speed, the alternator voltage may be too low.

CHARGING SYSTEM DIAGNOSIS

When diagnosing a charging system, one of the most important tests is the **alternator output test**. During the output test, the alternator is forced to produce full output. If the alternator passes this test, the alternator is satisfactory and this proves the problem must be in some other part of the charging system.

> *NOTE:* The alternator output test is often called *full fielding* the alternator.

Alternator Output Test

If the battery is discharged, the alternator may not be producing the specified voltage and current. An alternator output test may be performed to determine if the alternator output in amperes is satisfactory. Follow this procedure to perform an alternator output test:

1. Inspect the alternator belt condition and measure the belt tension. Adjust or replace the alternator belt as necessary.
2. Connect the volt-amp tester to the battery terminals with the correct polarity **(Figure 20-7)**.
3. Install the ammeter inductive clamp over the negative battery cable.

4. Turn off all the electrical accessories on the vehicle.
5. Start the engine and increase the engine rpm to 2,000.
6. Turn the load control clockwise on the volt-amp tester until the voltmeter indicates 13 volts, and read the ammeter. The ammeter reading should be within 10 amperes of the rated alternator output if the alternator output is satisfactory. If the alternator output is not satisfactory, the alternator should be replaced.
7. Turn the tester load control fully counterclockwise and allow the engine to idle. Shut the engine off and disconnect the tester cables.

Did You Know? **When performing an alternator output test, do not lower the charging system voltage below 12.6 volts. This action causes excessive current flow from the alternator and battery through the carbon pile load in the volt-amp tester. Under this condition, the carbon pile load and other tester components may be damaged.**

Full fielding the alternator involves connecting a jumper wire or making a connection to the alternator field circuit that bypasses the voltage regulator and allows maximum current to flow through the field coil in the rotor. This results in maximum voltage and current output in the stator windings.

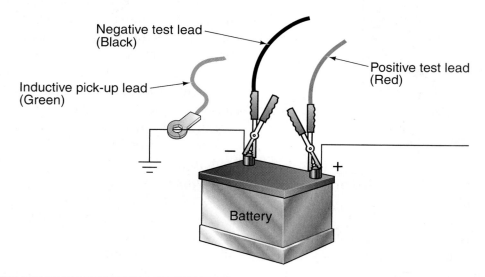

Figure 20-7 Meter test connections for an alternator output test.

A carbon pile load connected across the battery terminals limits the voltage to a safe value, such as 15 volts, during this test. Caution should be used when full fielding an alternator; allowing the alternator to run in the full field mode for long will cause it to overheat and possibly damage it.

> ***Did You Know?*** *On some older vehicles, the manufacturer's recommended testing alternator output is by full fielding the alternator. Vehicle manufacturers have discontinued this procedure because of the possibility of high voltage damaging electronic systems and components on the vehicle.*

Charging System Voltage Test

In most charging systems, the voltage regulator limits the charging system voltage to 13.8 volts to 14.8 volts. If the charging system voltage is continually below 13.8 volts, the battery may become discharged. A charging system voltage above 14.8 volts may overcharge the battery, resulting in excessive battery gassing. A charging system voltage above 14.8 volts may also damage electrical/ electronic components on the vehicle.

> ***Did You Know?*** *Electronic and mechanical voltage regulators are temperature compensated, which allows the regulator to provide a higher charging system voltage when the atmospheric temperature and the regulator are cold. This higher voltage compensates for the higher resistance that naturally occurs in a cold battery. This higher charging system voltage maintains some charging rate through a cold battery.*

Follow this procedure to test the charging system voltage:

1. Connect the amp-volt tester to the battery terminals as described for the alternator output test.
2. Start the engine and turn off all the electrical accessories.
3. Increase the engine speed to 1,500 to 2,000 rpm. Observe the ammeter reading.
4. The charging rate on the ammeter should be about 10 amperes. If the charging rate is high,

the charging system voltage may not be at peak. Allow the engine to run until the charging rate is about 10 amperes.
5. Observe the charging voltage on the voltmeter. If the charging voltage is above or below the vehicle manufacturer's specifications, replace the voltage regulator.

Scan Tool Diagnosis of Charging Systems

On many vehicles, **Diagnostic trouble codes (DTCs)** are set in the PCM memory if there is a defect detected in the charging system by the PCM. DTCs are set in the PCM memory when charging system defects occur on systems with the voltage regulator in the PCM and also on some systems with the voltage regulator in the alternator.

SAFETY TIP *When connecting or disconnecting a scan tool, the ignition switch must be off. If the ignition switch is on during this connection or disconnection, electronic systems on the vehicle or the scan tool may be damaged.*

On some systems with the voltage regulator in the alternator, the PCM senses the charging system voltage and commands the regulator to turn the alternator field on and off. If the PCM senses a charging system defect the required number of times and sets a DTC, the PCM illuminates the MIL in the instrument panel. A scan tool is connected to the DLC to read the DTCs **(Figure 20-8)**. On 1996 and newer

Figure 20-8 A scan tool.

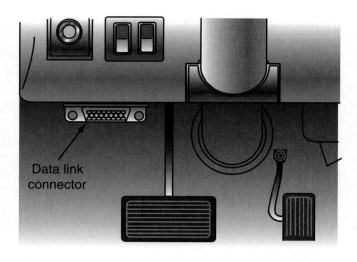

Figure 20-9 A data link connector (DLC) in an on-board diagnostic II (OBD II) system.

vehicles with OBD II systems, the DLC is a 16-terminal connector located under the left side of the dash **(Figure 20-9)**. After the scan tool is connected, turn the ignition switch on and complete the initial entries on the scan tool. These entries may include vehicle make, model year, and engine code. After the initial entries, select READ CODES on the scan tool. DTCs vary depending on the vehicle make and model year. Always use the DTCs and diagnostic procedure in the vehicle manufacturer's service manual. Typical OBD II system DTCs related to charging system faults are the following:

1. P0562 Battery voltage low—The PCM sets this DTC if the battery-sensed voltage is 1 volt below the specified charging voltage range for 13.47 seconds. The PCM senses the battery voltage and then turns off the alternator field circuit and senses the voltage again. When the two sensed voltages are the same, the PCM sets the DTC. The PCM sets this DTC the first time the fault is sensed.

2. P0563 Battery voltage high—The PCM sets this DTC if the battery voltage is 1 volt higher than the specified charging system voltage range. The PCM sets this DTC the first time the fault is sensed.

3. P0622 Alternator field control circuit—The PCM sets this DTC if the PCM tries to regulate the field current with no result. The PCM sets this DTC the first time the fault is sensed.

4. P2503 Charging system voltage low—The conditions required to set this DTC are the same as the conditions for P0562.

> **Tech Tip** *There are some standards, such as SAE J2012, established for the DTCs on OBD II vehicles. The P in the codes indicates they are related to powertrain systems. If the second digit is a zero or two in the powertrain DTCs, the DTC is mandated by the SAE. When the second digit is a one or a three, the DTC is established by the vehicle manufacturer. The third digit in the DTC indicates the subgroup to which the DTC belongs. For example, if the third digit is a five, the DTC is in the vehicle speed, idle control, and auxiliary inputs subgroup. When the third digit is a six, the DTC is in the computer and auxiliary outputs subgroup.*

Charging Circuit Voltage Drop Testing

Excessive resistance in the charging circuit between the alternator battery terminal and the positive battery terminal reduces the voltage supplied to the battery. For example, a higher than normal resistance in this circuit may cause a 1-volt drop across the circuit. Under this condition, if the alternator voltage is 14.5 volts, the voltage supplied to the battery is 13.5 volts. When this condition is present, the current flow through the battery is reduced and the battery may become discharged. To measure the voltage drop across the charging circuit, connect the positive voltmeter lead to the alternator battery terminal and connect the negative meter lead to the positive battery terminal.

Select the lowest voltmeter scale. If necessary, connect a carbon pile load across the battery terminals and adjust the load to obtain 10 to 15 amperes charging rate with the engine running. A voltage drop above 0.5 volt indicates excessive charging circuit resistance. To measure voltage drop across the alternator ground circuit, connect the negative voltmeter lead to the negative battery terminal and attach the positive meter lead to the alternator case. With the engine running and a 10- to 15-ampere charging rate, the maximum voltage drop should be 0.2 volt.

Alternator Service and Inspection

If an alternator output test indicates the alternator is defective, the common procedure in the automotive service industry is to replace the alternator. Because of high labor costs, it is usually more economically feasible to replace rather than repair an alternator. Always disconnect the negative battery cable before attempting to remove the alternator. Some replacement alternators are supplied without a pulley, so the pulley must be removed from the old alternator and installed on the replacement alternator. If the alternator has a press-on pulley, the proper pulling and installation tools must be used to remove and replace the pulley. On many alternators, the pulley is retained on the shaft with a nut and lock washer. These alternators usually have a hex opening in the center of the rotor shaft. An Allen wrench may be installed in this opening to hold the shaft while the pulley nut is loosened. After the pulley nut and lock washer are installed, be sure the nut is tightened to the specified torque. If the mounting holes in the alternator bracket are worn, replace the alternator bracket.

Summary

- The alternator belt must have the specified tension to provide proper alternator output.

- The friction surfaces are on the sides of a V-belt.

- Serpentine belts usually have an automatic tensioner pulley.

- Some tensioner pulleys have a belt length scale that indicates used belt range and new belt range.

- Defects in the alternator diodes or stator may cause a whining noise from the alternator with the engine running.

- The alternator output may be tested by lowering the charging system voltage to 13 volts with a carbon pile load connected across the battery.

- The specified voltage regulator operating range is 13.8 volts to 14.8 volts.

- On many late-model charging systems, the PCM will store DTCs when a defect occurs in the charging system.

- A scan tool is connected to the DLC to obtain the DTCs.

- The resistance in the charging system is measured by testing the voltage drop in the system.

Review Questions

1. Technician A says a loose alternator belt may cause a discharged battery. Technician B says a V-belt that is contacting the bottom of the alternator pulley may cause a discharged battery. Who is correct?

   A. Technician A

   B. Technician B

   C. Both Technician A and Technician B

   D. Neither Technician A nor Technician B

2. An alternator has a whining noise with the engine running. Technician A says the alternator may have an open field circuit. Technician B says one of the alternator diodes may be defective. Who is correct?

   A. Technician A

   B. Technician B

   C. Both Technician A and Technician B

   D. Neither Technician A nor Technician B

3. Technician A says during an alternator output test, the engine should be running at 2,000 rpm. Technician B says during an alternator output test, the charging system voltage should be lowered to 13 volts. Who is correct?

A. Technician A

B. Technician B

C. Both Technician A and Technician B

D. Neither Technician A nor Technician B

4. A vehicle has a charging system voltage of 15.5 volts with the engine at normal operating temperature. Technician A says this voltage may cause a discharged battery. Technician B says this voltage may damage electrical/electronic components on the vehicle. Who is correct?

A. Technician A

B. Technician B

C. Both Technician A and Technician B

D. Neither Technician A nor Technician B

5. All of these statements about testing alternator output are true *except*:

A. A slipping alternator belt may cause low alternator output.

B. Low alternator output may be caused by worn brushes and slip rings.

C. Low alternator output may cause a discharged battery.

D. Low alternator output may cause damaged electronic equipment on the vehicle.

6. An alternator produces a squealing noise when the vehicle is accelerated. When the alternator field wire is disconnected, the noise is not present. The most likely cause of this problem is:

A. an open alternator field winding.

B. a defective alternator diode.

C. a loose alternator drive belt.

D. a defective alternator voltage regulator.

7. When testing alternator output, a carbon pile load may be used to lower the alternator voltage to:

A. 14.2 volts.

B. 13.8 volts.

C. 13.6 volts.

D. 13.0 volts.

8. When discussing charging circuit diagnosis, Technician A says some charging circuits may be diagnosed with a scan tool. Technician B says when disconnecting a scan tool from the DLC, the ignition switch should be in the ON position. Who is correct?

A. Technician A

B. Technician B

C. Both Technician A and Technician B

D. Neither Technician A nor Technician B

9. With a 10-ampere charging rate, the maximum voltage drop between the alternator battery terminal and the positive battery terminal is:

A 0.1 volt.

B. 0.2 volt.

C. 0.5 volt.

D. 1.2 volts.

10. The most likely result of high resistance in the wire between the alternator battery terminal and the positive battery terminal is:

A. excessive battery gassing.

B. an undercharged battery.

C. damage to electronic components.

D. damage to the electronic voltage regulator.

11. In a charging system that has the voltage regulator in the PCM, DTCs related to charging system defects may be retrieved with a(n) _____.

12. When testing the resistance between the alternator battery terminal and the positive battery terminal, connect the positive voltmeter lead to the _____.

13. When testing the resistance in the alternator ground circuit, connect one voltmeter lead to the negative battery terminal and connect the other voltmeter lead to the _____.

14. Describe one precaution to be observed when connecting and disconnecting a scan tool from the DLC.

15. Describe the results of higher than specified charging system voltage.

16. Describe the result of higher than specified resistance in the charging system between the alternator battery terminal and the positive battery terminal.

17. If the PCM senses charging system defects, list four DTCs that may be stored in the PCM memory.

18. Explain the reason why vehicle manufacturers no longer recommend full fielding an alternator during an output test.

CHAPTER

21 Ignition Systems

Learning Objectives

After you have read, studied, and practiced the contents of this unit, you should be able to:

- Describe DI and EI ignition system operation.
- Describe briefly how the PCM controls spark advance.
- Describe basic operating principles of a COP system.

Key Terms

Hall effect switch

Heat range

Multiport fuel injection (MFI)

Sequential fuel injection (SFI)

Spark plug electrodes

Waste spark

INTRODUCTION

Proper ignition system operation is extremely important to obtain satisfactory engine performance and fuel economy. Ignition system defects, such as misfiring and incorrect timing reduce engine performance and economy. Understanding ignition system operating principles provides the necessary background information so the ignition diagnostic procedures described in subsequent chapters can be easily understood.

SPARK PLUGS

In gasoline internal combustion engines, spark plugs provide a critical air gap for the spark to jump across and ignite the fuel mixture. The purpose of the ignition system is to create a spark across the **spark plug electrodes** at the correct point in time in relation to piston movement. The spark at the plug electrodes ignites the compressed air-fuel mixture in the combustion chamber.

A spark plug is contained in a metal shell and the lower end of this shell has threads that match the threads in the cylinder head spark plug opening. A metal center electrode is positioned in the center of the spark plug and a ceramic insulator surrounds this electrode. The center electrode and insulator assembly is mounted in the metal shell. The upper end of the shell is crimped over the insulator preventing cylinder pressure from leaking up through the spark plug. A ground electrode is attached to the lower edge of the shell and the outer end of the ground electrode is positioned over the top of the center electrode. The spark plug gap is the distance between the center and ground electrodes **(Figure 21-1)**. This gap may vary from 0.035 inch (.889 millimeter) to 0.080 inch (2.032 millimeter) depending on the engine and ignition system design. Most spark plugs have a resistor in the center electrode to reduce radio static. The terminal at the top of the insulator contacts the center electrode. The spark plug wire is connected to the terminal on top of the spark plug.

The spark plug reach is the length of the threaded area. The spark plug reach usually varies from 3/8 inch (9.52 millimeters) to 3/4 inch (19.050 millimeters) depending on cylinder head and combustion chamber design. Replacement spark plugs must have reach specified by the vehicle manufacturer. If the spark plug reach is longer than specified, the spark plug extends too far into the combustion chamber **(Figure 21-2)**. Under this condition, the piston may strike the spark plug electrodes, resulting in severe piston damage. If the spark plug reach is too short, the electrodes are not positioned properly in the combustion chamber. This may result in improper combustion. The spark plug shell is sealed to the cylinder head

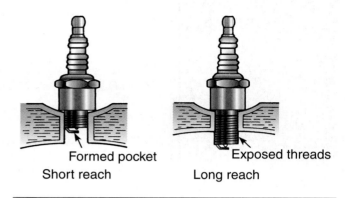

Figure 21-2 Spark plug reach.

opening by a metal gasket or a tapered seat on the shell. The spark plug **heat range** indicates the ability of the spark plug to dissipate heat from the center electrode through the spark plug insulator to coolant circulated through a cylinder head passage surrounding the spark plug. Hot spark plugs are designed with a long heat path so the heat has to travel further up through the insulator before it is dissipated through the shell into the cylinder head and coolant. In a spark plug with a colder heat range, the distance is shorter from the center electrode to the point where the insulator contacts the shell. This design allows faster heat dissipation through the spark plug so the electrode temperature is reduced **(Figure 21-3)**. On many spark plugs, the number on the spark plug indicates the heat range; a higher number represents a hotter range spark plug **(Figure 21-4)**. In recent years, many engines have platinum-tipped electrodes **(Figure 21-5)**. The platinum coating on the electrodes resists erosion and high temperatures to provide longer spark plug life.

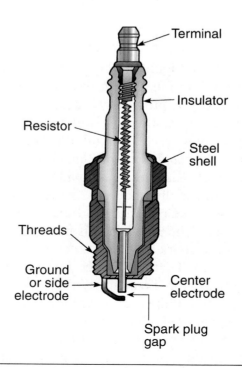

Figure 21-1 A typical spark plug design.

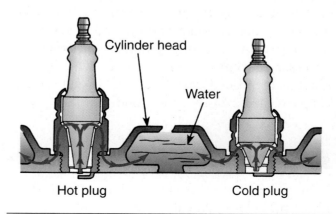

Figure 21-3 Spark plug heat range.

Figure 21-4 Heat range designators on spark plugs.

Figure 21-5 A platinum-tipped spark plug.

Manufacturers commonly specify platinum-tipped spark plug replacement at 100,000 miles (160,000 kilometers).

Most vehicles are equipped with spark plug cables that contain a resistance to provide television/radio suppression (TVRS). These wires have a conductor in the center that contains

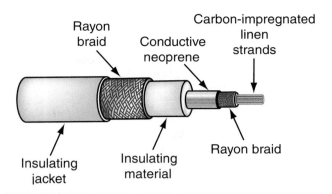

Figure 21-6 A suppression-type spark plug wire.

carbon-impregnated linen strands **(Figure 21-6)**. The center conductor is surrounded by a double braid and insulating material. Because spark plug wires conduct high voltage, they must have heavy insulation made from hypalon or silicone. Metal ends on the spark plug cable connect the center conductor to the spark plug terminal and the distributor cap or coil. Most spark plug cables have an outer diameter of approximately 0.312 inch (8 millimeters).

DISTRIBUTOR IGNITION (DI) SYSTEMS

As the name suggests, distributor ignition systems contain a distributor **(Figure 21-7)**. The main purpose of the distributor is to distribute the spark from the ignition coil to the spark plugs. This is accomplished by the distributor cap and rotor. The distributor often contains the primary ignition circuit controls. Modern distributors have a pickup coil and an ignition module to provide this function **(Figure 21-8)**. In some ignition systems, the module is mounted externally from the distributor. Distributors on vehicles built before computer controls have mechanical and vacuum advance mechanisms to provide spark advance **(Figure 21-9)**. Late-model engines often do not have a distributor because the computer provides for these functions. Regardless of the distributor design, this component plays a very important role in providing proper ignition operation and maintaining engine performance, economy, and emission levels.

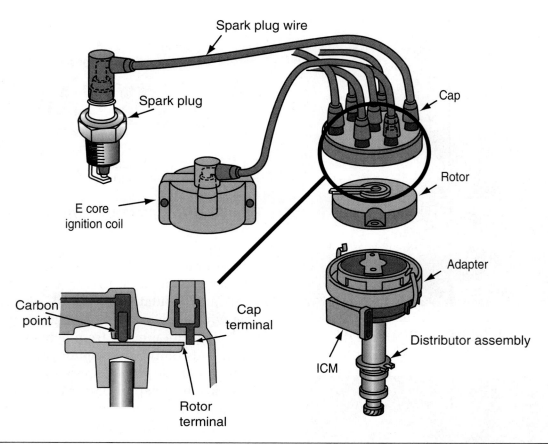

Figure 21-7 Ignition coil designs.

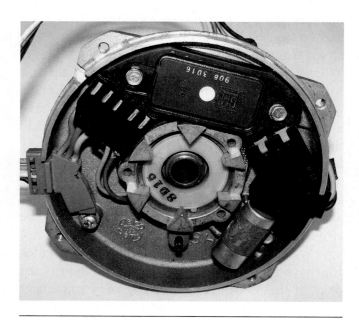

Figure 21-8 Pickup coil and module.

Figure 21-9 Mechanical and vacuum advance mechanisms.

Figure 21-10 Ignition coils.

Ignition Coils

A conventional ignition coil has a laminated soft iron core in the center of the coil **(Figure 21-10)**. This core contains many strips of metal. This type of core magnetizes and de-magnetizes faster than a solid metal core. A secondary winding contains thousands of turns of very fine wire wound around the core. One end of the secondary winding is connected to a terminal in the coil tower and the other end of this winding is usually connected to one of the primary terminals in the tower. Insulation is applied to the wire in the secondary winding so the turns of wire do not contact each other, and insulating paper is positioned between each layer of secondary turns. The primary winding contains a few hundred turns of heavier wire and this winding is wound on top of the secondary winding. The ends of the primary winding are connected to the primary terminals in the coil tower **(Figure 21-11)**. Some metal sheathing is placed around the outside of the primary winding and the winding assembly is installed in a metal can. The metal sheathing concentrates the magnetic field on the outside of the coil windings. In the manufacturing process, the coil is filled with oil through the screw opening in the coil tower. Then the screw is installed in the tower. The oil helps to cool the coil and also prevents air space in the coil that would allow moisture formation because of temperature change. If moisture forms in a coil, arcing will occur because water vapor has less resistance than air or oil.

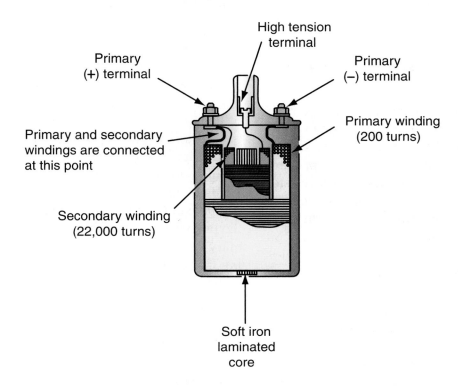

Figure 21-11 Ignition coil design.

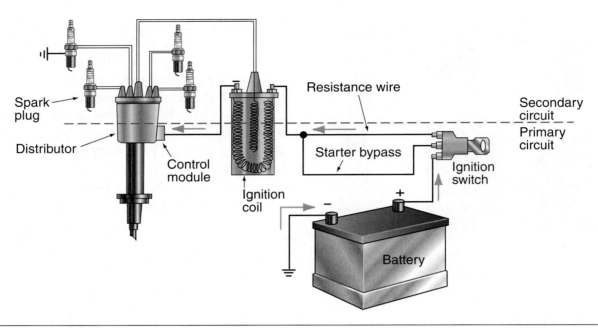

Figure 21-12 Primary and secondary ignition circuits.

Tech Tip *An ignition coil is a step-up transformer. Battery or charging system voltage is supplied to the primary winding and the ignition coil steps this voltage up to a very high voltage to fire the spark plugs in the secondary circuit.*

Primary and Secondary Ignition Systems

The primary ignition system contains the ignition switch, primary coil winding, ignition module, and pickup assembly. The primary ignition system is the control portion of the circuit. Some primary ignition circuits on older vehicles have a resistor connected in series between the ignition switch and the coil primary winding. The wire from the ignition switch is connected to the positive primary coil terminal and the negative primary coil terminal is connected to the ignition module, which is often mounted in the distributor. The primary circuit operates on battery or charging system voltage.

The secondary ignition circuit contains the secondary coil winding, coil secondary wire, distributor cap and rotor, spark plug wires, and spark plugs **(Figure 21-12)**. The secondary circuit operates on a very high voltage produced in the coil secondary winding. The voltage induced into the secondary coil windings is in the thousands of volts and extreme caution should be used when working around secondary ignition systems.

Pickup Assembly

The pickup assembly is usually mounted in the distributor. Many pickup assemblies contain a coil of wire surrounding a magnet. The ends of the pickup winding are connected to the ignition module. A reluctor attached to the distributor shaft has a high point for each engine cylinder. A small, specified clearance is provided between the reluctor high points and the pickup coil. Each time a reluctor high point rotates past and breaks the magnetic fields of the magnet, a voltage is inducted in the pickup coil **(Figure 21-13)**. This induced voltage in the pickup coil signals the module to open the primary circuit, causing the field that has been building in the primary side of the coil to collapse and induce a voltage in the secondary coil windings.

NOTE: Some vehicle manufacturers refer to the distributor reluctor as a *timer core* or an *armature*.

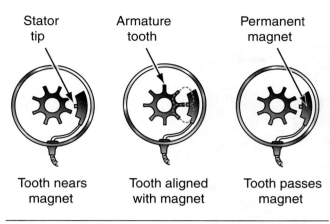

Figure 21-13 Pickup coil operation.

Figure 21-15 Hall effect switch.

Another method commonly used to control the opening and closing of the primary circuit is the **Hall effect switch (Figure 21-14)**. In this type of pickup, a metal blade assembly attached to the underside of the rotor has an individual blade for each engine cylinder. As these blades rotate through the pickup, the Hall element signal goes from low voltage to high voltage **(Figure 21-15)**. The Hall element is connected to the ignition module.

Ignition Modules

The ignition module contains many electronic components such as diodes, transistors, capacitors, and resistors. The ignition module is non-serviceable. Some ignition modules are mounted inside the distributor where they are bolted to the distributor housing, whereas other modules are bolted on the

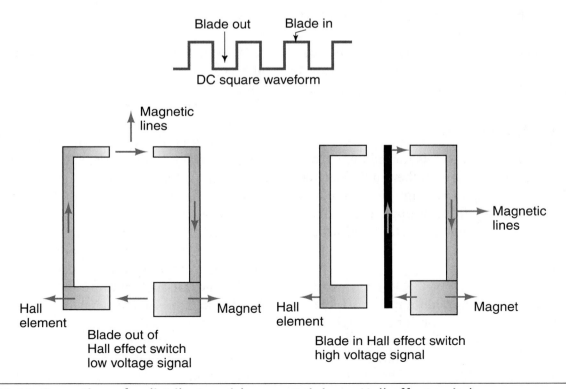

Figure 21-14 Operation of a distributor pickup containing a Hall effect switch.

Figure 21-16 Ignition modules.

outside of the distributor. Some modules are mounted externally from the distributor **(Figure 21-16)**. If the ignition module is mounted in or on the distributor, the module is grounded to the distributor housing. Most of these modules require a heat dissipating grease on the module surface that is in contact with the distributor housing to prevent module overheating. Externally mounted modules usually have a ground wire connected from the module to the vehicle ground. The purpose of the ignition module is to open and close the primary circuit at the right instant.

Ignition Operation

When the ignition switch is turned on, current flows from the battery through the ignition switch, coil primary winding, and the ignition module to ground. This current flow through the primary winding creates a strong magnetic field around both coil windings.

NOTE: The ignition module must turn the primary ignition circuit on long enough to allow the magnetic field to build up in the ignition coil. This buildup time is called *dwell time.* The dwell time is determined by the module design and it is not adjustable on

DI systems. If the module does not supply enough dwell time, the magnetic field will not have time to build up in the coil. This condition causes a weak magnetic field and reduced secondary voltage to fire the spark plug, resulting in spark plug misfiring.

When the engine begins cranking and a reluctor high point rotates past the pickup coil, a voltage signal is sent from the pickup coil, the module opens the primary circuit, causing the magnetic field in the coil to collapse rapidly across both coil windings. As the magnetic field collapses across the thousands of turns in the secondary winding, a very high voltage is induced in the secondary winding. This high voltage in the secondary winding forces current to flow through the coil secondary wire, distributor cap and rotor, and spark plug wire to the appropriate spark plug. As the secondary current arcs across the spark plug gap, it ignites the air-fuel mixture in the combustion chamber and the engine starts. When the magnetic field collapses in the coil, a low voltage is induced in the primary winding. This voltage is absorbed by protective circuitry in the module.

Each time a reluctor high point rotates past the pickup coil, the module opens the primary circuit

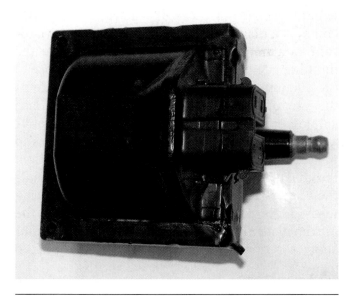

Figure 21-17 Laminated core coil.

and the magnetic field collapses in the coil, resulting in high induced voltage in the secondary winding and spark plug firing. The distributor cap and rotor distribute the secondary current to the appropriate spark plug.

The maximum secondary coil voltage may be over 35,000 volts on a typical coil. Late-model ignition systems often have laminated core coils that can generate over 100,000 volts **(Figure 21-17)**. However, if the spark plug gaps are set to specifications and the engine is operating at idle, the average voltage required to fire the spark plugs is 10,000 volts. The difference between the voltage required to fire the spark plugs and the maximum

available coil voltage is called reserve coil voltage. Reserve secondary coil voltage is necessary to compensate for extra resistance in the secondary circuit as the spark plug gaps erode and become wider. Reserve secondary coil voltage is also necessary to compensate for high cylinder pressures when the engine is operating under heavy load. For example, if the engine is operating at wide open throttle, resulting in high cylinder pressures, the voltage required to fire the spark plugs may be 18,000 volts.

Distributor Advances

When the distributor is installed in the engine, the engine must be positioned with number 1 piston at TDC on the compression stroke and the timing marks on the crankshaft pulley or flywheel in the 0 degree position. The distributor is then installed with the rotor under number 1 spark plug wire terminal in the distributor cap and one of the reluctor high points lined up with the pickup coil. The other spark plug wires must be installed in the distributor cap in the direction of distributor shaft rotation and in the proper engine firing and routing order **(Figure 21-18)**.

With the engine operating at idle speed, the distributor is rotated to provide the specified timing. This specified timing is usually between 2 degrees and 12 degrees BTDC depending on the engine. When the engine speed increases, the pistons move up and down much faster and the ignition system must advance the spark timing, so the air-fuel mixture has to start to burn before the piston

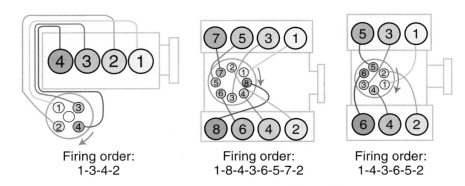

Figure 21-18 Proper spark plug wire installation in relation to firing order and distributor shaft rotation.

is at TDC **(Figure 21-19)**. This action maintains maximum downward force on the piston from the combustion process. If the ignition system does not advance the spark timing in relation to engine speed, the piston will be too far down in the power stroke by the time the air-fuel mixture completes its burn. Under this condition, engine power is greatly reduced. Some distributors have a centrifugal and a vacuum advance to control spark timing. The centrifugal advance contains pivoted weights that fly outward against a spring tension as the distributor shaft speed increases. This outward weight movement rotates the reluctor ahead of the distributor shaft to provide spark advance in relation to

engine speed **(Figure 21-20)**. The vacuum advance contains a diaphragm in a sealed chamber. Manifold vacuum or ported vacuum from above the throttle plate is supplied through a vacuum hose to the vacuum advance chamber. The other side of the vacuum advance chamber is open to atmospheric pressure and a rod on this side of the diaphragm is connected to the distributor pickup plate. When manifold vacuum is high during light throttle operation, the high manifold vacuum pulls the vacuum advance diaphragm against the spring tension. As the diaphragm moves, it rotates the pickup plate in the opposite direction to distributor shaft rotation. This action causes the pickup

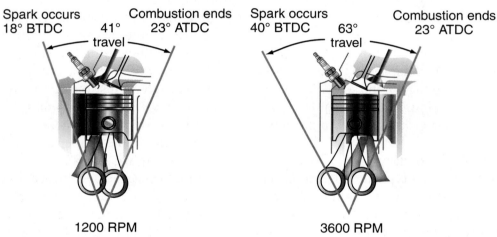

Spark timing must be advanced
as engine speed increases

Spark occurs
18° BTDC

41°
travel

Combustion ends
23° ATDC

Spark occurs
40° BTDC

63°
travel

Combustion ends
23° ATDC

1200 RPM

3600 RPM

Figure 21-19 Spark timing in relation to engine speed.

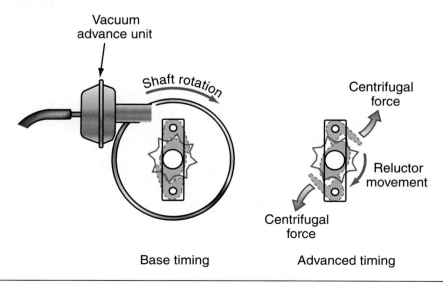

Vacuum
advance unit

Shaft rotation

Centrifugal
force

Reluctor
movement

Centrifugal
force

Base timing

Advanced timing

Figure 21-20 Centrifugal advance operation.

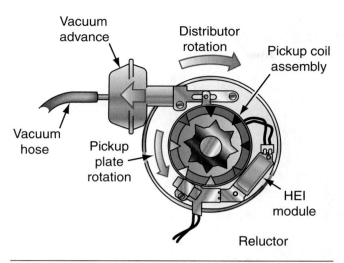

Figure 21-21 Vacuum advance operation.

coil to be aligned with the reluctor high points sooner. This results in additional spark advance **(Figure 21-21)**. When the throttle is 75 percent

or more open, manifold vacuum decreases. Under this condition, the spring on the vacuum advance diaphragm moves the diaphragm back to the retarded position. The vacuum advance controls spark advance in relation to engine load. During part throttle, light load operation, the vacuum advance provides additional spark advance to improve fuel economy and engine performance. If the engine is operating at wide throttle under heavy load conditions, the vacuum advance retards to prevent engine detonation.

Computer-Controlled Spark Advance

In many ignition systems, the spark advance is controlled by the PCM and the distributor advances are not required. The ignition module is connected to the PCM and the negative coil primary terminal is connected to the ignition module **(Figure 21-22)**.

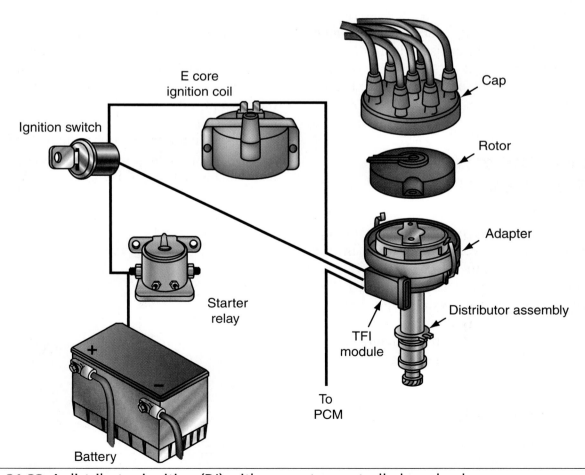

Figure 21-22 A distributor ignition (DI) with computer-controlled spark advance.

Figure 21-23 A crankshaft position sensor.

On the basis of the pickup coil signals and signals from other sensors, such as the ECT sensor, the PCM determines the precise spark advance required by the engine. The PCM signals the ignition module to open the primary ignition circuit and fire each spark plug at the correct instant to provide the necessary spark advance.

ELECTRONIC IGNITION (EI) SYSTEMS

The SAE Standard J1930 is an attempt to standardize automotive powertrain terminology. This standard was developed in 1991 and has since been revised. In the J1930, the DI refers to ignition systems with a distributor and EI refers to distributorless ignition systems. For instance, in the J1930 terminology the term for an engine computer is PCM. Most vehicle manufacturers use the J1930 terminology in their service publications.

Crankshaft Position Sensor

A distributor is not required in the EI system and maximum secondary coil voltage is 20 percent higher compared to previous ignition systems. The

crankshaft position sensor is mounted in one of three places, in an opening in the transaxle bell housing **(Figure 21-23)**, at the front of the crank shaft behind the belt pulley, or on the side of the block often behind the coil and module assembly. The inner end of this sensor is positioned near a ring with a series of notches and slots such as the ones on this transaxle flexplate **(Figure 21-24)**. A group of four slots is located on the transaxle drive plate for each pair of engine cylinders; thus a total of twelve slots are positioned around the drive plate on a V6 engine.

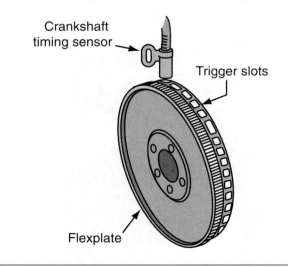

Figure 21-24 Types of crankshaft position sensors.

Figure 21-25 A camshaft reference sensor.

The slots in each group are positioned 20 degrees apart. Often there will be one slot or space between slots that will appear to not match the proper order; however, this is a reference slot and the PCM uses this to identify cylinder number 1. Once cylinder number 1 is identified, the PCM then knows the proper order of all cylinders after that since it was designed and built that way and does not change.

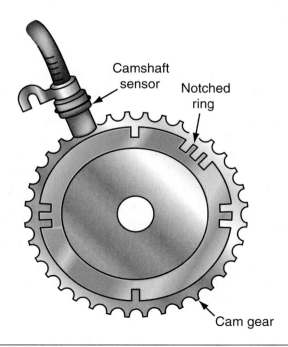

Figure 21-26 A notched ring on a camshaft gear.

When the slots on the transaxle drive plate rotate past the crankshaft position sensor, the voltage signal from the sensor changes from 0 volt to 5 volts. This varying voltage signal informs the PCM regarding crankshaft position and speed and the PCM calculates spark advance from this signal. The PCM also uses the crankshaft position sensor signal along with other inputs to determine the air-fuel ratio. Base timing is fixed at the factory and is not externally adjustable.

Camshaft Reference Sensor

The camshaft reference sensor is mounted in the top of the timing gear cover **(Figure 21-25)**. A notched ring on the camshaft gear rotates past the end of the camshaft reference sensor. This ring contains two single slots, two double slots, and a triple slot **(Figure 21-26)**.

When a camshaft gear notch rotates past the camshaft reference sensor, the signal from the sensor commonly changes from 0 volt to 5 volts. The single, double, and triple notches provide different voltage signals from the camshaft reference sensor. These voltage signals are sent to the PCM. The PCM determines the exact camshaft and crankshaft position from the camshaft reference sensor signals. The PCM uses these signals to sequence the

Figure 21-27 An EI system coil assembly.

coil primary windings and each pair of injectors at the correct instant. The PCM supplies the voltage reference signal to both the crankshaft position sensor and the camshaft reference sensor. The ground side of the circuit also returns through the PCM and not through a more common frame type ground.

Coil Assembly

The coil assembly contains three ignition coils on a V6 engine **(Figure 21-27)**. The ends of each secondary winding are connected to the dual secondary terminals on each coil. Two spark plug wires are connected from the secondary terminals on each coil to the spark plugs. When the ignition switch is turned on, 12 volts are supplied through the ignition switch to the positive primary terminal on each ignition coil. The negative primary terminal on each coil is connected to the PCM.

IGNITION SYSTEM OPERATION

When the engine starts cranking, the spark plugs fire and the injectors discharge fuel within one crankshaft revolution. The spark plug wires

from coil number 1 are connected to cylinders 1 and 4 and the spark plug wires from coil number 2 go to cylinders 2 and 5. The spark plug wires from coil number 3 are connected to cylinders 3 and 6. The firing order of V6 engines is typically 1-2-3-4-5-6. These cylinders are paired together to take advantage of the fact that when one cylinder of the pair is at TDC the other cylinder is beginning the exhaust stroke and firing a spark plug on a cylinder that has already had its fuel burned will have no adverse affect.

> **Tech Tip** *In the EI system being explained, the ignition module is in the PCM. Some other EI systems have an ignition module that is mounted externally to the PCM, but the operating principle of both systems is similar.*

When the single camshaft reference sensor slot rotates past the camshaft reference sensor, one high and one low-digital voltage signal are received by the PCM. The PCM is programmed to understand these pulses as indicators of where the camshaft is positioned in its rotation. After these signals are

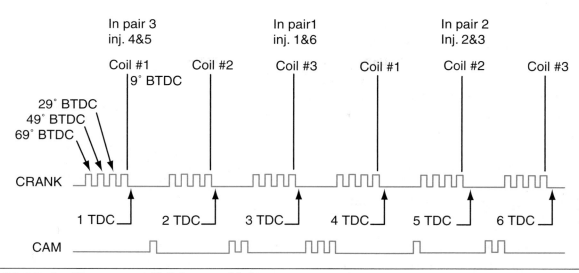

Figure 21-28 An EI system coil firing and injector sequencing.

received, the PCM receives the digital signals from the crankshaft position sensor as cylinder number 2 is approaching TDC on the compression stroke **(Figure 21-28)**. When the engine is cranking and the last decreasing voltage signal is received from the crankshaft position sensor at 9 degrees BTDC, the PCM opens the primary circuit on the number 2 coil and fires spark plugs 2 and 5.

The coils in many EI systems operate on a **waste spark** principle of firing two spark plugs simultaneously with one spark plug being in a cylinder on the compression stroke and the other spark plug located in a cylinder on the exhaust stroke **(Figure 21-29)**. When a coil fires two spark plugs in a waste spark system, the secondary current flows down through one pair of spark plug electrodes and up through the other pair of spark plug electrodes.

When the engine is cranking, the PCM fires all cylinders at a predetermined advance BTDC. When the PCM fires spark plugs 2 and 5, it also sequences the injectors in to provide fuel when the intake valves open. In a **multiport fuel injection (MFI)** system, the PCM opens two or more injectors simultaneously. These injectors can be opened in pairs, all together, or by sides in a V8 depending on the particular model and manufacturer. The reference signal to open injector typically comes from the camshaft position sensor.

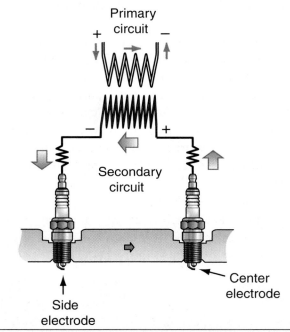

Figure 21-29 Spark plugs firing in an EI system.

In order to provide the precise spark advance required by the engine, the PCM receives crankshaft position sensor signals at predetermined intervals that correspond to BTDC on each cylinder. By counting the time interval between these crankshaft position sensor signals, the PCM can fire any cylinder at the precise spark advance required by the engine based upon the conditions at which the engine is operating. The PCM determines this precise spark advance requirement from all the input data it receives from other input sensors. Some engines have a detonation sensor located in the cowl side of the engine block. If the engine detonates, this sensor signals the PCM to reduce the spark advance.

Late-model engines are now equipped with **sequential fuel injection (SFI)**, in which the PCM opens each injector individually, just prior to a cylinder's intake valve or valves opening. The basic inputs are the same as MFI.

Coil-On-Plug and Coil-Near-Plug Ignition Systems

Many engines are presently equipped with coil-on-plug ignition systems. On these systems, the coil secondary terminals are connected directly to the spark plugs and secondary plug wires are not required **(Figure 21-30)**. This design eliminates the possibility of high voltage leakage from secondary spark plug wires. Early coil-on-plug ignition systems often had two spark plugs connected to each coil as described previously in the EI system **(Figure 21-31)**. Some coil-on-plug ignition systems have individual coils connected to each spark plug. In these systems, 12 volts are supplied to each positive primary terminal when the ignition switch is turned on and each negative primary terminal is connected to the ignition module. The crankshaft position and camshaft reference sensors are similar to those on the EI system. Coil-near-plug ignition systems have individual coils for each spark plug and a short spark plug wire is connected from each coil secondary terminal to the spark plug **(Figure 21-32)**.

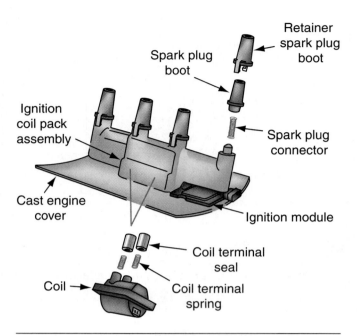

Figure 21-31 Early model coil-on-plug system.

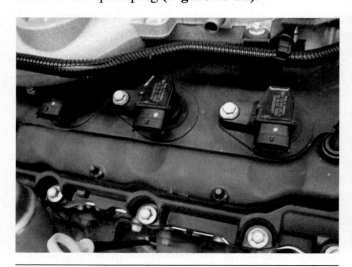

Figure 21-30 Coil-on-plug ignition system.

Figure 21-32 A coil-near-plug ignition system.

Summary

- Spark plug reach is the distance from the lower edge of the plug shell to the shoulder on the shell that fits against the cylinder head.

- Spark plug heat range is determined by the length of the heat path from the center electrode to the location where the insulator contacts the plug shell.

- The recommended replacement interval for platinum-tipped spark plugs is 100,000 miles (160,000 kilometers).

- The primary winding in an ignition coil contains a few hundred turns of heavy wire and the secondary winding contains thousands of turns of very fine wire.

- The ignition switch, primary coil winding, ignition module, and pickup coil are in the primary ignition circuit.

- The secondary coil winding, coil wire, distributor cap and rotor, spark plug wires, and spark plugs are in the secondary ignition circuit.

- A voltage is produced in the distributor pickup coil each time a reluctor high point rotates past the pickup coil.

- The ignition module opens the primary ignition circuit each time it receives a voltage signal from the pickup coil.

- The ignition module must turn the primary circuit on long enough to allow the magnetic field to build up in the ignition coil. This buildup time is referred to as dwell time.

- The centrifugal advance controls spark advance in relation to engine speed.

- The vacuum advance controls spark advance in relation to engine load.

- In DI systems, the PCM controls the spark advance and the centrifugal and vacuum advances are not required.

- In an EI system, the distributor is eliminated and a coil is provided for each pair of cylinders.

- In an EI system, each coil fires two spark plugs simultaneously. One of these spark plugs is in a cylinder on the compression stroke and the other spark plug is in a cylinder on the exhaust stroke.

- In an EI system, the ignition module may be located in the PCM or may be mounted externally to the PCM.

- In a coil-on-plug ignition system, the spark plug wires are eliminated and the coil secondary terminals are connected directly to the spark plugs.

- A coil-near-plug ignition system has an individual coil mounted near each spark plug and a short secondary wire is connected from the coil secondary terminal to the spark plug.

Review Questions

1. Technician A says that spark plugs with a hotter heat range have a higher number compared to spark plugs with a colder heat range. Technician B says spark plugs with a hotter heat range have a longer heat path compared to spark plugs with a colder heat range. Who is correct?

   A. Technician A

   B. Technician B

   C. Both Technician A and Technician B

   D. Neither Technician A nor Technician B

2. Technician A says the secondary coil winding is connected to the ignition module. Technician B says the primary coil winding contains many turns of very fine wire. Who is correct?

   A. Technician A

   B. Technician B

   C. Both Technician A and Technician B

   D. Neither Technician A nor Technician B

3. Technician A says in some EI systems, each coil fires two spark plugs simultaneously. Technician B says in some EI systems, the camshaft reference sensor can be mounted in the transaxle bell housing. Who is correct?

   A. Technician A

   B. Technician B

   C. Both Technician A and Technician B

   D. Neither Technician A nor Technician B

4. Technician A says in an MFI system, the PCM fires each injector individually. Technician B says in an EI system, the PCM uses the camshaft reference sensor signal to fire the coils in the proper order. Who is correct?

   A. Technician A

   B. Technician B

   C. Both Technician A and Technician B

   D. Neither Technician A nor Technician B

5. A typical service interval on platinum-tipped spark plugs is:

   A. 35,000 miles.

   B. 50,000 miles.

   C. 80,000 miles.

   D. 100,000 miles.

6. All of these statements about DI systems are true *except*:

   A. The distributor pickup coil is part of the secondary circuit.

   B. The pickup coil leads are connected to the ignition module.

   C. The ignition module opens and closes the primary circuit.

   D. The ignition module may be mounted externally from the distributor.

7. In a DI system, the high voltage in the secondary winding is produced:

   A. when the reluctor high points are not aligned with the pickup coil.

   B. when the magnetic field begins to build up in the coil.

   C. when there is no voltage signal from the pickup coil to the module.

   D. when the magnetic field collapses in the coil.

8. Technician A says secondary reserve voltage in an ignition system is necessary to compensate for high combustion chamber pressures at heavy engine loads. Technician B says secondary reserve voltage in an ignition system is necessary to compensate for wear at spark plug electrodes. Who is correct?

   A. Technician A

   B. Technician B

   C. Both Technician A and Technician B

   D. Neither Technician A nor Technician B

9. When discussing DI systems, Technician A says the spark plug wires must be installed in the distributor cap in the opposite direction to distributor shaft rotation. Technician B says the spark plug wires must be installed in the distributor cap in the engine firing order. Who is correct?

   A. Technician A

   B. Technician B

   C. Both Technician A and Technician B

   D. Neither Technician A nor Technician B

10. All of these statements about EI systems are true *except*:

   A. The crankshaft position sensor informs the PCM regarding crankshaft position and speed.

   B. On many EI systems, the PCM uses the camshaft position sensor signals to properly sequence the injectors and ignition coils.

   C. On many EI systems, each coil fires two spark plugs at the same time.

   D. On some EI systems, each coil fires one spark plug on the compression stroke and another spark plug on the intake stroke.

11. The vacuum advance controls ignition spark advance in relation to engine _____.

12. When the centrifugal advance weights move outward, they move the _____ in relation to the distributor shaft.

13. In a DI system, the pickup coil leads are connected to the _____.

14. In a primary ignition circuit, the dwell time is the time that the module keeps the primary circuit _____.

15. Explain the importance of reserve voltage in the secondary ignition system.

16. Explain the waste spark principle in an EI system.

17. Describe the difference between MFI and SFI systems.

18. Explain the importance of dwell time in the primary ignition circuit.

CHAPTER

22

Ignition System Maintenance, Diagnosis, and Service

Learning Objectives

After you have read, studied, and practiced the contents of this unit, you should be able to:

- Describe fouled spark plug conditions.
- Diagnose normal spark plug wear.
- Perform a DI ignition no-start diagnosis.
- Diagnose and test coils.
- Time a distributor to an engine and adjust ignition timing.
- Test crankshaft and camshaft position sensors.

Key Terms

Graphing voltmeter

Lab scope

Open pickup coil

Preiginition

Shorted pickup coil

Short to ground

INTRODUCTION

Proper ignition system operation is extremely important to obtain satisfactory engine performance and fuel economy. For example, a defective spark plug or spark plug wire that results in cylinder misfiring will greatly reduce engine performance and fuel economy. A defective ignition coil causes cylinder misfiring on some or all cylinders depending on the ignition system. This misfiring is most noticeable during hard engine acceleration and results in a significant reduction in engine power and fuel economy. Proper ignition system operation is maintained by performing correct ignition system maintenance, diagnosis, and service.

SPARK PLUG MAINTENANCE, DIAGNOSIS, AND SERVICE

In a four-cylinder engine running at 3,000 rpm, each spark plug is firing twenty-five times per second. A considerable amount of intermittent spark plug misfiring can go undetected because of the very rapid spark plug firing. This intermittent misfiring causes increased exhaust emissions and reduced fuel economy. Spark plug maintenance, diagnosis, and service are extremely important to provide the intended fuel economy and emission levels designed into the vehicle by the manufacturer.

The condition of an engine's spark plugs indicate what is happening in the cylinder. Correctly interpreting the type and amount of deposits on a

Tech Tip *In DI systems, the center spark plug electrodes erode faster than the ground electrodes because current is flowing from the center electrode to the ground electrode. In some EI systems, each coil fires two spark plugs, with current flowing down through one set of spark plug electrodes and up through the other set of electrodes. On the spark plugs with current flowing up through the electrodes, the ground electrode erodes faster than the center electrode. When center spark plug electrodes become eroded and rounded instead of having a flat surface, a higher secondary voltage is required to fire the spark plugs.*

Figure 22-1 Spark plug sockets.

used spark plug gives a good indicator of the conditions that are happening in that particular cylinder while operating. In some cases, spark plug conditions indicate a cylinder problem. If this problem is corrected when it is first detected, it may prevent a more expensive repair bill and/or tow bill later on.

Spark Plug Removal

Spark plugs should be removed at the vehicle manufacturer's recommended service interval. This service interval may be from 20,000 to 100,000 miles (32,000 to 160,000 kilometers) depending on the type of spark plug and the operating conditions. For example, many engines are presently equipped with platinum-tipped spark plugs with a specified replacement interval of 100,000 miles (160,000 kilometers). As a general rule all spark plug service is done when the engine is cold. Aluminum engine heads are particularly susceptible to damage. If the spark plugs are removed when the engine is hot, often the threads are damaged if extreme care is not taken.

Before removing the spark plugs, grasp the spark plug wire boot and twist it back and forth on the spark plug before pulling the wire off the spark plug. Applying excessive pulling force to the spark plug wire may damage the wire.

When loosening and removing spark plugs, use a special spark plug socket and keep the ratchet or breaker bar as close as possible to a 90-degree angle to the socket **(Figure 22-1)**. This procedure helps to prevent the socket from cocking to one side and breaking the spark plug insulator. A spark plug socket has an internal rubber grommet that reduces the possibility of cracking the spark plug insulator. Before the spark plugs are loosened, use a shop air gun to blow any debris out of the spark plug recesses in the cylinder head. This prevents debris from entering the cylinder when the spark plug is removed. When the spark plugs are removed, keep them in order so you can identify the spark plug removed from each cylinder.

Spark Plug Maintenance and Diagnosis

If normal combustion is occurring in all the cylinders, the lower end of each spark plug insulator should have a minimum amount of light tan or gray carbon deposits. Inspect the spark plug electrodes. The current arcing across the electrodes slowly erodes the electrodes, but this electrode erosion should not exceed 0.001 inch (.0254 millimeter) for every 10,000 miles (16,000 kilometers). Excessive electrode burning may be caused by too hot a spark plug heat range, improper spark plug torque and cooling, a lean air-fuel mixture, or engine detonation. Compare the spark plug part number with the part specified by the manufacturer. Severely damaged spark plug electrodes indicate **preignition** damage **(Figure 22-2)**. Inspect the upper and lower ends of the insulator for cracks. Spark plugs with burned electrodes or cracked insulators must be replaced.

Preignition is the ignition of the air-fuel mixture in the combustion chamber by something other than the spark at the spark plug electrodes. This is usually caused by hot carbon deposits or spark plugs with a heat range that is too hot.

NOTE: Preignition or detonation causes an engine noise similar to stones rattling around inside a metal container. The shop term for this noise is pinging

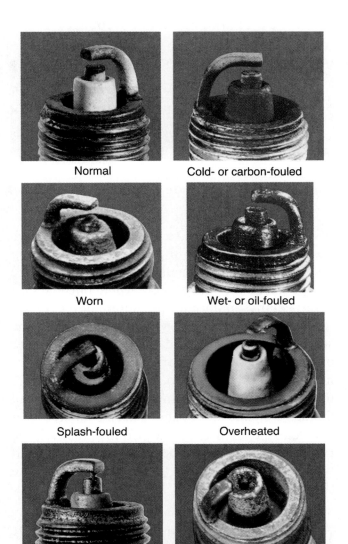

Normal

Cold- or carbon-fouled

Worn

Wet- or oil-fouled

Splash-fouled

Overheated

Gap bridging

Preignition damage

Figure 22-2 Defective spark plug conditions. *(Courtesy of Champion Spark Plug Company)*

Carbon-fouled spark plugs have dry, black carbon deposits. Any type of carbon deposits on the spark plug insulator causes the high secondary ignition voltage to arc from the center electrode across the carbon deposits to the spark plug shell. Carbon-fouled plugs will misfire because it is easier for electricity to conduct along the carbon than to cross the spark gap This results in a cylinder misfiring and the air-fuel mixture in the combustion chamber is not ignited. Carbon-fouled spark plugs may becaused by a rich air-fuel mixture, secondary ignition misfiring on that spark plug, a colder than-specified spark plug heat range, or low cylinder compression.

Oil-fouled spark plugs are recognized by wet oil deposits on the spark plug insulator and electrodes. Oil-fouled spark plugs have the same effect as carbon-fouled spark plugs. Oil-fouled spark plugs are caused by oil entering the combustion chamber. The causes of this problem may be worn piston rings or cylinder walls, worn valve guide seals or valve guides, leaking intake manifold gaskets on some V8 or V6 engines, a leaking transmission modulator, or a completely restricted PCV system.

Splash-fouled spark plugs have hard carbon deposits splashed on the spark plug insulators. A splash-fouled spark plug is usually caused by combustion chamber deposits that loosen and stick to the spark plug insulator. This condition may occur after a cylinder has been misfiring, allowing some carbon deposits to build up in the combustion chamber.

Tech Tip *Fouled spark plugs may not cause misfiring at idle speed because secondary ignition voltage required to fire is low under this condition. During hard acceleration, the secondary ignition voltage must increase because of the increase in combustion chamber pressure. This higher voltage increases the possibility of arcing across the carbon, oil, or splash-fouling spark plug deposits.*

When the cause of the misfiring is corrected, these carbon deposits may loosen and cause a splash-fouled spark plug. Occasionally hot combustion chamber deposits loosen and become lodged between the spark plug electrodes. This spark plug condition is called gap bridging. Spark plugs that show any of these defective conditions must be replaced. The cause of the defective spark plug condition must also be corrected. For example, if spark plugs indicate oil fouling, the cause of this problem must be corrected before new spark plugs are installed.

Spark Plug Service and Installation

When installing a new set of spark plugs, always use the spark plugs with the number specified by the vehicle manufacturer. Spark plug electrode gaps must be set to the vehicle manufacturer's specifications using a round wire feeler gauge or a gapping tool **(Figure 22-3)**. The specific spark plug gap and part number is often located on the vehicle's under-hood tag **(Figure 22-4)**.

Tech Tip *Several years ago it was a recommended practice to clean spark plugs in a sand blaster. In modern engines, this procedure is not practical because spark plugs are inexpensive and the labor cost is high to remove and replace them on many engines. Attempting to clean platinum-tipped spark plugs in a sand blast-type cleaner will destroy the platinum coating on the electrodes.*

Figure 22-3 **Spark plug gapping tools.**

Figure 22-4 **Typical under-hood emission tag.**

If spark plugs require a metal gasket, always be sure this gasket is in place before installing the plugs. Place a small amount of anti-seize compound on the first two spark plug threads. This is especially important on aluminum cylinder heads to prevent electrolytic action between the steel spark plug shell and the aluminum cylinder head. Always start the spark plug into the cylinder head threads by hand to avoid damaging these threads. Turn the spark plug into the cylinder head threads by hand until the gasket or tapered seat contacts the cylinder head. Follow the vehicle manufacturer's recommended procedure and tighten the spark plugs to the specified torque **(Figure 22-5)**.

SAFETY TIP *When spark plug torque is less than specified, the spark plug electrodes are not properly cooled. Under this condition, the excessive electrode temperature may cause engine detonation and severe engine damage.*

Torque Specifications

Description	Nm	Lb/Ft	Lb/In
Oil Level Indicator Tube Bolt	10–13	–	72–99
Intake Manifold Runner Control Actuator Bolts	10–13	–	72–99
Front Engine Accessory Drive Bracket Nut	40–55	30–40	–
Speed Control Cable Bracket Bolts	10–12	–	72–90
Front Engine Accessory Drive Bracket Bolts	42–57	32–42	–
Intake Manifold Nuts	22–30	17–24	–

Figure 22-5 Torque tightening table.

Tech Tip *If spark plug wires from cylinders that fire one after the other are placed beside each other, the magnetic field from current flow through one wire may build up and collapse across the adjacent wire. This action induces a voltage in the adjacent wire that fires the spark plug connected to this wire when the piston in this cylinder is approaching TDC on the compression stroke. This action may cause preignition, resulting in spark plug electrode and possible engine damage.*

DI SYSTEM MAINTENANCE, DIAGNOSIS, AND SERVICE

If DI system maintenance, diagnosis, and service is ignored or is not provided at the vehicle manufacturer's recommended interval, the result may be an expensive tow bill and a considerable amount of inconvenience. When maintenance, diagnosis, and service are ignored, minor ignition defects may become major problems that cause the engine to quit running on a busy highway or perhaps in an area where there is very little traffic and no service facilities nearby.

Ignoring DI system maintenance, diagnosis, and service is more expensive in the long term than having the vehicle's service completed at the manufacturer's recommended service interval.

DI System Maintenance

At the vehicle manufacturer's recommended service interval, the components in the DI system should be inspected, diagnosed, and serviced. Spark plug wires should be inspected for corroded terminals on both ends. These wires should also be inspected for insulation cracks and heat damage. Test the spark plug wires with an ohmmeter on the X1,000 scale **(Figure 22-6)**. Most TVRS spark plug wires should have a maximum resistance of 10,000 ohms per foot. Generally, if one wire needs replacement it is best to replace the entire set, unless one was damaged by unusual circumstances such as being cut. Normally, if one is bad the rest will be in similar condition due to age and it is in the customer's best interest to replace as a set to avoid a return for a similar problem in the near future.

Spark plug wires should be placed in retaining clips so they are positioned away from exhaust manifolds and engine drive belts. When spark plug wires are placed in retaining clips, the wires from cylinders that fire one after the other should not be positioned beside each other. Spark plug wires should be routed away from computers and computer wiring harnesses. Be sure the spark plug wires are installed completely into the distributor cap terminals. If the spark plug wires are to be reused, it is a good idea to put a small dab of dielectric grease

Ohmmeter

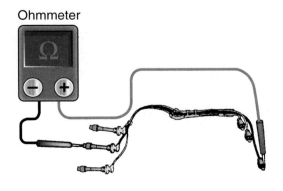

Figure 22-6 Testing spark plug wires with an ohmmeter.

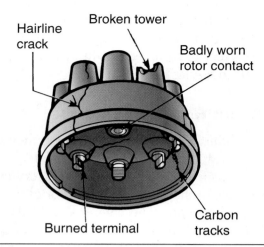

Figure 22-7 Distributor cap inspection.

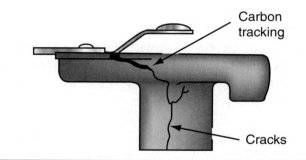

Figure 22-8 Rotor inspection.

inside the boot; not only will it help seal the older boot but will help keep the boot from sticking to the spark plug for future maintenance.

When performing basic ignition system maintenance, the distributor should be inspected. Inspect the distributor cap for corroded or worn terminals and cracks or carbon tracking **(Figure 22-7)**. Cracked distributor caps must be replaced. Inspect the rotor for cracks and corroded or worn terminals **(Figure 22-8)**. If the terminals are corroded or damaged, the cap should be replaced. If the distributor has advances, inspect the centrifugal advance for free movement and wear on the weight pivots. Use a vacuum hand pump to apply a vacuum to the vacuum advance diaphragm. If this diaphragm is leaking, the vacuum will slowly decrease. When this vacuum is applied to the diaphragm, be sure the pickup plate rotates freely. If the vacuum advance diaphragm is leaking, replacement is necessary.

Worn distributor shaft bushings result in excessive sideways shaft movement. Check to see if this movement is excessive; most vehicle manufacturers do not supply replacement bushings, and so

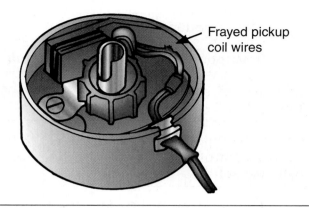

Figure 22-9 Inspecting pickup coil leads.

distributor replacement is required. Inspect the pickup coil and module leads for worn insulation and loose terminals **(Figure 22-9)**.

DI System No-Start Diagnosis

If the engine fails to start, use a voltmeter to measure the voltage from the positive primary ignition coil terminal to ground with the ignition switch on. On most ignition systems, 12 volts should be available at this terminal. Some older ignition systems have a resistor in the primary circuit that reduced the voltage at the positive primary terminal to approximately 6 volts. When the specified voltage is not available at the positive primary ignition coil terminal, test the ignition switch and the wire from this switch to the coil. If the specified voltage is available at the positive primary coil terminal, connect a 12-volt test light from the negative primary terminal to ground and crank the engine. If the test light flashes on and off, the primary circuit is being triggered on and off by the module, indicating that the pickup coil and module are satisfactory. When the test light does not flutter, one of these components is defective and individual component testing is necessary to determine the defective component. If the test light flashes on and off, connect the proper test spark tester from the coil secondary lead to ground and crank the engine **(Figure 22-10)**. If the spark tester does not fire, the ignition coil is defective. When the test spark plug fires normally, the coil is satisfactory. Connect the spark tester from several spark plug wires to ground and crank the engine, if the spark tester does not fire when connected to the spark plug wires, the secondary voltage is leaking in the distributor cap and rotor and they will need to be replaced.

Figure 22-10 Spark testers.

PICKUP COIL DIAGNOSIS

An electromagnetic-type pickup coil may be tested with an ohmmeter. Connect the ohmmeter leads to the pickup. An infinite ohmmeter reading indicates an open pickup coil, whereas a reading below the specified value indicates a shorted pickup coil.

An electromagnetic pickup coil contains a permanent magnet surrounded by a coil of wire. An infinite ohmmeter reading is an ohm reading beyond measurement, causing an OL display on a digital ohmmeter. An **open pickup coil** has an unwanted break in the winding. In a **shorted pickup coil**, the turns of wire are touching each other, thereby reducing the effective number of turns and the coil resistance.

> **Tech Tip** A non-magnetic feeler gauge must be used to measure the pickup coil gap because a metal gauge sticks to the pickup magnet and provides an inaccurate reading. Some pickup coils are riveted to the breaker plate and the gap on these pickup coils is not adjustable.

Connect the ohmmeter leads from one of the pickup coil leads to ground. An infinite reading indicates the pickup coil is not grounded, whereas a low reading indicates a grounded pickup coil. If either of the pickup coil tests are unsatisfactory, the pickup coil must be replaced. A pickup coil with a **short to ground** has an unwanted connection between the pickup coil winding or lead wires and ground on the distributor housing.

The gap between the head on the pickup coil and the reluctor high points may be measured with a non-magnetic feeler gauge. On some pickup coils, the mounting screws may be loosened and the pickup coil moved to adjust the pickup coil gap. Always check for excessive side-to-side distributor shaft movement, which indicates worn distributor bushings. Often when the bushings are worn, the pickup will be damaged from the reluctor hitting it as it turns.

> **Tech Tip** A variety of testers are available to test ignition modules. These testers usually check the ability of the module to switch the primary ignition circuit on and off.

Testing Ignition Coil Windings

> **Tech Tip** *A shorted primary winding causes excessive primary current, which may damage other components in the primary circuit, such as the ignition module.*

Place an ohmmeter on the lowest scale and connect the meter leads to the primary coil terminals **(Figure 22-11)**. A satisfactory primary winding provides the specified resistance, which is usually between 0.5 and 2.0 ohms. If the ohmmeter reading is less than specified, the primary winding is shorted. An infinite ohmmeter reading indicates an open primary winding.

Switch the ohmmeter to a higher scale and connect the ohmmeter leads from the secondary coil terminal in the coil tower to one of the primary terminals **(Figure 22-12)**. A satisfactory secondary winding provides the specified ohmmeter reading, which is usually 8,000 to 20,000 ohms. An infinite reading indicates an open secondary winding and a

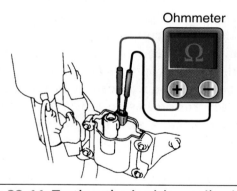

Figure 22-11 Testing the ignition coil primary winding.

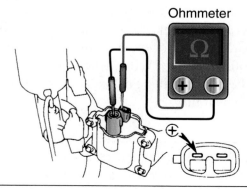

Figure 22-12 Testing the secondary ignition coil winding.

lower-than-specified reading indicates a shorted secondary winding. Keep in mind that it is not uncommon for a back coil to pass an ohm test, particularly a shorted one, so if all other test results are good and still point toward the coil, it is most likely the fault.

> **Tech Tips** *Insulation leakage around the secondary coil winding will cause low maximum secondary coil voltage and the coil winding tests with an ohmmeter may be satisfactory.*
>
> *The values that are given for tests in this book are generalities. Always look up the exact specifications for the vehicle being worked on.*

TIMING THE DISTRIBUTOR TO THE ENGINE

If the distributor is removed, it must be properly timed to the engine when it is re-installed. Follow this procedure to time the distributor to the engine:

1. Remove the spark plug from cylinder number 1 and place a compression gauge hose fitting in the spark plug hole.

2. Crank the engine a small amount at a time until compression pressure appears on the gauge.

3. Crank the engine a very small amount at a time until the 0 degree position on the timing marks is aligned with the timing indicator.

4. Locate the number 1 spark plug wire position in the distributor cap. The wire terminals in some caps are marked (the manufacturer's service manual provides this information). If number 1 is not marked, use a pen to mark it so the rotor can be easily lined up later.

5. Install the distributor in the block with the rotor positioned under the number 1 plug wire position in the distributor cap and the vacuum advance in the original position. The distributor drive gear easily meshes with the camshaft gear, but many distributors also drive the oil pump with a hex-shaped or slotted drive in the lower end of the distributor gear or shaft. It may be necessary to turn the oil pump drive to get mesh with the distributor.

6. Rotate the distributor a small amount until a high point on the reluctor is aligned with the head on the pickup coil.

7. Install the distributor hold down clamp and bolt, leaving this bolt loose enough so the distributor can be turned by hand.

8. Install the spark plug wires in the distributor cap in the direction of distributor shaft rotation and in the cylinder firing order **(Figure 22-13)**.

9. Connect the distributor wiring connectors. The vacuum advance hose is usually left disconnected and plugged until the ignition timing is set with the engine running.

> **Tech Tip** *The distributor shaft rotates in the opposite direction to which the vacuum advance pulls the pickup plate.*

CHECKING AND ADJUSTING IGNITION TIMING

The ignition timing procedure varies depending on the make and year of vehicle and the type of ignition system. Ignition timing specifications and instructions are included on the under-hood emissions label. More detailed instructions are provided in the vehicle manufacturer's service manual. The ignition timing procedure recommended by the vehicle manufacturer must be followed. On distributors with advance mechanisms, manufacturers usually recommend disconnecting and plugging the vacuum advance hose while checking the ignition timing. On carbureted engines, the manufacturer usually specifies a certain engine rpm while checking ignition timing. The timing light pickup is connected to number 1 spark plug wire and the power supply wires on the light are connected to the battery terminals with the correct polarity. The following is a general guide for setting most systems:

1. If the spark advance is controlled by the PCM, disconnect the in-line timing connector. The under-hood emissions label usually provides the location of this connector. When the timing connector is disconnected, the PCM cannot provide any spark advance.

2. Connect the timing light and start the engine.

3. The engine must be idling at the specified rpm and all other timing procedures must be followed.

4. Aim the timing light at the timing indicator and observe the timing marks **(Figure 22-14)**.

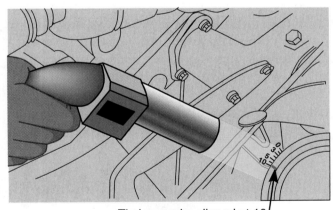

Timing marks aligned at 10

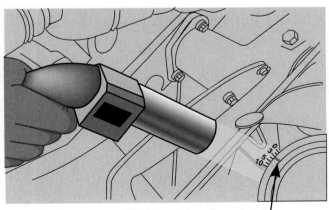

Timing marks aligned at 3

Figure 22-14 Checking ignition timing.

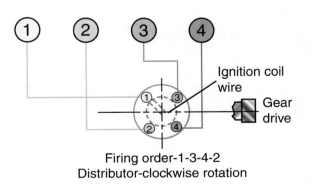

Firing order-1-3-4-2
Distributor-clockwise rotation

Figure 22-13 Installation of spark plug wires in the firing order and direction of distributor shaft rotation.

5. If the timing mark is not at the specified position, rotate the distributor a small amount until this mark is at the specified location.

6. Tighten the distributor hold down bolt to the specified torque and recheck the timing mark position.

7. Connect the vacuum advance hose, the in-line timing connector, and any other connectors, hoses, or components that were disconnected for the timing procedure.

EI MAINTENANCE, DIAGNOSIS, AND SERVICE

Spark plug and spark plug wire inspection, diagnosis, and service is basically the same on DI and EI systems. Since EI systems do not have a distributor, servicing of this component is eliminated.

On most EI systems, if the crankshaft position sensor or camshaft reference sensor are defective, the engine will not start. On some vehicles, if the camshaft reference sensor becomes defective when the engine is running, the engine continues to run but after the engine is shut off it will not restart. On other EI systems, if the camshaft reference sensor is defective, the PCM will allow the engine to start because it estimates the camshaft position

after the engine cranks a few times. There is no maintenance for these sensors.

If the crankshaft position sensor or camshaft reference sensor is defective, a DTC is stored in the PCM memory. These DTCs may be read by connecting a scan tool **(Figure 22-15)** to the data link connector, which is usually located under the dash **(Figure 22-16)**. A DTC indicates a problem in a specific area. For example, a DTC representing a crankshaft position sensor fault indicates a defective crankshaft position sensor, faulty wires between the sensor and the PCM, or a problem

Figure 22-15 A scan tool.

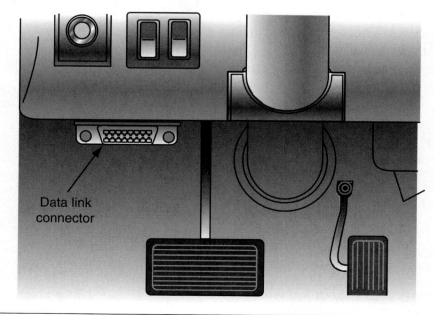

Figure 22-16 A data link connector (DLC).

with the PCM receiving this signal. It is best to think of DTCs as guides to the fault rather than telling you what the fault is. Often many things can give the same code. The crankshaft position sensor and camshaft reference sensor may be tested using the procedure in the vehicle manufacturer's service manual. If the EI system has electromagnetic crankshaft and camshaft sensors, they may be tested with an ohmmeter using the same procedure as described previously for testing electromagnetic pickup coils. When the EI system has Hall effect crankshaft and camshaft sensors, these sensors may be tested with a digital voltmeter, graphing voltmeter, or lab scope. Hall effect sensors have three wires: a voltage supply wire, a signal wire, and a ground wire. When testing this type of sensor, always be sure the specified voltage is available on the voltage supply wire with the ignition switch on. This voltage is usually between 5 volts and 12 volts depending on the system. Connect a digital voltmeter to the Hall effect sensor signal wire and crank the engine. The signal voltage should vary from a very low voltage to a high voltage if the sensor is satisfactory. On a graphing voltmeter or lab scope, a satisfactory Hall effect switch should provide a square waveform as indicated in

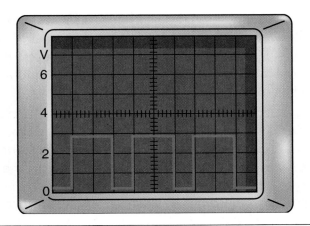

Figure 22-17 A voltage waveform from a Hall effect type sensor.

(Figure 22-17). If the Hall effect crankshaft and camshaft sensor signals are satisfactory, it may be necessary to test the wires between these sensors and the PCM to locate the cause of the DTC representing one of these sensors.

A **graphing voltmeter** or **lab scope** provides waveforms or voltage traces representing the voltages in electronic circuits **(Figure 22-18)**. However, the lab scope scans voltage signal much faster. For this reason, the lab scope waveform will indicate momentary defects in electronic components that other types of test equipment may fail to display.

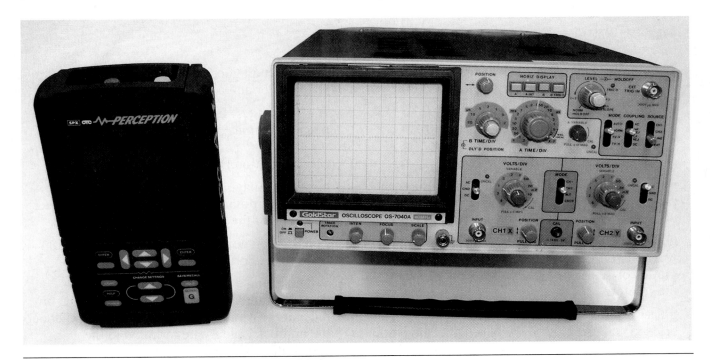

Figure 22-18 A graphing multimeter and lab scope.

A spark tester may be used to test for secondary voltage in EI systems. If the spark plugs are not firing, always be sure 12 volts are supplied to the primary winding of each coil with the ignition switch on.

> **Tech Tip** *In some EI systems, the crankshaft position sensor is mounted at the front of the crankshaft and a notched ring on the crankshaft pulley rotates near the tip of this sensor. On some EI systems, a gap adjustment is necessary between the crankshaft position sensor and the notched ring. The crankshaft sensor position may be moved to correct this gap.*

> **Tech Tip** *On most Daimler-Chrysler vehicles, voltage is supplied to the primary coil windings through an automatic shutdown (ASD) relay. On most systems, this relay also supplies voltage to the fuel injectors and the oxygen sensor heaters. If the ASD relay is defective, the primary ignition circuit and the fuel injectors are inoperative.*

Ignition coils in these systems may be tested with an ohmmeter as discussed previously. Timing adjustments are not possible on EI systems.

COIL-ON-PLUG AND COIL-NEAR-PLUG IGNITION SYSTEMS MAINTENANCE, DIAGNOSIS, AND SERVICE

Spark plug and spark plug wire maintenance, diagnosis, and service on coil-on-plug and coil-near-plug ignition systems is basically the same as the service on these components in DI systems described previously. Since coil-on-plug ignition systems do not have spark plug wires, servicing of these items is not required. Distributor service and timing procedures are not required on these systems because the distributor is eliminated.

The crankshaft position sensor and camshaft reference sensor on coil-on-plug and coil-near-plug systems may be diagnosed using the same procedure as described previously for the EI systems. An ohmmeter may be used to test the coil windings as explained previously for DI systems.

Summary

- Platinum-tipped spark plugs usually have a service interval of 100,000 miles (160,000 kilometers).

- Spark plug conditions are an indicator of cylinder operating conditions.

- Carbon-fouled spark plugs may be caused by a rich air fuel mixture, too cold a spark plug heat range, cylinder misfiring, or low cylinder compression.

- Oil-fouled spark plugs may be caused by worn piston rings or cylinders, worn valve guides or seals, or a defective transmission modulator.

- Splash-fouled spark plugs or spark plugs with bridged gaps are caused by loosened combustion chamber deposits that adhere to the spark plug insulator or electrodes.

- A no-start diagnosis may be performed with a spark tester and a 12-volt test light.

- If a 12-volt test light connected from the negative primary coil terminal to ground does not flash when cranking the engine, the pickup coil or ignition module is defective.

- An electromagnetic-type pickup coil may be tested for grounds by connecting an ohmmeter from one of the pickup coil leads to ground.

- The ignition coil windings may be tested for opens and shorts with an ohmmeter.

- When timing a distributor to the engine, cylinder number 1 must be located at TDC on the compression stroke, the rotor must be under the number 1 plug wire terminal in the distributor cap, and one of the reluctor high points must be aligned with the pickup coil.

- On many DI systems with PCM-controlled spark advance, an in-line timing connector must be disconnected when checking ignition timing.

- Hall effect-type sensors may be tested with a digital voltmeter, a graphing voltmeter, or a scope.

- Timing adjustments are not possible on EI, coil-on plug, or coil-near-plug systems.

Review Questions

1. While diagnosing an EI system with a scan tool, a DTC is obtained representing the crankshaft position sensor. Technician A says this DTC proves the crankshaft position sensor is defective. Technician B says the wires from the crankshaft sensor to the PCM may be defective. Who is correct?

   A. Technician A

   B. Technician B

   C. Both Technician A and Technician B

   D. Neither Technician A nor Technician B

2. Technician A says a Hall effect-type sensor may be tested with a lab scope. Technician B says this type of sensor may be tested with a digital voltmeter. Who is correct?

   A. Technician A

   B. Technician B

   C. Both Technician A and Technician B

   D. Neither Technician A nor Technician B

3. Technician A says a defective crankshaft position sensor causes a no-start condition on an EI system. Technician B says if the spark plugs are not firing on an EI system, check the voltage supply to the positive primary terminal on each ignition coil. Who is correct?

   A. Technician A

   B. Technician B

   C. Both Technician A and Technician B

   D. Neither Technician A nor Technician B

4. Technician A says when installing spark plug wires in a distributor cap, they should be installed in the cylinder firing order and in the direction of distributor shaft rotation. Technician B says in a DI system, the distributor shaft rotates in the same direction as the vacuum advance pulls the pickup plate. Who is correct?

   A. Technician A

   B. Technician B

   C. Both Technician A and Technician B

   D. Neither Technician A nor Technician B

5. When the spark plug is removed from cylinder number 3, the spark plug is wet with oil. The most likely cause of this problem is:

   A. excessive valve clearance.

   B. excessive low engine operating temperature.

   C. worn piston rings in cylinder number 3.

   D. a rich air-fuel ratio.

6. Spark plug misfiring is most likely to occur:

   A. when the engine is at normal operating temperature and idle speed.

   B. when it is at or near the wide-open position.

   C. when the engine is decelerating and the throttle is closed.

   D. at a constant cruising speed of 60 mph.

7. All of these conditions must be present to properly time a distributor to an engine *except*:

   A. The pickup coil must be aligned with the mark on the distributor housing.

   B. The piston in cylinder number 1 must be positioned at TDC on the compression stroke with the timing marks in the 0-degree position.

   C. The distributor rotor must be under the number 1 spark plug wire terminal in the distributor cap.

   D. One of the reluctor high points must be aligned with the pickup coil.

8. When diagnosing a no-start complaint on a DI system, a test spark plug fires when connected from the coil secondary lead wire to ground and the engine is cranked, but the test spark plug does not fire when connected from several spark plug wires to ground and the engine is cranked. Technician A says the ignition coil is defective. Technician B says the pickup coil is defective. Who is correct?

   A. Technician A

   B. Technician B

   C. Both Technician A and Technician B

   D. Neither Technician A nor Technician B

9. All of these statements about crankshaft position sensors and camshaft reference sensors in EI systems are true *except*:

   A. A defective crankshaft position (CP) sensor causes a no-start condition.

   B. A Hall effect CP sensor has three wires.

   C. When tested with a lab scope, a satisfactory Hall effect CP sensor produces a square-wave voltage signal.

   D. When tested with a lab scope, a satisfactory electromagnetic CP sensor produces a digital voltage signal.

10. An engine with a coil-on-plug ignition system has a no-start condition and the spark plugs are not firing. Technician A says to test the voltage supply to each coil positive primary terminal with the ignition switch on. Technician B says to test the CP sensor signal. Who is correct?

    A. Technician A

    B. Technician B

    C. Both Technician A and Technician B

    D. Neither Technician A nor Technician B

11. Carbon-fouled spark plugs have _____ _____ carbon deposits.

12. Normal spark plug carbon deposits should be _____ or _____ colored.

13. TVRS spark plug wires should have a maximum resistance of _____ ohms per foot.

14. When testing an electromagnetic-type pickup coil, the ohmmeter leads are connected to the pickup leads and an infinite meter reading is obtained. This reading indicates the pickup coil is _____ .

15. List the causes of oil-fouled spark plugs.

16. Define preignition and detonation in the combustion chambers and explain the causes of these problems.

17. Describe a no-start diagnosis on a DI system using a 12-volt test light and a test spark plug.

18. Explain the proper procedures for diagnosing and testing crankshaft position and camshaft reference sensors.

CHAPTER

23 Brake System Design and Operation

Learning Objectives

After you have read, studied, and practiced the contents of this unit, you should be able to:

- Explain the basic brake principle related to pressurized fluid in a confined space.

- Describe three different classifications of brake fluid.

- Explain master cylinder design and operation.

- Describe drum brake operation.

- Describe disc brake operation, including the advantages of disc brakes.

- Explain the purpose of metering, proportioning, and pressure differential valves.

- Describe the operation of a vacuum brake booster.

- Explain the operation of the parking brakes on a drum brake system.

- Describe the advantages of an ABS.

- Describe the operation of an ABS system during the anti-lock function.

- Explain the operation of the amber and red brake warning lights.

- Explain the difference between a three-channel and a four-channel ABS.

Key Terms

Brake fade

Department of Transportation (DOT)

Diagonally split brake system

Electronic control unit (ECU)

Equilibrium boiling point (ERBP)

Federal Motor Vehicle Safety Standards (FMVSS)

Floating caliper

Four-channel ABS

Front/rear split brake system

Hydraulic control unit (HCU)

International Standard Organization (ISO)

Kinetic energy

Metering valve

One-channel ABS

Pressure differential valve

Proportioning valve

Society of Automotive Engineers (SAE)

Servo action

Three-channel ABS

Vacuum brake booster

INTRODUCTION

The brake system is one of the most important systems on the vehicle. The brake system must slow and stop a vehicle in a short distance and it must perform this function without causing steering pull or premature wheel lock-up. Technicians who service brake systems must be highly skilled experts because the work they perform can affect the safety and lives of those who travel in vehicles.

HYDRAULIC PRINCIPLES

One of the basic principles of hydraulics is that liquids are not compressible. A second principle of hydraulics used in brake systems is that pressure on a confined fluid is transmitted equally in all directions and acts with equal force on equal areas. Most brake systems make use of a brake pedal to exert a mechanical advantage on the master cylinder. If the pedal length is 10 inches and the master cylinder attaches 2 inches from the

fulcrum point, then the mechanical advantage is 5:1 (10 in/2 in = 5 × advantage). When 50 pounds of force is applied to the pedal, then 250 pounds of force are exerted against the master cylinder **(Figure 23-1)**.

NOTE: Pascal's Law states that changes in pressure at any point in an enclosed fluid at rest are transmitted undiminished to all points in the fluid and act in all directions.

When 250 pounds of force is applied to the master cylinder piston and the master cylinder piston has an area of .8 square inch, the force exerted by the fluid in the master cylinder is 250 lb/.8 square in. = 312.5 psi. If the wheel cylinders also have an area of 1 square inch, the 312.5 psi from the master cylinder exerts a pressure of 312.5 psi in each wheel cylinder (312.5 psi × 1 square in = 312.5 lb. Thus the force of 50 lb at the pedal may be transferred using hydraulic principles, to

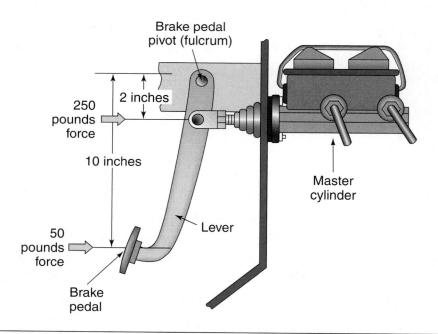

Figure 23-1 The brake pedal assembly uses mechanical leverage to increase the force applied to the master cylinder.

a force of 312.5 lb at the wheels. Brake system principles that must be understood include the following:

- If the master cylinder piston and bore size are decreased, the pressure exerted by the master cylinder increases for a given pressure on the brake pedal.

- If the master cylinder piston and bore size are increased, the pressure exerted by the master cylinder decreases for a given pressure on the brake pedal.

- A smaller diameter master cylinder requires more piston travel to displace the same amount of fluid as a larger piston.

- The force on the brake pedal and the diameter of the master cylinder piston determine the pressure in the brake system.

- The diameter of the wheel cylinders or calipers determines the force against the brake shoes or pads. A larger diameter wheel cylinder or caliper piston exerts more force on the brake shoes or pads.

BRAKE FLUIDS

Brake fluid quality is extremely important to provide proper brake system operation. Brake fluid must have these qualities:

- A controlled amount of swell. Brake fluids must provide a controlled amount of swell in cups and seals to provide adequate sealing in calipers, wheel cylinders, and master cylinders. Excessive swelling of cups and seals causes brake drag and inadequate brake response.

- Temperature extremes. Brake fluid must operate at temperatures from −104°F (−75°C) to 500°F (260°C).

- Compatibility with rubber. Brake fluid must be compatible with rubber in master cylinder cups, wheel cylinder cups, caliper seals, and brake hoses.

- Lubricating ability. Brake fluid must have a satisfactory lubricating ability to provide smooth operation of brake components.

- Resistance to evaporation. Brake fluid must resist evaporation at high temperatures.

- Antirust and anticorrosion. Brake fluid must combat rust and corrosion in brake system components.

Tech Tip *DOT 3, DOT 4, and DOT 5.1 brake fluids are hygroscopic, which means they attract moisture from the air. Therefore, containers containing these types of brake fluid must be kept tightly closed. DOT 5 brake fluid is non-hygroscopic, which indicates it does not attract moisture from the air. DOT 5 brake fluid must not be mixed with DOT 3, DOT 4, or DOT 5.1 brake fluids.*

Classification Standards

The **Department of Transportation (DOT)** designates that brake fluids must meet classification standards established by the SAE. The following brake fluid classifications are used at the present time:

- **DOT 3**—A brake fluid classified as DOT 3 has a minimum dry **equilibrium boiling point (ERBP)** of 401°F (205°C) and a minimum wet ERBP of 284°F (140°C).

- **DOT 4**—A brake fluid classified as DOT 4 has a minimum dry ERBP of 446°F (230°C) and a minimum wet ERBP of 311°F (155°C).

- **DOT 5.1**—A brake fluid classified as DOT 5.1 has a minimum dry ERBP of 500°F (270°C) and a minimum wet ERBP of 365°F (185°C).

- **DOT 5**—A brake fluid classified as DOT 5 is silicone based and this fluid has a minimum dry ERBP of 500°F (260°C) and a minimum wet ERBP of 356°F (180°C). DOT 5 has a purple color added for identification purposes. Always use the brake fluid specified by the vehicle manufacturer.

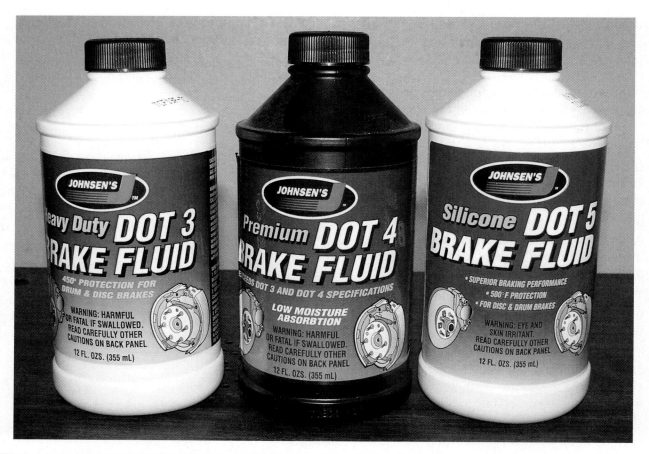

Figure 23-2 The two most commonly used brake fluids are DOT 3 and DOT 4.

Brake fluids in use today are DOT 3, DOT 4, DOT 5, and DOT 5.1 and are shown in **Figure 23-2**.

MASTER CYLINDERS

The purpose of the master cylinder is to supply brake fluid pressure to the wheel cylinders or calipers during a brake application **(Figure 23-3)**. Dual-master cylinders have been mandated since 1967 by **Federal Motor Vehicle Safety Standards (FMVSS)**. If fluid pressure is not available in one section of the master cylinder, fluid pressure in the other section is applied at two wheels to stop the vehicle.

Tech Tip *When the brakes are applied, a switch operated by the brake pedal illuminates the brake lights on the rear of the vehicle.*

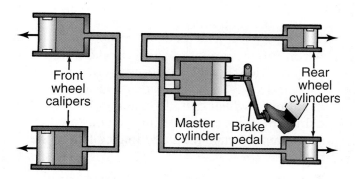

Figure 23-3 The master cylinder supplies brake fluid pressure to the calipers and wheel cylinders.

The master cylinder may be manufactured from cast iron or aluminum. Cast iron master cylinders have an integral reservoir and a removable plastic reservoir is used on aluminum master cylinders. A cover bail is used to retain the cover and gasket on top of the master cylinder. The primary and secondary pistons are mounted in the master cylinder bore. Rubber cups are used to seal these pistons in the master cylinder bore. A spring behind each piston holds the pistons in the released position **(Figure 23-4)**. Steel brake lines and brake hoses are connected from the master cylinder to the wheel cylinders or calipers. On rear-wheel drive vehicles, the primary piston outlet is connected to the front brakes and the secondary piston outlet is connected to the rear brakes. This is a **front/rear split brake system (Figure 23-5)**. On many front-wheel drive vehicles, the brake system is diagonally split so the primary master cylinder piston supplies fluid to the right front and left rear brakes and the secondary piston supplies fluid to the left front and right rear brakes. This is a **diagonally split brake system (Figure 23-6)**.

When the brake pedal is depressed, the primary piston moves down the master cylinder bore and the primary cup seals the vent port. This action seals the primary piston cylinder and forces fluid from the primary outlet port. Pressure in the primary bore moves the secondary piston down its bore and the vent port is sealed by the secondary piston cup. Further movement of the secondary piston forces fluid from the secondary outlet to the wheel calipers or cylinders.

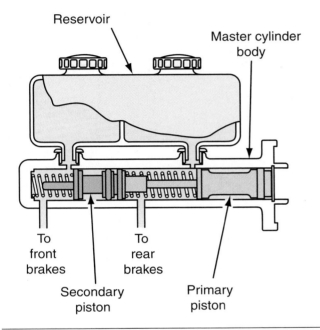

Figure 23-4 Master cylinder design.

SAFETY TIP *In past years, asbestos was used in brake lining material. Because asbestos has been proven to contribute to lung cancer, asbestos has been eliminated from brake linings currently manufactured in North America. However, brake dust is still considered to he a hazardous material.*

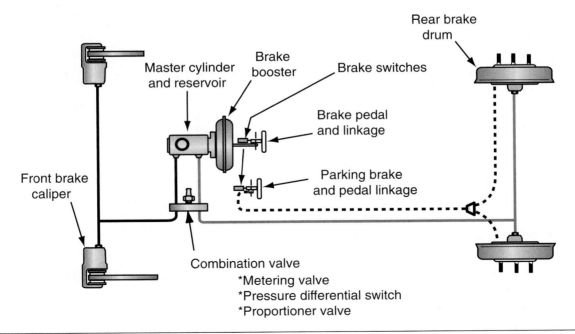

Figure 23-5 A front-to-rear split brake system.

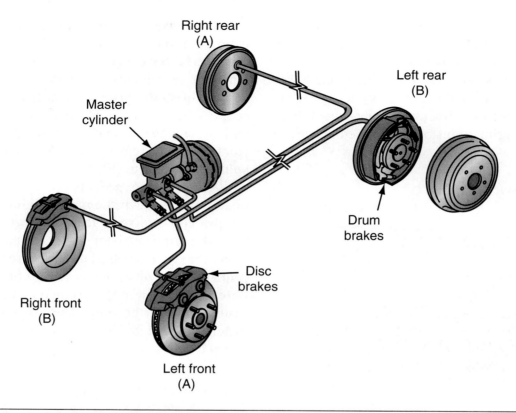

Figure 23-6 A diagonally-split brake system.

When fluid is forced from both outlets, the brakes are applied on all four wheels. If the brakes are applied repeatedly, the brake fluid may become very hot. Under this condition, the fluid expands and some fluid flows back through the vent ports when the brakes are released. The vent ports prevent excessive pressure buildup in the brake system. If the brake shoe adjustment is not correct or air enters the brake system, excessive brake pedal movement is required to apply the brakes. Under this condition, the driver may pump the brake pedal rapidly to apply the brakes. This pumping action may cause a pressure drop in the master cylinder because the fluid cannot flow back from the wheel cylinders or calipers as quickly as the master cylinder pistons can move when the brakes are released. When pressure drops in the master cylinder, atmospheric pressure on top of the fluid in the reservoir forces fluid past the cup seals into the piston bores.

On a cast iron master cylinder, the rubber cover gasket seals the cover to the reservoir preventing moisture from entering the system. A wire cover retainer holds the cover on the master cylinder (**Figure 23-7**). Atmospheric vents are located

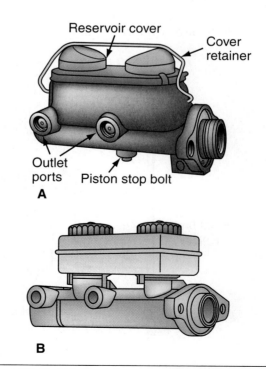

Figure 23-7 (A) Cast-iron master cylinder, and (B) aluminum/composite master cylinder.

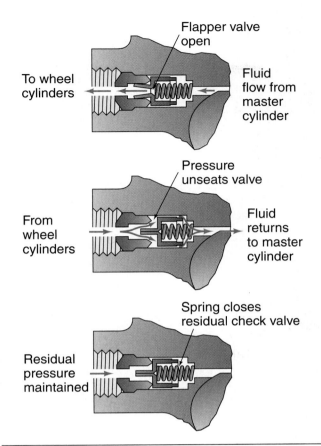

Flapper valve open

To wheel cylinders

Fluid flow from master cylinder

Pressure unseats valve

From wheel cylinders

Fluid returns to master cylinder

Spring closes residual check valve

Residual pressure maintained

Figure 23-8 A residual pressure valve used with drum brakes. This valve typically holds 5 to 8 PSI pressure in the drum braking system.

between the cover and the gasket. If the fluid level goes down in the reservoir, the bellows on the cover gasket expand into the reservoir and air flows through the vents into the area between the cover and the gasket. Some drum brake master cylinders have a residual valve at the outlet connected to the drum brakes **(Figure 23-8)**. This residual valve is only used for drum brakes and maintains a slight pressure in the brake system that keeps the wheel cylinder cups expanded outward to provide improved cup sealing. Other brake systems have wheel cylinder cup expanders in place of the residual valve.

DRUM BRAKES

In a drum brake system, the brake shoes are forced outward against the drum by the wheel cylinder pistons **(Figure 23-9)**. Each wheel cylinder contains two pistons and cup seals with expanders **(Figure 23-10)**. The expanders prevent air from

entering the system when the pistons and cups are moving. A spring is positioned between the piston cups. A rubber dust boot is mounted on each end of the wheel cylinder. Pushrods are located between the pistons and the brake shoes. The pushrods fit snugly in the dust boot openings. The dust boots keep contaminants out of the wheel cylinder. The wheel cylinder is bolted to the backing plate and a bleeder screw is mounted in the back of the wheel cylinder. The bleeder screw allows air to be bled from the wheel cylinder and brake system.

Drum brake shoes are made from stamped steel. The brake linings are riveted or bonded to the brake shoes. The curvature of the brake shoes and linings matches the contour of the brake drums. Brake linings must be able to withstand extreme heat. The brake linings may be organic, semi-metallic, metallic, or synthetic. Organic brake linings are made from non-metallic fibers bonded together to form a composite material. Organic brake linings contain friction modifiers that may include graphite and powdered metals. Fillers, binders, and curing agents are also used on organic brake linings.

Drum brakes are either servo-type or leading/ trailing. In a servo-type, the brake shoe return springs hold the shoes against the anchor pin mounted in the top of the backing plate. An adjusting mechanism is located between the lower ends of the brake shoes, but the lower ends of the brake

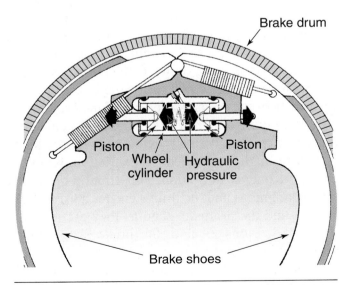

Brake drum

Piston

Piston

Wheel cylinder

Hydraulic pressure

Brake shoes

Figure 23-9 Hydraulic pressure moves the two pistons outward to force the brake shoes against the drum.

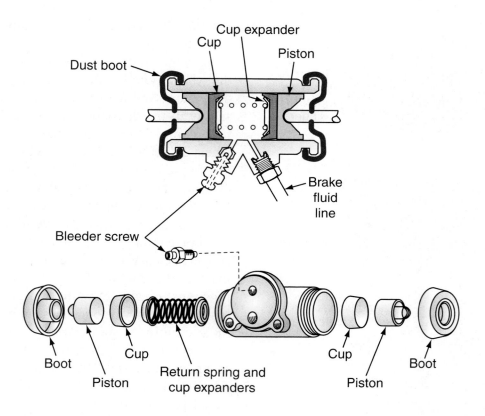

Figure 23-10 A typical drum wheel cylinder design.

shoes are not connected to the backing plate. The edges of the brake shoes contact the support pads on the backing plate **(Figure 23-11)**. The primary brake shoe is positioned toward the front of the vehicle and the secondary shoe is positioned toward the rear of the vehicle **(Figure 23-12)**. When the brakes are applied, pressure from the master cylinder forces the wheel cylinder pistons outward. As these pistons move outward, the pushrods force the brake shoes outward against the brake drums. Friction between the brake shoes and the drum provides a braking action. The brake system changes the kinetic energy of the moving vehicle into heat energy through the application of friction. **Kinetic energy** is energy in motion. This friction between the brake linings and the drums creates a great deal of heat and the drums expand as they are heated. When the drums expand, the brake shoes must move farther outward to provide braking action. This additional shoe movement causes increased brake pedal movement and brake fade. **Brake fade** occurs on drum brakes when the drums become very hot during hard, repeated braking.

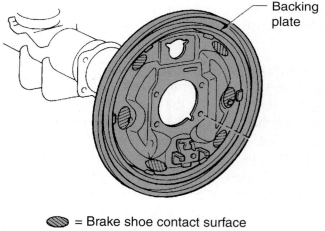

= Brake shoe contact surface

Figure 23-11 Backing plate with anchor pin and support pads.

Tech Tip *Brake fade is defined as the reduction in braking due to loss of friction between the brake shoes and drum or pads and disc. Brake fade often occurs from heat buildup due to repeated or prolonged brake application.*

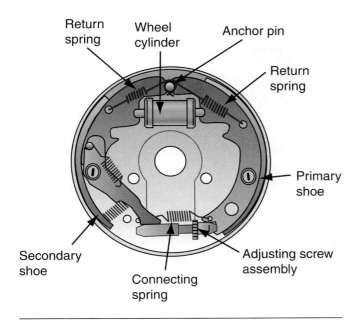

Figure 23-12 A servo-type drum brake.

Under these operating conditions, the drums expand and the shoes must travel further outward to provide braking action. Increased shoe movement requires greater pedal movement.

When the primary shoe is forced against the drum surface, the wheel rotation tends to transfer movement to the secondary shoe by servo action and force this shoe against the brake drum surface **(Figure 23-13)**. **Servo action** occurs when the operation of the primary shoe applies mechanical force to the secondary shoe to assist in its application. Therefore, the secondary shoe is forced against the brake drum by movement from the primary shoe and brake drum and also by fluid pressure in the wheel cylinder. In a servo-type brake, approximately 75 percent of the braking force is from the secondary shoe. Therefore, the secondary shoe has a longer lining compared to the primary shoe. The adjusting mechanism between the lower end of the brake shoes contains a threaded star wheel with two extensions that fit over the lower ends of the brake shoes.

Tech Tip *If a drum brake does not have a self-adjusting mechanism, a tool may be inserted through an opening in the backing plate to rotate the star wheel and adjust the brakes manually.*

NOTE: When the brake shoe is forced against the rotating drum, the rotation of the drum forces the shoe against its anchor point. This movement, called *self-energizing action*, forces the shoes tighter against the drum.

In a leading/trailing brake system there is less servo action **(Figure 23-14)**. When the leading shoe is forced against the drum surface, the wheel rotation tends to force this shoe against the brake

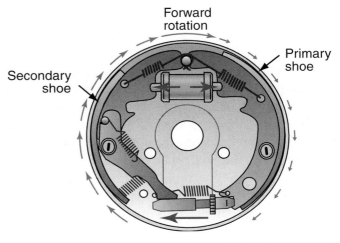

Figure 23-13 Servo-type braking forces.

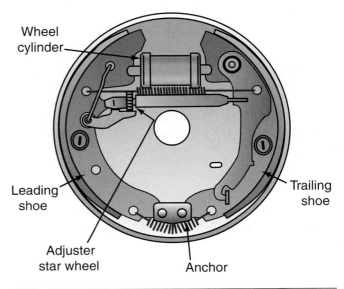

Figure 23-14 A leading-trailing brake assembly.

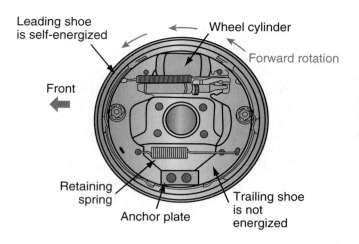

Figure 23-15 Leading-trailing type braking forces.

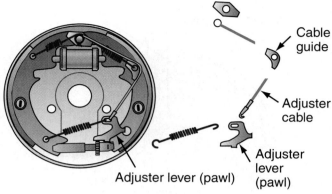

Figure 23-16 A cable-type self-adjusting mechanism for a servo-type brake.

drum surface **(Figure 23-15)**. Some servo action occurs when the leading shoe presses against the anchor pin at the bottom of the backing plate and applies some mechanical force itself to assist in its application. However, the trailing shoe is forced against the brake drum by movement from fluid pressure in the wheel cylinder alone. In a leading/trailing type brake, more of the braking force is from the leading shoe during a forward stop and more of the braking force is from the trailing shoe during a reverse stop. Both shoes have the same length linings. The adjusting mechanism is between the upper ends of the brake shoes and contains a threaded star wheel with two extensions that fit over the upper ends of the brake shoes.

NOTE: On some brake systems, the brake shoes have a backing plate anchor between the lower ends of the brake shoes. This type of brake does not provide servo action and may be referred to as a *leading/trailing brake*.

Most drum brakes have a self-adjusting mechanism **(Figure 23-16)**. The self-adjusting mechanism for the servo-type brake contains a cable attached to the anchor pin. The adjusters are usually opposite the anchor pin. The lower end of this cable is connected to a pivoted adjusting lever

mounted above the star wheel. A spring holds the adjusting lever downward against the star wheel. As the brake lining wears, the outward shoe movement increases when the brakes are applied. During a brake application in reverse, the secondary shoe moves away from the anchor pin. If the lining is worn enough to allow sufficient shoe movement, the spring pulls the adjusting lever down enough to rotate the star wheel to the next notch. This action lengthens the star wheel assembly and moves the brake shoes outward so the linings are closer to the drum surface. The adjuster for the leading/trailing brake is very similar to that of the servo-type brake except it is found near the wheel cylinder **(Figure 23-17)**.

DISC BRAKES

A disc brake has a cast iron disc or rotor mounted on the wheel hub. Both sides of the rotor have machined surfaces. During a brake application, the brake pad linings contact both sides of the rotor. Many front brake rotors have ventilating slots between the two sides of the rotor to act as a fan and dissipate heat from the rotor **(Figure 23-18)**. The caliper and brake pad assembly is mounted over the top of the rotor and the caliper is bolted to the steering knuckle **(Figure 23-19)**. Many calipers contain a single piston mounted in the caliper bore on the inboard side of the caliper. This type of disc brake has a floating caliper. A **floating caliper** is mounted so it is free to slide sideways on the mounting bolts. The brake hose is threaded into the back of the caliper and a bleeder screw is also positioned in the rear of

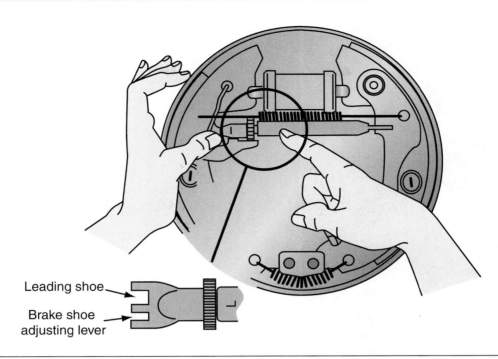

Leading shoe

Brake shoe
adjusting lever

Figure 23-17 A self-adjusting mechanism is being installed in a leading/trailing type brake.

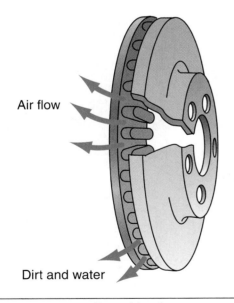

Air flow

Dirt and water

Figure 23-18 A ventilated brake rotor.

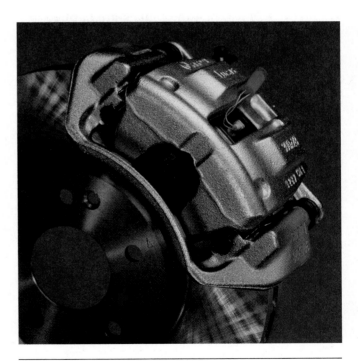

Figure 23-19 Disc brake assembly.

the caliper. The brake linings are bonded or riveted to the brake pads. Similar materials are used in the brake linings on drum and disc brakes. Metal locating tabs on most brake pads retain the pads to the caliper. The caliper piston has a seal around the outer diameter of the piston. A dust boot is mounted between the caliper piston and the housing to keep contaminants out of the caliper bore **(Figure 23-20)**.

When the brakes are applied, the caliper piston is forced outward. This movement pushes the inboard brake pad lining against the rotor. This action forces the floating caliper inward so the outer brake pad lining is forced against the outside of the rotor

Figure 23-20 A caliper piston dust boot.

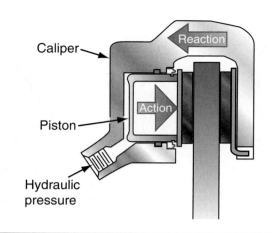

Figure 23-21 Floating caliper action during a brake application.

(Figure 23-21). When the brakes are applied, the caliper piston seal twists. After the brakes are released, the seal returns to its original shape. This seal action pulls the caliper piston and brake pad lining away from the rotor surface **(Figure 23-22).** As the brake pad linings wear, the caliper pistons move outward. This eliminates the need for adjusting disc brakes. Disc brakes do not experience brake fade because as the rotors are heated they expand and move slightly closer to the brake pad linings.

> **NOTE:** During severe braking the friction material may vaporize and produce a gas that will greatly reduce friction. Performance brake rotors have special holes and slots to vent this gas and improve braking performance.

Tech Tip *Some calipers contain two pistons, one on each side of the rotor. Other calipers contain two pistons on each side of the rotor. Some disc brakes have an electronic sensor that illuminates a warning light in the instrument panel when the brake linings are worn a specific amount.*

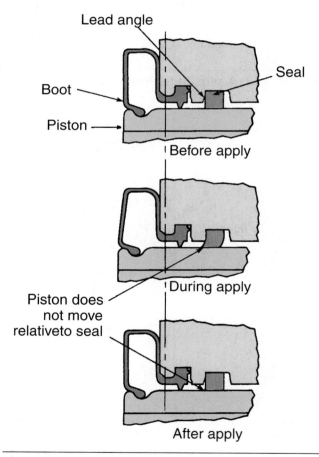

Figure 23-22 Caliper piston seal action during a brake application and brake release.

Disc brakes do not provide any servo action. Many disc brakes have a wear indicator attached to one of the brake pads. When the brake pad lining wears a specific amount, the wear indicator contacts

NEW PAD   **WORN PAD**

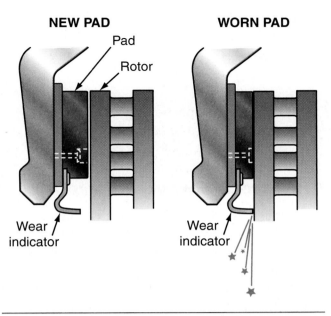

Figure 23-23 Brake lining wear indicator.

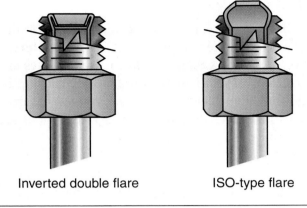

Inverted double flare   ISO-type flare

Figure 23-24 Inverted double flare and ISO flare on brake tubing.

the rotor causing a scraping noise that alerts the driver about the brake problem **(Figure 23-23)**.

BRAKE LINES, HOSES, AND VALVES

Brake lines are made from seamless steel tubing that is coated with zinc or tin for corrosion protection. Brake lines must conform to **Society of Automotive Engineers (SAE)** standard J1047 which requires that an 18-inch section of tubing must withstand an internal pressure of 8,000 psi. Brake lines may have an inverted double flare or an **International Standard Organization (ISO)** flare. The inverted double flare has the end of the tubing

flared out and then it is formed back onto itself **(Figure 23-24)**. An ISO flare has a bubble-shaped end formed on the tubing. Different fittings are required with each type of flare and these fittings are not interchangeable.

Brake tubing is specially bent to fit a specific location on the vehicle. When a brake line is leaking, the complete line must be replaced, not repaired. It is preferable to replace brake lines with the OEM pre-shaped brake line. Brake hoses are a flexible connection between the chassis and the wheel calipers or cylinders or between the rear axle and the chassis. Brake hoses must conform to the SAE J1401 standard that requires the hose to withstand 4,000-psi pressure for 2 minutes without bursting. Brake hoses contain an inner liner surrounded by metal fabric plies, a rubber separator layer, and an outer jacket **(Figure 23-25)**. The fittings are attached to the hose during the manufacturing process.

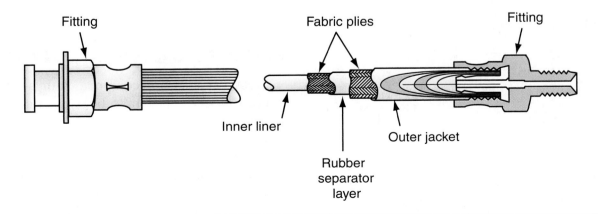

Figure 23-25 Construction details of a brake hose.

Metering Valve

The metering valve is connected in the brake line to the front brakes **(Figure 23-26)**. The **metering valve** is used on vehicles with front disc brakes and rear drum brakes. During a brake application, the metering valve delays the fluid pressure to the front brakes momentarily until the rear brake shoes are forced outward against the drums.

NOTE: The metering valve is sometimes called a *hold-off valve.*

Proportioning Valve

During a brake application, vehicle inertia and momentum tend to shift the vehicle weight forward. This weight shift is proportional to braking force and the rate of deceleration. This weight shift reduces traction between the rear tires and the road surface which may result in rear wheel lock-up. To provide maximum braking, equal friction must be maintained between the front and rear tires and the road surface. During a brake application, the **proportioning valve** modulates pressure to the rear brakes to prevent rear wheel lock-up. The pro-

portioning valves may be integral in the master cylinder or they may be external to the master cylinder.

NOTE: The proportioning valve is sometimes referred to as a *balancing valve.*

Pressure Differential Valve

A **pressure differential valve** is connected in the brake lines from the master cylinder to the front and rear brakes. A wire from the brake warning light in the instrument panel is connected to a switch in the pressure differential valve. If the brake fluid pressure is equal in both the primary and secondary sections of the master cylinder, the switch in the pressure differential valve remains open. If one section of the master cylinder is low on brake fluid and the pressure in this section decreases, the piston in the pressure differential valve moves toward the section in the pressure differential valve with the low pressure. This piston movement pushes the stem upward in the brake-warning switch and closes the switch contacts **(Figure 23-27)**. Under this condition, current flows through the brake warning light, the switch contacts to ground, and the brake warning light is illuminated to warn the driver that a brake problem exists. Some brake systems have the pressure differential valve combined with the metering and proportioning valves in a combination valve assembly. Pressure differential, metering, and proportioning valves are non-serviceable components.

VACUUM BRAKE BOOSTERS

Most vehicles are equipped with a **vacuum brake booster**. The brake booster is connected between the master cylinder and the brake pedal. A pushrod is connected from the booster to the brake pedal, and another pushrod is mounted between the brake booster and the master cylinder. The brake booster operates in three modes, released, applying, and hold. When the engine is running, manifold vacuum is supplied through a

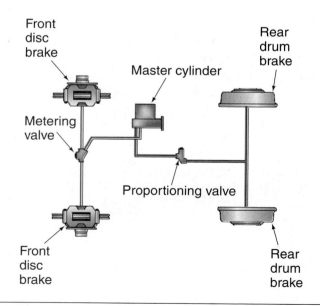

Figure 23-26 A metering valve and proportioning valve in a brake system.

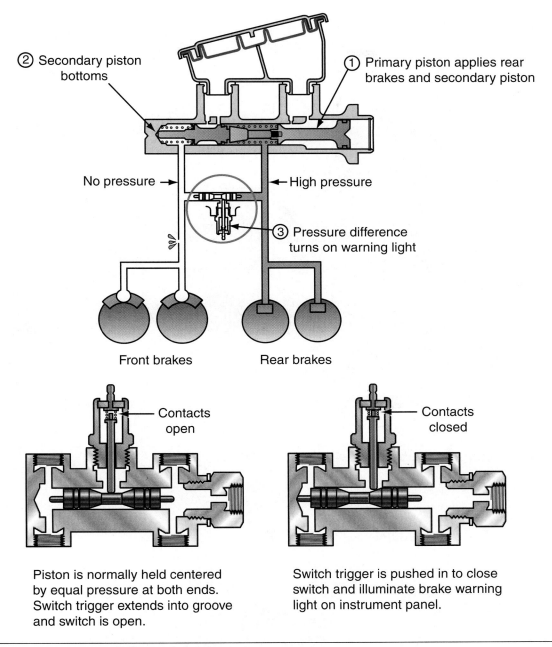

① Primary piston applies rear brakes and secondary piston

② Secondary piston bottoms

No pressure → ← High pressure

③ Pressure difference turns on warning light

Front brakes   Rear brakes

Contacts open

Contacts closed

Piston is normally held centered by equal pressure at both ends. Switch trigger extends into groove and switch is open.

Switch trigger is pushed in to close switch and illuminate brake warning light on instrument panel.

Figure 23-27 A pressure differential valve.

hose to the booster. If the brakes are released, manifold vacuum is supplied to both sides of the booster diaphragm. This diaphragm does not exert any force on the master cylinder pistons **(Figure 23-28)**. This is the release mode.

When the brakes are applied, the brake pedal movement closes the vacuum passage to the rear of the booster diaphragm and opens an atmospheric pressure port to allow air pressure on the rear side of the diaphragm **(Figure 23-29)**. Because

manifold vacuum is supplied to the front side of the diaphragm and atmospheric pressure is provided on the rear side of the diaphragm, the diaphragm is moved toward the master cylinder with considerable force. This is the apply mode. Under this condition, the diaphragm moves the pushrod and supplies force to the master cylinder pistons to provide brake assist.

When the brakes are being held at the desired stopping force, both valves are closed to trap

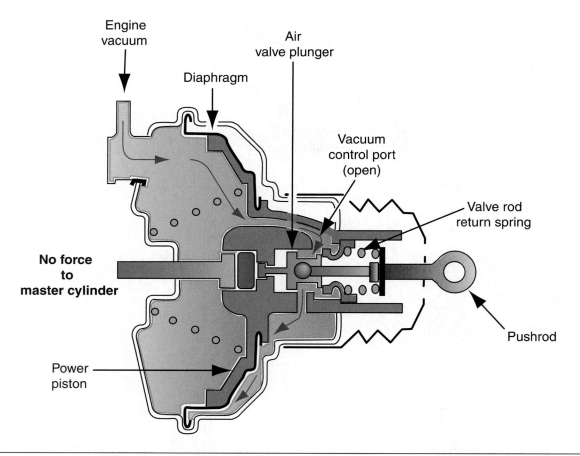

Figure 23-28 A vacuum brake booster with brakes released.

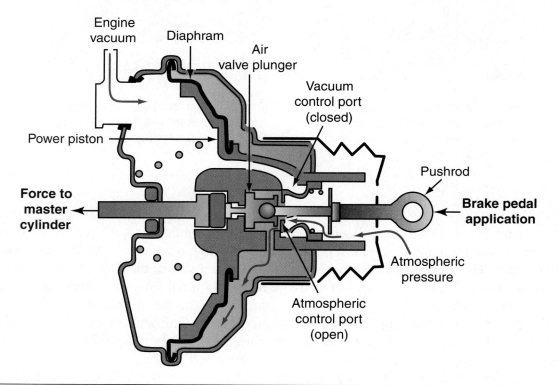

Figure 23-29 A vacuum brake booster with brakes applied.

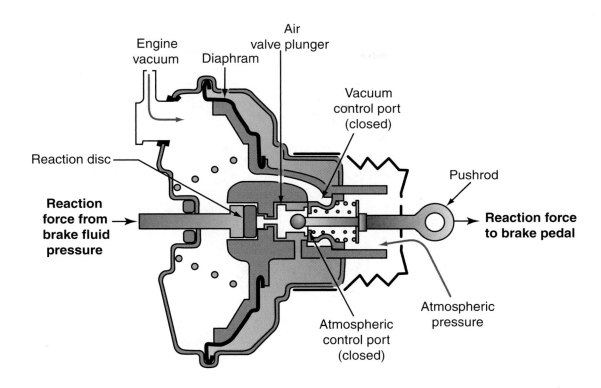

Figure 23-30 A vacuum brake booster with brakes holding.

vacuum and atmospheric pressure and keeps the diaphragm in a hold position. A reaction disc between the booster diaphragm and the master cylinder pushrod provides brake pedal feel to the driver **(Figure 23-30)**. This is the hold mode.

HYDRO-BOOST BRAKE SYSTEM

Some vehicles have a hydro-boost brake system with a hydraulic brake booster. These systems are often used on diesel engines because of the low manifold vacuum in these engines. In a hydro-boost brake system, a hydraulic brake booster is mounted between the master cylinder and the brake pedal. The power-steering pump supplies fluid pressure to the hydro-boost unit and the power-steering gear **(Figure 23-31)**. Some vehicles have a separate pump for the hydro-boost system. When the brakes are released, power-steering pump pressure forces fluid through the hydro-boost unit, but this pressure is not applied to the master cylinder pistons. If the brakes are applied, brake pedal movement operates a lever in the hydro-boost unit and opens a valve that supplies power-steering pump fluid pressure to the master cylinder pistons and provides brake assist.

PARKING BRAKES

Parking brakes are applied by a foot- or hand-operated lever in the passenger compartment **(Figure 23-32)**. When the parking brakes are applied, a switch on the parking brake lever grounds the red brake warning light in the instrument panel. A release handle on the parking brake lever is pulled to release the parking brakes. Some parking brakes have a vacuum-operated release mechanism that releases the parking brake when vacuum is supplied to the vacuum diaphragm in the parking brake release actuator. This type of parking brake also has a mechanical release lever to release the parking brake if the vacuum actuator is not operating. A cable is connected from the parking brake lever to an equalizer under the vehicle. Dual cables are connected from the equalizer to the rear brakes **(Figure 23-33)**.

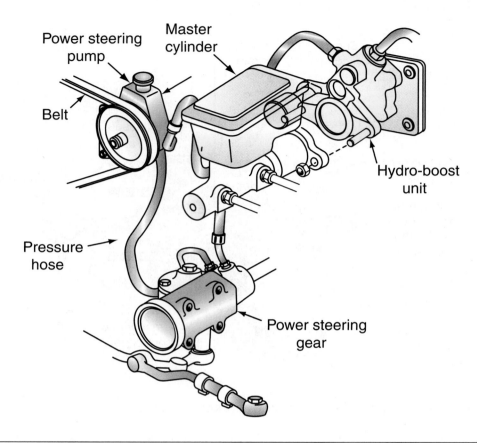

Figure 23-31 A hydro-boost power brake system.

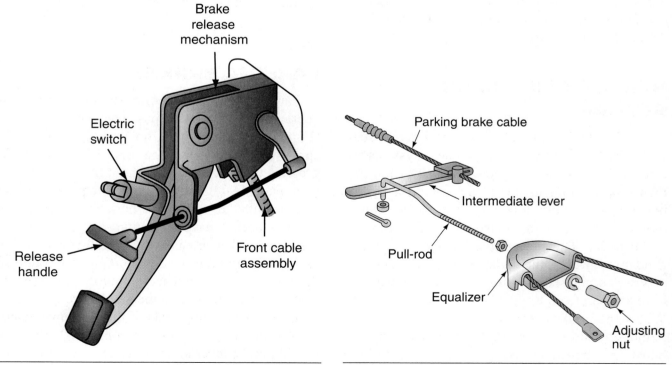

Figure 23-32 A foot-operated parking brake mechanism.

Figure 23-33 Parking brake equalizer and cables.

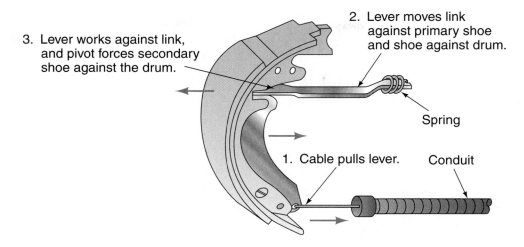

3. Lever works against link, and pivot forces secondary shoe against the drum.

2. Lever moves link against primary shoe and shoe against drum.

Spring

1. Cable pulls lever.

Conduit

Figure 23-34 Parking brake lever and strut operation.

When the parking brakes are applied, the cable in each rear wheel pulls a lever that forces the secondary and primary brake shoes outward against the drum to apply the brakes **(Figure 23-34)**. Some rear disc brake systems have a small drum brake inside the rotor that is used for the parking brake. This drum in a disc rotor is sometimes called a *drum in hat*.

If the vehicle has rear disc brakes, the parking brake cables are connected to a lever on the rear brake calipers. When the parking brakes are applied, the cable and lever turns a threaded actuator screw inside a nut with matching threads. As the actuator screw is rotated by the parking brake cable and lever, the nut moves inward and pushes the caliper piston so it forces the brake pad lining against the rotor to apply the brakes.

ANTILOCK BRAKE SYSTEM PRINCIPLES

Today many vehicles use antilock brakes. Antilock brake systems (ABS) are designed to prevent wheel lock-up under heavy braking conditions on any type of road condition. The main purpose of ABS is to help the driver retain directional stability or control of steering, stop the vehicle as quickly as possible given the road conditions, and retain maximum control of the vehicle.

The antilock brake system can be thought of as an add-on system to the normal braking system

of the vehicle. Vehicles equipped with ABS still require conventional brake service work to be performed for normal brake concerns. Brake lines, hoses, calipers, drums, and friction components still require the regular attention of the technician. ABS systems are usually one of three types: **one channel ABS** in which both rear wheels are controlled together, **three channel ABS** in which each front wheel is controlled individually and the rear wheels are controlled together, and **four channel ABS** in which both front wheels are controlled individually and both rear wheels are controlled individually.

The ABS system should operate only when wheel lock-up is eminent. The antilock brake system can be thought of as an electronic hydraulic pumping of the brakes or stopping of the vehicle under panic conditions or under adverse road conditions. It is a myth for one to think or believe that the antilock brake system will prevent the wheel from skidding. Although it does prevent complete wheel lock-up, it does allow some wheel slip in order to achieve the best braking. During ABS operation the slip will be between 10 percent to 20 percent. Slip rate of 15 percent means that the velocity of a wheel spent less than that of a free rolling wheel at the same vehicle speed. In the event of an ABS system failure the brake system should operate as a conventional braking system without the benefits of ABS. The components that comprise the ABS include an **electronic control unit (ECU)**,

hydraulic control unit (HCU), wheel speed sensors, as well as the brake booster and master cylinder **(Figure 23-35).**

The purpose of the ECU is to receive signals from the wheel speed sensors and to operate the hydraulic control unit to control the brake pressure at the wheel according to the data analyzed. The hydraulic control unit function is to modulate the pressure to each of the wheels. It receives the signals necessary to do this from the electronic control unit or ECU to apply or release the brakes under ABS conditions

If a wheel locks up during a brake application, the tire exhibits a loss of traction. Conversely, if enough braking force is supplied so the tire slips

Tech Tip *When the module detects a defect in the ABS, the module shuts down the antilock function, but normal power-assisted braking is maintained.*

to a certain extent without wheel lock-up, traction increase is experienced. This traction increase is highest at 10 percent to 20 percent tire slip on the road surface. If there is 20 percent tire slip, the wheel is turning at 80 percent of the vehicle speed. An ABS is designed to provide approximately 10 percent to 20 percent tire slip without experiencing wheel lock-up during the antilock

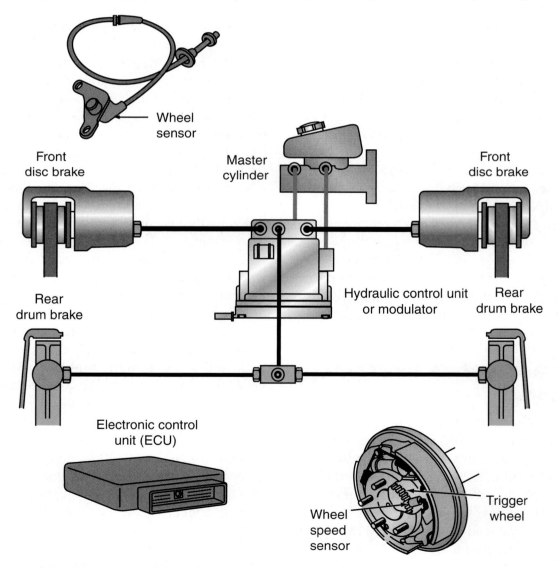

Figure 23-35 Main components of an ABS system.

brake function. This action provides tire traction while braking, and stopping distance may be reduced depending on the road surface and other variables.

Cornering tire traction also decreases significantly if the wheel locks up during a brake application. Wheel lock-up and reduced cornering traction during a brake application may result in loss of steering stability and control. If the tire slips about 15 percent without wheel lock-up, cornering traction is greatly improved. Since the ABS provides about 15 percent tire slip without wheel lock-up even during panic stops, steering control and vehicle stability are significantly improved. If a wheel or wheels lock-up during a panic stop, the vehicle may swerve sideways and the driver may lose steering control.

An amber ABS warning light is mounted in the instrument panel or in the roof console. When the ignition switch is turned on, the control

> **Tech Tip** *In most ABS, the warning lights operate in the same way.*

module performs a check of the ABS electrical system. This check requires 3 to 4 seconds and during this time the ABS warning light is on. If the ABS warning light is on with the engine running, the module has detected a defect in the ABS.

Many ABS systems also have a red brake warning light. If this light is illuminated with the engine running, the parking brake is applied or one section of the master cylinder is low on brake fluid caused by a pressure differential between the two sections of the master cylinder. In some ABS, the control module illuminates both the amber and red brake-warning lights when certain serious defects occur. Under these conditions the ABS is disabled and the system will operate as a conventional power brake system.

WHEEL SPEED SENSORS

Many vehicles have a four-wheel ABS system with a wheel speed sensor at each wheel. A toothed ring on the wheel hub rotates past the tip of the wheel speed sensor. These can be mounted on the half-shaft outer joint. On some front-wheel drive cars, the wheel speed sensors are integral with the wheel hubs and cannot be serviced separately **(Figure 23-36)**.

As the wheel rotates, a toothed sensor ring passes by a sensor, which is made with a coil and magnet. As the high and low spots of the teeth pass the sensor an AC voltage is induced in the coil **(Figure 23-37)**. The different signals from each wheel are analyzed by the ECU and appropriate action will be taken by the ECU to modulate brake pressure if necessary **(Figure 23-38)**.

ABS WITH LOW-PRESSURE ACCUMULATORS

Some vehicles have an ABS with the solenoid valves, low-pressure accumulators, pressure pump,

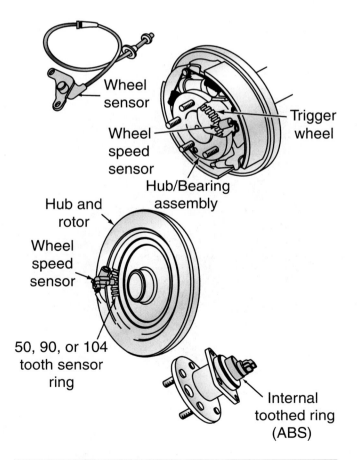

Figure 23-36 Various wheel speed sensors.

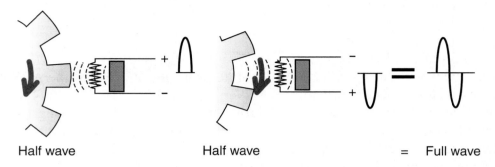

Figure 23-37 Wheel speed sensor output.

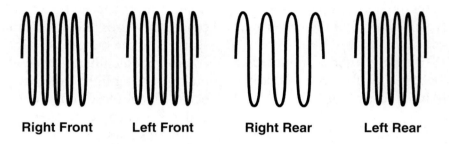

Right Front   **Left Front**   **Right Rear**   **Left Rear**

Figure 23-38 The ECU compared the signals from each of the wheel speed sensors.

and the control module designed into one unit **(Figure 23-39)**. This type of ABS has a conventional vacuum brake booster to provide brake assist. The solenoids in this system are referred to as isolation and dump valves. The VSS in the transmission extension housing also acts as a rear wheel speed sensor and sends voltage signals to the ABS control module and the PCM **(Figure 23-40)**. One pair of solenoids controls the brake fluid pressure to both rear wheels in this three-channel system.

Many vehicles are equipped with four-channel ABS. All the solenoid valves are mounted in a valve block attached to the master cylinder. A three-channel ABS has a pair of solenoids at each front wheel, but only one pair of solenoids for both rear wheels. These systems cannot modulate the brake fluid pressure to each rear wheel individually. A four-channel ABS has a pair of solenoids at each wheel, which can modulate the brake fluid pressure.

During a brake application if a wheel speed sensor signal indicates to the control module that a

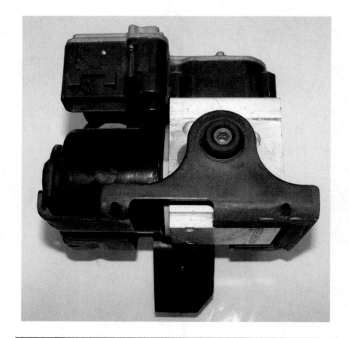

Figure 23-39 ABS with low-pressure accumulator.

wheel is quickly approaching a lock-up condition, the module closes the normally open solenoid con-

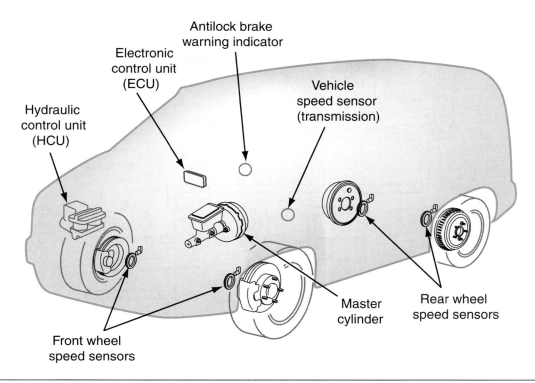

Figure 23-40 ABS system with low-pressure accumulator and related components.

nected to that wheel caliper. This action prevents any further increase in fluid pressure to the wheel caliper. In a few milliseconds of the wheel speed sensor still indicating that wheel lock-up is about to occur, the module opens the normally closed solenoid and allows some brake fluid out of the brake caliper back into the master cylinder reservoir. This action reduces pressure in the caliper and prevents wheel lock-up. The module pulses the solenoids on and off to maintain maximum braking force without allowing wheel lock-up. When the module pulses the solenoids on and off during the antilock function, the driver may feel pedal pulsations. **Figure 23-41** shows the hydraulic schematic of a Bosch three-channel ABS system.

When a wheel speed sensor signal indicates wheel lock-up is about to occur, the control module energizes the isolation valve. This valve closes the fluid passage between the master cylinder and the wheel that is about to lock up. If wheel lock-up is still imminent, the control module energizes the dump valve that allows some brake fluid out of the wheel cylinder or caliper back into the low-pressure accumulator. The control module pulses

the dump valve on and off very quickly to apply maximum braking force without allowing wheel lock-up. During a prolonged brake application, the repeated cycling of the isolation and dump valves takes fluid out of the master cylinder and places it in the low-pressure accumulators. For this reason, the control module starts the ABS pump motor when the system enters the antilock mode. This pump forces brake fluid back against the master cylinder pistons and to the isolation valves. This action maintains brake pedal height during the antilock mode. In the antilock mode, the driver may feel pedal pulsations and a limited brake pedal fade followed by upward brake pedal movement.

> **Tech Tip** *Some light-duty trucks have a rear-wheel antilock (RWAL) system, which has ABS only on the rear wheels. If the vehicle has four-wheel drive and the four-wheel drive is engaged, then the ABS system is turned off.*

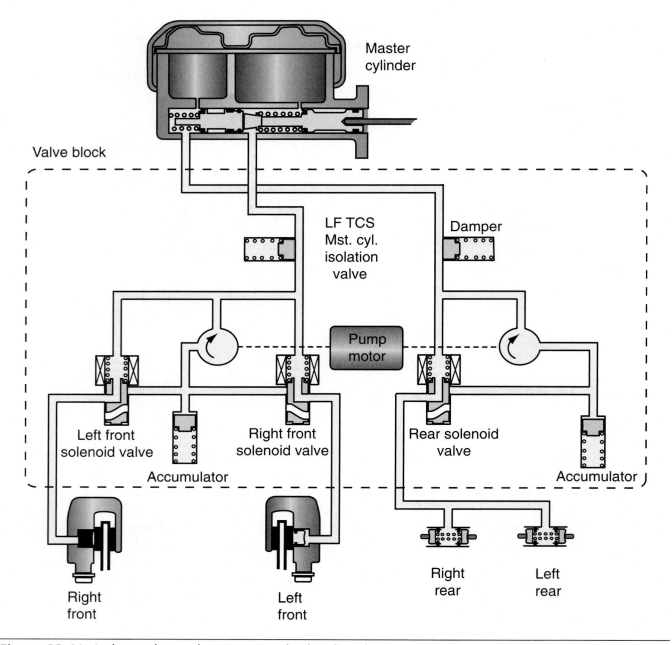

Figure 23-41 A three-channel ABS system hydraulic schematic.

FOUR-WHEEL ABS WITH HIGH PRESSURE ACCUMULATOR

Many vehicles have a four-wheel ABS system with a wheel speed sensor at each wheel. A toothed ring on the wheel hub rotates past the tip of the wheel speed sensor. On some front-wheel drive cars, the wheel speed sensors are integral with the wheel hubs and cannot be serviced separately. Some four-wheel ABS have a high-pressure accumulator mounted on the master cylinder **(Figure 23-42)**. The high-pressure accumulator

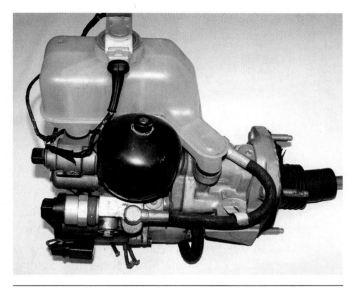

Figure 23-42 ABS with high-pressure accumulator.

contains a heavy diaphragm in the center of the accumulator. A nitrogen gas charge is permanently sealed in the upper accumulator chamber above the diaphragm. A pump integral with the master cylinder pumps brake fluid into the lower accumulator chamber. The pump maintains a brake fluid pressure of 2,000 to 2,600 psi (14,000 to 16,000 kPa) in the accumulator. If the accumulator pressure drops below 2,000 psi (14,000 kPa), a pressure switch on the master cylinder signals the control module to start the pump motor and increase the pressure.

Summary

- When pressure is applied to a confined liquid, the pressure is transmitted equally in all directions and acts with equal force on equal areas.

- Brake fluids may be classified as DOT 3, DOT 4, or DOT 5.

- Dual master cylinders contain a primary and a secondary piston and two wheels on the vehicle are supplied with pressure from each piston.

- Many drum brakes are servo-type in which the operation of the primary shoe applies mechanical force to the secondary shoe.

- Many drum brakes are self-adjusting.

- Many disc brakes have floating calipers with a single piston on the inside of the caliper.

- Disc brakes do not provide a servo action. This type of brake does not experience brake fade.

- A metering valve delays fluid pressure to the front calipers until the rear brake shoes are moved outward against the drums.

- A proportioning valve reduces pressure to the rear brakes to prevent rear wheel lock-up caused by weight shift to the front of the vehicle while braking.

- A pressure differential valve illuminates a brake warning light in the instrument panel if there is unequal pressure between the two sections in the master cylinder.

- Brake boosters may be vacuum operated or hydraulically operated.

- Parking brakes apply the rear brakes through a lever and cable system.

- ABS provides maximum braking force with approximately 20 percent tire slip without wheel lock-up.

- A wheel speed sensor produces an AC voltage signal in relation to wheel speed.

- On some ABS, the high-pressure accumulator supplies fluid pressure to the master cylinder pistons that acts as a brake booster.

- In many ABS, a pair of solenoids operated by the ABS module control fluid pressure in the wheel caliper or cylinder during the antilock brake function.

- In a three-channel ABS, a pair of solenoids controls the brake fluid pressure to both rear wheels.

■ A four-channel ABS has a pair of solenoids for each wheel.

■ In place of a high-pressure accumulator, some ABS have two low-pressure accumulators and a pump motor.

■ If the ABS is disabled the system will operate as a conventional power brake system.

Review Questions

1. Technician A says a larger diameter wheel cylinder bore requires more fluid than a smaller wheel cylinder and will travel a greater distance. Technician B says a smaller diameter master cylinder bore requires more piston travel to displace the same amount of fluid as a larger piston. Who is correct?

   A. Technician A

   B. Technician B

   C. Both Technician A and Technician B

   D. Neither Technician A nor Technician B

2. Technician A says a DOT 5 brake fluid is silicone-based. Technician B says a DOT 5 brake fluid has a higher dry ERBP boiling point compared to a DOT 4 brake fluid. Who is correct?

   A. Technician A

   B. Technician B

   C. Both Technician A and Technician B

   D. Neither Technician A nor Technician B

3. Technician A says the primary piston is positioned toward the brake pedal end of the master cylinder bore. Technician B says if the master cylinder has a residual valve, a slight pressure is maintained in the drum brake system with the brakes released. Who is correct?

   A. Technician A

   B. Technician B

   C. Both Technician A and Technician B

   D. Neither Technician A nor Technician B

4. Technician A says in a duo-servo drum brake system, the primary lining has a longer lining compared to the secondary lining. Technician B says in this type of brake system, the primary shoe applies most of the braking force. Who is correct?

   A. Technician A

   B. Technician B

   C. Both Technician A and Technician B

   D. Neither Technician A nor Technician B

5. In a duo-servo drum brake assembly:

   A. The primary shoe is positioned toward the rear of the vehicle.

   B. The lower ends of the brake shoes are connected to the backing plate.

   C. Drum heat and expansion can cause brake fade.

   D. The primary and secondary shoes provide equal braking force.

6. All of these statements about floating caliper disc brakes are true *except*:

   A. The brake caliper is mounted so it can slide sideways.

   B. The floating caliper usually contains a single piston.

   C. A piston return spring moves the lining away from the rotor surface.

   D. This type of brake does not require periodic adjustment.

7. While discussing vacuum brake boosters, Technician A says when the brakes are released, vacuum is supplied to both sides of the booster diaphragm. Technician B says when the brakes are applied, atmospheric pressure is supplied to the front side of the booster diaphragm. Who is correct?

   A. Technician A

   B. Technician B

   C. Both Technician A and Technician B

   D. Neither Technician A nor Technician B

8. All of these statements about high-pressure accumulators in ABS are true *except*:

   A. A high-pressure accumulator has an upper and lower chamber separated by a diaphragm.

   B. A high-pressure accumulator is fully charged at 2,000 psi.

   C. A high-pressure accumulator supplies fluid pressure that acts as a brake booster.

   D. A high-pressure accumulator continously supplies fluid pressure to the rear wheel calipers.

9. In an ABS with low-pressure accumulators:

   A. The module starts the motor pump when the brake pedal is depressed.

   B. The driver may feel pedal pulsations during the antilock mode.

   C. The driver should not feel any brake pedal fade in the antilock mode.

   D. The pump motor forces brake fluid back into the master cylinder reservoir.

10. During the antilock mode, the percentage of wheel slip supplied by an ABS is:

    A. 20 percent.

    B. 30 percent.

    C. 40 percent.

    D. 50 percent.

11. In a drum brake system, the self-adjusting mechanism adjusts the brakes when the brakes are applied during _____ vehicle motion.

12. A metering valve delays fluid pressure to the _____ brakes.

13. A proportioning valve reduces pressure to the _____ brakes.

14. A pressure differential valve illuminates a light in the instrument panel if there is

    _____ _____

    in the master cylinder sections.

15. Explain the operation of a vacuum brake booster with the brakes released and applied.

16. Describe the basic operation of a hydro-boost brake booster.

17. Describe the operation of the parking brake system on a vehicle with rear drum brakes.

18. Describe the purpose of the vent ports in the master cylinder bores.

Brake System Maintenance, Diagnosis, and Service

Learning Objectives

After you have read, studied, and practiced the contents of this unit, you should be able to:

- Perform brake system inspections.
- Diagnose brake system problems.
- Adjust pedal free-play and drum brakes.
- Bleed brake systems.
- Service drum brake and disc brake systems.
- Observe the ABS warning light and determine the ABS condition.
- Check the master cylinder fluid level on ABS.
- Obtain flash DTCs on various ABS.
- Obtain ABS DTCs using a scan tool.
- Perform all the ABS diagnostic functions using a scan tool.
- Diagnose ABS wheel speed sensors.
- Bleed ABS using various methods, including the automated bleed procedure with a scan tool.

Key Terms

Air gap

Brake pedal free-play

Brake pressure modulator valve (BPMV)

Data link connectors (DLCs)

Diagnostic trouble codes (DTCs)

Infinite ohmmeter reading

Non-directional finish (NDF)

Out-of-round

INTRODUCTION

The brake system is the most important safety system on the vehicle. Brake technicians have the customer's life in their hands. You must never compromise on brake safety. When brake components are in questionable condition, they should be replaced and quality brake components must always be installed. For example, if brake drum diameter is equal to the discard diameter, replace the brake drum. Do not take a chance on a worn-out brake drum that may crack and cause a brake failure. All brake components must be serviced using the vehicle manufacturer's recommended procedures.

ABS are more likely to experience basic brake problems than problems in the ABS control circuitry and components. Therefore, you must be familiar with conventional brake service because many of the same basic service procedures, such as changing brake shoes or pads, are required on ABS. You must also be familiar with the maintenance, diagnostic, and service procedures required for the electronic circuits and hydraulic systems in ABS. When you are knowledgeable regarding these maintenance, diagnostic, and service procedures, you will be able to solve ABS problems quickly and accurately.

BRAKE SYSTEM MAINTENANCE

During a brake inspection, one of the first items to check is the fluid level in both sections of the master cylinder. On disc brake systems, fluid level in the master cylinder becomes lower as the linings wear and the caliper pistons move outward. Fill the master cylinder to the proper level with the vehicle manufacturer's specified brake fluid **(Figure 24-1)**. Check the fluid for contamination, and flush if necessary. If the brake fluid is a light amber color, then it should be checked for moisture contamination with an approved brake fluid moisture tester. If the color is darker than a light amber, then it should be flushed.

Check the brake pedal for proper free-play. Inadequate brake pedal free-play may cause the master cylinder piston cups to cover the vent ports in the master cylinder. This action causes dragging brakes and a hard brake pedal. Excessive brake pedal free-play causes too much pedal movement and a low brake pedal during a brake application. The

Tech Tip *Always use brake fluid from a sealed container to avoid moisture contamination.*

brake pedal free-play is the amount of pedal movement before the booster pushrod contacts the master cylinder piston **(Figure 24-2)**. This should generally be approximately $\frac{1}{8}$ inch to $\frac{1}{4}$ inch (3–6 mm). Also check the brake pedal reserve. Pedal reserve is the distance between the pedal and the floor after the brakes are fully applied. This should generally not be less than 1–2 inches (24–50 mm).

Tech Tip *Brake fluid is very damaging to painted surfaces. Be careful not to spill brake fluid on the vehicle when servicing the brake system.*

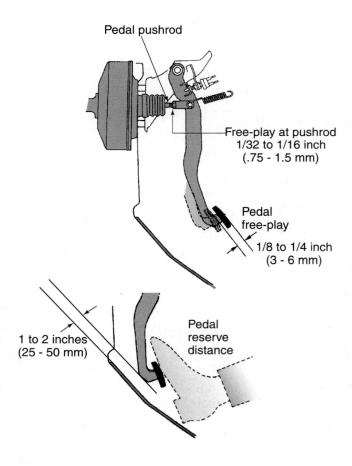

Pedal pushrod

Free-play at pushrod 1/32 to 1/16 inch (.75 - 1.5 mm)

Pedal free-play

1/8 to 1/4 inch (3 - 6 mm)

1 to 2 inches (25 - 50 mm)

Pedal reserve distance

Figure 24-1 Maintain the fluid level in the master cylinder at the proper level.

Figure 24-2 Brake pedal free-play and brake pedal reserve.

Check all the brake lines and fittings under the hood for leaks and unwanted contact with other components. Raise the vehicle on a lift and inspect all the brake lines and hoses under the vehicle. Check the brake flexible hoses for cracks, leaks, bulges, and unwanted contact with other components. Inspect all the brake tubing for rust and corrosion, leaks, flat spots, kinks, and damage. Inspect each wheel for evidence of brake fluid or grease leaks.

On many vehicles it is possible to partially inspect the condition of the rotor and brake linings on disc brakes without removing the caliper. Inspect the condition of the parking brake cables and pull on the cables to determine if they are moving freely. Check each wheel for free rotation. If one of the wheels is hard to rotate, the brakes are likely not releasing properly. Lower the vehicle onto the shop floor and road test the vehicle. Check the brake pedal feel during a brake application. If the brake pedal feels spongy, there may be air in the brake system or the master cylinder may be very low on brake fluid. If the fluid is very low, it may be possible to draw air into the brake system. When the pedal feels hard and requires excessive pedal effort, the brake linings may be oil soaked or the caliper or wheel cylinder pistons may be seized. Check the pedal for the correct amount of pedal travel during a brake application.

During a brake application, check for pedal pulsations and noises in the brake system such as scraping or rattling. Check for brake grabbing, wheel lock-up, and steering pull during a brake application. These problems are often the result of contaminated brake linings. To perform a complete inspection of the brake system components, the wheels and drums or calipers have to be removed. Park the vehicle and apply the parking brake. Start the engine and place the transmission in drive. Accelerate the engine slightly and determine if the parking brake holding capability is adequate.

....

NOTE: Brake fade is the reduction in braking effectiveness over a number of subsequent brake applications. Pedal fade is when the pedal slowly sinks to the floor when holding the brake pedal with steady pressure.

....

BRAKE SYSTEM DIAGNOSIS

The causes of various brake system problems are listed in the following diagnostic summary:

1. A low spongy pedal and excessive pedal travel with the red brake warning light on.
   a. Low fluid level in the master cylinder.
   b. Air in the hydraulic system.
2. Noise while braking.
   a. Worn brake linings causing a squealing or scraping noise.
   b. Worn bushings on floating calipers.
   c. Interference between brake linings and splash shield or backing plate.
   d. Chirping noise caused by lack of lubrication on backing plate supports.
3. Pedal pulsations during brake applications.
   a. Excessive thickness variation or runout on rotors or excessive out-of-round on brake drums.
4. Grabbing brakes during a brake application.
   a. Contaminated front brake pad linings.
   b. Contaminated rear brake shoe or pad linings.
5. Steering pulls to one side while braking.
   a. Contaminated brake linings.
   b. Seized caliper piston or wheel cylinder pistons on the opposite front wheel from the direction of steering pull.
6. Excessive pedal effort on non-power brakes.
   a. Caliper or wheel cylinder pistons seized.
   b. Contaminated or glazed brake linings.
7. Excessive pedal effort on power brakes.
   a. Defective brake booster.
   b. Low manifold vacuum supplied to the booster.
   c. Seized caliper or wheel cylinder pistons.
   d. Glazed or contaminated brake linings.
8. The rear brakes drag.
   a. Improper parking brake adjustment.
   b. Seized parking brake cables.

c. Improper rear brake adjustment.

d. Weak brake shoe return springs or seized caliper or wheel cylinder pistons.

9. The brakes on all wheels drag.

a. Seized caliper or wheel cylinder pistons.

b. Improperly adjusted brake light switch preventing pedal return.

c. Binding pedal linkage.

d. Contaminated brake fluid.

e. Corrosion in the master cylinder that plugs the vent ports.

10. Premature rear wheel lock-up.

a. A defective proportioning valve.

b. Rear brake linings contaminated.

c. Front caliper pistons seized.

11. Excessive pedal travel, or a low, firm pedal.

a. The drum brakes require adjusting.

b. One master cylinder section is defective.

12. The brakes release slowly and the pedal does not return fully.

a. Weak shoe return springs.

b. A defective booster or boost check valve.

c. Contaminated or wrong brake fluid.

d. The brake light switch is interfering with pedal return.

e. Pedal linkage binding.

13. Brake pedal fade while brakes are applied firmly.

a. A leaking brake line or bulging brake hose.

b. Leaking master cylinder piston cups.

> **Tech Tip** *A noise from the brakes without pressing on the brake pedal could be caused by the wear indicators on the brake pads.*

BRAKE SERVICE
Brake Pedal Free-Play Adjustment

Brake pedal free-play is very important to provide proper brake system operation. If this distance is not within specifications, lengthen or shorten the pedal pushrod to obtain the proper free-play.

Brake Shoe Adjustment

To determine if the drum brakes require adjusting, pump the brake pedal twice with the engine running. If the pedal is higher on the second pump, the drum brakes require adjustment. Raise the vehicle on a lift and remove the rubber plugs in the adjustment openings in the backing plate. Some drums may have the access opening in the face of the drum. Insert the end of a brake-adjusting tool through the backing plate opening until it contacts the star wheel on the brake adjuster. Move the tool up and down to rotate the star wheel in the proper direction to move the brake shoes outward while rotating the tire and wheel. Continue rotating the adjuster until the wheel is hard to turn. If the vehicle has self-adjusting brakes, insert a small screwdriver through the backing plate opening to move the self-adjuster lever away from the star wheel **(Figure 24-3)**. Use the brake-adjusting tool to rotate the star wheel to shorten the brake adjuster and move the brake shoes inward until the tire and wheel rotate freely. The shoes should be adjusted until they are just off the surface of the drum. Reinstall the rubber plug in the backing plate opening.

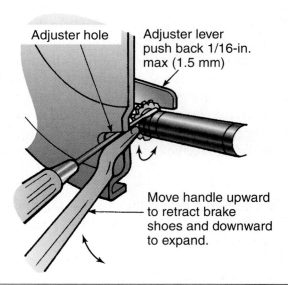

Adjuster hole

Adjuster lever push back 1/16-in. max (1.5 mm)

Move handle upward to retract brake shoes and downward to expand.

Figure 24-3 Brake shoe adjustment.

Brake System Bleeding

Brake fluid may be replaced in the brake system using five different methods:

- Manual bleeding
- Pressure bleeding
- Vacuum bleeding
- Gravity bleeding
- Reverse bleeding

Tech Tip *Special procedures must be followed on vehicles equipped with a high pressure accumulator and/or with ABS. It may be possible to disarm the ABS for the purposes of bleeding the brakes using a scan tool. Consult the service manual for the specific procedures for these vehicles.*

Air may be bled manually from the brake system. To use the manual bleeding procedure, fill the master cylinder to the required level with the specified brake fluid and raise the vehicle on a lift. Have a co-worker apply and hold the brake pedal. Connect a hose from the bleeder screw on the caliper or wheel cylinder in the right-rear wheel, which is furthest from the master cylinder **(Figure 24-4)**. Place the other end of the hose in a glass container. Open the bleed screw and allow the brake fluid to flow into the container. Close the bleeder screw and

inform your co-worker to release the brake pedal. Repeat this procedure until a clear stream of brake fluid with no air bubbles is discharged when the bleeder screw is opened. Refill the master cylinder and repeat the bleeding procedure on the other wheels in this order: left rear, right front, and left front. A pressure brake bleeder may be used to bleed the brakes.

The pressure bleeder is a storage tank with two compartments separated by a diaphragm. The lower half of this tank contains brake fluid, and 15 to 20 psi shop air pressure is applied above the diaphragm. Do not exceed the pressure bleeder's manufacturer's recommended air pressure as damage to the equipment or brake system may result. A gauge on the pressure bleeder indicates the air pressure in the bleeder tank. A hose is connected from the brake fluid reservoir in the bleeder to a special adapter installed on top of the master cylinder. A shut-off valve in the pressure bleeder hose allows you to open and close this hose. If the vehicle has a metering valve, then it will need to be manually opened to bleed that portion of the brake system. The pressure bleeder will likely not be able to force brake fluid past the metering valve as it is operating on only 15 to 20 psi of air pressure.

When using the pressure bleeder, be sure there is adequate brake fluid in the lower reservoir and close the valve in the brake fluid hose. Supply the proper air pressure to the upper reservoir in the bleeder. Install the adapter securely on top of the master cylinder and connect the brake fluid hose to this

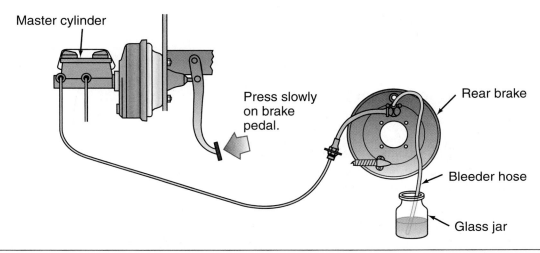

Figure 24-4 Manual brake bleeding procedure.

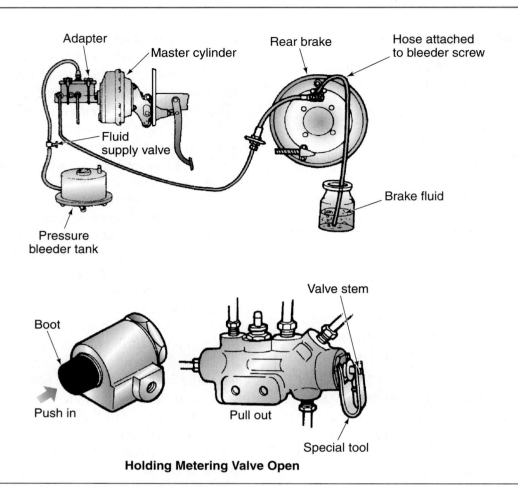

Figure 24-5 Pressure brake bleeding procedure.

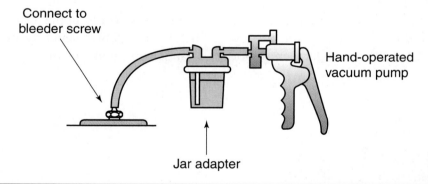

Figure 24-6 A basic setup for vacuum bleeding.

adapter **(Figure 24-5)**. Open the valve in the brake fluid hose and be sure there are no leaks between the adapter and the master cylinder. Bleed air from each wheel cylinder or caliper in the same sequence used for manual bleeding. After the bleeding procedure is completed, be sure the master cylinder has the proper brake fluid level.

A hand-operated or air-operated suction pump may be used to pull some brake fluid from the master cylinder to obtain the proper fluid level **(Figure 24-6)**. If the vehicle has a metering valve, then it will need to be manually opened to bleed that portion of the brake system. You may notice air bubbles even after bleeding the brake system repeatedly. This may be air leaking past the bleeder screw threads. One solution is to coat the threads of the bleeder screws with silicon grease. Do not use any other type of grease as contamination and serious damage can be done to the rubber components of the brake system. The thin coating of silicon grease will help seal the air from leaking around the threads. You may also notice that the bubbles are much smaller than those from air trapped in the brake system. Simply close the bleeder screw and test the brake pedal for proper operation.

Gravity bleeding is similar to manual bleeding except the brake pedal is not depressed. The same equipment is used. The bleeder screw is opened and the brake fluid is allowed to flow through the brake system solely under the influence of gravity. This method is not very efficient and may not be suitable for many brake systems, particularly those with ABS.

Reverse bleeding is a relatively recent development. The concept is simple in theory but requires specialized equipment. Brake fluid is forced into the brake bleeders and flows through the brake system as if it were brake fluid traveling back to the master cylinder when the brakes were released. The machine forces brake fluid into the bleeder screws and removes the excess or contaminated brake fluid from the master cylinder **(Figure 24-7)**. This same method is popular when flushing the brake system of contaminated fluid. It is important to consider that the most contaminated brake fluid

is at the wheel cylinders and calipers. If rust or other hard contaminants are present, then it is possible to push these contaminants to the ABS HCU or the master cylinder. It may be wise to inspect a sample of the brake fluid from the wheel cylinders before using the brake-flushing machine, especially on an older vehicle.

Drum Brake Service

After the wheel and brake drum are removed, inspect the brake linings. A depth micrometer may be used to measure the thickness of the linings above the rivet heads. This thickness should be a minimum of 0.030 inch (0.75 millimeter). Use a low-pressure wet cleaning system to wash the brake components **(Figure 24-8)**. Place a retaining clip on the wheel cylinder to prevent the pistons from moving outward in this cylinder **(Figure 24-9)**. Use a brake spring tool to remove the shoe return springs and remove the brake shoe hold-down clips and springs **(Figure 24-10)**.

Remove the brake shoes and then remove the lower retracting spring and adjuster and the self-

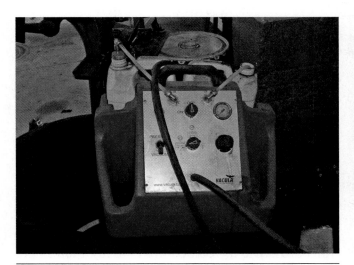

Figure 24-7 A brake flushing machine. *(Courtesy of Steffy Ford, Columbus NE)*

Figure 24-8 Wet brake cleaning equipment.

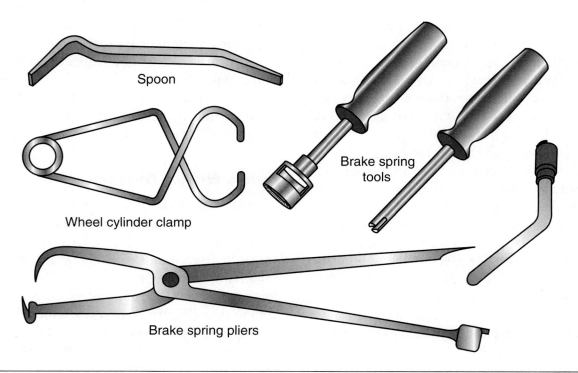

Figure 24-9 Common brake tools.

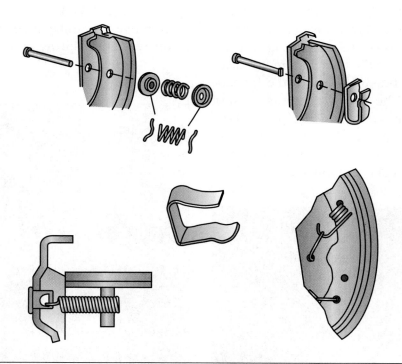

Figure 24-10 Various hold-down springs can be used on drum brakes.

adjusting mechanism (**Figure 24-11** and **Figure 24-12**). Replace the shoe return springs if they are damaged, distorted, bent, or corroded. Inspect the backing plates for cracks, distortion, or worn brake shoe supports and anchors. If any of these conditions are present, replace the backing plate. Peel the rubber boots back on the wheel cylinder and check for brake fluid and contaminants behind

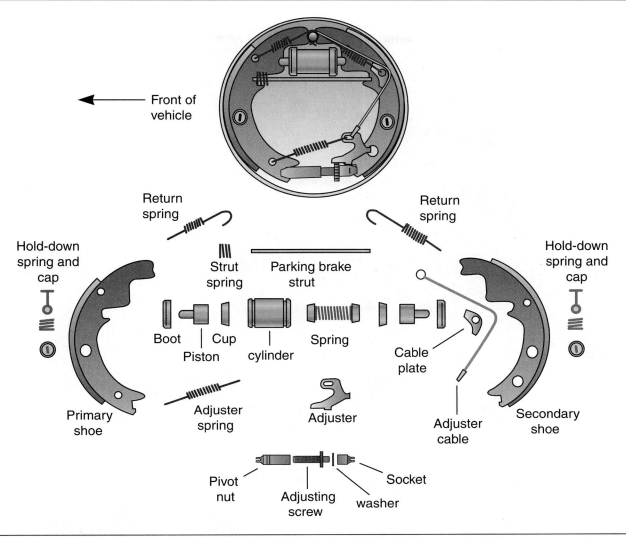

Figure 24-11 Drum brake components (servo type).

these boots. If brake fluid or contaminants are present behind the boots, wheel cylinder overhaul or replacement is required.

Inspect the brake drums for scoring, cracks, heat checks, or burned spots. Drums with heat checks, cracks, or severe burned spots must be replaced. If the burned spots are removed after turning the drum, the drum may be reused. Use a brake drum micrometer to measure the drum diameter at several locations around the drum **(Figure 24-13)**. Take four measurements 45° apart and record the results. Insert the micrometer in the drum. Measure the drum diameter at the inner and outer edges of the machined surface and also on the center of this surface. Hold the fixed anvil firmly against the inside surface of the drum and carefully rock the dial indicator until you get the highest reading. If drum taper or out-of-round exceeds 0.006 inch (0.152 millimeter), the drum must be machined or replaced. Brake drum **out-of-round** is a variation in drum diameter at various locations around the drum. Some common defective drum conditions are illustrated in **Figure 24-14**. The drum's discard diameter is stamped on the drum

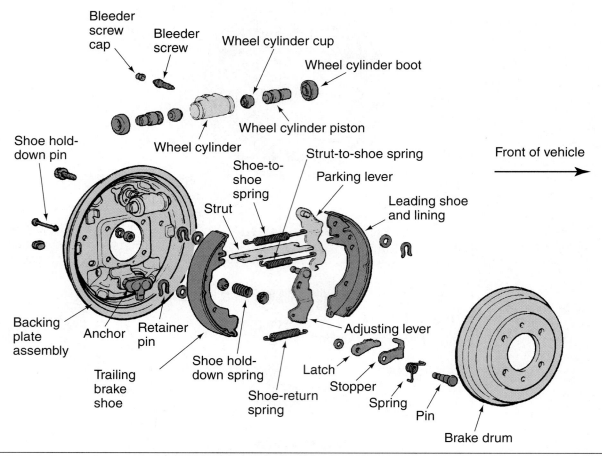

Bleeder screw cap

Bleeder screw

Wheel cylinder cup

Wheel cylinder boot

Wheel cylinder piston

Wheel cylinder

Shoe hold-down pin

Shoe-to-shoe spring

Strut-to-shoe spring

Strut

Parking lever

Leading shoe and lining

Front of vehicle

Backing plate assembly

Anchor

Retainer pin

Shoe hold-down spring

Trailing brake shoe

Shoe-return spring

Latch

Stopper

Adjusting lever

Spring

Pin

Brake drum

Figure 24-12 Drum brake components (leading/trailing type).

Figure 24-13 Measuring brake drum diameter with a drum micrometer.

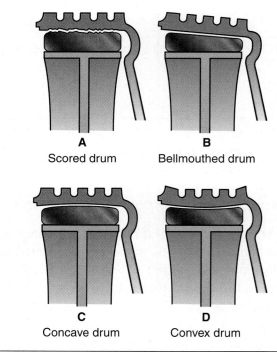

A
Scored drum

B
Bellmouthed drum

C
Concave drum

D
Convex drum

Figure 24-14 Defective brake drum conditions.

(Figure 24-15). After turning the drum, there must be 0.030 inch (0.762 millimeter) left for wear.

> **Tech Tip** *Brake drums should generally be within .010 inch of each other.*

> **Tech Tip** *If metal particles are not removed from a brake drum after turning the drum, these particles become embedded in the brake linings and cause rapid lining and drum wear.*

Mount the brake drum securely in the drum lathe and position a vibration band around the outer drum diameter **(Figure 24-16)**. Most brake lathe manu-

Figure 24-15 Brake drum discard diameter.

facturers recommend taking a very light cut that just scratched the inside surface of the drum. This is called a scratch cut. The drum is then loosened and rotated 180° on the arbor of the brake lathe and then retightened. A second scratch cut is made. If both scratch cuts are next to each other then the drum is mounted properly centered. If they are not, then the drum is not mounted properly and should be removed, cleaned, and remounted on the lathe.

Set the cutter on the lathe to remove a small amount of metal from the drum, usually 0.002 to 0.005 inch (0.05 to 0.15 millimeter). Follow the equipment manufacturer's recommended procedure when using the brake drum lathe. Several cuts may be required to remove scoring and other defective drum surface conditions. After turning the drum, the drum should be re-measured. If the drum is serviceable, it should be thoroughly cleaned with hot water and wiped with a lint-free shop towel to remove small metal particles from the drum.

Wheel cylinders are usually replaced rather than repaired. Wheel cylinders should be replaced during the servicing of the drums. Install the wheel cylinder in the backing plate. Lubricate the brake shoe supports on the backing plate and install the brake

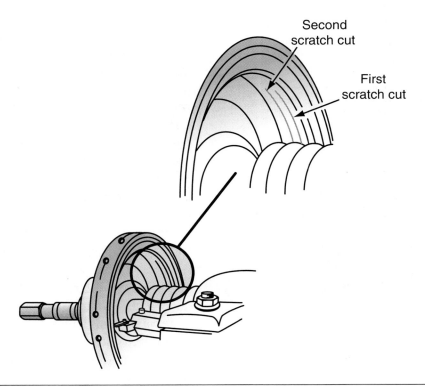

Figure 24-16 Brake drum with vibration damper band mounted in a drum lathe.

shoes with the adjuster and lower retracting spring. Install the self-adjusting mechanism and brake shoe hold-down clips. Use a brake spring tool to install the shoe return springs. Be sure all brake components are installed in their original position. Install one end of a brake shoe setting gauge in the drum and set the brake shoe setting gauge to the drum diameter **(Figure 24-17)**. Install the other side of the brake shoe setting gauge over the brake shoes and adjust the brake shoes so they lightly contact the brake shoe setting gauge jaws **(Figure 24-18)**. Using the brake shoe setting gauge sets the proper clearance between the brake shoes and the drum. Verify the brake shoe adjustment after the drum had been installed and make any necessary final adjustments.

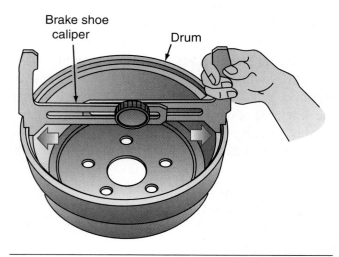

Figure 24-17 A brake shoe caliper is set to the inside diameter of the brake drum.

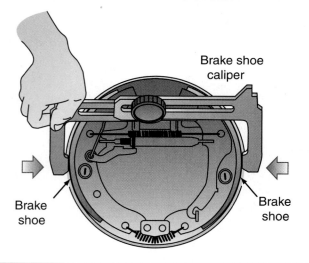

Figure 24-18 A brake shoe caliper is set to the inside diameter of the brake drum.

Tech Tip *If wheel cylinders are not replaced during a brake overhaul, the wheel cylinder pistons and cups are repositioned when the new brake linings are installed. Under this condition, the wheel cylinder cups may be positioned on top of accumulated debris in the bottom of the wheel cylinder. This action results in brake fluid leaking from the wheel cylinder.*

Disc Brake Caliper and Brake Pad Removal

The brake pad linings may be inspected through inspection holes in the caliper or through the outer ends of the caliper. If the lining thickness is equal to or less than the minimum thickness specified by the vehicle manufacturer, the brake pads must be replaced. Some floating calipers are removed by removing the top bolts, retainer clip, and antirattle spring **(Figure 24-19)**. Some floating calipers have two caliper pins **(Figure 24-20)**.

Tech Tip *The caliper's pistons will need to be pushed back into their bores to allow for the installation of the new pads as they will be thicker than the worn pads. It is recommended that the bleeder screws be open during this procedure to avoid pushing the used/contaminated fluid back up to the master cylinder or the ABS unit. This also eliminates the need to drain the excess fluid from the master cylinder should the pistons be compressed without opening the bleeder screws. Also consult the service manual for any specific procedures if the vehicle is equipped with ABS.*

On a fixed caliper, remove the bolts holding the caliper to the steering knuckle. Lift the caliper straight up off the rotor. When replacing only the brake pads, remove the brake pads from the caliper and hang the caliper from a suspension component with a length of wire. After the new brake pads and/or calipers are installed, the brake pedal should be applied several times to properly position the caliper pistons before attempting to drive the vehicle.

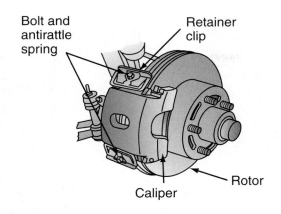

Figure 24-19 Floating caliper using retaining clips and springs.

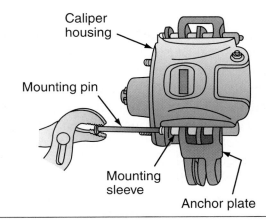

Figure 24-20 Floating caliper using pins.

> **Tech Tip** *Never allow the caliper to hang by the brake hose, as this will damage the brake hose. The hose can withstand pressure from the brake fluid but not stretching.*

Rotor Service

Inspect the rotor friction surfaces for scoring or grooving, cracks, broken edges, heat checking, and hard spots. Replace the rotor if it has cracks, broken edges, heat checking, or excessive scoring or grooving. Use a dial indicator to measure rotor runout and measure the rotor thickness variation with a micrometer at 8 to 12 locations around the rotor **(Figure 24-21)**. Be sure the wheel bearings are properly adjusted before measuring the rotor runout. If the rotor runout or thickness is not within vehicle manufacturer's specifications, the rotor must be machined or replaced. The rotor discard thickness is stamped on the rotor. After machining the rotor, the rotor thickness must be

> **Tech Tip** *Excessive rotor runout causes brake pedal pulsations and increased pedal travel.*

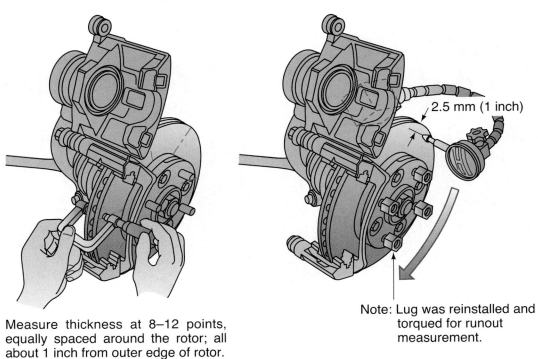

Measure thickness at 8–12 points, equally spaced around the rotor; all about 1 inch from outer edge of rotor.

2.5 mm (1 inch)

Note: Lug was reinstalled and torqued for runout measurement.

Figure 24-21 Measuring rotor thickness variation and lateral runout.

0.015 to 0.030 inch (0.40 to 0.75 millimeter) thicker than the discard dimension to allow for wear.

If the rotor is to be resurfaced, remove the rotor and hub from the vehicle. Clean the hub and install the hub and rotor on the lathe using the proper cones to support the hub. Install the vibration damper ring on the outside diameter of the rotor **(Figure 24-22)**. Rotors may also be turned using an on-the-car brake lathe. This has several advantages. The rotor need not be removed from the vehicle. Some vehicles require the removal of the wheel bearing when the rotor is removed. Also the rotor will be turned true to the axis of the bearings. Some vehicle manufacturers recommend this procedure **(Figure 24-23)**. Follow the equipment manufacturer's recommended procedure to machine the rotor surfaces. Adjust the cutting depth on the rotor lathe to remove a small amount of metal from the rotor surface with each cut. Several cuts are usually necessary to restore the rotor friction surfaces.

> **Tech Tip** *Rotor lathes are available that machine the rotor friction surfaces with the rotor and hub installed on the vehicle.*

After the rotor surfaces are machined, a sander may be used to apply a **non-directional finish**

Figure 24-23 Rotors may be turned using an on-the-car brake lathe.

(NDF) on the rotor surfaces **(Figure 24-24)**. The non-directional finish on the rotor friction surfaces does not follow the arc of rotor rotation. This type of finish reduces brake noise and provides improved brake lining seating. After the rotor machining is completed, thoroughly clean the rotor and hub to remove any metal particles. The rotor should be thoroughly washed with soap and water and then immediately air-dried. This will remove any metal particles left from the machining process.

Service, repack, or replace the wheel bearings and seals as necessary and install the hub and rotor on the steering knuckle. Install a new washer on the brake hose and install and tighten the brake

Figure 24-22 Rotor and hub mounted in lathe.

Figure 24-24 A technician is using a sander to apply a non-directional finish to the surface of the rotor.

hose in the caliper. Install the caliper and brake pads on the steering knuckle. On floating calipers, replace the mounting sleeves or bushings and the mounting pins. Be sure all the caliper hardware such as clips and antirattle springs are in good condition. Tighten all fasteners to the specified torque. Bleed the brakes as explained previously in this chapter.

Master Cylinder Service

Master cylinders are usually replaced rather than overhauled. However, repair kits are available to rebuild master cylinders. When the master cylinder piston cups are leaking, the brake pedal gradually fades downward during steady pedal pressure. If the brake fluid is contaminated or the reservoir contains excessive corrosion, master cylinder replacement is necessary. On aluminum master cylinders, the plastic reservoir may be replaced separately from the master cylinder. To remove a master cylinder, disconnect and cap the brake lines and remove the two mounting bolts. When installing a new master cylinder, it may be bled on the bench before installation **(Figure 24-25)**. Install plastic hoses from the master cylinder outlets into the reservoir. Use a wooden dowel to push on the primary piston. Continue stroking the primary piston until the brake fluid flowing through the plastic hoses is free from bubbles. Remove the plastic hoses and cap the fluid outlets.

Figure 24-25 A technician is bench bleeding a master cylinder.

Brake Booster Service

Check the brake pedal free-play before diagnosing brake booster problems. Adjust the free-play if necessary. To check the vacuum booster operation, repeatedly pump the brake pedal with the engine not running to remove all vacuum from the booster. Hold the brake pedal down and start the engine. If the booster is working properly, the pedal should move downward slightly and then stop moving. If the pedal does not move downward, the manifold vacuum is not supplied to the booster or the booster is defective. Inspect the vacuum hose to be sure it is connected tightly to the intake manifold and to the booster. Remove the vacuum hose from the booster. The engine should stall or run very roughly. If there is not much difference in engine operation, the vacuum hose is restricted. Be sure the one-way check valve in the vacuum hose allows airflow through it in one direction only.

Tech Tip *A leaking seal in the front of a vacuum booster causes brake fluid to be drawn from the master cylinder into the booster and intake manifold. If a booster has this problem, the brake fluid level goes down continually in the master cylinder with no indication of external leaks.*

Parking Brake Service

To check the parking brake, raise the vehicle on a lift. Pull on the parking brake cable to each rear wheel while rotating the rear wheel. When the parking brake cable is pulled by hand, the rear wheel should stop. When the parking brake cable is released, the cable should move freely back to its original position. If the parking brake cables are sticking, they should be replaced. To adjust the parking brake cables, be sure the parking brake lever in the passenger compartment is released. Adjust the nut on the front parking brake cable at the equalizer until the rear wheels begin to drag when

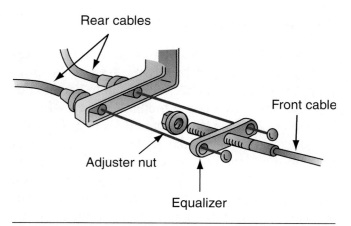

Figure 24-26 A parking brake equalizer and adjuster nut.

they are rotated **(Figure 24-26)**. Back off the adjusting nut on the front cable until the rear wheels rotate freely. Lower the vehicle onto the shop floor and apply the parking brake to be sure it operates properly.

ANTILOCK BRAKE SYSTEM MAINTENANCE

ABS maintenance includes a complete brake system inspection. During an ABS inspection, check all the ABS wiring harness for damage, worn insulation, or corroded connections. The wiring most likely to be damaged is the wheel speed sensor wiring. Inspect all the lines, fittings, and high-pressure accumulator for leaks. Start the engine and observe the amber ABS warning light. This light should be on when the ignition switch is turned on. It should remain on for approximately 4 seconds after the engine starts. During these few seconds, the ABS module completes a check of the ABS electrical system. If the ABS control module does not detect any electrical defects during the system check, the module turns the amber ABS warning light off. Any time the ABS module detects an electrical defect in the ABS system, the module turns on the ABS warning light with the engine running. When the ABS module turns on the ABS warning light with the engine running, the module usually cancels the ABS function. However, conventional, power-assisted braking is still available. If the ABS has a high-pressure accumulator, follow this procedure to check the fluid level in the master cylinder:

1. Turn the ignition switch on and pump the brake pedal until the hydraulic pump motor starts.
2. Wait until the hydraulic pump motor shuts off (the accumulator will be fully charged).
3. Check the fluid level in the master cylinder. If the fluid level is below the max fill line on the reservoir, bring the fluid level up to this line with the vehicle manufacturer's recommended brake fluid.

> **Tech Tip** *On some ABS while the module performs a check of the system, the module operates all the solenoids momentarily. This may cause audible clicking noises.*

ABS Diagnosis OBD I Vehicles

On OBD I vehicles prior to 1996, a variety of **data link connectors (DLCs)** were used by various vehicle manufacturers for scan tool connection. These DLCs were usually located under the hood or under the dash. Many OBD I vehicles had a separate DLC for ABS diagnosis. OBD II systems were mandated in 1996 and these systems have a standard 16-terminal DLC mounted under the dash near the steering column **(Figure 24-27)**. On vehicles with OBD II systems, data links interconnect the various computers on the vehicle. These data links are also connected to DLC terminals. Therefore, the different computer systems on

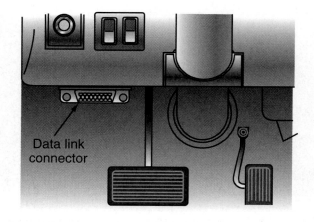

Figure 24-27 Data link connector (DLC) on an OBD II vehicle.

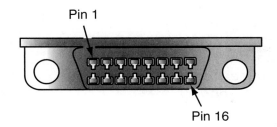

Pin 1   Secondary UART 8192 baud serial data
         Class B 160 baud serial data
Pin 2   J1850 Bus + line on 2 wire systems, or single wire (class 2)
Pin 3   Ride control diagnostic enable
Pin 4   Chassis ground pin
Pin 5   Signal ground pin
Pin 6   PCM/VCM Diagnostic enable
Pin 7   K line for International Standards Organization (ISO) application
Pin 8   Keyless entry enable, or MRD theft diagnostic enable
Pin 9   Primary UART
Pin 10   J1850 Bus-line for J1850-2 wire applications
Pin 11   Electronic Variable Orifice (EVO) steering
Pin 12   ABS diagnostic, or CCM diagnostic enable
Pin 13   SIR diagnostic enable
Pin 14   E&C bus
Pin 15   L line for International Standards Organization (ISO) application
Pin 16   Battery power from vehicle unswitched (4 AMP MAX)

Figure 24-28 The purpose of each of the 16 pins in a DLC for an OBD II system.

Tech Tip *On some OBD I vehicles with ABS, the ABS provides scan tool communication from the ABS computer to the DLC. These systems display ABS DTCs or diagnostic messages on the scan tool.*

Tech Tip *The ignition switch must be off when connecting or disconnecting the scan tool.*

the vehicle, including the ABS computer, may be diagnosed by connecting a scan tool to the DLC **(Figure 24-28)**.

The first step in any ABS diagnosis is to check the operation of the amber ABS warning light and the red brake warning light as explained previously. When diagnosing ABS on OBD 1 vehicles, connect a jumper wire between two terminals in the DLC or ABS DLC to retrieve **diagnostic trouble codes (DTCs)** from

the ABS computer. On General Motors vehicles with the ignition switch off, a special key may be used to connect terminals A and H or A and G in the DLC under the dash **(Figure 24-29)**. When the ignition switch is turned on, the ABS warning light flashes the DTCs. For example, three light flashes followed by a short pause and four

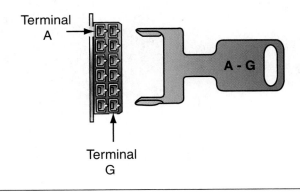

Figure 24-29 Connecting terminals A to G in the DLC for ABS diagnosis on a GM OBD I system.

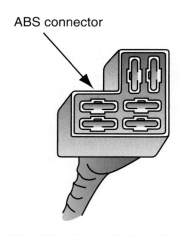

Figure 24-31 DaimlerChrysler ABS DLC.

light flashes indicates DTC 34. The vehicle manufacturer's service information provides the DTC interpretation. A DTC indicates an electrical problem in a certain area, such as a wheel speed sensor. Voltmeter or ohmmeter tests are usually necessary to determine if the defect is in the wheel speed sensor or the wires from this sensor to the ABS computer.

Two different DLC are used on Daimler-Chrysler ABS on OBD I vehicles **(Figure 24-30** and **Figure 24-31)**. A scan tool may be connected to these DTCs to retrieve ABS DTCs and perform some other diagnostic functions.

On Ford OBD I vehicles with ABS, the DLC is the same shape as the DLC for the PCM, but the ABS DLC is red. The ABS DLC may be located in the trunk or under the hood. To retrieve DTCs on a Ford ABS, turn the ignition switch off and

connect a jumper wire between the trigger self-test input terminal and the ground terminal in the DLC **(Figure 24-32)**. Connect an analog voltmeter from the self-test output terminal in the DLC to ground. When the ignition switch is turned on, the voltmeter needle sweeps out the DTCs. For example, two needle sweeps followed by a pause and one needle sweep indicates DTC 21. On some Ford ABS, a scan tool may be connected to the DLC to retrieve DTCs and perform other diagnostic tests.

To erase the DTCs on many ABS on OBD I vehicles, turn the ignition switch off and disconnect the wiring harness connector from the ABS computer. Reconnect the wiring harness. On some General Motors RWAL systems, disconnect the STOP/HAZARD fuse to erase the DTCs on the RWAL computer. Always follow the vehicle manufacturer's instructions when erasing ABS DTCs because special DTC erasing procedures are required on some ABS

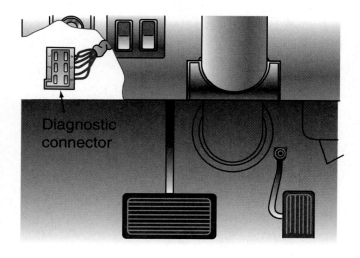

Figure 24-30 DaimlerChrysler DLC for body computers and ABS. Note this connector is blue.

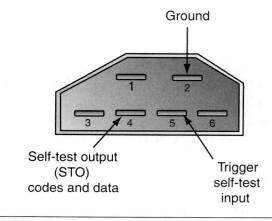

Figure 24-32 Ford ABS DLC.

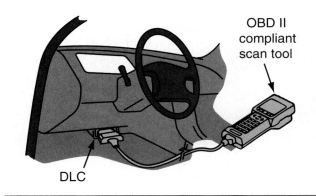

Figure 24-33 Attach the scan tool to the OBD II DLC when the ignition key is OFF.

on OBD I vehicles. On 1996 and newer vehicles, the scan tool is used to erase ABS DTCs.

ABS Diagnosis OBD II Vehicles

To diagnose the ABS on OBD II vehicles, connect the scan tool to the DLC with the ignition switch off. Be sure the scan tool contains the proper module for the ABS diagnosis on the vehicle being tested **(Figure 24-33)**. The following is a typical ABS diagnostic procedure on an OBD II vehicle:

1. Turn the ignition switch on, and select ABS diagnosis on the scan tool display. This establishes communication between the ABS computer and the scan tool.

2. Select ABS DTCs on the scan tool display. If there are any DTCs with a U prefix, there is a defect in the data links connected between the computers. This problem must be repaired before proceeding with the diagnosis.

3. Record any other DTCs displayed on the scan tool. Interpret the DTCs from the vehicle manufacturer's service information and repair the causes of the DTCs. Use the scan tool to erase the DTCs.

4. Select Functional Test on the scan tool. In this mode, the EBCM commands the ABS relay, solenoids, and pump motor on and off. If there is a defect in any of these systems, a DTC is set in the EBCM.

5. Select Automated Bleed on the scan tool. In this mode, the EBCM commands the valve solenoids and pump motor on and off in a special sequence to bleed air from the ABS system.

6. Select ABS Motor on the scan tool. This selection allows you to command the pump motor on and off. If a defect is present in the motor circuit, a DTC is set in the EBCM.

7. Select System Identification on the scan tool. The scan tool displays the hardware and software revision of the EBCM.

8. Select Tire Size Calibration on the scan tool. This calibration must be performed after an EBCM is replaced or when different size tires are installed on the vehicle.

9. Select Lamp Tests on the scan tool. This mode allows you to command the ABS amber warning light and the red brake warning light on and off.

10. Select Solenoid Tests on the scan tool. This test mode allows you to command a selected ABS solenoid valve on and off. The vehicle must be raised on a lift so the wheels and tires are about 1 foot off the shop floor. Command a specific solenoid on with the scan tool and depress the brake pedal. Have an assistant turn the wheel to which the solenoid is connected. The wheel should spin because the solenoid blocks fluid flow to the wheel. If the wheel does not turn, the solenoid is likely leaking. Repeat this procedure at each wheel.

11. Select ABS Relay on the scan tool. This mode allows you to turn the ABS relay on and off. If a malfunction is detected in the relay circuit, a DTC is set in the EBCM.

Tech Tip *The ignition switch must be off when connecting or disconnecting the scan tool.*

Tech Tip *OBD II systems have many standardized features including DTCs, monitors in the PCM to monitor various engine-related systems, and data links between computers on the vehicle.*

Wheel-Speed Sensor Diagnosis

If a DTC indicates a defect in a wheel-speed sensor and related circuit, an ohmmeter may be connected to the wheel-speed sensor terminals.

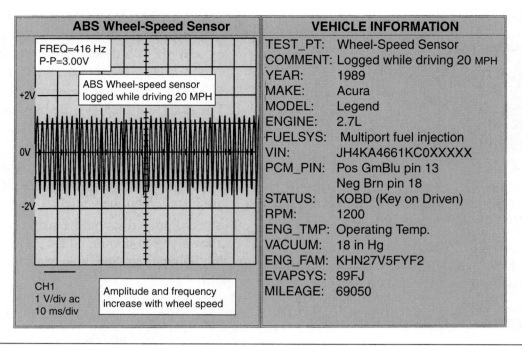

ABS Wheel-Speed Sensor	VEHICLE INFORMATION

FREQ=416 Hz
P-P=3.00V

ABS Wheel-speed sensor
logged while driving 20 MPH

+2V

0V

-2V

CH1
1 V/div ac
10 ms/div

Amplitude and frequency
increase with wheel speed

TEST_PT: Wheel-Speed Sensor
COMMENT: Logged while driving 20 MPH
YEAR: 1989
MAKE: Acura
MODEL: Legend
ENGINE: 2.7L
FUELSYS: Multiport fuel injection
VIN: JH4KA4661KC0XXXXX
PCM_PIN: Pos GmBlu pin 13
                   Neg Brn pin 18
STATUS: KOBD (Key on Driven)
RPM: 1200
ENG_TMP: Operating Temp.
VACUUM: 18 in Hg
ENG_FAM: KHN27V5FYF2
EVAPSYS: 89FJ
MILEAGE: 69050

Figure 24-34 Normal wheel-speed sensor voltage waveform.

The sensor winding should have the specified resistance. An open winding provides an infinite ohmmeter reading and an ohmmeter reading that is lower than specified indicates a shorted winding. An **infinite ohmmeter reading** is a resistance reading beyond the measurement capability of the ohmmeter.

When the ohmmeter is connected from one of the wheel-speed sensor terminals to ground, an infinite reading indicates the sensor winding is not grounded, whereas a low reading indicates a grounded sensor winding. A lab scope or graphing voltmeter may be used to obtain a waveform from a wheel-speed sensor **(Figure 24-34)**. If the wheel-speed sensor waveform is higher, lower, or erratic compared to the one in Figure 24-34, the sensor is defective.

ABS SERVICE

When a wheel-speed sensor is removed, always check the toothed ring for damage. If damage is present, the ring must be replaced. A paper shim must be installed on some wheel-speed sensor tips before the new sensor is installed. Push the sensor into place until the paper shim lightly contacts the toothed

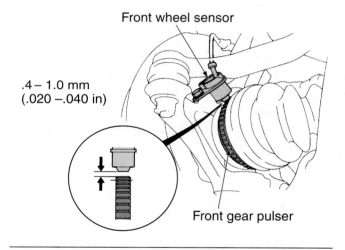

Front wheel sensor

.4 – 1.0 mm
(.020 – .040 in)

Front gear pulser

Figure 24-35 Proper air gap for a wheel-speed sensor.

ring and tighten the sensor-mounting bolt. The proper **air gap** must be maintained **(Figure 24-35)**.

ABS Bleeding

Always follow the vehicle manufacturer's recommended brake bleeding procedure. Some systems require the removal of power to the ABS system before bleeding the brakes while others may require no special steps. Always consult the service manual.

> **Tech Tip** *Many ABS are bled using the manual or pressure bleeder procedures used on conventional brake systems.*

On ABS with a high-pressure accumulator, if the accumulator pressure acts as a brake booster and also supplies fluid pressure to the rear brakes, the rear brakes may be bled with a fully charged accumulator. The front brakes may be bled manually or with a pressure bleeder. To bleed the rear brakes with a fully charged accumulator, follow this procedure:

1. Turn the ignition switch on and pump the brake pedal until the hydraulic pump motor starts.

2. Wait until the hydraulic pump motor shuts off.

3. With the ignition switch on, have an assistant apply the brake pedal.

4. Connect a hose from the right-rear bleeder screw into a glass container and loosen the right-rear bleeder screw until the brake pedal goes down to the floor.

5. Close the bleeder screw and release the brake pedal.

6. Repeat the bleeding procedure until a clear, bubble-free stream of brake fluid flows into the glass jar when the bleeder screw is opened.

7. Fill the master cylinder reservoir as required and repeat the bleeding procedure on the left-rear wheel.

> **SAFETY TIP** *On ABS with a high-pressure accumulator, always relieve the accumulator pressure before loosening any brake line or component in the hydraulic system. To relieve the accumulator pressure, turn the ignition switch off and pump the brake pedal twenty-five times. If this precaution is not followed, high-pressure brake fluid may cause personal injury or damage to vehicle paint surfaces if any brake line or component is loosened in the hydraulic system.*

Automated Bleeding Procedure

If the **brake pressure modulator valve (BPMV)** has been replaced on some ABS, the vehicle manufacturer recommends a manual bleeding procedure followed by an automated bleeding procedure using a scan tool. The BPMV is the assembly containing the solenoid valves connected to each wheel. Complete the manual brake bleeding procedure as explained previously in this chapter. Select ABS and AUTOMATED BLEED PROCEDURE on the scan tool. Follow any instructions provided on the scan tool. Apply the brake pedal and repeat the bleeding procedure on each wheel. Refill the master cylinder reservoir as required. Do not reuse the brake fluid that is bled from each wheel.

Summary

- Brake pedal free-play is necessary for proper brake system operation.

- A low, spongy brake pedal may be caused by a low brake fluid level in the master cylinder or air in the hydraulic system.

- Brake pedal pulsations may be caused by excessive drum out-of-round or rotor lateral runout.

- Grabbing brakes may be caused by contaminated brake linings.

- Excessive brake pedal effort on a power brake system may be caused by lack of vacuum at the brake booster.

- During a brake application, brake pedal fade may be caused by leaking piston seals in the master cylinder.

- Brake system bleeding may be done manually or with a pressure bleeder.

- Brake drums must not be machined so their diameter exceeds the discard diameter.

- There must be no free-play between the caliper fingers and the brake pad flanges.

- Rotor thickness must exceed the discard thickness after machining the rotors.

- Master cylinders may be bled on the bench before installation.

- A restricted vacuum hose to the brake booster causes excessive brake pedal effort.

- The amber ABS warning light should be on for about 4 seconds after the engine starts. Then it should remain off with the engine running.

- If the ABS computer detects an electrical defect in the ABS, the computer illuminates the amber ABS warning light.

- On an ABS with a high-pressure accumulator, the brake fluid level in the master cylinder should be checked with a fully-charged accumulator.

- In OBD I vehicles prior to 1996, the various DLCs were used for ABS diagnosis and other computer system diagnosis.

- OBD II systems on 1996 and newer vehicles have a standard 16-terminal DLC and data links connected between the various computers on the vehicle. The data links are also connected to the DLC.

- On some ABS on OBD I vehicles, the ABS warning light flashes DTCs when the two terminals are connected in the diagnostic connector.

- In some ABS on OBD I vehicles, the ABS DTCs are erased by disconnecting the ABS computer wiring harness with the ignition switch off.

- On OBD II vehicles, the scan tool will display DTCs and perform many other diagnostic functions.

- Wheel-speed sensor voltage waveforms may be displayed on a lab scope.

- Wheel-speed sensor windings may be tested with an ohmmeter.

- On ABS with a high-pressure accumulator, the accumulator pressure must be relieved before loosening or disconnecting any brake line or component in the hydraulic system.

- The accumulator pressure may be relieved by pumping the brake pedal twenty-five times with the ignition switch off.

Review Questions

1. Technician A says inadequate brake pedal free-play may cause dragging brakes. Technician B says excessive brake pedal free-play may cause a low brake pedal during a brake application. Who is correct?

   A. Technician A

   B. Technician B

   C. Both Technician A and Technician B

   D. Neither Technician A nor Technician B

2. Technician A says grabbing brakes may be caused by contaminated brake linings. Technician B says brake pedal pulsations may be caused by lateral wheel or tire runout. Who is correct?

   A. Technician A

   B. Technician B

   C. Both Technician A and Technician B

   D. Neither Technician A nor Technician B

3. Technician A says brake pedal fade during a brake application may be caused by contaminated brake linings. Technician B says a defective metering valve may cause brake pedal fade during a brake application. Who is correct?

   A. Technician A

   B. Technician B

   C. Both Technician A and Technician B

   D. Neither Technician A nor Technician B

4. Technician A says when a pressure bleeder is used to bleed the brakes, the left-front wheel should be bled first. Technician B says when using a pressure bleeder, the air pressure in the bleeder tank should be 75 psi (517 kPa). Who is correct?

   A. Technician A

   B. Technician B

   C. Both Technician A and Technician B

   D. Neither Technician A nor Technician B

5. All of these statements about the amber ABS warning light are true *except*:

   A. The amber ABS warning light should be on when the ignition switch is turned on.

   B. The amber ABS warning light should be on for about 4 seconds after the engine starts.

   C. The amber ABS warning light should remain off when the engine is running.

   D. If a serious ABS electrical defect occurs with the engine running, the amber ABS warning light begins flashing.

6. In an ABS with a high-pressure accumulator, the brake fluid level in the master cylinder should be checked with the high-pressure accumulator:

   A. discharged.

   B. charged at 500 psi.

   C. half charged.

   D. fully charged.

7. In some ABS on OBD I vehicles, the amber ABS warning light flashes DTCs when:

   A. the brake pedal is applied five times in a 10-second interval with the ignition switch on.

   B. the ABS fuse is removed and replaced.

   C. two terminals are connected in the DLC with the ignition switch on.

   D. when the ignition switch is cycled on and off three times in a 5-second interval.

8. When diagnosing an ABS, a DTC is obtained representing a wheel-speed sensor. Technician A says to replace the wheel-speed sensor. Technician B says to perform ohmmeter tests to determine if the problem is in the wheel-speed sensor or connecting wires. Who is correct?

   A. Technician A

   B. Technician B

   C. Both Technician A and Technician B

   D. Neither Technician A nor Technician B

9. All of these statements about diagnosing typical ABS on OBD II vehicles are true *except*:

   A. A scan tool must be used to perform a Tire Size Calibration each time an ABS is diagnosed.

   B. A scan tool may be used to perform a Functional Test that commands the EBCM to turn on and off various ABS relays, solenoids, and motors.

   C. A scan tool may be used to perform a lamp test that commands the EBCM to turn the amber and red brake warning lights on and off.

   D. A scan tool may be used to perform Solenoid Tests that allow the technician to command a specific ABS solenoid on and off.

10. With the wheel-speed sensor terminals disconnected, a pair of ohmmeter leads is connected from one of the sensor terminals to ground and a low ohmmeter reading is obtained. This ohmmeter reading indicates:

    A. a wheel-speed sensor with a shorted winding.

    B. a satisfactory wheel-speed sensor winding.

    C. an open circuit in the wheel-speed sensor winding.

    D. a shorted to ground wheel-speed sensor winding.

11. A convex-shaped brake drum surface has the largest diameter in the _____ of the drum friction surface.

12. After turning a brake drum in a lathe, the drum diameter must be at least _____ to _____ inch less than the discard diameter.

13. If the ohmmeter leads are connected to the two wheel speed sensor terminals, the sensor winding is being tested for _____ and _____ circuits.

14. If the ABS computer senses an electrical defect in the ABS, the computer _____ the ABS function.

15. Explain the manual brake bleeding procedure.

16. Describe the procedure for performing a basic test on a vacuum brake booster.

17. Describe the normal operation of the amber ABS warning light.

18. Explain the DTC erasing procedure on ABS used with OBD II systems having a 16-terminal DLC.

CHAPTER
25

Steering Columns, Steering Linkages, and Power Steering Pumps: Maintenance, Diagnosis, and Service

Learning Objectives

After you have read, studied, and practiced the contents of this unit, you should be able to:

- Describe two methods of steering column movement to protect the driver in a frontal collision.

- Explain the purpose of the pitman arm and the idler arm.

- Explain the rack-and-pinion steering linkage design.

- Describe the purpose of the worm shaft preload adjustment.

- Define gear ratio.

- Explain the purpose of the rack bearing and adjuster plug.

- Describe the fluid movement in a rack-and-pinion steering gear during a right turn.

- Explain the operation of the spool valve and rotary valve.

Key Terms

Ball socket

Castellated nut

Center link ends

Idler arm

Independent front
suspension

Parallelogram steering

Pitman arm

Rack-and-pinion steering

Steering damper

Tie-rods

INTRODUCTION

Steering columns play a significant part in directional control, safety, and driver convenience. The steering column connects the steering wheel to the steering gear. Collapsible steering columns provide some driver protection in a collision. Steering linkage mechanisms are used to connect the steering gear to the front wheels. Steering linkages help to position the front wheels and tires properly to provide proper steering control. Power steering systems have contributed to reduced driver fatigue and have made driving a more pleasant experience. Nearly all power steering systems at the present time use fluid pressure to assist the driver in turning the front wheels. New power steering systems using electrical power are currently in development.

Since driver effort required to turn the front wheels is reduced, driver fatigue is decreased. The advantages of power steering have been made available on many vehicles and safety has been maintained in these systems.

STEERING COLUMN DESIGN

Steering columns may be classified as non-tilting, tilting, and tilting/telescoping. Tilt steering columns facilitate driver entry and exit to and from the front seat. These columns also allow the driver to position the steering wheel to suit individual comfort requirements **(Figure 25-1)**. A tilting/telescoping steering column allows the driver to tilt and extend or retract the steering wheel. In this type of steering column, the driver has more steering wheel position choices.

Many steering columns contain a two-piece steering shaft connected by universal joints

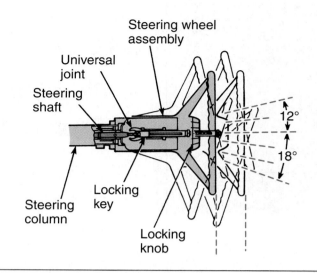

Figure 25-1 A tilt steering column.

(Figure 25-2). A jacket and shroud surround the steering shaft and the upper shaft is supported by two bearings in the jacket. In some steering columns a toe plate, seal, and silencer surround the lower steering shaft and cover the opening where the shaft extends through the floor **(Figure 25-3)**. A shield underneath the toe plate surrounds the lower steering

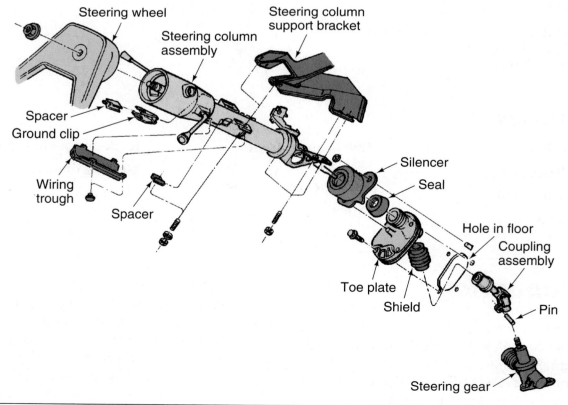

Figure 25-2 Typical steering column components.

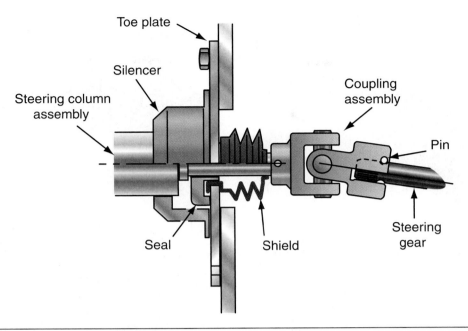

Figure 25-3 Toe plate, seal, and silencer surrounding the lower end of the steering column.

shaft. The lower universal joint couples the lower shaft to the stub shaft in the steering gear. In some steering columns, a flexible coupling is used in place of the lower universal joint **(Figure 25-4)**.

Studs and nuts retain the steering column bracket to the instrument panel support bracket. The steering column is designed to protect the driver if the vehicle is involved in a frontal collision. An energy-absorbing lower bracket and lower plastic adapter are used to connect the steering column to the instrument panel mounting bracket. This bracket allows the column to slide forward if the driver is thrown forward into the wheel in a frontal collision. The mounting bracket is also designed to prevent rearward movement toward the driver in a collision.

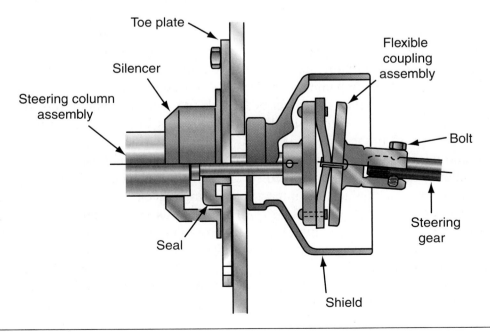

Figure 25-4 A flexible coupling.

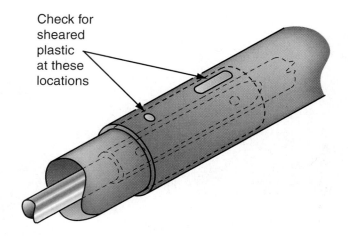

Figure 25-5 Injection plastic in a collapsible outer steering column jacket.

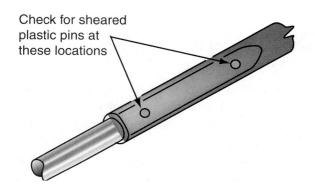

Figure 25-6 Injection plastic in a collapsible lower steering shaft.

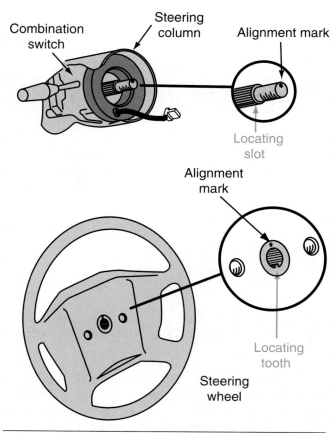

Figure 25-7 The steering wheel is indexed to the steering shaft.

In some steering columns, the outer column jacket is a two-piece unit retained with plastic pins **(Figure 25-5)**. In this type of column the lower steering shaft is also a two-piece sliding unit retained with plastic pins **(Figure 25-6)**. When the driver is thrown against the steering wheel in a frontal collision, the plastic pins shear off in the lower steering shaft and outer column jacket. The shearing action of the plastic pins allows the steering column to collapse away from the driver, which reduces the impact as the driver hits the steering wheel.

The steering wheel splines fit on matching splines on the top of the upper steering shaft. A nut retains the wheel on the shaft. Most steering wheels and shafts have matching alignment marks that must be aligned when the steering wheel is installed **(Figure 25-7)**.

An ignition switch cylinder is usually mounted in the upper-right side of the column housing.

The ignition switch is bolted on the lower side of the housing **(Figure 25-8)**. An operating rod connects the ignition switch cylinder to the ignition switch. Some ignition switches are integral with the lock cylinder in other steering columns. A lock plate is attached to the upper steering shaft and a lock pin engages with slots in this plate to lock the steering wheel and gearshift when the gearshift is in park and the ignition switch is in the lock position.

The turn signal switch and hazard-warning switch are mounted on the steering column near the steering wheel. Lugs on the bottom of the steering wheel are used to cancel the signal lights after a turn is completed. On many vehicles, the signal light lever also operates the wiper/washer switch and the headlight dimmer switch **(Figure 25-9)**. If the gearshift is mounted in the steering column, then a tube extends from the gearshift housing to the shift lever at the lower end of the steering column. This shift lever is connected through a linkage to the transaxle or transmission shift lever.

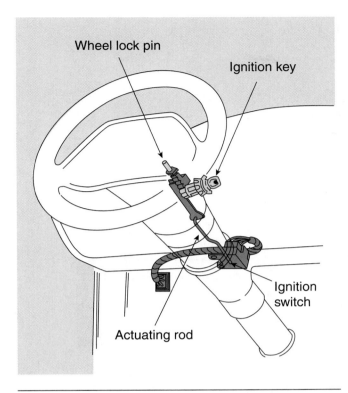

Figure 25-8 Ignition switch, ignition switch cylinder, and lock plate mounted in steering column.

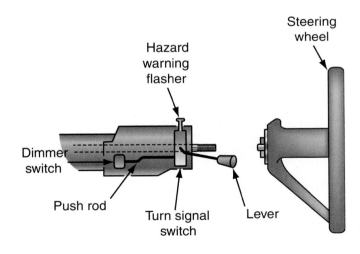

Figure 25-9 Turn signal switch, hazard warning switch, dimmer switch, and wipe/wash switch mounted in steering column.

NOTE: The signal light switch, wipe/wash switch, dimmer switch, and hazard warning switch may be called a *combination switch*, *multi-function switch*, or *smart switch*.

PARALLELOGRAM STEERING LINKAGES

Parallelogram steering linkages may be mounted behind the front axle centerline suspension **(Figure 25-10A)** or ahead of the front suspension **(Figure 25-10B)**. The parallelogram steering linkage must not interfere with the engine oil pan or chassis components. Regardless of the parallelogram steering linkage mounting position, this type of steering linkage contains the same components. The main components in this steering linkage mechanism are the pitman arm, center link, idler arm, tie-rods with sockets, adjusting sleeves, and tie-rod ends.

Parallelogram steering linkages are usually found on **independent front suspension** systems. In a parallelogram steering linkage, the tie-rods are connected parallel to the lower control arms. Road vibration and shock are transmitted from the tires and wheels to the steering linkage. These forces tend to wear the linkages and cause steering looseness. If the steering linkage components are worn, steering control is reduced. Since loose steering linkage components cause intermittent toe changes, this problem increases tire wear. The wear points in a parallelogram steering linkage are the **tie-rods**, **pitman arm**, **idler arm**, and **center link ends**.

Did You Know? *Always remember that a customer's life may depend on the condition of the steering linkages on his or her vehicle. During under-car service, always make a quick check of steering linkage condition.*

SAFETY TIP *Never attempt to straighten steering linkage components. This action may weaken the metal and cause sudden component failure, vehicle damage, or personal injury.*

Tie-Rods

The tie-rod assemblies connect the center link to the steering arms that are attached to the front steering knuckles. In most front suspensions, the steering arms are part of the steering knuckle

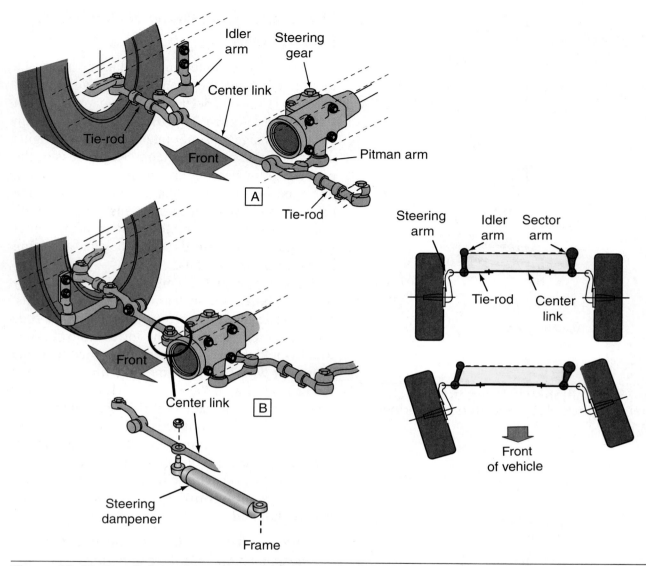

Figure 25-10 Parallelogram steering linkage (A) behind the suspension and (B) in front of the suspension.

while in some front suspension systems the steering arm is bolted to the knuckle. A **ball socket** is mounted on the inner end to each tie-rod and a tapered stud on this socket is mounted in a center link opening. A **castellated nut** and cotter pin retains the tie-rods to the center link. A threaded sleeve is mounted on the outer end of each tie-rod and a tie-rod end is threaded into the outer end of this sleeve **(Figure 25-11)**. Some tie-rod ends have a hardened steel upper bearing and a high-strength polymer lower bearing for increased durability **(Figure 25-12)**. Other tie-rod ends have a rubber-encapsulated ball stud **(Figure 25-13)**.

Each tie-rod sleeve contains a left-hand and a right-hand thread where they are threaded onto the tie-rod end and the tie-rod. Therefore, sleeve rotation changes the tie-rod length and provides a toe adjustment. Clamps are used to tighten the tie-rod sleeves. The clamp opening must be positioned away from the slot in the tie-rod sleeve. The design of the steering linkage mechanism allows multi-axial movement since the front suspension moves vertically and horizontally.

Ball and socket-type pivots are used on the tie-rod assemblies and center link. If the front wheels hit a bump, the wheels move up and down

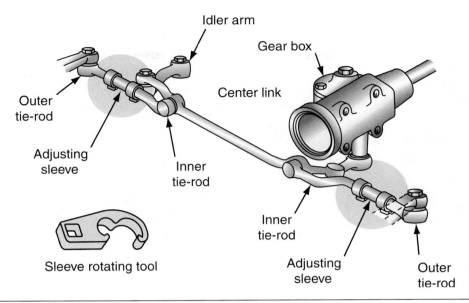

Figure 25-11 A threaded tie-rod adjuster sleeve.

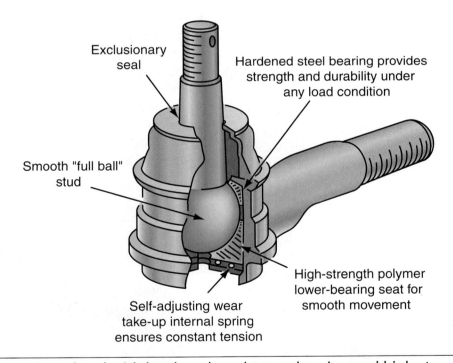

Figure 25-12 Outer tie-rod end with hardened steel upper bearing and high-strength polymer lower bearing.

and the control arms move through their respective arcs. Since the tie-rods are connected to the steering arms, these rods must move upward with the wheel. Under this condition, the inner end of the tie-rod acts as a pivot and the tie-rod also moves through an arc. This arc is almost the same as the lower control arm arc because the tie-rod is parallel to the lower control arm. Maintaining the same arc between the lower control arm and the tie-rod minimizes toe change on the front wheels during upward and downward wheel movement. This action improves the directional stability of the vehicle and reduces tread wear on the front tires.

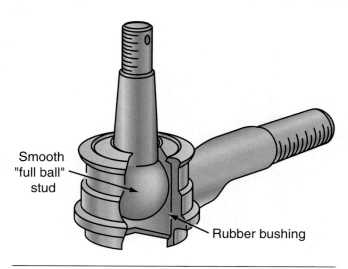

Figure 25-13 Outer tie-rod end with rubber-encapsulated ball stud.

Pitman Arm

The pitman arm connects the steering gear to the center link. This arm also supports the left side of the center link. Motion from the steering wheel and steering gear is transmitted to the pitman arm. This arm transfers the movement to the steering linkage. This pitman arm movement forces the steering linkage to move to the right or left and the linkage moves the front wheels in the desired direction. The pitman arm also positions the center link at the proper height to maintain the parallel relationship between the tie-rods and the lower control arms.

Wear-type pitman arms have ball sockets and studs at the outer end. This stud fits into the center link opening (Figure 25-14). The ball stud and socket are subject to wear and pitman arm replacement is

necessary if the ball stud is loose. A non-wear pitman arm has a tapered opening in the outer end. A ball stud in the center link fits into this opening. The non-wear pitman arm needs replacing only if the arm is damaged, bent, or in a collision. The opening in the inner end of both types of pitman arms have serrations that fit over matching serrations on the steering gear shaft. A nut and lock washer retains the pitman arm to the steering gear shaft.

SAFETY TIP *A binding idler arm may suddenly break off and cause complete loss of steering control, vehicle damage, or personal injury.*

Idler Arm

An idler arm support is bolted to the frame or chassis on the opposite end of the center link from the pitman arm. The idler arm is connected from the support bracket to the center link. Two bolts retain the idler arm bracket to the frame or chassis. In some idler arms, a ball stud on the outer end of the arm fits into a tapered opening in the center link, whereas in other idler arms a ball stud in the center link fits into a tapered opening in the idler arm **(Figure 25-15)**.

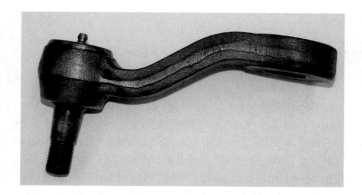

Figure 25-14 Pitman arm design.

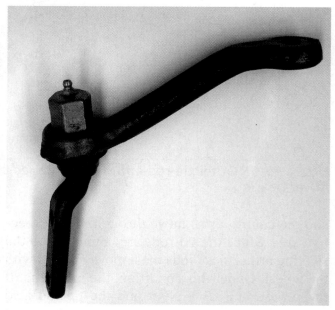

Figure 25-15 Idler arm design.

The idler arm supports the right side of the center link and helps to maintain the parallel relationship between the tie-rods and the lower control arms. The outer end of the idler arm is designed to swivel on the idler arm bracket. This swivel is subject to wear. A worn idler arm swivel causes excessive vertical steering linkage movement away from the vehicle centerline. This action results in excessive steering wheel freeplay with reduced steering control and front tire wear, erratic toe, and may also result in bump steer.

Center Links

The center link controls the sideways steering linkage movement. The center link, together with the pitman arm and idler arm, provide the proper height for the tie-rods, which is very important to minimize toe change on road irregularities which will help to minimize bump steer. Some center links have tapered openings in each end and the studs on the pitman arm and idler arm fit into these openings. This type of center link may be called a taper end or non-wear link. Other wear-type center links have ball sockets in each end with tapered studs extending from the sockets **(Figure 25-16)**. These tapered studs fit into openings in the pitman arm and idler arm and they are retained with castellated nuts and cotter pins.

NOTE: Bump steer is caused by steering changes introduced by the steering and suspension system as the vehicle travels over rough or irregular road surfaces.

RACK-AND-PINION STEERING LINKAGES

The **rack-and-pinion steering** linkage is used with rack-and-pinion steering gears **(Figure 25-17)**. In this type of steering gear, the rack is a rod with teeth on one side. This rack slides horizontally on bushings inside the gear housing. The rack teeth are meshed with teeth on a pinion gear and this pinion gear is connected to the steering column. When the steering wheel is turned, the pinion rotation moves the rack sideways. Tie-rods are connected directly from the ends of the rack to the steering arms. The tie-rods are similar to those found on parallelogram steering systems. An inner tie-rod end connects each tie-rod to the rack. Bellows boots are clamped to the gear housing and the tie-rods keep dirt out of these joints. The inner tie-rod end contains a spring-loaded ball socket and the outer tie-rod ends connected to the steering arms are basically the same as those in parallelogram steering linkages **(Figure 25-18)**. Some inner tie-rod ends contain a bolt and bushing. These tie-rod ends are threaded onto the rack. Since the rack is connected directly to the tie-rods, the rack replaces the center link in a parallelogram steering linkage.

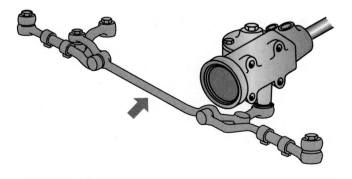

Figure 25-16 Center link design.

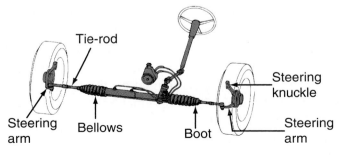

Figure 25-17 Rack-and-pinion steering gear.

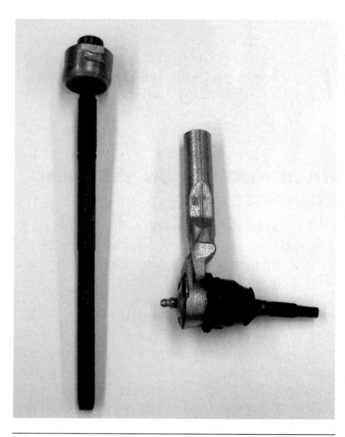

Figure 25-18 Inner tie-rod and outer tie-rod end, rack-and-pinion steering.

Tech Tip *The steering knuckle inner portion of the spindle is attached to and pivots on upper and lower ball joints. The steering knuckle may also be attached to a MacPherson strut and pivots on the upper strut bearing and lower ball joints. The steering arms are bolted to or manufactured as an integral part of the steering knuckles.*

The steering arm transmits the steering force from tie-rod to the knuckles causing the wheels to turn. The rack and pinion is mounted either ahead or behind the axle centerline. The rack-and-pinion steering gear may be mounted on the front crossmember **(Figure 25-19)** or attached to the firewall behind the engine **(Figure 25-20)**. Rubber insulating bushings surround the steering gear. These bushings are clamped to the crossmember or firewall. The rack-and-pinion steering gear is mounted

Figure 25-19 Rack-and-pinion steering gear mounting on front crossmember.

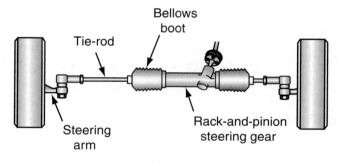

Figure 25-20 Rack-and-pinion steering gear mounted on firewall.

at the proper height to position the tie-rods and lower control arms parallel to each other. The number of friction points is reduced in a rack-and-pinion steering system. This system is light and compact. Many vehicles have rack-and-pinion steering. Since the rack is linked directly to the steering arms, this type of steering linkage provides good road feel.

STEERING DAMPER

A **steering damper** or stabilizer may be found on some rack-and-pinion or parallelogram steering linkages. The steering damper is similar to a shock absorber. This component is connected from one of the steering links to the chassis or frame **(Figure 25-21)**. When a front wheel strikes a road irregularity, a shock is transferred from the front wheel to the steering linkage, steering gear, and steering wheel. The steering damper helps to

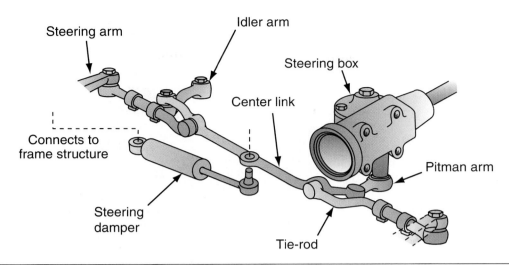

Figure 25-21 Steering damper and linkage.

absorb this road shock and prevents it from reaching the steering wheel. Heavy-duty steering dampers are available for severe road conditions, such as those sometimes encountered by four-wheel drive vehicles.

POWER STEERING PUMP DESIGN

The power steering pump is driven by a V-belt or multi-ribbed belt. Vehicle manufacturers have used various types of power steering pumps. Many vane-type power steering pumps have flat vanes, which seal the pump rotor to the elliptical pump cam ring **(Figure 25-22)**. Other vane-type power steering pumps have rollers to seal the rotor to the cam ring. In some pumps, inverted, U-shaped slippers are used for this purpose.

The major difference in these pumps is in the rotor design and the method of sealing the pump rotor in the elliptical pump ring. The operating principles of all three types of pumps are similar. A balanced pulley is pressed on the steering pump drive shaft. This pulley and shaft is belt-driven by the engine. The oblong pump reservoir is made from steel or plastic. A large O-ring seals the front of the reservoir to the pump

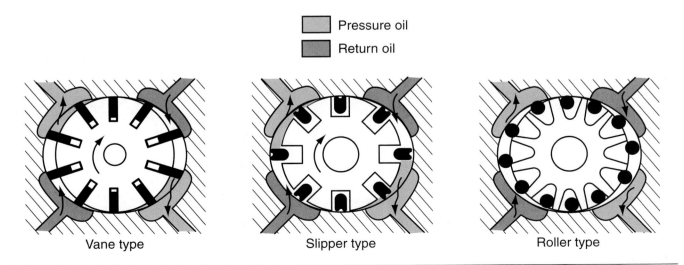

Figure 25-22 Power steering pumps are made in three types: rotor and vanes type, slipper type, and roller type.

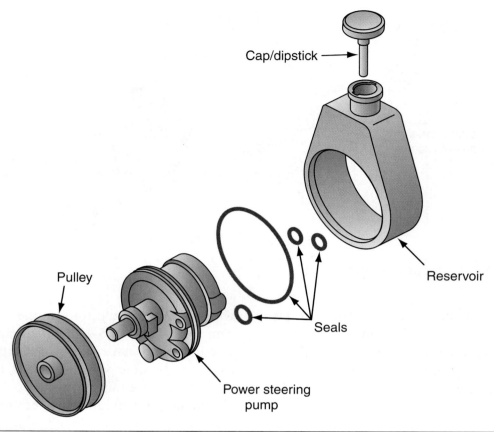

Figure 25-23 Power steering pump housing and reservoir.

housing **(Figure 25-23)**. Smaller O-rings seal the bolt fittings on the back of the reservoir. The combination cap and dipstick keep the fluid reserve in the pump and vents the reservoir to the atmosphere.

The rotating components inside the pump housing include the shaft and rotor with the vanes mounted in the rotor slots. As the pulley drives the pump shaft, the vanes rotate inside an elliptical-shaped opening in the cam ring. The cam ring remains in a fixed position inside the pump housing. As the vanes rotate and move toward the narrowest part of the cam ring, the space between the vanes becomes smaller. This action pressurizes the fluid between the vanes. When the rotating vanes move into the wider part of the cam ring, the fluid pressure decreases and fluid flows from the pump reservoir into the area between the vanes. A seal between the drive shaft and the housing prevents oil leaks around the shaft.

The flow control valve is a precision-fit valve controlled by spring pressure and fluid pressure.

Any dirt or roughness on the valve results in erratic pump pressure. The flow control valve contains a pressure relief ball **(Figure 25-24)**. High-pressure fluid is forced past the control valve

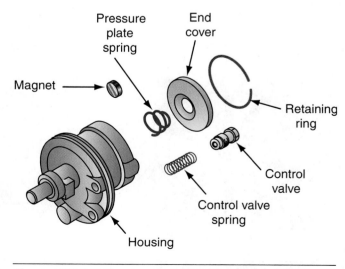

Figure 25-24 Power steering pump housing assembly with end cover, flow control valve, and magnet.

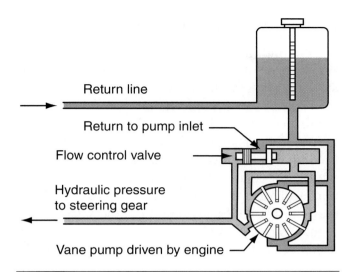

Figure 25-25 Hydraulic circuit of a typical power steering pump.

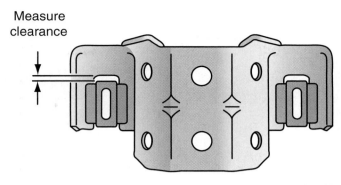

Figure 25-26 Capsules in steering column bracket.

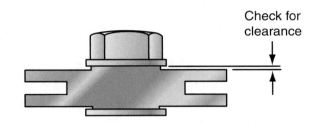

Figure 25-27 Bolt head to bracket clearance.

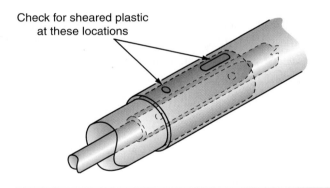

Figure 25-28 Inspecting for sheared plastic in jacket openings.

to the outlet fitting. The flow control valve controls pump pressure to provide adequate pressure for steering assist and also protect system hoses from excessive pressure. A high-pressure hose connects the outlet fitting to the inlet fitting on the steering gear, while a low-pressure hose returns the fluid from the steering gear to the inlet fitting in the pump reservoir. **Figure 25-25** shows a typical hydraulic schematic.

STEERING COLUMN MAINTENANCE AND DIAGNOSIS

Proper steering column maintenance and diagnosis is very important to locate minor problems in the column before they become a dangerous safety concern. For example, when a worn flexible coupling causes a rattling noise in the steering column and this noise is ignored for a period of time, the flexible coupling may break completely and result in a loss of steering control. This loss of steering control may cause a collision resulting in personal injury and vehicle damage.

Collapsible Steering Column Inspection

Since steering column design varies depending on the vehicle, the collapsible steering column inspection procedure in the vehicle manufacturer's service manual should be followed:

1. Measure the clearance between the capsules and the slots in the steering column bracket **(Figure 25-26)**. If this measurement is not within specifications, replace the bracket.

2. Check the contact between the bolt head and the bracket **(Figure 25-27)**. If the bolt head contacts the bracket, the shear load is too high and the bracket must be replaced.

3. Check the steering column jacket for sheared injected plastic in the openings on the side of the jacket **(Figure 25-28)**. If sheared plastic is present, the column is collapsed.

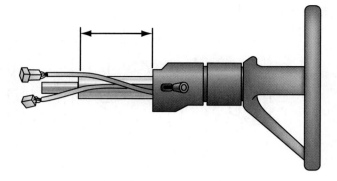

Figure 25-29 Measuring distance from the end of the bearing assembly to the upper steering column jacket.

4. Measure the distance from the end of the bearing assembly to the lower edge of the upper steering column jacket **(Figure 25-29)**. If this distance is not within the vehicle manufacturer's specification, a new jacket must be installed.

5. Check the steering column jacket for damage to the collapsible mesh of the steering column **(Figure 25-30)**. If any damage is present, the column is considered to have been collapsed and should be replaced.

Steering Column Noise Diagnosis

If a binding condition is present while turning the steering wheel, the problem may be in the steering column, the steering linkage, or suspension components. To determine the source of the binding condition, disconnect the flexible coupling or lower universal joint in the steering shaft. If the binding condition is still present, the steering column is the source of the problem. A worn upper universal joint or spherical bearing on a tilt column may cause a binding condition in the steering column. A binding problem may also be caused by interference between the lower steering shaft and the toe plate or silencer. A worn flexible coupling or loose universal joints may cause a rattling noise in the steering column.

Checking Steering Wheel Free-Play

With the engine stopped and the front wheels in the straight-ahead position, move the steering wheel in each direction with light finger pressure. Measure the amount of steering wheel movement before the front wheels begin to turn **(Figure 25-31)**. This movement is referred to as steering free-play. On some vehicles, this measurement should not exceed 1.2 inches (30 millimeters**)**. Always refer to the

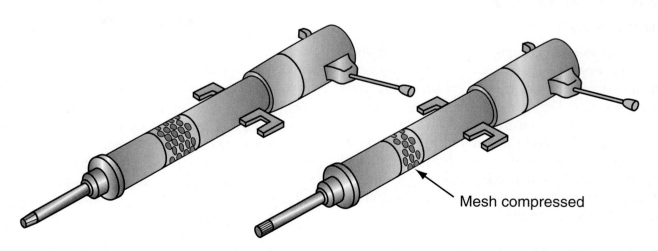

Mesh compressed

Figure 25-30 This figure shows the collapsible mesh of a steering column in its normal and collapsed positions.

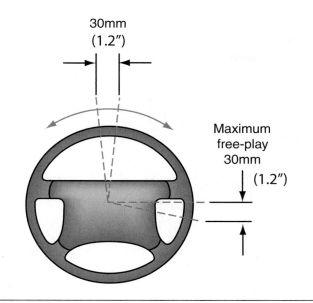

Figure 25-31 Measuring steering wheel free-play.

vehicle manufacturer's specifications. Worn steering shaft universal joints or a worn flexible coupling may cause excessive steering free-play. Other causes of excessive steering wheel free-play include worn steering linkage mechanisms or a worn or out-of-adjustment steering gear. With the normal vehicle weight resting on the front suspension, observe the flexible coupling or universal joint as an assistant turns the steering wheel ½ turn in each direction. If the vehicle has power steering, the engine should be running with the gear selector in park. The flexible coupling or universal joint must be replaced if there is free-play in this component.

Flexible Coupling Replacement

If the flexible coupling must be replaced, loosen the coupling-to-steering gear stub shaft bolt. Disconnect the steering column from the instrument panel and move the column rearward until the flexible coupling can be removed from the steering column shaft. Remove the coupling-to-steering shaft bolts and disconnect the coupling from the shaft. When installing the new coupling and the steering column, always use the vehicle manufacturer's specifications in the service manual.

STEERING LINKAGE MAINTENANCE AND DIAGNOSIS

When servicing or replacing steering wheels, columns, or linkages, you actually have the customer's life in your hands. If steering components are not serviced properly or are not tightened to the specified torque, the steering may become disconnected, resulting in a complete loss of steering control. This condition may cause a collision resulting in vehicle damage, personal injury, and an expensive lawsuit for you and the shop where you are employed. Therefore, when performing any automotive service, always be sure the vehicle manufacturer's recommended service procedures and torque specifications are followed.

Diagnosis of Center Link, Pitman Arm, and Tie-Rod Ends

The vehicle should be raised and safety stands must be positioned under the lower control arms to support the vehicle weight. Use vertical hand force to check for looseness in all the pivots on the tie-rod ends and the center link **(Figure 25-32)**. Check the seals on each tie-rod end and pivot on the center link or pitman arm for damage and cracks. Cracked seals allow dirt to enter the pivoted joints which results in rapid wear. If looseness or damaged seals are found on any pivoted joint on the tie-rods and center link, these components must be replaced. Inspect rubber-encapsulated tie-rod ends for looseness of the ball stud in the rubber capsule and looseness of the stud and rubber capsule in the outer housing. If either of these conditions is present, replace the tie-rod end.

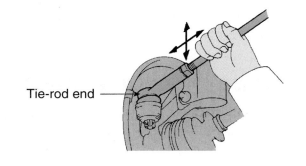

Tie-rod end

Figure 25-32 Checking for tie-rod wear.

The second part of this diagnosis is done with the front wheels resting on the shop floor. If the vehicle is equipped with power steering, start the engine and allow the engine to idle with the transmission in park and the parking brake applied. While someone turns the steering wheel one-quarter turn in each direction from the straight-ahead position, observe all the pivoted joints on the tie-rod ends and center link. This test allows you to check the steering linkage pivots under load. If any of the pivoted joints show a slight amount of play, they must be replaced.

Worn tie-rod ends result in excessive steering wheel free-play, incorrect front wheel toe setting, tire squeal on turns, tread wear on front tires, front wheel shimmy, and a rattling noise on road irregularities. Front wheel toe is the distance between the front edges of the front tires compared to the distance between the rear edges of these tires. Front wheel shimmy is a rapid oscillation of the front wheels to the right and the left.

Idler Arm Diagnosis

Grasp the center link firmly near the idler arm and apply a 25-pound vertical load to the idler arm. If the total idler arm vertical movement measured at the outer end of the arm exceeds ¼ inch, idler arm replacement is necessary **(Figure 25-33)**. If idler arm vertical movement is excessive, the tie-rod is not parallel to the lower control arm. Excessive idler arm vertical movement causes these steering problems: excessive toe change and front tire tread wear, excessive steering wheel free-play and reduced steering control, and front end shimmy. Binding idler arm bushings results in these complaints: hard steering, a squawking noise when the front wheels are turned, and poor steering wheel returnability.

STEERING LINKAGE SERVICE

Proper steering linkage service is extremely important to maintain vehicle safety and provide normal tire life. For example, if the retaining nut on a tie-rod end is not tightened to the specified torque and retained with a cotter pin, the tie-rod may become disconnected, resulting in a loss of

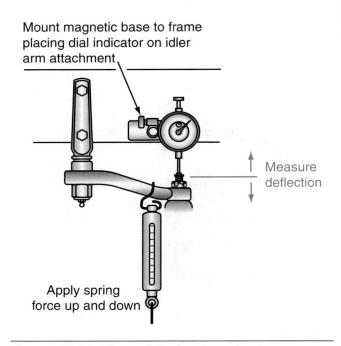

Mount magnetic base to frame placing dial indicator on idler arm attachment

Measure deflection

Apply spring force up and down

Figure 25-33 Tie-rod end removal using a puller.

steering control. This action may result in a collision causing vehicle damage and personal injury.

Tie-Rod End Replacement

The cotter pin and nut must be removed prior to tie-rodend replacement. A puller may be used to remove the tie-rod end from the steering arm **(Figure 25-34** and **Figure 25-35)**. Tie-rods may

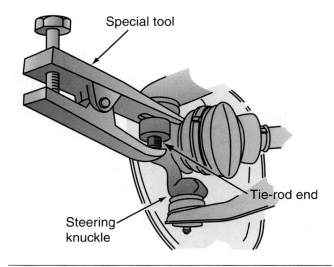

Special tool

Tie-rod end

Steering knuckle

Figure 25-34 Tie-rod end removal using a puller.

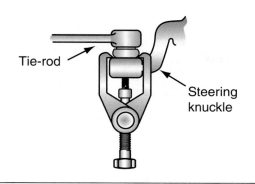

Figure 25-35 Tie-rod end removal using a "pickle fork."

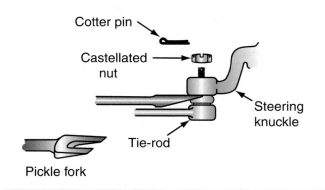

Figure 25-36 Tie-rod end removal using a puller.

also be removed using a "pickle fork." These pullers wedge between the steering knuckle and the tie-rod joint and are operated with either a hammer or an air hammer **(Figure 25-36)**. The pickle fork is a quick and efficient method to remove tie-rods and larger sizes are available for ball joint removal. It is important to note that the pickle fork usually damages the rubber boot of the tie-rod, which is not a concern if the tie-rod joint is to be replaced, but if the joint is being removed to perform other service, then the puller methods may be more suitable. The tie-rod clamp must be loosened before the tie-rod is removed from the sleeve. Count the number of turns required to remove the tie-rod end from the sleeve and install the new tie-rod with the same number of turns. Even when this procedure is followed, the toe may not be in proper alignment and the vehicle will need to have a wheel alignment.

Before the new tie-rod end is installed, center the stud in the tie-rod end. When the tie-rod stud is installed in the steering arm opening, only the

threads should be visible above the steering arm surface. If the machined surface of the tie-rod stud is visible above the steering arm surface or the stud fits loosely in the steering arm opening, this opening is worn or the tie-rod end is not correct for that application **(Figure 25-37)**. The tie-rod nut must be torqued to manufacturer's specifications and the cotter pin must be installed through the tie-rod end and nut openings **(Figure 25-38)**.

When rubber-encapsulated tie-rod ends are installed and tightened, the front wheels must be in the straight-ahead position. This helps to keep the bushings in their normal straight-ahead position. If the wheels are turned when these are tightened then the vehicle may tend to steer back to the position that the bushings were tightened. The tie-rod nut must never be loosened from the specified

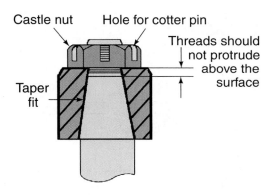

Figure 25-37 Steering linkage taper connection.

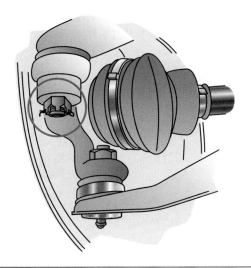

Figure 25-38 Tie-rod nut and cotter pin installation.

torque to install the cotter pin. Another method of positioning replacement tie-rod ends is to measure the distance from the center of the tie-rod stud to the end of the sleeve prior to removal **(Figure 25-39)**. When the new tie-rod is installed, be sure this measurement is the same. The slots in the tie-rod sleeve must be positioned away from the opening in the sleeve clamps **(Figure 25-40)**. Leave the sleeve clamps loose until the front wheel toe is checked and then tighten the sleeve clamp bolts to the specified torque. A special tool is available

Figure 25-41 Tie-rod sleeve adjusting tool.

to rotate the tie-rod sleeves and set the front wheel toe **(Figure 25-41)**.

Pitman Arm Diagnosis and Replacement

Some pitman arms contain a ball socket joint on the outer end. The threaded extension on this ball socket fits into the center link. On other steering linkages, the ball socket joint is in the center link and the threaded extension fits into the pitman arm opening. If the pitman arm is bent, it must be replaced. If the pitman arm is bent, the tie-rod is not parallel to the lower control arm. Under this condition, excessive front wheel toe change occurs on road irregularities and front tire wear may be excessive. The following is a typical pitman arm replacement procedure:

1. Position the front wheels straight ahead and remove the cotter pin and nut from the ball socket joint on the outer end of the pitman arm.

2. Remove the ball socket extension from the pitman arm or center link with a tie-rod end puller.

3. Loosen the pitman arm to pitman shaft nut.

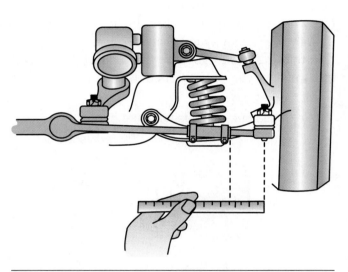

Figure 25-39 Measure the distance between the center of the tie rod and the end of the sleeve. This allows the preliminary toe setting when installing the new tie rod. The technician may also count the number of turns required to remove the tie-rod end from the sleeve and install the new tie rod with the same number of turns.

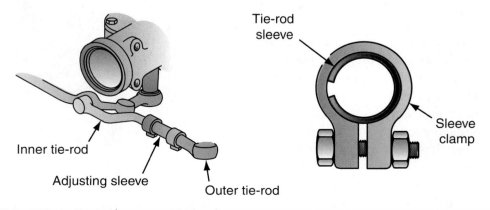

Figure 25-40 Proper slot and clamp position on tie-rod sleeves.

4. Use a puller to pull the pitman arm loose on the shaft.

5. Remove the nut, lock washer, and pitman arm.

6. Check the pitman shaft splines. If the splines are damaged or twisted, the shaft must be replaced.

7. Reverse steps 1 through 5 to install the pitman arm.

The pitman arm-to-shaft nut and the ball-socket-extension nut must be tightened to the manufacturer's specified torque. Be sure the pitman arm is installed in the correct position on the shaft splines. Install the cotter pin in the ball socket extension.

Center Link Replacement

Follow these steps for a typical center link replacement:

1. Remove the cotter pins from the tie-rod to center link nuts and the idler arm and pitman arm to center link nuts.

2. Remove the nuts on the tie-rod inner ends, idler arm to center link ball socket extension, and the pitman arm to center link ball socket extension.

3. Use a tie-rod end puller to pull the inner tie-rods from the center link. Follow the same procedure to remove the center link to pitman arm ball socket extension.

4. Remove the idler arm from the center link and remove the center link.

5. Reverse steps 1 through 4 to install the center link.

Tighten all the ball socket nuts to the manufacturer's specified torque and install cotter pins in all the nuts. If the ball sockets have grease fittings, lubricate the ball sockets with a grease gun and chassis lubricant. The following is a typical idler arm removal and replacement procedure:

1. Remove the idler arm to center link cotter pin and nut.

2. Remove the center link from the idler arm.

3. Remove the idler arm bracket mounting bolts and remove the idler arm.

4. If the idler arm has a steel bushing, thread the bracket into the idler arm bushing until the

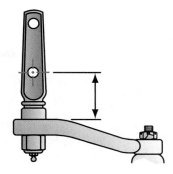

Figure 25-42 Specified clearance between center of lower bracket bolt hole and upper idler arm surface.

specified clearance is obtained between the center of the lower bracket bolt hole and the upper idler arm surface **(Figure 25-42)**.

5. Install the idler arm bracket to frame bolts and tighten the bolts to the specified torque. Be sure that lock washers are installed on the bolts.

6. Install the center link into the idler arm and tighten the mounting nut to the specified torque. Install the cotter pin in the nut.

7. If the idler arm steel bushing or bushings contain a grease fitting, lubricate as required. The idler arm adjustment is very important. If this adjustment is incorrect, front wheel toe is affected. After idler arm replacement, the front wheel toe should be checked.

Steering Damper Diagnosis and Replacement

Some steering systems have a damper connected between the center link and the chassis. A damper is similar to a small shock absorber. The purpose of the damper is to prevent the transmission of steering shock and vibrations to the steering wheel. A worn-out steering damper may cause excessive steering shock and vibration on the steering wheel, especially on irregular road surfaces. A rattling noise occurs if the damper's mounting bolts or brackets are loose.

Steering Arm Diagnosis

If the front rims have been damaged, the steering arms should be checked for a bent condition.

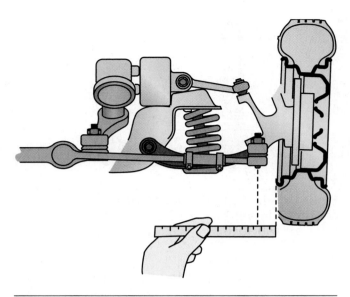

Figure 25-43 Measure the distance between the center of the tie-rod and the end of the adjusting sleeve.

Measure the distance from the center of the tie-rod stud to the edge of the rim on each side. Make sure the bent rim has been replaced and the new rim checked to ensure it has lateral runout that is within the manufacturer's specifications. Unequal readings may indicate a bent steering arm **(Figure 25-43)**. Bent steering arms can also be diagnosed during a wheel alignment and will show as incorrect toe-out-on-turn. Bent steering arms must be replaced.

POWER STEERING PUMP MAINTENANCE AND DIAGNOSIS

Proper power steering pump maintenance and diagnosis is very important to maintain vehicle safety. A sudden loss of power steering pump pressure results in a large increase in steering effort. When turning a corner, a sudden loss of power steering pump pressure may cause the driver to lose steering control, resulting in a collision. When performing power steering pump maintenance and diagnosis, one of the first steps is to inspect the pump drive belt and measure the belt tension. A loose power steering belt causes low pump pressure, hard steering, and belt chirping or squealing, especially on acceleration.

Fluid Level Checking

Most vehicle manufacturers recommend power steering fluid or automatic transmission fluid in power steering systems. Always use the type of fluid recommended in the vehicle manufacturer's service manual. If the power steering fluid level is low, steering effort is increased and may be erratic. A low fluid level may cause a growling noise in the power steering pump. Some vehicle manufacturers now recommend checking the power steering pump fluid level with the fluid at an ambient temperature of 176°F (80°C). Follow these steps to check the power steering fluid level:

1. With the engine idling at 1,000 rpm or less, turn the steering wheel slowly and completely in each direction several times to boost the fluid temperature.
2. If the vehicle has a remote power steering fluid reservoir, check for foaming in the reservoir, which indicates a low fluid level, or air in the system.
3. Observe the fluid level in the remote reservoir. This level should be at the hot full mark. Shut the engine off and remove dirt from the neck of the reservoir with a shop towel. If the power steering pump has an integral reservoir, the level should be at the hot level on the dipstick. When an external reservoir is used, the dipstick is located in the external reservoir **(Figure 25-44)**.
4. Pour the required amount of the vehicle manufacturer's recommended power steering fluid into the reservoir to bring the fluid level to the hot full mark on the reservoir or dipstick with the engine idling.

> **Tech Tip** *Some power steering pumps have an integral reservoir, but other pumps have a remote reservoir.*

Power Steering Pump Oil Leak Diagnosis

The possible sources of power steering pump oil leaks are the drive shaft seal, reservoir O-ring

Figure 25-44 Power steering reservoir and dipstick.

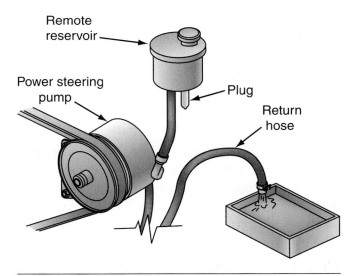

Check oil level. If leakage persists with the right level and cap tight, replace the cap.

Replace O-ring seal and tighten hose fitting nut to 35 N·m (25 ft. lb.).

Replace drive shaft seal.

Replace reservoir O-ring.

Tighten fitting to 55 N·m (40 ft. lb.). If leakage persists, replace O-ring seal.

Figure 25-45 Power steering pump oil leak diagnosis.

seal, high-pressure outlet fitting, and the dipstick cap. If leaks occur at any of the seal locations, seal replacement is necessary. When a leak is present at the high-pressure outlet fitting, tighten this fitting to the specified torque **(Figure 25-45)**. If the leak still occurs, replace the O-ring seal on the fitting and retighten the fitting.

Power Steering System Draining and Flushing

If the power steering fluid is contaminated with moisture, dirt, or metal particles, the system must be drained and new fluid installed. Follow these steps to drain and flush the power steering system:

1. Lift the front of the vehicle with a floor jack and install jack stands under the suspension. Lower the vehicle onto the jack stands and remove the floor jack.
2. Remove the return hose from the remote reservoir that is connected to the steering gear. Place a plug on the reservoir outlet and position the return hose in an empty drain pan **(Figure 25-46)**.

Remote reservoir

Power steering pump

Plug

Return hose

Figure 25-46 Return hose installed in drain pan for power steering draining and flushing.

3. With the engine idling, turn the steering wheel fully in each direction and stop the engine.
4. Fill the reservoir to the hot full mark with the manufacturer's recommended fluid.
5. Start the engine and run the engine at 1,000 rpm while observing the return hose in the

drain pan. When fluid begins to discharge from the return hose, shut the engine off.

6. Repeat steps 4 and 5 until there is no air in the fluid discharging from the return hose.

7. Remove the plug from the reservoir and reconnect the return hose. Bleed the power steering system.

Bleeding Air from the Power Steering System

When air is present in the power steering fluid, a growling noise may be heard in the pump and steering effort may be increased or erratic. When a power steering system has been drained and refilled, follow this procedure to remove air from the system:

1. Fill the power steering pump reservoir as outlined previously.

2. With the engine running at 1,000 rpm, turn the steering wheel fully in each direction three or four times. Each time the steering wheel is turned fully to the right or left, hold it there for 2 to 3 seconds before turning it in the other direction.

3. Check for foaming of the fluid in the reservoir. When foaming is present, repeat steps 1 and 2.

4. Check the fluid level with the engine running and be sure it is at the hot full mark. Shut the engine off and make sure the fluid level does not increase more than 0.020 inch (0.5 millimeter).

Another method to remove air is to use a vacuum pump to purge the air from the power steering system. This method is very similar to that of purging the air from the cooling system. Follow the manufacturer's recommendation when using equipment.

Power Steering Pump Pressure Test

Since there are some variations in power steering pump pressure test procedures and pressure specifications, the vehicle manufacturer's test procedures and specifications must be used. If the power steering pump pressure is low, steering effort is increased. Erratic power steering pump pressure causes variations in steering effort and the steering wheel may jerk as it is turned. Since a

power steering pump will never develop the specified pressure if the belt is slipping, the belt tension must be checked and adjusted if necessary prior to a pump pressure test. The following is a typical power steering pressure test procedure:

1. With the engine stopped, disconnect the pressure line from the power steering pump and connect the gauge side of the pressure gauge to the pump outlet fitting. Connect the valve side of the gauge to the pressure line.

2. Start the engine and turn the steering wheel fully in each direction two or three times to bleed air from the system. Be sure the fluid level is correct and the fluid temperature is at least 176°F (80°C). A thermometer may be inserted in the pump reservoir fluid to measure the fluid temperature.

3. With the engine idling, close the pressure gauge valve for no more than 10 seconds and observe the pressure gauge reading **(Figure 25-47)**. Turn the pressure gauge valve to the fully open

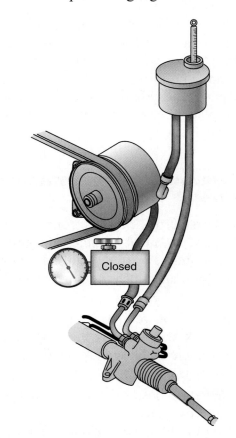

Figure 25-47 Power steering pump pressure test with gauge valve closed.

position. If the pressure gauge reading, when the valve was closed, did not equal the vehicle manufacturer's specifications, repair or replace the power steering pump.

4. Check the power steering pump pressure with the engine running at 1,000 rpm and 3,000 rpm and record the pressure difference between the two readings. If the pressure difference between the pressure readings at 1,000 rpm and 3,000 rpm does not equal the vehicle manufacturer's specifications, repair or replace the flow control valve in the power steering pump.

5. With the engine running, turn the steering wheel fully in one direction and observe the steering pump pressure while holding the steering wheel in this position. If the pump pressure is less than the vehicle manufacturer's specifications, the steering gear housing has an internal leak and should be repaired or replaced.

6. Be sure the front tire pressures are correct and center the steering wheel with the engine idling. Raise the wheels off the ground. Connect a spring scale to the steering wheel and measure the steering effort in both directions **(Figure 25-48)**. If the power steering pump pressure is satisfactory and the steering effort is more than the vehicle manufacturer's specifications, the power steering gear should be repaired.

Figure 25-48 Steering effort measurement.

SAFETY TIP *During the power steering pump pressure test, if the pressure gauge valve is closed for more than 10 seconds, excessive pump pressure may cause power steering hoses to rupture resulting in personal injury. Do not allow the fluid to become too hot during the power steering pump pressure test. Excessively high fluid temperature reduces pump pressure. Wear protective gloves and always shut the engine off before disconnecting gauge fittings because the hot fluid may cause burns.*

POWER STEERING PUMP SERVICE

If a growling noise is present in the power steering pump after the fluid level is checked and air has been bled from the system, the pump bearings or other components are defective and pump replacement is required. When the power steering pump pressure is lower than specified and the flow control valve is operating normally, pump replacement is necessary. When the power steering pump is replaced, proceed as follows:

1. Disconnect the power steering return hose from the remote reservoir or pump and allow the fluid to drain from this hose into a drain pan. Discard the used fluid.

2. Loosen the bracket or belt tension adjusting bolt and the pump mounting bolts.

3. Loosen the belt tension until the belt can be removed. On some cars it is necessary to lift the vehicle on a hoist to gain access to the power steering pump from underneath the vehicle.

4. Remove the hoses from the pump and cap the pump fittings and hoses.

5. Remove the belt tension adjusting bolt and the mounting bolt and remove the pump.

6. Check the pump bolts and bolt holes for wear. Worn bolts must be replaced. If the bolt mounting holes in the pump housing are worn, pump replacement is necessary.

7. Reverse steps 1 through 5 to install the power steering pump. Tighten the belt as described

previously and tighten the pump mounting and bracket bolts to the manufacturer's specifications. If O-rings are used on the pressure hose, replace the O-ring. Be sure the hoses are not contacting the exhaust manifold, catalytic converter, or exhaust pipe during or after pump replacement.

8. Fill the pump reservoir with the manufacturer's recommended power steering fluid and bleed air from the power steering system as previously described.

Power Steering Pump Pulley Replacement

If the pulley wobbles while the pulley is rotating, the pulley is probably bent and pulley replacement is necessary. Worn pulley grooves also require pulley replacement. Always check the pulley for cracks. If this condition is present, pulley replacement is essential. A pulley that is loose on the pump shaft must be replaced. Never hammer on the pump drive shaft during pulley removal or replacement. This action will damage internal pump components. If the pulley is pressed onto the pump shaft, a special puller is required to remove the pulley **(Figure 25-49)** and a pulley installation tool is used to install the pulley **(Figure 25-50)**.

If the power steering pump pulley is retained with a nut, mount the pump in a vise. Always tighten

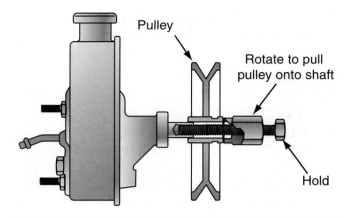

Figure 25-50 Press-on power steering pump pulley installation.

the vise on one of the pump mounting bolt surfaces and do not tighten the vise with excessive force. Use a special holding tool to keep the pulley from turning and loosen the pulley nut with a box-end wrench. Remove the nut, pulley, and woodruff key. Inspect the pulley, shaft, and woodruff key for wear. Be sure the key slots in the shaft and pulley are not worn. Replace all worn components.

Checking Power Steering Lines and Hoses

Power steering lines should be checked for leaks, dents, sharp bends, cracks, or contact with other components. Lines and hoses must not rub against other components. This action could wear a hole in the line or hose. Many high-pressure power steering lines are made from high-pressure steel-braided hose with molded steel fittings on each end.

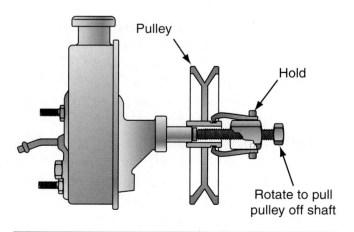

Figure 25-49 Press-on power steering pump pulley removal.

Summary

- Steering columns help to provide steering control, driver convenience, and driver safety.

- Many steering columns provide some method of energy absorption to protect the driver protection during a frontal collision.

- Tilt steering columns provide increased driver convenience while driving and getting in or out of the driver's seat.

- The ignition switch, dimmer switch, signal light switch, hazard switch, and the wipe/wash switch may be mounted in the steering column.

- When the ignition switch is in the lock position, a locking plate and lever in the upper steering column locks the steering wheel and the gearshift.

- In a parallelogram steering linkage, the tie-rods are parallel to the lower control arms.

- The parallelogram steering linkage minimizes toe change as the control arms move up and down on road irregularities.

- A rack-and-pinion steering linkage has reduced friction points and it is lightweight and compact compared to a parallelogram steering linkage.

- A power steering pump may have a vane-, roller-, or slipper-type rotor assembly, but all three types of pumps operate on the same basic principle.

- The flow control valve in a power steering pump controls pump pressure.

- A loose power steering belt causes low pump pressure, hard steering, and belt chirping or squealing, especially on acceleration.

- Most vehicle manufacturers recommend power steering fluid or automatic transmission fluid in the power steering system.

- Air may be bled from the power steering system by turning the steering wheel fully in each direction several times with the engine running. Each time the wheel is turned fully to the right or left, it should be held in that position for 2 to 3 seconds.

- A pressure gauge and manual valve are connected in series in the power steering pump pressure hose to check pump pressure. Closing the manual valve for less than 10 seconds checks maximum pump pressure.

- After the power steering pump pressure is tested and proven to be satisfactory, turning the steering wheel fully in one direction and observing the pressure reading may test steering gear leakage. If the power steering pump pressure is less than specified, the steering gear has an internal leak.

- With the engine idling, a spring scale may be attached at the outer end of the steering wheel crossbar to measure steering effort.

Review Questions

1. While discussing energy absorbing or collapsible steering columns, Technician A says on some energy absorbing steering columns, the column-to-instrument panel mount is designed to allow column movement if the driver is thrown against the steering wheel in a frontal collision. Technician B says in some collapsible steering columns, steel pins shear off in the two-piece jacket and steering shaft to allow column collapse if the driver is thrown against the steering wheel in a frontal collision. Who is correct?

   A. Technician A

   B. Technician B

   C. Both Technician A and Technician B

   D. Neither Technician A nor Technician B

2. While discussing parallelogram steering linkages, Technician A says the tie-rods are parallel to the upper control arms. Technician B says the idler arm helps to maintain the proper center link and tie-rod height. Who is correct?

A. Technician A

B. Technician B

C. Both Technician A and Technician B

D. Neither Technician A nor Technician B

3. While discussing parallelogram steering linkages, Technician A says loose steering linkage components may cause excessive tire wear. Technician B says the tie-rod sleeves provide a method of toe adjustment. Who is correct?

A. Technician A

B. Technician B

C. Both Technician A and Technician B

D. Neither Technician A nor Technician B

4. While discussing types of steering linkages, Technician A says compared to a parallelogram steering linkage, a rack-and-pinion linkage has more friction points. Technician B says in a rack-and-pinion steering linkage, the tie-rods are parallel to the lower control arms. Who is correct?

A. Technician A

B. Technician B

C. Both Technician A and Technician B

D. Neither Technician A nor Technician B

5. All these statements about parallelogram steering linkages are true *except*:

A. Parallelogram steering linkages may be mounted in front of or behind the front suspension.

B. Parallelogram steering linkages have the tie-rods mounted parallel to the lower control arms.

C. Parallelogram steering linkages have a pitman arm that connects the center link to the tie-rods.

D. Parallelogram steering linkages have an idler arm that supports one end of the center link.

6. On a vehicle with a parallelogram steering linkage, when one front wheel strikes a bump on the road surface, the front wheel and suspension move upward. During this upward wheel action:

A. The lower control arm moves through an arc.

B. The tie-rod moves through an arc that is much different compared to the control arm arc.

C. There is a considerable amount of front wheel toe change.

D. During upward and downward wheel movement, toe change causes tire tread wear.

7. In a rack-and-pinion steering gear and linkage:

A. The tie-rods are connected to the rack through inner tie-rod ends.

B. The tie-rods are not parallel to the lower control arms.

C. The pinion gear is connected to the sector shaft.

D. The mounting position of the steering gear does not affect vehicle steering.

8. All these statements about power steering pump pressure testing are true *except*:

A. The pressure gauge and valve should be connected in the power steering pump return hose to check pump pressure.

B. The pressure gauge valve should be closed for 10 seconds during the power steering pump pressure test.

C. The pressure test should be performed when the power steering fluid is cold.

D. The engine should be running at 2,500 rpm during the pressure test.

9. While discussing power steering pump pressure, Technician A says if the maximum pump

pressure is satisfactory with the pressure gauge valve closed but the pressure is lower than specified with the steering wheel turned fully in one direction, the steering gear may have an internal leak. Technician B says if the pump pressure is satisfactory and the front tires are properly inflated, if the steering effort is higher than specified, the steering gear may require replacing or repairing. Who is correct?

A. Technician A

B. Technician B

C. Both Technician A and Technician B

D. Neither Technician A nor Technician B

10. A loose power steering belt may cause all of these problems *except*:

A. squealing during engine acceleration.

B. low power steering pump pressure.

C. damage to the power steering pump bearings.

D. excessive steering effort.

11. In some collapsible steering columns _____ _____ in the outer column jacket and steering shaft shear off if the driver is thrown against the steering wheel in a frontal collision.

12. In a parallelogram steering linkage, the center link connects the pitman arm to the _____ _____.

13. A worn flexible coupling may cause a _____ noise in the steering column.

14. Power steering fluid level should be checked with the fluid _____.

15. Describe the proper clamp position on a tie-rod sleeve.

16. Describe the proper front wheel position when a rubber-encapsulated tie-rod end is installed and tightened.

17. Explain the results of a worn idler arm.

18. Explain the advantages of a rack-and-pinion steering linkage compared to a parallelogram steering linkage.

CHAPTER

26

Manual and Power Steering Gear Maintenance, Diagnosis, and Service

Learning Objectives

After you have read, studied, and practiced the contents of this unit, you should be able to:

- Diagnose steering columns.
- Inspect collapsible steering columns for damage.
- Diagnose and service steering linkage mechanisms.
- Check power steering fluid level.
- Drain, flush, and bleed air from the power steering system.
- Perform power steering pump pressure test.

- Remove and replace power steering pump pulleys.
- Diagnose manual and power recirculating ball steering gear problems.
- Adjust manual and power steering gears.
- Diagnose manual and power steering systems.
- Diagnose oil leaks in manual and power steering gears.

Key Terms

Constant ratio

Electrical power steering (EPS)

Helical gear

Pulse width modulated (PWM)

Rotary valve

Spur gear

Variable effort steering (VES)

Variable ratio

INTRODUCTION

During the late 1970s and 1980s, the domestic automotive industry shifted much of their production from larger RWD cars to smaller, lightweight, and more fuel-efficient FWD cars. These FWD cars required smaller, lightweight components wherever possible. Manual and power rack-and-pinion steering gears are lighter and more compact compared to the recirculating ball steering gears and parallelogram steering linkages used on most RWD cars. Therefore, rack-and-pinion steering gears are ideally suited to these compact FWD cars **(Figure 26-1)**.

Steering systems have not escaped the electronics revolution. Many cars are presently equipped with a **variable effort steering (VES)** that provides greater power assist during low-speed cornering and parking for increased driver convenience. These systems also reduce the assist during cruising speeds when little or no assist is needed.

Steering gear maintenance, diagnosis, and service are extremely important to maintain vehicle safety. If a vehicle has excessive steering wheel free-play, this problem should be diagnosed and corrected as soon as possible. If worn tie-rod ends or loose steering gear mounting bolts causes this problem, these conditions create a safety hazard. If a worn tie-rod end becomes disconnected, the result is a complete loss of steering control. This condition may cause a collision resulting in vehicle damage and/or personal injury.

STEERING GEARS

In a recirculating ball steering gear, the steering wheel and steering shaft are connected to the worm shaft. Ball bearings support both ends of the worm shaft in the steering gear housing. A seal above the upper worm shaft bearing prevents oil leaks and an adjusting plug is provided on the lower worm shaft bearing to adjust worm shaft bearing preload. Proper preloading of the worm shaft bearing is necessary to eliminate worm shaft endplay and prevent steering gear free-play and vehicle wander. A ball nut is mounted over the

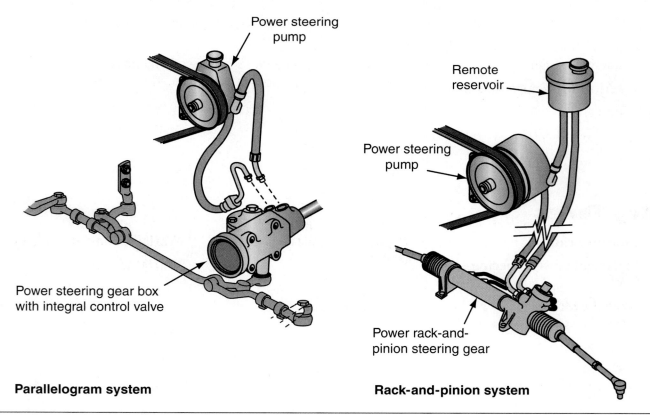

Parallelogram system   **Rack-and-pinion system**

Figure 26-1 A parallelogram and a rack-and-pinion steering systems.

worm shaft and internal threads or grooves on the ball nut match the grooves on the worm shaft. Ball bearings run in the ball nut and worm shaft grooves **(Figure 26-2)**. The recirculating balls operate much like the balls in a ball bearing. The rotation of the balls within the operating members reduces friction **(Figure 26-3)**.

When the worm shaft is rotated by the steering wheel, the ball nut is moved up or down on the worm shaft. The gear teeth on the ball nut are meshed with matching gear teeth on the pitman sector shaft. Therefore, ball nut movement causes pitman sector shaft rotation. Since the pitman sector shaft is connected through the pitman arm and steering linkage to the front wheels, the front wheels are turned by the pitman sector shaft. A bushing or a needle bearing in the steering gear

housing usually supports the lower end of the pitman sector shaft. A bushing in the sector cover supports the upper end of this shaft.

NOTE: Rotation of the steering wheel from extreme left to extreme right is called *lock-to-lock* or *stop-to-stop*.

When the front wheels are straight ahead, an interference fit exists between the sector shaft and ball nut teeth. This interference fit eliminates gear tooth lash when the front wheels are straight ahead and provides the driver with a positive feel of the road. Proper axial adjustment of the sector shaft is necessary to obtain the necessary interference

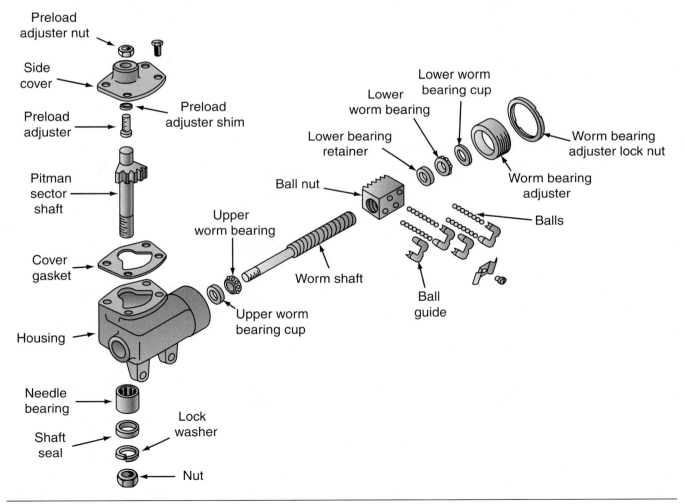

Figure 26-2 An exploded view of a manual recirculating ball steering gear.

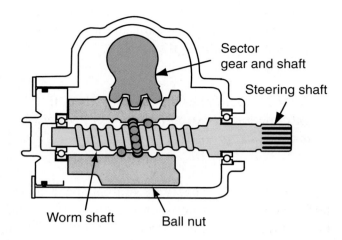

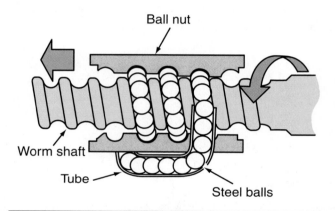

Figure 26-3 Operation of the recirculating balls in a manual recirculating ball steering gear.

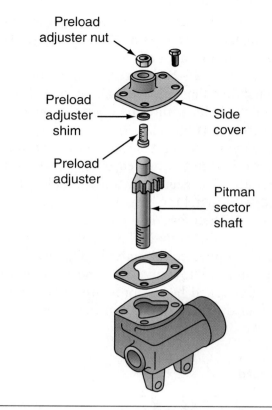

Figure 26-4 A sector shaft adjusting nut.

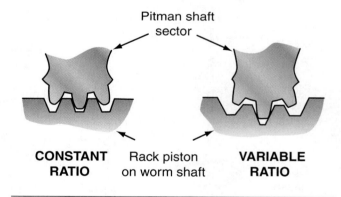

Figure 26-5 Constant and variable ratio steering gears.

fit between the sector shaft and worm shaft teeth. A sector shaft adjuster screw is threaded into the sector shaft cover to provide axial sector shaft adjustment **(Figure 26-4)**.

Manual recirculating ball steering gears have sector gear teeth and are usually designed to provide a **constant ratio**, whereas power recirculating ball steering gears usually have sector gear teeth with a **variable ratio (Figure 26-5)**. The sector gear teeth have equal lengths in a constant ratio steering gear, but the center sector gear tooth is longer compared to the other teeth in a variable ratio gear. The variable ratio steering gear varies the amount of mechanical advantage provided by the steering gear in relation to steering wheel position. This variable ratio provides "faster" steering. The steering gear ratio in a constant ratio manual steering gear is usually 15:1 or 16:1, whereas the

average variable ratio steering gear ratio may be 13:1. When the same types of steering gears are compared, a higher numerical ratio provides reduced steering effort and increased steering wheel movement in relation to the amount of front wheel movement.

Many recirculating ball steering gears are bolted to the frame with hard steel bolts **(Figure 26-6)**. These bolts must be tightened to the vehicle manufacturer's specified torque.

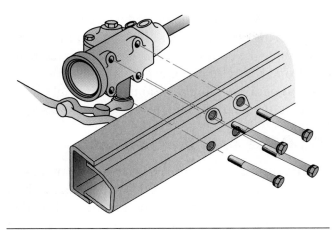

Figure 26-6 Steering gear mounting on the vehicle frame.

Tech Tip *Hard steel bolts may be used for steering gear mounting. Ribs on the bolt head indicate bolt hardness. Harder bolts have five, six, or seven ribs on the bolt heads. Never substitute softer steel bolts in place of the original harder bolts because these softer bolts may break allowing the steering box to detach from the frame, which results in loss of steering control.*

MANUAL RACK-AND-PINION STEERING GEARS

The rack is a toothed bar that slides back and forth in a metal housing. The steering gear housing is mounted in a fixed position on the front cross-member or on the firewall. The rack takes the place of the idler and pitman arms in a parallelogram steering system and maintains the proper height of the tie-rods so they are parallel to the lower control arms. The rack may be compared to the center link in a parallelogram steering linkage. Bushings support the rack in the steering gear housing. Sideways movement of the rack pulls or pushes the tie-rods and steers the front wheels **(Figure 26-7)**.

NOTE: In most rack-and-pinion steering gears, the tie-rods are attached to the ends of the rack. This type of steering gear may be called an *end take-off steering gear.*

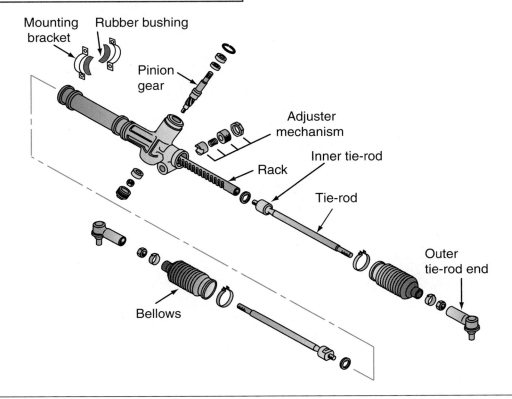

Figure 26-7 Manual rack-and-pinion steering gear components.

NOTE: The metal partition between the passenger compartment and the engine compartment is called the *firewall*. A bulkhead is a structural partition that separates compartments. This is generally a metal wall that extends from one side of a vehicle to the other such as the radiator bulkhead or trunk bulkhead.

Pinion

The pinion is a toothed shaft mounted in the steering gear housing so the pinion teeth are meshed with the rack teeth. The upper end of the pinion shaft is connected to the steering shaft from the steering column. Therefore, steering wheel rotation moves the rack sideways to steer the front wheels. The pinion is supported on ball bearings in the steering gear housing. Teeth on the rack and pinion may be spur or helical. **Spur gear** teeth are positioned parallel to the gear's rotational axis. **Helical gear** teeth are positioned at an angle in relation to the gear's rotational axis **(Figure 26-8)**.

Tie-Rods and Tie-Rod Ends

The tie-rods are similar to those used on parallelogram steering linkages. A spring-loaded ball socket on the inner end of the tie-rod is threaded onto the rack. When these ball sockets are torqued to the vehicle manufacturer's specification, a preload is placed on the ball socket. A bellows boot is clamped to the housing and tie-rod on each side of the steering gear. These boots keep contaminants out of the ball socket and rack. A tie-rod end is threaded onto the outer end of each tie-rod. These

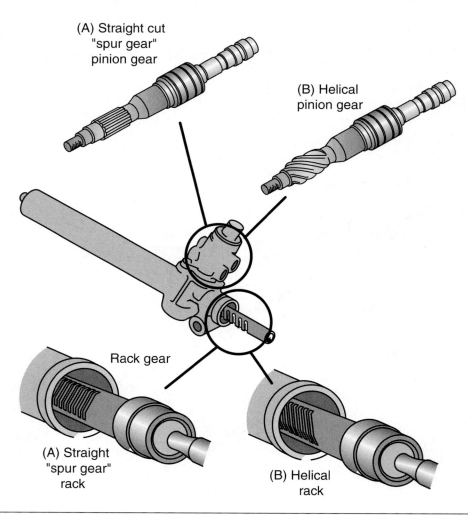

Figure 26-8 Examples of (A) a spur cut pinion and rack and (B) a helical cut pinion and rack.

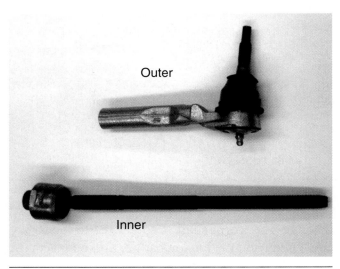

Figure 26-9 Inner and outer tie-rod ends for a rack and pinion steering.

tie-rod ends are similar to those used on parallelo-gram steering linkages. A jamb nut locks the outer tie-rod end to the tie-rod **(Figure 26-9)**.

> **NOTE:** On some rack-and-pinion steering gears, the tie-rods are attached to a moveable sleeve on the center of the gear. This type of gear may be called a *center take-off gear*.

Rack Adjustment

A rack bearing is positioned against the smooth side of the rack. A spring is located between the rack bearing and the rack adjuster plug that is threaded into the housing. This adjuster plug is re-tained with a locknut. The rack bearing adjustment sets the preload between the rack-and-pinion teeth, which affects steering harshness, noise, and feedback. The manufacturer's procedure should be followed when adjusting the rack bearing **(Figure 26-10)**.

Steering Gear Ratio

When the steering wheel is rotated from lock to lock, the front wheels turn about 30° each in each direction from the straight-ahead position. There-fore, the total front wheel movement from left to

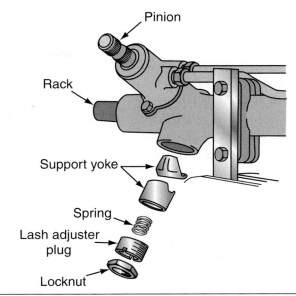

Figure 26-10 A rack bearing, spring, adjuster plug, and locknut.

right is approximately 60°. With a steering ratio of 1:1, 1° of steering wheel rotation would turn the front wheels 1°, and 30° of steering wheel rotation in either direction would result in lock-to-lock front wheel movement. This steering ratio is much too extreme because the slightest steering wheel movement would cause the vehicle to swerve. The steering gear must have a ratio that allows more steering wheel rotation in relation to front wheel movement.

A steering ratio of 15:1 is acceptable. This ratio provides 1° of front wheel movement for every 15° of steering wheel rotation. To calculate the steering ratio, divide the lock-to-lock steering wheel rotation in degrees by the total front wheel movement in degrees. For example, if the lock-to-lock steering wheel rotation is 3.5 turns, or 1,260°, and the total front wheel movement is 60°, the steering ratio is 1260 ÷ 60 = 21:1. As a general rule, large, heavy cars will have higher numerical steering ratios compared to small, lightweight cars.

> **NOTE:** A steering gear with a lower numer-ical ratio may be called a *faster steering gear* compared to a steering gear with a higher numerical gear ratio.

Manual Rack-and-Pinion Steering Gear Mounting

Large rubber insulating grommets are positioned between the steering gear and the mounting brackets. These bushings help prevent the transfer of road noise and vibration from the steering gear to the chassis and passenger compartment. The rack-and-pinion steering gear may be attached to the front crossmember or the firewall as shown in the previous chapter. Proper steering gear mounting is important to maintain the parallel relationship between the tie-rods and the lower control arms. The firewall is reinforced at the steering gear mounting locations to maintain the proper steering gear position.

Advantages and Disadvantages of Rack-and-Pinion Steering

As mentioned previously, the rack-and-pinion steering gear is lighter and more compact compared to a recirculating ball steering gear and parallelogram steering linkage. Therefore, the rack-and-pinion steering gear is most suitable for FWD unibody vehicles.

Since there are fewer friction points in the rack-and-pinion steering compared to the recirculating ball steering gear with a parallelogram steering linkage, the driver has a greater feeling of the road with rack-and-pinion steering gear. However, fewer friction points reduce the steering system's ability to isolate road noise and vibration. Therefore, the rack-and-pinion steering system may have more complaints of road noise and vibration transfer to the steering wheel and passenger compartment.

POWER RECIRCULATING BALL STEERING GEARS

The ball nut and pitman sector shaft are similar in manual and power recirculating ball steering gears. In the power steering gear, a torsion bar is connected between the steering shaft and the worm shaft. Since the front wheels are resting on the road surface, they resist turning and the parts attached to the worm shaft also resist turning. This turning resistance causes torsion bar deflection when the wheels are turned. This deflection is limited to a predetermined

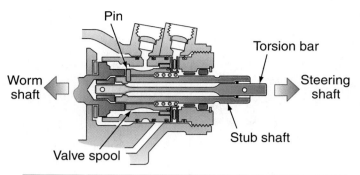

Figure 26-11 A torsion bar and stub shaft.

amount. The worm shaft is connected to the rotary valve body and the torsion bar pin also connects the torsion bar to the worm shaft. The upper end of the torsion bar is attached to the steering shaft and wheel. A stub shaft is mounted inside the rotary valve and a pin connects the outer end of this shaft to the torsion bar. The pin on the inner end of the stub shaft is connected to the spool valve in the center of the rotary valve **(Figure 26-11)**.

Rotary Valve and Spool Valve Operation

When the car is driven with the front wheels straight ahead, oil flows from the power steering pump through the spool valve, **rotary valve**, and low-pressure return line to the pump inlet **(Figure 26-12)**. In the straight-ahead steering gear position, oil pressure is equal on both sides of the recirculating ball piston and the oil acts as

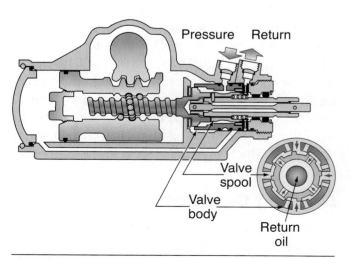

Figure 26-12 Power steering fluid flow with the wheels straight ahead.

a cushion that prevents road shocks from reaching the steering wheel. If the driver makes a left turn, torsion bar deflection moves the spool valve inside the rotary valve body so that oil flow is directed through the rotary valve to the left turn holes in the spool valve **(Figure 26-13)**. Since power steering fluid is directed from these left turn holes to the upper side of the recirculating ball piston **(Figure 26-14)**, this hydraulic pressure on the piston assists the driver in turning the wheels to the left.

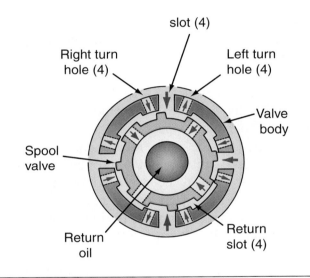

Figure 26-13 Spool valve position during a left turn.

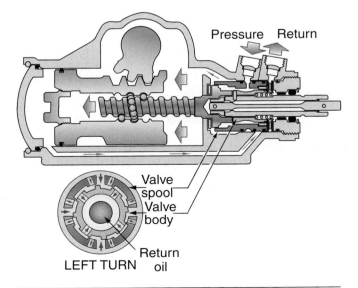

Figure 26-14 Power steering gear fluid flow during a left turn.

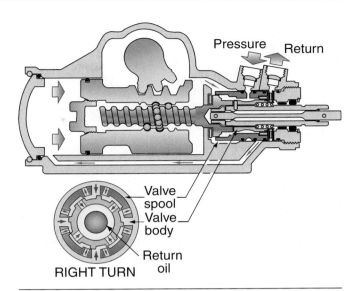

Figure 26-15 Power steering gear fluid flow during a right turn.

When the driver makes a right turn, torsion bar deflection moves the spool valve so that oil flows through the spool valve, rotary valve, and a passage in the housing to the pressure chamber at the lower end of the ball nut piston **(Figure 26-15)**. During a right turn, hydraulic pressure applied to the lower end of the recirculating ball piston helps the driver to turn the wheels. If a front wheel strikes a bump during a turn and the front wheels are driven in the direction opposite to the turning direction, the recirculating ball piston tends to move against the hydraulic pressure and force oil back out the pressure inlet port. This action would create a kickback on the steering wheel. Under this condition, a poppet valve in the pressure inlet fitting closes and prevents kickback action.

POWER RACK-AND-PINION STEERING GEARS

A power assisted rack-and-pinion steering gear uses the same basic parts as a manual rack-and-pinion steering gear. It uses hydraulic fluid pressure from the power steering pump to reduce steering effort. A rack piston is integral with the rack. This piston is located in a sealed chamber in the steering gear housing. Hydraulic fluid lines are connected to each end of this chamber and rack seals are positioned

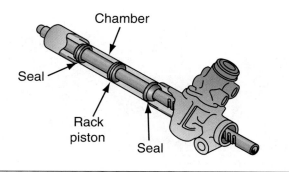

Figure 26-16 Hydraulic chambers in a power rack-and-pinion steering gear.

in the housing at ends of the chamber. A seal is also located on the rack piston (**Figure 26-16**).

Fluid direction in the steering gear is controlled by a spool valve attached to the pinion assembly (**Figure 26-17**). A stub shaft on the pinion assembly is connected to the steering shaft and wheel.

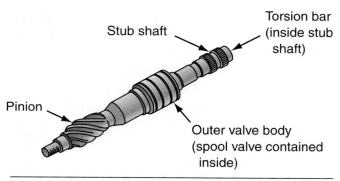

Figure 26-17 Pinion assembly for a power rack-and-pinion steering gear.

The pinion is connected to the stub shaft through a torsion bar that twists when the steering wheel is rotated and springs back to the center position when the wheel is released. A rotary valve body contains an inner spool valve that is mounted over the torsion bar on the pinion assembly.

Teflon rings, or O-rings, seal the rotary valve ring lands to the steering gear housing. A great deal of force is required to turn the pinion and move the rack because of the vehicle weight on the front wheels. When the driver turns the wheel, he or she forces the stub shaft to turn. However, the pinion resists turning because it is in mesh with the rack, which is connected to the front wheels. This resistance of the pinion to rotation results in torsion bar twisting. During this twisting action, a pin on the torsion bar moves the spool valve with a circular motion inside the rotary valve. If the driver makes a left turn, the spool valve movement aligns the inlet center rotary valve passage with the outlet passage to the left side of the rack piston. Therefore, hydraulic fluid pressure applied to the left side of the rack piston assists the driver in moving the rack to the right. When a right turn is made, twisting of the torsion bar moves the spool valve and aligns the center rotary valve passage with the outlet passage to the right side of the rack piston (**Figure 26-18**). Under this condition, hydraulic fluid pressure applied to the rack piston helps the driver to move the rack to the left.

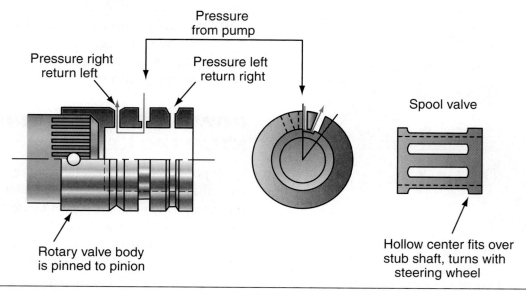

Figure 26-18 Rack movement during right and left turns.

The torsion bar provides a feel of the road to the driver. When the steering wheel is released after a turn, the torsion bar centers the spool valve and power assist stops. If hydraulic fluid pressure is not available from the pump, the power steering system operates like a manual system, but steering effort is higher. When the torsion bar is twisted to a designed limit, tangs on the stub shaft engage with drive tabs on the pinion. This action mechanically transfers motion from the steering wheel to the rack and front wheels. Since hydraulic pressure is not available on the rack piston, greater steering effort is required.

The right and left side of a vehicle is determined from the driver's seat. If a left turn is completed, fluid is pumped into the left side of the fluid chamber and exhausted from the right chamber area. This hydraulic pressure on the left side of the rack piston helps the pinion to move the rack to the right **(Figure 26-19 A)**.

When a right turn is made, fluid is pumped into the right side of the fluid chamber and fluid flows out of the left end of the chamber. Thus hydraulic pressure is exerted on the right side of the rack piston, which assists the pinion gear in moving the rack to the left **(Figure 26-19 B)**. Since the steering gear is mounted behind the front wheels, rack movement to the left is necessary for a right turn, while rack movement to the right causes a left turn. Power rack-and-pinion steering gears may be classified as end take-off or center take-off.

When the front wheels are in the straight-ahead position, fluid flows from the pump through the

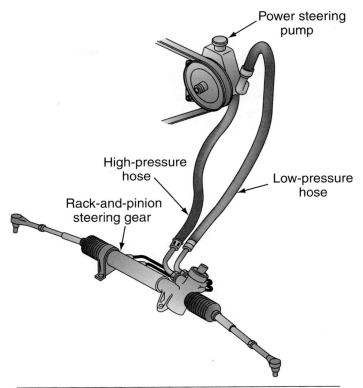

Figure 26-20 Spool valve movement inside the rotary valve.

high-pressure hose to the center rotary valve body passage. Fluid is then routed through the valve body to the low-pressure return hose and the pump reservoir **(Figure 26-20)**.

The rack boots are clamped to the housing and the rack. Because the boots are sealed and air cannot be moved through the housing, a breather

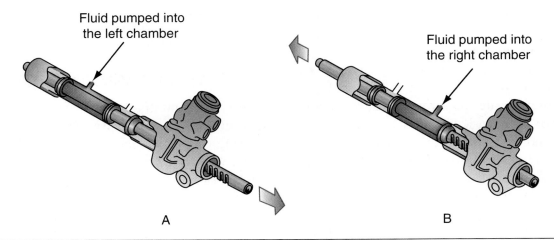

Figure 26-19 Power rack-and-pinion steering gear with connecting hoses.

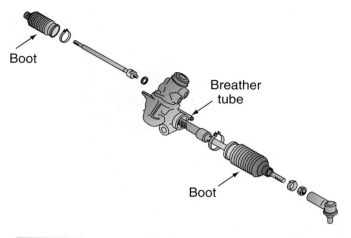

Figure 26-21 Breather tube and boot.

tube is necessary to move air from one boot to the other when the steering wheel is turned (**Figure 26-21**). This air movement through the vent tube prevents pressure changes in the bellows boots during a turn.

VARIABLE EFFORT POWER STEERING SYSTEMS

Some vehicles have an electronically controlled power steering system. These are usually referred to as a variable effort steering (VES). In this system, the vehicle speed sensor input is sent to a power steering controller. This controller supplies a **pulse width modulated (PWM)** voltage to the actuator solenoid in the power steering pump. The controller also provides a ground connection for the actuator solenoid (**Figure 26-22**).

When the vehicle is operating at low speeds or the steering wheel is being turned quickly, the power steering controller supplies a PWM signal to position the actuator solenoid plunger so the power steering pump pressure is higher. Under this condition, greater power assist is provided for cornering or parking. If the vehicle is operating at higher speed and the steering wheel is being turned slowly or very little, the controller changes the PWM signal to the actuator solenoid and the solenoid plunger is positioned to reduce power steering pump pressure. This action reduces power steering assist to provide improved road feel for the driver (**Figure 26-23**).

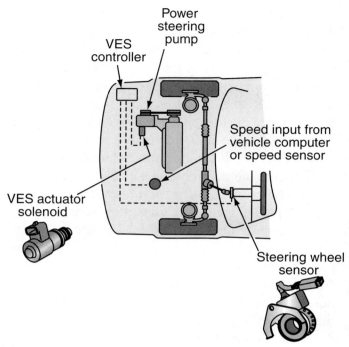

Figure 26-22 The main components of a variable effort steering (VES) system.

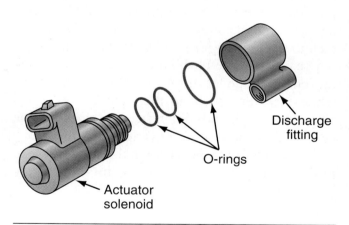

Figure 26-23 An actuator solenoid in a variable effort steering (VES) system.

ELECTRIC POWER STEERING

A significant number of vehicles are equipped with **electrical power steering (EPS)**. In these steering gears, an electric motor provides steering assist when the front wheels are turned (**Figure 26-24**). Because the electronic power steering system does not have a power steering pump, there is no engine power required to drive this pump and fuel economy is increased slightly; the unit uses power only as needed.

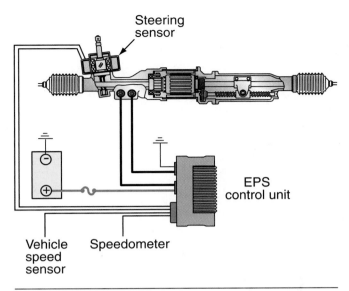

Figure 26-24 An electronic power steering (EPS) rack-and-pinion unit.

EPS is also a cheaper and more flexible solution than a hydraulic pump. Besides providing a 5 percent improvement in fuel efficiency, EPS is lighter (4 Kg to 6 Kg or 8 lbs. to 13 lbs.), and mechanically simpler (no pump, fan belt, or fluid reservoir). The same EPS system can provide performance driving for a sports car and smooth, power-assisted steering for the family sedan simply by changing the controller's software. **Figure 26-25**

shows a block diagram for an EPS system. The motor may be mounted in a variety of locations depending on the needs of the design engineers and available space (**Figure 26-26**).

MANUAL STEERING GEAR MAINTENANCE AND DIAGNOSIS

Manual steering gear maintenance is very important to provide normal steering gear life and proper steering gear operation. For example, if a manual steering gear is operated for a period of time with a low lubricant level, steering gear life is shortened.

Manual Recirculating Ball Steering Gear Maintenance and Diagnosis

When the oil level plug is removed from the steering gear, the lubricant should be level with the bottom of the plug opening. If the oil level is low, fill the steering gear with the manufacturer's specified steering gear lubricant. If a steering gear provides binding and hard, uneven steering effort, the oil level is very low. If the oil level is low, visually check the

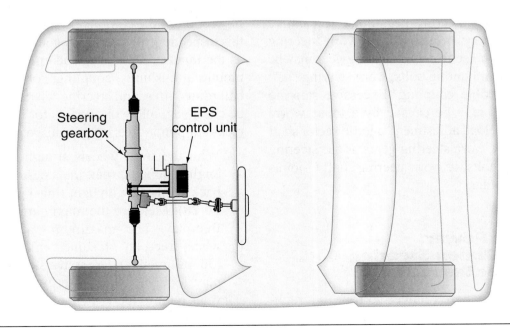

Figure 26-25 An electronic power steering (EPS) system components.

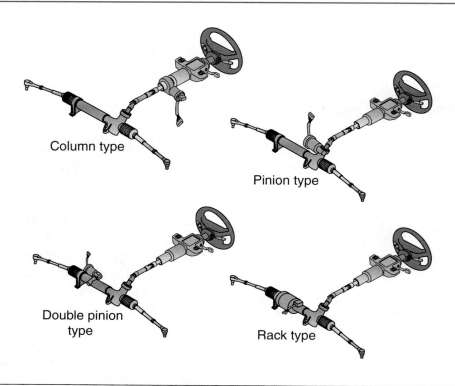

Column type

Pinion type

Double pinion
type

Rack type

Figure 26-26 Various motor locations are possible and include column, pinion, rack, and double pinion.

sector shaft seal and the worm shaft seal area for leaks. Leaking seals must be replaced. Hard steering effort may also be caused by wheel alignment problems. Defective worm shaft bearings cause uneven turning effort and steering gear noise. Excessive steering effort may be caused by worn steering gears. A rattling noise from the steering gear may be caused by loose mounting bolts, worn steering shaft U-joints, or flexible coupling. Excessive steering wheel free-play may be caused by a loose worm shaft bearing preload adjustment, a loose sector shaft lash adjustment, worn steering gears, loose steering gear mounting bolts, or worn steering shaft U-joints or flexible coupling.

Manual or Power Rack-and-Pinion Steering Gear On-Car Inspection

The wear points are reduced to four in a rack-and-pinion steering gear. These wear points are the inner and outer tie-rod ends on both sides of the rack-and-pinion assembly **(Figure 26-27)**.

The first step in manual or power rack-and-pinion steering gear diagnosis is a very thorough inspection of the complete steering system. During this inspection, all steering system components, such as the inner and outer tie-rod ends, bellows boots, mounting bushings, couplings, or universal joints, ball joints, tires, and steering wheel free-play, must be checked. Follow these steps for manual or power rack-and-pinion steering gear inspection:

1. With the front wheels straight ahead and the engine stopped, rock the steering wheel gently back and forth with light finger pressure **(Figure 26-28)**. Measure the maximum steering wheel free-play. The maximum specified steering wheel free-play on some vehicles is 1.2 inches (30 millimeters). Excessive steering wheel free-play indicates worn steering components. Always refer to the vehicle manufacturer's specifications in the service manual.

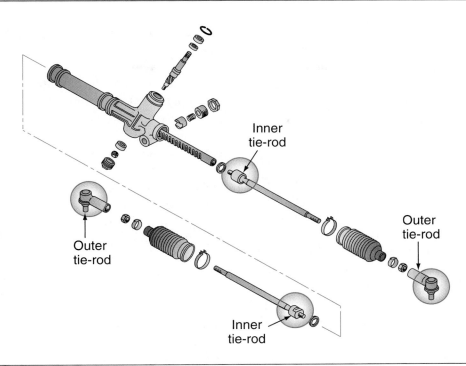

Figure 26-27 Wear points at the inner and outer tie-rod ends in a rack-and-pinion steering gear.

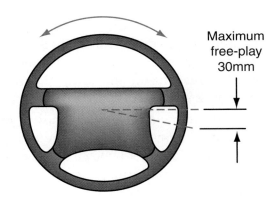

Figure 26-28 Measuring steering wheel free-play.

2. With the vehicle sitting on the shop floor and the front wheels straight ahead, have an assistant turn the steering wheel about one-quarter turn in both directions. Watch for looseness in the flexible coupling in the steering shaft. If looseness is observed, replace the coupling or universal joint.

3. While an assistant turns the steering wheel about one-half turn in both directions, watch

for movement of the steering gear housing in the mounting bushings. If there is any movement of the housing in these bushings, replace the bushings. Oil soaking, heat, or age may cause deteriorated steering gear mounting bushings.

4. Grasp the pinion shaft extending from the steering gear and attempt to move it vertically. If there is steering shaft vertical movement, a pinion bearing preload adjustment may be required. When the steering gear does not have a pinion bearing preload adjustment, replace the necessary steering gear components.

5. Road test the vehicle and check for excessive steering effort. A bent steering rack, tight rack bearing adjustment, or damaged front drive axle joints in a front wheel drive car may cause excessive steering effort.

6. Visually inspect the bellows boots for cracks, splits, leaks, and proper clamp installation. Replace any boot that indicates any of these conditions. If the boot clamps are loose or

improperly installed, tighten or replace the clamps as necessary. Since the bellows boots protect the inner tie-rod ends and the rack from contamination, boot condition is extremely important. Boots should be inspected each time undercar service such as oil and filter change or chassis lubrication is performed.

7. Loosen the inner bellows boot clamps and move each boot toward the outer tie-rod end until the inner tie-rod end is visible. Push outward and inward on each front tire and watch for movement in the inner tie-rod end. If any movement or looseness is present, replace the inner tie-rod end. An alternate method of checking the inner tie-rod ends is to squeeze the bellows boots and grasp the inner tie-rod end socket. Movement in the inner tie-rod end is then felt as the front wheel is moved inward and outward. Hard plastic bellows boots may be found on some applications. With this type of bellows boot, remove the ignition key from the switch to lock the steering column and push inward and outward on the front tire while observing any lateral movement in the tie-rod. When lateral movement is observed, replace the inner tie-rod end.

8. Grasp each outer tie-rod end and check for vertical movement. While an assistant turns the steering wheel one-quarter turn in each direction, watch for looseness in the outer tie-rod ends. If any looseness or vertical movement is present, replace the tie-rod end.

9. Check the outer tie-rod end seals for cracks and proper installation of the nuts and cotter pins. Cracked seals must be replaced. Inspect the tie-rods for a bent condition. Bent tie-rods or other steering components must be replaced. Do not attempt to straighten these components.

SAFETY TIP *Bent steering components must be replaced. Never straighten steering components because this action may weaken the metal and result in sudden component failure, serious personal injury, and vehicle damage.*

MANUAL STEERING GEAR SERVICE

Proper manual steering gear service is essential to supply normal steering control and vehicle safety. For example, loose manual steering gear worm shaft bearing preload and sector shaft adjustments cause vehicle wander and reduced steering control.

SAFETY TIP *Recirculating ball steering gears are often mounted near the exhaust manifold, which may be extremely hot. Wear protective gloves and use caution when inspecting, adjusting, and servicing the steering gear.*

Manual Recirculating Ball Steering Gear Removal and Replacement

Follow these preliminary steps prior to steering gear removal or adjustment:

1. Disconnect the battery ground cable.

2. Raise the vehicle with the front wheels in the straight-ahead position. If the vehicle is lifted with a floor jack, place jack stands under the chassis or suspension and lower the vehicle onto the jack stands.

3. Remove the pitman arm nut and washer. Mark the pitman arm position in relation to the pitman shaft with a center punch and use a puller to remove the pitman arm.

4. Loosen the sector shaft backlash adjuster lock nut and back the adjuster screw off one-quarter turn.

5. Turn the steering wheel gently in one direction until it is stopped by the gear and then count the number of turns as the steering wheel is rotated in the opposite direction until it is stopped by the gear. Then turn the steering wheel back one-half the total number of turns toward the center position.

6. Remove the center steering wheel cover and place a socket and inch-pound torque wrench on the steering wheel nut. Do not use a torque wrench with a maximum scale reading above 50 in. lb.

7. Rotate the steering wheel through a 90-degree arc and record the turning torque, which indicates the worm shaft bearing preload.

Follow these steps for manual steering gear removal and replacement:

1. Disconnect the flexible coupling from the worm shaft.

2. Remove the steering gear-to-frame mounting bolts.

3. Remove the steering gear from the chassis. The steering gear may be cleaned externally with an approved cleaning solution.

4. Reverse steps 1 through 3 to re-install the steering gear. All bolts must be tightened to the specified torque. Be sure the steering gear is filled with the manufacturer's specified steering gear lubricant.

Tech Tip *When the steering linkage is disconnected from the gear, do not turn the steering wheel hard against the stops. This action may damage the ball guides in the steering gear.*

Manual or Power Rack-and-Pinion Steering Gear Removal and Replacement

The replacement procedure is similar for manual or power rack-and-pinion steering gears. This removal and replacement procedure varies depending on the vehicle. On some vehicles, the front crossmember or engine support cradle must be lowered to remove the rack-and-pinion steering gear. Always follow the vehicle manufacturer's recommended procedure in the service manual.

The following is a typical rack-and-pinion steering gear removal and replacement procedure:

1. Place the front wheels in the straight-ahead position and remove the ignition key from the ignition switch to lock the steering wheel. Place the driver's seat belt through the steering wheel to prevent wheel rotation if the ignition switch is turned on. This action maintains the clockspring electrical connector or spiral cable in

the centered position on air bag-equipped vehicles.

2. Lift the front end with a floor jack and place jack stands under the vehicle chassis. Lower the vehicle onto the jack stands. Remove the left and right fender apron seals if equipped **(Figure 26-29)**.

3. Place punch marks on the lower universal joint and the steering gear pinion shaft so they may be reassembled in the same position **(Figure 26-30)**. Loosen the upper universal joint bolt and remove the lower universal joint bolt and disconnect this joint.

4. Remove the cotter pins from the outer tie-rod ends. Use a tie-rod end puller to loosen the outer tie-rod ends in the steering arms. Remove the tie-rod ends from the arms.

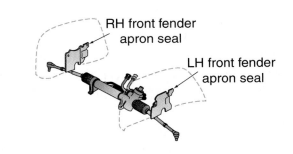

Figure 26-29 Left and right fender apron seals.

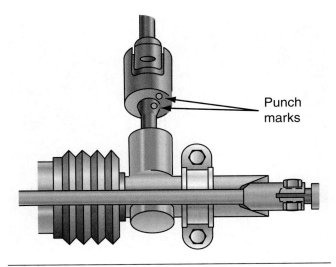

Figure 26-30 Punch marks on universal joint and pinion shaft.

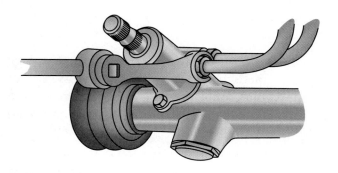

Figure 26-31 Removing high-pressure and return hoses from the steering gear.

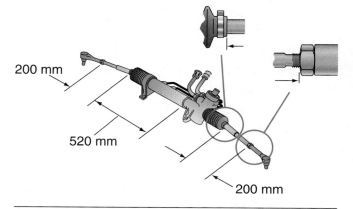

Figure 26-33 Right and left tie-rod position in relation to the steering gear housing prior to installation.

5. Use the proper wrenches to disconnect the high pressure hose and return hose from the steering gear (**Figure 26-31**). This step is not required on a manual steering gear.

6. Remove the four steering stabilizer bar mounting bolts (**Figure 26-32**).

7. Remove steering gear mounting bolts.

8. Remove the steering gear assembly from the car.

9. Position the right and left tie-rods the specified distance from the steering gear housing (**Figure 26-33**). Install the steering gear through the right fender apron.

10. With the punch marks aligned, install the pinion shaft in the universal joint. Tighten the upper and lower universal joint bolts to the specified torque.

11. Install the steering gear mounting bolts and tighten these bolts to the specified torque.

12. Install the stabilizer bar mounting bolts and torque these bolts to specifications.

13. Install and tighten the high-pressure and return hoses to the specified torque. This step is not required on a manual rack-and-pinion steering gear.

14. Install the outer tie-rod ends in the steering knuckles and tighten the nuts to the specified torque. Install the cotter pins in the nuts.

15. Check the front wheel toe and adjust as necessary. Tighten the outer tie-rod end jam nuts to the specified torque and tighten the outer bellows boot clamps.

16. Install the left and right fender apron seals and lower the vehicle with a floor jack.

17. Fill the power steering pump reservoir with the vehicle manufacturer's recommended power steering fluid and bleed air from the power steering system. This step is not required on a manual rack-and-pinion steering gear.

18. Road test the vehicle and check for proper steering gear operation and steering control.

Manual Recirculating Ball Steering Gear Worm Shaft Bearing Preload Adjustment

If the worm shaft bearing preload adjustment is loose, steering wheel free-play is excessive. This results in steering wander when the vehicle is driven

Figure 26-32 Removing steering stabilizer bar mounting bolts.

straight ahead. Steering effort is increased if the worm shaft bearing preload adjustment is too tight.

> **Tech Tip** *Applying force to the worm shaft at either of the stops may damage the steering gear.*

When the worm shaft bearing preload is adjusted, use the following procedure:

1. Follow steps 1 and 2 listed in the previous section.

2. Loosen the worm shaft adjuster plug locknut with a brass punch and a hammer. Tighten the adjuster plug until all the worm shaft endplay is removed and then loosen the plug one-quarter turn.

3. Turn the worm shaft fully to the right with a socket and an inch-pound torque wrench. Then turn the worm shaft one-half turn toward the center position.

4. Tighten the adjuster plug until the specified bearing preload is indicated on the torque wrench as the worm shaft is rotated **(Figure 26-34)**. The specification on some steering gears is 5 to 8 in. lb.

Figure 26-34 Adjusting worm shaft bearing preload, manual recirculating ball steering gear.

(0.56 to 0.896 N·m). Always use the vehicle manufacturer's specified preload.

5. Tighten the adjuster plug locknut to 85 ft. lb. (114 N·m).

Manual Recirculating Ball Steering Gear Sector Lash Adjustment

When the sector shaft lash adjustment is too loose, steering wheel free-play is excessive and vehicle wander occurs when the vehicle is driven straight ahead. A loose sector shaft lash adjustment decreases driver road feel. If the sector lash adjustment is too tight, steering effort is increased, especially with the front wheels in the straight-ahead position and steering wheel returnability can be diminished.

The following procedure may be used when the pitman sector shaft lash is adjusted:

1. Turn the pitman backlash adjuster screw outward until it stops and then turn it in one turn.

2. Rotate the worm shaft fully from one stop to the other stop and carefully count the number of shaft rotations.

3. Turn the worm shaft back exactly one-half the total number of turns from one of the stops.

4. With the steering gear positioned as it was in step 3, connect an inch-pound torque wrench and socket to the worm shaft and note the steering gear turning torque while rotating the worm shaft 45° in each direction.

5. Turn the pitman backlash adjuster screw until the torque wrench reading is 6 to 10 in. lb. (0.44 to 1.12 N·m) more than the worm shaft bearing preload torque in step 4 **(Figure 26-35)**.

6. Tighten the pitman backlash adjuster screw locknut to the specified torque.

POWER STEERING GEAR MAINTENANCE AND DIAGNOSIS

If the steering gear is noisy, check these items:

1. A loose pitman shaft lash adjustment may cause a rattling noise when the steering wheel is turned.

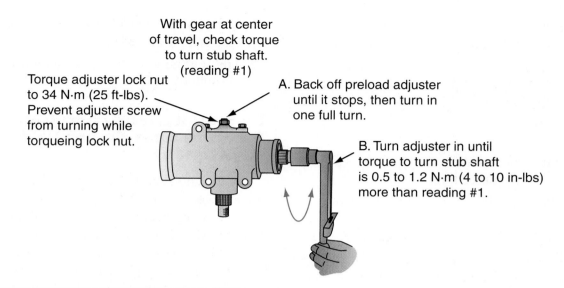

With gear at center of travel, check torque to turn stub shaft. (reading #1)

Torque adjuster lock nut to 34 N·m (25 ft-lbs). Prevent adjuster screw from turning while torqueing lock nut.

A. Back off preload adjuster until it stops, then turn in one full turn.

B. Turn adjuster in until torque to turn stub shaft is 0.5 to 1.2 N·m (4 to 10 in-lbs) more than reading #1.

Figure 26-35 Adjusting pitman backlash adjuster screw, manual recirculating ball steering gear.

2. Cut or worn dampener O-ring on the valve spool. When this defect is present, a squawking noise is heard during a turn.

3. Loose steering gear mounting bolts.

4. Loose or worn flexible coupling or steering shaft U-joints.

A hissing noise from the power steering gear is normal if the steering wheel is at the end of its travel or when the steering wheel is rotated with the vehicle standing still. If the steering wheel jerks or surges when it is turned with the engine running, check the power steering pump belt condition and tension. When excessive kickback is felt on the steering wheel, check the poppet valve in the steering gear.

SAFETY TIP *If the engine has been running for a length of time, power steering gears, pumps, lines, and fluid may be very hot. Wear eye protection and protective gloves when servicing these components.*

When the steering is loose, check these defects:

1. Air in the power steering system. To remove the air, fill the power steering pump reservoir and rotate the steering wheel fully in each direction several times.

2. Loose pitman lash adjustment.

3. Loose worm shaft thrust bearing preload adjustment.

4. Worn flexible coupling or universal joint.

5. Loose steering gear mounting bolts.

6. Worn steering gears.

A complaint of hard steering while parking could be caused by one of these defects:

1. Loose or worn power steering pump belt.

2. Low oil level in the power steering pump.

3. Excessively tight steering gear adjustments.

4. Defective power steering pump with low pressure output.

5. Restricted power steering hoses.

6. Defects in the steering gear such as:

   a) Pressure loss in the cylinder because of scored cylinder, worn piston ring, or damaged back-up O-ring.

   b) Excessively loose spool in valve body.

   c) Defective or improperly installed gear check poppet valve.

Power Recirculating Ball Steering Gear Oil Leak Diagnosis

Five locations where oil leaks may occur in a power steering gear are the following:

1. Side cover O-ring seal.

2. Adjuster plug seal.

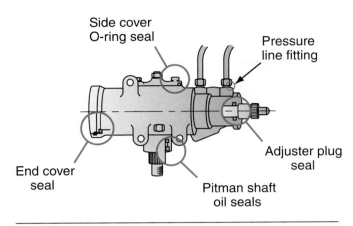

Figure 26-36 Power recirculating ball steering gear oil leak locations.

3. Pressure line fitting.

4. Pitman shaft oil seals.

5. Top cover seal.

If an oil leak is present at any of these areas, complete or partial steering gear disassembly and seal or O-ring replacement is necessary **(Figure 26-36)**.

Power Rack-and-Pinion Steering Gear Oil Leak Diagnosis

If power steering fluid leaks from the cylinder end of the power steering gear, the outer rack seal is leaking **(Figure 26-37)**.

The inner rack seal is defective if oil leaks from the pinion end of the housing when the rack reaches the left internal stop **(Figure 26-38)**. An

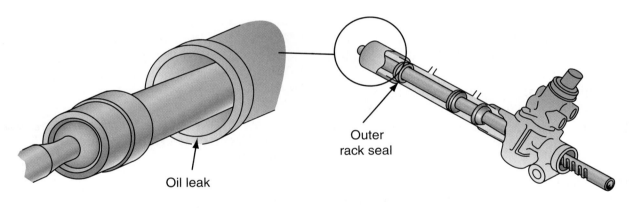

If a leak is detected at the housing cylinder end, the origin of the leak is the outer rack seal.

Figure 26-37 Oil leak diagnosis at outer rack seal.

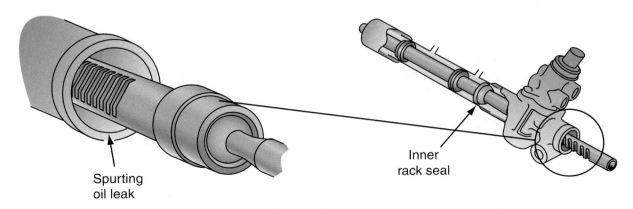

If the leak at the pinion end of the housing spurts when the rack reaches the left internal stop, the inner rack seal is at fault.

Figure 26-38 Inner rack seal leak diagnosis.

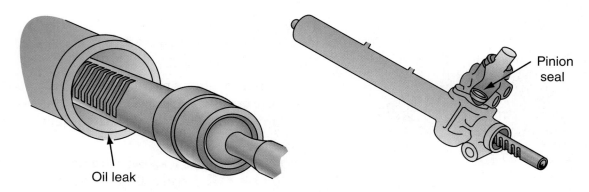

Oil leak

If you detect a leak at the pinion end of the housing and it is not influenced by the direction of the turn, the origin of the leak is the pinion seal.

Pinion seal

Figure 26-39 Pinion seal leak diagnosis.

oil leak at one rack seal may result in oil leaks from both boots because the oil may travel through the breather tube between the boots.

If an oil leak occurs at the pinion end of the housing and this leak is not influenced when the steering wheel is turned, the pinion seal is defective **(Figure 26-39)**.

If an oil leak occurs in the pinion coupling area, the input shaft seal is leaking **(Figure 26-40)**. This seal and the pinion seal will require replacement because the pinion seal must be replaced if the pinion is removed.

When the rack is removed from the housing, the inner and outer rack seals and the pinion seal must be replaced **(Figure 26-41)**.

If oil leaks occur at the hydraulic fittings, these fittings must be torqued to the manufacturer's specifications. If the leak is still present, the line

and fitting should be replaced. Leaks in the lines or hoses require line or hose replacement.

Tech Tip *To provide equal turning effort in both directions, the rack must be centered with the front wheels straight ahead.*

Power Rack-and-Pinion Steering Gear Turning Imbalance Diagnosis

The same amount of effort should be required to turn the steering wheel in either direction. A pressure gauge connected to the high-pressure hose should indicate the same pressure when the steering wheel is turned in each direction. Defective rack seals may cause steering effort

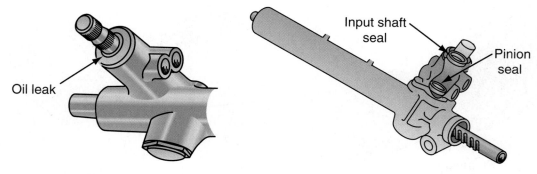

Oil leak

Input shaft seal

Pinion seal

If you discover a leak at the pinion coupling area, you will have to replace both the input shaft seal and the lower pinion seal.

Figure 26-40 Oil leak diagnosis in pinion coupling area.

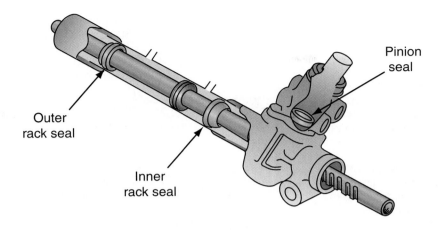

Figure 26-41 Rack seals and pinion seal.

imbalance or lower power assist in each direction **(Figure 26-42)**. Defective rotary valve rings and seals or restricted hoses and lines may cause steering effort imbalance or low power assist in both directions may also be caused by defective rack seals **(Figure 26-43)**.

POWER STEERING GEAR SERVICE

Proper power steering gear service procedures are critical to maintain vehicle safety and normal steering control. For example, if steering gear

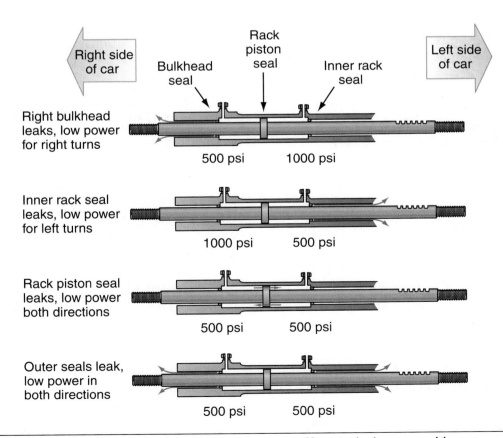

Figure 26-42 Effect of defective rack seals on steering effort imbalance and low power assist.

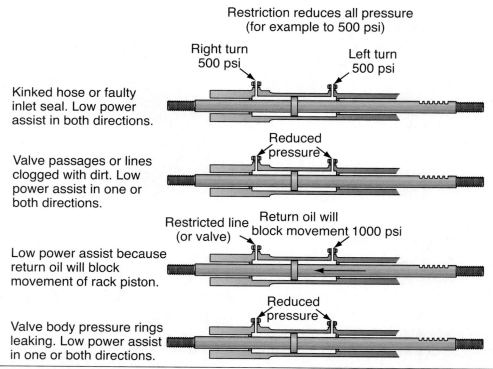

Restriction reduces all pressure
(for example to 500 psi)

Right turn
500 psi

Left turn
500 psi

Kinked hose or faulty inlet seal. Low power assist in both directions.

Reduced pressure

Valve passages or lines clogged with dirt. Low power assist in one or both directions.

Restricted line (or valve)   Return oil will block movement 1000 psi

Low power assist because return oil will block movement of rack piston.

Reduced pressure

Valve body pressure rings leaking. Low power assist in one or both directions.

Figure 26-43 Effect of worn rotary valve rings and seals or restricted lines or hoses on steering effort.

mounting bolts are not tightened to the specified torque, these bolts may loosen and fall out resulting in a complete loss of steering control. This action may cause a collision and personal injury.

Power Recirculating Ball Steering Gear Replacement

When the power steering gear is replaced, proceed as follows:

1. Disconnect the hoses from the steering gear and cap the lines and fittings to prevent dirt from entering the system.
2. Remove the pitman arm nut and washer and mark the pitman arm in relation to the shaft with a center punch. Use a puller to remove the pitman arm.
3. Disconnect the steering shaft from the worn shaft.
4. Remove the steering gear mounting bolts and remove the steering gear from the chassis.
5. Reverse steps 1 through 4 to install the steering gear. All bolts must be tightened to the specified

torque. Be sure the pitman arm is installed in the original position.

Power Recirculating Ball Steering Gear Adjustment

A loose worm shaft thrust bearing preload adjustment or sector lash adjustment causes excessive steering freeplay and steering wander. The power recirculating ball steering gear adjustment procedures may vary depending on the vehicle's make and model year. Always follow the vehicle manufacturer's recommended procedure in the service manual.

The following is a typical worm shaft thrust bearing preload adjustment:

1. Remove the worm shaft thrust bearing adjuster plug locknut with a special tool.
2. Turn this adjuster plug inward or clockwise until it bottoms and tighten the plug to 20 ft. lb. (27 N·m).
3. Place an index mark on the steering gear housing next to one of the holes in the adjuster plug **(Figure 26-44)**.

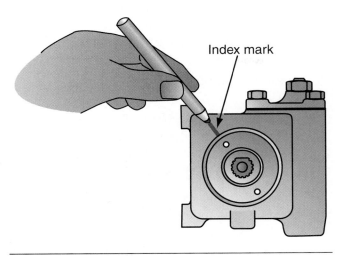

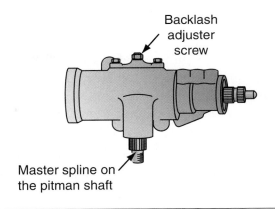

Figure 26-46 Pitman shaft master spline aligned with pitman backlash adjuster screw.

Figure 26-44 Placing index mark on steering gear housing opposite one of the adjuster plug holes.

4. Measure 0.50 inch (13 millimeters) counterclockwise from the index mark and place a second index mark at this position **(Figure 26-45)**.

5. Rotate the adjuster plug counterclockwise until the hole in the adjuster plug is aligned with the second index mark placed on the housing.

6. Install and tighten the adjuster plug locknut to the specified torque.

The following is a typical pitman sector shaft lash adjustment:

1. Rotate the stub shaft from stop to stop and count the number of turns.

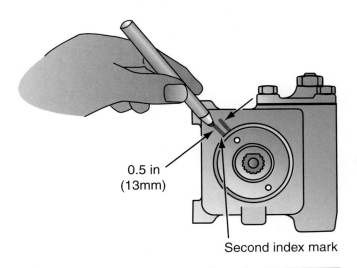

Figure 26-45 Measuring 0.50 inch (13 millimeters) counterclockwise from the index mark on the steering gear housing.

2. Starting at either stop, turn the stub shaft back two thirds of the total number of turns. In this position, the flat on the stub shaft should be facing upward and the master spline on the pitman shaft should be aligned with the pitman shaft backlash adjuster screw **(Figure 26-46)**.

3. Turn the pitman shaft backlash adjuster screw fully counterclockwise and then turn it clockwise one turn.

4. Use an inch-pound torque wrench to turn the stub shaft through a 45-degree arc on each side of the position in step 2. Read the over-center torque as the stub shaft turns through the center position.

5. Continue to adjust the pitman shaft adjuster screw until the torque is 6 to 10 in. lb. (0.6 to 1.2 N·m) more than the torque in step 3.

6. Hold the pitman shaft adjuster screw in this position and tighten the locknut to the specified torque.

Power Recirculating Ball Steering Gear Side Cover O-Ring Replacement

Prior to any disassembly procedure, clean the steering gear with an approved cleaning solution. The steering gear service procedures vary depending on the make of gear. Always follow the vehicle manufacturer's recommended procedure in the service manual.

The following is a typical side cover O-ring replacement procedure:

1. Loosen the pitman backlash adjuster screw locknut and remove the side cover bolts. Rotate the pitman backlash adjuster screw clockwise to remove the cover from the screw.

2. Discard the O-ring and inspect the side cover matching surfaces for metal burrs and scratches.

3. Lubricate a new O-ring with the vehicle manufacturer's recommended power steering fluid and install the O-ring.

4. Rotate the pitman backlash adjuster screw counterclockwise into the side cover until the side cover is properly positioned on the gear housing. Turn this adjuster screw fully counterclockwise and then one turn clockwise. Install and tighten the side cover bolts to the specified torque. Adjust the pitman sector shaft lash as explained previously.

Power Recirculating Ball Steering Gear End Plug Seal Replacement

The following is a typical end plug seal replacement procedure:

1. Insert a punch into the access hole in the steering gear housing to unseat the retaining ring and remove the ring.

2. Remove the end plug and seal.

3. Clean the end plug and seal contact area in the housing with a shop towel.

4. Lubricate a new seal with the vehicle manufacturer's recommended power steering fluid and install the seal.

5. Install the end plug and retaining ring.

Power Recirculating Ball Steering Gear Worm Shaft Bearing Adjuster Plug Seal and Bearing Replacement

The following is a typical worm shaft bearing adjuster plug seal and bearing service:

1. Remove the adjuster plug locknut and use a special tool to remove the adjuster plug.

2. Use a small pry bar to pry at the raised area of the bearing retainer to remove this retainer from the adjuster plug **(Figure 26-47)**.

3. Place the adjuster plug face down on a suitable support, and use the proper driver to remove the needle bearing, dust seal, and lip seal.

4. Place the adjuster plug outside face up on a suitable support and use the proper driver to install the needle bearing dust seal and lip seal.

5. Install the bearing retainer in the adjuster plug and lubricate the bearing and seal with the vehicle manufacturer's recommended power steering fluid.

6. Install the adjuster plug and locknut and adjust the worm shaft bearing preload as discussed previously.

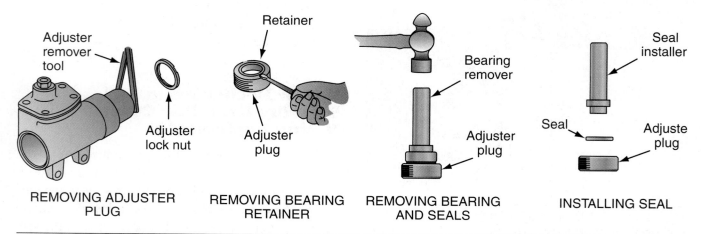

Figure 26-47 Removing and replacing worm shaft adjuster plug, bearing, and seal.

Tech Tip *The bearing identification number must face the driving tool to prevent bearing damage during installation.*

Power Rack-and-Pinion Steering Gear Adjustment

The following is a typical rack bearing adjustment on a power rack-and-pinion steering gear:

1. Use the proper tool to rotate the rack spring cap 12° counterclockwise.
2. Turn the pinion shaft fully in each direction and repeat this action.
3. Loosen the rack spring cap until there is no tension on the rack guide spring.
4. Place the proper turning tool and an inch-pound torque wrench on top of the pinion shaft.
5. Tighten the rack spring cap while rotating the pinion shaft back and forth (**Figure 26-48**). Continue tightening the rack spring cap until the specified turning torque is indicated on the torque wrench.
6. Install and tighten the rack spring cap locknut.

Removing and Replacing Tie-Rod Ends

The following is a typical method to replace the inner tie-rod ends on a power rack-and-pinion steering gear:

1. Use a hammer and chisel to remove the claw washers from the inner tie-rod socket (**Figure 26-49** and **Figure 26-50**).
2. Use a wrench to hold the rack from turning and loosen the inner tie-rod socket with the proper tool (**Figure 26-51**).

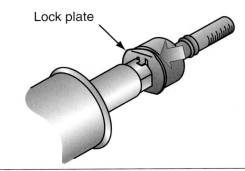

Figure 26-49 A lock plate is used to retain the inner tie-rod.

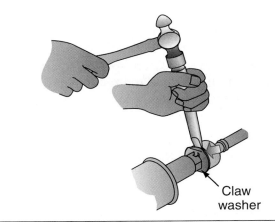

Figure 26-50 Removing claw washers from the inner tie-rod sockets.

Figure 26-48 Adjusting pinion turning torque.

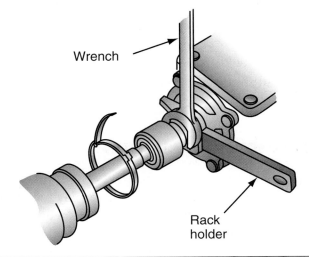

Figure 26-51 Loosening inner tie-rod sockets.

3. Remove the inner tie-rod socket from the rack and install the new inner tie-rod socket and claw washer on the rack.

4. Hold the rack from turning and tighten the inner tie-rod socket to the specified torque.

5. Use a hammer and punch to stake the claw washer on the inner tie-rod socket.

> **Tech Tip** *The inner tie-rod end removal and replacement procedure varies depending on the method of tie-rod attachment to the rack.*

Summary

- With the front wheels straight ahead in a recirculating ball steering gear, an interference fit exists between the ball nut teeth and the sector shaft teeth.

- When the front wheels are turned in, a power recirculating ball steering gear torsion bar deflection moves the spool valve inside the rotary valve. This valve movement directs the power steering fluid to the appropriate side of the recirculating ball piston to provide steering assist.

- Manual or power rack-and-pinion steering systems are lighter and more compact compared to recirculating ball steering gears and parallelogram steering linkages.

- The rack-and-pinion steering gear must be mounted so the rack maintains the tie-rods in a parallel position in relation to the lower control arms.

- The inner tie-rods are connected to the rack through spring-loaded inner ball sockets.

- The rack bearing and adjuster plug maintains proper preload between the rack-and-pinion teeth.

- Steering ratio is the relationship between steering wheel movement and front wheel movement to the right or left.

- A rattling noise and excessive steering free-play may be caused by loose gear mounting bolts,

worn steering shaft U-joints, or worn flexible coupling in a recirculating ball steering gear system.

- A loose worm shaft bearing preload adjustment or sector lash adjustment on a manual or power recirculating ball steering gear causes excessive steering wheel free-play, steering wander, and reduced feel of the road.

- On a manual or power recirculating ball steering gear, the sector lash adjusting screw is tightened until the correct worm shaft turning torque is obtained as the worm shaft is rotated through the center position.

- Because bellows boots protect the inner tie-rod ends from contamination, boot condition should be checked each time undercar service such as oil and filter change or chassis lubrication is performed.

- Bellows boots that are cracked, split, leaking, or deteriorated must be replaced.

- Never straighten bent steering components such as tie-rods.

- Always hold the rack while loosening the inner tie-rod ends.

- The rack adjuster plug must be adjusted until the correct turning torque is obtained on the pinion shaft.

Review Questions

1. While discussing steering wheel free-play, Technician A says steering free-play refers to the amount of steering wheel rotation before the front wheels begin to move. Technician B says excessive steering free-play causes vehicle wander when the vehicle is driven straight ahead. Who is correct?

   A. Technician A
   B. Technician B
   C. Both Technician A and Technician B
   D. Neither Technician A nor Technician B

2. While discussing interference fit between the sector shaft teeth and ball nut teeth in a recirculating ball steering gear, Technician A says the interference fit between the sector shaft teeth and ball nut teeth is present when the front wheels are straight ahead. Technician B says the interference fit between the sector shaft teeth and the ball nut teeth provides the driver with a positive feel of the road. Who is correct?

A. Technician A

B. Technician B

C. Both Technician A and Technician B

D. Neither Technician A nor Technician B

3. While discussing steering ratio, Technician A says steering ratio is the relationship between the degrees of steering wheel rotation and the degrees of front wheel movement to the right or left. Technician B says if the lock-to-lock steering wheel rotation is 2.5 turns and the total front wheel movement is 60°, the steering ratio is 17:1. Who is correct?

A. Technician A

B. Technician B

C. Both Technician A and Technician B

D. Neither Technician A nor Technician B

4. While discussing power rack-and-pinion steering gear operation, Technician A says fluid is directed to the appropriate side of the rack piston by the position of the rotary valve inside the spool valve. Technician B says fluid is directed to the appropriate side of the rack piston by the position of the rack in relation to the pinion. Who is correct?

A. Technician A

B. Technician B

C. Both Technician A and Technician B

D. Neither Technician A nor Technician B

5. Excessive steering wheel free-play may be caused by all of these problems *except*:

A. a worn tie-rod end.

B. a tight rack bearing adjustment.

C. a loose steering gear mounting.

D. a worn flexible coupling.

6. A vehicle with a power recirculating ball steering gear has a complaint of steering wander and reduced road feel. The most likely cause of this problem is:

A. a loose sector lash adjustment.

B. low power steering fluid level.

C. a loose idler arm.

D. a bent pitman arm.

7. A vehicle with a power recirculating ball steering gear requires excessive steering effort. All of these defects may be the cause of this complaint *except*:

A. low power steering pump pressure.

B. a restricted high-pressure power steering hose.

C. under inflated front tires.

D. loose outer tie-rod ends.

8. A power rack-and-pinion steering gear has a fluid leak at the pinion end of the housing and this leak is worse when the steering wheel is turned with the engine running. The most likely cause of this leak is:

A. a worn input shaft seal.

B. a leaking inner rack seal.

C. a leaking outer rack seal.

D. a leaking pinion seal.

9. A power steering gear has an excessive steering effort complaint and the power steering pump pressure is satisfactory. The most likely cause of this problem is:

A. a leaking rack piston seal.

B. a leaking inner rack seal.

C. a leaking outer rack seal.

D. a worn inner tie-rod end.

10. A manual rack-and-pinion steering gear has a loose mounting bushing. The most likely complaint caused by this problem is:

A. reduced steering effort.

B. reduced road feel.

C. excessive steering wheel free-play.

D. a rattling noise.

11. When the steering has free-play, there is some _____ _____ movement before the front wheels start to turn.

12. The worm shaft end plug provides a worm shaft bearing _____ adjustment.

13. During a left turn, the power steering pump forces fluid into the _____ side of the rack chamber and fluid is removed from the _____ side of this chamber.

14. When adjusting a power rack-and-pinion steering gear, the rack _____ _____ is rotated while measuring the pinion _____ _____.

15. Explain the procedure for performing a rack bearing adjustment on a power rack-and-pinion steering gear.

16. Define the term "faster steering."

17. Explain the purpose of the bellows boots in a rack-and-pinion steering gear.

18. Explain how the spool valve is moved inside the rotary valve in a power rack-and-pinion steering gear.

CHAPTER

27 Tires, Wheels, and Hubs

Learning Objectives

After you have read, studied, and practiced the contents of this unit, you should be able to:

- Describe general tire purposes.
- Describe three types of tire ply and belt designs.
- Explain the purpose of the tire performance criteria (TPC) rating.
- Define tire contact area, free tire diameter, and rolling tire diameter.
- Describe the tire motion forces while a tire and wheel assembly is rotating on a vehicle.
- Define wheel tramp and wheel shimmy.
- Describe three different types of bearing loads.
- Explain the advantage of tapered roller bearings over other types of bearings.

Key Terms

Angular load

Aspect ratio

Axial load

Conicity

Direct pressure monitoring

Drop center

Dynamic wheel imbalance

Indirect pressure monitoring

National Highway Traffic Safety Administration (NHTSA)

Radial load

Rim offset

Static imbalance

Thrust load

Tire Performance Criteria (TPC)

Uniform Tire Quality Grading (UTQG)

Wheel shimmy

Wheel tramp

INTRODUCTION

Although tires are often taken for granted, they contribute greatly to the ride and steering quality of a vehicle. Tires also play a significant role in vehicle safety. Improper types of tires, incorrect inflation pressure, and worn out tires create a safety hazard. When tires and wheels are out of balance, tire wear and driver fatigue are increased, which may create a driving hazard. The purposes of tires may be summarized as follows:

1. Tires cushion the vehicle to provide a comfortable ride for the occupants.
2. The vehicle weight must be supported firmly by the tires.
3. Tires must develop traction to drive and steer the vehicle under a wide variety of road conditions.

4. The tires contribute to directional stability of the vehicle and they must absorb all the stresses of accelerating, braking, and centrifugal force in turns.

TIRE DESIGN

Tire construction varies depending on the manufacturer and the type of tire. The three main areas of a tire are the beads, carcass and casing, and the tread **(Figure 27-1)**. A typical modern tire contains many components **(Figure 27-2)**, including: bead wire, bead filler, liner, steel reinforcement in the sidewall, sidewall with hard side compound, rayon carcass plies, steel belts, belt cover, hard

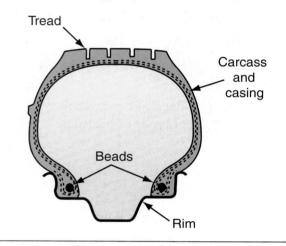

Figure 27-1 Three main sections of a tire.

under-tread compound, and hard high-grip tread compound.

The tire bead contains several turns of bronze-coated steel wire in a continuous loop. This bead is molded into the tire at the inner circumference and is wrapped in the cord plies. The bead anchors the tire to the wheel. The bead filler above the bead reinforces the sidewall and acts as a rim extender.

Tire sidewalls are made from a blend of rubber that absorbs shocks and impacts from road irregularities and prevents damage to the plies. A lettering and numbering arrangement provides for tire identification and is located on the outside of the sidewall **(Figure 27-3)**. The sidewall material contains antioxidants and other chemicals that are gradually released to the surface of the sidewall during the life of the tire. These antioxidants help to keep the sidewall from cracking, and protect it from ultraviolet radiation and ozone attack. Because the sidewall must be flexible to provide ride quality, minimum thickness of this component is essential. Tire manufacturers have reduced sidewall thickness by 40 percent in recent years to reduce weight, heat buildup, and improve ride quality.

Some tires may be found with sidewall reinforcements to improve their performance if the tire loses air pressure. These tires are sometimes referred to as "run flat" tires. They may be driven for a short time and distance without air pressure without serious damage to the tire and rim. They

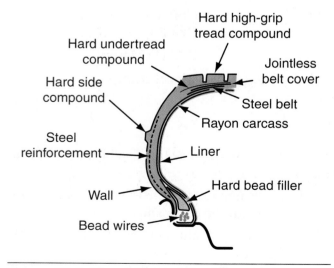

Figure 27-2 Tire designed for improved steering and handling with a nylon bead reinforcement and a hard bead filler with slim, tapered profile.

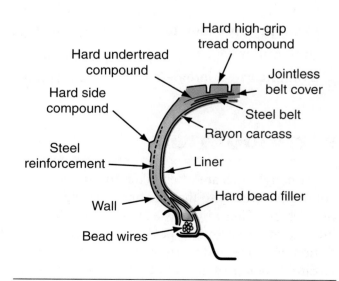

Figure 27-3 Tire sidewalls contain identification information.

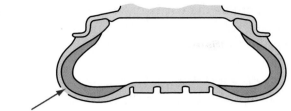

Sidewall reinforcement:
• Flexible low-hysteresis rubber
• Thermal-resistive material
• Metallic and/or textile tissues

Figure 27-4 A tire with sidewall reinforcement for enhanced performance during air pressure loss.

are not meant to run for long periods of time in this fashion, but rather allow the driver to drive the car to a suitable location for changing or repairing the tire **(Figure 27-4)**.

The cord plies surround both beads and extend around the inner surface of the tire to enable the tire to carry its load. The plies are molded into the sidewalls. Each ply is a layer of rubber with parallel cords imbedded in its body. The load capacity of a tire may be increased by adding more cords in each ply or by installing additional plies. The most common materials in tire plies are polyester, rayon, and nylon. Passenger car tires usually have two cord plies, whereas many trucks and recreation vehicles are equipped with six- or eight-ply tires to carry the heavier loads of these vehicles. In general, tires with more plies have stiffer sidewalls, which provide less cushioning and reduced ride quality.

Steel is the most common material in tire belts, although other belt materials such as polyester have been used to some extent. Many tires contain two steel belts. The tire belts restrict ply movement and provide tread stability and resistance to deformation. This belt action provides longer tread wear and reduces heat buildup in the tire. Steel belts expand as wheel speed and tire temperature increase. Centrifugal force and belt expansion tend to tear the tire apart at high speeds and temperatures. Therefore, high-speed tires usually have a nylon belt cover. This nylon belt cover contracts as it is heated and helps to hold the tire together, thereby providing longer tire life, improved stability, and better handling.

Tire treads are made from a blend of rubber compounds that are very resistant to abrasion wear. Spaces between the tire treads allow tire distortion on the road without scrubbing that accelerates wear. Modern automotive tires contain two layers of tread materials. The first tread layer is designed to provide cool operation, low rolling resistance, and durability. The outer layer, or tread, is designed for long life and maximum traction. Tread rubber is a blend of many different synthetic and natural rubbers. Tire manufacturers may use up to thirty different synthetic rubbers and eight natural rubbers in their tires. The manufacturers blend these synthetic and natural rubbers in both tread layers to provide the desired traction and durability. Tire treads must provide traction between the tire and the road surface when the vehicle is accelerating, braking, and cornering. This traction must be maintained as much as possible on a wide variety of road surfaces. For example, on wet pavement, tire treads must be designed to drain off water between the tire and the road surface. This draining action is extremely important to maintain adequate acceleration, braking, and directional control. Lines cut across the tread provide a wiping action that helps to dry the tire/road contact area.

The synthetic gum rubber liner is bonded to the inner surface of the tire to seal the tire. Nearly all passenger car and light truck tires are tubeless-type. In these tires, the tire bead must provide an airtight seal on the rim and both the tire and the wheel rim must be completely sealed. Some heavy-duty truck tires have inner tubes mounted inside the tire. On tube-type tires, the air is sealed in the inner tube and the sealing qualities of the tire and wheel rim are not important. Designing tires is a very complex engineering operation and the average all-season tire contains these components by weight:

- Synthetic rubber—5.5 lb. (2.49 kg)
- Carbon black—5 lb. (2.27 kg)
- Natural rubber—4.5 lb. (2.04 kg)
- Chemicals, waxes, oils, and pigments—3 lb. (1.36 kg)
- Steel cord for belts—1.5 lb. (0.68 kg)

- Polyester and nylon—1 lb. (0.45 kg)
- Bead wire—0.5 lb. (0.23 kg)
- **Total weight**—21 lb. (9.52 kg)

TIRE PLY AND BELT DESIGN

The most commonly used tire designs are bias, belted bias, and belted radial. In bias-ply or belted bias-ply tires, the cords crisscross each other. These cords are usually at an angle of 25° to 45° to the tire centerline. The belt ply cord angle is usually 5° less than the cord angle in the tire casing. Two plies and two belts are most commonly used, but four plies and four belts may be used in some tires. Compared to a bias-ply tire, a belted bias-ply tire has greater tread rigidity. The belts reduce tread motion during road contact. This action provides extended tread life compared to a bias-ply tire.

In radial tires, the ply cords are arranged radially at a right angle to the tire centerline **(Figure 27-5)**. Steel belts are the most common in radial tires, but other belt materials such as fiberglass, nylon, and rayon have been used. The steel or fiberglass cords in the belts are crisscrossed at an angle of 10° to 30° in relation to the tire centerline. Many radial tires have two plies and two belts. Radial tires provide less rolling resistance, improved steering characteristics, and longer tread life compared to bias-ply tires.

Regardless of the type of tire construction, the tire must be uniform in diameter and width. Radial runout refers to variations in tire diameter. A tire with excessive radial runout causes a tire-thumping problem as the car is driven. When a tire has excessive variations in width, this condition is called lateral runout. A tire with excessive lateral runout causes the chassis to "waddle" when the car is driven.

The tire plies and belts must be level across the tread area. If the plies and/or belts are not level across the tread area, the tire is cone-shaped. This condition is referred to as tire **conicity**. When a front tire has conicity, the steering may pull to one side as the car is driven straight ahead. A rear tire with conicity will not affect the steering as much as a front tire with conicity **(Figure 27-6)**.

> **Tech Tip** *A defective tire or a tire worn by incorrect camber settings can still cause a vehicle to pull even after the camber problem has been corrected.*

SAFETY TIP *If different sizes or types of tires are combined on the same axle or front to rear on a vehicle, handling and braking could be seriously affected. This action may result in vehicle damage and/or personal injury. If a vehicle is equipped with an ABS, installing different size tires than recommended by the vehicle manufacturer may result in braking defects and cause vehicle damage and/or personal injury. If replacement tires have any ratings of lower value than the original tires, a safety hazard may be created, which could result in vehicle damage and/or personal injury.*

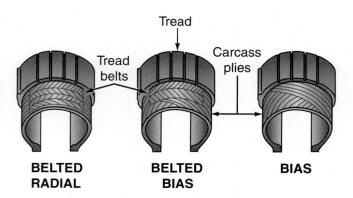

Figure 27-5 Three types of tire construction: belted radial-ply, belted bias-ply, and bias-ply.

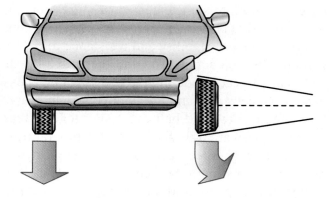

Figure 27-6 Tire conicity can cause a vehicle to pull to one side.

TIRE RATINGS

A great deal of important information is molded into the sidewall of the average passenger car or light-truck tire. The tire rating is part of the information located on the sidewall. The tire rating is a group of letters and numbers that identify the tire type, section width, aspect ratio, construction type, rim diameter, load capacity, and speed symbol. For example, when a tire has a P215/65R15 89H rating on the sidewall, the P indicates a passenger car tire **(Figure 27-7)**. The number 215 is the size of the tire in millimeters measured from sidewall to sidewall with the tire mounted on the recommended rim width.

The number 65 indicates the **aspect ratio**, which is the ratio of the height to the width. With a 65 aspect ratio, the tire's height is 65 percent of its width. The letter R indicates a radial-ply tire design.

The number 15 is the rim diameter in inches. The number 89 represents the load index. This load rating indicates the tire has a load capacity of 1,279 pounds. Various numbers represent different maximum loads.

Some tire manufacturers use the letters B, C, or D to indicate the load rating. The letter B indicates the lowest load rating and the letter C represents a higher load rating. A tire with a D load rating is designed for light-duty trucks.

Speed Ratings

Many tires sold in the United States are speed rated with various letters that indicate the maximum

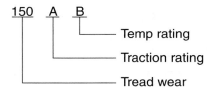

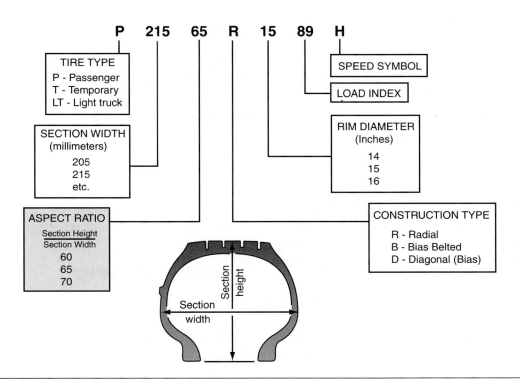

Figure 27-7 Tire sidewall information.

Speed Symbols						
Speed Symbol	**Speed (km/h)**	**Speed (mph)**		**Speed Symbol**	**Speed (km/h)**	**Speed (mph)**
A1	5	3		K	110	68
A2	10	6		L	120	75
A3	15	9		M	130	81
A4	20	12		N	140	87
A5	25	16		P	150	94
A6	30	19		Q	160	100
A7	35	22		R	170	106
A8	40	25		S	180	112
B	50	31		T	190	118
C	60	37		U	200	124
D	65	40		H	210	130
E	70	43		V*	above 210	above 130
F	80	50		V	240	149
G	90	56		W	270	168
J	100	62		Y	300	186
					above 300	above 186

*For Unlimited V tires without the Service Description, the speed category is over 210 km/h (130 mph).

Figure 27-8 Tire speed ratings.

speed capabilities of the tire **(Figure 27-8)**. The letter designation for the speed rating is included on the sidewall markings. Although tires may be speed rated, tire manufacturers do not endorse the operation of a vehicle in an unlawful or unsafe manner. Speed ratings are based on laboratory tests and these ratings are not valid if tires are worn out, damaged, altered, under-inflated, or overloaded. Tire speed ratings do not suggest that vehicles can be driven safely at the designated speed rating because many different road and weather conditions may be encountered. Also, the condition of the vehicle may affect high-speed operation.

Tread Wear Rating

Some other tire ratings available from the manufacturers include tread wear, traction, and temperature resistance **(Figure 27-9)**. Tread wear ratings allow consumers to compare tire life expectancies. When tires are tread-wear rated, they are installed on a vehicle and driven on a test

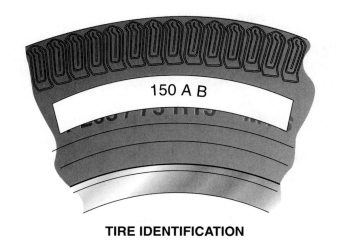

TIRE IDENTIFICATION

Figure 27-9 A tire sidewall showing the tread wear, traction, and temperature rating for the tire.

course for 7,200 miles (11,587 kilometers). Tread wear is calculated after this test course run and the mileage to tread wear out is determined from this calculation. The projected mileage is adjusted for test condition variations and is compared to 30,000 miles (48,279 kilometers) on the test course. This calculation provides a number, or ratio, to compare various tires. For example, a tire with a 150 tread wear rating will provide 50 percent more tire wear mileage than a tire with a 100 tread wear rating. To obtain maximum tire tread life, most vehicle manufacturers recommend tire rotation at specific mileage intervals.

Traction Rating

Traction ratings indicate the braking capabilities of the tire to the consumer. To determine the traction rating, ten skid tests are completed on wetted asphalt and concrete surfaces. Test conditions are carefully controlled to maintain uniformity. The results of the ten skid tests are averaged and the traction rating is designated AA, A, B, or C, with an A rating having the best traction and C having the lowest traction.

Temperature Rating

Temperature resistance ratings indicate the tire's ability to withstand heat generated during tire operation. The National Highway Traffic Safety Administration (NHTSA) has established controlled procedures on a laboratory test wheel for temperature resistance testing of tires. The tire's temperature rating indicates how long the tire can last on the test wheel. Temperature ratings are A, B, or C. An A rating has the best temperature resistance. Tires must have a minimum letter C temperature rating to meet **National Highway Traffic Safety Administration (NHTSA)** standards in the United States.

Uniform Tire Quality Grading (UTQG) and Department of Transportation (DOT) Designations

Department of Transportation (DOT) requirements specify that tires must have the **Uniform Tire Quality Grading (UTQG)** system designations molded into the sidewall. The UTQG system designations include the tread wear, traction, and temperature ratings. Typical UTQG designations are Tread wear 160, Traction B, and Temperature B.

Some tire manufacturers use a DOT designation and indicates that the tire has met specific quality tests approved by the DOT. Federal law in the United States requires tire manufacturers to place these designations on tire sidewalls:

- Size
- Load range
- Maximum load
- Maximum pressure in pounds per square inch (psi)
- Number of plies under the tread and in the sidewalls
- Manufacturer's name
- Tubeless or tube construction
- Radial construction (if a radial tire)
- DOT approval number which includes manufacturer's code number, size, type, date of manufacture, and tubeless or tube type

TPC Number

A **Tire Performance Criteria (TPC)** specification number is molded into the sidewall near the

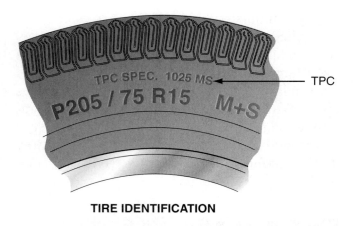

TIRE IDENTIFICATION

Figure 27-10 Tire performance criteria (TPC) number.

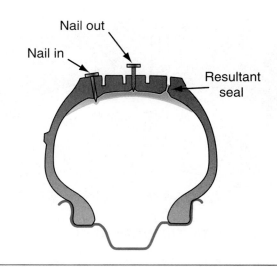

Figure 27-11 A puncture-sealing tire.

tire rating **(Figure 27-10)**. The TPC number assures that the tire meets the car manufacturer's performance standards for traction, endurance, dimensions, noise, handling, and rolling resistance. Most car manufacturers assign a different TPC number to each tire size. When replacement tires are selected, these tires should have the same size, load capacity, and construction as the original tires. The replacement tires should have the same TPC number to assure that these tires meet the same performance standards as the original tires.

ALL-SEASON AND SPECIALTY TIRES

Many tires sold at present are classified as all-season radial tires. These tires have 37 percent higher average snow traction than non-all-season tires. All-season tires may have slightly improved performance in areas such as wet traction, rolling resistance, tread life, and air retention. Improvements in tread design and tread compounds provide the superior qualities in all-season tires. These all-season tires are identified by an M+S suffix after the TPC number.

Puncture Resistant Tires

Puncture-sealing tires are available as an option on certain car lines and some rubber companies sell these tires in the replacement tire market. These tires contain a special rubber-sealing compound applied under the tread area during the

manufacturing process. When a nail or other object up to 3/16 inch in diameter punctures the tread area, it picks up a coating of sealant. If the object is removed, the sealant sticks to the object and is pulled into the puncture. This sealant completely fills the puncture and forms a permanent seal to maintain tire inflation pressure **(Figure 27-11)**. Puncture-sealing tires usually have a special warranty and these tires can be serviced with conventional tire changing and balancing equipment.

Mud and Snow Tires

Mud and snow tires are available in various ply and belt designs. These tires provide increased traction in snow or mud compared to conventional tires. Mud and snow tires are identified with an MS suffix after the TPC number on the tire sidewall. When snow tires are installed on a vehicle, these tires should be the same size and type as the other tires on the vehicle. In areas where snow is encountered, all-season tires have replaced snow tires to a large extent. Studded tires provide improved traction on ice, but laws in many states may prohibit these tires because their use resulted in road surface damage. Many states that allow studded tires have strict guidelines as to their usage for certain times of the year.

Run-Flat Tires

Some luxury and sport-type vehicles are equipped with run-flat tires. Run-flat tires eliminate the need

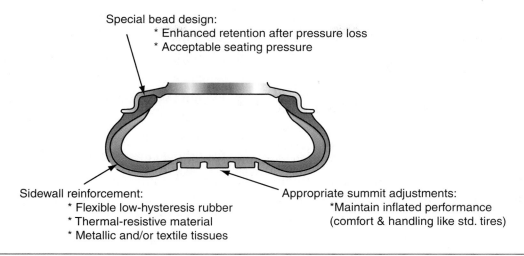

Special bead design:
* Enhanced retention after pressure loss
* Acceptable seating pressure

Sidewall reinforcement:
* Flexible low-hysteresis rubber
* Thermal-resistive material
* Metallic and/or textile tissues

Appropriate summit adjustments:
*Maintain inflated performance
(comfort & handling like std. tires)

Figure 27-12 A run-flat tire with sidewall reinforcement.

for a spare tire and a jack on these cars. This provides a weight and space savings. Run-flat tires must minimize the difference between run-flat tires and conventional tires and provide sufficient zero-pressure durability so the vehicle can be driven a reasonable distance to a repair facility. Some run-flat tires have stiffer sidewalls that partially support the vehicle weight without air pressure in the tire **(Figure 27-12)**. Other run-flat tires have a flexible rubber support ring mounted on a special rim to support the vehicle weight if deflation occurs **(Figure 27-13)**.

REPLACEMENT TIRES

Most tires have tread wear indicators built into the tread. When the tread wears a specific amount, the wear indicators appear as bands across the tread **(Figure 27-14)**. Some car manufacturers recommend tire replacement when the wear indicators appear in two or more tread grooves at three locations around the tire.

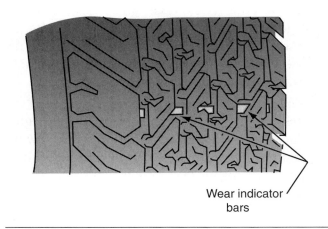

Wear indicator bars

Figure 27-14 Tread wear indicators.

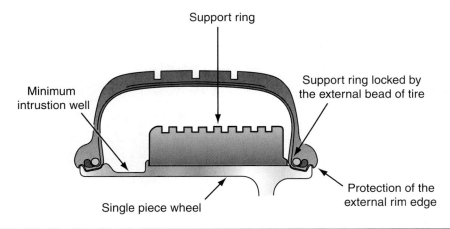

Support ring

Minimum intrustion well

Support ring locked by the external bead of tire

Single piece wheel

Protection of the external rim edge

Figure 27-13 A run-flat tire with support ring.

If replacement tires have a different size or construction type than the original tires, vehicle handling, ride quality, and speedometer/odometer calibration may be seriously affected. When replacement tires are a different size than the original tires, the vehicle ground clearance and tire-to-body clearance may be altered. Steering and braking quality may be seriously affected if different sizes or types of tires are installed on a vehicle. The temporary spare tire can also affect the vehicle handling and the driver must keep this in mind when the use of the temporary spare is required. Many vehicles manufactured in recent years are equipped with ABS. When different-sized tires are installed on these vehicles, the ABS operation can be compromised, which may result in serious braking defects.

When selecting replacement tires, these precautions must be observed to maintain vehicle safety:

1. Replacement tires must be installed in pairs on the same axle. Never mix tire sizes or designs on the same axle. If it is necessary to replace only one tire, it should be paired with the tire having the most tread to equalize braking traction.

2. The tire load rating must be adequate for the vehicle on which the tire is installed. Light-duty trucks, station wagons, and trailer-towing vehicles are examples of vehicles that require tires with higher load ratings compared to passenger car tires.

3. Snow tires should be the same type and size as the other tires on the vehicle.

4. Four-wheel drive vehicles should have the same type and size of tires on all four wheels.

5. Do not install tires with a load rating less than the car manufacturer's recommended rating.

6. Replacement tire ratings should be equivalent to the original tire ratings in all rating designations.

7. When combining different tires front to rear, check the car manufacturer's or tire manufacturer's recommendations.

TIRE VALVES

The tire valve allows air to flow into the tire and it is also used to release air from the tire. The core in the center of the valve is spring-loaded and allows air to flow inward while the tire is inflated **(Figure 27-15)**. This valve is very similar to the Schrader valve used in the air conditioning and fuel systems. Once the tire is inflated, the valve core seats and prevents airflow out of the tire. The small pin on the outer end of the valve core may be pushed to unseat the valve core and release air from the tire. An airtight cap on the outer end of the valve keeps dirt out of the valve and provides an extra seal against air leakage. A deep groove is cut around the inner end of the tire valve. When the valve assembly is pulled into the wheel opening, this groove seals the valve in the opening.

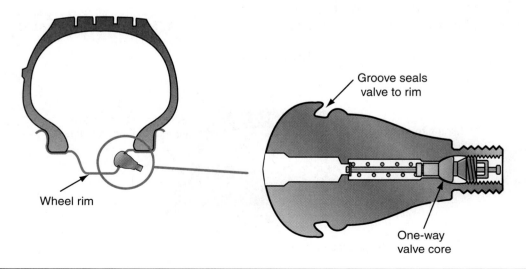

Figure 27-15 A tire valve.

COMPACT SPARE TIRES

Since cars have been downsized in recent years, space and weight have become major concerns for vehicle manufacturers. For this reason, many car manufacturers have marketed cars with compact spare tires to provide a weight and space savings. The high-pressure mini spare tire is the most common type of compact spare **(Figure 27-16)**. This compact spare rim is usually 4 inches wide. The compact spare rim should not be used with standard tires, snow tires, wheel covers, or trim rings. Any of these uses may result in damage to these items or other parts of the vehicle. The compact spare should be used only on vehicles that offered it as original equipment. Inflation pressure in the compact spare should be maintained at 60 psi (415 kPa) or at the manufacturer's recommendation. The compact spare tire is designed for temporary use until the conventional tire can be repaired or replaced. The driver should limit driving speed to 50 mph (80 kph) and distance to 50 miles (80 km) when the high-pressure mini spare is installed on a vehicle.

The space-saver spare tire must be inflated with a special compressor. Battery voltage is supplied to the compressor from the cigarette lighter. This type of compact spare should be inflated to 35 psi (240 kPa). After the tire is inflated, be sure there are no folds in the sidewalls. The lightweight skin spare tire is a bias-ply tire with a reduced tread depth to provide an estimated 2,000 miles (3,200 km) of tread life. This type of spare tire is also designed for emergency use only. Driving speed must be limited to 50 mph (80 kph) when this tire is installed on a vehicle. Always inflate the lightweight skin spare to the pressure specified on the tire placard.

***Did You Know?** The "50/50 rule" stated that the tire should not be driven for more than 50 miles at 50 MPH to minimize wear to the differential should the tire be installed on the driving axle and to minimize the risk of running a smaller tire than normal.*

TIRE CONTACT AREA

The tire contact area refers to the area of the tire that is in contact with the road surface when the tire is supporting the vehicle weight. The free diameter of a tire is the distance of a horizontal line through the center of the spindle and wheel to the outer edges of the tread. The rolling diameter of a tire is the distance of a perpendicular straight line through the center of the spindle to the outer edges of the tread when the tire is supporting the vehicle weight. The rolling diameter is always less than the free diameter. The difference between the free diameter and the rolling diameter is referred to as deflection. Tire tread grooves take up excess rubber and prevent scrubbing as the tire deflects in the contact area **(Figure 27-17)**.

- Temporary use only
- Inflate to 60 psi

Figure 27-16 A compact, high-pressure mini spare tire.

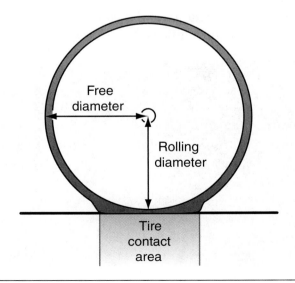

Figure 27-17 Tire rolling diameter, free diameter, and contact area.

TIRE PLACARD AND INFLATION PRESSURE

The correct air pressure exerted evenly against the entire interior tire surface, which produces tension in the tire carcass, supports the vehicle weight. Therefore, tire pressure is extremely important. Under-inflation decreases the rolling diameter and increases the contact area, which results in excessive sidewall flexing and tread wear. Over-inflation decreases the contact area, increases the rolling diameter, and stiffens the tire. This action results in excessive wear on the center of the tread. Tire pressure should be checked when the tires are cool. Since tire pressure normally increases at high tire temperatures, air pressure should not be released from hot tires. Excessive heat buildup in a tire may be caused by under-inflation. This condition may lead to severe tire damage.

On many vehicles, the tire placard is permanently attached to the rear face of the driver's door **(Figure 27-18)**. This placard provides tire information such as maximum vehicle load, tire size including spare, and cold inflation pressure including spare **(Figure 27-19)**. The vehicle manufacturer carefully calculates tire pressure to provide satisfactory tread life, handling, ride, and load-carrying capacity. Most vehicle manufacturers recommend that tire pressures be checked cold once a month or prior to any extended trip. The manufacturer considers the tires to be cold after the vehicle has sat for three hours or when the vehicle has been driven less than one mile. The tires should be inflated to the pressure indicated

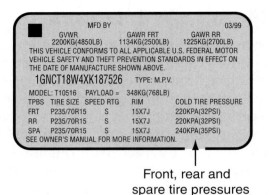

Front, rear and
spare tire pressures

Figure 27-19 A tire placard lists tire inflation data.

on the tire placard. Tire pressures may be listed in metric or USC system values.

> **Tech Tip** *Tire pressure changes about 1 pound for every 10° of temperature change.*

TIRE PRESSURE MONITORING SYSTEMS

The U.S. government has passed legislation that requires all passenger cars and light trucks under 10,000 pounds of gross vehicle weight to eventually be equipped with tire pressure monitoring systems. The main purpose of these systems is to warn the driver if their tires are losing air pressure, leaving the tires under-inflated and dangerous.

Direct Pressure Monitoring

Direct pressure monitoring systems have a pressure sensor strapped to the drop center in each rim **(Figure 27-20)**. **Indirect pressure monitoring** systems have a pressure sensor threaded onto the end of the valve stem.

Direct systems will have a pressure sensor/ transmitter inside the tire. Many systems attach their air pressure sensor/ transmitter to a special tire valve that clamps into the wheel. Because rubber valves are used for most wheels that do not have a direct system, the presence of a metal clamp-in valve typically identifies the presence of a direct tire pressure monitoring system. The transmitter's

Figure 27-18 A tire placard is usually located on the door pillar.

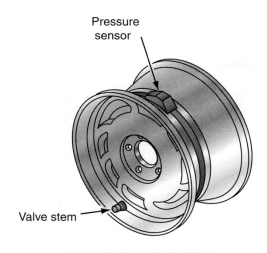

Location	Color code
Right front	Blue
Left front	Green
Right rear	Orange
Left rear	Yellow

WARNING: Pressure sensor inside tire. Avoid contacting sensor with tire changing equipment tools or tire bead.

Service note: Pressure sensor must be mounted directly across from valve stem.

Figure 27-20 A direct tire pressure sensor.

signal is broadcast to the in-car receiver and the information is displayed to the driver.

The pressure sensors send radio frequency (RF) signals to the module in the tire pressure monitoring system. These RF signals change if the tire is deflated a specific amount. When the module senses a tire with low air pressure, the module illuminates a warning light in the instrument panel.

Because direct systems have a sensor in each wheel, they usually generate more accurate warnings and can alert the driver quickly if the pressure in any one tire falls below a predetermined level due to air loss caused by a puncture. In addition, a gradual air loss over time can also be detected and the driver warned of under-inflated tires.

Indirect Pressure Monitoring

Indirect tire pressure monitoring systems use the wheel speed sensor signals in the ABS to monitor tire inflation pressure. Therefore, the wheel speed sensor signals may be used to indicate low tire pressure.

A lower-cost solution is the use of the indirect tire pressure monitoring system. This was developed by vehicle manufacturers wishing to comply with the law while minimizing development time and cost. Indirect systems use the vehicle's antilock braking system's wheel speed sensors to compare the rotational speed of one tire versus the other three tires on the vehicle. When a tire is deflated to some extent, the tire diameter is smaller and wheel speed increases. Reading the same signals used to control ABS systems, the vehicle's onboard computer can warn the driver when a single tire is running at a reduced inflation pressure compared to the other three.

Unfortunately, indirect tire pressure monitoring systems have several disadvantages. Indirect systems cannot tell the driver which tire is low on pressure, and cannot warn the driver if all four tires are losing pressure at the same rate. Some indirect systems cannot operate reliably at highway speeds or during extremes in ambient temperature. Indirect systems can generate false warnings. An example of a condition that can cause a false warning is when the vehicle is driven around a long curve that causes the outside tires to rotate faster than the inside tires, or when the tires spin on ice and snow-covered roads. False warnings could lead the driver to ignore the tire pressure monitoring system's warnings.

RIMS

Rims are manufactured from stamped or pressed steel discs that are riveted or welded together to form the circular rim. Some rims are cast aluminum alloy wheel rims or cast magnesium alloy wheel rims. Sometimes these wheels are referred to as "mag" wheels. These wheels are lighter and generally more accurately designed than pressed steel wheel rims.

The **rim offset** is the distance between the rim centerline and the mounting face of the disc. If a rim is designed with positive offset, it has the mounting face outboard of the rim centerline

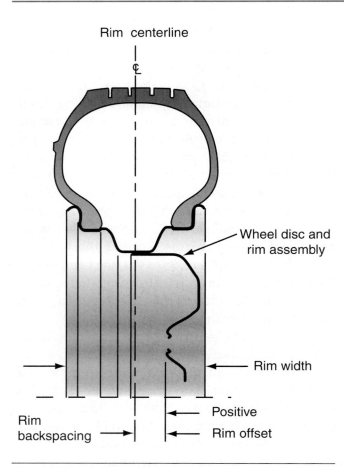

Figure 27-21 Wheel rim design.

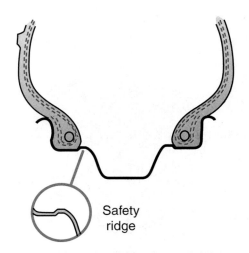

Figure 27-22 A wheel rim with drop center and safety ridges.

(Figure 27-21). A rim with negative offset has the mounting face inboard of the rim centerline. The rim offset affects front suspension loading and operation.

The width of the wheel is measured between the rim flanges. Rim diameter is determined by measuring across the wheel from the top to the bottom. Rim backspacing is the inside depth of the rim. It is measured from the backside of the rim to the mounting surface of the rim.

A large hole in the center of the rim fits over a flange on the mounting surface and the rim has a small hole for the valve stem. The wheel stud mounting holes in the rims are tapered to match the taper on the wheel nuts.

A **drop center** in the rim makes tire changing easier **(Figure 27-22)**. Rims have safety ridges behind the tire bead locations, which help to prevent the beads from moving into the drop center area if the tire blows out. If the tire blows out and a bead enters the drop center area, the tire may come off the

wheel. Replacement wheel rims must be the same as the original equipment wheels in load capacity, offset, width, diameter, and mounting configuration. An incorrect wheel can affect tire life, steering quality, wheel bearing life, vehicle ground clearance, tire clearance, and speedometer/odometer calibrations.

STATIC WHEEL BALANCE THEORY

When a wheel and tire have proper static balance, the weight is equally distributed around the axis of rotation and gravity will not force the wheel to rotate from its rest position. If a vehicle is raised off the floor and a wheel is rotated in 120-degree intervals, a statically balanced wheel will remain stationary at each interval. When wheel and tire are statically imbalanced, the tire has a heavy portion at one location. The force of gravity acting on this heavy portion will cause the wheel to rotate when the heavy portion is located near the top of the tire **(Figure 27-23)**.

Results of Static Imbalance

Centrifugal force may be defined as the force that tends to move a rotating mass away from its axis of rotation. As we have explained previously, a tire and wheel are subjected to very strong acceleration and deceleration forces when a vehicle is in motion. The heavy portion of a statically imbalanced wheel is influenced by centrifugal force. This

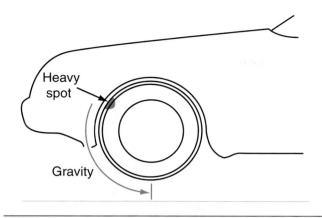

Figure 27-23 Static wheel imbalance.

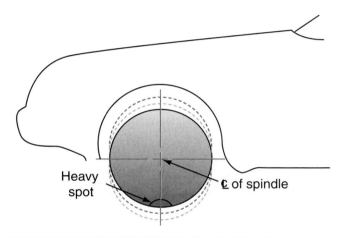

Figure 27-24 Effects of static imbalance, this causes wheel tramp.

influence attempts to move the heavy spot on a tangent line away from the wheel axis. This action tends to lift the wheel assembly off the road surface. The wheel lifting action caused by **static imbalance** may be referred to as **wheel tramp (Figure 27-24)**. This wheel tramp action allows the tire to slip momentarily when it is lifted vertically.

When the wheel and tire move downward as the heavy spot decelerates, the tire strikes the road surface with a pounding action. This repeated slipping and pounding action causes severe tire scuffing and cupping **(Figure 27-25)**. The vertical wheel motion from static imbalance is transferred to the suspension system and is then absorbed by the chassis and body. This action causes rapid wear on suspension and steering components. The wheel tramp action resulting from static imbalance is also transmitted to the passenger compartment and can cause passenger discomfort and driver fatigue.

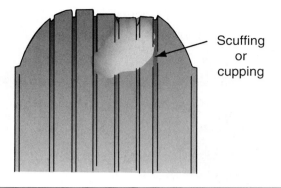

Figure 27-25 Cupping tire wear caused by static imbalance.

When a vehicle is traveling at normal highway cruising speed, the average wheel speed would be 850 revolutions per minute (rpm). A statically imbalanced tire and wheel assembly is an uncontrolled mass of weight in motion. When a vehicle is traveling at 60 miles per hour (97 kilometers per hour), if a tire has 2 ounces (57 grams) of static imbalance, the resultant pounding force is approximately 15 pounds (6.8 kilograms) against the road surface.

DYNAMIC WHEEL BALANCE THEORY

When a wheel and tire assembly have correct dynamic balance, the weight of the assembly is distributed equally on both sides of the wheel center viewed from the front. **Dynamic wheel balance** may be explained by dividing the tire into four sections **(Figure 27-26)**.

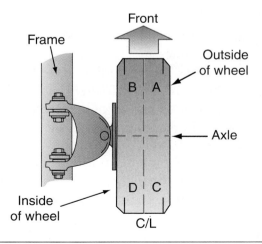

Figure 27-26 Dynamic wheel balance theory.

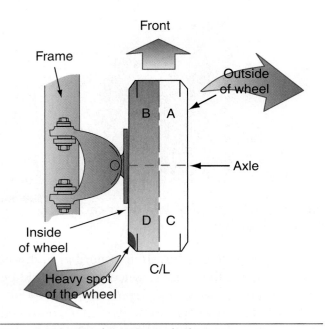

Figure 27-27 Dynamic imbalance.

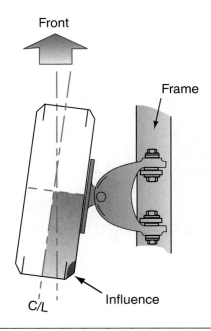

Figure 27-28 Dynamic wheel imbalance with heavy spot at the rear of the left front wheel.

In Figure 27-27, if sections A and C have the same weight and sections B and D also have the same weight, the tire has proper dynamic balance. If a tire has dynamic imbalance, section D may have a heavy spot, and thus sections B and D have different weights (Figure 27-27).

From our discussion of dynamic balance, we can understand that a tire and wheel assembly may be in static balance but have dynamic imbalance. Therefore, wheels must be in balance statically and dynamically.

Results of Dynamic Wheel Imbalance

When a dynamically imbalanced wheel is rotating, centrifugal force moves the heavy spot toward the tire centerline. The centerline of the heavy spot arc is at a 90-degree angle to the spindle. This action turns the true centerline of the left front wheel inward when the heavy spot is at the rear of the wheel (Figure 27-28).

When the wheel rotates until the heavy spot is at the front of the wheel, the heavy spot movement turns the left front wheel outward (Figure 27-29).

From these explanations we can understand that dynamic wheel imbalance causes **wheel shimmy** (Figure 27-30). This action causes steering wheel

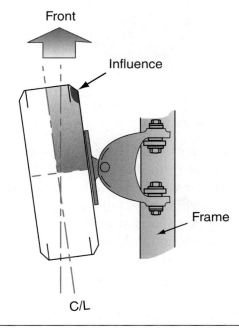

Figure 27-29 Dynamic wheel imbalance with heavy spot at the front of the left front wheel.

oscillations at medium and high speeds with resultant driver fatigue and passenger discomfort. Wheel shimmy and steering wheel oscillations also cause unstable directional control of the vehicle. Wheel shimmy is the rapid, repeated lateral wheel movement.

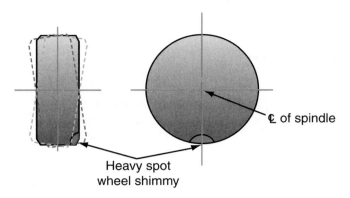

Figure 27-30 Dynamic wheel imbalance causes wheel shimmy.

In a previous discussion of wheel rotation in this chapter, we mentioned that a tire stops momentarily where it contacts the road surface. A wheel with dynamic imbalance is forced to pivot on the contact area, which results in excessive tire scuffing and wear. Dynamic wheel imbalance causes premature wear on steering linkage and suspension components. Therefore, dynamic wheel balance is extremely important to provide normal tire life, reduce steering and suspension component wear, increase directional control, and decrease driver fatigue. The main purposes of proper wheel balance are to maintain normal tire tread life, provide extended life of suspension and steering components, help to provide directional control of the vehicle, reduce driver fatigue, increase passenger comfort, and help to maintain the life of body and chassis components.

WHEEL BEARINGS

Bearings are precision-machined assemblies that provide smooth operation and long life. When bearings are properly installed and maintained, bearing failure is rare. When you understand different bearing loads, various types of wheel bearings, and the loads these bearings are designed to withstand, diagnosing wheel bearing problems becomes much easier.

NOTE: If a wheel does not have the same width, diameter, offset, load capacity, and mounting configuration as the original wheel,

steering quality, vehicle control, tire life, and wheel bearing life may be adversely affected. An incorrect wheel may create a safety hazard and cause vehicle damage and/or personal injury.

Bearing Loads

When a bearing load is applied perpendicular to the axis or shaft, it is called a **radial load**. If the vehicle weight is applied upward on a bearing, this weight is a radial load on the bearing. An **axial load** is applied in the same direction as the axis or shaft **(Figure 27-31)**. Axial loads are sometimes referred to as **thrust loads**. For example, while a vehicle is turning a corner, horizontal force is applied to the front wheel bearings. This is an axial load. When **angular load** is applied to a bearing, the angle of the applied load is somewhere between the radial and axial and is actually a combination of the radial and axial loads.

Front and rear wheel bearings may be cylindrical ball bearings or roller bearings. Either type of bearing contains these basic parts: an inner race, or cone; a separator, also called a cage or retainer; rolling elements, balls or rollers; and an outer race, or cup.

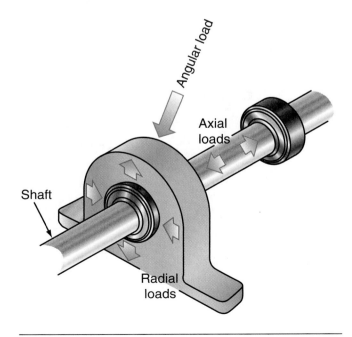

Figure 27-31 Types of bearing loads.

The inner race is an accurately machined component and the inner surface of the race is mounted on the shaft with a precision fit. The rolling elements are mounted on a smooth machined surface on the inner race. Positioned between the inner and outer races, the separator retains the rolling elements and keeps them evenly spaced. The rolling elements have precision-machined surfaces. These elements are mounted between the inner and outer races. The outer race is the bearing's exterior ring and both sides of this component have precision-machined surfaces. The outer surface of this race supports the bearing in the housing and the inner surface is in contact with the rolling elements.

A single-row ball bearing has a crescent-shaped machined surface in the inner and outer races in which the balls are mounted **(Figure 27-32)**. When a ball bearing is at rest, the load is distributed equally through the balls and races in the contact area. When one of the races and the balls begin to rotate, the bearing load causes the metal in the race to bulge out in front of the ball and flatten out behind the ball **(Figure 27-33)**. This action creates a small amount of friction within the bearing and the same action is repeated for each ball while the bearing is rotating. If metal-to-metal contact were allowed between the balls and races, these components would experience faster wear. Therefore, bearing lubrication is extremely important to

Figure 27-33 When a load is applied to a ball bearing, the metal in the race bulges out in front of the ball and flattens out behind the ball.

eliminate metal-to-metal contact in the bearing and thus reduce wear.

A cylindrical ball bearing is designed primarily to handle radial loads. However, this type of bearing can also withstand a considerable amount of thrust load in either direction even at high speeds. A maximum capacity ball bearing has extra balls for greater radial load-carrying capacity. Ball bearings are available in many different sizes for various applications.

Double-row ball bearings contain two rows of balls side by side. As in the single-row ball bearing, the balls in the double-row bearing are mounted in crescent-shaped grooves in the inner and outer races. The double-row ball bearing can support heavy radial loads and this type of bearing can also withstand thrust loads in either direction ball **(Figure 27-34)**.

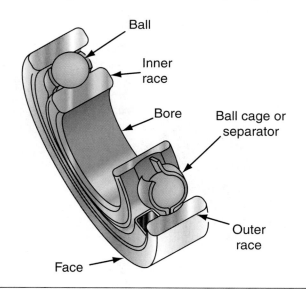

Figure 27-32 Parts of a cylindrical ball bearing.

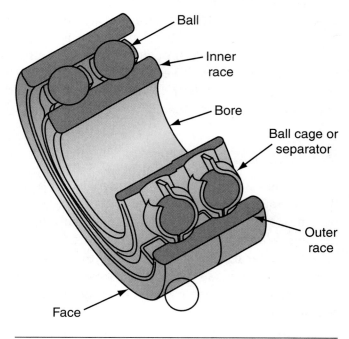

Figure 27-34 Parts of a double roller ball bearing.

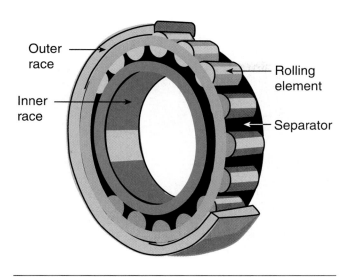

Figure 27-35 Parts of a cylindrical roller bearing.

Cylindrical Roller Bearings

A cylindrical roller bearing contains precision-machined rollers that have the same diameter at both ends. These rollers are mounted in square-cut grooves in the outer and inner races **(Figure 27-35)**. In the cylindrical roller bearing, the races and rollers run parallel to one another. Cylindrical roller bearings are designed primarily to carry radial loads, but they can withstand some thrust load.

Tapered Roller Bearings

In a tapered roller bearing, the inner and outer races are cone-shaped. If imaginary lines extend through the inner and outer races, these lines taper and eventually meet at a point extended through the center of the bearing **(Figure 27-36)**.

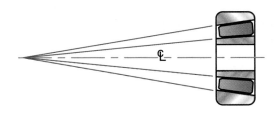

Figure 27-36 Imaginary lines extending from the tapered roller bearing races eventually meet at a point extending from the bearing center.

Figure 27-37 Tapered roller bearings.

The most important advantage of the tapered roller bearing compared to other bearings is an excellent capability to carry radial, axial, and angular loads, especially when used in pairs. In the tapered roller bearing, the rollers are mounted on cone-shaped precision surfaces in the outer and inner races. The bearing separator has an open space over each roller **(Figure 27-37)**. Grooves cut in the side of the separator roller openings match the curvature of the roller. This design allows the rollers to rotate evenly without interference between the rollers and the separator. Lubrication and proper adjustment are critical on tapered roller bearings.

WHEEL BEARING SEALS

Seals are designed to keep lubricant in the bearing and prevent dirt particles and contaminants from entering the bearing. Wheel bearing seals are mounted in front and rear wheel hubs and in rear-axle housings on rear-wheel drive cars. The metal seal case has a surface coating that resists corrosion and rust and acts as a bonding agent for the seal material. Seals have many different designs including single lip, double lip, and fluted. The seal material is usually made of a synthetic rubber compound such as nitrile, silicon, polyacrylate, or other elastomers. The actual seal material depends on the lubricant and contaminants that the seal encounters.

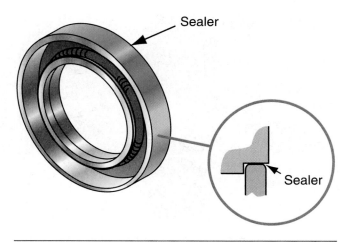

Figure 27-38 Sealer painted on the seal case prevents leaks between the case and the housing.

Some seals have a sealer painted on the outside surface of the metal seal housing. When the seal is installed, this sealer prevents leaks between the seal case and the housing (**Figure 27-38**). Some rear axle bearings are an example of a bearing with seals attached to each side of the bearing (**Figure 27-39**).

If a seal must direct oil back into a housing, the seal lip is fluted. This seal design provides a pumping action to redirect the oil back into the housing (**Figure 27-40**).

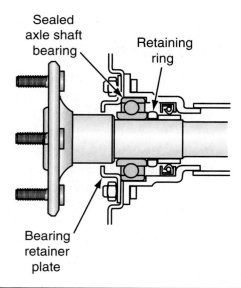

Figure 27-39 Some rear axle bearings are sealed on both sides and retained on the axle with a retainer ring.

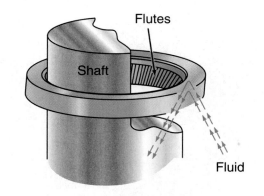

Figure 27-40 Fluted seal lip redirects oil back into the housing.

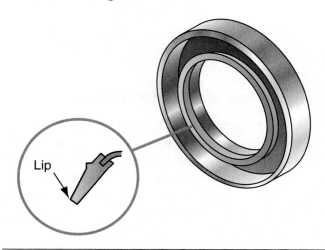

Figure 27-41 A spring-less seal.

Spring-Less Seals

All seals may be divided into two groups, spring-less and spring-loaded. Spring-less seals are used in some front- or rear-wheel hubs where they seal a heavy lubricant into the hub (**Figure 27-41**).

Spring-Loaded Seals

In a spring-loaded seal, the garter spring behind the seal provides additional force on the seal lip to compensate for lip wear, shaft movement, and bore eccentricity (**Figure 27-42**).

WHEEL BEARING HUB ASSEMBLIES

Some front-wheel drive vehicles have a type of bearing containing two rows of ball bearings with an angular contact angle of 32° (**Figure 27-43**).

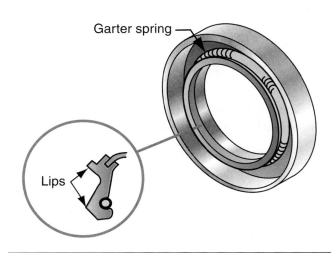

Figure 27-42 A spring-loaded seal.

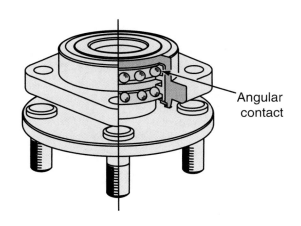

Figure 27-43 A double-row, sealed wheel bearing hub unit.

These front-wheel bearing and hub assemblies are bolted to the steering knuckles **(Figure 27-44)**. The bearings are lubricated and sealed and the complete bearing and hub assembly are replaced as a unit. The bearing and hub unit are more compact than other types of wheel bearings mounted in the wheel hub.

The inner bearing assembly bore is splined and the inner ring extends to the outside to form a flange and spigot. The flange attached to the outer ring contains boltholes. Bolts extend through these holes into the steering knuckle. This type of bearing attachment allows the bearing to become a structural member of the front suspension.

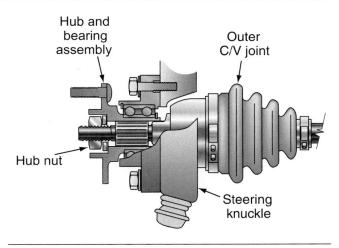

Figure 27-44 Wheel bearing and hub assembly.

Since the bearing outer ring is self-supporting, the main concern in knuckle design is fatigue strength rather than stiffness. The drive axle shaft transmits torque to the inner bearing race. This shaft is not designed to hold the bearing together. This type of wheel bearing is designed for mid-sized front-wheel drive (FWD) vehicles.

Each front drive axle has splines that fit into matching splines inside the bearing hubs **(Figure 27-45)**. A hub nut secures the drive axle into the inner bearing race.

Front Steering Knuckles with Two Separate Tapered Roller Bearings

Some front-wheel drive vehicles have a one-piece, dual-tapered roller bearing assembly mounted in the steering knuckles **(Figure 27-46)**.

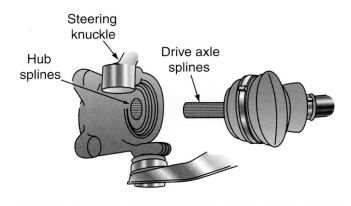

Figure 27-45 A front drive axle installed in a wheel-bearing hub.

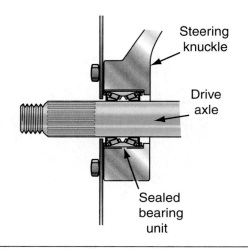

Figure 27-46 A steering knuckle with a pressed-in bearing.

Other front-wheel drive vehicles have separate tapered roller bearings with individual races pressed into the steering knuckle and seals are located in the knuckle on the outboard side of each bearing **(Figure 27-47)**. The hub nut torque supplies correct bearing endplay adjustment. The wheel hub is pressed into the inner bearing races and the drive axle splines are meshed with matching splines in the wheel hub.

Wheel Hubs with Two Separate Tapered Roller Bearings

Many rear-wheel drive cars have two tapered roller bearings in the front hubs that support the hubs and wheels on the spindles. This type of front wheel bearing has the bearing races pressed into the hub. A grease seal is pressed into the inner end of the hub to prevent grease leaks and to keep contaminants out of the bearings. The hub and bearing assemblies are retained on the spindle with a washer, adjusting nut, nut lock, and cotter pin. The adjusting nut must be properly adjusted to provide the correct bearing endplay. A grease cap is pressed into the hub. Some front-wheel drive cars have two tapered roller bearings in the rear-wheel hubs that are very similar to the front wheel bearings in a rear-wheel drive vehicle **(Figure 27-48)**.

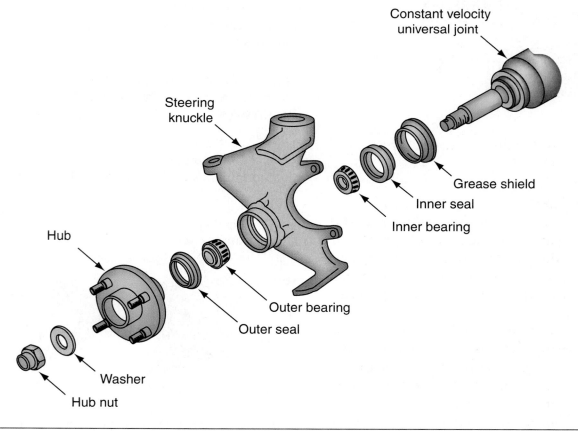

Figure 27-47 A steering knuckle with two separate tapered roller bearings.

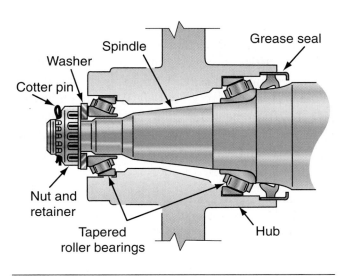

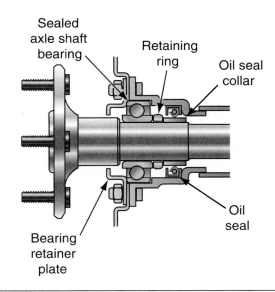

Figure 27-48 A front-wheel bearing assembly, rear-wheel drive car.

Figure 27-50 Rear axle bearing and retainer.

REAR AXLE BEARINGS

On many rear-wheel drive vehicles, the rear axles are supported by roller bearings mounted near the outer ends of the axle housing. The outer bearing race is pressed into the housing. A machined surface on the axle contacts the inner roller surface. A seal is mounted in the axle housing on the outboard side of each bearing **(Figure 27-49)**. This type of axle bearing is usually not sealed and lubricant in the differential and rear axle housing provides axle-bearing lubrication. The seals prevent lubricant leaks from the outer ends of the axle housing and keep dirt out of the bearings. Other rear axle bearings on rear-wheel drive vehicles

have sealed roller bearings pressed onto the rear axles. These axle bearings are sealed on both sides, and a retainer ring is pressed onto the axle on the inboard side of the bearing **(Figure 27-50)**. The outer bearing race is mounted in the rear axle housing with a light press fit. A seal is positioned in the housing on the inboard side of the bearing and adaptor ring. A retainer plate is mounted between the bearing and the outer end of the axle. This retainer plate is bolted to the axle housing to retain the axle in place.

BEARING LUBRICATION

Proper bearing lubrication is extremely important to maintain bearing life. Bearing lubricant reduces friction and wear, dissipates heat, and protects surfaces from dirt and corrosion. Sealed or shielded bearings are lubricated during the manufacturing process. No attempt should be made to wash these bearings or pack them with grease. Bearings that are not sealed or shielded require cleaning and repacking at intervals specified by the vehicle manufacturer. Always use the bearing grease specified by the vehicle manufacturer. Bearing lubricants may be classified as greases or oils. Many wheel-bearing greases are lithium or sodium based.

New bearings usually have a protective coating to prevent rust and corrosion. This coating

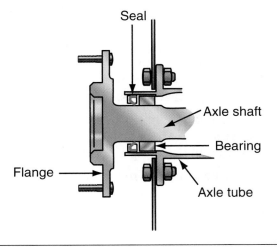

Figure 27-49 A rear axle bearing, rear-wheel drive car.

should not be washed from the bearing. When rear axle bearings are lubricated from the differential housing, the type and level of oil in the housing is important.

Vehicle manufacturers usually recommend an SAE 75W-90 or SAE 140 hypoid gear oil in the differential. In very cold climates, the manufacturer may recommend SAE 80 differential gear oil. The API classifies gear lubricants as GL-1, GL-2, GL-3, GL-4, and GL-5. The GL-4 lubricant is used for hypoid gears under normal conditions. The GL-5 lubricant is used in heavy-duty hypoid gears. Always use the vehicle manufacturer's specified differential gear oil. The differential should be filled until the lubricant is level with the bottom of the filler plug opening in the differential housing. If the differential is overfilled, excessive lubricant may be present at the bearings and seals. Under this condition, the lubricant may leak past the seal. When the lubricant level is low in the differential, the lubricant may not be available in the axle housings. When this condition exists, the bearings do not receive enough lubrication and bearing life is shortened.

> **Tech Tip** *If a bearing is operated without proper lubrication, bearing life will be very short.*

Summary

- Tires are extremely important because they provide ride quality, support the vehicle weight, provide traction for the drive wheels, and contribute to steering quality and directional stability.

- Tires may be bias-ply, belted bias-ply, or belted radial-ply.

- The TPC number assures that the tire meets the car manufacturer's performance standards for traction, endurance, dimensions, noise, handling, and rolling resistance.

- The UTQG designation includes tread wear, traction, and temperature ratings.

- The tire placard provides valuable information regarding the tires on the vehicle.

- To maintain vehicle safety, wheel rims must have the same width, diameter, offset, load capacity, and mounting configuration as the original rims.

- Static wheel imbalance causes wheel tramp and severe tire cupping.

- Dynamic wheel imbalance causes wheel shimmy, increased tire wear, unstable directional control, driver fatigue, and increased wear on suspension and steering components.

- A bearing reduces friction, carries a load, and guides certain components such as pivots, shafts, and wheels.

- Radial bearing loads are applied in a vertical direction.

- Thrust bearing loads are applied in a horizontal direction.

- A cylindrical ball bearing, or roller bearing, is designed primarily to withstand radial loads, but these bearings can handle a considerable thrust load.

- Bearing seals keep lubricant in the bearing and prevent dirt from entering the bearing.

- Tapered roller bearings have excellent radial, thrust, and angular load-carrying capabilities.

- Bearing hub units are compact compared to the previous bearings in the wheel hub. This compactness makes hub bearing units suitable for front-wheel drive cars.

- Some bearing hub units are bolted to the steering knuckle and other bearing hub units are pressed into the steering knuckle.

- Rear axle bearings are mounted between the drive axles and the housing on rear-wheel drive cars.

Review Questions

1. While discussing tire design and operation, Technician A says a belted radial-ply tire provides improved steering characteristics compared to a belted bias-ply tire. Technician B says a belted radial-ply tire provides longer tread life than a belted bias-ply tire. Who is correct?

   A. Technician A

   B. Technician B

   C. Both Technician A and Technician B

   D. Neither Technician A nor Technician B

2. While discussing tires in motion, Technician A says when a vehicle is traveling at normal cruising speed, centrifugal and belt expansion forces tend to tear the tire apart. Technician B says steel belts expand as wheel speed and tire temperature increase. Who is correct?

   A. Technician A

   B. Technician B

   C. Both Technician A and Technician B

   D. Neither Technician A nor Technician B

3. While discussing wheel balance, Technician A says static imbalance causes wear on the center of the tire tread. Technician B says static imbalance causes cupped tire wear. Who is correct?

   A. Technician A

   B. Technician B

   C. Both Technician A and Technician B

   D. Neither Technician A nor Technician B

4. While discussing wheel balance, Technician A says dynamic wheel imbalance causes lateral wheel shimmy. Technician B says dynamic wheel imbalance causes vertical wheel tramp. Who is correct?

   A. Technician A

   B. Technician B

   C. Both Technician A and Technician B

   D. Neither Technician A nor Technician B

5. All these statements about tire design are true *except*:

   A. The tire sidewalls are made from a blend of rubber that absorbs shocks and impacts from road irregularities.

   B. Antioxidants in the sidewall material help to prevent sidewall cracking.

   C. Increased sidewall thickness improves ride quality.

   D. The bead filler above the bead reinforces the tire sidewall and acts as a rim extender.

6. When manufacturing tires, a typical tire may contain:

   A. up to five different synthetic rubbers and one type of natural rubber.

   B. up to eight different synthetic rubbers and two different natural rubbers.

   C. up to fourteen different synthetic rubbers and three different natural rubbers.

   D. up to thirty different synthetic rubbers and eight different natural rubbers.

7. While discussing run-flat tires, Technician A says some run-flat tires have stiffer sidewalls. Technician B says some run-flat tires have a flexible rubber support ring mounted on special rims. Who is correct?

   A. Technician A

   B. Technician B

   C. Both Technician A and Technician B

   D. Neither Technician A nor Technician B

8. While discussing tapered roller bearings, Technician A says a tapered roller bearing can withstand high radial, thrust, and angular loads. Technician B says lubrication and proper end-play adjustment are critical on tapered roller bearings. Who is correct?

   A. Technician A

   B. Technician B

   C. Both Technician A and Technician B

   D. Neither Technician A nor Technician B

9. A front-wheel bearing hub assembly

   A. is not serviceable.

   B. is more compact compared to bearings in the wheel hub.

   C. is bolted to the steering arm.

   D. may contain two rows of ball bearings.

10. All of these statements about roller bearing-type rear axle bearings in rear-wheel drive vehicles are true *except*:

    A. The outer bearing race is pressed into the rear axle housing.

    B. The bearing seal is on the inboard side of the wheel bearing.

    C. The bearing rollers contact a machined surface on the rear axle.

    D. Lubricant in the differential provides rear axle-bearing lubrication.

11. To calculate the tire aspect ratio, the tire section height is divided by the _____.

12. Car manufacturers recommend that tire inflation pressures should be checked when the tires are _____.

13. The rim offset is the vertical distance between the rim centerline and the _____ _____ of the disc.

14. A spring-less seal may be used to seal a _____ lubricant into a hub.

15. Define a radial bearing load.

16. Define a thrust-bearing load.

17. Describe the advantage of a bearing hub unit compared to the previous bearings mounted in the wheel hub.

18. Explain the purpose of the wheel rim drop center and safety ridges.

CHAPTER

28

Tires, Wheels, and Hubs: Maintenance, Diagnosis, and Service

Learning Objectives

After you have read, studied, and practiced the contents of this unit, you should be able to:

- Diagnose steering pull problems related to tire condition.

- Rotate tires according to the vehicle manufacturer's recommended procedure.

- Demount, inspect, repair, and remount tires.

- Diagnose problems caused by excessive radial, lateral, wheel, or tire runout.

- Perform off-car static and dynamic wheel balance procedures.

- Diagnose tire wear problems caused by tire and wheel imbalance.

- Clean, repack, reassemble, and adjust wheel bearings.

- Diagnose wheel bearings on the vehicle.

- Remove and replace rear axle bearings on rear-wheel drive cars.

Key Terms

Dynamic wheel balance

Lateral tire runout

Match mount

Radial runout

Static wheel balancing

INTRODUCTION

Proper servicing of tires and wheels is extremely important to maintain vehicle safety and provide normal tire life. Improperly serviced and/or balanced tires and wheels cause wheel vibration and shimmy problems, resulting in excessive tire tread wear, increased wear on suspension and steering components, and decreased vehicle stability and steering control.

TIRE MAINTENANCE

Proper tire maintenance is essential to provide normal tire life and maintain vehicle safety. When tires are under-inflated, tire operating temperatures can increase, tire life is shortened, and worn-out tires are a safety concern. A lack of tire rotation can also reduce tire life.

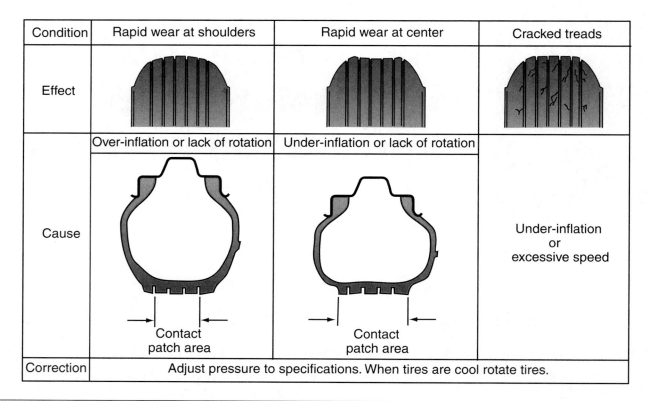

Condition	Rapid wear at shoulders	Rapid wear at center	Cracked treads
Effect			
Cause	Over-inflation or lack of rotation — Contact patch area	Under-inflation or lack of rotation — Contact patch area	Under-inflation or excessive speed
Correction	Adjust pressure to specifications. When tires are cool rotate tires.		

Figure 28-1 Tire tread wear caused by under-inflation and over-inflation.

Tire Inflation Pressure

A tire depends on correct inflation pressure to maintain its correct shape and support the vehicle weight. Excessive inflation pressure causes excessive center tread wear **(Figure 28-1)**, hard ride, and damage to the tire carcass.

When tires are under-inflated, these problems may be evident:

- Excessive wear on each side of the tread
- Hard steering, wheel damage
- Excessive heat buildup in the tire which can lead to tire failure
- Possible severe tire damage with resultant hazardous driving

Tread Wear Measurement

On most tires, the tread wear indicators appear as wide bands across the tread when tread depth is worn to 1/16 inch (1.6 millimeters). Most tire manufacturers recommend tire replacement when the wear indicators appear across two or more tread grooves at any three locations around the tire **(Figure 28-2)**. If tires do not have wear indicators, a

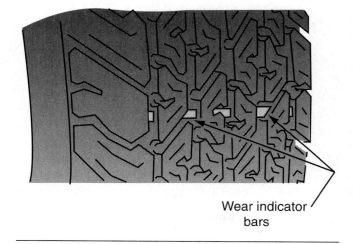

Wear indicator bars

Figure 28-2 Tire tread wear indicators.

tread depth gauge may be used to measure the tread depth **(Figure 28-3)**. The tread depth gauge reads in 32nds of an inch, and tires with 2/32 tread depth or less should be replaced **(Figure 28-4)**.

Tire Rotation

To a large extent, driving habits determine tire life. Severe brake applications, high-speed driving, turning at high speeds, rapid acceleration and

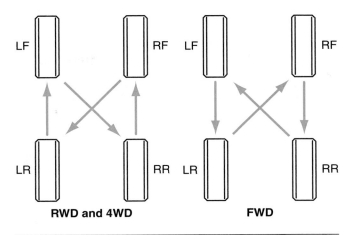

Figure 28-5 Tire rotation procedure.

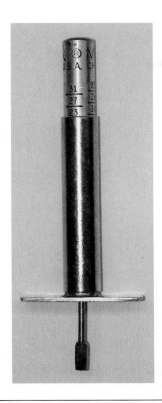

Figure 28-3 Tread depth gauge. These are marked in 1/32nds of an inch or in millimeters.

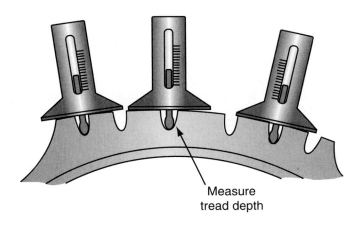

Measure tread depth

Figure 28-4 Measure the tread in several places across the tread.

deceleration, and striking curbs are just a few driving habits that shorten tire life. Most car manufacturers recommend tire rotation at specified intervals to obtain maximum tire life. The exact tire rotation procedure depends on the model year, the type of tires, and whether the vehicle has a conventional spare or a compact spare. Tire rotation procedures do not include the compact spare. The vehicle

manufacturer provides tire rotation information in the owner's manual and service manual.

Vehicle manufacturers usually recommend different tire rotation procedures for bias-ply tires and radial tires **(Figure 28-5)**:

- On front-wheel drive vehicles, the tires should be moved from the front straight to the rear (LF to LR and RF to RR). The rear tires should be crossed to the front (LR to RF and RR to LF).

- On rear-wheel drive, front-wheel driver, and all-wheel drive vehicles, the front tires should be crossed to the rear (LF to RR and RF to LR) and the rear tires should be moved straight to the front (LR to LF and RR to RF).

> **Tech Tip** *A rule of thumb to follow is that the main driving axle tires are moved straight to the opposite axle and the other axles tires are crossed to the main driving axle.*

When tires and wheels are installed on a vehicle, it is very important that the wheel nuts are torqued to the manufacturer's specifications in the proper sequence **(Figure 28-6)**. Do not use an impact wrench when tightening wheel nuts to the specified torque.

TIRE DIAGNOSIS

Uneven tread surfaces may cause tire noises that seem to originate elsewhere in the vehicle. These noises may be confused with differential noise. Differential noise usually varies with acceleration

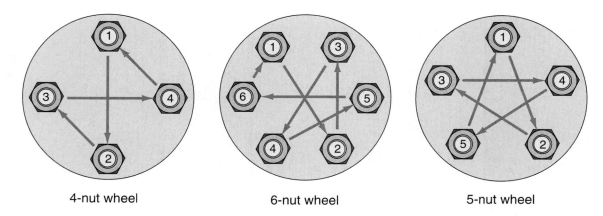

4-nut wheel   6-nut wheel   5-nut wheel

Figure 28-6 Wheel nut tightening sequence.

and deceleration, while tire noise remains more constant in relation to these forces. Tire noise is most pronounced on smooth asphalt road surfaces at speeds of 15 to 45 miles per hour (24 to 72 kilometers per hour).

> **Tech Tip** *Tire noise varies with road surface conditions, whereas differential noise is not affected when various road surfaces are encountered.*

When tire thump and vibration is present, check for cupped tire treads; excessive tire radial runout; manufacturing defects such as heavy spots, weak spots, or tread chunking; and incorrect wheel balance.

A vehicle should maintain the straight-ahead forward direction on smooth, straight road surfaces without excessive steering wheel correction by the driver. If the steering gradually pulls to one side on a smooth, straight road surface, a tire, steering, or suspension defect is present. Tires of different types, sizes, designs, or inflation pressures on opposite sides of a vehicle cause steering pull.

Sometimes a tire-manufacturing defect occurs in which the belts are wound off-center on the tire. This condition is referred to as conicity **(Figure 28-7)**. A cone-shaped object rolls in the direction of its smaller diameter. Similarly, a tire with conicity tends to lead or pull to one side, which causes the vehicle to follow the action of the tire. Tire conicity cannot be diagnosed by a visual inspection. Tire conicity is

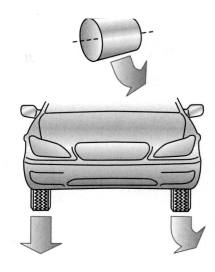

Figure 28-7 Tire conicity can cause a vehicle to pull to one side.

diagnosed by switching the two front tires and reversing the front and rear tires **(Figure 28-8)**. Incorrect front suspension alignment angles also cause steering pull.

WHEEL AND TIRE SERVICE

Proper wheel and tire service are extremely important to maintain vehicle safety and provide nor-

> **Tech Tip** *Tire conicity is not visible and can be diagnosed only by changing the tire and wheel position.*

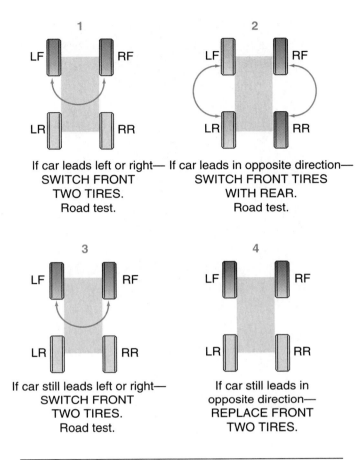

Figure 28-8 Tire conicity diagnosis.

1. Remove the wheel cover. If the vehicle is equipped with anti-theft locking wheel covers, the lock bolt for each wheel cover is located behind the ornament in the center of the wheel cover. A special key wrench is supplied to the owner for ornament and lock bolt removal. If the customer's key wrench has been lost, a master key is available from the vehicle dealer.

2. Loosen the wheel lug nuts about one-half turn, but do not remove the wheel nuts. Some vehicles are equipped with anti-theft wheel nuts. A special lug nut key is supplied to the vehicle owner. This lug nut key has a hex nut on the outer end and a special internal projection that fits in the wheel nut opening. Install the lug nut key on the lug nuts and connect the lug nut wrench on the key hex nut to loosen the lug nuts.

3. Raise the vehicle on a lift or with a floor jack to a convenient working level.

4. Remove the lug nuts and remove the wheel and tire assembly. If the wheel is rusted and will not come off, hit the inside of the wheel with a large rubber mallet. Do not hit the wheel with a steel hammer because this action could damage the wheel. Do not heat the wheel.

Tech Tip *A defective tire or a tire worn by incorrect camber settings can still cause vehicle pull even after the camber problem has been corrected.*

---

mal tire life with adequate driver and passenger comfort. Improperly balanced tires and wheels greatly reduce tire life and the vibrations from this problem result in driver discomfort and fatigue.

Wheel and Tire Removal

When it is necessary to remove a wheel and tire assembly, follow these steps:

SAFETY TIP *Before the vehicle is raised on a lift, be sure the lift is contacting the car manufacturer's recommended lift points. If the vehicle is lifted with a floor jack, place safety stands under the suspension or frame and lower the vehicle onto the safety stands. Then remove the floor jack from under the vehicle.*

Tech Tip *Some wheel covers have fake plastic lug nuts that must be removed to access the lug nuts. Be careful not to break the fake lug nuts.*

> **Tech Tip** *If heat is used to loosen a rusted wheel, the wheel and/or wheel bearings may be damaged.*

Tire and Wheel Service Precautions

There are many different types of tire-changing equipment in the automotive service industry. However, specific precautions apply to the use of any tire-changing equipment:

1. Before you operate any tire-changing equipment, always be absolutely certain that you are familiar with the operation of the equipment.

2. When operating tire-changing equipment, always follow the equipment manufacturer's recommended procedure.

3. Always deflate a tire completely before attempting to demount the tire.

4. The bead seats on the wheel rim must be clean before the tire is mounted on the wheel rim.

5. The outer surface of the tire beads should be lubricated with rubber lubricant prior to mounting the tire on the wheel rim.

6. When the tire is mounted on the wheel rim, be sure the tire is positioned evenly on the wheel rim.

7. While inflating a tire, do not stand directly over the tire. An air hose extension allows you to stand back from the tire during the inflation process.

8. Do not over-inflate tires.

9. When mounting tires on cast aluminum alloy wheel rims or cast magnesium alloy wheel rims, always use the tire changing equipment manufacturer's recommended tools and procedures.

Tire Demounting

Always use a tire changer to demount tires. Rim clamp machines are the most common type of tire changers used **(Figure 28-9)**. Do not use hand tools or tire irons for this purpose. Always follow

Figure 28-9 A rim clamp tire changer.

the manufacturer's procedure. The following is a typical tire demounting procedure:

1. Remove the valve core and be sure the tire is completely deflated.

2. Place the wheel and tire on the side of the tire. Follow the operating procedure recommended by the manufacturer of the tire changer to force the tire bead inward and separate it from the rim on both sides **(Figure 28-10)**. Turn the wheel around and repeat the loosening procedure on the other side of the wheel. This should be the long side of the drop center.

3. Apply approved tire lubricant to the bead area of the tire.

4. Place the wheel on the platform of the tire changer and clamp the wheel to the changer **(Figure 28-11)**.

5. Move the swing arm into position. Pull the locking handle forward to release the slide. Push down on the top of the vertical slide to move the head into contact with the rim edge.

Figure 28-10 Separating the tire bead from the rim.

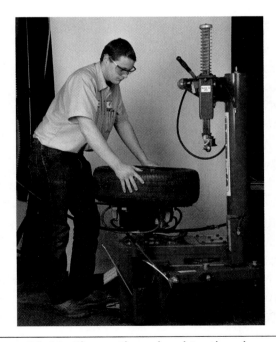

Figure 28-11 Clamp the wheel to the changer.

Push the locking handle back to lock the slide into place and the head should move upward approximately 1/8 inch from the rim edge. The mount/demount head roller should be in contact with the rim edge. Turn the arm-adjusting knob to move the roller away from the rim approximately 1/8 inch **(Figure 28-12)**.

Figure 28-12 Adjusting the head for proper clearance.

6. Push one edge of the top bead into the drop center of the rim.

7. Place the tire changer's bar or lever between the bead and the rim on the opposite side of the rim from where the bead is in the drop center **(Figure 28-13)**.

Figure 28-13 Operating the changer to remove the tire from the rim.

Figure 28-14 Inside bead is being removed from the rim.

Figure 28-15 Inspect and clean the rim before mounting the tire.

8. Operate the tire changer to rotate the bar or lever and move the bead over the top of the rim.

9. Repeat steps 4, 5, and 6 to move the lower bead over the top of the rim **(Figure 28-14)**.

> **Tech Tip** *If hand tools or tire irons are used to demount tires, tire bead and wheel rim damage may occur.*

Tire Mounting

The following is a typical tire mounting procedure.

1. Inspect the wheel closely for corrosion and damage. Clean the wheel and remove any light corrosion or rubber residue **(Figure 28-15)**. Do not attempt to use corroded or damaged rims.

2. Inspect the tire for damage. Closely inspect the beads. Make sure the tire and wheel are the correct size.

3. Lubricate tire beads and rim with approved tire lubricant.

4. Secure the wheel rim on the tire changer with the narrow bead ledge facing upward.

5. Place the tire over the wheel and move the swing arm into position. Position the tire so that the lower bead is above the rear extension of the head and below the front of the head.

6. Rotate the wheel to mount the lower bead. Use the drop center of the wheel to reduce the force on the bead by pressing down on the tire directly across from the head. Rotate table until lower bead is fully mounted.

7. For top bead, rotate the rim until the valve stem is directly across from the mount head. Lift the upper bead up and over the rear of the mount head. Press down on the tire between the mount head and the valve stem to hold the tire in the drop center. Rotate the tire until the bead is mounted **(Figure 28-16)**.

8. Follow the tire changer's manufacturer's procedure for seating the beads.

Figure 28-16 Mounting the tire on rim.

Figure 28-17 Check the air pressure after the beads have been seated.

9. Inflate the tire to the proper pressure. Inspect the tire stem valve for leakage **(Figure 28-17)**.

SAFETY TIP *When a bead expander is installed around the tire, never exceed 10-psi (69 kPa) pressure in the tire. A higher pressure may cause the expander to break and fly off the tire, causing serious personal injury or property damage. When a bead expander is not used, never exceed 40-psi (276 kPa) tire pressure to move the tire beads out tightly against the wheel rim. A higher pressure may blow the tire bead against the rim with excessive force. This action could burst the rim or tire, resulting in serious personal injury or property damage. While inflating a tire, do not stand directly over a tire. In this position, serious injury could occur if the tire or wheel rim flies apart.*

Tire Inspection and Repair

To find a leak in a tire and wheel, inflate the tire to the pressure marked on the sidewall and then submerge the tire and wheel in a tank of water. An alternate method of leak detection is to sponge soapy water on the tire and wheel. Bubbles will appear wherever the leak is located in the tire or wheel. Mark the leak location in the tire or wheel rim with a crayon and mark the tire at the valve stem location so the tire can be reinstalled in the same position on the wheel.

Tech Tip *The tire should be reinstalled in the same position on the wheel as the valve stem is typically the low spot of the rim. The tire should have been mounted so the high spot of the tire corresponds with the low spot of the rim. Marking the tire helps maintain this optimum mounting.*

Tech Tip *The tire will need to be balanced after the tire leak is repaired.*

A puncture is the most common cause of a tire leak. Many punctures can be repaired satisfactorily. Do not attempt to repair punctures over 1/4-inch diameter. Punctures in the sidewalls or on the tire

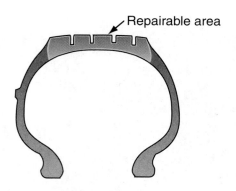

Figure 28-18 Repairable area on bias-ply and belted bias-ply tires.

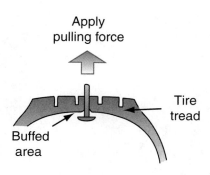

Figure 28-19 Plug installation procedure.

shoulders should not be repaired. The repairable area in belted bias-ply tires is approximately the width of the belts **(Figure 28-18)**. The belts in radial tires are wider compared to bias-ply tires and the repairable area in radial tires is also the width of the belts. Since compact spare tires have thin treads, do not attempt to repair these tires.

Inspect the tire and do not repair a tire with any of these defects, signs of damage, or excessive wear: tires with the wear indicators showing, tires worn until the fabric or belts are exposed, bulges or blisters, ply separation, broken or cracked beads, or cuts or cracks anywhere in the tire Since most vehicles are equipped with tubeless tires, we will discuss this type of tire repair. If the cause of the puncture, such as a nail, is still in the tire, remove it from the tire. Most punctures can be repaired from inside the tire with a service plug or vulcanized patch service kit. Some tire manufacturers may recommend both the plug and cold patch method as the plug helps to seal the belts inside of the tire and the cold patch helps to maintain the integrity of the liner. The instructions from the tire service kit manufacturer should be followed, but we will discuss three common tire repair procedures.

Plug Installation Procedure

The following is a typical plug installation procedure.

1. Buff the area around the puncture with a wire brush or wire buffing wheel.
2. Select a plug slightly larger than the puncture opening and insert the plug in the eye of the insertion tool.

3. Wet the plug and the insertion tool with vulcanizing fluid.
4. While holding and stretching the plug, pull the plug into the puncture from the inside of the tire **(Figure 28-19)**. The head of the plug should contact the inside of the tire. If the plug pulls through the tire, repeat the procedure.
5. Cut the plug off 1/32 inch from the tread surface. Do not stretch the plug while cutting.

Cold Patch Installation Procedure

The following is a typical cold patch installation procedure.

1. Buff the area around the puncture with a wire brush or buffing wheel.
2. Apply vulcanizing fluid to the buffed area and allow it to dry until it is tacky.
3. Peel the backing from the patch and apply the patch over the puncture. Center the patch over the puncture.
4. Run a stitching tool back and forth over the patch to improve bonding.

Tech Tip *Radial tire patches should have arrows that must be positioned parallel to the radial plies.*

Hot Patch Installation Procedure

The area around the puncture must be buffed with a wire brush or buffing wheel. Many hot patches require the application of vulcanizing fluid on the buffed area. Peel the backing from the patch and install the patch so it is centered over the puncture on the inside of the tire. Many hot patches are heated with an electric heating element clamped over the patch. This element should be clamped in place for the amount of time recommended by the equipment or patch manufacturer. After the heating element is removed, allow the patch to cool for a few minutes and be sure the patch is properly bonded to the tire.

WHEEL RIM SERVICE

Steel rims should be spray cleaned with a water hose. Aluminum or magnesium wheel rims should be cleaned with a mild soap and water solution and rinsed with clean water. The use of abrasive cleaners, alkaline-base detergents, or caustic agents may damage aluminum or magnesium wheel rims. Clean the rim bead seats on these wheel rims thoroughly with the mild soap and water solution. The rim bead seats on steel wheel rims should be cleaned with a wire brush or coarse steel wool.

SAFETY TIP *Steel wheel rims must not be welded, heated, or peened with a ball peen hammer. These procedures may weaken the rim and create a safety hazard. Installing an inner tube to correct leaks in a tubeless tire or wheel rim is not an approved procedure.*

Steel wheel rims should be inspected for excessive rust and corrosion, cracks, loose rivets or welds, bent or damaged bead seats, and elongated lug nut holes. Aluminum or magnesium wheel rims should be inspected for damaged bead seats, elongated lug nut holes, cracks, and porosity. If any of these conditions are present on either type of wheel rim, replace the wheel rim.

Many shops always replace the tire valve assembly when a tire is repaired or replaced. This policy helps to prevent future problems with tire valve leaks. The inner end of the valve may be cut off with a pair of diagonal pliers and then the outer end may be pulled from the rim. Coat the new valve with rubber tire lubricant and pull it into the rim opening with a special puller screwed onto the valve threads.

Wheel Rim Leak Repair

A wheel rim leak may be repaired if the leak is caused by minor rust or corrosion and the rim is in satisfactory condition. Follow these steps for wheel rim leak repair:

1. Use #80 grit sandpaper to thoroughly clean the area around the leak on the tire side of the rim.
2. Wash the rim to remove any grit from the leak area.

Tire and Wheel Runout Measurement

Ideally, a tire and wheel assembly should be perfectly round. However, this condition is rarely achieved and most tires have a certain amount of radial runout. **Radial runout** is the amount of diameter variation in a tire. If the radial runout exceeds manufacturer's specifications, a vibration may occur because the radial runout causes the spindle to move up and down **(Figure 28-20)**. A defective tire with a variation in stiffness may also cause this up and down spindle action.

Did You Know? **Radial runout is at a right angle to the axis of rotation.**

A dial indicator gauge may be positioned against the center of the tire tread as the tire is rotated

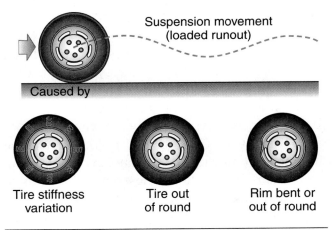

Figure 28-20 Vertical tire and wheel vibrations caused by radial tire or wheel runout or variation in tire stiffness.

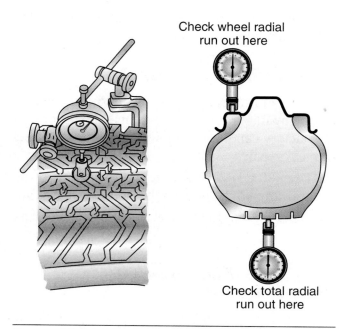

Figure 28-21 Measuring tire radial runout.

Figure 28-22 A setup for measuring tire radial runout.

slowly to measure radial runout **(Figure 28-21)**. Radial runout of more than 0.060 inch (1.5 millimeters) will cause vehicle shake. If the radial runout is between 0.045 inch to 0.060 inch (1.1 millimeters to 1.5 millimeters), vehicle shake may occur. These are typical radial runout specifications. Always consult the vehicle manufacturer's specifications. Mark the highest point of radial runout on the tire with a crayon and mark the valve stem position on the tire.

If the radial tire runout is excessive, demount the tire and check the runout of the wheel rim with a dial indicator positioned against the lip of the rim while the rim is rotated **(Figure 28-22)**. The use of the wheel balancer for this procedure can be helpful. Mount the rim on the balancer, rotate the rim, and measure the runout. Use a crayon to mark the highest point of radial runout on the wheel rim. Radial wheel runout should not exceed 0.035 inch (0.9 millimeter), whereas the maximum lateral wheel runout is 0.045 inch (1.1 millimeter). If the highest point of wheel radial runout coincides with the chalk mark from the highest point of maximum tire radial runout, the tire may be rotated 180° on the wheel to reduce radial runout. Tires or wheels with excessive runout are usually replaced.

Lateral tire runout may be measured with a dial indicator located against the sidewall of the

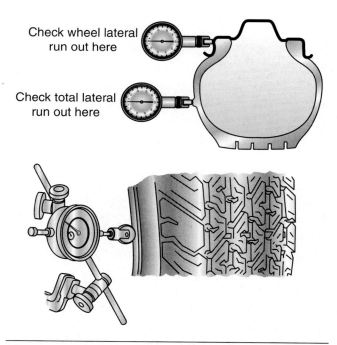

Figure 28-23 Measuring wheel lateral runout.

tire **(Figure 28-23)**. Lateral tire runout is the amount of sideways wobble in a rotating tire. Excessive lateral runout causes the tire to waddle as it turns. This waddling sensation may be transmitted

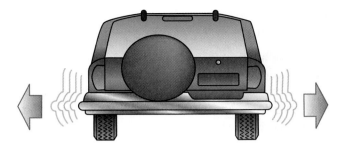

Tire waddle often caused by

- Steel belt not straight within tire
- Excessive lateral runout

Figure 28-24 Chassis waddle caused by lateral tire or wheel runout or a defective tire with a belt that is not straight.

Figure 28-25 A hand spin electronic wheel balancer.

to the passenger compartment **(Figure 28-24)**. A defective tire in which the belts are not straight may also cause chassis waddle. If the lateral runout exceeds 0.080 inch (2.0 millimeters), wheel shake problems will occur on the vehicle. Use chalk to mark the tire and wheel at the highest point of radial runout. When the tire runout is excessive, the tire should be removed from the wheel and the wheel lateral runout should be measured with a dial indicator positioned against the edge of the wheel as the wheel is rotated. Wheels or tires with excessive lateral runout should be replaced.

Did You Know? *Lateral runout is at parallel to the axis of rotation.*

TIRE AND WHEEL BALANCING

The electronic wheel balancer is the most common type in use at the present time **(Figure 28-25)**. There are many different types of electronic wheel balancers in the automotive service industry. Some balancers even have the capability of measuring the rolling resistance of the tire with a special roller **(Figure 28-26)**. These balancers are very effective in diagnosing customer concerns that stem from tires with variations in stiffness or rolling resistance. With an electronic balancer, the operator must enter the wheel diameter, width, and offset. The balancer must have this information to perform its calculations.

Figure 28-26 An electronic wheel balancer capable of measuring the rolling resistance of the tire.

These preliminary checks should be completed before a wheel and tire are balanced:

1. Check for objects in tire tread.
2. Check for objects inside tire.
3. Inspect the tread and sidewall.
4. Check inflation pressure.
5. Measure tire and wheel runout.
6. Check wheel bearing adjustment.
7. Check for mud collected on the inside of the wheel.

All of the items on the preliminary checklist influence wheel balance or safety. Therefore, it is extremely important that the preliminary checks are completed. Since a tire and wheel assembly is rotated at high speed during the **dynamic wheel balance** procedure, it is very important that objects such as stones are removed from the treads. Centrifugal force may dislodge objects from the treads and cause serious personal injury. For this reason, it is also extremely important that the old wheel weights are removed from the wheel prior to balancing and the new weights are attached securely to the wheel during the balance procedure. Follow these preliminary steps before attempting to balance a wheel and tire:

1. When most types of wheel balancers are used, the wheel and tire must be removed from the vehicle and installed on the balancer. All mud, dust, and debris must be washed from the wheel after it is removed.

2. Objects inside a tire, such as balls of rubber, liquid sealants, water, or other foreign material make balance impossible. When the wheel and tire assembly is mounted on the balancer, be absolutely sure that the wheel is securely tightened on the balancer. As the tire is rotated slowly, listen for objects rolling inside the tire. If objects are present, they must be removed prior to wheel balancing.

3. The tire should be inspected for tread and sidewall defects before the balance procedure. These defects create safety hazards and may influence wheel balance. For example,

tread chunking makes the wheel balancing difficult.

The tires should be inflated to the car manufacturer's recommended pressure prior to the balance procedure. Loose wheel bearings allow lateral wheel shaking and simulate an imbalanced wheel condition when the vehicle is in motion. Therefore, wheel bearing adjustments should be checked when wheel balance conditions are diagnosed.

> **Tech Tip** *Always check the vehicle placard for the recommended tire inflation pressure. Do not inflate the tire using the maximum tire pressure rating of the tire on the sidewall of the tire.*

Static Wheel Balance Procedures

Many different types of wheel balancers are available in the automotive service industry and the exact wheel balance procedure may vary depending on the type of wheel balancer. **Static wheel balancing** is seldom used and dynamic balancing is used instead.

On some types of wheel balancers, the wheel is allowed to rotate by gravity during the static balance procedure. A heavy spot rotates the tire until this spot is at the bottom. The necessary static balance weights are then added at the top of the wheel 180° from the heavy spot. When the wheel and tire assembly is balanced statically, gravity does not rotate the wheel from the at-rest position. Rotate the tire by hand and check the static balance at 120-degree intervals.

> **Tech Tip** *On-car wheel balancers are available that use an electrically driven roller to spin the tire and wheel on the vehicle.*

Dynamic Wheel Balance Procedure

Dynamic imbalance is corrected with lead weights. Many wheel weights have a spring clamp

Figure 28-27 Wheel weight pliers are used to remove and install wheel weights.

that holds the weight on the edge of the rim. A special pair of wheel weight pliers is used to tap the wheel weight onto the rim and remove the weight from the rim (**Figure 28-27**). Magnesium or aluminum wheels may require the use of stick-on wheel weights. This type of weight must also be used on some wheels on which the wheel covers interfere with the conventional weights.

The balancer may have different balancing modes depending on where the weights are to be placed. On some alloy rims it would not be desirable to have the weights on the outside of the wheel, as it would affect the visual appeal of the rim. Other special modes include the ability to **match mount**. This balancing procedure determines the best positioning of the tire on the rim so that a minimum amount of additional weight is required to balance the wheel. This procedure requires repositioning of the tire on the rim using a special balancing procedure. You can use the match mount procedure when excessive radial runout is noticed in the tire and wheel during balancing. It may also be used when the customer complains of ride problems, or when the balancer calls for weight in excess of 2 ounces on either plane of passenger car tires during normal dynamic balancing. Regardless of the wheel weight type, they must be attached securely to the wheel.

Many electronic wheel balancers have an electric drive motor that spins the tire. On some electronic wheel balancers, the tire is spun by hand during the balance procedure. The electronic-type balancer performs static and dynamic balance calculations simultaneously and indicates the correct weight size and location to the operator. The tire and wheel assembly must be fastened securely on the balancer with the correct size adapters.

Off-car electronic high-speed wheel balancers have a safety hood that must be positioned over the tire prior to dynamic balancing. Hand-spun electronic balancers generally do not require a safety hood. The preliminary checks mentioned previously in this chapter also apply to dynamic balancing. Wheel balancers must always be operated according to the manufacturer's instructions.

When the heavy spot is located on the inside edge of the tread, the correct size weight installed on the inside of the wheel 180° from the heavy spot provides proper dynamic balance (**Figure 28-28**).

When the heavy spot is on the outside edge of the tread, the correct size wheel weight installed 180° from the heavy spot on the outside

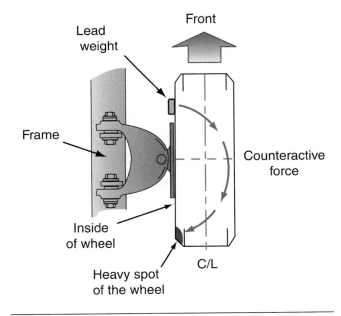

Figure 28-28 Dynamic wheel balance with heavy spot on the inside edge of the wheel.

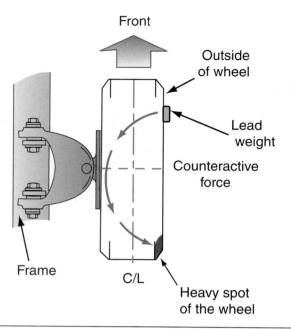

Figure 28-29 Dynamic wheel balance with heavy spot on the outside edge of the tread.

of the wheel provides proper dynamic balance **(Figure 28-29)**.

SAFETY TIP *On many wheel balancers, the tire and wheel are spun at high speed during the dynamic balance procedure. Be sure that all wheel weights are attached securely and check for other loose objects on the tire and wheel, such as stones in the tread. If loose objects are detached from the tire or wheel at high speed, they may cause serious personal injury or property damage. On the type of wheel balancer that spins the tire and wheel at high speed during the dynamic balance procedure, always attach the tire and wheel assembly securely to the balancer. Follow the equipment manufacturer's recommended wheel mounting procedure. If the tire and wheel assembly ever becomes loose on the balancer at high speed, serious personal injury or property damage may result. Prior to spinning a tire and wheel at high speed on a wheel balancer, always lower the protection shield over the tire. This shield provides protection in case anything flies off the tire or wheel.*

WHEEL BEARING MAINTENANCE

Technicians must accurately diagnose wheel bearing problems to avoid repeat bearing failures and thus provide customer satisfaction. Accurate wheel bearing service procedures are essential to maintain vehicle safety! Improper wheel bearing service may cause brake problems, steering complaints, and premature bearing failure. Improper wheel bearing service may even cause a wheel to fly off a vehicle resulting in personal injury and vehicle damage. You must also understand rear-axle bearing service procedures on rear-wheel drive vehicles.

Two separate tapered roller bearings are used in the front wheel hubs of many rear-wheel drive cars and the rear wheel hubs on some front-wheel drive cars have the same type of bearings. Similar service and adjustment procedures apply to these tapered roller bearings. These bearings should be cleaned, inspected, and packed with wheel bearing grease at the vehicle manufacturer's recommended service intervals. Remove the grease seal out of the inner hub opening with a seal puller and discard the seal. This seal should always be replaced when the bearings are serviced. Do not attempt to wash sealed bearings or bearings that are shielded on both sides. If a bearing is sealed on one side, it may be washed in solvent and repacked with grease.

Bearings may be placed in a tray and lowered into a container of clean solvent. A brush may be used to remove old grease from the bearing **(Figure 28-30)**.

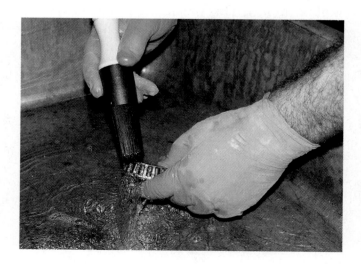

Figure 28-30 Cleaning a bearing.

The bearings may be dried with compressed air after the cleaning operation. Be sure the shop air supply is free from moisture, which causes rust formation in the bearing. Do not spin the bearing with the air pressure, as damage to the bearing will occur.

After all the old grease has been cleaned from the bearing, rinse the bearing in clean solvent and dry it thoroughly with compressed air. When bearing cleaning is completed, bearings should be inspected for the defects illustrated in **Figure 28-31** and

TAPERED ROLLER BEARING DIAGNOSIS

Consider the following factors when diagnosing bearing condition:
1. General condition of all parts during disassembly and inspection.
2. Classify the failure with the aid of the illustrations.
3. Determine the cause.
4. Make all repairs following recommended procedures.

ABRASIVE STEP WEAR

Pattern on roller ends caused by fine abrasives. Clean all parts and housings, check seals and bearings, and replace if leaking, rough or noisy.

GALLING

Metal smears on roller ends due to overheating, lubricant failure, or overload. Replace bearing, check seals, and check for proper lubrication.

BENT CAGE

Cage damaged due to improper handling or tool usage. Replace bearing.

ABRASIVE ROLLER WEAR

Pattern on races and rollers caused by fine abrasives. Clean all parts and housings, check seals and bearings, and replace if leaking, rough or noisy.

ETCHING

Bearing surfaces appear gray or grayish black in color with related etching away of material usually at roller spacing. Replace bearings, check seals, and check for proper lubrication.

BENT CAGE

Cage damaged due to improper handling or tool usage. Replace bearing.

INDENTATIONS

Surface depressions on race and rollers caused by hard particles of foreign material. Clean all parts and housings. Check seals, and replace bearings if rough or noisy.

MISALIGNMENT

Outer race misalignment due to foreign object. Clean related parts and replace bearing. Make sure races are properly sealed.

Figure 28-31 Bearing failures and corrective procedures.

FATIGUE SPALLING

Flaking of surface metal resulting from fatigue. Replace bearing, clean all related parts.

STAIN DISCOLORATION

Discoloration can range from light brown to black caused by incorrect lubricant or moisture. Re-use bearings if stains can be removed by light polishing or if no evidence of overheating is observed. Check seals and related parts for damage.

CAGE WEAR

Wear around outside diameter of cage and roller pockets caused by abrasive material and inefficient lubrication. Clean related parts and housings. Check seals and replace bearings.

HEAT DISCOLORATION

Heat discoloration can range from faint yellow to dark blue, resulting from overload or incorrect lubricant. Excessive heat can cause softening of races or rollers. To check for loss of temper on races or rollers, a simple file test may be made. A file drawn over a tempered part will grab and cut metal, whereas, a file drawn over a hard part will glide readily with no metal cutting. Replace bearings if overheating damage is indicated. Check seals and other parts.

FRETTAGE

Corrosion set up by small relative movement of parts with no lubrication. Replace bearings. Clean related parts. Check seals and check for proper lubrication.

BRINELLING

Surface indentations in raceway caused by rollers either under impact loading or vibration while the bearing is not rotating. Replace bearing if rough or noisy.

SMEARS

Smearing of metal due to slippage. Slippage can be caused by poor fits, lubrication, overheating, overloads, or handling damage. Replace bearings, clean related parts and check for proper fit and lubrication.

CRACKED INNER RACE

Race cracked due to improper fit, cocking, or poor bearing seats. Replace bearing and correct bearing seats.

Figure 28-32 Bearing failures and corrective procedures, continued.

Figure 28-32. If any of these conditions are present on the bearing, replacement is necessary.

Tapered roller bearings and their matching outer races must be replaced as a set. If the bearing installation is not done immediately, cover the bearings with a protective lubricant and wrap them in waterproof paper (**Figure 28-33**). Be sure to identify the bearings or lay them in order so you reinstall them in their original location. Use care when cleaning bearings or races with paper towels as the paper towel may leave lint.

> **Tech Tip** *Lint from shop towels or paper towels may contaminate the bearing. Be sure to check for any contamination after the cleaning process is completed.*

Figure 28-33 Wrapping a bearing in waterproof paper.

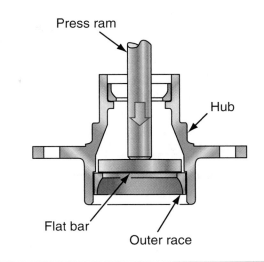

Figure 28-34 Bearing race removal.

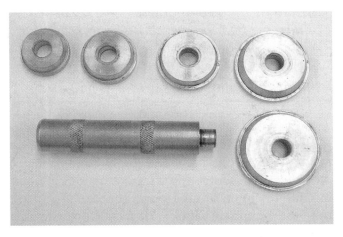

Figure 28-35 Correct bearing driver selection for bearing race installation.

Bearing races and the inner part of the wheel hub should be thoroughly cleaned with solvent and dried with compressed air. Inspect the seal mounting area in the hub for metal burrs. Remove any burrs with a fine round file. Bearing races must be replaced if they indicate any of the defects described in Figure 28-31 and Figure 28-32. The proper bearing race driving tool must be used to remove the bearing races **(Figure 28-34)**. If a driver is not available for the bearing races, a long punch and hammer may be used to drive the races from the hub. When a hammer and punch are used for this purpose, be careful not to damage the hub's inner surface with the punch.

SAFETY TIP *Do not spin the bearing at high speed with compressed air; bearing damage or disintegration may result. Bearing disintegration may cause serious personal injury. Never strike a bearing with a ball peen hammer. This action will damage the bearing and the bearing may shatter causing severe personal injury.*

The new bearing races should be installed in the hub with the correct bearing race driving tool **(Figure 28-35** and **Figure 28-36)**. When bearings

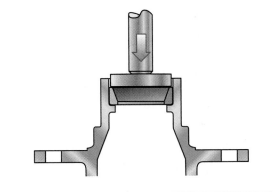

Figure 28-36 Bearing race installation.

and races are replaced, be sure they are the same as the original bearings. The part numbers should be the same on the old bearings and the replacement bearings.

Inspect the bearing and seal mounting surfaces on the spindle. Small metal burrs may be removed from the spindle with a fine-toothed file. If the spindle is severely scored in the bearing or seal mounting areas, spindle replacement is necessary.

Bearing Lubrication and Assembly

After the bearings and races have been cleaned and inspected, the bearings should be packed with grease.

> **Tech Tip** *Cleanliness is very important during wheel bearing service. Always maintain cleanliness of hands, tools, work area, and all related bearing components. One small piece of dirt in a bearing will cause bearing failure. Always keep grease containers covered when not in use. Uncovered grease containers are easily contaminated with dirt and moisture.*

Always use the vehicle manufacturer's specified wheel bearing grease. Place a lump of grease in the palm of one hand and grasp the bearing in the other hand. Force the widest edge of the bearing into the lump of grease and squeeze the grease into the bearing **(Figure 28-37)**. Continue this process until grease is forced into the bearing around the entire bearing circumference. Place a coating of grease around the outside of the rollers and apply a light coating of grease to the races. A bearing packing tool may be used to force grease into the bearings rather than the hand method **(Figure 28-38)**.

> **Tech Tip** *Always use the vehicle manufacturer's specified wheel bearing grease.*

Figure 28-37 A hand method for packing wheel bearings.

Figure 28-38 Mechanical wheel bearing packer.

Figure 28-39 Wheel bearing lubrication.

Place some grease in the wheel hub cavity and position the inner bearing in the hub **(Figure 28-39)**. Check the fit of the new bearing seal on the spindle and in the hub. The seal lip must fit snugly on the spindle and the seal case must fit properly in the hub opening. The part number on the old seal and the replacement seal should be the same. Be sure the seal is installed in the proper direction with the garter spring and higher part of the lip toward the lubricant in the hub. The new inner bearing seal must be installed in the hub with a suitable seal driver **(Figure 28-40)**. Place a light coating of wheel bearing grease on the spindle and slide the hub assembly onto the spindle. Install the outer wheel bearing and be sure there is adequate

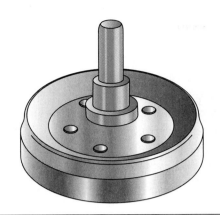

Figure 28-40 Seal installation.

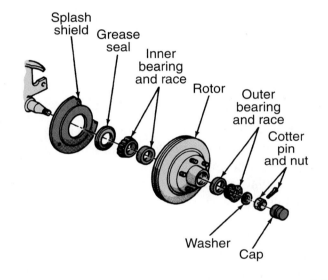

Figure 28-41 Installation of wheel bearings and related components.

lubrication on the bearing and race. Be sure the washer and nut are clean and install these components on the spindle **(Figure 28-41)**. Tighten the nut until it is finger tight.

Wheel Bearing Adjustment with Two Separate Tapered Roller Bearings in the Wheel Hub

Loose front wheel bearing adjustment results in lateral front wheel movement and reduced directional stability. If the wheel bearing adjusting nut is tightened excessively, the bearings may overheat resulting in premature bearing failure. The bearing adjustment procedure will vary depending on the make of vehicle. Always follow the procedure in the vehicle manufacturer's service manual. The following is a typical bearing adjustment procedure:

1. With the hub and bearings assembled on the spindle, tighten the adjusting nut to 17 to 25 ft. lb. (23 to 34 N·m) while the hub is rotated **(Figure 28-42)**.

2. Loosen the adjusting nut one-half turn and retighten it to 10 to 15 in. lb. (1.0 to 1.7 N·m). This specification varies depending on the make of vehicle. Always use manufacturer's specifications.

3. Position the adjusting nut retainer over the nut so the retainer slots are aligned with the cotter keyhole in the spindle.

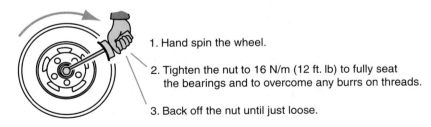

1. Hand spin the wheel.

2. Tighten the nut to 16 N/m (12 ft. lb) to fully seat the bearings and to overcome any burrs on threads.

3. Back off the nut until just loose.

Bend end of cotter pin legs flat against nut. Cut off extra length.

4. Snug up the nut by hand.

5. Loosen the nut until a hole in the spindle lines up with a slot in the nut. Insert cotter pin.

6. When the bearing is properly adjusted there will be from 0.03 to 13 mm (0.001" to 0.005") end play.

Figure 28-42 Typical wheel bearing adjustment.

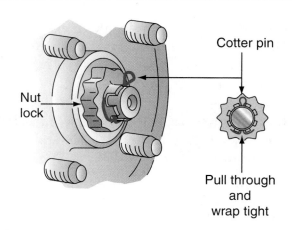

Figure 28-43 Nut lock and cotter pin installation.

4. Install a new cotter pin and bend the ends around the retainer flange **(Figure 28-43)**.

5. Install the grease cap and make sure the hub and drum rotate freely.

WHEEL BEARING DIAGNOSIS

Bearings are designed to provide long life, but there are many causes of premature bearing failure. If a bearing fails, you must decide if the bearing failure was caused by normal wear or if the bearing failed prematurely. For example, if a front wheel bearing fails on a car that is 1 year old with an original odometer reading of 15,000 miles (24,000 kilometers), experience tells us the bearing failure is premature because front wheel bearings normally last for a much longer mileage period. Always listen to the customer's complaints and obtain as much information as possible from the customer. Ask the customer specific questions about abnormal or unusual vehicle noises and operation. If a bearing fails prematurely, there must be some cause for the failure. The causes of premature bearing failure are lack of lubrication, improper type of lubrication, incorrect end-play adjustment (where applicable), misalignment of related components such as shafts or housings, excessive bearing load, improper installation or service procedures, excessive heat, or dirt or contamination.

When a bearing fails prematurely, you must correct the cause of this failure to prevent the new bearing from failing. The first indication of bearing failure is usually a howling noise while the bearing is rotating. The howling noise will likely vary depending on the bearing load. A front wheel bearing usually provides a more noticeable howl when the vehicle is turning a corner because this places additional thrust load on the bearing. A defective rear axle bearing usually provides a howling noise that is more noticeable at lower speeds. The howling noise is more noticeable when driving on a narrow street with buildings on each side because the noise resonates off the nearby buildings. A rear axle bearing noise is present during acceleration and deceleration because the vehicle weight places a load on the bearing regardless of the operating condition. The rear axle bearing noise may be somewhat more noticeable during deceleration because there is less engine noise at that time.

WHEEL HUB UNIT DIAGNOSIS

When wheel bearings and hubs are an integral assembly, the bearing endplay should be measured with a dial indicator stem mounted against the hub. If the endplay exceeds 0.005 inch (0.127 millimeter) as the hub is moved in and out, the hub and bearing assembly should be replaced. This specification is typical, but the vehicle manufacturer's specifications must be used. Hub and bearing replacement is also necessary if the bearing is rough or noisy. Integral-type bearing and hub assemblies are used on the front and rear wheels on some front-wheel drive cars. When the front wheel bearings are mounted in the steering knuckle, the wheel bearings may be checked with the vehicle raised on the hoist and a dial indicator positioned against the outer wheel rim lip as pictured in **Figure 28-44**. When the wheel is moved in and out, the maximum bearing movement on the dial indicator should be as follows:

- 0.020 inch (0.508 millimeter) for 13-inch (33 centimeter) wheels

- 0.023 inch (0.584 millimeter) for 14-inch (35.5 centimeter) wheels

- 0.025 inch (0.635 millimeter) for 15-inch (38 centimeter) wheels

If the bearing movement is excessive, check the hub nut torque before replacing the bearing. When this torque is correct and bearing movement is excessive, the bearing should be replaced.

Figure 28-44 Wheel bearing diagnosis on vehicle.

REAR AXLE BEARING AND SEAL SERVICE, REAR-WHEEL DRIVE VEHICLES

Rear axle bearing noise may be diagnosed with the vehicle raised on a lift. Be sure the hoist safety mechanism is engaged after the vehicle is raised on the lift. With the engine running and the transmission in drive, operate the vehicle at moderate speed (35 to 45 miles per hour) (56 to 72 kilometers per hour) and listen with a stethoscope placed on the rear axle housing directly over the axle bearings. If grinding or clicking noises are heard, bearing replacement is necessary.

Some rear axle shafts have the bearing mounted on the axle and retained by a wheel bearing retainer plate **(Figure 28-45)**. Many other axle shafts in rear-wheel drive cars have a roller bearing and seal at the outer end **(Figure 28-46)**. These axle shafts are often retained in the differential with C locks that must be removed before the axles.

The rear axle bearing removal and replacement procedure varies depending on the vehicle make and model year. Always follow the rear axle bearing removal and replacement procedure in the manufacturer's service manual.

The following is a typical rear axle shaft removal and replacement procedure on a rear-wheel drive car with C lock axle retainers:

1. Loosen the rear wheel nuts and chalk mark the rear wheel position in relation to the rear axle **(Figure 28-47)**.

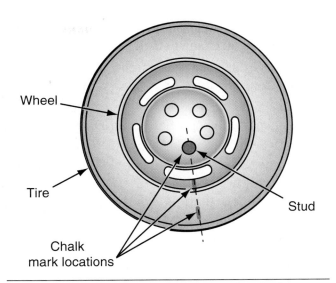

Figure 28-45 Chalk marking on wheel, tire, and stud.

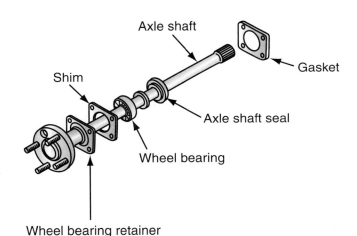

Figure 28-46 A rear axle bearing retained by a plate.

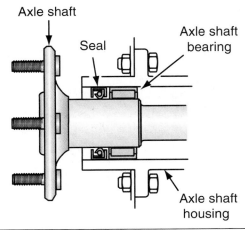

Figure 28-47 Rear axle roller bearing and seal, rear-wheel drive car.

2. Raise the vehicle on a lift and make sure the lift safety mechanism is in place.

3. Remove the rear wheels and brake drums or calipers and rotors.

4. Place a drain pan under the differential and remove the differential cover. Discard the old lubricant.

5. Remove the differential lock bolt, pin, and shaft **(Figure 28-48)**.

6. Push the axle shaft inward and remove the axle C lock.

7. Pull the axle from the differential housing.

Reverse the axle removal procedure to reinstall the axle. Always use a new differential cover gasket or sealant and fill the differential to the bottom of the filler plug opening with the manufacturer's recommended lubricant. Be sure all fasteners, including the wheel nuts, are tightened to the specified torque.

The following is a typical axle bearing and seal removal procedure:

1. Remove the axle seal with a seal puller.

2. Use the proper bearing puller to remove the axle bearing **(Figure 28-49)**.

3. Clean the axle housing seal and bearing mounting area with solvent and a brush. Clean this area with compressed air.

4. Check the seal and bearing mounting area in the housing for metal burrs and scratches.

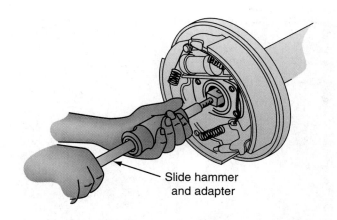

Figure 28-49 Rear axle bearing puller.

Remove any burrs or irregularities with a fine-toothed round file.

5. Wash the axle shaft with solvent and blow it dry with compressed air.

6. Check the bearing contact area on the axle for roughness, pits, and scratches. If any of these conditions are present, axle replacement is necessary.

7. Be sure the new bearing fits properly on the axle and in the housing. Install the new bearing with the proper bearing driver **(Figure 28-50)**. The bearing driver must apply pressure to the outer race that is pressed into the housing.

8. Be sure the new seal fits properly on the axle shaft and in the housing. Make sure the garter spring on the seal faces toward the differential.

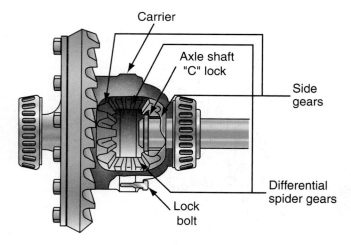

Figure 28-48 Rear axle "C" lock, lock bolt, and pinion gears.

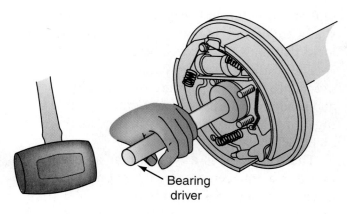

Figure 28-50 Rear axle bearing driver.

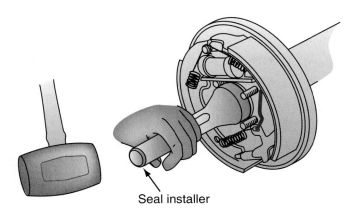

Seal installer

Figure 28-51 Installing rear axle seal.

Tech Tip *When a lip seal is installed, the garter spring should always face toward the flow of lubricant.*

Use the proper seal driver to install the new seal in the housing **(Figure 28-51)**.

9. Lubricate the bearing, seal, and bearing surface on the axle with the manufacturer's specified differential lubricant.

10. Reverse the rear axle removal procedure to reinstall the rear axle.

11. Be sure all fasteners are tightened to the specified torque.

Some rear axles have a sealed bearing that is pressed onto the axle shaft and held in place with a retainer plate. These rear axles usually do not have C locks in the differential. A retainer plate is mounted on the axle between the bearing and the outer end of the axle. This plate is bolted to the outer end of the differential housing. After the axle retainer plate bolts are removed, a slide hammer-type puller is attached to the axle studs to remove this type of axle. When this type of axle bearing is removed, the retainer ring must be split with a hammer and chisel while the axle is held in a vise. Do not heat the retainer ring or the bearing with an acetylene torch during the removal or installation process. After the retainer ring is removed, the bearing must be pressed from the axle shaft and the bearing must not be reused. A new bearing and retainer ring must be pressed onto the axle shaft. The bearing removal and replacement procedure is illustrated in **Figure 28-52**.

SAFETY TIP *Never use an acetylene torch to heat axle bearings or retainer rings during the removal and replacement procedure. The heat may cause fatigue in the steel axle and the axle may break suddenly causing the rear wheel to fall off. This action will likely result in severe vehicle damage and personal injury.*

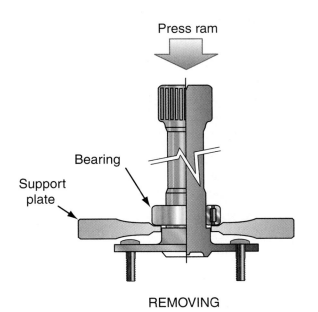

REMOVING

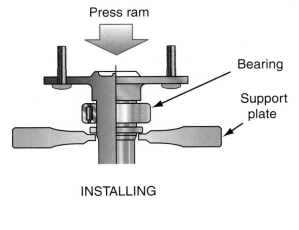

INSTALLING

Figure 28-52 Axle bearing and retainer plate removal and replacement.

Summary

- Tire noises vary with road conditions, but differential noise is not affected by road conditions.
- Steering pull may be caused by defects in the suspension or steering systems and tire conicity.
- Tire conicity occurs when the belt in a tire is wound off center during the manufacturing process.
- Tire conicity cannot be diagnosed by a visual inspection.
- The tire rotation procedure varies depending on the model year, type of tires, and whether the vehicle has a compact spare or a conventional spare.
- Hand tools and tire irons should not be used to demount tires.
- Tire punctures over 1/4-inch diameter and sidewall punctures should not be repaired.
- Compact spare tires should not be repaired.
- Do not use caustic agents, alkaline-based detergents, or abrasive cleaners to clean aluminum or magnesium wheels.
- When a bead expander is used to mount a tire, do not increase tire pressure above 10 psi (69 kPa).
- When a bead expander is not used to mount a tire, do not increase tire pressure above 40 psi (276 kPa) to move the tire beads out against the wheel rim.
- While mounting a tire, be sure that the tire beads and wheel rim bead seats are coated with rubber tire lubricant.
- Radial tire and wheel runout cause tire thump.
- Lateral tire and wheel runout cause tire and chassis waddling.
- Balls of rubber or other objects inside a tire make proper balance impossible.
- Aluminum or magnesium wheels require the use of stick-on wheel weights.
- Electronic wheel balancers perform the necessary wheel weight calculations and indicate the exact locations where these weights should be installed.
- When a wheel bearing fails prematurely, you must determine the cause of the failure and correct this cause to prevent a second bearing failure.
- Before a tire and wheel assembly is removed from a car, the wheel and tire assembly must be chalk-marked in relation to the hub or axle to maintain wheel balance when the wheel is reinstalled.
- Tapered roller bearings and races must be replaced as a matched set.
- When cleaning bearings, never spin the bearings with compressed air.
- When drying bearings with compressed air, be sure the shop air supply is free from moisture.
- Do not wipe bearings with paper towels or shop towels with lint on them.
- Always inspect bearing and seal mounting areas for metal burrs and scratches. Burrs may be removed with a fine-toothed file.
- While servicing wheel bearings, always keep hands, tools, and work area clean.
- Wheel bearing hub units must be checked for endplay with a dial indicator.
- Rear axle bearings and retainer rings that are pressed onto the axle shaft should not be heated with an acetylene torch.
- On some rear-wheel drive cars, the rear axle C locks inside the differential must be removed prior to rear axle removal.

Review Questions

1. While discussing a tire-thumping problem, Technician A says this problem may be caused by cupped tire treads. Technician B says a heavy spot in the tire may cause this complaint. Who is correct?

   A. Technician A

   B. Technician B

   C. Both Technician A and Technician B

   D. Neither Technician A nor Technician B

2. While discussing a vehicle that pulls to one side, Technician A says that excessive radial runout on the right front tire may cause this problem. Technician B says that tire conicity may be the cause of this complaint. Who is correct?

   A. Technician A

   B. Technician B

   C. Both Technician A and Technician B

   D. Neither Technician A nor Technician B

3. While discussing tire wear, Technician A says that static imbalance causes feathered tread wear. Technician B says that dynamic imbalance causes cupped wear and bald spots on the tire tread. Who is correct?

   A. Technician A

   B. Technician B

   C. Both Technician A and Technician B

   D. Neither Technician A nor Technician B

4. While discussing tire noise, Technician A says that tire noise varies with road surface conditions. Technician B says that tire noise remains constant when the vehicle is accelerated and decelerated. Who is correct?

   A. Technician A

   B. Technician B

   C. Both Technician A and Technician B

   D. Neither Technician A nor Technician B

5. Chassis waddle may occur if the lateral tire runout exceeds:

   A. 0.010 inch.

   B. 0.025 inch.

   C. 0.050 inch.

   D. 0.080 inch.

6. Radial wheel runout should not exceed:

   A. 0.005 inch.

   B. 0.015 inch.

   C. 0.020 inch.

   D. 0.035 inch.

7. While driving a vehicle straight ahead on a smooth road surface, the steering pulls to the right. All of these defects may be the cause of the problem *except*:

   A. two tires of different sizes on the front wheels.

   B. different inflation pressures in the front tires.

   C. improper static wheel balance on one front wheel and tire assembly.

   D different tire designs on the two front wheels.

8. The growling noise produced by a defective front wheel bearing is most noticeable when:

   A. turning a corner at low speeds.

   B. driving straight ahead at low speeds.

   C. driving at the freeway speed limit.

   D. the vehicle is just starting off after a stop.

9. On a typical sealed wheel bearing hub assembly, the maximum end play should be:

   A. 0.005 inch.

   B. 0.008 inch.

   C. 0.010 inch.

   D. 0.018 inch.

10. When servicing rear axle bearings that are pressed onto the axle shaft and held in place with a retainer ring:

    A. Remove the C locks from the inner end of the axles.

    B. Cut the retainer ring off the axle with an acetylene torch.

    C. Use a hydraulic press to remove and install the bearing on the axle shaft.

    D. Heat the retainer ring with an acetylene torch before installing it on the axle.

11. The maximum allowable radial tire runout is _____ inch(es).

12. A seal should be installed with the garter spring facing toward the _____ _____.

13. After wheel bearings are cleaned and repacked with grease, the bearing-adjusting nut should be tightened to 17 to 25 ft. lbs, loosened 1/2 turn, and retightened to _____ to _____ in. lb.

14. Balls of rubber inside a tire make wheel-balancing _____.

15. Explain front wheel shimmy and describe the causes and results of this problem.

16. Describe the wheel bearing adjustment procedure on a wheel hub containing two tapered roller bearings.

17. List six causes of premature wheel bearing failure.

18. Describe the conditions when a defective front wheel bearing noise is most noticeable.

CHAPTER

29 Suspension Fundamentals

Learning Objectives

After you have read, studied, and practiced the contents of this unit, you should be able to:

- Describe shock absorber operation during wheel jounce.

- Describe the advantages of nitrogen gas-filled shock absorbers and struts.

- Explain shock absorber ratios.

- Explain basic torsion bar action as a front wheel strikes a road irregularity.

- Describe two types of load carrying ball joints and explain the location of the control arm in each type.

- List the two steering knuckle pivot points in a Macpherson strut front suspension system.

- Describe the rear axle housing movement during vehicle acceleration.

- Explain the difference between a semi-independent and an independent rear suspension system.

- Describe how individual rear wheel movement is provided in a semi-independent rear suspension system.

- Describe the basic operation of an electronic air suspension system.

- Explain the basic operation of a continuously variable road sensing suspension (CVRSS) system.

Key Terms

Analog voltage signals

Axle tramp

Continuously variable road sensing suspension (CVRSS) system

Electronic air suspension system

Jounce travel

Load carrying ball joint

Non-load carrying ball joint

Rebound travel

Sprung body weight

Stabilizer bar

Trim height

Unsprung weight

INTRODUCTION

The front and rear suspension systems must provide several extremely important functions to maintain vehicle safety and owner satisfaction. The suspension system must supply steering control for the driver under all road conditions. Vehicle owners expect the suspension system to provide a comfortable ride. The suspension, together with the chassis, must maintain proper vehicle tracking and directional stability. Another important purpose of the suspension system is to provide proper wheel alignment and minimize tire wear.

Computer-controlled suspension systems now provide a soft ride during normal freeway driving and then instantly switch to a firm ride during hard cornering, braking, fast acceleration, and high-speed driving. Therefore, the computer-controlled suspension system allows the same car to meet the demands of the driver who wants a soft ride and the driver who wants a firm ride. Since computer-controlled suspension systems reduce body sway during hard cornering, these systems provide improved steering control.

Some computer-controlled suspension systems also supply a constant vehicle riding height regardless of the vehicle passenger or cargo load. This action maintains the vehicle's cosmetic appearance as the passenger and/or cargo load is changed. Maintaining a constant riding height also supplies more consistent suspension alignment angles, which provides improved steering control.

SHOCK ABSORBERS AND STRUTS

The lower half of a shock absorber is a twin tube steel unit filled with hydraulic oil **(Figure 29-1)**. Some shock absorbers may use nitrogen gas. A relief valve is located in the bottom of the unit and a circular lower mounting is attached to the lower tube. This mounting contains a rubber isolating bushing, or grommets. A piston and rod assembly is connected to the upper half of the shock absorber. This upper portion of the shock absorber has a dust shield that surrounds the lower twin tube

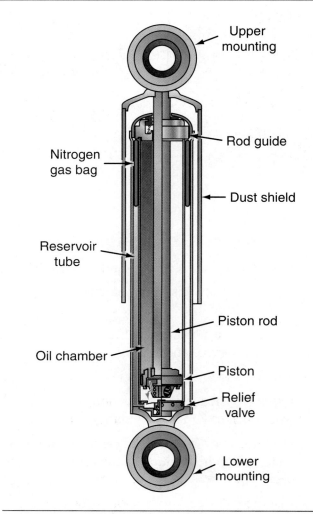

Figure 29-1 Shock absorber design.

unit. The piston is a precision fit in the inner cylinder of the lower unit. A piston rod guide and seal are located in the top of the lower unit. A circular upper mounting with a rubber bushing is attached to the top of the shock absorber. A shock absorber is mounted between the wheel assembly and the chassis.

Shock Absorber Operation

Shock absorbers are usually mounted between the lower control arms and the chassis. When the vehicle wheel strikes a bump, the wheel and suspension move upward in relation to the chassis. Upward wheel movement is referred to as **jounce travel**. This jounce action causes the spring to deflect or compress. Under this condition, the spring stores energy and springs back downward with all the energy absorbed when it deflected upward. This downward spring and wheel action are called **rebound travel**. If this spring action is not controlled, the wheel would strike the road with a strong downward force and the wheel jounce would occur again. Therefore, some device must be installed to control the spring action or the wheel would bounce up and down many times after it hit a bump, thereby causing passenger discomfort, directional instability, and suspension component wear.

Shock absorbers are installed on suspension systems to control spring action. Two basic designs of shocks are used. One design has the jounce and rebound valves located on the piston and the other has the rebound valve in the piston and the jounce valve in the bottom of the shock.

In the first design, when a wheel strikes a bump and jounce travel occurs, the shock absorber lower tube unit is forced upward. This action forces the piston downward in the lower tube unit. Because oil cannot leak past the piston, the oil in the lower unit is forced through the piston valves to the upper oil chamber.

When a wheel moves downward in rebound travel, the shock absorber piston moves upward in the lower half of the shock absorber. Under this condition, oil is forced to flow through the piston valve from the lower part of the unit to the area below the piston (**Figure 29-2**).

In the second design, when a wheel strikes a bump and jounce travel occurs, the shock absorber lower tube unit is forced upward. This action forces the piston downward in the lower tube unit.

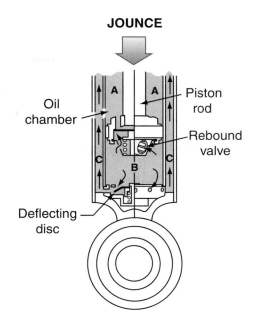

Figure 29-3 Shock absorber jounce operation.

Since oil cannot leak past the piston, the oil in the lower unit is forced through the lower tube valves to the upper oil chamber around the outside of the shock absorber (**Figure 29-3**).

During rebound the wheel moves downward and the shock absorber piston moves upward in the lower half of the shock absorber. Because the rebound valve is located in the piston, oil is forced to flow through the piston valve from the lower part of the unit to the area below the piston (**Figure 29-4**).

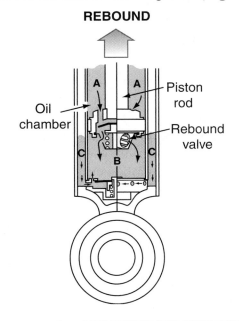

Figure 29-4 Shock absorber rebound operation.

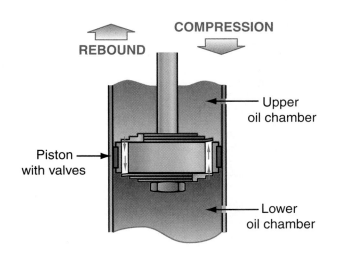

Figure 29-2 Shock absorber operation.

Shock absorber valves provide precise oil flow control and control the upward and downward action of the wheel and suspension. Regardless of the piston orifice and valve design, the shock absorber must be precisely matched to absorb the spring's energy.

Gas-Filled Struts and Shock Absorbers

During fast upward wheel movement on the compression stroke, excessive pressure in the lower oil chamber forces the base valve open, thus allowing oil to flow through this valve to the reservoir. The nitrogen gas provides a compensating space for the oil that is displaced into the reservoir on the compression stroke and when the oil is heated. Since the gas exerts pressure on the oil, cavitation or foaming of the oil is eliminated. When oil bubbles are eliminated in this way, the shock absorber provides continuous damping for wheel deflections as small as 0.078 inch (2.0 millimeters).

A rebound rubber is located on top of the piston. If the wheel drops downward into a hole, the shock absorber may become fully extended. Under this condition, the rebound rubber provides a cushioning action. Gas-filled units are identified with a warning label. If a gas-filled shock absorber is removed and compressed to its shortest length, it should re-extend when it is released. Failure to re-extend indicates that shock absorber or strut replacement is necessary.

NOTE: New gas-filled shock absorbers are wired in the compressed position for shipping purposes. Exercise caution when removing the shipping wire because the shock extends when this strap is cut. After the upper attaching bolt is installed on the shock absorber, the wire may be cut to allow the unit to extend. Front gas-filled struts have an internal catch that holds them in the compressed position. This catch is released when the rod is held and the strut is rotated 45° counterclockwise.

Heavy-Duty Shock Absorber Design

Some heavy-duty shock absorbers have a dividing piston in the lower oil chamber. The area below this position is pressurized with nitrogen gas to 360 psi (2482 kPa). Hydraulic oil is contained in the oil chamber above the dividing piston. The other main features of the heavy-duty shock absorber are a high-quality seal for longer life, a single tube design to prevent excessive heat buildup, and a rising rate valve that provides precise spring control under all conditions.

The operation of the heavy-duty shock absorber is similar to the conventional type **(Figure 29-5)**.

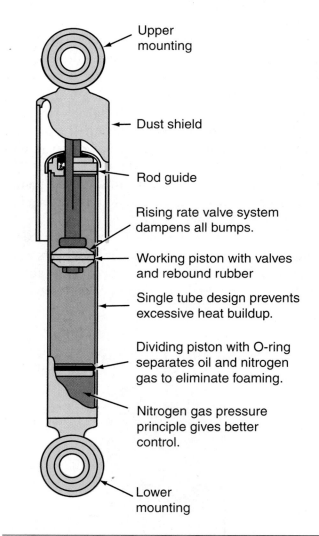

Upper mounting

Dust shield

Rod guide

Rising rate valve system dampens all bumps.

Working piston with valves and rebound rubber

Single tube design prevents excessive heat buildup.

Dividing piston with O-ring separates oil and nitrogen gas to eliminate foaming.

Nitrogen gas pressure principle gives better control.

Lower mounting

Figure 29-5 A heavy duty shock absorber design using pressurized nitrogen.

Shock Absorber Ratios

Most automotive shock absorbers are a double-acting type that controls spring action during jounce and rebound wheel movements. The piston and valves in many shock absorbers are designed to provide more extension control than compression control.

Sprung body weight is the chassis weight supported by the springs. **Unsprung weight** is the weight of the wheel and suspension components that are not supported by the springs. Shock absorbers are mounted between the sprung and unsprung weights or the wheel assembly and the chassis **(Figure 29-6)**.

An average shock absorber may have 70 percent of the total control on the extension cycle; thus 30 percent of the total control is on the compression cycle. Shock absorbers usually have this type of design because they must control the heavier sprung body weight on the extension cycle. The lighter unsprung weight of the axle, wheel, and tire is controlled by the shock absorber on the compression cycle. A shock absorber with this type of design is referred to as a 70/30 type. Shock absorber ratios vary from 50:50 to 90:10.

Strut Design

A strut-type front suspension is used on most front-wheel drive cars and some rear-wheel drive cars. Internal strut design is very similar to shock absorber design and struts perform the same functions as shock absorbers. Some struts have a replaceable cartridge. In many strut-type suspension systems, the coil spring is mounted on the strut. The coil spring is largely responsible for proper curb ride height. A weak or broken coil spring reduces curb ride height and provides harsh riding.

The lower end of the front suspension strut is bolted to the steering knuckle **(Figure 29-7)**. An upper strut mount is attached to the strut. This mount is bolted into the chassis strut tower. A lower spring seat is part of the strut assembly and a lower insulator is positioned between the coil spring and the spring seat on the strut. Another insulator is located between the coil spring and the upper strut mount. The two insulators prevent metal-to-metal contact between the spring and the strut or mount. These isolators reduce the transmission of noise and harshness from the suspension to the chassis. A rubber spring bumper is positioned around the strut piston rod. When a front wheel strikes a large road irregularity and the strut is fully compressed, the spring bumper provides a cushioning action between the top of the strut and the upper support.

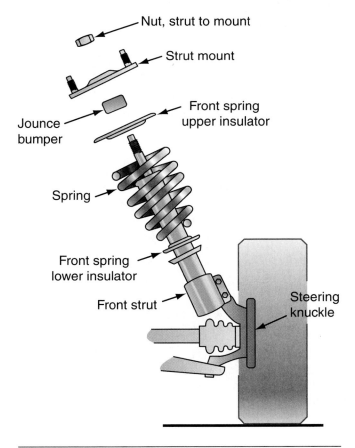

Figure 29-7 Front strut mounting.

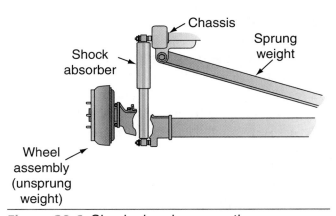

Figure 29-6 Shock absorber mounting.

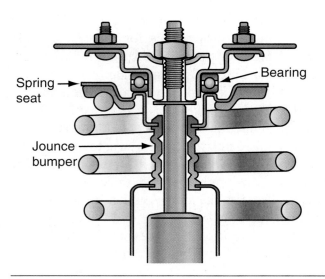

Figure 29-8 Upper strut mount.

The upper strut mount contains a bearing, an upper spring seat, and a jounce bumper **(Figure 29-8)**. When the front wheels are turned, the front strut and coil spring rotate with the steering knuckle. The strut and spring assembly rotates on the upper strut mount bearing.

Some vehicles may use a modified Macpherson strut design. In this design the coil spring has been separated from the strut and moved to the frame area of the vehicle. This design allows for a smaller strut tower **(Figure 29-9)**.

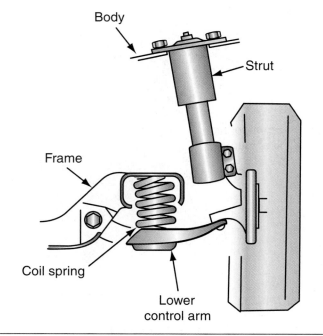

Figure 29-9 Modified Macpherson strut mounting.

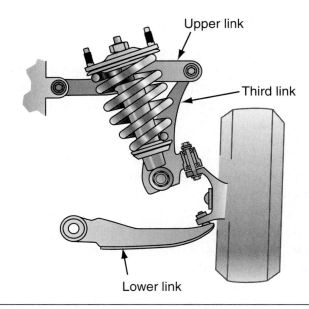

Figure 29-10 Multi-link front suspension with strut connected between the upper link and the strut tower.

Some cars have a multi-link front suspension with an upper link connected from the chassis to the steering knuckle. The strut is connected from the upper link to the strut tower **(Figure 29-10)**. A bearing is mounted between the upper link and the steering knuckle and the wheel and knuckle turn on this bearing and the lower ball joint. Therefore, the coil spring and strut do not turn when the front wheels are turned and a bearing in the upper strut mount is not required. Rear struts are similar to front struts, but rear struts do not require any steering action.

Load-Leveling Shock Absorbers

Load-leveling rear struts (or shock absorbers) are used with an electronic height control system. An on-board air compressor pumps air into the rear shocks to raise the rear of the vehicle and an electric solenoid releases air from the shocks to lower the rear chassis. An electromagnetic height sensor may be contained in the shock absorber, or an external sensor may be used **(Figure 29-11)**. This sensor sends a signal to an electronic control module in relation to the rear suspension height. The module controls the air compressor and the exhaust solenoid to control air pressure in the shock absorbers. This action maintains a specific rear suspension trim height regardless of the load on the

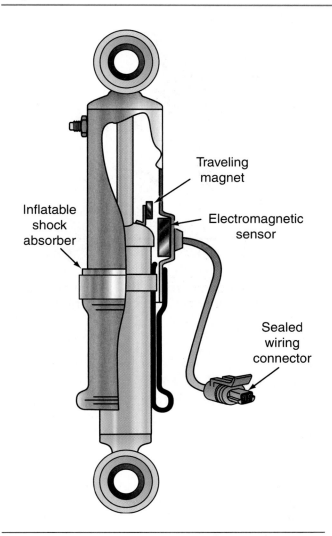

Figure 29-11 Load-leveling shock absorber.

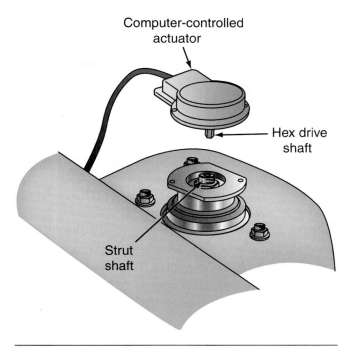

Figure 29-12 Computer-controlled shock absorber.

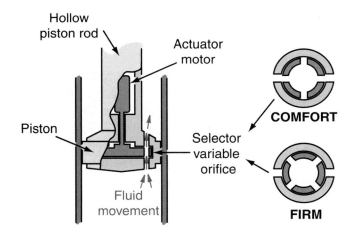

Figure 29-13 A computer-controlled actuator can change ride quality from comfort to firm.

rear suspension. If a heavy package is placed in the trunk, the vehicle chassis is forced downward. However, the load-leveling shock absorbers extend to restore the original rear suspension height.

Electronically Controlled Shock Absorbers and Struts

Many cars are now equipped with computer-controlled suspension systems. In these systems, a computer-controlled actuator is positioned in the top of each shock absorber or strut (**Figure 29-12**). The shock absorber or strut actuators rotate a shaft inside the piston rod. This shaft is connected to the shock valve.

Many of these systems have two modes, soft and firm. In the soft mode, the actuators position the shock absorber valves so there is less restriction to the movement of oil. When the computer changes

the actuators to the firm mode, the actuators position the shock valves so they provide more restriction to oil movement, which provides a firmer ride (**Figure 29-13**).

BALL JOINTS

Ball joints must be in satisfactory condition to provide proper steering control and vehicle safety. Worn ball joints may cause steering wander, inadequate steering control, and excessive tire tread wear.

Load Carrying Ball Joints

The ball joints act as pivot points that allow the front wheels and spindles (or knuckles) to turn between the upper and lower control arms. Ball joints may be grouped into two classifications, **load carrying** and **non-load carrying**. Ball joints may be manufactured with forged, stamped, cold-formed, or screw-machined housings. A load-carrying ball joint supports the vehicle weight. The coil spring is seated on the control arm to which the load-carrying ball joint is attached.

When the coil spring is mounted between the lower control arm and the chassis, the lower ball joint is a load-carrying joint **(Figure 29-14)**. In a torsion bar suspension, the load-carrying ball joint is mounted on the control arm to which the torsion bar is attached. When the spring is mounted above the upper control arm and the chassis, the upper ball joint is a load-carrying joint **(Figure 29-15)**.

In a load-carrying ball joint, the vehicle weight forces the ball stud into contact with the bearing surface in the joint. Load-carrying ball joints may be compression loaded or tension loaded.

When the control arm is mounted above the lower end of the knuckle and rests on the knuckle, the ball joint is compression loaded. In this type of ball joint, the vehicle weight is pushing downward on the control arm and this weight is supported on the tire and wheel that are attached to the steering knuckle. Since the ball joint is mounted between the control arm and the steering

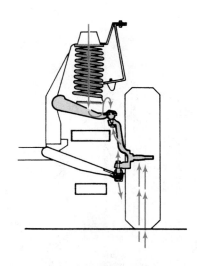

Figure 29-15 Load-carrying ball joint mounted on the upper control arm on which the spring is seated.

knuckle, the vehicle weight squeezes the ball joint together **(Figure 29-16)**. In this type of ball joint mounting, the ball joint is mounted in the lower control arm and the ball joint stud faces downward.

NOTE: The upper strut mount is the load-carrying joint in a Macpherson strut suspension.

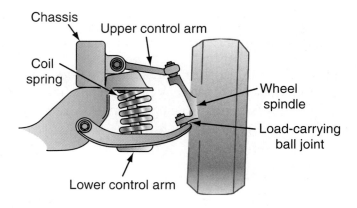

Figure 29-14 Load-carrying ball joint mounted on the lower control arm on which the spring is seated.

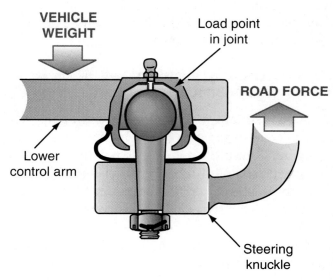

Figure 29-16 Compression-loaded ball joint.

When the lower control arm is positioned below the steering knuckle, the vehicle weight is pulling the ball joint away from the knuckle. This type of ball joint mounting is referred to as tension loaded and is mounted in the lower control arm with the ball joint stud facing upward into the knuckle **(Figure 29-17)**.

Because the load-carrying ball joint supports the vehicle weight, this ball joint wears faster compared to a non-load carrying ball joint. Many load-carrying ball joints have built-in wear indicators. These ball joints have an indicator on the grease nipple surface that recedes into the housing as the joint wears. If the ball joint is in good condition,

the grease fitting shoulder extends a specified distance out of the housing. If the grease fitting shoulder is even with or inside of the ball joint housing, the ball joint is worn and replacement is necessary **(Figure 29-18)**.

Non-Load Carrying Ball Joints

A non-load carrying ball joint may be referred to as a stabilizing ball joint. A non-load carrying ball joint is designed with a preload, which provides damping action **(Figure 29-19)**. This ball joint preload provides improved steering quality and vehicle stability.

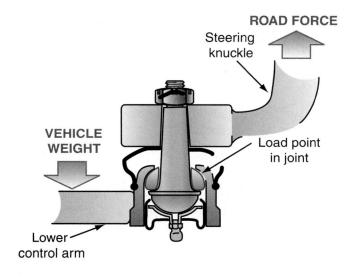

Figure 29-17 Tension-loaded ball joint.

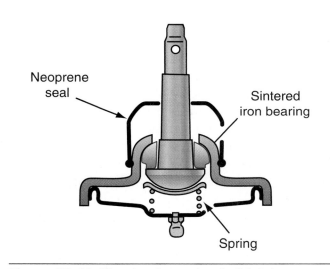

Figure 29-19 Non-load carrying ball joint.

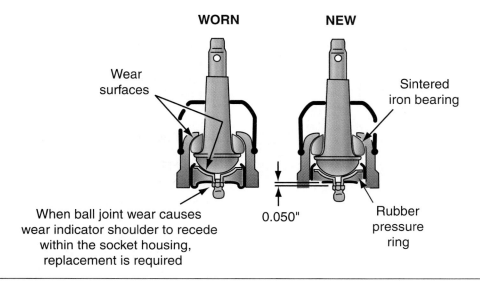

Figure 29-18 Ball joint wear indicator.

SHORT, LONG-ARM (SLA) FRONT SUSPENSION SYSTEMS

An SLA front suspension system has coil springs between the lower control arms and the chassis. Since wheel jounce or rebound movement of one front wheel does not directly affect the opposite front wheel, the control arm suspension is an independent system. Many rear-wheel drive cars have SLA front suspension systems.

Upper and Lower Control Arms

In SLA front suspension systems, the upper control arm is shorter than the lower control arm. During wheel jounce and rebound travel in this suspension system, the upper control arm moves in a shorter arc compared to the lower control arm. This action moves the top of the tire in and out slightly, but the bottom of the tire remains in a more constant position **(Figure 29-20)**. This SLA front suspension system provides little tire tread wear, good ride quality, and good directional stability.

The inner end of the lower control arm contains large rubber insulating bushings and the ball joint is attached to the outer end of the control arm. The lower control arm is bolted to the front crossmember and the attaching bolts are positioned in the center of the lower control arm bushings **(Figure 29-21)**.

The ball joint may be riveted, bolted, pressed, or threaded into the control arm. A spring seat is

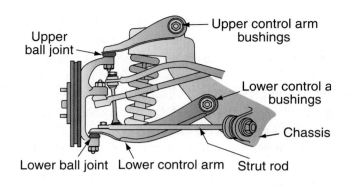

Figure 29-21 Short, long-arm front suspension system with control arm bushings, strut rod, and retaining bolts.

located in the lower control arm. An upper control arm shaft is bolted to the frame and rubber insulators are located between this shaft and the control arm. On some SLA front suspension systems, the coil spring is positioned between the upper control arm and the chassis.

Steering Knuckle

The upper and lower ball joint studs extend through openings in the steering knuckle. Nuts are threaded onto the ball joint studs to retain the ball joints in the knuckle and the nuts are secured with cotter keys. The wheel hub and bearings are positioned on the steering knuckle extension and the wheel assembly is bolted to the wheel hub. When the steering wheel is turned, the steering gear and linkage turn the steering knuckle. During this turning action, the steering knuckle pivots on the upper and lower ball joints. The upper and lower control arms must be positioned properly to provide correct tracking and wheelbase between the front and rear wheels. The control arm bushings must be in satisfactory condition to position the control arms properly.

Coil Spring and Shock Absorber

Coil springs are actually a coiled spring steel bar. When a vehicle wheel strikes a road irregularity, the coil spring compresses to absorb shock and then recoils back to its original installed height. Many coil springs contain a steel alloy that contains different types of steel mixed with other elements such as silicon or chromium. The coil spring is positioned between the lower control arm and the spring seat in the frame. A spring seat is located in the lower

Figure 29-20 Short, long-arm front suspension system.

control arm and an insulator is positioned between the top of the coil spring and the spring seat in the frame. The shock absorber is usually mounted in the center of the coil spring and the lower shock absorber bushing is bolted to the lower control arm. The top of the shock absorber extends through an opening in the frame above the upper spring seat. Washers, grommets, and a nut retain the top of the shock absorber to the frame.

Stabilizer Bar

The stabilizer bar is attached to the crossmember and interconnects the lower control arms. Rubber insulating bushings are used at all stabilizer bar attachment points **(Figure 29-22)**. When jounce and rebound wheel movements affect one front wheel, the stabilizer bar transmits part of this movement to the opposite lower control arm and wheel, which reduces and stabilizes body roll. A **stabilizer bar** may be called a sway bar or an anti-roll bar.

Strut Rod

On some front suspension systems, a strut rod is connected from the lower control arm to the chassis. The strut rod is bolted to the control arm and a large rubber bushing surrounds the strut rod in the chassis opening. The outer end of the strut rod is threaded and steel washers are positioned on each side of the strut rod bushing. Two nuts tighten the strut rod into the bushing **(Figure 29-23)**. The strut rod prevents fore and aft movement of the lower control arm. In some suspension systems, the strut rod nut position provides proper front wheel adjustment **(Figure 29-24)**.

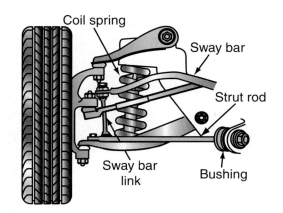

Figure 29-23 A strut rod.

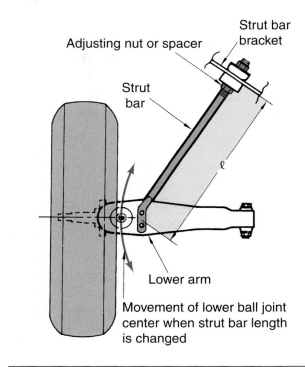

Figure 29-24 The strut rod adjusting nuts maintain proper ball joint position.

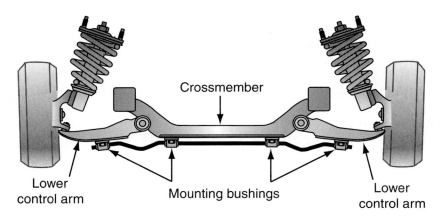

Figure 29-22 A stabilizer bar.

MACPHERSON STRUT-TYPE FRONT SUSPENSION SYSTEM DESIGN

When smaller front-wheel drive cars became popular, most of these cars had Macpherson strut-type front suspension systems. Since the upper control arm is not required in these suspension systems, they are more compact and therefore very suitable for smaller cars.

Lower Control Arms and Support

On some Macpherson strut front suspension systems, a steel support is positioned longitudinally on each side of the front suspension. These supports are bolted to the unitized body. The inner ends of the lower control arms contain large insulating bushings with a bolt opening in the bushing center. The control arm retaining bolts extend through the center of these bushings and openings in the support **(Figure 29-25)**.

Road irregularities cause the tire and wheel to move up and down vertically. The lower control arm bushings pivot on the mounting bolts during this movement. When the vehicle is driven over road irregularities, vibration and noise are applied to the tire and wheel. The control arm bushings help to prevent the transfer of this noise and vibration to the support, unitized body, and passenger compartment. Proper location of the support and lower control arm is important to provide correct vehicle tracking.

NOTE: The supports also carry the engine and transaxle weight. Large rubber mounts are positioned between the engine and transaxle and the supports to absorb engine vibration.

Stabilizer Bar

The stabilizer bar in a Macpherson strut front suspension is similar to the stabilizer bar in a short, long-arm front suspension.

Lower Ball Joint

The lower ball joint is attached to the outer end of the lower control arm. Methods of ball joint to control arm attachment include bolting, riveting, pressing, and threading.

A threaded stud extends from the top of the lower ball joint. This stud fits snugly into a hole in the bottom of the steering knuckle. When the ball joint stud is installed in the steering knuckle opening, a nut and cotter pin retain the ball joint **(Figure 29-26)**.

Steering Knuckle and Bearing Assembly

The front wheel bearing assembly is bolted to the outer end of the steering knuckle and the brake rotor and wheel rim are retained on the studs in the wheel bearing assembly. This front wheel bearing assembly is a complete non-serviceable sealed unit. The front drive shaft is splined into the center

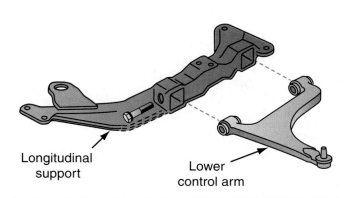

Figure 29-25 Lower control arm and support.

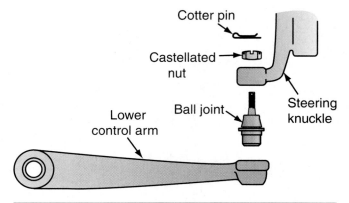

Figure 29-26 A lower ball joint.

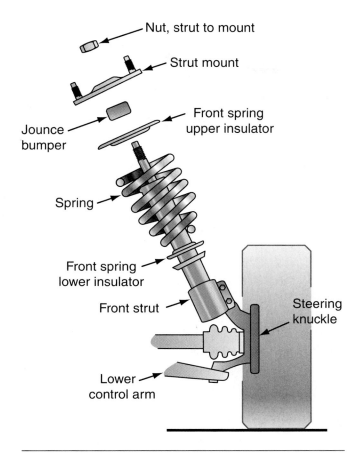

Figure 29-27 Complete Macpherson strut suspension system.

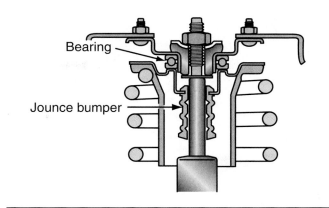

Figure 29-28 Strut bearing and jounce bumper.

of the wheel bearing hub, thus drive axle torque is applied to the front wheel. A tie-rod end connects the steering linkage from the steering gear to the steering knuckle. The top end of the steering knuckle is bolted to the lower end of the strut **(Figure 29-27)**.

Strut and Coil Spring Assembly

The strut is the shock absorber in the front suspension and the lower spring seat is attached near the center area of the strut. An insulator is located between the lower spring seat and the bottom of the coil spring. An upper strut mount is retained on top of the strut with a nut threaded onto the upper end of the strut rod. The upper strut mount contains a bearing and upper spring seat and an insulator is positioned between the top of the coil spring and the seat. The upper and lower insulators help to prevent the transfer of noise and vibration from the spring to the strut and body.

A bumper is located on the upper end of the strut rod. This bumper reduces harshness while driving on severe road irregularities. During upward wheel movement, the bumper strikes the lower spring seat before the coils in the spring hit each other **(Figure 29-28)**. Therefore, this bumper reduces harshness when the wheel and suspension move fully upward. The spring tension is applied against the upper and lower spring seats and insulators. However, the nut on top of the upper mount holds the spring in the compressed position between the upper and lower spring seats. When the steering wheel is turned, the steering linkage turns the steering knuckles to the right or left. During this front wheel turning action, the strut and spring assembly pivot on the lower ball joint and the upper strut mount bearing. All the suspension-to-chassis mounting devices, such as the lower control arm bushings, and the upper strut mount must be positioned properly and be in satisfactory condition to provide correct vehicle tracking and the same wheelbase on both sides of the vehicle.

TORSION BAR SUSPENSION

In some front suspension systems, torsion bars replace the coil springs. During wheel jounce, the torsion bar twists. During wheel rebound, the torsion bar unwinds back to its original position. Torsion bar front suspension systems are used on some four-wheel drive light-duty trucks. Each torsion bar is anchored to the front crossmember

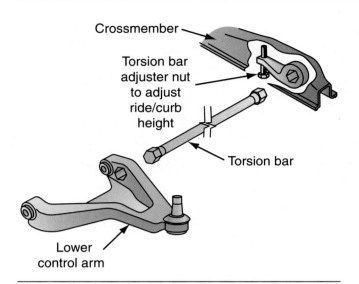

Figure 29-29 Torsion bar mounting.

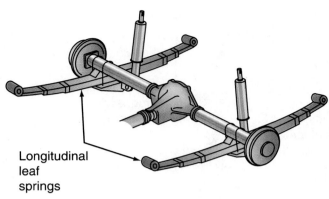

Figure 29-30 Live-axle, leaf-spring rear suspension system.

and the lower control arm **(Figure 29-29)**. A pivot cushion bushing is mounted around the torsion bar. This bushing is bolted to the crossmember opposite to the torsion bar anchor. An insulating bushing is positioned on the end of the torsion bar where it is connected to the lower control arm. The torsion bar anchor adjusting bolts in the crossmember controls vehicle ride height. Front suspension heights must be within specifications for correct wheel alignment, tire wear, satisfactory ride, and accurate bumper heights. A conventional stabilizer bar is connected between the lower control arms and the crossmember. Ball joints are located in the upper and lower control arms. Both ball joints are bolted into the steering knuckle. The shock absorbers are connected between the lower control arms and the crossmember support and the inner ends of the lower control arms are bolted to the crossmember through an insulating bushing.

LIVE-AXLE REAR SUSPENSION SYSTEMS

In a live-axle rear suspension with leaf springs, a leaf spring is mounted longitudinally on each side of the rear suspension rear-wheel drive cars and trucks **(Figure 29-30)**. Leaf springs may be multiple leaf or mono leaf and may be constructed from steel or composite materials. Multiple-leaf springs usually have a series of flat steel leaves of varying lengths that are clamped together. A center bolt extends through all the leaves to maintain the leaf position in the spring. The upper leaf is called the main leaf and this leaf has an eye on each end. An insulating bushing is pressed into each main leaf eye. The front bushing is attached to the frame and the rear bushing is connected through a shackle to the frame. The shackle provides fore and aft movement as the spring compresses. These relatively flat springs provide excellent lateral stability and reduce side sway, which contributes to a well-controlled ride with very good handling characteristics.

However, leaf spring rear suspension systems have a great deal of unsprung weight and leaf springs require a considerable amount of space. A live-axle rear suspension is one in which the differential housing, wheel bearings, and brakes act as a unit. Since the differential axle housing is a one-piece unit, jounce and rebound travel of one rear wheel affects the position of the other rear wheel. This action increases tire wear and decreases ride quality and traction.

The differential axle housing is mounted above or below the springs and a spring plate with an insulating clamp and U-bolts retain the springs to the rear axle housing **(Figure 29-31)**. The shock absorbers are mounted between the spring plates and the frame. The vehicle weight is supported by the springs through the rear axle housing and wheels.

When the vehicle is accelerated, the rear wheels are turning counterclockwise viewed from the left vehicle side. One of Newton's Laws of Motion states that for every action, there is an equal and

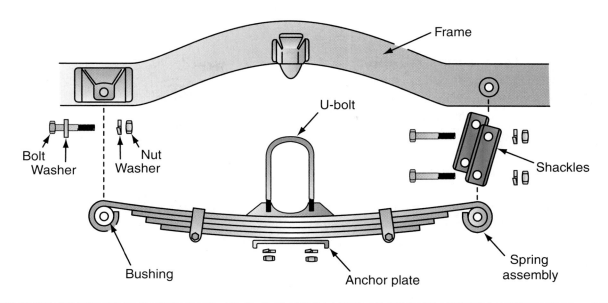

Figure 29-31 Rear leaf springs, shackles, and U-bolts.

opposite reaction. Therefore, when the wheels are turning counterclockwise viewed from the left, the rear axle housing tries to rotate clockwise. This rear axle torque action is absorbed by the rear springs and the chassis moves downward **(Figure 29-32)**. Engine torque supplied through the drive shaft to the differential tends to twist the differential housing and the springs. This twisting action may be referred to as axle windup. Many leaf springs have a shorter distance from the center bolt to the front of the spring compared to the distance from the center bolt to the rear of the spring. This type of leaf spring is referred to as asymmetrical. The shorter distance from the center bolt to the front of the spring resists axle windup. A symmetrical leaf spring has the same distance from the center bolt to the front and rear of the spring.

When braking and decelerating, the rear axle housing tries to turn counterclockwise. This rear axle torque action applied to the springs lifts the chassis. During hard acceleration, the entire powertrain twists in the opposite direction to engine crankshaft and drive shaft rotation. The engine and transmission mounts absorb this torque. However, the twisting action of the drive shaft and differential pinion shaft tends to lift the rear wheel on the passenger's side of the vehicle. Extremely hard acceleration may cause the rear wheel on the passenger's side to lift off the road surface. Once this rear wheel slips on the road surface, engine torque is reduced and the leaf spring forces the wheel downward. When this rear tire contacts the road surface, engine torque increases and the cycle repeats. This repeated lifting of the differential housing is called

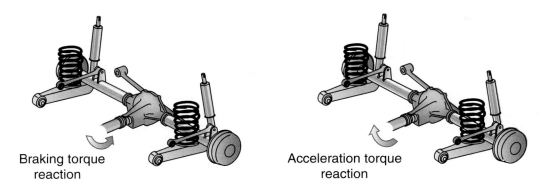

Braking torque reaction

Acceleration torque reaction

Figure 29-32 Rear axle torque action during acceleration and braking.

axle tramp. This action is provided on live-axle rear suspension systems.

Axle tramp is the repeated lifting of one rear wheel off the road surface during hard acceleration. Axle tramp is more noticeable on live-axle leaf spring rear suspension systems in which the springs have to absorb all the differential torque. For this reason, only engines with moderate horsepower were used with this type of rear suspension. Rear suspension and axle components such as spring mounts, shock absorbers, and wheel bearings may be damaged by axle tramp. Mounting one rear shock absorber in front of the rear axle and the other rear shock behind the rear axle helps to reduce axle tramp.

Coil Spring Rear Suspension

Some rear-wheel drive vehicles have a coil spring rear suspension. Upper and lower suspension arms with insulating bushings are connected between the differential housing and the frame **(Figure 29-33)**. The upper arms control lateral movement and the lower trailing control arms absorb differential torque. In some rear suspension systems, the upper arms are replaced with strut rods. The front of the upper and lower arms contain large rubber bushings. When strut rods are used in place of the upper arms, both ends of these rods contain large rubber bushings to prevent

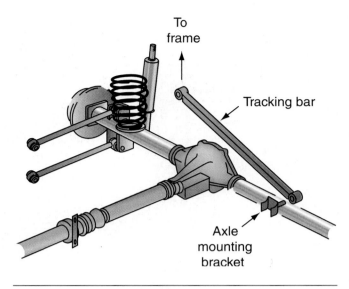

Figure 29-34 Tracking bar.

noise and vibration transfer from the suspension to the chassis. The coil springs are usually mounted between the lower suspension arms and the frame, while the shock absorbers are mounted between the back of the suspension arms and the frame.

Some rear suspension systems have a track bar connected from one side of the differential housing to the chassis to prevent lateral chassis movement **(Figure 29-34)**. Large rubber insulating bushings are positioned in each end of the tracking bar.

NOTE: A tracking bar may be referred to as a *Panhard* or *Watts rod*.

SEMI-INDEPENDENT REAR SUSPENSION SYSTEMS

Many front-wheel drive vehicles have a semi-independent rear suspension that has a solid axle beam connected between the rear trailing arms **(Figure 29-35)**. Some of these rear axle beams are fabricated from a transverse inverted U-section channel.

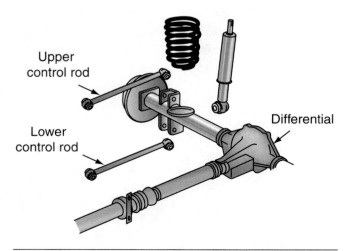

Figure 29-33 Coil-spring rear suspension system with upper and lower control rods.

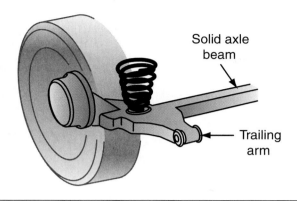

Figure 29-35 Semi-independent rear suspension system.

In some rear suspension systems, the inverted U-section channel contains an integral tubular stabilizer bar. When one rear wheel strikes a road irregularity and the wheel moves upward, the inverted U-section channel twists, which allows some independent rear wheel movement. The trailing arms are connected to chassis brackets through rubber insulating bushings. In some semi-independent rear suspension systems, the coil springs are mounted on the rear struts and the lower spring seat is located on the strut with the upper spring seat positioned on the upper strut mount.

In other semi-independent rear suspension systems, the coil springs are mounted separately from the shock absorbers. Coil spring seats are located on the trailing arms and the shock absorbers are connected from the trailing arms to the chassis. A crossmember connected between the trailing arms provides a twisting action and some independent rear wheel movement.

Some semi-independent rear suspension systems have a track bar connected from a rear axle bracket to a chassis bracket. In some applications, an extra brace is connected from this chassis bracket to the rear upper crossmember. The track bar prevents lateral rear axle movement.

INDEPENDENT REAR SUSPENSION SYSTEMS

In an independent rear suspension system, each rear wheel can move independently from the opposite rear wheel. Independent rear suspension systems may be found on front-wheel drive and rear-wheel drive vehicles. When rear wheel movement is independent, ride quality, tire life, steering control, and traction are improved.

Macpherson Strut Independent Rear Suspension System

In a Macpherson strut rear suspension system, the coil springs are mounted on the rear struts. A lower spring seat is located on the strut and the upper spring seat is positioned on the upper strut mount. This upper strut mount is bolted into the inner fender reinforcement. Dual lower control arms on each side of the suspension are connected from the chassis to the lower end of the spindle **(Figure 29-36)**. The lower end of each strut is bolted to the spindle. Two strut rods are connected forward from the spindles to the chassis. Rubber insulating bushings are located in both ends of the strut rods. A stabilizer bar is mounted in rubber bushings connected to the chassis and the ends of this bar are linked to the struts.

Insulators are mounted between the lower end of the coil spring and the lower spring seat and the

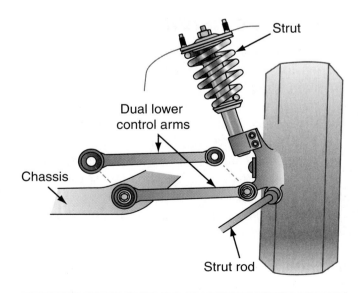

Figure 29-36 Macpherson strut independent rear suspension system.

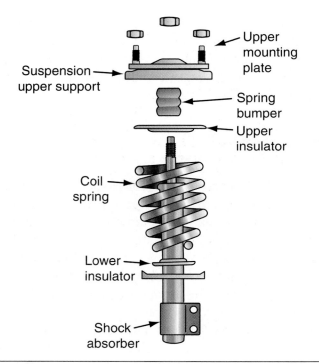

Figure 29-37 Upper and lower spring insulators, Macpherson strut independent rear suspension.

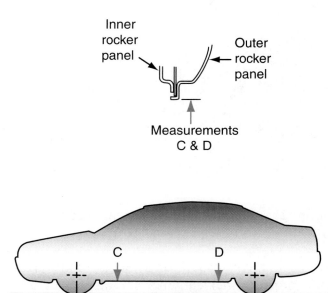

Figure 29-38 Curb ride height measurement.

top of the coil spring and the upper spring support **(Figure 29-37)**. These insulators help to prevent the transfer of spring noise and vibration to the chassis and passenger compartment.

CURB RIDE HEIGHT

Regular inspection and proper maintenance of suspension systems is extremely important to maintain vehicle safety. The curb ride height is determined mainly by spring condition. Other suspension components, such as control arm bushings, will affect curb height if they are worn. Because incorrect curb ride height affects most of the other suspension angles, this measurement is critical.

The curb ride height must be measured at the vehicle manufacturer's specified location, which varies depending on the type of suspension system. When the vehicle is on a level floor or an alignment rack, measure the curb ride height from the floor to the manufacturer's specified location on the chassis on the front and rear suspension **(Figure 29-38)**.

SPRING SAG, CURB RIDE HEIGHT, AND CASTER ANGLE

Sagged springs cause insufficient curb ride height. Therefore, the distance is reduced between the strikeout bumper and its stop. This distance reduction causes the bumper to hit the stop frequently with resulting harsh ride quality.

When both rear springs are sagged, the caster angle tilts excessively toward the rear of the vehicle. This type of angle is called positive caster **(Figure 29-39)**. Rear spring sag and excessive positive caster increases steering effort and causes rapid steering wheel return after a turn is completed.

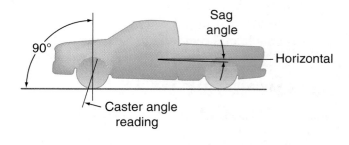

Figure 29-39 Effects of rear spring sag on caster angle.

ELECTRONIC AIR SUSPENSION SYSTEM

In an **electronic air suspension system**, the air springs replace the coil springs in a conventional suspension system. These components are used in conjunction with air valves, an air compressor, sensors, and a module to control the ride height of the vehicle **(Figure 29-40)**. These air springs have a composite rubber and plastic membrane that is clamped to a piston located in the lower end of the spring. An end cap is clamped to the top of the membrane and an air spring valve is positioned in the end cap. The air springs are inflated or deflated to provide a constant vehicle trim height. Front air springs are mounted between the control arms and the crossmember **(Figure 29-41)**. The lower end of these air springs is retained in the control arm with a clip and the upper end is positioned in a crossmember spring seat. The rear air springs are the same as the front air springs except for their mounting. The lower ends of the rear springs are bolted to the rear suspension arms and the upper ends of these springs are attached to the frame.

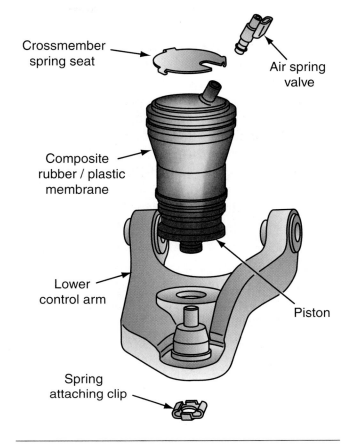

Figure 29-41 Front air spring.

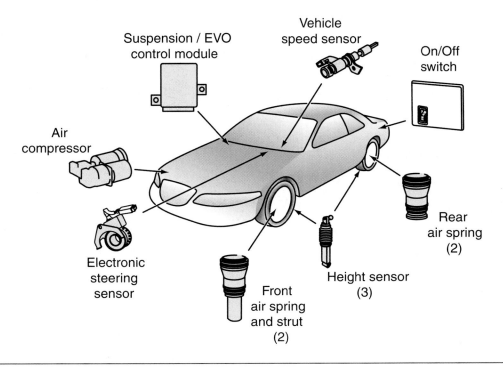

Figure 29-40 Components of an air suspension system.

Air Spring Valves

An air spring valve is mounted in the top of each air spring. These are electric solenoid-type valves that are normally closed. When the valve winding is energized, plunger movement opens the air passage to the air spring. Under this condition, air may enter or be exhausted from the air spring. Two O-ring seals are located on the end of the valves to seal them into the air spring cap. The valves are installed in the air spring cap with a two-stage rotating action similar to a radiator pressure cap.

> *NOTE:* Never rotate an air spring solenoid valve to the release slot in the cap fitting until all the air is released from the spring. If one of these solenoid valves is loosened with air pressure in the spring, the air pressure drives the solenoid out of the spring with extreme force. This action may result in personal injury.

Air Compressor

A single piston in the air compressor is moved up and down in the cylinder by a crankshaft and connecting rod **(Figure 29-42)**. The armature is connected to the crankshaft and the rotating action of the armature moves the piston up and down. Armature rotation occurs when 12 volts are supplied to the compressor input terminal. Intake and discharge

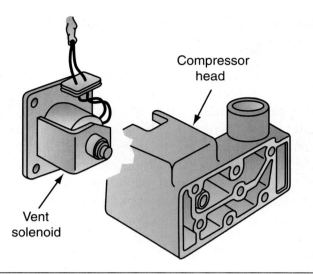

Figure 29-43 An air vent valve.

valves are located in the cylinder head. An air dryer that contains a silica gel dessicant is mounted on the compressor. This silica gel removes moisture from the air as it enters the system. Nylon air lines are connected from the compressor outlets to the air spring valves. The compressor operates when it is necessary to force air into one or more air springs to restore the vehicle ride height. On these computer-controlled suspension systems, the curb ride height may be called the **trim height**.

An air vent valve is located in the compressor cylinder head **(Figure 29-43)**. This normally closed electric solenoid valve allows air to be vented from the system. When it is necessary to exhaust air from an air spring, the air spring valve and vent valve must be energized at the same time with the compressor shut off. Air exhausting is necessary if the vehicle trim height is too high.

> *NOTE:* The on/off switch must be in the off position before the car is hoisted, jacked, towed, or raised off the ground. If this precaution is not observed, it may result in personal injury, component damage, or vehicle damage.

Compressor Relay

When the compressor relay is energized, it supplies 12 volts through the relay contacts to the

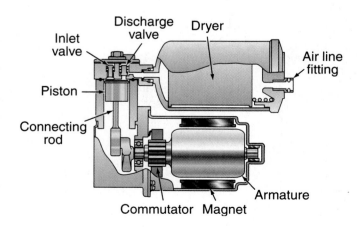

Figure 29-42 An air compressor.

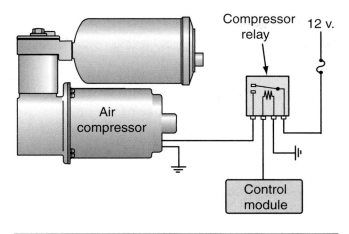

Figure 29-44 A compressor relay.

compressor input terminal **(Figure 29-44)**. The relay contacts open the circuit to the compressor if the relay is de-energized.

Control Module

The control module is a microprocessor that operates the compressor, vent valve, and air spring valves to control the amount of air in the air springs and maintain the trim height. The control module is usually located in the trunk **(Figure 29-45)**. The control module turns on the service indicator to alert the driver when a system defect is present. Diagnostic capabilities are also designed into the control module to flash codes or a scan tool may be used.

On/Off Switch

The on/off switch opens the 12-volt supply circuit to the control module. This switch is located in the trunk near the control module. When the air suspension system is serviced or some other vehicle service is performed, the switch must be in the off position.

Height Sensors

In the air suspension system, there are two front height sensors located between the lower control arms and the crossmember. A single rear height sensor is positioned between the suspension arm and the frame. Each height sensor contains a magnet slide that is attached to the upper end of the sensor. This magnet slide moves up and down in the lower sensor housing as changes in vehicle trim height occur **(Figure 29-46)**. The lower sensor housing contains two electronic switches that are connected through a wiring harness to the control module. When the vehicle is at trim height, the switches remain closed and the control module receives a trim height signal. If the magnet slide moves upward, the above trim switch opens. When the module receives the signal it opens the appropriate air spring valve and the vent valve. This action exhausts air from the air spring and corrects the above trim height condition. Downward magnet slide movement closes the above trim switch and opens the below trim switch. When the control module receives this signal, it energizes the compressor relay and starts the compressor. The control module opens the appropriate air

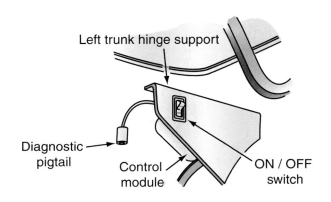

Figure 29-45 A control module and its on/off switch.

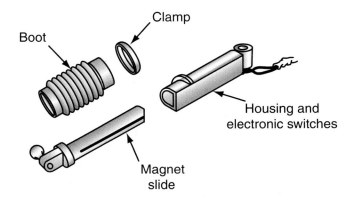

Figure 29-46 Rear height sensor mounting.

spring valve. This action forces air into the air spring to correct the below trim height condition. The height sensors are serviced as a unit.

Tech Tip *Never attempt to probe or inspect the electronic switches in the height sensor since this may cause sensor damage.*

GENERAL SYSTEM OPERATION

If a door is opened with the brake pedal released, raise-vehicle commands are completed immediately, but lower-vehicle requests are serviced after the door is closed. This action prevents an open door from catching on curbs or other objects.

If the doors are closed and the brake pedal is released, all commands are serviced by a 45-second averaging method to prevent excessive suspension height corrections on irregular road surfaces.

If the brake is applied and a door is open, raise-vehicle commands are completed immediately, but lower-vehicle requests are ignored.

When the doors are closed and the brake pedal is applied, all requests are ignored by the control module. If a request to raise the rear suspension is in progress under these conditions, this request will be completed. This action prevents correction of front end jounce while braking.

Warning Lamp

When the control module senses a system defect, the module turns the air suspension warning lamp which informs the driver that a problem exists. If the air suspension system is working normally, the warning lamp will be on for 1 second when the ignition switch is turned from the off to the run position. After this time, the warning lamp should remain off. This lamp does not operate with the ignition switch in the start position. The warning lamp is used during the self-diagnostic procedure and the spring fill sequence.

CONTINUOUSLY VARIABLE ROAD SENSING SUSPENSION (CVRSS) SYSTEM

The **CVRSS system** controls shock absorber and strut firmness very precisely to provide improved ride quality, steering control, and directional stability.

General Description

The CVRSS system may be referred to as a real-time damping (RTD) system in the on-board diagnostics. The CVRSS controls damping forces in the front struts and rear shock absorbers in response to various road and driving conditions. The CVRSS system changes shock and strut damping forces in 10 to 15 milliseconds, whereas other suspension damping systems require a much longer time interval to change damping forces. It requires about 200 milliseconds to blink your eye—this gives us some idea how quickly the CVRSS system reacts. The CVRSS module receives inputs regarding vertical acceleration, wheel-to-body position and speed of wheel movement, vehicle speed, and lift/dive **(Figure 29-47)**. The CVRSS module evaluates these inputs and controls a solenoid in each shock or strut to provide suspension damping control. The solenoids in the shocks and struts can react much faster compared to the strut actuators explained previously in some systems.

Position Sensors

A position sensor is mounted at each corner of the vehicle between a control arm and the chassis. These sensor inputs provide analog voltage signals to the CVRSS module regarding relative wheel-to-body movement and the velocity of wheel movement **(Figure 29-48)**. **Analog voltage signals** vary continually within a specific voltage range.

The rear position sensor inputs also provide rear suspension height information to the CVRSS module. This information is used by the module to control the rear suspension trim height. All four position sensors have the same design.

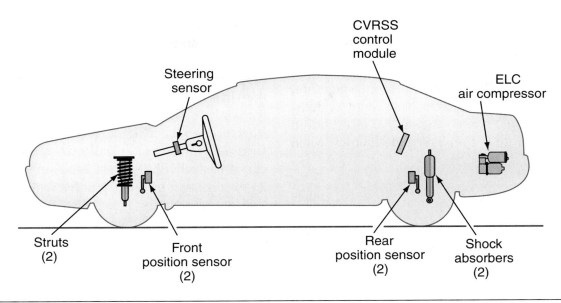

Figure 29-47 Continuously variable road sensing suspension (CVRSS) system components.

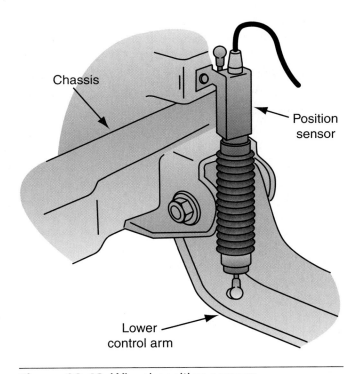

Figure 29-48 Wheel position sensor.

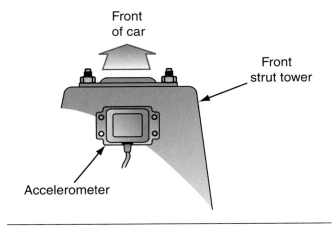

Figure 29-49 Front accelerometer mounting location.

the strut towers **(Figure 29-49)** and the rear accelerometers are located on the rear chassis near the rear suspension support. All four accelerometers are similar in design; they send analog voltage signals to the CVRSS module.

Accelerometer

An accelerometer is mounted on each corner of the vehicle. These inputs send information to the RSS module in relation to vertical acceleration of the body. The front accelerometers are mounted on

Vehicle Speed Sensor

The VSS is mounted in the transaxle. This sensor sends a voltage signal to the PCM in relation to vehicle speed. The VSS signal is transmitted from the PCM to the CVRSS module.

Lift/Dive Input

The lift/dive input is sent from the PCM to the CVRSS module. Suspension lift information is obtained by the PCM from the throttle position, vehicle speed, and transaxle gear input signals. The PCM calculates suspension dive information from the rate of vehicle speed change when decelerating.

CVRSS Module

The CVRSS module is mounted on the right side of the electronics bay in the trunk. Extensive self-diagnostic capabilities are programmed into the CVRSS module.

Damper Solenoid Valves

Each strut or shock damper contains a solenoid that is controlled by the CVRSS module. Each damper solenoid provides two levels of damping, firm or soft. In the soft mode, the CVRSS module switches on the solenoid. This causes the oil in the shock or strut to bypass the main damper valving. Voltage is supplied through an RSS damper relay

to each strut or shock damper solenoid and the RSS module energizes each damper solenoid by providing a ground for the solenoid winding **(Figure 29-50)**. When the system switches to the firm mode, the CVRSS module switches off the damper solenoid. This action causes the oil to flow through the main damper valving to provide a firm ride. Each strut or shock damper solenoid circuit is basically the same. Each damper solenoid is an integral part of the damper assembly and is not serviced separately. The CVRSS system operates automatically without any driver-controlled inputs. The fast reaction time of the CVRSS system provides excellent control over ride quality and body lift or dive, which provides improved vehicle stability and handling. Since the position sensors actually sense the velocity of upward and downward wheel movements and the damper solenoid reaction time is 10 to 15 milliseconds, the RSS module can react to these position sensor inputs very quickly. For example, if a road irregularity drives a wheel upward, the CVRSS module switches the damper solenoid to the firm mode before that wheel strikes the road again during downward (rebound) movement.

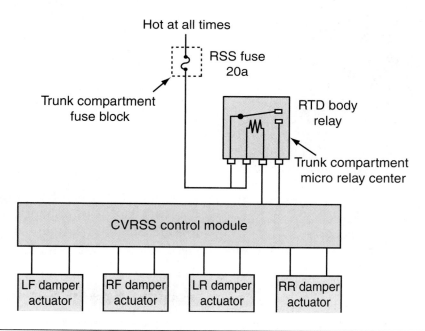

Figure 29-50 Strut damper solenoids and damper relay wiring connections.

Summary

- Wheel and tire jounce travel occur when a tire strikes a hump in the road surface and the wheel and tire move upward.

- Rebound wheel and tire travel occur when the tire and wheel move downward after jounce travel.

- The shock absorbers control spring action and prevent excessive wheel and tire oscillations.

- Shock absorber valves are matched to the amount of energy that may be stored in the spring.

- A nitrogen gas charge is located in the oil reservoir of many shock absorbers and struts to prevent oil cavitation and foaming, which provides more positive shock absorber action.

- Shock absorber ratio refers to the difference between the shock absorber control on the compression and extension cycle. Many shock absorbers provide more control on the extension cycle.

- Most front struts are connected between the steering knuckle and the upper strut mount.

- In a torsion bar suspension system, the torsion bars replace the coil springs.

- A load-carrying ball joint may be compression loaded or tension loaded.

- In an SLA suspension system, the coil springs may be mounted between the lower control arm and the frame or between the upper control arm and the chassis.

- In a live-axle leaf spring rear suspension system, the leaf springs absorb differential torque and provide lateral control.

- In a live-axle coil spring rear suspension system, the lower control arms absorb differential torque and the upper arms control lateral movement.

- A semi-independent rear suspension has a limited amount of individual rear wheel movement provided by a steel U-section channel or crossmember.

- In an independent rear suspension, each rear wheel can move individually without affecting the opposite rear wheel.

- Compared to a live-axle rear suspension system, an independent rear suspension provides improved ride quality, steering control, tire life, and traction.

- An electronic air suspension system maintains a constant vehicle trim height regardless of passenger or cargo load.

- The air spring valves are retained in the air spring caps with a two-stage rotating action, much like a radiator cap.

- An air spring valve must never be loosened until the air is exhausted from the spring.

- The on/off switch in an electronic air suspension system supplies 12 volts to the control module. This switch must be off before the car is hoisted, jacked, towed, or raised off the ground.

- In an electronic air suspension system, if the doors are closed and the brake pedal is released, all requests to the control module are serviced by a 45-second averaging method.

- If the control module in an electronic air suspension system cannot complete a request from a height sensor in 3 minutes, the control module illuminates the suspension warning lamp.

- The continuously variable road sensing suspension system changes shock and strut damping forces in 10 to 15 milliseconds.

Review Questions

1. When discussing a torsion bar front suspension system, Technician A says the one end of the torsion bar is attached to the upper control arm. Technician B says the suspension height adjustment is positioned on the end of the torsion bar connected to the control arm. Who is correct?

   A. Technician A

   B. Technician B

   C. Both Technician A and Technician B

   D. Neither Technician A nor Technician B

2. While discussing semi-independent rear suspensions, Technician A says some individual rear wheel movement is provided by the trailing arms. Technician B says some independent rear wheel movement is provided by the struts. Who is correct?

   A. Technician A

   B. Technician B

   C. Both Technician A and Technician B

   D. Neither Technician A nor Technician B

3. While discussing rear axle tramp, Technician A says rear axle tramp is the repeated lifting of the passenger's side tire off the road surface during hard acceleration. Technician B says rear axle tramp occurs because of the engine torque transmitted through the drive shaft. Who is correct?

   A. Technician A

   B. Technician B

   C. Both Technician A and Technician B

   D. Neither Technician A nor Technician B

4. All these statements about shock absorber and strut design and operation are true *except*:

   A. Jounce wheel travel occurs when a wheel moves upward.

   B. Rebound wheel travel occurs when a wheel moves downward.

   C. Some shock absorbers and struts have a nitrogen gas charge in place of hydraulic oil.

   D. Shock absorbers and struts control the spring action.

5. A typical shock absorber ratio is:

   A. 20:60.

   B. 40:80.

   C. 70:30.

   D. 45:90.

6. While discussing strut design, Technician A says some struts have a replaceable cartridge. Technician B says on many front suspension systems when the front wheels are turned, the strut and coil spring assembly rotate on the upper strut mount. Who is correct?

   A. Technician A

   B. Technician B

   C. Both Technician A and Technician B

   D. Neither Technician A nor Technician B

7. All these statements about ball joints are true *except*:

   A. A non-load carrying ball joint wears faster than a load carrying ball joint.

   B. If the lower control arm is mounted above the lower end of the steering knuckle, the ball joint between these components is compression loaded.

   C. A load carrying ball joint supports the vehicle weight.

   D. In a ball joint with wear-indicating capabilities, the grease fitting shoulder should be extended from the joint housing.

8. The electronic air suspension switch should be in the off position if the:

   A. vehicle is boosted with a booster battery.

   B. vehicle is diagnosed with a scan tool.

C. vehicle is jacked up on one corner to change a tire.

D. battery is being changed.

9. All these statements about CVRSS are true *except*:

   A. Position sensors are mounted at each corner of the vehicle.

   B. The position sensors send digital voltage signals to the CVRSS module.

   C. The position sensor signals are in relation to the amount and speed of wheel-to-body movement.

   D. An accelerometer is mounted on each corner of the vehicle.

10. In a CVRSS, the PCM sends a lift/dive signal to the CVRSS module. The PCM calculates the lift/dive signal by using information from:

    A. the engine coolant temperature sensor and engine speed signals.

    B. the mass airflow sensor and intake air temperature sensor signals.

    C. the crankshaft position sensor and camshaft position sensor signals.

    D. the throttle position sensor, vehicle speed sensor, and transaxle gear position sensor signals.

11. If the lower control arm is mounted above the steering knuckle and rests on the knuckle, the lower ball joint is _____ _____ .

12. When one front wheel strikes a road irregularity, the stabilizer bar reduces _____ _____ .

13. The strut rod prevents _____ and _____ lower control arm movement.

14. When the front springs are sagged, the front wheel caster becomes more _____ .

15. Explain the purpose of the nitrogen gas charge in shock absorbers and struts.

16. Explain the differential torque action during acceleration and describe how this torque is absorbed in a live axle coil spring rear suspension.

17. List the conditions when the on/off switch in an electronic air suspension system must be turned off.

18. Explain why the control module in an electronic air suspension system services all commands by a 45-second averaging method when the doors are closed and the brake pedal is released.

CHAPTER

30

Shocks and Suspension Service

Learning Objectives

After you have read, studied, and practiced the contents of this unit, you should be able to:

- Maintain, diagnose, and service shock absorbers.

- Maintain, diagnose, and service front and rear suspension systems.

- Measure ball joint wear.

- Diagnose electronic air suspension systems.

- Adjust front and rear trim height on electronic air suspension systems.

- Diagnose continuously variable road sensing suspension systems.

Key Terms

Bounce test

Compression stroke

Rebound stroke

Strut chatter

INTRODUCTION

Proper front and rear suspension system maintenance, diagnosis, and service are extremely important to provide adequate vehicle safety and maintain ride comfort and normal tire life. If worn or loose front suspension system components are ignored, steering control may be adversely affected, which may result in loss of steering control and an expensive collision. Defective front suspension system components, such as worn-out shock absorbers and broken springs, may cause rough riding that results in driver and passenger discomfort. Other worn front suspension components, such as worn ball joints and control arm bushings, can cause improper alignment angles that cause excessive front tire wear. Therefore, you must be familiar with front and rear suspension maintenance, diagnosis, and service.

Each year, more vehicles are equipped with computer-controlled suspension systems. These systems are becoming increasingly complex. Therefore, you must understand the correct procedures for diagnosing and servicing these systems. When you understand computer-controlled suspension systems and the proper diagnostic procedures for these systems, diagnosis becomes faster and more accurate.

SHOCK ABSORBER MAINTENANCE

Shock absorbers should be inspected for loose mounting bolts and worn mounting bushings. If these components are loose, rattling noise is evident, and replacement of the bushings and bolts is necessary.

In some shock absorbers, the bushing is permanently mounted in the shock and the complete unit must be replaced if the bushing is worn. When the mounting bushings are worn, the shock absorber will not provide proper spring control.

Shock absorbers and struts should be inspected for oil leakage. A slight oil film on the lower oil chamber is acceptable. Any indication of oil dripping is not acceptable and the unit should be replaced. Replacement of shock absorbers and struts should always be done in pairs **(Figure 30-1)**. Struts and shock absorbers should be inspected visually for a bent condition and severe dents or punctures **(Figure 30-2)**.

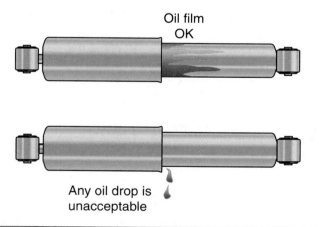

Oil film
OK

Any oil drop is
unacceptable

Figure 30-1 Shock absorber and strut oil leak diagnosis.

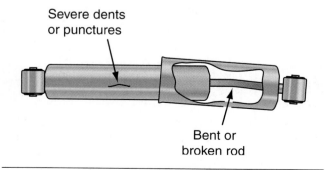

Severe dents
or punctures

Bent or
broken rod

Figure 30-2 Damaged shock absorber inspection.

SHOCK ABSORBER DIAGNOSIS

A **bounce test** may be performed to determine shock absorber condition. When the bounce test is performed, the bumper is pushed downward with considerable weight applied on each corner of the vehicle. The bumper is released after this action and one free upward bounce should stop the vertical chassis movement if the shock absorber or strut provides proper spring control. Shock absorber replacement is required if more than one free upward bounce occurs.

A manual test may be performed on shock absorbers. When this test is performed, disconnect the lower end of the shock and move the shock up and down as rapidly as possible. A satisfactory shock absorber should offer a strong, steady resistance to movement on the entire compression and rebound strokes. The amount of resistance may be different on the **compression stroke** compared to the **rebound stroke**. If a loss of resistance is experienced during either stroke, shock replacement is essential.

Did You Know? *The compression stroke occurs when the wheel travels or moves upward toward the body. The rebound stroke occurs when the wheel moves downward or away from the vehicle body.*

Did You Know? *The rebound stroke of the shock absorber occurs when the shock absorber is extending.*

Some defective shock absorbers or struts may have internal clunking, clicking and squawking noises, or binding conditions. When these shock absorber noises or conditions are experienced, shock absorber or strut replacement is necessary. Strut chatter may be heard when the steering wheel is turned with the vehicle not moving or moving at low speed. **Strut chatter** is a repeated clicking noise when the front wheels are turned to the right or left. To verify the location of this chattering noise, place one hand on a front coil spring while someone turns the steering wheel. If strut chatter is present, the spring binds and releases as it turns. The upper spring seat binding against the strut bearing mount causes this condition. A noise that occurs on sharp turns or during front suspension jounce may be caused by interference between the upper strut rebound stop and the upper mount or strut tower, the coil spring and tower, or the coil spring and the upper mount.

On some models, these coil spring interference problems may be corrected by installing upper coil spring spacers on top of the coil spring. Spring removal from the strut is required to install these spacers.

> **Tech Tip** *Gas-filled shock absorbers will extend when disconnected.*

SHOCK ABSORBER SERVICE

When shock absorber replacement is necessary, follow this procedure:

1. Prior to rear shock absorber replacement, raise the vehicle on a lift and support the rear axle on jack stands so the shock absorbers are not fully extended.

2. When a front shock absorber is changed, lift the front end of the vehicle with a floor jack and then place jack stands under the lower control arms. Lower the vehicle onto the jack stands and remove the floor jack.

3. Disconnect the upper shock mounting nut and grommet.

4. Remove the lower shock mounting nut or bolts and remove the shock absorber.

5. Reverse steps 1 through 3 to install the new shock absorber and grommets. Do not overtighten the rubber bushings. The bushing should not protrude past the retaining washers.

6. With the full vehicle weight supported on the suspension, tighten the shock mounting nuts to the specified torque.

> **SAFETY TIP** *Never apply heat to the lower shock absorber or strut chamber with an acetylene torch. Excessive heat may cause a shock absorber or strut explosion and result in personal injury.*

Strut Removal and Replacement

Before a front strut and spring assembly is removed, the strut must be removed from the steering knuckle and top strut mount bolts must be removed from the strut tower.

If the strut uses an upper plate for wheel alignment, always mark the position of the retaining bolts and other fasteners for correct position during reassembly **(Figure 30-3)**. If an eccentric camber bolt is used to attach the strut to the knuckle, always mark the eccentric and bolt head in relation to the strut and reinstall the bolt in the same position **(Figure 30-4)**. Always follow the vehicle manufacturer's recommended procedure in the service manual for removal of the strut and spring assembly.

The following is a typical procedure for strut and spring assembly removal:

1. Raise the vehicle on a lift or floor jack. If a floor jack is used to raise the vehicle, lower the vehicle

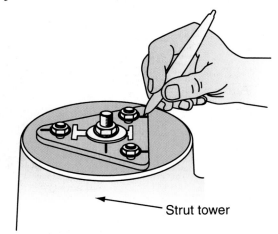

Strut tower

Figure 30-3 **Upper plate bolt marking for strut removal.**

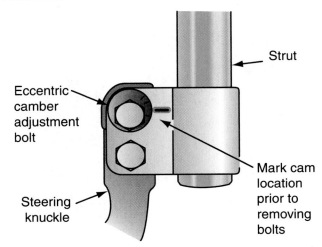

Strut

Eccentric camber adjustment bolt

Mark cam location prior to removing bolts

Steering knuckle

Figure 30-4 **Camber bolt marking for strut removal.**

onto jack stands placed under the chassis so the lower control arms and front wheels drop downward. Remove the floor jack from under the vehicle.

2. Remove the brake line and ABS wheel speed sensor wire from clamps on the strut **(Figure 30-5)**. In some cases, these clamps may have to be removed from the strut.

3. Remove the strut-to-steering knuckle retaining bolts and remove the strut from the knuckle **(Figure 30-6)**.

4. Remove the upper strut mounting bolts on top of the strut tower and remove the strut and spring assembly **(Figure 30-7)**.

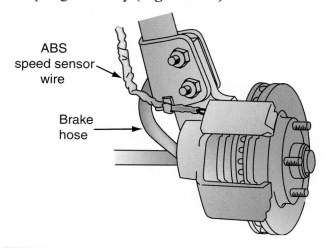

Figure 30-5 Brake hose and ABS wheel-speed sensor wire removal from the strut.

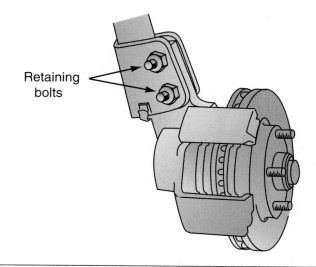

Figure 30-6 Removing strut-to-knuckle retaining bolts.

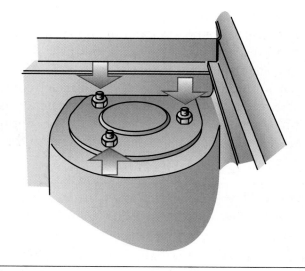

Figure 30-7 Removing upper mounting bolts on top of strut tower.

Removal of Strut from Coil Spring

The coil spring must be compressed with a special tool before the strut can be removed. All the tension must be removed from the upper spring seat before the upper strut piston rod nut is loosened. Many different spring compressing tools are available and they must always be used according to the manufacturer's recommended procedure. If the coil spring has an enamel-type coating, tape the spring where the compressing tool contacts the spring. The spring may break prematurely if this coating is chipped. A typical procedure for removing a strut from a coil spring is the following:

1. Install the strut assembly into the spring compressing tool according to the tool or vehicle manufacturer's recommended procedure.

2. Turn the handle on top of the compressing tool until all the spring tension is removed from the upper strut mount **(Figure 30-8)**.

3. Loosen the nut on the upper strut mount. Be sure all the spring tension is removed from the upper strut mount before loosening and removing this nut **(Figure 30-9)**.

4. Remove the upper strut mount, upper insulator, and spring bumper. Drop the strut from inside of the coil spring. The coil spring will stay inside the spring compressor. Remove the lower insulator from the spring **(Figure 30-10)**.

Figure 30-8 The spring compressing tool is tightened until all the spring tension is removed from the upper strut mount.

Figure 30-9 Removal of nut from strut piston rod.

Figure 30-10 Removal of upper strut mount, upper insulator, spring bumper, and lower insulator.

SAFETY TIP *Always use a coil spring compressing tool according to the tool or vehicle manufacturer's recommended service procedure. Be sure the tool is properly installed on the spring. If a coil spring slips off the tool when the spring is compressed, severe personal injury or property damage may occur. Never loosen the upper strut mount retaining nut on the end of the strut rod unless the spring is compressed enough to remove all spring tension from the upper strut mount. If this nut is loosened with spring tension on the upper mount, this mount becomes a dangerous projectile that may cause serious personal injury or property damage.*

Installation of Coil Spring on Strut

The following is a typical procedure for installing a coil spring on a strut:

1. Install the lower insulator on the lower strut spring seat and be sure the insulator is properly seated. **(Figure 30-11)**.

2. Install the spring bumper on the strut rod **(Figure 30-12)**. With the coil spring compressed in the spring compressing tool, install the spring on

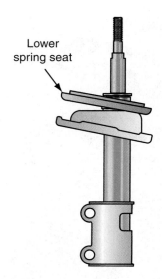

Figure 30-11 Insulator installation on lower spring seat.

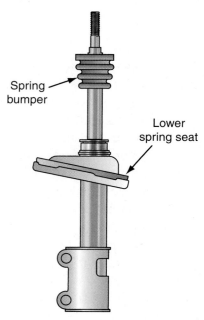

Figure 30-12 Spring bumper installation on strut piston rod.

Figure 30-13 Installing strut into spring compressing tool.

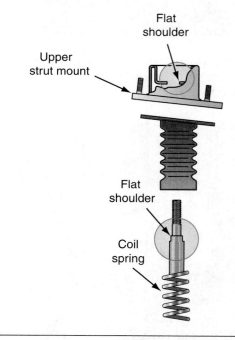

Figure 30-14 Upper strut mount properly positioned on strut piston rod.

the strut **(Figure 30-13)**. Be sure the spring is properly seated on the lower insulator spring seat.

3. Be sure the strut piston rod is fully extended and install the upper insulator on top of the coil spring.

4. Install the upper strut mount on the upper insulator. Be sure the spring, upper insulator, and upper strut mount are properly positioned and

seated on the coil spring and strut piston rod **(Figure 30-14)**.

5. Install a bolt in the upper strut-to-knuckle retaining bolt and clamp this bolt in a vise to hold the strut, spring, and compressing tool as in the disassembly procedure.

Figure 30-15 Tightening nut on strut piston rod.

6. Use the compressing tool bar to hold the strut and spring from turning and tighten the strut piston rod nut to the specified torque **(Figure 30-15)**.

7. Rotate the upper strut mount until the mount is properly aligned with the lower spring seat.

8. Gradually loosen the nut on the compressing tool until all the spring tension is released from the tool and remove the tool from the spring.

SAFETY TIP *Never clamp the lower strut or shock absorber chamber in a vise with excessive force. This action may distort the lower chamber and affect piston movement in the strut or shock absorber. Always follow the vehicle manufacturer's recommended procedure for strut disposal. Do not throw gas-filled struts or shock absorbers in a dumpster or in a fire of any kind. If the vehicle manufacturer recommends drilling the strut to release the gas charge, drill the strut only at the manufacturer's recommended location.*

Tech Tip *It is possible to install replacement cartridges in some struts rather than replacing the strut.*

FRONT AND REAR SUSPENSION SYSTEM MAINTENANCE

When performing under-car service, inspect the suspension system components. Inspect the steering knuckle for looseness where the tie-rod end fits into the knuckle. There should be no looseness in the tie-rod end and the tie-rod end must be properly secured to the steering knuckle with a nut and cotter pin.

Upper and lower control arms should be inspected for cracks, bent conditions, or worn bushings. If the control arm bushings are worn, steering is erratic especially on irregular road surfaces. Worn control arm bushings may cause a rattling noise while driving on irregular road surfaces. Dry or worn control arm bushings may cause a squeaking noise on irregular road surfaces. Caster and camber angles on the front suspension are altered by worn upper and lower control arm bushings. Incorrect caster or camber angles may cause the vehicle to pull to one side and tire wear may be excessive when the camber angle is not within specifications.

Leaf springs should be inspected for a sagged condition that causes the curb riding height to be less than specified. Leaf springs should also be visually inspected for broken leafs, broken center bolts, and worn shackles or bushings. Weak or broken leaf springs affect front suspension alignment angles and cause excessive tire tread wear, reduced directional stability, and harsh riding. A rattling noise while driving on irregular road surfaces may be caused by worn shackles or bushings. Worn shackles and bushings lower curb riding height and cause a rattling noise on road irregularities.

Many leaf springs have plastic silencers between the spring leafs. If these silencers are worn out, creaking and squawking noises are heard when the vehicle is driven over road irregularities at low speeds.

When the silencers require checking or replacement, lift the vehicle with a floor jack and support the frame on jack stands so the rear suspension moves downward. With the vehicle weight no longer applied to the springs, the spring leafs may be pried apart with a pry bar to remove and replace the silencers.

Worn shackle bushings, brackets, and mounts cause excessive chassis lateral movement and rattling noises. With the normal vehicle weight resting on the springs, insert a pry bar between the rear outer end of the spring and the frame. Apply downward pressure on the bar and observe the rear shackle for movement. Shackle bushings, brackets, or mounts must be replaced if there is movement in the shackle. The same procedure may be followed to check the front bushing in the main leaf. A broken spring center bolt may allow the rear axle assembly to move rearward on one side. This movement changes rear wheel tracking resulting in handling problems, tire wear, and reduced directional stability.

Sagged rear springs reduce the curb riding height. Spring replacement is necessary if the springs are sagged. Rebound bumpers are usually bolted to the lower control arm or to the chassis. Inspect the rebound bumpers for cracks, wear, and flattened conditions **(Figure 30-16)**. Damaged rebound bumpers may be caused by sagged springs and insufficient curb riding height or worn-out shock absorbers and struts. If the rebound bumpers must be replaced, remove the mounting bolts and the bumper. Install the new bumper and tighten the mounting bolts to the specified torque.

Worn stabilizer bar mounting bushings, grommets, or mounting bolts cause a rattling noise as the vehicle is driven on irregular road surfaces. A weak stabilizer bar or worn bushings and grommets cause harsh riding and excessive body sway while driving on irregular road surfaces. Worn or very dry stabilizer bar bushings may cause a squeaking noise on irregular road surfaces. All stabilizer bar components should be visually in-

spected for wear. Inspect the strut rod for a bent condition or a loose bushing.

A bent strut rod or loose bushing may cause steering pull. Rear tie-rods should be inspected for worn grommets, loose mountings, or bent conditions. Loose tie-rod bushings or a bent tie-rod will change the rear wheel tracking and result in reduced directional stability. Worn tie-rod bushings also cause a rattling noise on road irregularities. When the rear tie-rod is replaced, remove the front and rear rod mounting nuts. The lower control arm or rear axle may have to be pried rearward to remove the tie-rod. Check the tie-rod grommets and mountings for wear and replace parts as required. When the tie-rod is reinstalled, tighten the mounting bolts to specifications and check the rear wheel toe. After replacement of front or rear suspension components such as control arms, control arm bushings, springs, and strut rods, front and rear suspension alignment should be checked.

CURB RIDING HEIGHT MEASUREMENT

Regular inspection and proper maintenance of suspension systems is extremely important to maintain vehicle safety. The curb riding height is determined mainly by spring condition. Other suspension components, such as control arm bushings, affect curb riding height if they are worn. Since incorrect curb riding height affects most of the other suspension angles, this measurement is critical. Reduced curb riding height on the front suspension may cause decreased directional stability. If the curb riding height is reduced on one side of the front suspension, the steering may pull to one side. Reduced rear suspension height increases steering effort and causes rapid steering wheel return after turning a corner. Harsh riding occurs when the curb riding height is less than specified. The curb riding height must be measured at the vehicle manufacturer's specified location, which varies depending on the type of suspension system.

Curb ride height is measured when the vehicle is on a level floor or an alignment rack. On some cars the curb ride height is measured from the floor

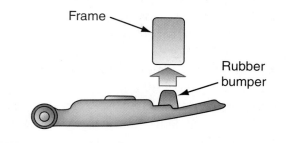

Figure 30-16 Rebound bumper.

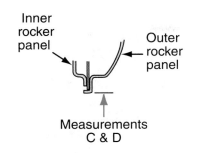

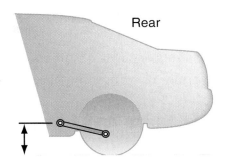

Figure 30-19 Curb riding height measurement, rear suspension.

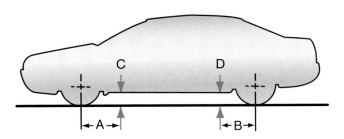

Figure 30-17 Curb riding height measurement, front and rear.

Tech Tip *Correcting curb height often corrects some concerns with alignment angles. If the curb height is not within specifications, then the vehicle cannot be properly aligned.*

---

to the rocker panel on both sides of the front and rear suspension **(Figure 30-17)**. On some vehicles the front curb ride height is measured from the floor to the center of the lower control arm-mounting bolt on both sides of the front suspension **(Figure 30-18)**. On other vehicles the rear curb ride height is measured from the floor to the center of the strut rod-mounting bolt **(Figure 30-19)**. Follow the manufacturer's procedure when making this measurement.

If the curb riding height is less than specified, the control arms and bushings should be inspected and replaced as necessary. When the control arms and bushings are in normal condition, the reduced curb riding height may be caused by sagged springs that require replacement.

FRONT AND REAR SUSPENSION DIAGNOSIS AND SERVICE

Proper suspension system diagnosis and service is very important to maintain ride quality, steering control, passenger comfort, and vehicle safety. Accurate suspension system diagnosis is extremely critical so the recommended repair corrects the customer's complaint. Some suspension problems may be confused with brake problems. For example, a worn, loose strut rod bushing may cause steering pull. This problem can also be caused by brake problems such as contaminated brake linings. Therefore, accurate suspension system diagnosis is extremely critical!

Noise Diagnosis

A squeaking noise in the front or rear suspension may be caused by a suspension bushing or a defective strut or shock absorber. If a rattling noise occurs in the rear suspension, check these components:

1. Worn or missing suspension bushings, such as control arm bushings, track bar bushings, stabilizer bar bushings, trailing arm bushings, and strut rod bushings.

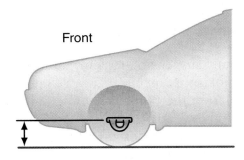

Figure 30-18 Curb riding height measurement, front suspension.

2. Worn strut or shock absorber bushings or mounts.

3. Defective struts or shock absorbers.

4. Broken springs or worn spring insulators.

Rear Body Sway and Lateral Movement Diagnosis

Excessive body sway or roll on road irregularities may be caused by a weak stabilizer bar or loose stabilizer bar bushings. If lateral movement is experienced on the rear of the chassis, the track bar, or track bar bushings, may be defective.

Torsion Bar Adjustment

On torsion bar front suspension systems, the torsion bars may be adjusted to correct the curb riding height. The curb riding height must be measured at the location specified by the vehicle manufacturer. If the curb riding height is not correct on a torsion bar front suspension, the torsion bar anchor adjusting bolts must be rotated until the curb riding height equals the vehicle manufacturer's specifications **(Figure 30-20)**.

Checking Ball Joints

Some ball joints have a grease fitting installed in a floating retainer. The grease fitting and retainer may be used as a ball joint wear indicator. With the vehicle weight resting on the wheels,

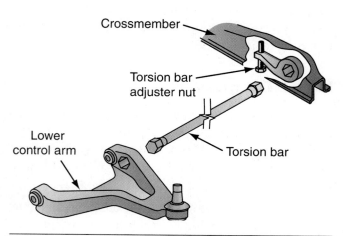

Figure 30-20 Curb riding height adjustment, torsion bar front suspension.

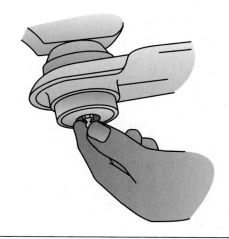

Figure 30-21 Ball joint grease fitting wear indicator.

grasp the grease fitting and check for movement **(Figure 30-21)**.

Tech Tip *Torsion bar adjustments are meant to correct ride height for a typical vehicle and normal loads. They are not meant to correct for excessive loads such as the addition of optional equipment such as snow plows.*

Some car manufacturers recommend ball joint replacement if any grease fitting movement is present. In some other ball joints, the grease-fitting retainer extends a short distance through the ball joint surface **(Figure 30-22)**. On this type of ball joint, the grease fitting extends .050 inch when the joint is new. On this type of joint, replacement is necessary if the grease fitting shoulder is flush with or below the ball joint cover surface.

Ball Joint Unloading

On many suspension systems, ball joint looseness is not apparent until the weight has been removed from the joint. When the coil spring is positioned between the lower control arm and the chassis, place a floor jack near the outer end of the lower control arm and raise the tire off the floor **(Figure 30-23)**. Be sure the rebound bumper is not in contact with the control arm or frame.

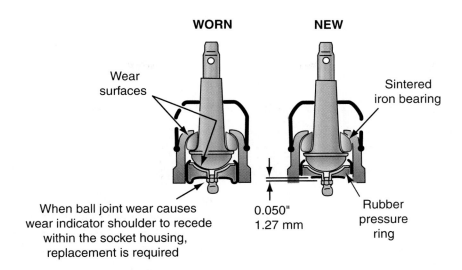

Figure 30-22 Ball joint wear indicator with grease fitting extending from ball joint surface.

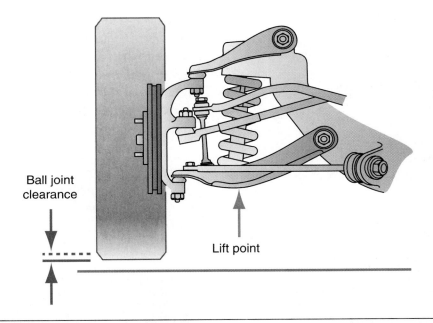

Figure 30-23 Floor jack position to check ball joint wear with spring between the lower control arm and the chassis.

When the coil springs are positioned between the upper control arm and the chassis, place a steel block between the upper control arm and the frame. With this block in place, raise the tire off the floor with a floor jack under the front crossmember **(Figure 30-24)**.

Ball Joint Axial Measurement

The vehicle manufacturer may provide ball joint axial (vertical) and radial (horizontal) tolerances.

A dial indicator is one of the most accurate ball joint measuring devices **(Figure 30-25)**. Always install the dial indicator at the vehicle manufacturer's recommended location for ball joint measurement.

Clean the top end of the lower ball joint stud and position the dial indicator stem against the top end of this stud **(Figure 30-26)**. Depress the dial indicator plunger approximately 0.250 inch.

Lift upward with a pry bar under the tire and observe the dial indicator reading. If the vertical

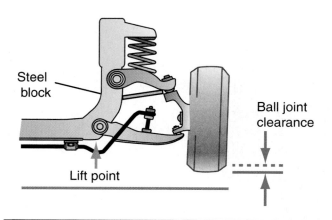

Figure 30-24 Floor jack position to check ball joint wear with spring between the upper control arm and the chassis.

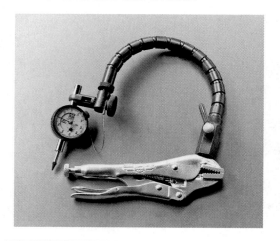

Figure 30-25 Dial indicator designed for ball joint measurement.

Figure 30-26 Dial indicator installed to measure vertical (axial) ball joint movement in a ball joint.

ball joint movement exceeds manufacturer's specifications, ball joint replacement is required.

Ball Joint Radial Measurement

Worn ball joints cause improper camber and caster angles, which result in reduced directional stability and tire tread wear. Connect the dial indicator to the lower control arm of the ball joint being checked and position the dial indicator stem against the edge of the wheel rim **(Figure 30-27)**. Be sure the front wheel bearings are adjusted properly prior to the ball joint radial measurement. While an assistant grasps the top and bottom of the raised tire and attempts to move the tire and wheel horizontally in and out, observe the reading on the dial indicator **(Figure 30-28)**. The lower ball joint on a Macpherson strut-type front suspension should be checked for radial movement with a dial indicator when the tire is lifted off the floor **(Figure 30-29)**. Since the spring load is carried by the upper and lower spring seats when the tire is lifted off the floor, it is not necessary to unload this type of ball joint.

> **Tech Tip** *The axial measurement is made along the centerline of the ball joint and the radial measurement is made at a right angle to the ball joint centerline.*

Figure 30-27 Dial indicator positioned to measure horizontal (radial) ball joint movement.

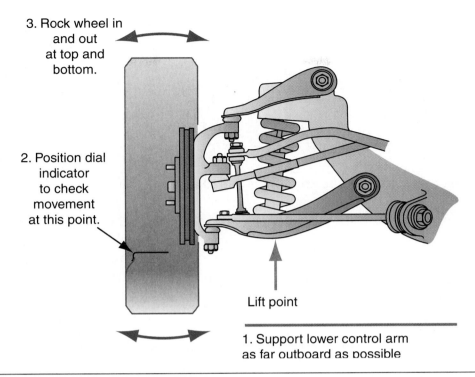

3. Rock wheel in
and out
at top and
bottom.

2. Position dial
indicator
to check
movement
at this point.

Lift point

1. Support lower control arm
as far outboard as possible

Figure 30-28 Measuring ball joint movement.

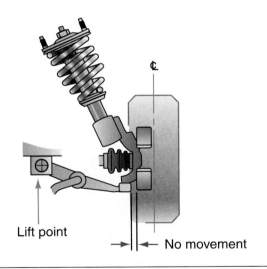

Lift point

No movement

Figure 30-29 Ball joint wear measurement on MacPherson strut front suspension.

Ball Joint Replacement

Ball joints are generally attached using one of three methods. These are press fit, rivet or bolt, and threaded **(Figure 30-30)**. The control arm may need to be removed from the vehicle to replace the ball joint. On some vehicles it may be possible to replace the ball joints while the control arms are still in the vehicle. Always follow the repair procedure recommended by the vehicle manufacturer.

Removal of the ball joint stud from the steering spindle will be necessary. The procedure for tapered studs is very similar to that of tie-rod end removal. Some vehicles will use a straight stud used in conjunction with a pinch bolt.

If the ball joint is fastened using a press fit, then it will be necessary to use a pressing tool to remove and reinstall the ball joint without damaging or distorting the control arm. Never hammer the ball joint to remove it if it is pressed in **(Figure 30-31)**.

On ball joints that are bolted in, remove the bolts and install the new ball joint using the new bolts supplied with the new ball joint. Riveted ball joints will require drilling the rivets. Generally a 1/8-inch twist drill is used to drill the center of the river to a depth of 1/4-inch. Do not drill through the rivet. Then use a 1/2-inch twist drill to remove the rivet head. Do not drill through the rivet or the control arm. With the head of the rivet gone, it is now possible to punch the rivet through the control arm and remove the ball joint. Install the new ball joint using the new bolts supplied with the new ball joint **(Figure 30-32)**.

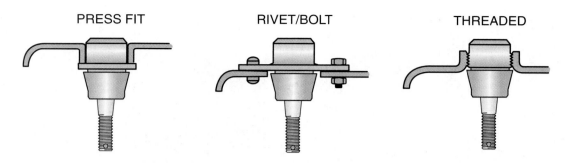

PRESS FIT   RIVET/BOLT   THREADED

Figure 30-30 Three methods of attaching ball joints to control arms.

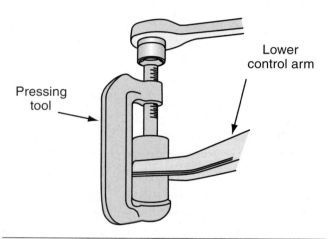

Pressing tool

Lower control arm

Figure 30-31 A pressing tool is used to service press fit ball joints.

Threaded ball joints are unscrewed using a special socket for the ball joint. Clean the control arm threads, lightly lubricate the new ball joint threads, and thread the new ball joint into the control arm. Make sure the ball joint starts straight and is not cross threading the control arm. Make sure the new ball joint is threaded all the way into the control arm. It is sometimes easy to confuse a threaded ball joint for a press fit ball joint. Examine the new ball joint to see if it is threaded. If the new ball joint is smooth then it is a press fit.

Reinstall the ball joint stud into the spindle and all components that were removed during service. It is strongly recommended that the vehicle have a wheel alignment performed to ensure proper driving, handling, and maximum tire life.

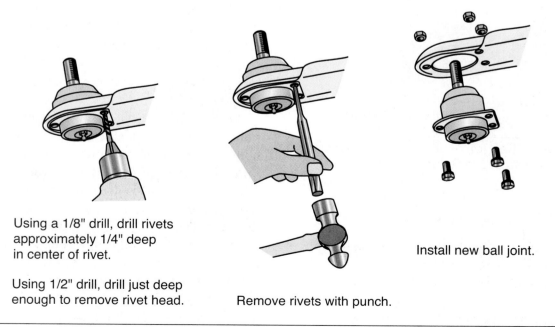

Using a 1/8" drill, drill rivets approximately 1/4" deep in center of rivet.

Using 1/2" drill, drill just deep enough to remove rivet head.

Remove rivets with punch.

Install new ball joint.

Figure 30-32 Ball joint rivet removal procedure.

ELECTRONIC AIR SUSPENSION SYSTEM MAINTENANCE

An electronic air suspension system requires a minimum amount of maintenance. When performing under-car service, inspect all the air lines for kinks, breaks, or contact with other components that would damage or wear the lines. Inspect all the wires in the system, including the module and switch wiring in the trunk, for worn insulation and unwanted contact with other components. The air springs should be inspected for worn spots on the membranes and proper retention to the chassis. Be sure the compressor mounting is secure. Most compressors have a rubber mount to reduce noise transmission to the passenger compartment. Inspect the height sensors for damage and loose mounting bushings.

Tech Tip *The system control switch must be in the off position when system components are serviced.*

ELECTRONIC AIR SUSPENSION DIAGNOSIS

If the air suspension warning lamp is illuminated with the engine running, the control module has detected a defect in the electronic air suspension system. The electronic air suspension diagnostic and service procedures vary depending on the vehicle. Always follow the vehicle manufacturer's recommended procedures in the service manual. When the air suspension warning lamp indicates a system defect, always follow the diagnostic procedure recommended by the vehicle manufacturer.

Tech Tip *During computer system diagnosis, use only the test equipment recommended in the vehicle manufacturer's service manual to prevent damage to computer system components.*

ELECTRONIC AIR SUSPENSION SYSTEM SERVICE

Many components in an electronic air suspension system such as control arms, shock absorbers, and stabilizer bars, are diagnosed and serviced in the same way as the components in a conventional suspension system. However, the air spring service procedures are different compared to coil spring service procedures on a conventional suspension system. Always follow the repair procedure recommended by the vehicle manufacturer.

SAFETY TIP *The system control switch must be turned off prior to hoisting, jacking, or towing the vehicle. If the front of the chassis is lifted with a bumper jack, the rear suspension moves downward. The electronic air suspension system will attempt to restore the rear trim height to normal. This action may cause the front of the chassis to fall off the bumper jack, resulting in personal injury or vehicle damage.*

Summary

- Shock absorber condition may be determined by performing a visual inspection, bounce test, or manual test.
- Front strut chatter may be caused by upper strut mount binding.
- Front strut noise on sharp turns or during suspension jounce may be caused by interference between the coil spring and the strut

tower or interference between the coil spring and the upper mount.

- If one of the front strut-to-steering knuckle bolts has an eccentric cam, the cam position should be marked on the strut prior to bolt removal.
- When a coil spring compressing tool is used, it is extremely important to follow the tool

manufacturer's and vehicle manufacturer's recommended spring compressing procedure. Never loosen the strut piston rod nut until the spring is compressed so all the tension is removed from the upper strut mount.

■ The curb riding height is critical because it affects most other front suspension alignment angles.

■ Bent control arms or worn control arm bushings affect curb riding height and front suspension alignment angles.

■ A weak stabilizer bar or worn stabilizer bar bushings cause harsh riding, excessive body sway, and suspension noise.

■ Excessive wear on strikeout bumpers may be caused by improper curb riding height or worn-out shock absorbers.

■ Bent strut rods or worn strut rod bushings cause improper front suspension alignment and reduced directional stability.

■ Excessive body sway or roll when one wheel strikes a road irregularity may be caused by a defective stabilizer bar or bushings.

■ Excessive lateral movement of the rear chassis may be caused by a defective track bar or bushings.

■ Sagged rear springs cause excessive steering effort and rapid steering wheel return after a turn.

■ If a coil spring has a vinyl coating, the spring should be taped in the compressing tool contact areas to prevent chipping the coating.

■ A broken leaf spring center bolt may allow the rear axle to move, resulting in improper tracking and reduced directional stability.

■ When servicing, hoisting, jacking, towing, hoisting, or lifting a vehicle equipped with an electronic air suspension system, always turn the air suspension switch off in the trunk.

■ If a defect occurs in an electronic air suspension system, the control module illuminates the suspension warning lamp with the engine running.

Review Questions

1. While discussing curb riding height, Technician A says worn control arm bushings reduce curb riding height. Technician B says incorrect curb riding height affects most other front suspension angles. Who is correct?

   A. Technician A

   B. Technician B

   C. Both Technician A and Technician B

   D. Neither Technician A nor Technician B

2. While discussing ball joint radial measurement, Technician A says the dial indicator should be positioned against the ball joint housing. Technician B says the front wheel bearing adjustment does not affect the ball joint radial measurement. Who is correct?

   A. Technician A

   B. Technician B

   C. Both Technician A and Technician B

   D. Neither Technician A nor Technician B

3. While discussing shock absorber and strut inspection, Technician A says an oily film is acceptable. Technician B says if oil drops are present, the shock absorber or strut must be replaced. Who is correct?

   A. Technician A

   B. Technician B

   C. Both Technician A and Technician B

   D. Neither Technician A nor Technician B

4. While discussing shock absorber and strut replacement, Technician A says shock absorbers and struts may be replaced individually or in pairs. Technician B says the shock absorbers or struts must be replaced as a set of four on the vehicle. Who is correct?

   A. Technician A

   B. Technician B

   C. Both Technician A and Technician B

   D. Neither Technician A nor Technician B

5. While discussing ball joint replacement, Technician A says to drill the rivets retaining the ball joint using a 3/8-inch twist drill all the way through the rivet and control arm. Technician B says the rivets on the ball joint can be removed using an air grinder. Who is correct?

   A. Technician A

   B. Technician B

   C. Both Technician A and Technician B

   D. Neither Technician A nor Technician B

6. While discussing rear wheel tracking, Technician A says a bent tie-rod could result in improper rear wheel tracking. Technician B says improper tracking occurs when both rear springs are sagged the same amount. Who is correct?

   A. Technician A

   B. Technician B

   C. Both Technician A and Technician B

   D. Neither Technician A nor Technician B

7. A vehicle with a Macpherson strut front suspension has a strut chatter problem when the front wheels are turned to the right or left. The most likely cause of this problem is:

   A. a defective lower ball joint.

   B. a defective upper strut mounts bearing.

   C. worn lower control arm bushings.

   D. a binding outer tie-rod end.

8. When discussing strut removal from a coil spring, Technician A says to tighten the coil spring compressor tool until all the spring tension is removed from the upper strut mount. Technician B says to loosen the strut rod nut before the coil spring and strut assembly is removed from the vehicle. Who is correct?

   A. Technician A

   B. Technician B

   C. Both Technician A and Technician B

   D. Neither Technician A nor Technician B

9. A vehicle with a Macpherson strut front suspension has severely damaged rebound bumpers on both sides of the front suspension. All of these defects may be the cause of this problem except:

   A. worn-out front struts.

   B. sagged front coil springs.

   C. worn lower control arm bushings.

   D. defective upper strut mount bearings.

10. When measuring ball joint movement on the right-hand side of a short arm, long arm front suspension with the coil spring mounted between the lower control arm and the chassis, Technician A says to position a floor jack under the lower control arm and raise the right front tire off the shop floor. Technician B says to lift the right front tire with a pry bar under the tire and measure the ball joint vertical movement with a dial indicator. Who is correct?

    A. Technician A

    B. Technician B

    C. Both Technician A and Technician B

    D. Neither Technician A nor Technician B

11. Worn ball joints may cause all of these problems except:

    A. harsh ride quality.

    B. excessive tire tread wear.

    C. steering wander.

    D. reduced directional stability on irregular road surfaces.

12. When the coil spring is positioned between the lower control arm and the chassis, place the floor jack under the _____ _____ _____ to unload the ball joint prior to ball joint wear measurement.

13. A false measurement of ball joint radial movement may be caused by a loose _____ _____ adjustment.

14. Explain the most likely cause of rear body lateral movement.

15. Describe the results of excessive front ball joint wear.

16. Explain why the load carrying ball joint must be unloaded before checking ball joint wear.

17. Explain why the lower ball joint on a Macpherson strut can be checked after lifting the vehicle on the frame.

18. Explain the difference between press fit and threaded ball joints and how to tell if the ball joint on the vehicle is press fit or threaded.

Wheel Alignment

Learning Objectives

After you have read, studied, and practiced the contents of this unit, you should be able to:

- Explain the importance of correct wheel alignment angles.

- Explain the benefits of wheel alignment.

- Describe the different functions of camber and caster.

- Explain the purposes of steering axis inclination.

- Explain why toe is the most critical tire wear factor of all the alignment angles.

- Identify the purposes of turning radius or toe-out on turns.

- Explain the condition known as tracking.

- Perform a pre-alignment inspection.

- Describe the equipment that can be used for alignment.

- Describe how alignment angles can be adjusted.

- Understand the importance of a four-wheel alignment.

- Know the difference between two-wheel and four-wheel alignment procedures.

Key Terms

Camber

Caster

Eccentric cam and bolt

Four-wheel alignment

Geometric centerline

Included angle

Ride height

Road crown

Scrub radius

Steering axis inclination (SAI)

Thrust angle

Tire contact patch

Toe

Toe-out on turns

Total toe

Two-wheel alignment

INTRODUCTION

Accurate steering is needed at all driving speeds. The wheels of the vehicle must be in proper alignment. Proper wheel alignment allows the wheels to roll on different types of roads and conditions with minimal wear and maximum control. Proper alignment of all wheels provides for the safety of the driver and passengers **(Figure 31-1)**. It also reduces fuel consumption, tire wear, and stress

Figure 31-1 A computerized wheel alignment machine and alignment rack.

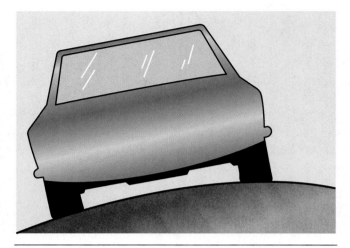

Figure 31-2 Roads are crowned to allow water to drain off them.

and strain on the steering and suspension systems of the vehicle.

There are a number of alignment angles and handling characteristics the automotive engineer considers when designing a vehicle. The engineer must balance the characteristics of the different alignment angles along with other considerations for the vehicle such as weight, drive train, and ride quality. You need not have an in-depth understanding of the engineering of the suspension system, but you must have an understanding of the alignment angles, the effect of these angles, and how to set the vehicle to the factory specifications.

A vehicle with a properly set wheel alignment will have the vehicle weight properly distributed across the chassis and suspension components. It is important to note that the factory specifications are designed for a vehicle that has not been modified, for example, larger tires or a suspension that has been lifted or lowered. The specifications are also for a vehicle with a specific set of conditions. These conditions may include driver weight, fuel load, ride height, and other considerations.

These alignment angles also take into consideration the operating environment of the vehicle and road surfaces. Most highway road surfaces are convex or crowned. This **road crown** allows water to drain away from the road surface **(Figure 31-2)**. Tires may have some positive camber to allow for the road crown.

> **Tech Tip** *The slight outward tilt of positive cambers allows for better contact with the road crown.*

ALIGNMENT GEOMETRY

Most wheel alignment angles are measured in degrees. Alignment angles may be measured in inches or millimeters. Is important to note that if the measurements are given in inches or millimeters, it is usually based on the standard rim diameter for that vehicle. If the vehicle has had different rims mounted on the vehicle, then inches or millimeters may not accurately be used to set the wheel alignment. Degrees is the preferred units when available. However, some manufacturers may give specs only in inches or millimeters.

When discussing alignment angles, it is important to understand the terms "more negative" and "more positive" **(Figure 31-3)**. Two readings of +1 and −1 would seem obvious. The −1 is more negative than the +1.

Now compare alignment readings of +1 and +1.5. Even though both numbers are positive,

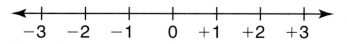

Figure 31-3 Wheel alignment readings may be compared to determine which are more positive or more negative.

the +1 reading is more negative or closer to the negative end of the number line. Conversely the +1.5 is more positive than the +1 reading.

Now compare alignment readings of −1 and −0.5. Even though both numbers are negative, the −0.5 reading is more positive or closer to the positive end of the number line. Conversely the −1 is more negative than the −0.5 reading.

The proper alignment of a vehicle suspension/steering system depends on the alignment and suspension angles discussed next.

Caster

Caster is the angle of the steering axis of a wheel compared to a vertical line, as viewed from the side of the vehicle. The forward or rearward tilt from the vertical line is caster **(Figure 31-4)**. If there is no tilt, then the caster is zero. If the line tilts forward at the top, then it is negative. If the caster tilts rearward at the top, then the caster is positive. When caster is positive, the steering axis intersects the road surface ahead of the tire contact patch. The **tire contact patch** is the area of the tire that is physically in contact with the road surface. The tire contact patch tends to follow the caster much as a caster on a bicycle, creeper wheel, or a furniture caster does **(Figure 31-5)**.

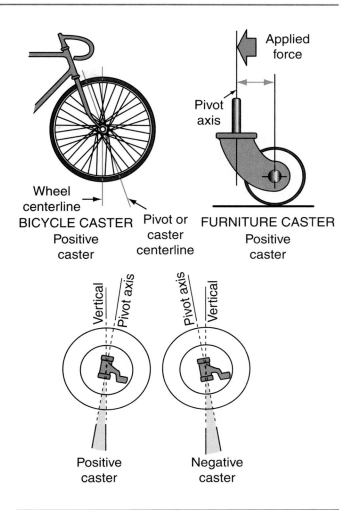

Figure 31-5 Caster on a bicycle, furniture caster, and on a vehicle.

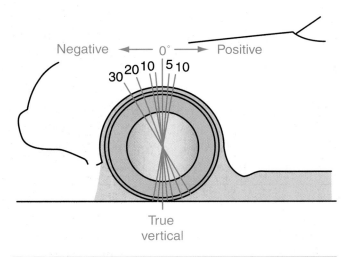

Figure 31-4 Caster is viewed from the side of the vehicle.

> **Tech Tip** *The rear wheels are aligned first and the front wheels are aligned last.*

Caster is usually the first angle adjusted when the front wheels are adjusted during a wheel alignment. Caster is a steering stability angle. When caster is positive, then very little effort is required to maintain the vehicle when driving straight ahead. When caster is negative, a greater effort will be required to maintain the vehicle when driving straight ahead and the vehicle will tend to wander more. The more positive caster will give a vehicle more steering stability, while more negative caster will give greater road feel to the driver.

The caster for each front wheel should be nearly the same. Caster angles that are too different from one another will cause the vehicle to pull toward the side with caster that is more negative. Caster is not a tire-wearing angle. If the tires are showing wear, it is not caused by caster.

> **Tech Tip** *Caster is not a tire-wearing angle. Tire wear cannot be caused by incorrect caster. Caster is not adjustable on the rear of most vehicles.*

Camber

Camber is the angle represented by the tilt of a wheel, inward or outward, compared to vertical as viewed from the front of the car **(Figure 31-6)**. Camber mainly compensates for road crown. Camber also compensates for vehicle weight transfer during cornering, vehicle weight, passenger weight, and normal cargo weight. If the top of the wheel is tilted inward at the top of the vehicle, then camber is negative. If the wheel is tilted outward, then the camber is positive.

The camber setting for each wheel is usually the same. A wheel will pull due to camber. This is because the wheel is tilted and acts like a cone and will roll in the direction of the most positive camber **(Figure 31-7)**. A vehicle will pull to the side that has the more positive camber. Camber also

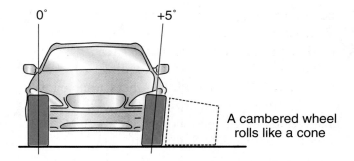

Figure 31-7 A wheel turns in the direction of camber.

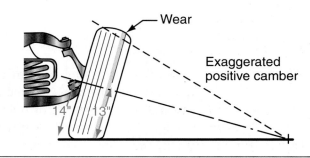

Figure 31-8 When camber is excessive tire wear will result.

properly distributes the vehicle on the bearings and steering system components. Too much camber will cause excess tire wear **(Figure 31-8)**.

Camber is usually the next angle to adjust during an alignment. Changes in vehicle ride height will almost always have an effect on camber. It is important to check the vehicle **ride height** before beginning the wheel alignment. Vehicles are designed to operate at proper ride height. Worn suspension components will also have detrimental effects on camber. Worn or damaged components must be replaced before an alignment.

> **Tech Tip** *Most suspension systems will experience camber change as the suspension travels up and down. If ride height is incorrect, then the camber too will be changed.*

> **Tech Tip** *Camber is a tire-wearing angle. Tire wear will occur on the side of the tire with the excessive camber.*

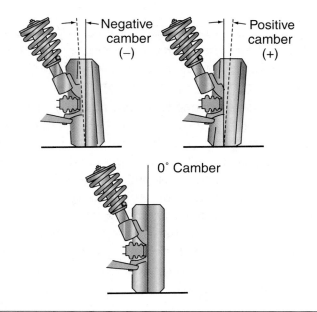

Figure 31-6 Positive and negative camber.

Toe

Toe is the angle represented by the position of a wheel, left or right, compared to the center of the vehicle as viewed from the top of the car **(Figure 31-9)**. Toe may be measured individually for the left and right tires. This measurement may be given in degrees, inches, or millimeters. Toe may be measured for both tires at the same time. This comparison of both tires is referred to as **total toe**.

If the front of the tire is closer to the center of the vehicle than the rear of the tire, then the wheel has toe-in and the toe is said to be positive. If the front of the tire points away from the centerline of the vehicle, then it has toe-out and is negative.

A wheel with an improper toe setting will effectively drag the tire sideways as it travels down the highway. A tire with excess toe-out will exhibit a feathered edge on the tire in the direction of excess toe **(Figure 31-10)**. A tire with excess toe-in will also show a feathered edge in the direction of the excess toe **(Figure 31-11)**.

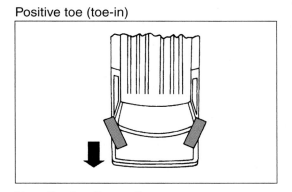

Positive toe (toe-in)

Negative toe (toe-out)

Figure 31-9 Toe is the difference between the centerline of the tires measured at the front and the back of the tire.

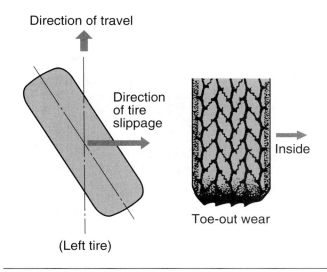

Figure 31-10 Toe-out wear.

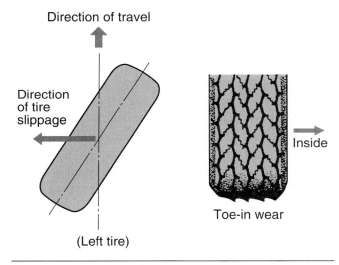

Figure 31-11 Toe-in wear.

Toe adjustments are made at the tie-rod **(Figure 31-12)**. The toe should be adjusted to the same setting on both sides of the car. If the settings are not the same, the driver may notice the steering wheel is off center. The toe should be adjusted when the steering wheel is centered. Toe is the last adjustment made during a wheel alignment.

Tech Tip *The rear wheels are aligned first and the front wheels are aligned last.*

Tech Tip *Toe is a tire-wearing angle. A tire with excess toe-in will show a feathered edge in the direction of the excess toe.*

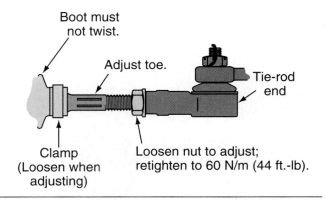

Boot must
not twist.

Adjust toe.

Tie-rod
end

Clamp
(Loosen when
adjusting)

Loosen nut to adjust;
retighten to 60 N/m (44 ft.-lb).

Figure 31-12 Toe is adjusted at the tie-rod end.

Thrust Line

A vehicle must run straight down the road. The rear tires should be aligned to the geometric centerline of the vehicle. The **geometric centerline** is a line that runs exactly down the centerline of the vehicle as viewed from the top of the vehicle. It divides the car into a left and right half. The vehicle should travel down the road with the geometric centerline traveling parallel to the road surface.

Should the rear toe be unequal from one side to another, then a thrust line or thrust angle is created that will cause the vehicle to run slightly sideways **(Figure 31-13)**. The thrust line is the algebraic sum of each individual rear wheel toe. If the total rear toe is toward the left side of the vehicle as compared to the geometric centerline, then the **thrust angle** is negative. If he total toe is to the right of the geometric centerline, then the thrust angle is positive. Ideally the thrust angle should be zero and parallel the geometric centerline.

Most alignment machines have the capability to check the rear alignment. The rear toe should be adjusted prior to adjusting the front wheels. Aligning the rear wheels will prevent the steering wheel

from having to be turned off-center to compensate for a vehicle with thrust angle. After rear toe has been properly adjusted, then the front toe can be set to the thrust line.

Performing this thrust angle alignment procedure is the main basis for a **four-wheel alignment**. If a **two-wheel alignment** is performed, then front toe is set to the geometric centerline of the vehicle and ignores the thrust line. It is important to perform a four-wheel alignment on today's vehicles.

A vehicle with a live rear axle should not need to have the thrust angle adjusted, but it should be checked. Should the thrust angle be out of specification, it might indicate a bent frame, chassis wear, rear axle offset, or chassis damage **(Figure 31-14)**. The problem will need to be diagnosed and corrected. Vehicles with independent rear suspensions will need to have the thrust angle checked and corrected as necessary.

Steering Axis Inclination

Steering axis inclination (SAI) is the angle of the steering axis of a wheel compared to a vertical line, as viewed from the front of the vehicle. The inward tilt from the vertical line is SAI **(Figure 31-15)**. When SAI tilts inward, it is positive. SAI is rarely negative. The upper and lower ball joints on a SLA suspension system create the steering axis. If a vehicle has a Macpherson strut suspension system, then the upper strut pivot and the lower ball joint create the steering axis.

SAI is an angle designed to project the pivot point of the wheel on the road surface near the center of the tire contact patch and uses the weight of the vehicle to enhance stability. SAI also helps return the steering to the straight-ahead position after completing a turn.

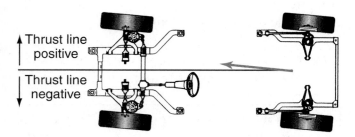

Thrust line
positive

Thrust line
negative

Figure 31-13 If toe is unequal then a thrust line is created.

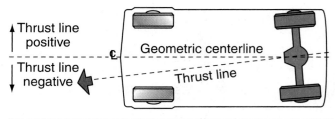

Thrust line
positive

Thrust line
negative

Geometric centerline

Thrust line

Figure 31-14 If a live axle is not perpendicular to the geometric centerline then a thrust line is created.

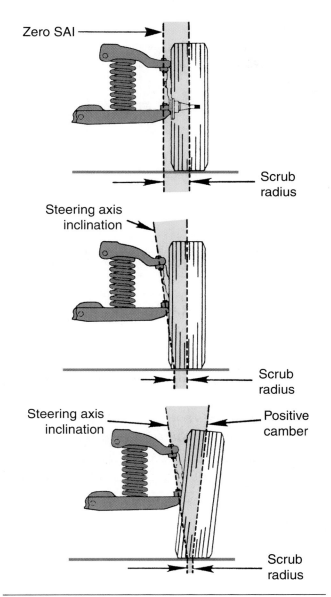

Figure 31-15 Steering axis is a line through the steering pivot points compared to vertical as viewed from the front of the vehicle.

this arc is when the wheels are straight ahead and lower when turned either left or right. The effect of this is to push the wheel away from the vehicle, which raises the vehicle slightly during a turn. This lifting force helps to center the steering after the completion of a turn.

Improperly adjusted or unequal SAI can result in torque steer, bump steer, and brake pull. SAI cannot be adjusted during a wheel alignment. Checking SAI is an important diagnostic tool that can identify a number of vehicle alignment and suspension problems. If SAI is found to be out of specification, then an out of position strut tower, bent lower control arm, shifted frame members, or other damaged or worn components could be the cause. If SAI is shifted equal amounts from one side to another, then a shifted engine cradle or racked body may be the cause.

Scrub Radius

Scrub radius is the difference between where SAI intersects the road surface compared to the center of the tire contact patch **(Figure 31-16)**. If the center of the tire contact patch is toward the outside of the vehicle compared to where the SAI intersects the road surface, then the scrub radius is positive. On a front-wheel drive vehicle the positive scrub radius tends to toe-in the wheel due to the torque applied to the wheel. A rear-wheel drive vehicle with positive scrub radius tends to toe-out the wheel due to the rolling resistance of the tire.

If the center of the tire contact patch is toward the inside of the vehicle compared to where the SAI intersects the road surface, then the scrub radius

Scrub radius is the difference between where SAI intersects the road surface compared to the center of the tire contact patch. If the upper ball joint were located directly over the lower ball joint, then the SAI would be zero. This would not be desirable, as the tire would likely have scrub radius. Undesirable results are increased steering effort, lack of steering wheel return after a turn, and lessened control.

As the wheel is turned, it travels through an arc. This arc is perpendicular to the SAI. Because SAI is positive, this arc too is tilted. The high point of

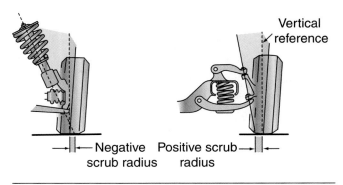

Figure 31-16 Negative and positive scrub radius.

is negative. Negative scrub radius tends to toe-out the wheel on a front-wheel drive vehicle due to the torque applied to the wheel. On a rear-wheel drive vehicle the negative scrub radius tends to toe-in the wheel due to the rolling resistance of the tire.

Tech Tip *Scrub radius is not adjustable.*

Included Angle

The **included angle** is the sum of the camber and SAI **(Figure 31-17)**. The included angle is not adjustable during a wheel alignment but is a diagnostic angle much like SAI. Diagnosing the included angle along with camber and SAI can point to a number of vehicle alignment and suspension problems. If SAI is correct but camber and the included angle are less, then a bent spindle may be the cause.

Toe-Out on Turns

When the vehicle travels straight ahead both front tires are nearly straight ahead. But when a vehicle turns a corner, then each wheel must travel a slightly different path. The wheel on the inside of the turn must travel in a smaller circle than the wheel on the outside of the turn **(Figure 31-18)**. During a turn the toe relationship between the front wheels must

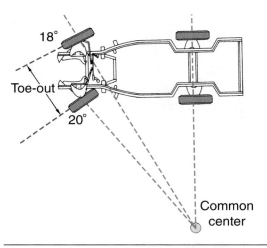

Figure 31-18 Tires must toe-out on during a turn.

change. The wheels must toe-out during a turn. This is referred to as **toe-out on turns**. This toe-out on turn minimizes tire scrub and wear by keeping each wheel pointed in the proper direction that they have to travel during a turn.

Like SAI and included angle, toe-out on turns is not adjustable but is an important diagnostic angle. If toe-out on turns is not correct, then worn or bent steering components will need to be replaced.

Tech Tip *Toe-out on turns is not adjustable.*

Setback

The amount by which one front wheel is further back from the front of the vehicle than the other is called setback **(Figure 31-19)**. It may also be

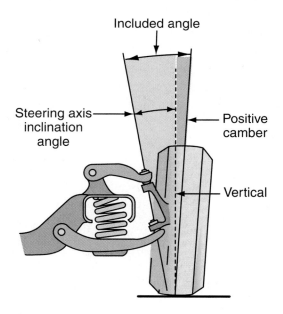

Figure 31-17 The included angle is the sum of the camber and SAI.

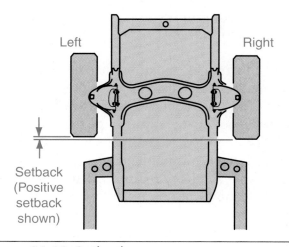

Figure 31-19 Setback.

measured as an angle formed by a line perpendicular to the axle centerline with respect to the vehicle's centerline. If the left wheel is further back than the right, setback is negative. If the right wheel is further back than the left, setback is positive.

Setback should usually be zero to less than a quarter inch or zero to less than one half a degree. Some vehicles may have an asymmetrical suspension design. Setback is typically measured with both wheels straight ahead and is usually done automatically by the alignment computer. Its main purpose during alignment is as a diagnostic angle. It can help to diagnose chassis misalignment or collision damage **(Figure 31-20)**. Excessive setback can also cause differences in toe-out on turn angle readings side-to-side.

> **Tech Tip** *Setback is not adjustable.*

PREALIGNMENT INSPECTION

An extremely important part of a wheel alignment is the preliminary inspection **(Figure 31-21)**. During this inspection if any parts are found to be defective, they should be replaced before proceeding with the alignment. The following are guidelines for the inspection:

- Road test the vehicle. Verify that the steering wheel is centered and note any vibration if present. Note any pull or wander in the vehicle. Also observe any unusual noise or vibration that may be present. Verify the customer's concern before beginning an alignment. Check for steering wheel play.

- Inspect the tires on the vehicle. Check all the tires for proper and matching sizes, proper inflation, abnormal wear, and any damage.

- Inspect the wheels for damage and runout. Inspect the wheel bearings.

- Load the vehicle for its expected service weight. Remove excess weight from the vehicle unless the vehicle will normally operate with these items in place.

- Check and correct the ride height of the vehicle **(Figure 31-22)**. Some wheel alignment machines may include an option for checking the ride height **(Figure 31-23)**. Replace or adjust springs as necessary or as recommended by the vehicle manufacturer.

- Inspect all the steering and suspension components such as control arms, ball joints, bushings, tie-rods, etc. Damaged or worn components must be replaced before performing a wheel alignment.

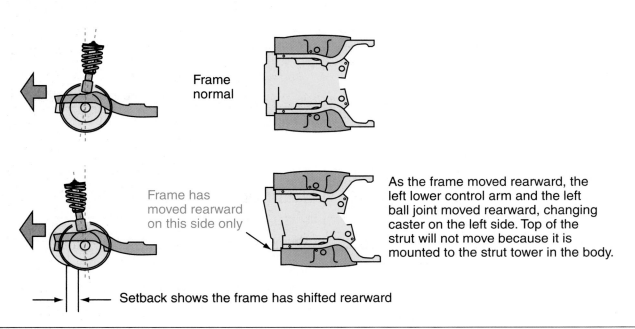

Frame normal

Frame has moved rearward on this side only

As the frame moved rearward, the left lower control arm and the left ball joint moved rearward, changing caster on the left side. Top of the strut will not move because it is mounted to the strut tower in the body.

Setback shows the frame has shifted rearward

Figure 31-20 Setback can show frame or chassis damage.

PREALIGNMENT INSPECTION CHECKLIST

Owner _____ Phone _____ Date _____

Adress _____ VIN _____

Make _____ Model _____ Year _____ Lic. number _____ Mileage_____

1. Road test results	Yes	No	Right	Left
Above 30 MPH				
Below 30 MPH				
Bump steer				
When braking				
Steering wheel movement Stopping from 2-3 MPH(Front)				
Vehicle steers hard				
Strg wheel returnability normal				
Strg wheel position				
Vibration	Yes	No	Frnt	Rear

2. Tire pressure | Specs Frnt ____ Rear ____

Record pressure found

RF _____ LF _____ RR _____ LR _____

3. Chassis height | Specs Frnt ____ Rear ____

Record height found

RF _____ LF _____ RR _____ LR _____

	Yes	No
Springs sagged		
Torsion bars adjusted		

4. Rubber bushings	OK
Upper control arm	
Lower control arm	
Sway bar / stabilizer link	
Strut rod	
Rear bushing	

5. Shock absorbers/struts	Frnt	Rear

6. Steering linkage	Frnt OK	Rear OK
Tie rod ends		
Idler arm		
Center link		
Sector shaft		
Pitman arm		
Gearbox/rack adjustment		
Gearbox/rack mounting		

7. Ball joints	OK
Load bearings	

Specs	Readings
Right ____ Left ____	Right ____ Left ____

Follower	
Upper strut bearing mount	
Rear	

8. Power steering	OK
Belt tension	
Fluid level	
Leaks/hose fittings	
Spool valve centered	

9. Tires/wheels	OK
Wheel runout	
Condition	
Equal tread depth	
Wheel bearing	

10. Brakes operating properly

11. Alignment	Spec		Initial reading		Adjusted reading	
	R	L	R	L	R	L
Camber						
Caster						
Toe						

Bump steer	Toe change right wheel		Toe change left wheel	
	Amount	Direction	Amount	Direction
Chassis down 3"				
Chassis up 3"				

	Spec		Initial reading		Adjusted reading	
	R	L	R	L	R	L
Toe-out on turns						
SAI						
Rear camber						
Rear total toe						
Rear indiv. toe						
Wheel balance						
Radial tire pull						

Figure 31-21 A prealignment and alignment checklist.

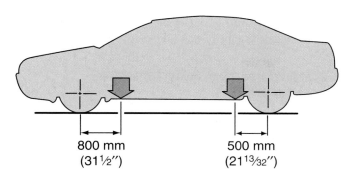

Figure 31-22 Typical measuring points for checking curb height of a vehicle.

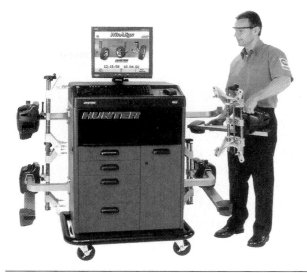

Figure 31-24 Computerized alignment machine.

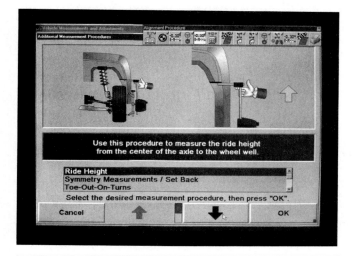

Figure 31-23 A wheel alignment machine with built in ride height sensors.

Tech Tip *Be sure to check the ride height of the vehicle. Correcting ride height to specification can often reduce concerns with other alignment angles. If the ride height is not correct, then the vehicle cannot be aligned properly.*

WHEEL ALIGNMENT EQUIPMENT

There are many different methods for aligning the wheels on a vehicle, but the method that provides the most accuracy is the computerized alignment machine **(Figure 31-24)**. Many of these machines can also display alignment adjustment points on the vehicle and have vehicle specifications available to the technician.

Most shops use an alignment machine to check all of the alignment angles. Normally an alignment rack is part of the alignment machine's package. The rack is best described as a limited purpose vehicle hoist (lift) equipped with turning radius plates. There are many varieties of alignment machines that have been used through the years. Some are equipped to measure alignment angles at all four wheels of the vehicle, and others measure the angles at only two wheels. Some alignment machines simply display the angle readings, whereas others display the readings plus give advice on how to correct the angles **(Figure 31-25)**.

Figure 31-25 Wheel alignment adjustment procedure given by the alignment machine.

Wheel Alignment Rack

The wheel alignment rack is usually part of a package of the alignment machine. The rack is a special fixture on which the vehicle is driven onto for the wheel alignment **(Figure 31-26)**. The rack is often mounted on a hoist and raised to various heights, as the technician needs access to various parts of the vehicle for adjustments.

The rack will have a set of turntables for the front wheels. These are plates that are held in place by locking pins while the vehicle is being driven on the rack. When the pins are removed the plates allow the wheel to freely slide sideways left to right, slide front to back, and turn left and right. The turntables also include a turning pointer to measure how many degrees each wheel is being turned left or right. This can be used to measure toe-out on turns. You turn the steering wheel to the right until the right tire turns 20° and reads the turn angle on the left turntable. You turn the steering wheel to the left until the left tire turns 20° and reads the turn angle on the right turntable. These readings are then compared to vehicle specification.

To perform a four-wheel alignment, the alignment rack should have rear slip plates. The slip plates are similar to the turntables but usually are larger to accommodate wheel base lengths of different vehicles. Slip plates also have a limited amount of movement compared to the front turntables. Locking pins hold the slip plate while the vehicle is driven on the alignment rack. When the pins are

removed, the plates allow the wheel to freely slide sideways left to right, and turn slightly left and right.

Computerized Alignment Equipment

Vehicle information is input into the computer and the alignment type is selected. The computer will then load the alignment specification for the vehicle. If the specifications are not in the computer, then you may enter them manually. The computer will then direct you to mount the wheel sensors as needed for the alignment type.

Sensors are mounted to each wheel after the vehicle is placed on the alignment rack. Runout compensation is then performed. This usually involves lifting the vehicle and rotating each tire a half turn, pressing the runout compensation button, rotating the tire another half turn, and pressing the runout compensation button again. Or left a quarter turn, pressing the runout compensation button, right a half turn, pressing the runout compensation button, left a quarter turn, and pressing the runout compensation button. Runout compensation will compensate for most problems with rim mounting or bent rims. It allows the computer to display alignment readings for the true rotational axis of the wheel.

Tech Tip *Runout compensation will vary depending on the equipment manufacturer's equipment and procedure.*

The computer will display the alignment after the compensation is completed **(Figure 31-27)**. You may need to steer the wheels left and right as directed by the computer to display the caster angles. The computer can display which alignment angles need adjustment, along with the tolerance for the adjustments. The display can include text and graphics and are often in color. The machine may also display where the adjustment points are located on the vehicle.

Alignment Tools

There are a number of specialty tools to assist in the adjustment of the alignment **(Figure 31-28)**. Some of these tools help speed the alignment

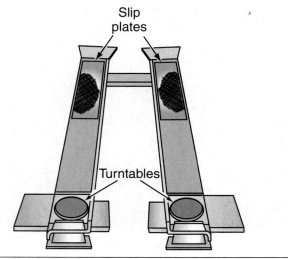

Figure 31-26 An alignment rack with turntables and slip plates.

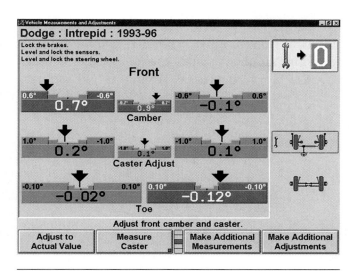

Figure 31-27 Alignment display.

Figure 31-28 Specialty tools available for steering and suspension work.

process and others are required. The service manual may specify which tools are needed for adjustment procedures. The aftermarket has developed many specialty tools for wheel alignment that are available to the technician.

Two-Wheel Alignment

Two-wheel alignment aligns the front wheels to the geometric centerline of the vehicle. The typical alignment order is to adjust:

1. front caster
2. front camber
3. front toe

This type of alignment does not account for the thrust line nor does it check the alignment of the rear wheels. As a result, proper steering and tracking of the vehicle may not be achieved. For these reasons four-wheel alignments are preferred. Consequently most alignment equipment is four-wheel alignment capable.

> **Tech Tip** *85 percent of vehicles today require a rear alignment as well as front alignment.*

Four-Wheel Alignment

A four-wheel alignment may be performed on a two-wheel alignment machine. The vehicle will need to be backed onto the rack and the rear aligned to the geometric centerline. There are several drawbacks to this method. You must reverse the toe reading. If the rear wheels are to have toe-in, then because the vehicle is on the rack backwards, the toe setting must be toed-out. You must also swap readings from right and left camber if there is an asymmetrical setting recommended by the manufacturer. This method is much more time consuming as the vehicle must then be driven off and back onto the rack to align the front end of the vehicle.

A four-wheel alignment aligns all of the wheels to the geometric centerline of the vehicle. While this may sound contradictory with the previous discussion of the importance of the thrust line, it is important to note that if the rear wheels are aligned to the geometric centerline, then the thrust line will be zero or nearly zero.

The typical alignment order is to adjust:

1. rear camber
2. rear toe
3. front caster
4. front camber
5. front toe

The main difference in the four-wheel alignment is the alignment of the rear tires and this is done before the front tires are adjusted. Adjusting the rear wheels will assure the quality and completeness of the alignment. The check and adjustment of the thrust line will help assure you that the alignment of the rear is in alignment with the vehicle chassis as the factory intended. The front toe then can be adjusted to the thrust line that will compensate for any variation in tolerances in the rear toe. It also serves to greatly reduce customer comeback due to an incomplete wheel alignment.

> **Tech Tip** *Four-wheel alignments are required for most vehicles.*

ADJUSTING WHEEL ALIGNMENT

Many of the alignment angles are related to one another. Adjusting one angle can change the setting of another angle. Caster and camber are two angles that can easily affect each other. Caster and camber are often adjusted at the same time. Changing caster and camber will change toe but not vice versa. This is why it is important to follow the alignment order: caster, camber, toe.

Vehicle adjustment methods are as varied as the suspension types. Some vehicles with Macpherson struts do not have factory adjustments for caster or camber while others may have these adjustments. Adjustment methods will vary from manufacturer to manufacturer, and will even vary from model to model, and even within the same model line for different model years. Consult the service manual or the alignment computer for these adjustment methods.

Caster/Camber Adjustment

Caster is a steering stability angle and assists steering wheel return. To compensate for road crown, most alignment specifications may allow for slightly more positive camber on the right wheel of the vehicle.

There are several methods used to adjust caster and camber.

Shims

Some vehicles may use shims for adjusting caster and camber. The shims are typically located between the control arm pivot shaft and the inside or outside of the frame. Caster and camber can both be adjusted at the same time. Shims are added or removed from between the front and rear control arm pivot shaft bolts to adjust caster **(Figure 31-29)**. Shims may also be moved from one shim pack to the other to change caster without effect-

ing camber. Equal amounts of shims are added or removed from both the front and rear shim packs to change camber without changing caster **(Figure 31-30)**. Most alignment computers will specify

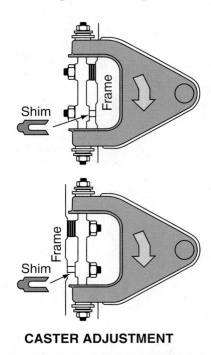

CASTER ADJUSTMENT

Figure 31-29 Adding or removing shims from one shim pack changes caster.

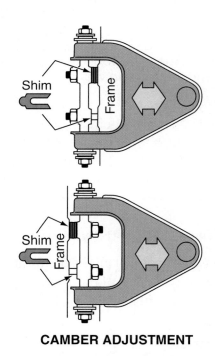

CAMBER ADJUSTMENT

Figure 31-30 Adding or removing equal shims from both shim packs changes camber.

how many shims to add and remove to the shim packs and reduce the need to use a trial and error method to figure out how many shims are needed.

Eccentric Cam and Bolt

An **eccentric cam and bolt** is essentially a bolt pressed through a heavy washer with an offset hole thus creating a cam. Some vehicles use an eccentric cam and bolt to adjust caster and camber. An eccentric cam and bolt on the upper control arm adjust both caster and camber. The bolts are loosened and turned to move the arm in and out.

When the eccentric cam and bolt is on the upper control arm and one bolt is adjusted, it will change caster and camber **(Figure 31-31)**. If one bolt is turned inward and the other bolt is turned an equal amount outward, then caster can be changed without affecting camber. If both bolts are turned an equal amount inward or outward, the camber is changed without affecting caster.

Some vehicles may have an eccentric cam and bolt on the spindle between the steering knuckle and upper control arm. This eccentric cam and bolt is used to adjust camber. Typically a strut rod is then used to adjust caster on these vehicles.

Some Macpherson struts may use an eccentric cam and bolt between the steering knuckle and strut. This eccentric cam and bolt is used to adjust camber. Caster may be adjusted on some of these vehicles but many times may not be adjustable **(Figure 31-32)**.

Other vehicles may use a bolt and offset washer that must be replaced with a kit that contains an eccentric cam and bolt that will then permit the adjustment of the vehicle alignment.

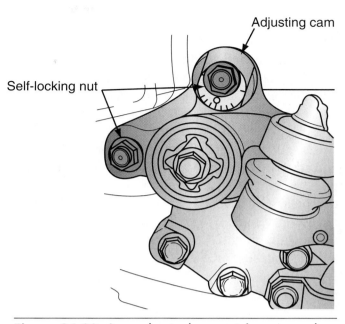

Figure 31-32 A graduated eccentric cam and bolt used to adjust camber.

Slotted Frame

Some vehicles may utilize a slot in the frame for caster and camber. A slot may be provided for adjusting camber and a strut rod for adjusting caster. Other vehicles may use slots for adjusting both caster and camber. These slots may be found on the upper or lower control arms. The pinch bolt in the slot is loosened and the control arm is moved inward or outward in the slot and then retightened.

Macpherson Strut Adjustments

Vehicle manufacturers usually allow for camber adjustment and may provide for adjustment of caster. Many of these adjustments are done with a plate at the top of the strut. The strut may have an adjustment that is riveted or spot welded at the factory to prevent movement of the adjustment plate **(Figure 31-33)**. Many of these adjustments are done with a

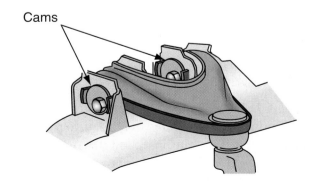

Figure 31-31 An eccentric cam and bolt used on an upper control arm.

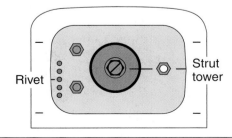

Figure 31-33 Caster and camber adjustment for a Macpherson strut.

plate at the top of the strut. The rivets or spot welds must be drilled out before adjustments can be made. After this is completed the plate can be loosened and the alignment can be completed. Tighten the plate when the caster and camber adjustment is complete. It is not necessary to re-rivet or re-weld the plate after the alignment is completed.

On other vehicles the plate is simply loosened and adjusted **(Figure 31-34)**. Be sure to tighten the plate when the caster and camber adjustment is complete. Other vehicles may require the use of an adapter plate that will permit the adjustment of the vehicle alignment. As previously mentioned, other vehicles may use an eccentric cam and bolt between the strut and steering knuckle **(Figure 31-35)**. This can be either the upper or the lower of the two bolts connecting the strut to the steering knuckle.

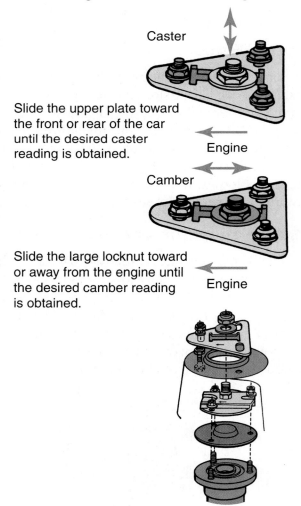

Figure 31-34 A plate for adjusting caster and camber for a Macpherson strut.

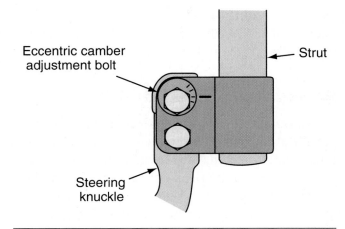

Figure 31-35 A camber adjustment bolt for a Macpherson strut.

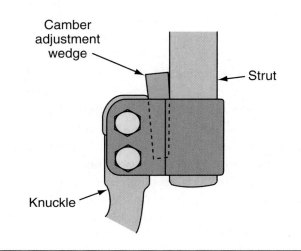

Figure 31-36 A wedge used to adjust camber on a Macpherson strut.

Both bolts must be loosened to permit the adjustment. The steering should be in the straight-ahead position when making this adjustment. Turn the cam bolt to the proper alignment and tighten the bolts to the manufacturer's torque specifications. Some vehicles may use a wedge to adjust camber **(Figure 31-36)**.

Rear-Wheel Adjustments

There are many methods used to adjust the rear suspension. Some vehicles may require the use of a tapered circular shim installed between the rear spindle and axle to change camber and toe **(Figure 31-37)**. The alignment computer can help calculate the specific tapered shim and the direction to install the shim **(Figure 31-38)**. Some vehicles may

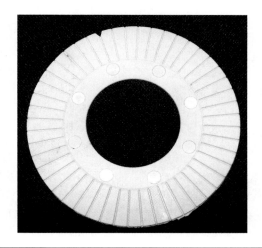

Figure 31-37 A tapered circular shim.

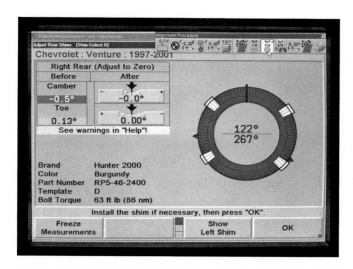

Figure 31-38 Shim selection screen on an alignment machine.

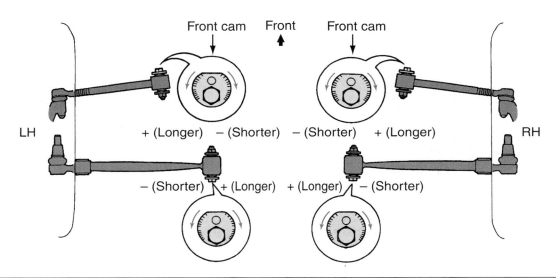

Figure 31-39 A typical eccentric cam and bolt adjustment for an independent rear suspension.

use a wedge type shim to make these adjustments. Other vehicles with independent rear suspensions may use eccentric cam and bolts **(Figure 31-39** and **Figure 31-40)**.

Toe Adjustment

Toe is the last adjustment to be performed during a wheel alignment. The steering wheel must always be centered before adjusting the toe **(Figure 31-41)**. The toe is usually adjusted one of two ways.

If the vehicle has tie-rod adjusting sleeves, then the sleeve clamp bolt will be loosened and the sleeve rotated to the desired toe setting **(Figure 31-42)**. The sleeve clamp bolts are then tightened to manufacturer specifications.

If the vehicle has a rack-and-pinion steering system, then the lock nut is loosened and the inner tie-rod shaft is rotated to achieve the desired toe setting **(Figure 31-43)**. The rack-and-pinion bellows boot must not be allowed to twist, as this will damage the boot. The clamp retaining the boot to the inner tie-rod shaft may need to be loosened or removed during this adjustment.

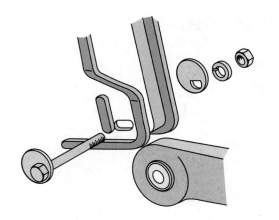

Figure 31-40 A typical eccentric cam and bolt adjustment for rear camber.

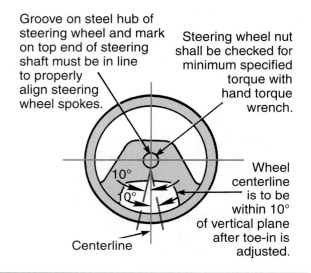

Groove on steel hub of steering wheel and mark on top end of steering shaft must be in line to properly align steering wheel spokes.

Steering wheel nut shall be checked for minimum specified torque with hand torque wrench.

10°
10°

Wheel centerline is to be within 10° of vertical plane after toe-in is adjusted.

Centerline

Figure 31-41 Acceptable range for centering the steering wheel.

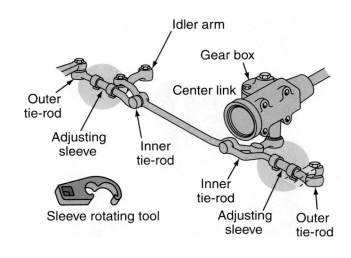

Idler arm

Gear box

Center link

Outer tie-rod

Adjusting sleeve

Inner tie-rod

Inner tie-rod

Sleeve rotating tool

Adjusting sleeve

Outer tie-rod

Figure 31-42 Adjust tie-rod sleeves to adjust toe.

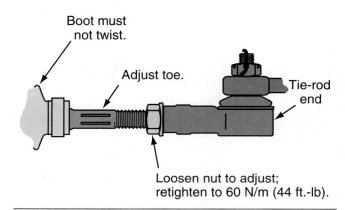

Boot must not twist.

Adjust toe.

Tie-rod end

Loosen nut to adjust; retighten to 60 N/m (44 ft.-lb).

Figure 31-43 Rotate the tie-rod shaft to adjust toe.

Summary

- Caster is the angle of the steering axis of a wheel compared to vertical, as viewed from the side of the vehicle. Tilting the steering axis forward at the top is negative caster. Tilting the steering axis backward at the top is positive.

- Camber is the angle represented by the tilt of either the front or rear wheels inward or outward from the vertical as viewed from the front of the car. The vehicle will pull to the side with the most positive camber.

- Toe is the distance comparison between the leading edge and trailing edge of the front tires.

- If the leading edge distance is less, then there is toe-in. If it is greater, there is toe-out.

- The difference of rear toe from the geometric centerline of the vehicle is called the thrust angle. The vehicle tends to travel in the direction of the thrust line, rather than straight ahead.

- The SAI is the angle between the true vertical and a line drawn between the steering pivots as viewed from the front of the vehicle. Steering axis inclination (SAI) angles located the vehicle weight to the inside or outside of the vertical centerline of the tire.

- Toe-out on turns is the amount of toe-out present on turns.

- In correct tracking, all suspensions and wheels are in their correct locations and conditions and are aligned so that the rear wheels follow directly behind the front wheels while moving in a straight line.

- It is important to remember that approximately 85 percent of today's vehicles not only undergo front-end alignment but require rear-wheel alignment as well.

- The primary objective of four-wheel or total-wheel alignment, whether front or rear drive, solid axle, or independent rear suspension, is to align all four wheels so that vehicle drives and tracks straight with the steering wheel centered. To accomplish this, the wheels must be parallel to one another and parallel to a common centerline.

Review Questions

1. Define camber.

2. Define caster.

3. What tire wear pattern may result from excessive positive camber?

4. Define toe-in.

5. Define thrust angle.

6. How will a vehicle handle if one wheel has more positive camber than the other?

7. How will a vehicle handle if one wheel has more positive caster than the other?

8. In what direction must the top of the wheel be moved to change toward the more positive camber?

9. Describe the reason for toe-out on turns.

10. Which of the following is a definition of SAI?

    A. Forward tilt of the top of the steering knuckle

    B. Inward tilt of the spindle steering arm

    C. Inward tilt of the top of the ball joint, kingpin, or Macpherson strut

    D. Outward tilt of the top of the ball joint, kingpin, or Macpherson strut

11. List the correct sequence for adjusting alignment angles during a four-wheel alignment.

12. Technician A says a wheel alignment rack should include rear slip plates. Technician B says a four-wheel alignment can be performed on a two-wheel alignment machine with some extra work. Who is correct?

    A. Technician A

    B. Technician B

    C. Both Technician A and Technician B

    D. Neither Technician A nor Technician B

13. Define the term *thrust line* as it applies to vehicle handling.

14. What checks should be made before undertaking a wheel alignment?

15. Technician A says positive camber provides directional stability. Technician B says the vehicle will pull toward the side with the more negative camber. Who is correct?

    A. Technician A

    B. Technician B

    C. Both Technician A and Technician B

    D. Neither Technician A nor Technician B

16. Technician A says the purpose of the caster adjustments is to make sure the steering wheel is centered after the alignment. Technician B says if the steering wheel is not centered after an alignment, camber should be re-adjusted. Who is correct?

    A. Technician A

    B. Technician B

    C. Both Technician A and Technician B

    D. Neither Technician A nor Technician B

17. Technician A says the presence of an improper thrust angle can cause a vehicle to travel down the highway sideways. Technician B says it can also increase tire wear as the front wheels fight the rear ones for steering control. Who is correct?

    A. Technician A

    B. Technician B

    C. Both Technician A and Technician B

    D. Neither Technician A nor Technician B

18. Technician A says camber changes as the suspension moves up and down. Technician B says camber changes as the vehicle is loaded and the suspension sags under the weight. Who is correct?

    A. Technician A

    B. Technician B

    C. Both Technician A and Technician B

    D. Neither Technician A nor Technician B

19. All of the following angles are adjustable during wheel alignment *except*:

    A. toe-in.

    B. toe-out on turns.

    C. caster.

    D. camber.

20. Technician A says if camber at both wheels is not equal, the vehicle will tend to pull toward the side with the most positive camber. Technician B says if caster at both wheels is not equal, the vehicle will tend to pull toward the side with the most positive caster. Who is correct?

    A. Technician A

    B. Technician B

    C. Both Technician A and Technician B

    D. Neither Technician A nor Technician B

CHAPTER

32

The Four-Stroke Cycle and Cylinder Arrangements

Learning Objectives

After you have read, studied, and practiced the contents of this unit, you should be able to:

- Define the four-stroke cycle theory.
- Describe the different cylinder arrangements in modern engines.
- Describe the different valve trains used in modern engines.
- Describe the power train differences in a hybrid vehicle.

Key Terms

Compression stroke

Dual overhead cam (DOHC)

Exhaust stroke

Horsepower

Intake stroke

Overhead cam (OHC)

Overhead valve (OHV)

Power stroke

Thermodynamics

Work

INTRODUCTION

One of the many laws of physics utilized within the automotive engine is thermodynamics. **Thermodynamics** is the relationship between heat energy and mechanical energy.

The driving force of the engine is the expansion of gases. Gasoline (a liquid fuel) will change states to a gas if it is heated or burned. Gasoline must be mixed with oxygen before it can burn. In addition, the air-fuel mixture must be burned in a confined area in order to produce power. Gasoline that is burned in an open container produces very little power, but if the same amount of fuel is burned in an enclosed container, it will expand with force. When the air-fuel mixture burns, it also expands as the molecules of the gas collide with each other and bounce apart. Increasing the temperature of the gasoline molecules increases their speed of travel, causing more collisions and expansion.

Heat is generated by compressing the air-fuel mixture in the combustion chamber. Igniting the compressed mixture causes the heat, pressure, and expansion to multiply. This process releases the energy of the gasoline so it can produce **work**. The igniting of the mixture is a controlled burn, not an explosion. The controlled combustion releases the fuel energy at a controlled rate in the form of heat energy. The heat, and consequential expansion of molecules, increases the pressure inside the combustion chamber. Typically, the pressure works on top of a piston that is connected by a connecting rod to the crankshaft. As the piston is driven, it causes the crankshaft to rotate. The engine produces torque, which is applied to the drive wheels. As the engine drives the wheels to move the vehicle, a certain amount of work is done. The rate of work being performed is measured in **horsepower**. Torque is a rotating force around a pivot point.

ENGINE CYCLES

A stroke is the amount of piston travel from top dead center (TDC) to bottom dead center (BDC) measured in inches or millimeters. For example, if the piston is at the top of its travel and is then moved to the bottom of its travel, one stroke has occurred. Another stroke occurs when the piston is moved from the bottom of its travel to the top again. A cycle is a sequence that is repeated. In the four-stroke engine, four strokes are required to complete one cycle. The internal combustion engine must draw in an air-fuel mixture, compress the mixture, ignite the mixture, and then expel the exhaust. This is accomplished in four piston strokes **(Figure 32-1)**. The process of drawing in the air-fuel mixture is actually accomplished by atmospheric pressure pushing it into a low-pressure area created by the downward movement of the piston.

The Intake Stroke

The first stroke of the cycle is the **intake stroke** **(Figure 32-1A)**. As the piston moves down from TDC, the intake valve is opened so the vaporized air-fuel mixture can be pushed into the cylinder by

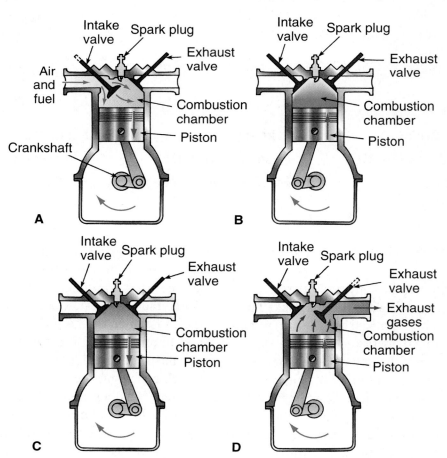

Figure 32-1 (A) Intake stroke. (B) Compression stroke. (C) Power stroke. (D) Exhaust stroke.

atmospheric pressure. At TDC, the piston is at the very top of its stroke. As the piston moves downward in its stroke, a vacuum is created (pressure that is lower than atmospheric pressure). Since high pressure moves toward a low pressure, the air-fuel mixture is pushed past the open intake valve and into the cylinder. After the piston reaches BDC, the intake valve is closed and the stroke is completed. At BDC, the piston is at the very bottom of its stroke. Closing the intake valve after BDC allows an additional amount of air-fuel mixture to enter the cylinder, increasing the volumetric efficiency of the engine. Even though the piston is at the end of its stroke and no more vacuum is created, the additional mixture enters the cylinder because it weighs more than air alone. Volumetric efficiency is the actual amount of air-fuel mixture entering the cylinders compared to the amount of air-fuel mixture that could enter the cylinders under ideal conditions.

The Compression Stroke

The **compression stroke** begins as the piston starts its travel back to TDC **(Figure 32-1B)**. The intake and exhaust valves are both closed, trapping the air-fuel mixture in the combustion chamber. The movement of the piston toward TDC compresses the mixture. As the molecules of the mixture are pressed tightly together, they begin to heat. When the piston reaches TDC, the mixture is fully compressed and a spark is induced in the cylinder by the ignition system. Compressing the mixture provides better burning and intense combustion.

> **Tech Tip** *Combustion chamber sealing is very important. If pressure can leak past the rings or valves during the compression stroke, cylinder pressure and engine power are reduced.*

The Power Stroke

When the spark occurs at the spark plug electrodes in the compressed mixture, the rapid burning causes the molecules to expand; this begins the **power stroke (Figure 32-1C)**. The expanding

molecules create a pressure above the piston and push it downward. The downward movement of the piston in this stroke is the only time the engine is productive concerning power output. During the power stroke, the intake and exhaust valves remain closed.

> **Tech Tip** *The effective power stroke is the shortest stroke in the four-stroke cycle. The extreme pressure created by combustion in the cylinder lasts for approximately only 25° of crankshaft rotation. Therefore, timing of the combustion event is very important. The spark at the plug electrodes must occur at just the right instant in relation to piston position so the combustion of the air-fuel mixture creates maximum downward force on the piston.*

NOTE: An engine that uses a spark at the spark plug electrodes to ignite the air-fuel mixture may be referred to as an *SI engine*.

NOTE: An engine that ignites the air-fuel mixture from the heat of combustion is called a *compression ignition (CI)* engine.

The Exhaust Stroke

The **exhaust stroke** of the cycle begins when the piston reaches BDC of the power stroke **(Figure 32-1D)**. Just prior to the piston reaching BDC, the exhaust valve is opened. The upward movement of the piston back toward TDC pushes out the exhaust gases from the cylinder past the exhaust valve and into the vehicle's exhaust system. As the piston approaches TDC, the intake valve opens and the exhaust valve is closed a few degrees after TDC. The degrees of crankshaft rotation when both the intake and exhaust valves are open is called valve overlap. During valve overlap, the incoming air-fuel mixture through the intake valve helps to purge any remaining exhaust gases from the cylinder. The cycle is then repeated again as the piston begins the intake stroke.

Most engines in use today are referred to as reciprocating. Power is produced by the up-and-down movement of the piston in the cylinder. This linear motion is then converted to rotary motion by a crankshaft.

> **Tech Tip** *The duration of valve overlap is very important to achieve maximum engine power, performance, and smooth engine operation. During valve overlap, the remaining burned exhaust gases are purged out of the cylinder and fresh air-fuel mixture must be swept into the combustion chamber.*

CYLINDER ARRANGEMENTS

One cylinder would not be able to produce sufficient power to meet the demands of today's vehicles. Most automotive and truck engines use three, four, five, six, eight, ten, or twelve cylinders. The number of cylinders used by the manufacturer is determined by the amount of work required from the engine.

Vehicle manufactures attempt to achieve a balance between power, economy, weight, and operating characteristics. An engine having more cylinders generally runs smoother than those having three or four cylinders because there is less crankshaft rotation between power strokes. However, adding more cylinders increases the weight of the vehicle and the cost of production. Engines are classified by the arrangement and number of cylinders. The cylinder arrangement used is determined by vehicle design and purpose. The most common engine designs are in-line and V-type.

The in-line engine places all of its cylinders in a single row **(Figure 32-2)**. Advantages of this engine design include ease of manufacturing and serviceability. In-line engines typically have higher torque output when compared to equal displacement engines of different cylinder arrangement. The disadvantages of this engine design are the block height and generally lower horsepower. Since the engine is tall, aerodynamic design of the vehicle is harder to achieve. Most manufacturers

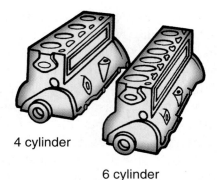

4 cylinder

6 cylinder

Figure 32-2 In-line engines.

overcome this disadvantage by mounting the engine transversely in the engine compartment of front-wheel drive vehicles **(Figure 32-3)**. A transversely mounted engine is mounted sideways on the vehicle. A variation of the in-line engine design is the slant cylinder **(Figure 32-4)**. This design is similar to other in-line engines except that the cylinders are placed at a slant. This design reduces the height of the engine and allows for more aerodynamic vehicle designs. The V-type engine has two rows of cylinders set generally 60° to 90° from each other **(Figure 32-5)**. The V-type design allows for a shorter block height and improved vehicle

Figure 32-3 A transverse mounted engine.

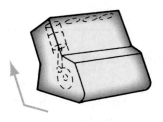

Figure 32-4 A slant-type engine.

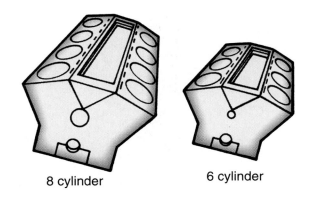

Figure 32-5 V-type engine design.

aerodynamics. The length of the block is also shorter than in-line engines with the same number of cylinders. V-type engines tend to produce higher horsepower compared to other designs of the same displacement. Recently Volkswagen has released a V-type engine that has only 15° between cylinder banks **(Figure 32-6)**. Having the banks so close allows it to share characteristics of both in-line and V-type engines. These engines have only one cylinder head because the banks are so close.

A common engine design used for rear-engine vehicles as well as the front-wheel drive Subaru is the opposed cylinder engine **(Figure 32-7)**. The opposing cylinder engine design has two rows of cylinders directly across from each other. An opposed cylinder engine has the piston banks mounted at 180° in relation to each other. The main advantage of this engine design is the very small vertical height, allowing the vehicle to have a lower center of gravity.

VALVE ARRANGEMENTS

Engines are also classified by their valve train. The three most commonly used valve trains are the:

- **Overhead valve (OHV).**—The intake and exhaust valves are located in the cylinder head while the camshaft and lifters are located in the engine block **(Figure 32-8)**. Valve train components of this design include lifters, pushrods, and rocker arms.

- **Overhead cam (OHC).**—The intake and exhaust valves are located in the cylinder head along with the camshaft **(Figure 32-9)**. The valves are oper-ated directly by the camshaft and a follower, thereby eliminating many of the moving parts required in the OHV engine. If a single camshaft is used in each cylinder head, the engine is classi-fied as a single overhead cam (SOHC) engine **(Figure 32-10)**. This designation is used even if the engine has two cylinder heads with one camshaft each.

Figure 32-6 A Volkswagen V-type engine.

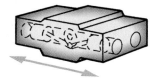

Figure 32-7 Opposing cylinder engine design.

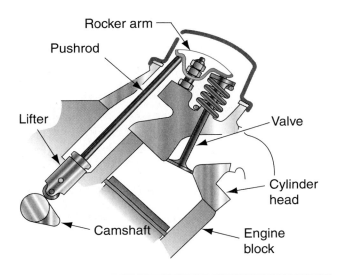

Figure 32-8 An overhead valve engine (OHV).

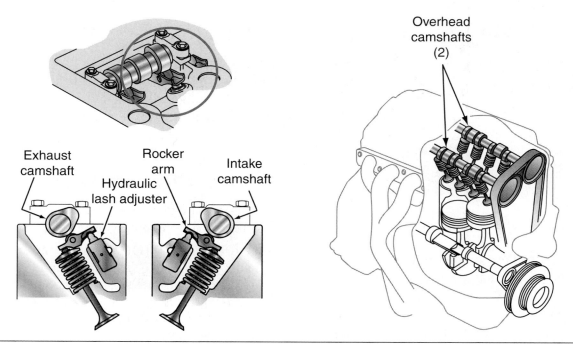

Figure 32-9 An overhead cam engine (OHC).

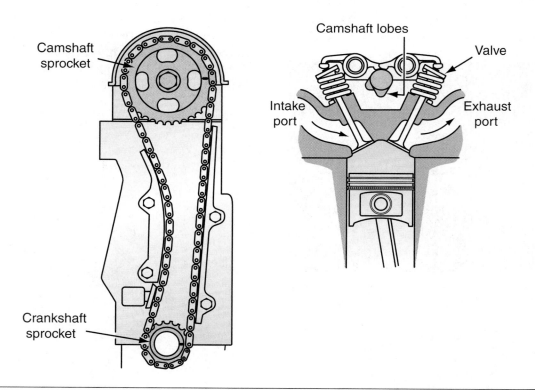

Figure 32-10 A single overhead cam engine (SOHC).

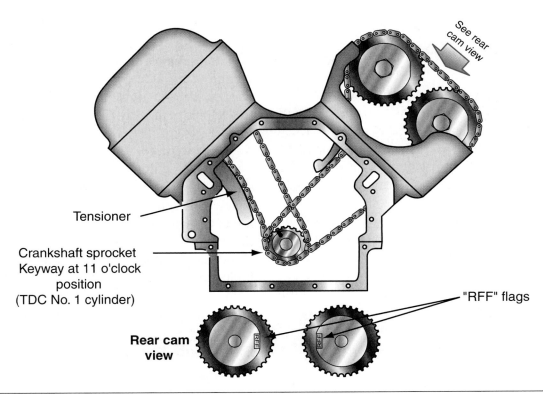

Tensioner

Crankshaft sprocket
Keyway at 11 o'clock
position
(TDC No. 1 cylinder)

"RFF" flags

**Rear cam
view**

Figure 32-11 A dual overhead cam engine (DOHC).

■ **Dual overhead cam (DOHC).**—The DOHC uses separate camshafts for the intake and exhaust valves. A DOHC V-8 engine is equipped with a total of four camshafts **(Figure 32-11)**. DOHC engines can be either in-line or V-type; they are generally high performance engines.

Diesel Engine Principles

Diesel and gasoline engines have many similar components and come in both in-line and V-type configurations. Instead of using a spark delivered by the ignition system, the diesel engine uses the heat produced by compressing air in the combustion chamber to ignite the fuel. Fuel injectors are used to supply fuel into the combustion chamber. The fuel is sprayed under pressure from the injector as the piston is completing its compression stroke. The temperature increase generated by compressing the air (approximately 1,000°F) is sufficient to ignite the fuel as it is injected into the cylinder. In order to attain such high combustion chamber pressures the diesel has very high compression ratios, well over 14:1. With this high compression the diesel will have over 600 lbs per square inch of pressure, compared to the typical gasoline engine that will have less than 10:1 compression ratios and generate only about 150 lbs per square inch.

Starting the diesel engine is dependent upon heating the intake air to a high enough level to ignite the fuel. A method of preheating the intake air is often required to start a cold engine. Some manufacturers use glow plugs to accomplish this. Another method includes using a heater grid in the air intake system. Glow plugs are small, round, electrical devices positioned in a precombustion chamber. When the engine is cold, voltage is supplied to the glow plugs and the resulting current flow heats the glow plugs, which heat the precombustion chambers. In addition to ignition methods, there are other differences between the gasoline and diesel engines. The combustion chambers of the diesel engine are designed to accommodate the different burning characteristics of diesel fuel.

Diesel engines can be either four-stroke or two-stroke designs. Two-stroke engines complete all cycles in two strokes of the piston, much like a gasoline two-stroke engine.

VEHICLES WITH ALTERNATE POWER SOURCES

Most vehicle manufacturers in the world are working on the development of vehicles that use alternate types of fuel to provide power. Some of these types are available in the North American market; however, the most common alternative fuel vehicles are hybrid vehicles. Hybrid vehicles operate on two power sources such as a gasoline engine and an electric motor.

Fuel cell vehicles are electrically driven and the fuel cells produce electricity for the electric drive motor(s). Fuel cells use hydrogen and oxygen to produce electricity, so these vehicles require hydrogen fuel. Fuel cell vehicles are zero-emission vehicles. Currently fuel cell vehicles are available for some fleets and as experimental vehicles. As the technology is being made more reliable and the cost is brought in line, vehicle manufacturers indicate they expect to be marketing fuel cell vehicles after 2010 to the general public.

Electric vehicles are driven with an electric motor only and rely upon battery packs for their power source. The battery packs need to be recharged by an external power source on a regular basis. Some are designed to be plugged into regular household outlets and some need special recharging stations. Due to the limited range and time to recharge, pure electric vehicles do not have the consumer demand that hybrid vehicles have.

Hybrid Vehicles

In the 1990s, most vehicle manufacturers began developing electric vehicles; however, pure electric vehicles have a few key shortcomings that limits their widespread use, including lack of both range and available recharging stations. The time it takes to recharge is also a drawback. The ability to use an already existing fuel infrastructure and no range limitations are just two of the driving factors behind the increasing popularity of hybrid vehicles.

A hybrid vehicle has two different power sources. In most hybrid vehicles, the power sources are a small displacement gasoline or diesel engine and an electric motor. Generally the electric motor is coupled between the engine and the transmission in

the flywheel housing. Currently the hybrid vehicles available for sale have two main types of operating principles. One design relies upon a small gasoline engine to provide power to drive the vehicle and uses an electric motor to provide additional power to accelerate quickly. The Honda Insight uses this type of system. The other hybrid design relies upon the electric motor for propulsion at lower speeds and the gas engine assists the electric when the speeds increase or extra power is required. In both cases the engine also runs a generator that keeps the battery packs charged. The Toyota hybrid uses this type of driving strategy. Currently Toyota and Honda account for over 90 percent of all hybrid vehicles sold in the United States. Currently all of the domestic makers are beginning to produce their own hybrid vehicles for domestic sale.

Fuel Cell-Powered Vehicles

A fuel cell-powered vehicle is an electric vehicle with some important differences. An electric motor supplies torque to the drive wheels, but the fuel cell produces and supplies electric power to the electric motor; in an electric vehicle the batteries supply this power. Most of the vehicle manufacturers, in cooperation with some independent laboratories and government funding, are involved in fuel cell research and development programs. A number of prototype fuel cell vehicles have been produced.

Fuel cells electrochemically combine oxygen from the air with hydrogen from a hydrocarbon fuel to produce electricity. The oxygen and hydrogen are fed to the fuel cell as "fuel" for the electrochemical reaction. There are different types of fuel cells, but the most common type is the proton exchange membrane (PEM) fuel cell. Each individual fuel cell contains two electrodes separated by a membrane.

Electrical power output from one individual fuel cell is very low, and so many fuel cells are contained in a fuel cell stack to supply enough electric energy to the electric drive motor.

An on-board reformer may be used to extract hydrogen from liquid fuels, such as gasoline or methanol. The main disadvantage to this system is the space required by the reformer. Using methanol would also require a new refueling system across the country. If a fuel cell-powered vehicle is fueled directly with hydrogen, the exhaust emissions are almost zero. A fuel cell-powered vehicle with an on-board reformer to extract hydrogen from some other fuel does create a very small amount of tailpipe emissions because of the reformer action. Recent developments include a multi-reformer that operates on different types of fuels. Hydrogen can be obtained by electrolysis from water. However, this process requires a great deal of electrical energy. With present technology, this method of obtaining hydrogen is not an option. It appears that the first fuel cell vehicles offered to customers will have on-board reformers that extract hydrogen from some well-known fuel. The two major obstacles to be overcome in the development of fuel cell vehicles are cost and the on-board space required by the system components. Research and development is taking place at a rapid pace, and components are quickly being downsized and improved. As production of fuel cell vehicle components increases, the price will be reduced.

Summary

- Most automotive engines operate on the four-stroke cycle principle.
- A stroke is the piston movement from TDC to BDC.
- During the intake stroke, the piston is moving downward, the intake valve is open, and the exhaust valve is closed.
- During the compression stroke, the piston is moving upward and both valves are closed.
- During the power stroke, the piston is forced downward by combustion pressure and both valves remain closed. The exhaust valve opens near BDC on the power stroke.

- During the exhaust stroke, the piston is moving upward, the exhaust valve is open, and the intake valve is closed. The intake valve opens a few degrees before TDC on the exhaust stroke, and the exhaust valve closes a few degrees after TDC on the exhaust stroke.
- Valve overlap is the number of degrees that the crankshaft rotates when both valves are open at the same time and the piston is near TDC on the exhaust stroke.
- Three of the most common valve train designs are overhead valve (OHV), overhead cam (OHC), and dual overhead cam (DOHC).

Review Questions

1. Technician A says that during the intake stroke, the intake and exhaust valves are open. Technician B says the spark plug fires when the piston is at BDC on the intake stroke. Who is correct?

   A. Technician A

   B. Technician B

   C. Both Technician A and Technician B

   D. Neither Technician A nor Technician B

2. Technician A says a stroke occurs when a piston moves from TDC to BDC. Technician B says the air is moved into the cylinder on the intake stroke by vacuum in the cylinder and atmospheric pressure. Who is correct?

   A. Technician A

   B. Technician B

   C. Both Technician A and Technician B

   D. Neither Technician A nor Technician B

3. Technician A says that during the compression stroke, the intake and exhaust valves are closed. Technician B says that during the compression stroke, the piston is moving upward in the cylinder. Who is correct?

   A. Technician A

   B. Technician B

   C. Both Technician A and Technician B

   D. Neither Technician A nor Technician B

4. Technician A says the intake valve opens when the piston is a few degrees before TDC on the exhaust stroke. Technician B says the exhaust valve closes when the piston is a few degrees after TDC on the exhaust stroke. Who is correct?

   A. Technician A

   B. Technician B

   C. Both Technician A and Technician B

   D. Neither Technician A nor Technician B

5. All of these statements about the power stroke in a four-cycle engine are true *except*:

   A. The power stroke is the shortest stroke in the four-stroke cycle.

   B. The extreme pressure created by combustion in the cylinder lasts for about 120° of crankshaft rotation.

   C. During the power stroke, the expanding molecules in the combustion chamber force the piston downward.

   D. Leaking piston rings decrease combustion chamber pressure and engine power.

6. In a four-stroke cycle engine during valve overlap:

   A. the intake valve is closing.

   B. the exhaust valve is beginning to open.

   C. both valves are open at the same time.

   D. the piston is at TDC on the power stroke.

7. All of these statements about diesel engine principles are true *except*:

   A. The heat developed by compressing the air in the cylinder ignites the air-fuel mixture.

   B. Compared to a gasoline engine, the diesel engine has a lower compression ratio.

   C. Fuel injectors spray fuel into the combustion chamber or precombustion chamber.

   D. Glow plugs supply additional combustion chamber heat when starting a cold engine.

8. While discussing engine design, Technician A says a V-type engine can have a 60-degree or a 90-degree angle between the two cylinder banks. Technician B says in an opposed cylinder the engine has a 180-degree angle between the two cylinder banks. Who is correct?

   A. Technician A

   B. Technician B

   C. Both Technician A and Technician B

   D. Neither Technician A nor Technician B

9. While discussing engine design, Technician A says in an OHV engine, the camshaft is mounted on top of the cylinder head. Technician B says in an OHV engine, the valve lifters are mounted in the cylinder head. Who is correct?

   A. Technician A

   B. Technician B

   C. Both Technician A and Technician B

   D. Neither Technician A nor Technician B

10. During valve overlap, both valves in a cylinder are_____.

11. A transverse-mounted engine is positioned _____ in the engine compartment.

12. In an OHC engine, the camshaft(s) are positioned _____ the _____.

13. Explain how an internal combustion engine produces power and torque.

14. Describe the piston and valve movement during the intake stroke.

15. Describe the piston and valve position during the power stroke.

16. Explain the purpose of valve overlap.

CHAPTER

33 Lubrication Systems

Learning Objectives

After you have read, studied, and practiced the contents of this unit, you should be able to:

- Describe the function of the lubrication system.

- Describe the basic types of oil additives.

- Explain the purpose of the SAE classifications of oil.

- Explain the purpose of the API classifications of oil.

Key Terms

American Petroleum Institute (API)

Cohesion

International Lubricant Standardization and Approval Committee (ILSAC)

Oxidation

Society of Automotive Engineers (SAE)

Viscosity

INTRODUCTION

When the engine is operating, the moving parts generate heat due to friction. If this heat and friction were not controlled, the components would weld together. In addition, heat is created from the combustion process. It is the function of the engine's lubrication system to supply oil to the high friction and wear locations in the engine and to dissipate heat away from them **(Figure 33-1)**. The purpose of engine oil is to lubricate, clean, seal, absorb shock, and cool.

It is impossible to eliminate all of the friction within an engine, but a properly operating lubrication system helps to reduce it. The oil must be at the correct **viscosity** or thickness to function properly.

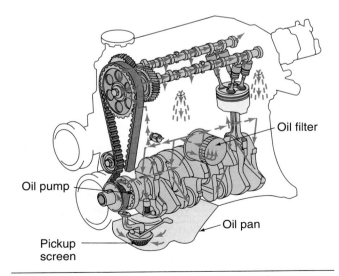

Figure 33-1 Oil flow through the engine lubrication system.

Oil molecules work as small bearings rolling over each other to reduce friction **(Figure 33-2)**. Lubrication systems provide an oil film to prevent moving parts from coming in direct contact with each other **(Figure 33-3)**.

Another function of the lubrication system is to act as a shock absorber between the connecting rod and crankshaft. Oil must also withstand high temperature and high shear conditions. This is achieved with a high temperature, high shear viscometer test. To qualify as a particular **Society of Automotive Engineers (SAE)** grade, the oil must meet minimum viscosity at high shear and high temperature conditions.

> ***Did You Know?*** *The SAE is a group of automotive engineers dedicated to the advancement of automotive and aerospace technology. SAE is responsible for establishing many automotive and aerospace standards, including oil viscosity ratings.*

Besides providing friction reduction, engine oil absorbs heat and transfers it to another area for cooling. As the oil flows through the engine, it conducts heat from the parts it comes in contact

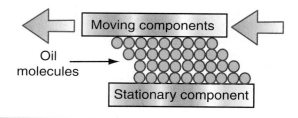

Figure 33-2 Oil molecules work as small bearings to reduce friction.

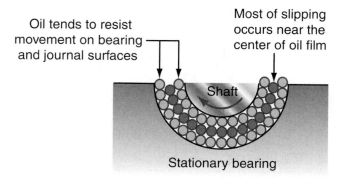

Figure 33-3 The oil film prevents metal-to-metal contact between engine compartments.

with. When the oil is returned to the oil pan, it is cooled by airflow over the pan. Other purposes of the oil include sealing the piston rings and washing away abrasive metal and dirt. To perform these functions, the engine lubrication system includes the following components:

- Engine oil
- Oil pan or sump
- Oil filter
- Oil pump
- Oil galleries

ENGINE OIL RATING AND CLASSIFICATION

Engine oil must provide a variety of functions under the entire extreme engine operating conditions. To perform these tasks, additives are mixed with natural oil. A brief description of these additives follows:

- Antifoaming agents—These additives are included to prevent aeration of the oil. Aeration will result in low oil pump pressure and insufficient lubrication to the engine parts.

- Anti-oxidation agents—Heat and oil agitation result in oxidation. These additives work to prevent the buildup of varnish and to prevent the oil from breaking down into harmful substances that can damage engine bearings. **Oxidation** occurs when some of the oil combines with oxygen in the air to form an undesirable compound.

- Detergents and dispersants—These are added to prevent deposit buildup resulting from carbon, metal particles, and dirt. Detergents break up larger deposits and prevent smaller ones from grouping together. A dispersant is added to prevent carbon particles from grouping together.

- Viscosity index improver—As the oil increases in temperature, it has a tendency to thin out. These additives prevent oil thinning.

- Pour point depressants—Prevent oil from becoming too thick to pour at low temperatures.

- Corrosion and rust inhibitors—Displace water from the metal surfaces and neutralize acid.

■ Cohesion agents—Maintain a film of oil in high-pressure points to prevent wear. Cohesion agents are deposited on parts as the oil flows over them and remain on the parts as the oil is pressed out. **Cohesion** of the engine oil is the tendency of the oil molecules to remain on the friction surfaces of engine components. Oil is rated by two organizations: the SAE and the API. The SAE has standardized oil viscosity ratings, while the API rates oil to classify its service or quality limits. These rating systems were developed to make proper selection of engine oil easier.

Through a licensing program, **American Petroleum Institute (API)** has standardized the labeling of engine oils by adopting the API donut logo **(Figure 33-4)**. The logo is designed to be placed in an easily seen position on quart containers of oil. API has also developed a starburst certification mark to select engine oils that meet the gasoline performance standards established by the **International Lubricant Standardization and Approval Committee (ILSAC) (Figure 33-5)**. This logo is displayed on the front of licensed oil products.

The API classifies engine oils by a two-letter system. The prefix letter is listed as either S-class or C-class to classify the oil usage. The S stands for a spark ignition (gasoline) engine and the C stands for a compression ignition (diesel) engine.

Did You Know? *The API is dedicated to advancement of the petroleum industry, and this organization is responsible for oil service classifications.*

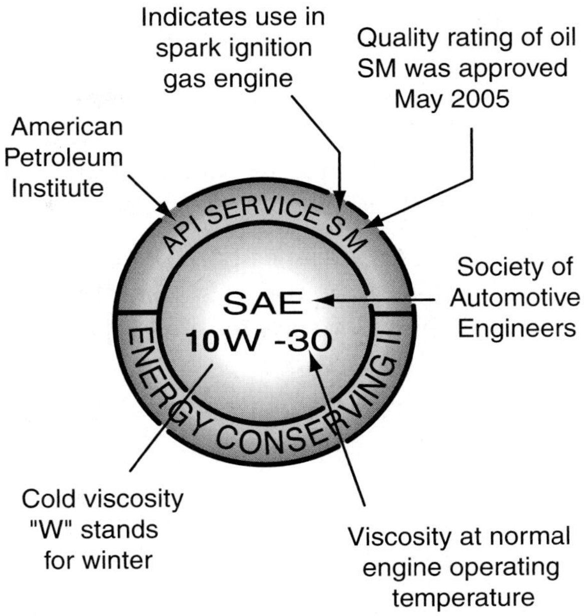

Figure 33-4 Oil containers have an API donut for API and SAE ratings.

The second letter denotes various grades of oil within each classification and denotes the oil's ability to meet the engine manufacturer's warranty requirements or its service rating **(Figure 33-6)**. The most recent classification is SM, and this oil meets the requirements for the latest automotive engines.

Did You Know? *Later classifications will also meet warranty requirements of new vehicles.*

SAE ratings provide a numeric system of determining the oil's viscosity. The higher the SAE

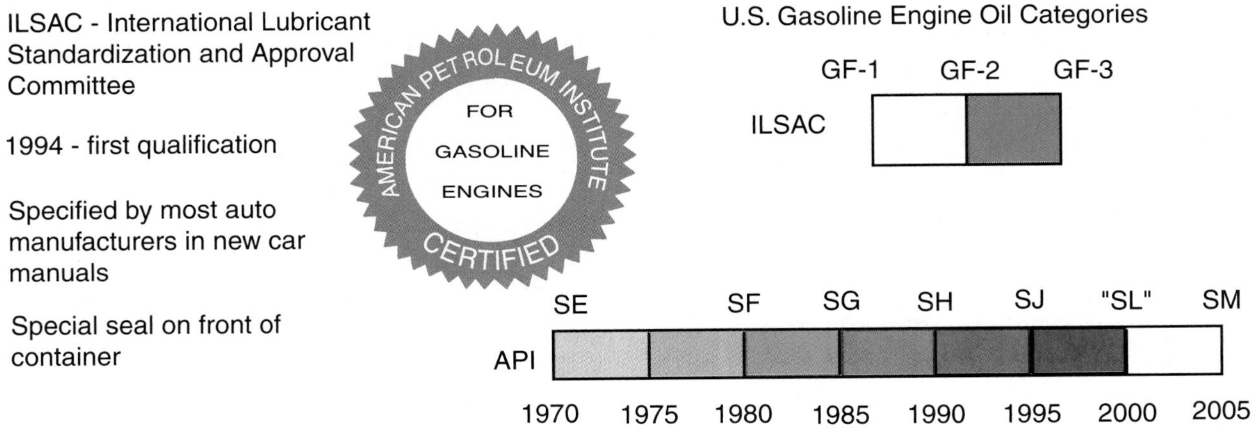

Figure 33-5 ILSAC symbol.

API GASOLINE ENGINE DESIGNATION

SF	Service typical of gasoline engines in automobiles and some trucks beginning with 1980. SF oils provide increased oxidation stability and improved antiwear performance over oils that meet API designation SE. It also provides protection against engine deposits, rust, and corrosion. SF oils can be used in engines requiring SC, SD, or SE oils.
SG	Service typical of gasoline automobiles and light-duty trucks, plus CC classification diesel engines beginning in the late 1980s. SG oils provide the best protection against engine wear, oxidation, engine deposits, rust, and corrosion. It can be used in engines requiring SC, SD, or SF oils.
SH	This classification was replaced in 1997. When compared to SG motor oil, this oil produces 19% less engine sludge, 20% less cam wear, 5% less engine varnish, 21% improved oxidation, 7% less piston varnish, 3% less engine rust, and 13% less bearing wear. As you can see, this is a superior product to anything used in the past and should prolong the life of any engine.
SJ	It meets the same minimum requirements in areas of deposit control, resistance to rust, and oxidation as the older classifications. It also allows the engine to run with less friction, which results in better fuel economy. This oil classification was introduced in 1997 and replaces all earlier API categories.
SL	This was released July of 2001.
SM	Released in May of 2005. SM is suitable for use in all automative engines presently in service.

Figure 33-6 The API rating system indicates the service rating of the oil and its ability to meet the manufacturer's requirements.

number, the thicker or heavier the weight of the oil. For example, oil classified as SAE 50 is thicker than SAE 20. The thicker the oil, the slower it will pour. Thicker oils should be used when engine temperatures are high, but they can cause lubrication problems in cooler climates. Thinner oils will flow through the engine faster and easier at colder temperatures but may break down under higher temperatures or heavy engine loads.

To provide a compromise between these conditions, multiviscosity oils have been developed. For example, an SAE 10W-30 oil has a viscosity of 10W at lower temperatures to provide easy flow. As the engine temperature increases, the viscosity of the oil actually increases to a viscosity of 30 to prevent damage resulting from excessively thin oil. The W in the designation refers to winter, which means the viscosity is determined at 0°F (−18°C). If there is no W, then the viscosity is determined at 210°F (100°C).

A common misconception is that multiviscosity oils somehow mysteriously get thicker when they are warmed up. Direct observation of pouring a cold quart container of oil as compared to a warm container of oil shows that the same oil when warm pours easier with less viscosity. An SAE 30 oil has a 30 rating when measured at 0°F and also gets a 30 rating when measured again at 210°F. The SAE 30 oil still thins out when warmed up **(Figure 33-7)**. A 10W-30 oil has the equivalent 10 rating when measured at 0°F and receives a 30 rating when measured at 210°F. The 10W-30 oil still thins out when warmed. It has the important benefit of having a lower viscosity at low temperatures for easier engine starting and better cold lubrication and the benefit of a 30 when the engine is warmed up, again providing better hot lubrication. The multiviscosity oils are more stable and do not thin out as fast as single viscosity oils.

SAE GRADES OF MOTOR OIL

Lowest Atmospheric Temperature Expected	Single-Grade Oils	Multigrade Oils
32°F (0°C)	20, 20W, 30	10W-30, 10W-40, 15W-50, 20W-40, 20W-50
0°F (−18°C)	10W	5W-30, 10W-30, 10W-40, 15W-40
−15°F (−26°C)	10W	10W-30, 10W-40, 5W-30
Below −15°F (−26°C)	5W*	*5W-20, 5W-30

Figure 33-7 The average atmospheric temperature in which the engine is operating determines the proper SAE oil grade.

Tech Tip *Many manufacturers may recommend 5W-20 oil for their vehicles. Use of a viscosity that is not recommended by the manufacturer will void the warranty.*

Oils have recently been developed that may be labeled "energy conserving" or "energy conserving II." These oils use friction modifiers to reduce friction and increase fuel economy. Selection of oil is based upon the type of engine and the conditions it will be running in. The API ratings must meet the requirements of the engine and provide protection under the normal expected running conditions. A main concern in selecting SAE ratings is ambient temperatures **(Figure 33-8)**. Always select oil that meets or exceeds the manufacturer's recommendations.

In recent years, synthetic oils have become increasingly popular. The advantage of synthetic oils is their stability over a wide temperature range. Before using these oils in an engine, refer to the vehicle's warranty information to confirm that use of this oil does not void the warranty.

Did You Know? **Energy conserving oils have produced a fuel economy improvement of 1.5 and energy conserving II oils produce an improvement of 2.7 percent or greater over standard reference oils.**

Did You Know? **Synthetic oils are produced either by a chemical reaction (synthesis), severe refining, or other complex chemical processes that yield molecular uniformity and purity that is impossible to achieve through normal refining processes.**

OIL PUMPS

There are two basic types of oil pumps: rotary and gear. Both types are positive displacement pumps. The rotor pump generally has a four-lobe inner rotor with a five-lobe outer rotor **(Figure 33-9)**. The outer rotor is driven by the inner rotor. As the lobes come out of mesh, a low pressure area is created and atmospheric pressures above the oil level in the pan forces it into the pickup tube. The

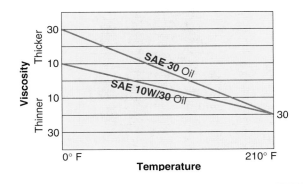

Figure 33-8 A comparison of a single and multiviscosity oil.

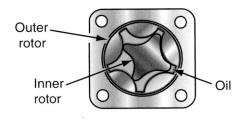

Figure 33-9 A typical rotor-type oil pump.

GEAR TYPE PUMPS

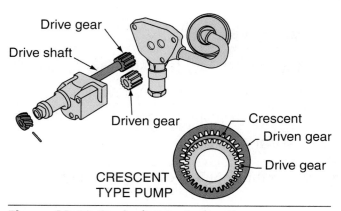

CRESCENT TYPE PUMP

Figure 33-10 Typical gear-type oil pumps.

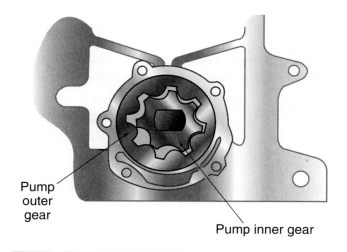

Figure 33-12 In many engines, the oil pump is located at the front of the engine block and driven directly by the crankshaft. Many distributorless engines use this type of pump.

oil is trapped between the lobes as it is directed to the outlet. As the lobes come back into mesh, the oil is pressurized and expelled from the pump.

Gear pumps use two gears riding in mesh with each other or use two gears and a crescent design **(Figure 33-10)**. Both types operate in the same manner as the rotor type pump. The advantage of the rotor-type pump is its capability to deliver a greater volume of oil since the rotor cavities are larger.

In the past, most oil pumps were driven off the camshaft by a drive shaft fitting into the bottom of the distributor shaft **(Figure 33-11)**. Many of today's engines do not use a distributor and may

drive the oil pump by the front of the crankshaft **(Figure 33-12)**.

Because oil pumps are positive displacement types, output pressures must be regulated to prevent excessive pressure buildup. A pressure relief valve opens to return oil to the sump if the specified pressure is exceeded **(Figure 33-13)**. A calibrated spring holds the valve closed, allowing pressure to increase. Once the oil pressure is great enough to overcome the spring pressure, the valve opens and returns the oil to the sump.

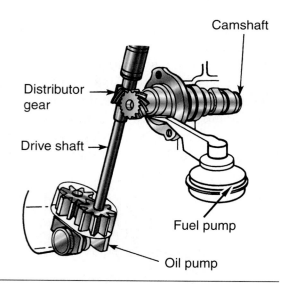

Figure 33-11 Many oil pumps are driven by a gear on the camshaft.

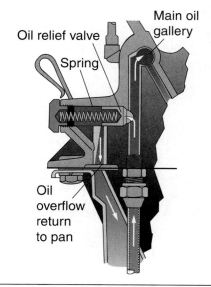

Figure 33-13 The oil pressure relief valve opens to return the oil to the sump if the oil pressure is excessive.

OIL FILTER

As the oil flows through the engine, it works to clean dirt and deposits from the internal parts. These contaminants are deposited into the oil pan or sump with the oil. Because the oil pump pickup tube draws oil from the sump, it can also pick up these contaminants. The pickup tube has a screen mesh to prevent larger contaminants from being picked up and sent back into the engine.

The finer contaminants must be filtered from the oil to prevent them from being sent with the oil through the engine **(Figure 33-14)**. Oil filter elements have a pleated paper or fibrous material designed to filter out particles between 20 and 30 microns **(Figure 33-15)**. A micron has a very small diameter. The thickness of a human hair is usually about 60 to 90 microns, and 100 microns is equal to 0.004 inch.

Most oil filters use a full filtration system that filters all the oil delivered by the pump. Under pressure from the pump, oil enters the filter on the outer areas of the element and works its way toward the center. If the filter becomes partially restricted and the pressure drop across the filter reaches a predetermined limit, a bypass valve opens and allows the oil to bypass the filter and enter the oil galleries. If this occurs, the oil will no longer be filtered **(Figure 33-16)**.

An inlet check valve is used to prevent oil drain back from the oil pump when the engine is shut off. The check valve is a rubber flap covering the

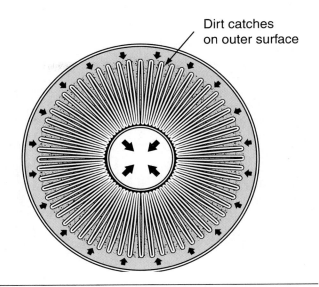

Figure 33-15 The filter catches contaminants on the surface of the pleated filter media.

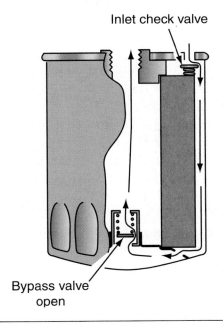

Figure 33-16 If the oil filter becomes plugged, the bypass valve opens to protect the engine.

inside of the inlet holes. When the engine is started and the pump begins to operate, the valve opens and allows oil to flow to the filter.

LUBRICATION SYSTEM PURPOSE AND OPERATION

After the oil leaves the filter, it is directed through galleries to various parts of the engine. Oil galleries are drilled passages in the cylinder block and

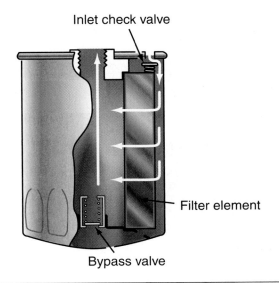

Figure 33-14 Oil flow through the oil filter.

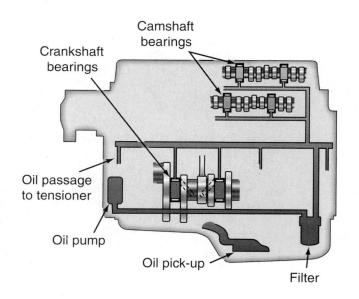

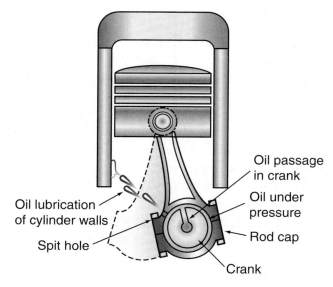

Figure 33-17 The lubrication system delivers oil to the high friction and wear areas of the engine.

Figure 33-19 A spit hole in the connecting rod sprays oil onto the cylinder walls and helps to cool the piston.

head. A main oil gallery is usually drilled the length of the block. From the main gallery, pressurized oil branches off to upper and lower portions of the engine (**Figure 33-17**). Oil is directed to the crankshaft main bearings through the main bearing saddles. Passages drilled in the crankshaft then direct the oil to the connecting rod bearings (**Figure 33-18**).

Some manufacturers drill a small spit or squirt hole in the connecting rod to spray pressurized oil delivered to the bearings out and onto the cylinder walls (**Figure 33-19**). When the spit hole aligns with the oil passage in the rod journal, it squirts the

oil onto the cylinder wall. This serves to help cool the underside of the piston and to assist in proper lubrication and sealing oil for the piston rings.

Each camshaft bearing (on OHV engines) also receives oil from passages drilled in the block (**Figure 33-20**). The valve train can receive oil through passages drilled in the block and then through a hollow rocker arm shaft. Engines using pushrods usually pump oil from the lifters through the pushrods to the valve train (**Figure 33-21**). Oil sent to the cylinder head is returned to the sump through drain passages cast into the head and block.

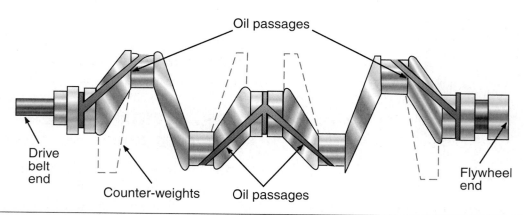

Figure 33-18 Oil passages drilled in the crankshaft supply lubrication to the bearings.

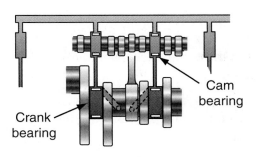

Figure 33-20 In an OHV engine, oil is supplied from the main bearings to the camshaft bearings.

Not all areas of the engine are lubricated by pressurized oil. The crankshaft rotates in the oil sump, creating some splash-and-throw effects. Oil that is picked up in this manner is thrown throughout the crankcase. This oil splashes onto the cylinder walls below the pistons. This provides lubrication and cooling to the piston pin and piston rings. In addition, oil that is forced out of the connecting rod bearings is thrown out to lubricate parts that are not fed pressurized oil. Valve timing chains are often

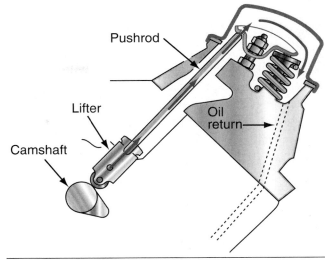

Figure 33-21 OHV engines often have hollow pushrods to supply oil from the valve lifters to the rocker arms and valve train.

lubricated by oil draining from the cylinder head being dumped onto the chain and gears.

Did You Know? *Oil cools most of the engine parts. The coolant mainly removes heat from the cylinder heads and cylinders.*

Summary

- The lubrication system provides an oil film to prevent moving parts from coming in direct contact with each other. Oil molecules work as small bearings rolling over each other to reduce friction. Another function is to act as a shock absorber between the connecting rod and crankshaft.

- Many different types of additives are used in engine oils to formulate a lubricant that will meet all of the demands in today's engines.

- Oil is rated by two organizations: the SAE and the API.

- The SAE has standardized oil viscosity ratings, while the API rates oil to classify its service or quality limits.

- There are two basic types of oil pumps: rotor and gear.

- Both types are positive displacement pumps.

- Oil filter elements have a pleated paper or fibrous material designed to filter out particles between 20 and 30 microns.

Review Questions

1. Technician A says engine oil in the lubrication system helps to clean engine components. Technician B says engine oil helps to provide a seal between the piston rings and the cylinder walls. Who is correct?

   A. Technician A

   B. Technician B

   C. Both Technician A and Technician B

   D. Neither Technician A nor Technician B

2. Technician A says pour point depressants prevent oil from becoming too thin at high temperatures. Technician B says cohesion agents in the engine oil help the oil to adhere to friction surfaces in the engine. Who is correct?

   A. Technician A

   B. Technician B

   C. Both Technician A and Technician B

   D. Neither Technician A nor Technician B

3. Technician A says an SAE 10W-30 engine oil has the viscosity of a 10W oil at low temperatures. Technician B says at high temperature the viscosity of a 10W-30 oil increases so this oil has the same viscosity as a SAE 30 oil. Who is correct?

   A. Technician A

   B. Technician B

   C. Both Technician A and Technician B

   D. Neither Technician A nor Technician B

4. Technician A says synthetic oil is made from crude oil. Technician B says synthetic oil maintains its stability over a wide temperature range. Who is correct?

   A. Technician A

   B. Technician B

   C. Both Technician A and Technician B

   D. Neither Technician A nor Technician B

5. All of these statements about oil additives are true *except*:

   A. Pour point depressants prevent oil from becoming thicker at high engine temperatures.

   B. Corrosion inhibitors displace water from metal surfaces.

   C. Dispersants prevent carbon particles from grouping together.

   D. Viscosity index improvers prevent oil thinning.

6. Technician A says an engine oil with an SAE 50 oil rating is thinner than an oil with an SAE 20 rating. Technician B says a typical car engine oil with an SAE 30 rating is suitable for use in extremely cold weather. Who is correct?

   A. Technician A

   B. Technician B

   C. Both Technician A and Technician B

   D. Neither Technician A nor Technician B

7. The viscosity rating of a 10W-30 engine oil is determined at:

   A. 32°F.

   B. 20°F.

   C. 10°F.

   D. 0°F.

8. All of these statements about engine oil pumps are true *except*:

   A. The oil pump may be driven off the front of the crankshaft.

   B. The oil pump may be rotor type or gear type.

   C. A calibrated spring holds the oil pressure relief valve open.

   D. The pressure relief valve prevents excessive oil pump pressure.

9. Technician A says most engine oil filters are connected so all the oil delivered by the pump flows through the filter. Technician B says if the oil filter becomes completely plugged, oil stops flowing to the engine oil system. Who is correct?

   A. Technician A

   B. Technician B

   C. Both Technician A and Technician B

   D. Neither Technician A nor Technician B

10. Technician A says in many OHV engines, oil is supplied through hollow pushrods to lubricate the rocker arms. Technician B says in many OHV engines, oil is supplied to the camshaft bearings through holes drilled in the engine block. Who is correct?

    A. Technician A

    B. Technician B

    C. Both Technician A and Technician B

    D. Neither Technician A nor Technician B

11. When the engine is running, oil is forced into the oil pump by _____ in the oil pump and _____ _____ above the oil level in the oil pan.

12. In many engines without a distributor, the oil pump is driven from the front of the _____ .

13. The oil pressure in the lubrication system is limited by the _____ _____ valve.

14. The oil filter bypass valve opens if the oil filter becomes_____ _____ .

15. Explain the oil flow through the lubrication system in an OHV engine.

16. Explain the functions of engine oil.

17. List and explain seven engine oil additives.

18. Describe the design of two different types of oil pumps.

34 Lubrication System Service

Learning Objectives

After you have read, studied, and practiced the contents of this unit, you should be able to:

- Change the oil and filter.
- List methods for oil leak detection.
- List potential locations for oil leaks.
- Diagnose oil leaks.
- List causes for oil consumption.
- Diagnose oil pressure indicators.
- Test oil pressure.

Key Terms

Blowby

Message center

Ultraviolet dye

INTRODUCTION

Proper lubrication system maintenance and service is extremely important to provide normal engine life. If the engine oil and filter are not changed at the vehicle manufacturer's recommended intervals, contamination in the engine oil may score engine components such as cylinder walls and piston rings. Oil contamination may also cause premature crankshaft main bearing and connecting rod bearing wear.

CHANGING ENGINE OIL AND FILTER

Lubrication system maintenance usually consists of periodic oil and filter changes. The oil pan is equipped with a drain plug to remove the oil from the crankcase. It is easier to remove all of the oil if the engine is warmed prior to draining. Contaminants are suspended in the oil when it is at operating temperature, and most of these contaminants are removed during an oil change.

SAFETY TIP *If the engine has been running for a period of time, the engine oil may be very hot. Always wear protective plastic-coated gloves when draining engine oil.*

The oil filter usually threads onto the engine block or adapter. A band-type wrench or special socket is used to remove the filter from the engine **(Figure 34-1)**. A rubber seal is located at the top of the oil filter to seal between the filter and block. Sometimes this seal may come off the filter and remain on the block. If this occurs, be sure to remove the seal from the block. Failure to remove the oil seal will result in oil leaks. Before installing the

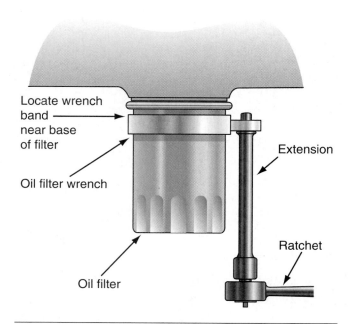

Figure 34-1 Using an oil filter wrench to remove an oil filter.

new oil filter, lubricate the seal with new engine oil and fill the filter with oil. Install the oil filter. Do not over-tighten the filter. Turn the filter 3/4 turn after the seal makes contact. Do not use the oil filter wrench to tighten the filter. Only hand pressure is required.

> **Tech Tip** *Over-tightening the oil filter may cause seal damage and result in oil leaks. Oil that is drained from the oil pan may contain dirt and grit. Using this oil to lubricate the seal may cause seal damage and an oil leak.*

After the proper amount of new oil is installed in the engine, start the engine and inspect the oil filter for leaks. Be sure the oil level on the dipstick is at the full mark. Used engine oil and filters are considered hazardous waste. Federal and state environmental regulations must be followed when disposing of these items. Used engine oil is usually collected and shipped to an oil recycling facility. Oil filters are usually drained, crushed, and recycled.

Some engines require pre-oiling before they are started after an oil change. This is especially true of engines equipped with turbochargers. The turbocharger rotates at a high rate of speed and requires lubrication to prevent damage to the turbocharger bearings. After the oil is drained and the oil filter replaced, it may take several moments before oil reaches the turbocharger. Pre-oiling the engine will assure all components receive oil before the engine is started. Use the following procedure to pre-oil the engine:

1. Make sure the oil filter and crankcase are filled with oil.
2. Disable the ignition system using the manufacturer's recommended procedure.
3. Crank the engine for 30 seconds, then allow the starter to cool, and crank the engine for another 30 seconds.
4. Repeat this procedure until oil pressure is indicated (oil pressure indicator light off or oil pressure gauge indications).
5. Enable the ignition system and start the engine.
6. Observe the oil pressure indicator. If oil pressure is not indicated, shut the engine off.
7. Recheck the oil level. Record the mileage in the vehicle log book on a maintenance reminder sticker.

> **Tech Tip** *Some customers feel they need to add additional treatments to their engine oil. A variety of aftermarket and manufacturer-supplied additives are available to remove carbon, improve ring sealing, loosen stuck valves, and so forth. Before using these additives, refer to the vehicle manufacturer's warranty. Use of some of these additives may void the warranty.*

OIL LEAK DIAGNOSIS

When performing an oil leak diagnosis, always check the oil level and condition. With the engine at normal operating temperature, the oil level on the dipstick should be at the full mark. If the crankcase is overfilled, excessive oil splash in the crankcase may cause oil leaks at the pan gasket and rear main bearing seal. If the oil is diluted with gasoline, it is excessively thin and this may aggravate oil leaks.

On a fuel-injected engine, a leaking fuel pressure regulator may cause gasoline to leak through the vacuum hose into the intake manifold resulting in diluted oil and an excessively rich air fuel ratio. A thorough visual inspection of the engine often locates the source of an oil leak.

When checking for oil leaks, inspect the positive crankcase ventilation (PCV) valve and hose. A restricted PCV valve or hose causes excessive pressure buildup in the engine, and this condition may cause oil leaks from engine gaskets or seals that are in normal condition **(Figure 34-2)**.

An **ultraviolet dye** can be added to the lubricating system to assist in detecting leaks. Simply add dye to the oil and allow the lubricating system to circulate the dye in the system and inspect for leaks with the ultraviolet light. The dye will show up with a fluorescent yellow color and is enhanced by using a specially tinted pair of safety glasses **(Figure 34-3)**. The UV kit can also help find coolant, A/C, and transmission leaks faster and easier.

Another cause of excessive pressure buildup in the engine is too much blowby. A restricted PCV

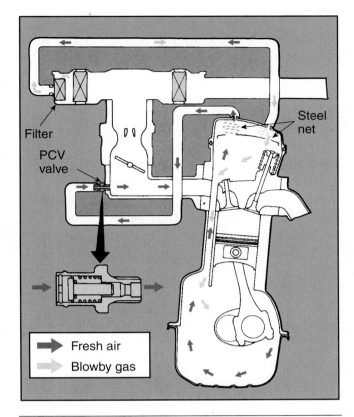

Figure 34-2 Normal PCV system operation.

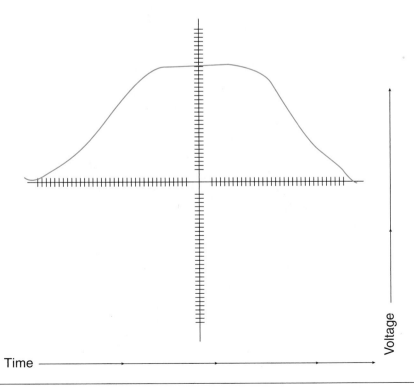

Figure 34-3 A UV leak detection kit.

system or excessive blowby builds up excessive pressure in the engine, and this condition may cause oil vapors to be forced from the engine through the clean air intake hose and into the air cleaner. Some of these oil vapors enter the air intake into the engine, resulting in excessive oil consumption. Excessive **blowby** refers to leakage between the piston rings and cylinder walls caused by worn rings or scored cylinder walls.

If the PCV valve is removed with the engine running and a great deal of oil vapor is escaping from the PCV valve opening, excessive blowby is indicated. Inspect the valve cover gasket area for indications of oil leaks. In many cases, the oil may be leaking around the valve cover gaskets and running down the sides and back of the block so it drips under the vehicle. Raise the vehicle on a lift and inspect the lower side of the engine for oil leaks **(Figure 34-4)**.

Always inspect the oil drain plug and oil filter for indications of oil leaks. Oil dripping from the oil drain plug may indicate a damaged gasket on this plug or stripped plug threads that do not allow sufficient torque on the plug. Oil leaking around the oil pan gasket area usually indicates a damaged oil pan gasket. Oil leaking from the flywheel housing usually indicates a leaking rear main bearing seal.

In an OHV engine, oil leaking from the flywheel housing may indicate a leaking oil gallery plug or a leaking plug in the rear camshaft-bearing opening. On some engines, a worn timing gear cover seal or gasket or a leaking oil pan gasket may cause oil leaks at the front of the engine.

On engines with a distributor, a damaged gasket between the distributor housing and the block may cause oil leaks. If oil is leaking from the valve cover gaskets or oil pan gasket, be sure the retaining bolts on these components are tightened to the specified torque. Some engines, such as light-duty diesels and gasoline engines in light-duty trucks with a trailer-hauling package, have an engine oil cooler. On these vehicles, always inspect the oil cooler and lines for leaks.

DIAGNOSING OIL PRESSURE INDICATORS

Most oil pressure warning light circuits use a normally closed switch threaded into the main oil

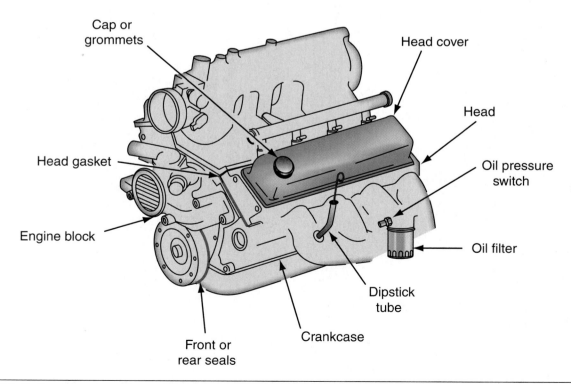

Figure 34-4 Common engine oil leak locations.

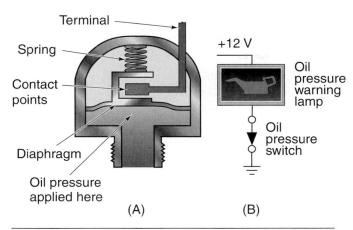

Figure 34-5 An oil pressure warning light circuit.

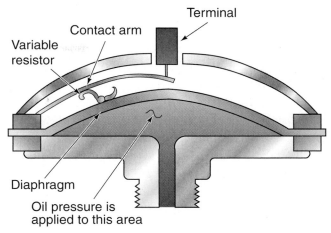

Figure 34-6 An oil pressure gauge sending unit containing a variable resistor.

gallery **(Figure 34-5)**. A diaphragm in the sending unit is exposed to the oil pressure. The movement of the diaphragm controls the switch contacts. When the ignition switch is turned to the on position with the engine not running, the oil warning light turns on. Since there is no pressure to the diaphragm, the contacts remain closed and the circuit is complete to ground.

When the engine is started, oil pressure builds and the diaphragm moves the contacts apart. This opens the circuit and the warning light goes off. The amount of oil pressure required to move the diaphragm is about 3 psi (20.6 kPa). If the oil warning light comes on while the engine is running, it indicates that the oil pressure has dropped below the 3-psi (20.6 kPa) limit.

When diagnosing this type of oil pressure warning light circuit, the first step is to be sure the warning light bulb is operating. Turn the ignition switch on, remove the wire from the oil pressure switch, and ground this wire to an engine ground. If the oil pressure warning light is not on, there is no power to the bulb, the bulb is burned out, or the wire from the bulb to the switch has an open circuit. If the oil pressure-warning light bulb operates normally, test the engine oil pressure. When the oil pressure is within specifications but the oil pressure warning light is on with the engine running, replace the oil pressure switch.

Oil pressure gauges are usually connected to a sending unit containing a variable resistance. Engine oil pressure causes the flexible diaphragm to move in the sending unit **(Figure 34-6)**. The diaphragm

movement is transferred to a contact arm that slides along the resistor. The position of the sliding contacts on the arm in relation to the resistance coil determines the resistance value and the amount of current flow through the gauge to ground. Always be sure the engine oil pressure is within specifications before diagnosing this system.

Connecting a special variable resistance test tool to the oil sending unit wire may test this type of oil pressure gauge circuit. With the ignition switch on, rotate the control knob on the special test tool and observe the oil pressure gauge. If this gauge operates normally when connected to the test tool but fails to operate properly when connected to the oil pressure-sending unit, replace the sending unit.

RESETTING CHANGE OIL WARNING MESSAGES

Many vehicles now have a computer-operated message center in the instrument panel that displays various warning messages. The **message center** is a digital readout in the instrument panel that provides warnings regarding abnormal operation of automotive systems, low fluid levels, or required system service.

Low engine oil pressure is indicated on the oil pressure gauge, but a LOW OIL PRESSURE warning is also displayed in the message center. When the engine oil requires changing, CHANGE OIL is displayed in the message center. The computer senses various parameters such as mileage, engine temperature, and air intake temperature. On the basis of this input information, the computer illuminates the CHANGE OIL message at the ideal oil change interval.

After the oil is changed, this message must be turned off. The procedure for turning off the CHANGE OIL message varies depending on the vehicle make, but this procedure is in the service manual and owner's manual. In some cases a special tool may be required to reset the oil change light or message.

DIAGNOSIS OF EXCESSIVE OIL CONSUMPTION

When diagnosing excessive oil consumption, always check the oil level and condition. If the crankcase is overfilled, excessive oil splash in the crankcase causes too much oil on the cylinder walls, which results in higher-than normal oil consumption. Diluted oil also results in excessive oil consumption.

When a customer complains about excessive oil consumption, always check the engine for oil leaks because this may be the cause of the problem. Worn valve guides or valve guide seals allow excessive amounts of oil to run down between the valve guides and stems into the combustion chamber. Worn valve guides and seals are often indicated by blue smoke from the tailpipe each time a warm engine is started. Worn piston rings and/or scored cylinder walls cause excessive blowby as explained previously. This condition also allows excessive oil to enter the combustion chamber resulting in abnormal oil consumption.

Tech Tip *An engine compression test is performed with the engine warmed up, all of the spark plugs removed, and the ignition and fuel systems disabled. A pressure gauge is installed in each spark plug hole in sequence and the engine is cranked through four compression strokes on the cylinder being tested.*

If blue smoke is escaping from the tailpipe during acceleration and deceleration, worn piston rings and/or scored cylinder walls may be indicated. An engine compression test with a compression gauge or a cylinder balance test with an engine analyzer will verify this condition.

Tech Tip *A cylinder balance test is performed with an engine analyzer. The analyzer prevents each cylinder from firing for a brief time period and the analyzer records the rpm drop during this time. All the cylinders should have nearly the same rpm drop. If a cylinder has low compression, it is not contributing to engine power output and that cylinder has very little rpm drop during the balance test.*

OIL PRESSURE DIAGNOSIS

Oil pressure is dependent upon oil clearances and proper delivery. If the clearance between a journal and the bearing becomes excessive, pressure is lost. Not all low oil pressure conditions are the result of bearing wear. Other causes include improper oil level, improper oil grade, sticking open pressure relief valve, and oil pump wear. Another common cause of low oil pressure is thinning oil as a result of excessive temperatures or gas dilution.

If low oil pressure is suspected, begin by checking the oil level. Too low a level will cause the oil pump to aerate and lose pressure. If the oil level is too high, it may be that gasoline is entering the crankcase as a result of a damaged mechanical fuel pump, improperly adjusted carburetor, sticking choke, or a leaking injector or fuel pressure regulator. If the oil level and condition are satisfactory, check the oil pressure using a shop gauge.

OIL PRESSURE TESTING

To perform an oil pressure test, remove the sending unit and install an oil pressure gauge with the appropriate adapter to determine the actual oil

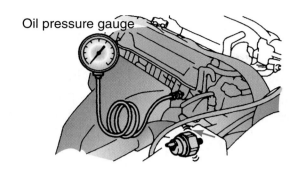

Oil pressure gauge

Figure 34-7 An oil pressure gauge is used to verify actual oil pressure in the lubricating system.

pressure in the system **(Figure 34-7)**. Using the correct size adapters, connect the oil pressure gauge to the oil passage. Start the engine and observe the gauge as the engine idles. Watch the gauge as the engine warms to note any excessive drops due to temperature. Increase the engine rpm to 2,000 while observing the gauge. Compare the test results with the manufacturer's specifications. After the test is complete, reinstall the oil pressure-sending unit, connect the sending unit wiring connector, start the engine, and confirm that there are no leaks.

> **Tech Tip** *Always verify the actual oil pressure before continuing diagnosis. A low oil pressure concern could be the result of oil that is too thin or a defective oil pressure sensor.*

No oil pressure indicates the oil pump drive may be broken, the pickup screen is plugged, the gallery plugs are leaking, or there is a hole in the pickup tube. The engine should be shut off immediately if it is suspected that there is no oil pressure. Lower-than-specified oil pressure indicates improper oil viscosity, a worn oil pump, a plugged oil pickup tube, a sticking or weak oil pressure relief valve, or worn bearings. An air leak in the oil pump pickup tube allows air to enter the oil pump. This lowers the oil pressure because oil is compressible. An air leak in the oil pickup tube also causes damage to engine components because of the air in the lubrication system. It is possible to have oil pressure that is too high. This can be caused by a stuck pressure relief valve or by a blockage in an oil gallery.

OIL PUMP SERVICE

The oil pump is usually replaced whenever the engine is rebuilt or when the oil pump fails. Since the oil pump is vital to proper engine operation, it is usually replaced along with the pickup tube and screen whenever the engine is rebuilt. The relatively low cost of an oil pump is considered cheap insurance by most technicians. However, there may be instances when the oil pump will have to be disassembled, inspected, and repaired. If the pump housing is not damaged, an oil pump rebuild kit containing new rotors or gears, a relief valve and spring, and seals may be available.

Because the oil is delivered to the pump before it goes through the oil filter, the pump is subject to wear and damage as contaminated oil passes through it. The particles passing through the gears or rotors of the pump wear away the surface area, resulting in a reduction of pump efficiency. If the particles are large enough, they may cause the metal of the rotor or gear surfaces to rise, resulting in pump seizure. In addition, these larger particles can form a wedge in the pump and cause it to lock up.

The pickup tube and screen should be replaced whenever the oil pump is replaced or rebuilt. If the tube attaches to the oil pump, install the new tube using light taps from a hammer. Make sure the pickup tube is properly positioned to prevent interference with the oil pan or crankshaft. Bolt on pickup tubes may use a rubber O-ring to seal them. Do not use a sealer in place of the O-ring. Lubricate the O-ring with engine oil prior to assembly then alternately tighten the attaching bolts until the specified torque is obtained. Prime the oil pump before installing it by submerging it into a pan of clean engine oil and rotating the gears by hand. When the pump discharges a stream of oil, the pump is primed.

If a gasket is installed between the pump and the engine block, check to make sure no gasket material blocks passages. Soak the gasket in oil to soften the material and allow for good compression. When installing the oil pump, make sure the drive shaft is properly seated and torque the fasteners to the specified value.

Summary

- The most common lubrication system maintenance is oil and filter changes.

- Used engine oil and filters must be disposed of according to hazardous material disposal regulations.

- Turbocharged engines require pre-oiling after an oil change or after some engine repairs.

- On fuel-injected engines, a leaking fuel pressure regulator diaphragm will cause the engine oil to be diluted with gasoline.

- A restricted PCV system or excessive engine blowby cause excessive pressure buildup in the engine and this condition may force oil vapors into the air cleaner.

- A leaking rear main bearing seal, a leaking main oil gallery plug, or a leaking rear camshaft bearing plug may cause oil leaking from the flywheel housing.

- An off/on switch located in the main oil gallery operates most oil pressure warning lights.

- Pushing the accelerator pedal to the floor three times in a 5-second interval with the ignition switch on turns off some CHANGE OIL messages in the message center.

- Excessive oil consumption may be caused by oil leaks, worn piston rings, scored cylinder walls, worn valve guides and seals, or excessive pressure buildup in the engine caused by a restricted PCV valve or excessive blowby.

- Low engine oil pressure may be caused by diluted oil, improper grade of oil, worn crankshaft or camshaft bearings, worn oil pump, or an air leak into the oil pump pickup.

- A test gauge is installed in the oil sender location to test oil pressure.

Review Questions

1. Technician A says a new oil filter should be tightened with an oil filter wrench. Technician B says used oil filters should be thrown in a dumpster. Who is correct?

   A. Technician A

   B. Technician B

   C. Both Technician A and Technician B

   D. Neither Technician A nor Technician B

2. Technician A says on fuel-injected engines, a leaking fuel pressure regulator diaphragm may cause oil dilution. Technician B says on a carbureted engine, a leaking fuel pump diaphragm may cause oil dilution. Who is correct?

   A. Technician A

   B. Technician B

   C. Both Technician A and Technician B

   D. Neither Technician A nor Technician B

3. Technician A says a restricted PCV valve may cause an air filter element saturated with oil. Technician B says an air filter element saturated

with oil may be caused by excessive blowby in the engine. Who is correct?

   A. Technician A

   B. Technician B

   C. Both Technician A and Technician B

   D. Neither Technician A nor Technician B

4. A low oil pressure warning light is on with the engine running. With the ignition switch on and the wire to the sending unit switch grounded, this warning light is on. All of these defects could cause the problem *except*:

   A. a defective sending unit switch.

   B. excessively low engine oil pressure.

   C. an air leak in the oil pump pickup pipe.

   D. an open circuit in the wire from the oil pressure warning light to the sending unit switch.

5. All of these statements about pre-oiling an engine are true *except*:

   A. The vehicle manufacturer may recommend pre-oiling on turbocharged engines.

B. Pre-oiling prevents damage to the oil pump.

C. During engine pre-oiling, the ignition system is disabled.

D. Cranking the engine until oil pressure is available performs pre-oiling.

6. Excessive engine blowby may cause all of these problems *except:*

A. oil in the air cleaner.

B. excessive crankcase pressure in the engine.

C. excessive engine oil pressure.

D. excessive oil vapors escaping from the PCV valve opening.

7. In an OHV engine, oil leaking from the bottom of the flywheel housing may be caused by a:

A. leaking timing gear cover seal.

B. leaking expansion plug in the rear of the engine block.

C. leaking pilot bearing seal.

D. leaking oil gallery plug.

8. An oil pressure warning light is illuminated with the engine running, but this light goes out when the wire is removed from the oil-sending unit. An oil pressure test indicates the engine oil pressure is within specifications. Technician A says the wire between the ignition switch and the oil-sending unit is grounded. Technician B says the oil-sending unit is satisfactory. Who is correct?

A. Technician A

B. Technician B

C. Both Technician A and Technician B

D. Neither Technician A nor Technician B

9. Technician A says the CHANGE OIL message in some message centers is erased by fully depressing the accelerator pedal three times in a 5-second interval with the ignition switch on. Technician B says in some vehicles the CHANGE OIL message is erased automatically when the oil is drained from the crankcase. Who is correct?

A. Technician A

B. Technician B

C. Both Technician A and Technician B

D. Neither Technician A nor Technician B

10. Excessive oil consumption may be caused by all of these problems *except:*

A. worn rocker arms.

B. worn valve guides.

C. worn piston rings.

D. scored cylinder walls.

11. Explain the proper oil filter tightening procedure.

12. List six causes of low oil pressure.

1. _____

2. _____

3. _____

4. _____

5. _____

6. _____

13. When testing engine oil pressure, a test gauge is installed in the _____ _____ _____ opening.

14. In some engines, oil jet valves squirt oil against the underside of the _____.

15. Explain how excessive engine blowby causes the air filter to become contaminated with oil.

16. Describe the operation of a low oil pressure warning light and sending unit switch with the ignition switch on and with the engine running.

17. Describe the operation of an oil pressure-sending unit containing a variable resistance.

18. Describe three data sources that an engine computer uses to determine when to illuminate the CHANGE OIL message.

35 Cooling Systems

Learning Objectives

After you have read, studied, and practiced the contents of this unit, you should be able to:

- Describe the purpose of the cooling system.
- Explain the operation of the thermostat.
- Describe the purpose of antifreeze.
- Describe the operation of an engine temperature warning light.

Key Terms

Compression ratio

Crossflow radiators

Detonation

Downflow radiators

Hybrid organic additive technology (HOAT)

Inorganic additive technology (IAT)

Organic additive technology (OAT)

INTRODUCTION

Without the cooling system, the heat created during the combustion process would quickly increase engine component temperature to a point where engine damage would occur. Engines may be cooled by air or liquid. Regardless of the type of cooling system, the purpose of this system is to disperse the heat from the engine to the atmosphere. Proper cooling system operation is extremely important to maintain engine component life. The components of a liquid cooling system include: coolant, recovery system, water pump, heater system, radiator, coolant jackets, thermostat, transmission cooler, radiator cap, hoses, cooling fan, and temperature indicator.

ENGINE COOLANT

Water by itself does not provide the proper characteristics needed to protect an engine. It works fairly well to transfer heat, but does not protect the engine in colder climates. The boiling point of water is 212°F (100°C) and it freezes at 32°F (0°C). Temperatures around the cylinders and in the cylinder head can reach above 500°F (250°C) and the engine will often be exposed to ambient temperatures well below 32°F (0°C). Water by itself does not make a suitable coolant. In addition, water reacts with metals to produce rust, corrosion, and electrolysis.

Most engine manufacturers use a coolant solution of water and an ethylene glycol-based antifreeze.

The proper coolant mixture will protect the engine under most conditions. Water expands as it freezes. If the coolant in the engine block freezes, the expansion could cause the block or cylinder head to crack. If the coolant boils, liquid is no longer in contact with the cylinder walls and combustion chambers. The vapors are not capable of removing the heat and the pistons may collapse or the head may warp from the excessive heat. Ethylene glycol by itself has a boiling point of 330°F (165°C) but it does not transfer heat very well. The freezing point of ethylene glycol is –8°F (–20°C).

> **Tech Tip** *Do not operate an engine with straight antifreeze. The poor heat transfer qualities of antifreeze can cause the engine to overheat.*

There are several types of antifreeze available. The antifreeze used for many years is the familiar green-colored antifreeze. This antifreeze has an ethylene glycol base and uses an **inorganic additive technology (IAT)** for corrosion prevention. This coolant generally has a two-year life span.

> **Tech Tip** *Polypropylene glycol has been available for a number of years and has been advertised as a less toxic antifreeze than ethylene glycol. However, with increasing performance demands on coolant and the release of newer coolant additive packages, its use has not been widespread. It should be used only if recommended by the vehicle manufacturer.*

In recent years most vehicle manufacturers now recommend extended life coolants. These extended life coolants are ethylene glycol-based but use two different additive packages: either **organic** additive technology **(OAT)** or **hybrid organic additive technology (HOAT) (Figure 35-1)**.

OAT coolant is usually orange in color and contains special additives to minimize corrosion in the cooling system and provide longer coolant life. OAT is based primarily on carbon-based molecules, typically organic acids, to protect the cooling system. These fluids do last up to five years in service,

Figure 35-1 Examples of coolant using OAT (left) and HOAT (right).

but may not protect quickly in fast corrosion conditions like boiling. If the coolant has boiled in the cooling system due to an overheating condition, then the coolant should be replaced. OAT is not backward compatible with older green coolant engines. OAT is not generally compatible with older vehicles particularly those using brass radiators.

Another type of coolant called HOAT was developed in recent years. HOAT uses both inorganic and organic, carbon-based, additives for long life protection. The HOAT coolant comes in a variety of colors depending on the manufacturer's request. The idea with hybrid antifreeze is to provide excellent all-around corrosion protection and extended service intervals. HOAT coolant generally can replace green coolant in older vehicles.

A cooling system with an OAT or HOAT additive usually has a 5-year/100,000-mile warranty. In most climates where below-freezing temperatures are encountered, the coolant is a mixture of 50 percent ethylene glycol and 50 percent water. This mixture provides a coolant freeze point of –35°F (–37°C). The coolant specified in the vehicle manufacturer's service manual must be used. It is important to not mix different types of engine coolants as each has a different formulation and additive package.

To improve the transfer of heat and lower the freezing point, water is added in a mix of about 50/50. At a 50/50 mix, the boiling point is 226°F

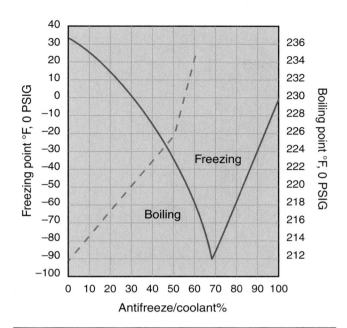

Figure 35-2 Changing the strength of the antifreeze solution changes its boiling and freezing points.

(108°C) and the freezing point is –34°F (–37°C). These temperature characteristics can be altered by changing the ratio of the mixture of antifreeze to water **(Figure 35-2)**. It may appear that lowering the boiling point by adding water is not desirable. In fact, this is required for proper engine cooling. The boiling point of the coolant is increased 3¼°F for every 1 psi increase. Under pressure, the boiling point of the coolant mix is about 263°F (128°C). The temperature next to the engine cylinder or cylinder head may be in excess of 500°F (250°C). As coolant droplets touch the metal walls, they are turned into a gas as they boil. The superheated gas bubbles are quickly carried away into the middle of the coolant flow where they cool and condense back into a liquid. If this nucleate boiling did not occur, the coolant would not be capable of removing heat fast enough to protect the engine.

> *Did You Know? Every pound per square inch (psi) of cooling system pressure increase raises the coolant boiling point by approximately 3.25°.*

The engine coolant should always be replaced whenever an engine is rebuilt. Special agents are added to antifreezes used in aluminum engines or with aluminum radiators. Always refer to the service manual for the type and mixture recommended. In addition, proper maintenance of the cooling system is required to maintain good operation. Over time, the coolant can become slightly acidic because of the minerals and metals in the cooling system. A small electrical current may flow between metals through the acid. The electrical current has a corrosive effect on the metal used in the cooling system. Anticorrosion agents are mixed into the antifreeze; however, these agents may be depleted over time. All vehicle manufacturers provide a maintenance schedule recommending cooling system flushing and refill based on time and mileage.

RADIATORS, PRESSURE CAPS, AND COOLANT RECOVERY SYSTEMS

The radiator is a series of tubes and fins that transfers the heat in the coolant to the air. As the coolant is circulated throughout the engine, it attracts and absorbs the heat within the engine, then flows into the radiator intake tank. The coolant then flows through the tubes to the outlet tank. As the heated coolant flows through the radiator tubes, the heat is dissipated into the airflow through the fins.

There are two basic radiator designs: downflow and crossflow. **Downflow radiators** have vertical fins that direct coolant flow from the top inlet tank to the bottom outlet tank. **Crossflow radiators** use horizontal fins to direct coolant flow across the core **(Figure 35-3)**. The core of the radiator can be constructed using a tube-and-fin or a cellular-type design. The fins on the tube-and-fin can be a flat plate fin or a serpentine fin **(Figure 35-4)**. The material used for the core is usually copper, brass, or aluminum. Aluminum-core radiators generally use nylon-constructed tanks **(Figure 35-5)**.

For heat to be transferred to the air effectively, there must be a difference in temperature between the coolant and the air. The greater the temperature difference, the more effectively heat is transferred. The radiator cap allows for an increase of pressure within the cooling system, increasing the boiling

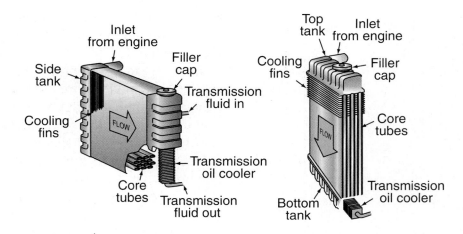

Figure 35-3 Crossflow and downflow radiators.

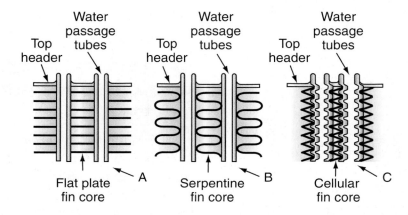

Figure 35-4 (A) Construction of a tube-and-fin (flat plate fin) radiator core. (B) Construction of a tube-and-fin radiator (serpentine fin) core. (C) Construction of a cellular radiator core.

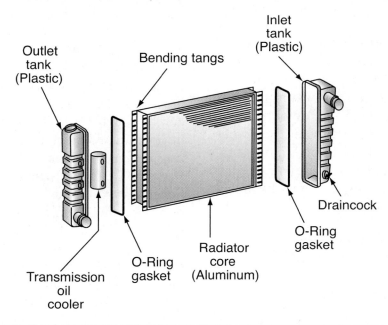

Figure 35-5 Construction of an aluminum core radiator with nylon tanks.

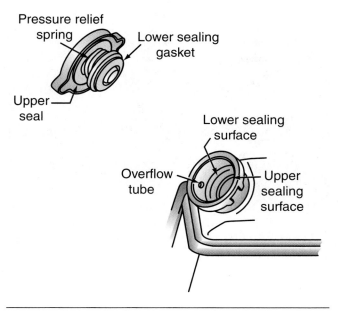

Figure 35-6 A radiator cap and sealing surfaces are shown.

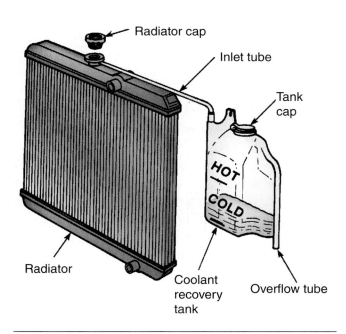

Figure 35-7 A typical coolant recovery system holds the coolant released from the radiator.

point of the coolant. The radiator cap uses a pressure valve to pressurize the radiator to between 14 and 18 pounds **(Figure 35-6)**.

If the pressure increases over the setting of the cap, the cap's seal will lift and release the pressure into a recovery tank **(Figure 35-7)**. Some vehicles may use a pressurized recovery tank called a degas tank. This system is designed to remove any air

bubbles from the coolant and keep the cooling system fed with a steady supply of coolant without any air bubbles **(Figure 35-8)**.

Vehicles with automatic transmissions usually have a transmission cooler mounted in the radiator outlet tank. As transmission fluid is circulated through this cooler, heat is transferred from the

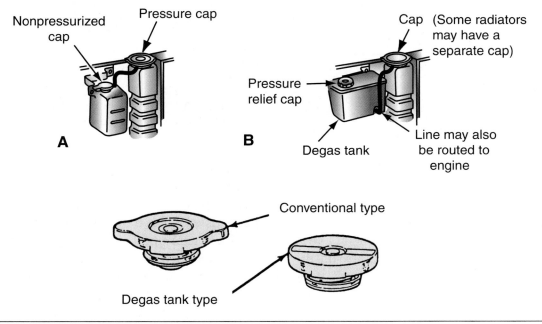

Figure 35-8 A cooling system with a (nonpressurized) recovery tank and a degas (pressurized) tank. Pressure caps for conventional and degas tank are also shown.

fluid to the coolant. The coolant recovery system contains a reservoir that is connected to the radiator by a small hose. The coolant expelled from the radiator during a high-pressure condition is sent to this reservoir.

When the engine cools, a vacuum is created in the radiator and the vacuum valve in the radiator cap opens. Under this condition, atmospheric pressure on the coolant in the overflow reservoir pushes it back into the radiator. The action of the vacuum valve in the radiator cap prevents a vacuum in the cooling system when the coolant temperature decreases as the engine cools off. Whenever an engine is rebuilt, the radiator should be thoroughly cleaned and then inspected by pressure testing it. The radiator cap should be replaced with a new one.

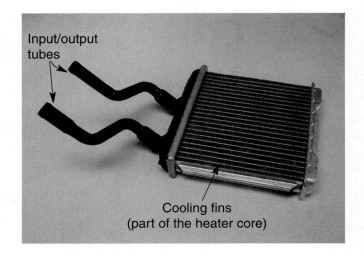

Figure 35-9 A heater core is basically a small radiator or heat exchanger.

SAFETY TIP *Never remove a pressure cap from the cooling system until the engine is allowed to cool. The removal of the cap will cause the reduction of pressure on the coolant, which in turn will lower the boiling point of the coolant. Should the lowered boiling point be lower than the temperature of the coolant, the coolant could boil violently and expel the hot coolant, causing severe burns and injury.*

> **Tech Tip** *If a vacuum occurs in the cooling system, radiator hoses may be collapsed.*

HEATER CORE

The heater core is similar to a small version of the radiator **(Figure 35-9)**. The heater core is located in a housing, usually in the passenger compartment of the vehicle **(Figure 35-10)**. Some of the hot engine coolant is routed to the heater core by hoses. The heat is dispersed to the air inside the vehicle, thus warming the passenger compartment. To aid in quicker heating of the compartment, a heater fan blows the radiated heat into the compartment.

HOSES

Hoses are used to direct the coolant from the engine into the radiator and back to the engine. In

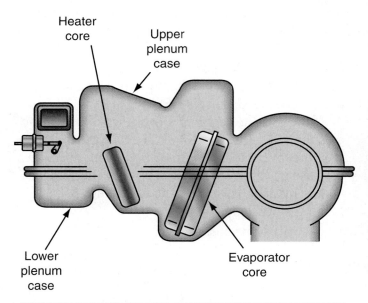

Figure 35-10 The heater core uses hot engine coolant to warm the passenger compartment.

addition, hoses are used to direct coolant from the engine into the heater core. On some engines, a bypass hose is used to bypass coolant around the thermostat when the thermostat is closed during engine warm-up. This action prevents excessive coolant pressure in the engine. Because radiator and heater hoses deteriorate from the inside out, most manufacturers recommend periodic replacement of the hoses as preventive maintenance. Whenever the engine is rebuilt, the hoses should always be replaced.

WATER PUMP

The water pump is the heart of the engine's cooling system. It forces the coolant through the engine block and into the radiator and heater core **(Figure 35-11)**. The water pump may be driven by a V-belt, ribbed V-belt, timing belt, or directly from the camshaft. Most water pumps are centrifugal design, using a rotating impeller to move the coolant **(Figure 35-12)**. When the engine is running, the impeller rotates, forcing coolant from the inside of the cavity outward toward the tips by centrifugal force **(Figure 35-13)**. Once inside the block, the coolant flows around the cylinders and into the cylinder heads, absorbing the heat from these components. If the thermostat is open, the coolant will then enter the radiator. The vacuum created by the empty impeller cavity allows the pressurized coolant to be pushed from the radiator to fill the cavity and repeat the cycle. When the thermostat is closed because coolant temperatures are too cold, the coolant is circulated through a bypass. This keeps the coolant circulating through

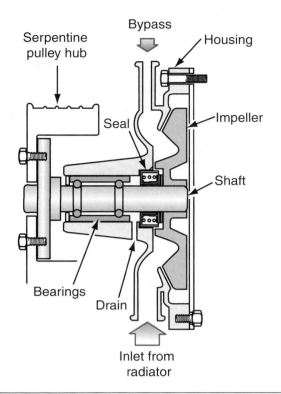

Figure 35-12 Most water pumps use an impeller to move the coolant through the system.

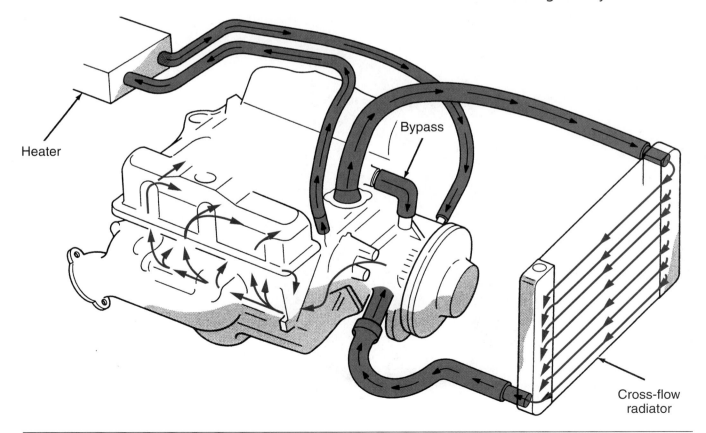

Figure 35-11 Coolant flow through the engine.

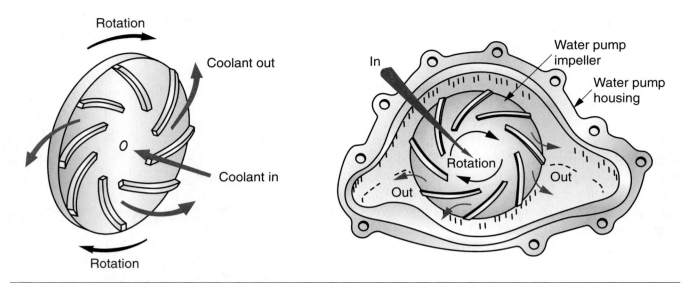

Figure 35-13 A centrifugal pump draws the fluid into the center of the vanes, and the centrifugal forces force the fluid outward.

the engine block until it becomes warm enough to open the thermostat.

WATER PUMP DRIVE BELTS

Many water pumps are driven by a V-belt that surrounds the crankshaft and water pump pulleys. This V-belt may also drive other components such as the alternator. The sides of a V-belt are the friction surfaces that contact the sides of the pulley groove and drive the water pump **(Figure 35-14)**. If the sides of the V-belt are worn and the lower edge of the belt is contacting the bottom of the pulley, the belt will slip. The drive pulleys on all

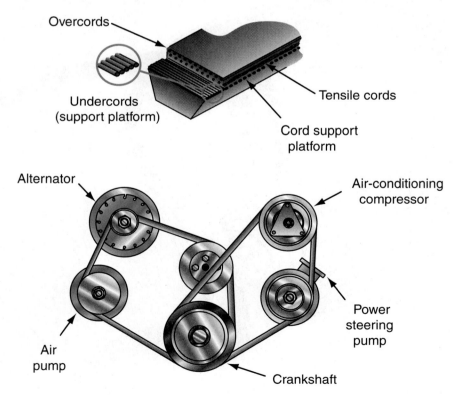

Figure 35-14 A conventional V-belt and typical V-belt routing.

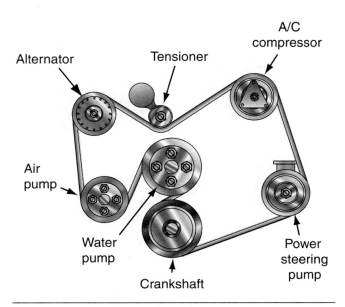

Figure 35-15 A multi-groove belt and typical multi-groove routing.

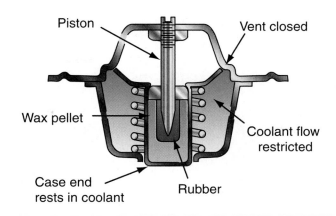

Figure 35-16 The wax pellet controls the flow of coolant through the thermostat.

the components driven by a V-belt must be properly aligned. Pulley misalignment causes V-belt edge wear.

A ribbed V-belt is used on many engines and this belt may be used to drive all the belt-driven components. Driving all the belt-driven components with one belt allows all these components to be placed on the same vertical plane, which saves a considerable amount of underhood space **(Figure 35-15)**. The smooth backside of a ribbed V-belt may be used to drive one of the components such as the water pump. The backside of the belt also contacts the idler and tensioner pulleys. The ribbed V-belt is much wider than a conventional V-belt and the underside of this belt has a number of small ribbed grooves. Regardless of the type of belt, belt tension is critical to prevent belt slipping. Most ribbed V-belts have an automatic spring-loaded belt tensioner that eliminates periodic belt tension adjustments.

Some engines may have water pumps driven by the timing belt, or by a belt from the rear of the camshaft. Others may also drive the water pump from the cam gear.

THERMOSTAT

Control of engine temperatures is the function of the thermostat. The thermostat is usually located at the outlet passage from the engine block

to the radiator. When the coolant is below normal operating temperatures, the thermostat is closed, preventing coolant from entering the radiator. In this case, the coolant flows through a bypass passage and returns directly to the water pump. The thermostat is rated at the temperature it opens in degrees Fahrenheit. If the rating is 195°F (90.5°C), this is the temperature at which the thermostat begins to open. Once the thermostat is open, it allows the coolant to enter the radiator to be cooled.

The thermostat cycles open and closed to maintain proper engine temperatures. Operation of the thermostat is controlled by a specially formulated wax and powdered metal pellet located in a heat-conducting copper cup **(Figure 35-16)**.

When the wax pellet is exposed to heat, it begins to expand. This causes the piston to move outward, opening the valve **(Figure 35-17)**. The thermostat should be replaced during an engine rebuild or scheduled cooling system service. Use a thermostat with a temperature rating recommended by the engine manufacturer. For proper operation, the thermostat must be installed in the correct direction. In most applications, the pellet is installed toward the engine block, but always refer to the appropriate service manual.

COOLING FANS

To increase the efficiency of the cooling system, a fan is mounted to direct the flow of air over the radiator cores. In the past, at high vehicle speeds, airflow through the grill and past the radiator core was

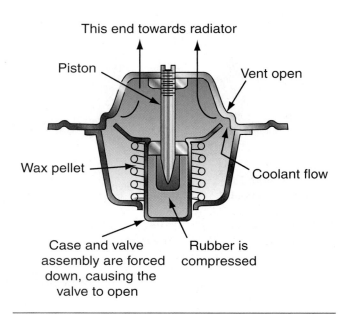

Figure 35-17 When the pellet expands, it opens the thermostat.

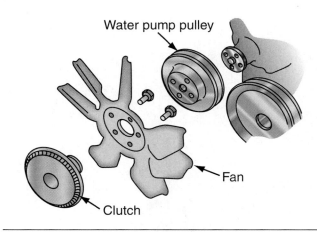

Figure 35-19 Some fans use a fan clutch to reduce noise and load on the engine.

sufficient to remove the heat. The cooling fan was really needed only for low-speed conditions when airflow was reduced. As modern cars have become more aerodynamic, airflow through the grill has declined; thus, proper operation of the cooling fan has become more critical.

Belt-driven fans are usually attached to the water pump pulley **(Figure 35-18)**. The cooling fan drive belt surrounds the crankshaft and water pump pulleys, and this belt may also drive other components such as the alternator. Some belt-driven fans may use a fan clutch to enhance performance **(Figure 35-19)**. The clutch helps to reduce

noise and reduce the power requirement to operate them as they only provide maximum airflow when needed. Two main types of clutches are the viscous-drive clutch and the thermostatic clutch.

The viscous-drive clutch operates the fan in relation to engine temperature by silicone oil **(Figure 35-20)**. If the engine is hot, the silicone oil in the clutch expands and locks the fan to the pump hub. The fan now rotates at the same speed as the water pump. When the engine is cold, the silicone oil contracts and the fan rotates at a reduced speed.

Some fan clutches use a thermostatic coil that winds and unwinds in response to the engine

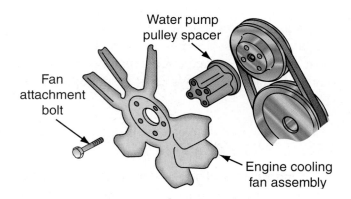

Figure 35-18 Belt-driven cooling fans are usually mounted to the water pump pulley.

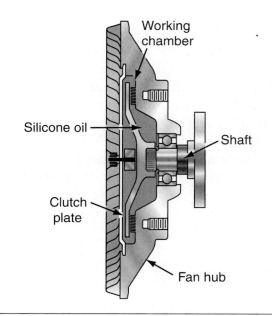

Figure 35-20 The viscous-drive clutch uses silicone oil to lock the fan to the hub.

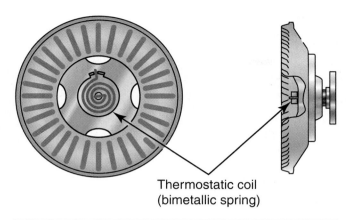

Thermostatic coil
(bimetallic spring)

Figure 35-21 A thermostatic spring connected to a piston is another common type of clutch fan.

temperatures **(Figure 35-21)**. The coil controls a piston located in a silicone-filled chamber. When it is hot, the coil unwinds and moves the piston into the chamber, which increases the pressure of the oil and locks the clutch.

ELECTRIC-DRIVE COOLING FANS

Electric drive fans are common on today's vehicles because they operate only when needed, thus reducing engine loads. Some of the earlier designs of electric cooling fans use a temperature switch in the radiator or engine block that closes when the temperature of the coolant reaches a predetermined value. With the switch closed, the electrical circuit is completed for the fan motor relay control circuit

> **Tech Tip** *Electric cooling fans are much more practical for transversely mounted engines as routing the drive belt to the fan would be impractical.*

and the fan turns on **(Figure 35-22)**. The voltage is supplied to the cooling fan motor from the battery through the relay contacts.

In recent years, most electric-drive cooling fans are operated by the PCM. The PCM receives engine coolant temperature inputs from the thermistor-type sensor. When the thermistor value indicates the temperature is hot enough to turn the fans on, the controller activates the fan control **(Figure 35-23)**. If the engine coolant is hot, the cooling fan may be turned on by the PCM with the ignition switch off. The PCM can also turn on the fans whenever air conditioning is turned on, regardless of the engine temperature. The PCM turns on a relay which in turn energizes the electric cooling fan.

> **SAFETY TIP** *Do not keep your hands near an electric cooling fan because it may start working at any time.*

COOLING SYSTEM OPERATION

In many cooling systems, the water pump forces coolant into the engine block where it flows around the cylinders. Heat is transferred from the pistons

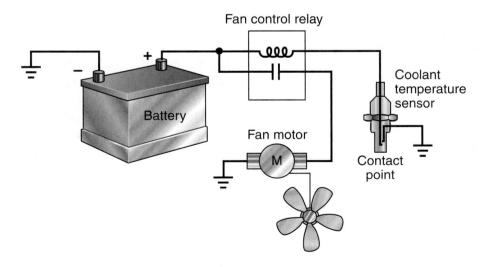

Figure 35-22 Simplified electric fan circuit.

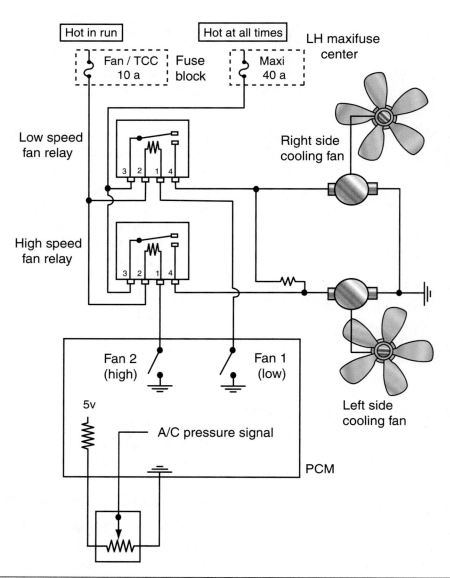

Figure 35-23 An example of the electric fan circuit controlled by the PCM.

and cylinders to the coolant **(Figure 35-24)**. The coolant then flows through passages in the head gasket and into the cylinder head where it flows around the valve seats and spark plug recesses. Heat is transferred from the combustion chamber, valves, valve seats, and spark plugs to the coolant. In many engines, coolant also flows through passages in the intake manifold. Coolant flows through the thermostat and the upper radiator hose into the radiator. When coolant flows through the radiator, heat is transferred from coolant to the air flowing through the air passages in the radiator core. Coolant flows from the radiator through the lower radiator hose to the water pump inlet. Many engines now have the

thermostat located in a housing on the inlet side of the water pump rather than on top of the intake manifold or cylinder head.

Some engines have a reverse-flow cooling system. In these systems, coolant flows from the water pump into the cylinder head and then it flows through the block **(Figure 35-25)**. This direction of coolant flow provides improved combustion chamber cooling. This action allows higher compression ratios without engine detonation.

The engine **compression ratio** is the relationship between the combustion chamber volume with the piston at TDC and the volume with the piston at BDC. Typical compression ratios are from 8:1 to 10:1.

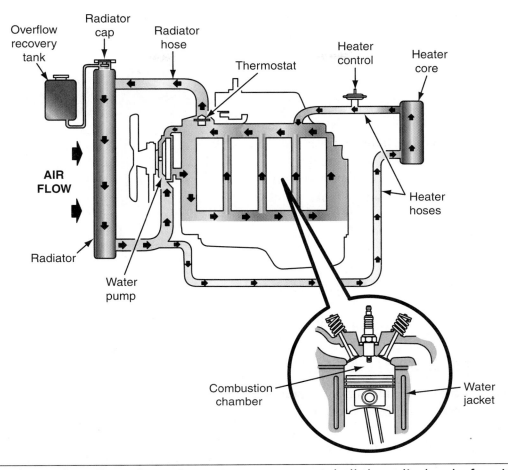

Figure 35-24 The coolant flows from the water pump around all the cylinders before it flows through the cylinder head.

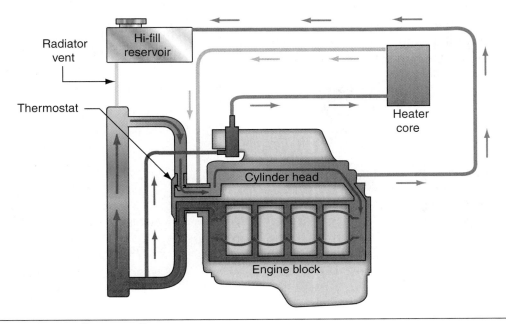

Figure 35-25 A reverse-flow cooling system.

Detonation occurs in a combustion chamber during the combustion process. The air-fuel mixture begins to burn normally, but glowing carbon or something ignites the remaining air-fuel mixture creating a second flame front that suddenly meets the original flame front. This action causes a sudden explosion of the remaining air-fuel mixture rather than a smooth burning action. This explosion suddenly drives the piston against the cylinder wall resulting in a rapping noise. Corrosion or contaminants can plug the coolant passages. This results in reduced heat transfer from the cylinder to the coolant. To prevent the formation of deposits, the cooling system should be flushed and coolant changed at regular intervals as specified by the manufacturer. Any time the engine is removed and rebuilt, it is a good practice to thoroughly clean the coolant passages in the block and cylinder head.

TEMPERATURE INDICATORS

Vehicle manufacturers provide some method of indicating cooling system problems to the driver. This is done by the use of an indicating gauge or by the illumination of a light. Some manufacturers control the operation of the gauge or light by a computer. Regardless of the type of control, sensors or switches provide the needed input.

Most coolant temperature warning light circuits use a normally open switch **(Figure 35-26)**. The temperature-sending unit consists of a fixed contact and a contact on a bimetallic strip. As the coolant temperature increases, the bimetallic strip bends. As

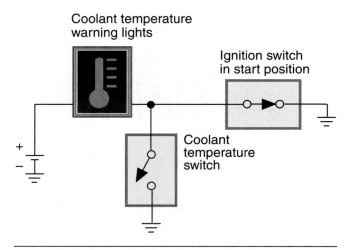

Figure 35-27 A prove-out circuit included in the temperature indicator light system.

the strip bends, the contacts move closer to each other. Once a predetermined temperature level has been exceeded, the contacts are closed and the circuit to ground is closed. When this happens, the warning light is turned on. With normally open switches, the contacts are not closed when the ignition switch is turned on. In order to perform a bulb check on normally open switches, a prove-out circuit is included that illuminates the bulb while cranking the engine **(Figure 35-27)**.

The most common sensor type used for monitoring the cooling system is a thermistor. In a simple coolant temperature-sensing circuit, current is sent from the gauge unit into the top terminal of the sending unit, through the variable resistor (thermistor), and to the engine block (ground). The resistance value of the thermistor changes in proportion to coolant temperature **(Figure 35-28)**.

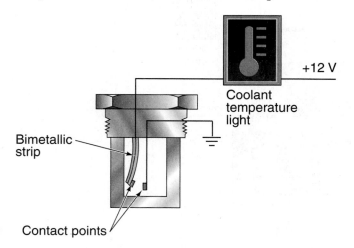

Figure 35-26 A temperature indicator light circuit.

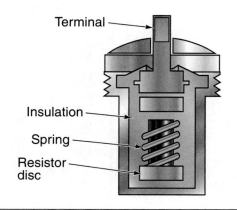

Figure 35-28 A thermistor used to sense engine temperature.

As the temperature rises, the resistance decreases and the current flow through the gauge increases. As the coolant temperature lowers, the resistance value increases and the current flow decreases.

Many vehicles have a computer-operated message center that displays various warning messages. In these systems, a thermistor-type engine coolant temperature (ECT) sensor is continually sending a voltage signal to the computer in relation to engine coolant temperature **(Figure 35-29)**. The computer senses the voltage drop across the ECT sensor as this sensor changes resistance. If the coolant temperature exceeds a predetermined value, the computer illuminates an ENGINE OVERHEATING message in the message center. Some message centers also display a LOW COOLANT message if the engine coolant level in the coolant recovery container is low. These systems have a coolant level sensor in the coolant recovery container.

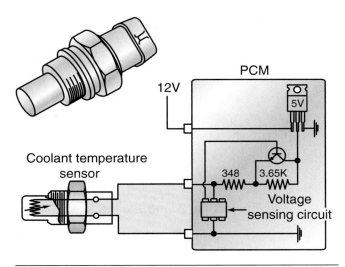

Figure 35-29 A coolant temperature sensor used in conjunction with the PCM.

Summary

- Many cooling systems are filled with a 50/50 mixture of ethylene glycol and water.

- A 50/50 mixture of ethylene glycol and water freezes at –34°F (–37°C) and boils at 265°F (130°C). Many radiator caps have a 15 psi (103 kPa) rating.

- If cooling system pressure exceeds the rating of the radiator cap, the cap pressure valve opens and releases coolant into the coolant recovery system.

- When the engine is shut off and the coolant temperature decreases, the vacuum valve in the radiator cap opens and allows coolant to flow from the recovery system into the radiator.

- The coolant is forced through the cooling system by pressure in the water pump.

- Water pump drive belts may be V-type or ribbed V-type, or the timing belt may drive the water pump.

- The thermostat controls engine temperature.

- Belt-driven cooling fans are usually mounted on the water pump pulley.

- The engine controller usually operates electric-drive cooling fans.

- In many cooling systems, the coolant flows through the engine block and then into the cylinder head.

- As the coolant flows through the radiator, heat is transferred from the coolant to the air flowing through the air passages in the radiator.

- Engine temperature indicator lights have a proving circuit that proves the indicator light bulb is working each time the engine is cranked.

- In many systems, a thermistor-type coolant temperature sensor sends a voltage signal to the engine controller in relation to engine temperature. If the engine overheats, the engine controller illuminates a warning light or displays a warning message in the message center.

Review Questions

1. Technician A says coolant boils at the same temperature regardless of the cooling system pressure. Technician B says the temperature next to the cylinder walls could be 500°F (250°C). Who is correct?

   A. Technician A

   B. Technician B

   C. Both Technician A and Technician B

   D. Neither Technician A nor Technician B

2. Technician A says some radiators have aluminum cores. Technician B says coolant is released into the recovery container if the radiator pressure exceeds the cap pressure rating. Who is correct?

   A. Technician A

   B. Technician B

   C. Both Technician A and Technician B

   D. Neither Technician A nor Technician B

3. Technician A says when the engine cools down, the vacuum valve in the radiator cap opens. Technician B says when the vacuum valve in the radiator cap opens, coolant flows from the recovery container into the radiator. Who is correct?

   A. Technician A

   B. Technician B

   C. Both Technician A and Technician B

   D. Neither Technician A nor Technician B

4. Technician A says a ribbed V-belt system allows the belt-driven components to be on the same vertical plane. Technician B says the friction surface on a V-belt is the underside of the belt. Who is correct?

   A. Technician A

   B. Technician B

   C. Both Technician A and Technician B

   D. Neither Technician A nor Technician B

5. All of these statements about ethylene glycol are true *except*:

   A. Pure ethylene glycol has a boiling point of 330°F.

   B. The freezing point of pure ethylene glycol is –20°F.

   C. The freezing point of a 50/50 ethylene glycol and water mixture is –34°F.

   D. The boiling point of a 50/50 ethylene glycol and water mixture is 226°F.

6. Every pound per square inch (psi) of cooling system pressure increases the coolant boiling point by:

   A. 1°F.

   B. 1.5°F.

   C. 2.0°F.

   D. 2.5°F.

7. While discussing cooling system operation, Technician A says a small electric current may flow between the metals in the cooling system through the acid that accumulates in the cooling system. Technician B says anticorrosion agents are mixed with most brands of antifreeze. Who is correct?

   A. Technician A

   B. Technician B

   C. Both Technician A and Technician B

   D. Neither Technician A nor Technician B

8. A typical radiator cap has a pressure rating of:

   A. 8 psi.

   B. 15 psi.

   C. 24 psi.

   D. 32 psi.

9. After a hot engine is shut down and allowed to cool off, the upper radiator hose gradually collapses. The most likely cause of this problem is:

   A. a defective vacuum valve in the radiator cap.

   B. a defective pressure relief valve in the radiator cap.

   C. an engine thermostat that is stuck open.

   D. a leaking water pump seal.

10. A typical engine thermostat has a temperature rating of:

    A. 150°F.

    B. 170°F.

    C. 195°F.

    D. 225°F.

11. A thermostat is opened by a _____ _____ in a copper cup.

12. The engine thermostat is usually installed with the pellet facing toward the _____.

13. Many fan clutches have a reservoir filled with _____ _____.

14. In a reverse-flow cooling system, the coolant flows from the water pump directly into the _____ _____.

15. Explain the operation of a coolant temperature warning light, including the prove-out circuit.

16. Explain the resistance change in a thermistor in relation to temperature.

17. Explain two messages that may be displayed in the message center in relation to engine temperature and discuss the cooling system conditions necessary to illuminate these warnings.

18. Describe the coolant flow through a conventional cooling system.

CHAPTER

36 Cooling System Service

Learning Objectives

After you have read, studied, and practiced the contents of this unit, you should be able to:

- Check belt tension.
- Inspect hoses.
- Clean and inspect the radiator.
- Inspect and diagnose the viscous fan coupling.
- Inspect and test the cooling system pressure cap.
- Drain, flush, and refill the cooling system.
- Diagnose and replace the thermostat.
- Test the operation of electric cooling fans.

Key Terms

Air purge tools

Bore scope

Chemical radiator flush

Coolant hydrometer

Coolant refractometer

Engine coolant temperature (ECT) sensor

Flex fan

Flushing gun

Pressure tester

Ultraviolet dye

INTRODUCTION

It is the function of the engine's cooling system to remove heat from the internal parts of the engine and dissipate it into the air. Proper engine operation and service life depend upon the cooling system functioning as designed.

COOLING SYSTEM MAINTENANCE

Proper cooling system maintenance is extremely important. If cooling system maintenance is not performed, engine overheating may occur, and this condition may require a tow truck to haul the vehicle to an automotive repair shop. This experience can be time consuming and expensive. If the engine overheating is severe or prolonged, engine damage may occur.

Checking Coolant Level

Checking the coolant level on older vehicles without a recovery system should be done only when the engine is cool. These vehicles require the radiator cap to be removed so you can see if the coolant level is above the radiator tubes.

Regardless of the system used, removing the radiator cap on a hot engine will release the pressure in the system, causing the coolant to boil immediately. This causes coolant to be expelled from the radiator at a high rate, which can result in severe burns.

SAFETY TIP *On some vehicles, the radiator cap is mounted on the recovery container rather than on the radiator. The recovery container is designed to withstand cooling system pressure. Never loosen the radiator cap on this type of recovery container with the engine warm or hot.*

Most vehicles are now equipped with a coolant recovery system, so the radiator cap usually does not need to be removed to check or add coolant to the system. The coolant recovery system provides quick visual checks of coolant level through the translucent bottle. Simply observe that the level is between the ADD and FULL marks while the engine is idling and warmed to normal operating temperatures. If coolant needs to be added, the coolant should be added directly to the coolant recovery tank. In the case of a degas tank (pressurized recovery tank) always allow the engine to cool before removing the pressure cap. The use of aluminum cylinder heads, blocks, and radiator cores requires the selection of proper coolant. Always use the manufacturer's recommended antifreeze.

Hose Inspection

Check all the radiator and heater hoses for soft spots, cracks, bulges, deterioration, and oil contamination **(Figure 36-1)**. Replace hoses that indicate any of these conditions. Inspect the radiator and heater hoses for contact with other components that could rub a hole in the hose. Reroute hoses away from other components if necessary. Be sure all the hose clamps are tight and in satisfactory condition.

Some new vehicles may make use of branched hoses, which consist of one or more hoses branching off the main hose. These can be found on some cooling systems and may be found in other areas

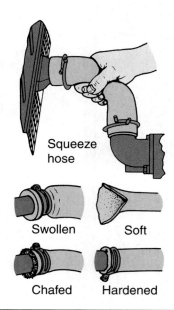

Figure 36-1 Common hose defects.

of the vehicle in the future. These hoses reduce the number of connections, thus reducing potential leak points **(Figure 36-2)**.

Drive Belt Inspection and Tension Testing

Inspect the water pump drive belt for cracks, missing chunks, fluid contamination, fraying, and bottoming in the pulley. If the belt indicates any of these conditions, belt replacement is necessary. Belt tension may be tested with the engine stopped and a belt tension gauge installed over the belt at the center of the longest belt span **(Figure 36-3)**. Measuring the specified belt deflection in the center of the longest belt span may also test belt tension.

When the belt tension is not within specifications, adjust the belt as necessary. A belt that is worn or loose will not turn the fan at a sufficient speed to draw the optimum mass of air over the radiator. Many ribbed V-belts have an automatic tensioner and do not require adjusting. Many automatic belt tensioners have a 1/2-inch drive opening in which a ratchet or flex-handle may be inserted to move the tensioner off the belt during belt replacement **(Figure 36-4)**. Some automatic belt tensioners have a wear indicator that indicates the amount of belt wear **(Figure 36-5)**.

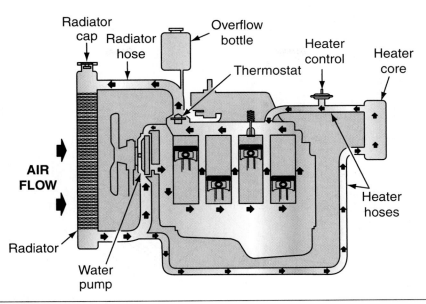

Figure 36-2 A branched cooling hose.

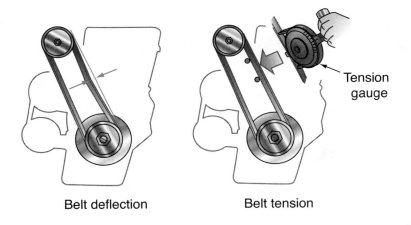

Belt deflection   Belt tension

Figure 36-3 Methods of checking bolt tension.

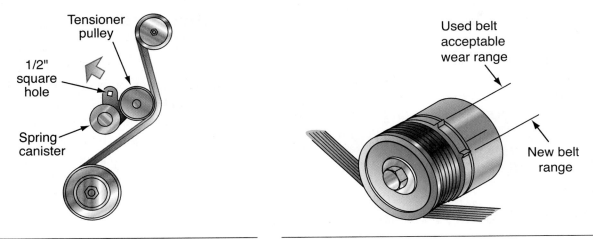

Figure 36-4 A one-half inch drive opening in the tensioner pulley.

Figure 36-5 A belt tension scale.

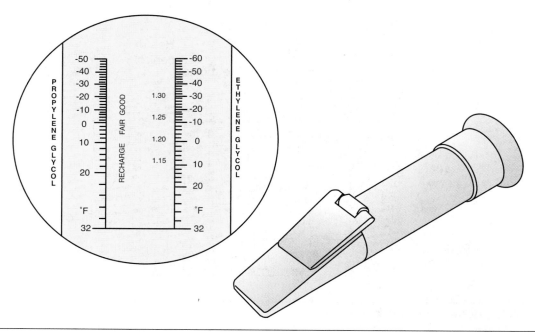

Figure 36-6 A coolant refractometer.

Coolant Testing

The antifreeze content in the coolant may be tested with a coolant hydrometer. A **coolant hydrometer** measures coolant weight or specific gravity to indicate the antifreeze content of the coolant.

The vehicle manufacturer's specified antifreeze content must be maintained in the cooling system because the antifreeze contains a rust inhibitor to protect the cooling system. Insert the hydrometer pickup tube into the coolant level in the recovery container and squeeze the hydrometer bulb to pull a coolant sample into the hydrometer to the specified level. The freezing point of the coolant is indicated on the hydrometer float.

A **coolant refractometer** may also be used to measure the antifreeze content of the coolant. Though these testers may be more expensive than a hydrometer, they are usually far more accurate. Some refractometers may also have graduations for coolants using propylene glycol or for testing battery electrolyte **(Figure 36-6)**.

COOLANT CONTAMINATION

Inspect the coolant for rust, scale, corrosion, and other contaminants such as engine oil or automatic transmission fluid. If the coolant is contaminated, the cooling system must be drained and flushed. If oil is visible floating on top of the coolant, the automatic transmission cooler may be leaking. This condition also contaminates the transmission fluid with coolant. If the vehicle has an external engine oil cooler, it may also be a source of oil contamination in the coolant. These coolers may be pressure tested to determine if they are leaking. A radiator specialty shop usually does testing or repairing of radiators and oil coolers.

COOLING SYSTEM DIAGNOSIS

Inspect the air passages through the radiator core and external engine oil cooler for contamination with debris or bugs. These may be removed from the radiator core with an air gun and shop air, or water pressure from a water hose. Visually inspect the radiator and all cooling system hoses for leaks. Leaking components must be replaced or repaired. Inspect the heater core area for signs of coolant leaks.

Some vehicles have the heater core mounted under the dash while other vehicles have this component under the hood. If the heater core is mounted under the dash, a leaking core may cause coolant to drip on the floor mat. Inspect the engine for coolant leaks at locations such as the thermostat housing, core plugs, and water pump. Coolant leaks at the water pump usually appear at the water pump drain hole or behind the pulley. A leaking water pump must be replaced.

DRAINING THE COOLING SYSTEM

Since most manufacturers recommend periodic coolant changes, a drain plug is usually provided at the bottom of the radiator **(Figure 36-7)**. The following is a typical procedure for draining the cooling system:

1. Start the engine.
2. Move the temperature selector of the heater control panel to the full heat position.
3. Shut the engine off before it begins to warm.
4. Allow the engine to cool. Never loosen the radiator cap or open the drain plug when the engine is hot.
5. Without removing the radiator cap, loosen the drain plug.
6. The coolant recovery tank should empty first. Then remove the radiator cap to drain the rest of the system. If the recovery tank does not

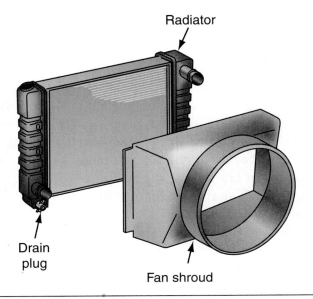

Figure 36-7 A drain plug in the lower radiator tank.

drain then it will be necessary to remove the coolant from the recovery tank. The tank can be drained using a VAC-U-FILL unit from Snap-On (or similar unit) or the recovery tank may be removed and emptied.

7. On most engines it is necessary to drain the block separate from the radiator. Remove the drain plug(s) from the side of the engine block to drain the block.

Currently, there are no federal requirements for the management and disposal of used antifreeze. Many states have regulations for the management and disposal of waste antifreeze, regardless of whether it is a hazardous waste. Always follow state and local regulations regarding the proper handling and disposal of used antifreeze. The antifreeze may be tested to determine if it has hazardous levels of lead and if present, the antifreeze may be considered a hazardous waste. Used antifreeze may be recycled offsite by an EPA-approved facility.

COOLING SYSTEM FLUSHING

The effectiveness of antifreeze and the additives mixed with it decreases over time. All manufacturers have a maintenance requirement for the cooling system. The recommended schedule for drain, flush, and refill ranges from once a year to once every five years. At the same time that the refill is performed, all cooling system components should be inspected.

Flushing of the cooling system is accomplished by using water forced through the cooling system in a reverse direction to the normal coolant flow. A special **flushing gun** mixes low air pressure with water. Reverse flushing causes the deposits to dislodge from the various components. The engine block and radiator should be flushed separately.

> **Tech Tip** Do not allow internal pressures in the radiator to increase over 20 psi (138 kPa), or damage to the radiator could result.

To flush the radiator, drain the radiator and disconnect the upper and lower radiator hoses. A long hose may be attached to the upper hose

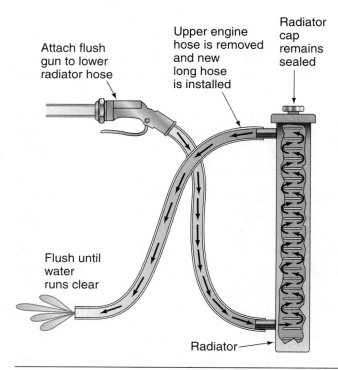

Figure 36-8 Reverse flushing the radiator.

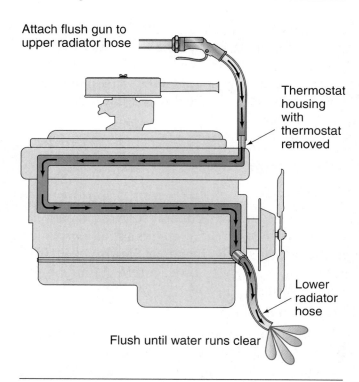

Figure 36-9 Reverse flushing the engine block.

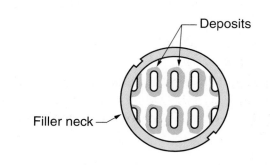

Figure 36-10 Deposits can be seen on the radiator tubes.

outlet to deflect the water. Disconnect and plug any heater hoses that are attached to the radiator. Fit the flushing gun to the lower hose opening. This causes the radiator to be flushed upward **(Figure 36-8)**. Fill the radiator with water and turn the gun on in short bursts. Continue to flush the radiator until the water being expelled from the upper hose outlet is clean.

To flush the engine block, disconnect the radiator upper and lower hoses. Also, remove the thermostat and reinstall the thermostat housing. Install the flushing gun to the thermostat housing hose **(Figure 36-9)**. Turn the water on until the engine is full. Then turn the air on in short bursts. Allow the engine to refill with water between blasts of air. Repeat this process until the water runs clean.

Water is usually sufficient to remove most contaminants from the cooling system. However, the cooling system may be contaminated by iron oxide, silicate and calcium deposits, and aluminum hydroxide, and these contaminants will reduce the flow of coolant and the effectiveness of heat transfer **(Figure 36-10)**. One of the major factors in the cause of these deposits is low coolant level. These contaminants will need to be removed using a **chemical radiator flush**.

Since heat is a catalyst, while flushing with the recommended chemical cleaner, be sure to allow the engine to achieve a normal operating temperature. The cleaner must be able to flow past the contamination; check to be sure that not more than the top two rows of tubes of the radiator are plugged. If the coolant will not flow through the radiator, then flushing will do no good; the radiator core must be replaced.

Aluminum hydroxide deposits require a two-part cleaner to remove. The two parts are an oxalic acid and a neutralizer. The acid is usually added to the system first. Then the engine is allowed to idle for the specified length of time. The cooling system is then flushed. After the flush is completed, the neutralizer is added to prevent the acid from damaging metal components.

If chemical cleaning fails to remove the deposits, the radiator will need to be removed and cleaned out or replaced. Internal radiator cleaning is usually performed in a radiator repair shop.

FILLING THE COOLING SYSTEM

The cooling system is usually filled with a mixture of 50/50 antifreeze to water. To fill the cooling system, make sure all hoses are installed and clamps are tight. Also, close the drain plug before filling. Look up the cooling capacity in the service manual. Fill the system to half of its capacity with 100 percent antifreeze through the radiator cap opening. Then complete filling with water.

Because many vehicles are designed with the radiator lower than the engine, a bleed valve is opened when filling the system **(Figure 36-11)**. Loosen the bleed valve while the radiator is being filled. Do not allow coolant from the open bleed valve to drip on the drive belts. Close the valve when coolant begins to flow out in a steady stream without bubbles. Leave the radiator cap off, start the engine, and let the engine warm up. Continue to fill the radiator as needed as the engine warms. When the radiator is full, install the radiator cap and fill the recovery container to the cold level.

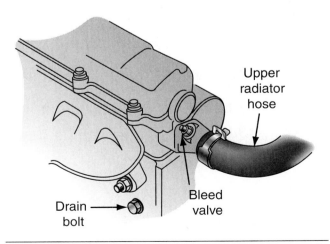

Figure 36-11 A bleed valve to remove trapped air from the cooling system.

Run the engine until it reaches normal operating temperature. It may be necessary to add additional antifreeze mix to the recovery tank. On some vehicles, it may take as many as four warm-up cycles before all of the air is removed from the system and the recovery tank equalizes.

Another method gaining rapid popularity is the use of a tool to place a vacuum on the cooling system. These **air purge tools** are designed to remove air from the cooling system. The tool creates negative pressure within the cooling system. By creating this vacuum, it removes the trapped air inside the system. When the vacuum is released, atmospheric pressure will push coolant into the spaces in the cooling system that previously held air.

These units reduce the time to fill the cooling system, eliminate airlocks, and can also be used to check for system leaks **(Figure 36-12)**. They are designed to work on most domestic and import passenger vehicles and light trucks.

DIAGNOSIS OF IMPROPER OPERATING TEMPERATURE

If the engine is operating at a lower-than-normal temperature, the customer may complain about insufficient heat from the heater in cold weather. On fuel-injected vehicles with this problem, the customer may complain about reduced fuel mileage. When the engine operating temperature is below normal, the engine coolant temperature (ECT) sensor sends a cold signal to the engine controller. When this signal is received, the computer provides a richer air-fuel ratio, resulting in reduced fuel mileage. If the engine operating temperature is below normal, the most common cause is a defective thermostat. The **engine coolant temperature (ECT) sensor** is a thermistor-type sensor that is usually mounted in the engine block or cylinder head. The ECT sensor sends a voltage signal to the engine controller in relation to coolant temperature.

When the engine operating temperature is above normal, the thermostat may not be opening properly or it may be improperly installed. Most thermostats are installed in a housing on top of the cylinder head or intake manifold. These thermostats are mounted on the outlet side of

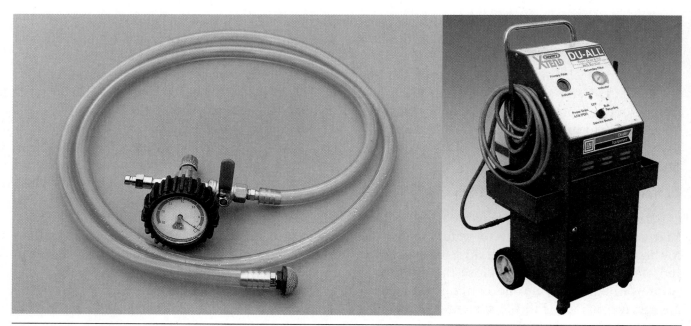

Figure 36-12 A small portable cooling system vacuum tool and a coolant exchanger/filler.

the water pump and the thermostat pellet must face toward the engine block. On some late-model engines, the thermostat is mounted on the inlet side of the water pump. Other causes of engine overheating are a loose water pump and fan drive belt, inoperative electric-drive cooling fan, a defective radiator pressure cap, a damaged radiator shroud, restricted air passages through the radiator, restricted coolant passages through the radiator, a defective thermostatic fan clutch, late ignition timing or insufficient spark advance on the engine, or a defective head gasket. Engine overheating causes a high coolant level in the recovery container. You must inspect and test these components or circuits to find the exact cause of the overheating problem.

THERMOSTAT TESTING

If a customer brings in a vehicle with an overheating problem, it is possible the thermostat is not opening. A faulty thermostat may also cause an engine that fails to reach operating temperature.

To check thermostat operation, it must be removed from the engine. Visually inspect the thermostat for rust or other contamination. Check the temperature rating of the thermostat and confirm it is the proper one for the engine application. Also

confirm it was installed in the right direction. In many engines, the thermostat pellet faces toward the engine block. The testing of the thermostat should be done as a diagnostic step. If the thermostat fails to open at its specified temperature or opens too early, this can help you confirm the thermostat was at fault. A new thermostat should always be used when servicing the cooling system thermostat.

To test the thermostat, submerge it in a container of water **(Figure 36-13)**. Use a thermometer so the temperature can be determined when the thermostat opens. Heat the water while observing the thermostat. At the rated temperature of the thermostat, it should begin to open.

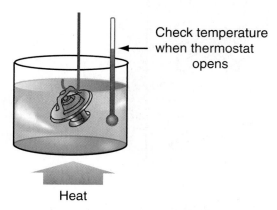

Check temperature when thermostat opens

Heat

Figure 36-13 Thermostat testing.

PRESSURE CAP DIAGNOSIS AND COOLING SYSTEM LEAK DIAGNOSIS

With the engine cool, remove the radiator cap and connect the **pressure tester** to the radiator filler neck. If the pressure cap is located on the degas tank, install the pressure tester on the filler neck on this container **(Figure 36-14)**. Pump the tester handle until the gauge on the tester indicates the rated cooling system pressure stamped on the pressure cap. Leave the pressure applied to the cooling system for 5 to 15 minutes. If the pressure does not drop on the tester gauge, the cooling system is not leaking. If the pressure on the tester gauge slowly drops, the cooling system is leaking. A visual inspection of the complete cooling system usually reveals the source of the coolant leak. Always remember to inspect the expansion plugs in the sides of the engine block for coolant leaks. These plugs may become rusted after a period of time. Rusted or leaking expansion plugs must be replaced.

If the pressure on the tester gauge dropped off but there is no evidence of an external leak, the leak could be in the head gasket, combustion chamber, or transmission cooler. To check head gasket and combustion chamber leaks, remove the

SAFETY TIP *Do not loosen the cap on a hot radiator. The radiator is under pressure and opening the cap will cause hot coolant to spray out of the filler tube.*

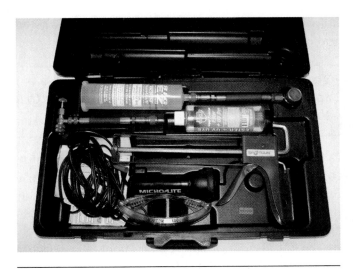

Figure 36-15 A UV dye leak detection kit.

spark plugs and check for indications of moisture in the cylinders when performing an engine compression test. The transmission cooler may be pressure tested to check it for leaks.

An **ultraviolet dye** can be added to the cooling system to assist in detecting leaks. Simply add dye to the system and allow the cooling system to circulate the dye in the system and inspect for leaks with the ultraviolet light. The dye will show up with a fluorescent yellow color and is enhanced by using a specially tinted pair of safety glasses **(Figure 36-15)**. A **bore scope** with UV capability can also be used to inspect for coolant leaks inside of the combustion chambers and other difficult to reach areas **(Figure 36-16)**. The UV kit can also

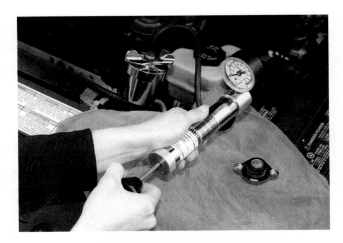

Figure 36-14 Pressure testing the cooling system.

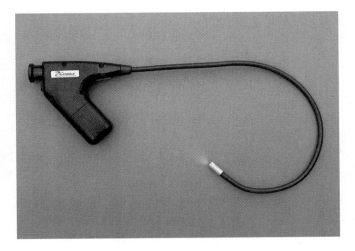

Figure 36-16 A bore scope with UV capability.

help find A/C, oil, and transmission leaks faster and easier.

> **Tech Tip** *A heater core leak may cause a coolant leak from the A/C/heater case onto the front floor mat.*

To test the radiator cap, use a special adaptor to connect the cap to the pressure tester **(Figure 36-17)**. Operate the tester pump and observe the pressure gauge. The cap should release the pressure in the tester at the specified pressure stamped on the cap. Radiator cap replacement is necessary if the cap does not release the pressure at the cap rating or fails to hold the pressure. Always check the vehicle manufacturer's specifications

Figure 36-17 Testing the radiator pressure cap.

to be sure the radiator cap has the correct pressure rating.

COOLING SYSTEM SERVICE

The fan is driven by a drive belt from the crankshaft or operated electrically **(Figure 36-18)**. Regardless of the design used, inspect the fan blades for stress cracks. The fan blades are balanced to prevent damage to the water pump bearings and seals. If any of the blades are damaged, replace the fan.

Engines with belt-driven fans use flex fans and/or a fan clutch. The flex fan changes its pitch in relation to engine speed. A **flex fan** has blades that are designed to flex or straighten out as engine and fan speed increases. This action moves more air through the radiator at low speed and less air through the radiator at high engine speed when vehicle motion is forcing air through the radiator. Flex fans are inspected for stress cracks as any other fan.

The viscous fan clutch should be visually inspected for silicone leakage indicated by black streaks radiating outward from the center of the fan clutch. Grasp the fan blades with your hands and move these blades back and forth while checking for excessive movement in the viscous clutch. Next, observe movement of the thermostatic spring coil and shaft. If the amount of movement is out of specification, replace the clutch assembly. Also, the shaft should rotate with the coil.

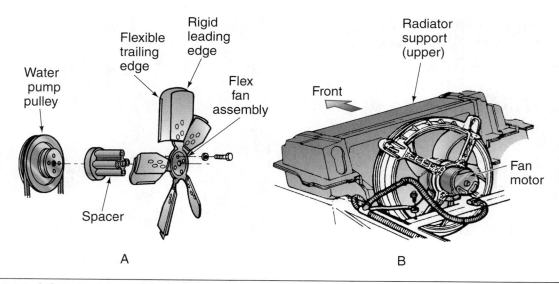

Figure 36-18 (A) Belt-driven fan. (B) An electric-driven fan.

Electric fans are also inspected for damage and looseness. If the fan fails to turn on at the proper temperature, the problem could be the temperature sensor, the fan motor, the fan control relay, the circuit wires, or the controller. On some computer-controlled systems, a scan tool may be used to activate the fan. If the fan operates, the problem is probably in the coolant sensor circuit. Also operating the A/C system should also cause the fans to operate, as the condenser will need airflow to function properly.

To isolate the cause of the malfunction, test all of the relays to verify their proper operation. A jumper wire can also be used to jump battery voltage directly to the cooling fan. If the fan motor fails to operate, check for proper ground connections before faulting the motor. Test the circuit wiring for proper voltages and resistance. Some vehicles use low speed, high speed, and fan brake relays and their proper functioning is important to the operation of the fan motors. These tests should be done in accordance with the manufacturer's test procedures.

To direct airflow more efficiently, many manufacturers use a shroud **(Figure 36-19)**. Proper location of the fan within the shroud is required for proper operation. Generally, the fan should be at least 50 percent inside the shroud. If the fan is outside the shroud, the engine may experience overheating due to hot under-hood air being drawn by the fan instead of the cooler outside air. If the shroud is broken, it should be repaired or replaced. Do not drive a vehicle without the shroud installed.

Inspect the idler pulley and belt tensioner, if so equipped **(Figure 36-20)**. Many manufacturers use idler pulleys so components such as generators, air pumps, A/C compressors, and so forth will be provided a greater area of belt contact on their pulleys. Tensioners are used to maintain proper drive

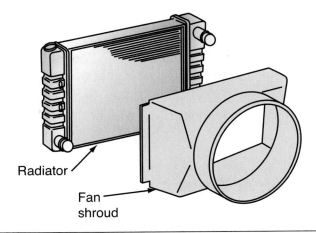

Figure 36-19 The fan shroud increases airflow through the radiator.

Figure 36-20 A typical belt tensioner with the tensioner mark within the proper tension range.

belt tension. Test the tensioner pulley for free rotation and inspect for any worn grooves or looseness. If the pulley fails inspection, it must be replaced.

Summary

- In most cooling systems, the coolant level may be checked by visually inspecting the level in the coolant recovery container.

- Belt tension may be measured with a belt tension gauge or by measuring the belt deflection in the center of the longest belt span.

- Antifreeze content in the coolant is measured with a coolant hydrometer.

- The radiator pressure cap must never be loosened with the engine warm or hot.

- In many cooling systems, a bleed valve is opened to allow air to escape from the system during the filling process.

- Cooling systems may be reverse flushed to remove contaminants.
- A defective thermostat may cause engine temperature to be lower or higher than normal.
- The engine thermostat may be removed and tested in a container filled with hot water.
- A pressure tester is used to leak test the cooling system and test the radiator pressure cap.
- Fan blade assemblies must be replaced if any of the blades have stress cracks.
- A damaged fan shroud causes engine overheating.

Review Questions

1. Technician A says the water pump drive belt tension may be tested with a belt tension gauge mounted on the shortest belt span. Technician B says measuring the belt deflection at the center of the shortest span may test belt tension. Who is correct?

   A. Technician A

   B. Technician B

   C. Both Technician A and Technician B

   D. Neither Technician A nor Technician B

2. Technician A says some automatic belt tensioners have a 1/2-inch drive opening in which a flex handle may be inserted to lift the tensioner off the belt. Technician B says some automatic belt tensioners have a wear indicator to show the amount of belt wear. Who is correct?

   A. Technician A

   B. Technician B

   C. Both Technician A and Technician B

   D. Neither Technician A nor Technician B

3. Technician A says if coolant is leaking from the water pump drain hole, replacing the water pump gasket will fix the leak. Technician B says coolant dripping on the front floor mat indicates a leaking air conditioning evaporator. Who is correct?

   A. Technician A

   B. Technician B

   C. Both Technician A and Technician B

   D. Neither Technician A nor Technician B

4. Technician A says when filling a cooling system, the bleed valve should be closed. Technician B says reverse cooling system flushing is done with water and low air pressure. Who is correct?

   A. Technician A

   B. Technician B

   C. Both Technician A and Technician B

   D. Neither Technician A nor Technician B

5. All of these statements about cooling system service are true *except*:

   A. The radiator cap should not be loosened when the engine is hot.

   B. In some cooling systems the radiator cap is mounted on the coolant recovery container.

   C. An engine with an aluminum block, cylinder heads, and/or radiator requires the selection of the proper antifreeze.

   D. When the radiator cap is mounted on the coolant recovery container, add coolant to the radiator.

6. When discussing a cooling system that is contaminated with oil, Technician A says the external engine oil cooler may be leaking. Technician B says the automatic transmission oil cooler may be leaking. Who is correct?

   A. Technician A

   B. Technician B

   C. Both Technician A and Technician B

   D. Neither Technician A nor Technician B

7. The most likely result of a loose cooling fan and water pump drive belt is:

   A. damage to the water pump bearing.

   B. engine overheating.

   C. excessive belt wear.

   D. excessive wear on the water pump pulley.

8. When discussing improper engine coolant temperature, Technician A says on a fuel-injected engine with a coolant temperature sensor, lower-than-normal coolant temperature causes a lean air-fuel ratio. Technician B says a defective head gasket may cause engine overheating. Who is correct?

   A. Technician A

   B. Technician B

   C. Both Technician A and Technician B

   D. Neither Technician A nor Technician B

9. During a cooling system pressure test, a pressure tester is used to pressurize the cooling system to 15 psi. After 15 minutes, the pressure decreased to 3 psi and there are no visible coolant leaks in the engine compartment or on the front floor of the vehicle. The most likely cause of the pressure decrease is:

   A. a leaking head gasket.

   B. a leaking heater core.

   C. a leaking coolant control valve.

   D. a loose water pump bearing.

10. An electric-drive cooling fan is inoperative at all engine temperatures. Technician A says on some cooling fan circuits, a scan tool may be used to activate the cooling fan. Technician B says the engine coolant temperature sensor may be defective. Who is correct?

    A. Technician A

    B. Technician B

    C. Both Technician A and Technician B

    D. Neither Technician A nor Technician B

11. List eight possible causes of engine overheating.

    1. _____

    2. _____

    3. _____

    4. _____

    5. _____

    6. _____

    7. _____

    8. _____

12. During a cooling system pressure test, the pressure gauge reading drops off and there are no external cooling system leaks. List three causes of these symptoms.

    1. _____

    2. _____

    3. _____

13. Silicone fluid leaking from a viscous fan clutch is indicated by _____ _____ radiating outward from the center of the fan clutch.

14. When diagnosing the electric-drive cooling fan on some fuel-injected engines, a _____ _____ may be used to activate the cooling fan.

15. Explain why a lower-than-normal engine operating temperature causes reduced fuel mileage on a fuel injected engine.

16. Explain the most likely cause of coolant dripping on the front floor mat.

17. Describe the correct installation position for a thermostat that is mounted on the outlet side of the water pump.

18. Describe the proper procedure for using a belt tension gauge to test belt tension on the water pump and cooling fan drive belt.

CHAPTER

37 Fuel System Fundamentals

Learning Objectives

After you have read, studied, and practiced the contents of this unit, you should be able to:

- Understand the importance of the fuel system.
- List the most common problems that fuel additives are used to control.
- List the basic carburetor circuits.
- List the basic fuel injection modes.
- Describe the similarities of the two fuel systems.

Key Terms

Atomize

Duty cycle

Octane

Vaporize

INTRODUCTION

There are three main things all internal combustion engines must have in order to run: compression, ignition, and fuel/air mixture. The next three chapters will explore fuel delivery as a necessary component of an internal combustion engine.

Of the many complex systems found on modern autos most have not drastically changed in the last fifty years in how they function. The fuel system is one of the systems that has changed the most dramatically in its capabilities and method of delivery. In the earliest days up until the mid 1980s nearly all gasoline powered vehicles had a carburetor to **atomize** fuel and mix it with the appropriate amount of air for delivery to the individual cylinders. The vehicles that were built just prior to the widespread use of fuel injection had some electronic controls, but even for these carburetors the computer had only minimal control. Once the throttle was opened past an idle the computer had little control. Today's autos have full computer control of fuel from startup until shutdown. Every little input is calculated to give the engine the best possible fuel mixture to control emissions and deliver optimum performance within the emission standards.

GASOLINE

As the fuel systems and engines have evolved in the past fifty years, so has the fuel that is fed

Figure 37-1 Fuel in the 1950s.

into it. Today's engines would run poorly at best and would soon stop running on the gasoline that was used just fifty years ago. The gasoline of today is a far more complex blend than was needed in the 1950s **(Figure 37-1)**. The fuel of that era was of low quality by modern standards; large amounts of materials were left over in the refining process that would ruin a modern fuel injection system in short order.

Modern gasoline fuels are blended to match the local climate and regulations. Fuels are seasonally blended to burn optimally according to local prevalent temperatures. If you were to drive across the country, the fuel purchased in New York is significantly different in its additive content than the fuel you would purchase in Denver or Los Angeles days later. Additives are added to clean the fuel system internally as the car is driven to help ensure optimum performance **(Figure 37-2)**. Numerous additives are used to control things like icing, rust, gum and varnish, with **octane** boosters to prevent engine knocking.

The octane rating number found displayed on gas pumps is a rating that describes a fuel's ability to resist detonation. Detonation is when fuel begins to burn before it is ignited by the ignition system from a combination of heat and high compression. The average octane rating of regular fuel is usually around 90 octane. This will provide adequate power and resist detonation in nearly all current production engines today. Generally the only time higher

Figure 37-2 Additives are used to control unwanted results of combustion.

octane is needed is for high performance motors that the manufacturer will state as needing "premium fuel." Because of the octane additives it is rare that fuel is the cause of detonation problems. Buying premium fuel is usually a waste of money for an engine that doesn't require it, as the additional octane does not improve performance and costs more.

MTBE was a common additive for increasing the available oxygen for burning until it was discovered that it does not break down in the burning process and eventually contaminates nearby water. Currently alcohol is the most common additive blended to gasoline to increase the oxygen content. In certain areas of the country it is blended in all fuels while in others it is optional and often goes under the E-10 label **(Figure 37-3)**. The "E" stands for ethanol, and the 10 represents the amount that is mixed in, 10 percent in this case. It is possible to currently get up to an 85 percent

Figure 37-3 Ethanol is commonly used to oxygenate fuel and is often made from corn.

concentration of ethanol, but this high an alcohol content requires modifications to the vehicle's fuel system and therefore E-85 is considered an alternative fuel.

Did You Know? Oxygenators are added to fuel to help supply the extra oxygen needed to complete the combustion process more effectively and reduce hydrocarbon emissions.

FUEL SYSTEMS

Whether the vehicle has an older carbureted fuel system or a more modern electronic fuel injected system the basic goal and function are the same. The main difference between the two is in how they do the job. Gasoline is a liquid and it must be reduced to a vapor and mixed with oxygen to burn. Carburetors rely upon moving air to create a vacuum to atomize the fuel; with fuel injection the fuel is under constant pressure and is sprayed into the intake manifold.

Carburetors

A basic carburetor system has changed little from the time it was invented until the present. While there were many improvements in design, the major components remained. All carburetors have six major delivery systems: fuel level control, cold enrichment, idle circuit, power enrichment circuit, main metering circuit, and the acceleration circuit **(Figure 37-4)**.

Fuel Level Control

Because fuel is stored inside the carburetor, it must be kept at a constant level to ensure that an adequate amount is available. This is accomplished

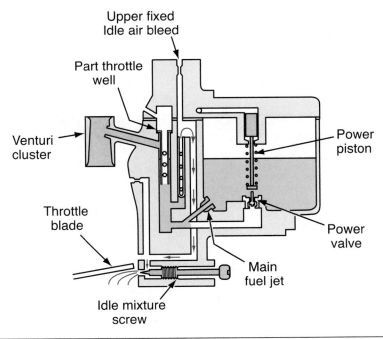

Figure 37-4 Carburetion circuits.

with a float that reaches a preset height and will shut off additional fuel from entering the bowl.

Cold Enrichment

Because a cold engine will not efficiently **vaporize** fuel, additional fuel is added to compensate and allow easy startup. The simplest way in a carburetor to add additional fuel is to reduce the amount of air in relation to fuel put into the intake manifold. By closing a valve (the choke) that was placed in front of the venturi the air would be sped up, increasing the vacuum pressure and pulling more fuel into the manifold than if it was fully open.

Idle Circuit

To control the engine at idle, a set of passages was cut to allow a preset amount of air past the main metering circuit. Often a small needle would be set to allow a preset amount of fuel to mix with the incoming air. Because the idle that would allow an engine to work properly was based upon optional equipment and engine type and design, idle speed was set to accommodate these changes. For example, idle speed would often be set differently if a car had an automatic transmission or a manual, and air conditioning or no air conditioning **(Figure 37-5)**.

Main Metering Circuit

As the name implies, the main metering circuit is the primary means for the carburetor to mix the

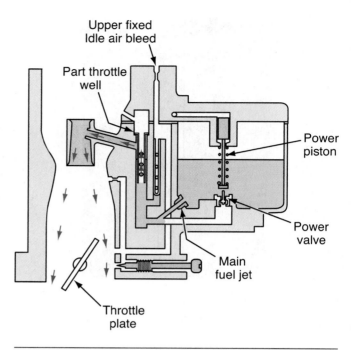

Figure 37-6 Throttle linkages and the throttle plates.

right amount of fuel with the incoming air. This is accomplished with a series of preset orifices called jets. The size of the jet determines the amount of fuel that is to be mixed with the incoming air. The main metering circuit is controlled by a set of throttle plates attached to the gas pedal with a cable and linkages **(Figure 37-6)**. As the throttle is opened, more air is allowed to pass through the carburetor body; more air flow increases the amount of fuel drawn into the venturi to be atomized with the air and sent to the intake manifold.

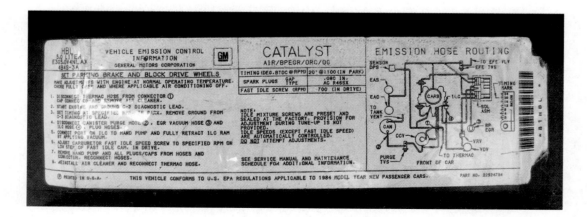

Figure 37-5 On carbureted vehicles, idle speed was determined by the options installed on the vehicle.

Power Enrichment Circuit

The power enrichment circuit's function is to provide additional fuel during heavy loads. Two main methods are employed to accomplish this: metering rods in the main jets and power valves. Both work on the principle that as an engine's load increases, the manifold vacuum would decrease. As the vacuum would decrease, it would no longer be able to maintain enough strength to hold back a valve with a spring. The spring opens additional fuel passages, allowing more fuel to flow.

Acceleration Circuit

The acceleration circuit is used to quickly add additional fuel when the throttle is quickly opened. Most commonly a small pump is built in the carburetor and attached to the throttle through the linkage. Because the carburetor is unable to distinguish a quickly opened throttle from a slow one the accelerator pump always squirts additional fuel every time the throttle is opened.

Fuel Injection

All new vehicles sold in this country for nearly the past two decades have had electronic fuel injection as the standard fuel delivery system. Initially this was brought about as a requirement to meet ever more stringent federal emission standards. As the emission requirements began to encompass more than just startup and idle, a means of precisely controlling all fuel delivery was essential. This could be met only with a computer precisely delivering the exact amount of fuel at the exact time it was needed.

As previously stated, the job of a fuel injection system is the same as a carbureted one: deliver the correct amount of fuel when needed. Even the modes are similar: cold start, idle, main metering, power, acceleration, and fuel control. The parts and method are different. However, if you understand that the requirements for an engine to run are similar, then repair and diagnosis are much easier.

The basic fuel injection system is made up of five main components: fuel pump, pressure regulator, fuel injectors, PCM and sensors, and idle air control **(Figure 37-7)**. These parts and their function and diagnosis will be covered in a later chapter; for now, understanding what the system does to deliver fuel is key. Generally all fuel injectors are cycled on and off to control fuel spray several

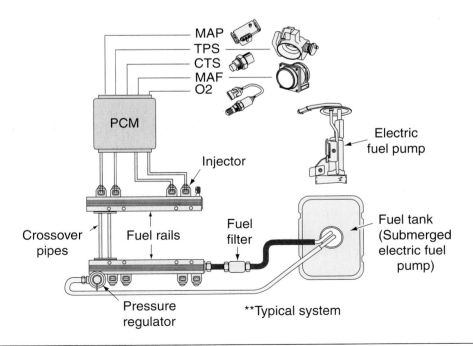

Figure 37-7 Fuel injection system.

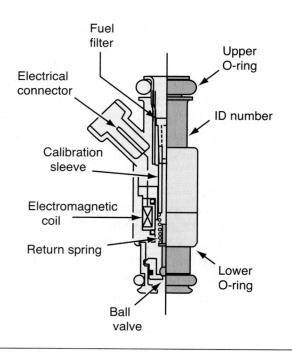

Figure 37-8 A fuel injector.

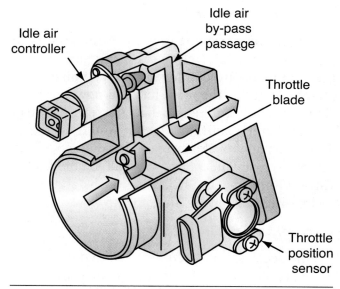

Figure 37-9 Idle air bypass.

times in one second. The amount of on time is compared to the amount of time the injector is off by the PCM (**Figure 37-8**). This cycle is often called the **duty cycle** or pulse width. To increase fuel delivery, the injector is cycled on more than off and to decrease fuel the opposite, hence more or less fuel is delivered.

Cold Start

As with carburetion all engines need a little extra fuel to aid in starting when the engine is cold. Instead of closing off air flow, the fuel injection system will spray extra fuel into the intake manifold. In earlier fuel injection systems it was common to have a separate injector just to aid in cold start.

Idle

To control the idle a small servomotor is generally used near the throttle plates to allow a certain amount of air around the throttle plates.

. . . .

NOTE: The servomotor used to control idle air bypass is commonly called the *IAC idle air control.*

. . . .

The motor opens and closes as directed by the PCM to maintain a steady rpm at idle (**Figure 37-9**). For instance, the PCM would recognize the A/C switch being turned on and would activate the IAC to open the idle air passage to let enough air by to maintain an optimum rpm. This often happens so fast that the driver does not hear the engine rpm change with the A/C cycling on and off.

Main Metering

As the vehicle is driven at a steady speed the fuel system constantly is making fuel ratio changes. This is being done for two reasons: the PCM is always trying to deliver the optimum power to fuel ratio and to meet emission standards. How the fuel injection system does this is covered in later chapters.

Power

When the load is increased on the engine and the throttle is not pushed down, the PCM will read changes in the data coming in from the engine sensors and determine the correct amount of duty cycle time to meet the demand.

Acceleration

Acceleration is handled in a similar fashion as power demand. The PCM will see that the throttle is being opened for more air.

Fuel Control

Unlike a carbureted system, for a fuel injection system to function properly the fuel must be kept under constant pressure. The amount of pressure is determined by the design of the system. This pressure is maintained by a high pressure electric fuel pump and a pressure regulator that keeps the pressure at a preset level.

Summary

- Numerous additives are used to control things like icing, rust, gum and varnish, with octane boosters to prevent engine knocking.

- Ethanol is the most common additive to add oxygen and boost octane in gasoline.

- Carburetors rely upon negative air pressure (vacuum) to meter fuel into the intake manifold.

- Carburetors have six major delivery systems: fuel level control, cold enrichment, idle circuit, power enrichment circuit, main metering circuit, and the acceleration circuit.

- The fuel injection system is made up of five main components: fuel pump, pressure regulator, fuel injectors, PCM and sensors, and idle air control.

- Fuel injectors are cycled on and off to control fuel spray several times in one second.

- For a fuel injection system to function properly the fuel must be kept under constant pressure.

Review Questions

1. Technician A says that a high octane fuel is needed to prevent detonation in a high compression engine. Technician B says that fuel additives are used to control icing and harmful deposits in the injectors. Who is correct?

   A. Technician A

   B. Technician B

   C. Both Technician A and Technician B

   D. Neither Technician A nor Technician B

2. Technician A says that one of the reasons fuel injection became standard equipment was to meet federal emission standards. Technician B says that some carburetion systems had minimal computer control. Who is correct?

   A. Technician A

   B. Technician B

   C. Both Technician A and Technician B

   D. Neither Technician A nor Technician B

3. Technician A says that during cold start fuel injection systems squirt more fuel to help starting. Technician B says that to aid cold startup more air is needed to over come increased vaporization. Who is correct?

   A. Technician A

   B. Technician B

   C. Both Technician A and Technician B

   D. Neither Technician A nor Technician B

4. Technician A says the job of the fuel system is to provide input to the vehicle's PCM. Technician B says that the duty cycle is adjusted to meet the engine's fuel demand by the PCM. Who is correct?

   A. Technician A

   B. Technician B

   C. Both Technician A and Technician B

   D. Neither Technician A nor Technician B

5. Technician A says that the job of the fuel injection system is to deliver the amount of fuel needed by the engine to meet run at its optimum power rating. Technician B says carburetors spray fuel into the intake because the fuel is kept under constant pressure.

   A. Technician A

   B. Technician B

   C. Both Technician A and Technician B

   D. Neither Technician A nor Technician B

6. Technician A says E-10 uses MTBE and is being phased out. Technician B says ethanol is used to control icing. Who is correct?

   A. Technician A

   B. Technician B

   C. Both Technician A and Technician B

   D. Neither Technician A nor Technician B

7. How is the fuel level in a carburetor maintained?

8. Technician A says that the idle air system in a fuel injected car is controlled by the driver. Technician B says that the PCM often adjusts idle speed so fast that it is often hard to tell it has happened. Who is correct?

   A. Technician A

   B. Technician B

   C. Both Technician A and Technician B

   D. Neither Technician A nor Technician B

9. The duty cycle is a comparison of an injector's on time compared to _____.

10. The PCM uses _____ inputs to adjust fuel delivery.

11. The PCM has to maintain control of fuel delivered at all times in order to meet _____ standards.

38 Fuel System Components

Learning Objectives

After you have read, studied, and practiced the contents of this unit, you should be able to:

- Understand how a TBI system works.
- Understand how ISC and IAC motors control idle speed.
- Understand how an MFI system works.
- Understand how an SFI system works.
- Understand how a CPI system works.
- Explain the similarities in the various fuel injection systems.
- Understand the effect that various PCM inputs have on a fuel injection system.

Key Terms

Central port injection (CPI)

Drive cycle

Idle speed control (ISC)

Injector driver

Injector pulse width

Multiport fuel injection (MFI)

Sequential fuel injection (SFI)

Stoichiometric air-fuel ratio

INTRODUCTION

A well-developed understanding of the various types of fuel injection systems is necessary to successfully diagnose these systems. For example, you must understand the relationship between the injectors and the PCM in various injection systems and how the PCM controls the air-fuel ratio in response to sensor inputs. You must also understand the various monitors in OBD II systems and the requirements to illuminate the MIL in these systems. An understanding of OBD terminology such as trip and drive cycle will aid in diagnostics and repair.

THROTTLE BODY INJECTION (TBI) SYSTEMS

In a TBI system, the throttle body assembly may contain one or two injectors depending on the engine displacement. Four-cylinder engines have a single injector in the throttle body assembly (Figure 38-1) and V6 or V8 engines have dual injectors. The injector body is sealed to the throttle body with O-ring seals. When the ignition switch is turned on, 12 volts are supplied through a fuse to the injector(s) (Figure 38-2). The other terminal on each injector is connected to the PCM. The PCM supplies a ground for the injector windings to open each injector. The

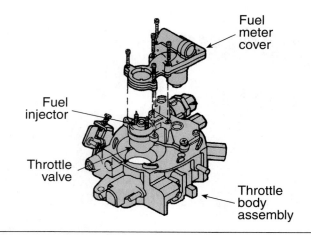

Figure 38-1 A single throttle body assembly.

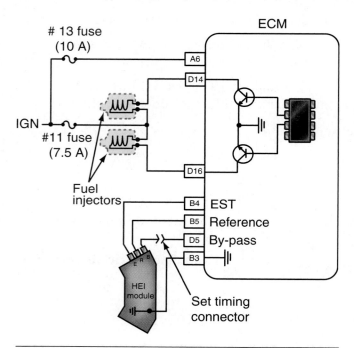

Figure 38-2 Dual throttle body injector wiring connections.

portion of the PCM that completes the circuit to ground and cycle the injector is called the **injector driver**. Grounding an injector winding allows current to flow through the winding and the resulting coil magnetism lifts the injector plunger. When the plunger tip is lifted off its seat, fuel is discharged from the injector orifice **(Figure 38-3)**. The amount of fuel discharged by the injector is determined by the injector pulse width. **Injector pulse width** is the length of time in milliseconds that the PCM keeps the injectors open.

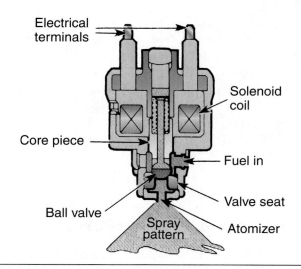

Figure 38-3 A throttle body injector.

> **NOTE:** Prior to OBDII each manufacturer had their own terms to describe their particular system and its components. It is quite common to have several different names for the same part depending on the manufacturer.

At idle speed, the PCM may supply an injector pulse width of 2 milliseconds, whereas at wide-open throttle the injector pulse width may be 12 milliseconds. The PCM always provides the correct injector pulse width to maintain the stoichiometric air-fuel ratio. The **stoichiometric air-fuel ratio** is the ideal air-fuel ratio at which the engine provides the best performance and economy and still meets emission standards. In a gasoline fuel system, this ratio is 14.7 pounds of air to every 1 pound of fuel. This ratio is expressed as 14.7:1. Although operating at higher ratios of air to fuel may improve fuel economy, this produces higher emissions of certain pollutants. This will be covered in greater detail in the emissions chapters that follow later in the book.

> **Tech Tip** *In some heated air inlet systems, the air control valve is operated by a wax-filled thermostatic pellet rather than a vacuum diaphragm.*

In a TBI system, the fuel is injected above the throttle plates and the intake manifold must be filled with fuel vapor. Because throttle body systems were among the first and most common systems introduced when carburetion was discontinued, they share several traits and operating principles. On many TBI systems some type of intake manifold or intake air heating is provided to keep the fuel vapor from condensing on the cool intake manifold passages. A heated air inlet system contains a vacuum-operated air control valve in the air cleaner inlet passage and a bimetal vacuum switch mounted in the lower part of the air cleaner housing **(Figure 38-4)**.

When the air cleaner temperature is cold and the engine is running, manifold vacuum is supplied through the bimetal vacuum switch to the vacuum diaphragm. Under this condition, the vacuum pulls the diaphragm upward and positions the air control valve so it blocks the cold air passage into the air cleaner inlet. With the air control valve in this position, air is drawn through a flexible hose and heat stove on the exhaust manifold into the air cleaner. This warm air from the exhaust manifold heat stove increases the temperature of the air-fuel mixture in the intake manifold to prevent fuel condensation in the intake manifold. When the air cleaner temperature reaches approximately 125°F (52°C), the bimetal vacuum switch begins bleeding off the manifold vacuum, causing the air control valve to open and allow cooler air to be drawn through the air cleaner inlet into the air cleaner. If the heated air inlet system does not pull hot air through the manifold stove into the air cleaner during engine warm-up, acceleration stumbles may occur. Some TBI systems have an **idle speed control (ISC)** motor mounted on the throttle body assembly

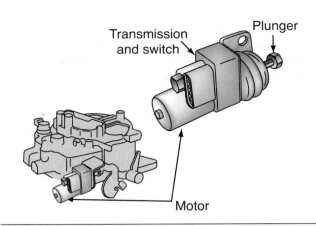

Figure 38-5 An idle speed control (ISC) motor.

(Figure 38-5). The ISC motor is a reversible motor that is controlled by the PCM. The stem of the ISC motor pushes against the throttle linkage. Two ISC motor windings are connected to the PCM. When one of these windings is activated, the motor stem extends and opens the throttle. Energizing the opposite winding retracts the ISC motor stem. The PCM operates the ISC motor to supply the proper idle speed at all engine temperatures. When the engine coolant is cold, the PCM operates the ISC motor to supply a faster idle speed. Some ISC motors have an idle switch in the ISC motor stem. When the throttle linkage contacts the ISC motor stem, the idle switch closes and sends a signal to the PCM. The PCM controls the ISC motor only when the throttle linkage is contacting the motor stem. Some throttle body assemblies have an idle air control (IAC) motor mounted on the throttle body **(Figure 38-6)**. The IAC motor is operated by the

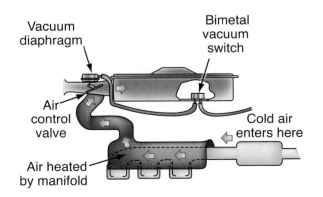

Figure 38-4 A heated air inlet system.

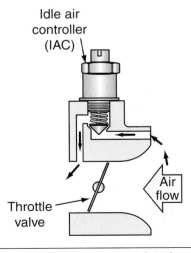

Figure 38-6 An idle air control (IAC) motor.

PCM. This motor contains a plunger that opens and closes an air passage. The air flow in the IAC motor passage bypasses the throttle. The IAC motor plunger controls idle speed by regulating the amount of air bypassing the throttle. When the engine is cold, the PCM moves the IAC motor plunger to provide more air flow through the IAC motor passage around the throttle. This action increases engine rpm. As the engine warms up, the PCM gradually moves the IAC motor plunger toward the closed position to reduce engine rpm. The IAC is used to control engine speed at idle any time a higher rpm is demanded by the PCM. For instance, while at idle the driver turns on the A/C, needing a higher rpm to prevent stalling the engine, the PCM will advance the IAC to keep the rpm's at the right speed until the driver manually takes control.

MULTIPORT FUEL INJECTION (MFI) SYSTEM

In a **multiport fuel injection (MFI)** system, the injectors are positioned in the intake manifold and the injector tips are located near the intake ports **(Figure 38-7)**.

An O-ring on the lower end of the injector seals the injector to the intake manifold and a second O-ring on the upper end of the injector seals the injector to the fuel rail. In some MFI systems, the injectors are connected in pairs or groups to the PCM and the PCM operates two to four injectors

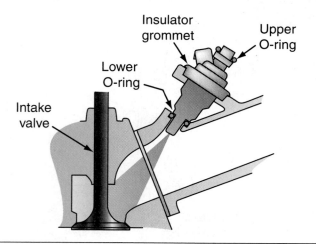

Figure 38-7 A multiport injector located in the intake port.

simultaneously **(Figure 38-8)**. The PCM always supplies the correct injector pulse width to maintain a 14.7:1 air-fuel ratio under part throttle engine operation. During cold engine operation, hard acceleration, or wide-open throttle conditions, the PCM supplies a richer air-fuel ratio to maintain engine performance. When the engine is decelerating, the PCM supplies a leaner air-fuel ratio to prevent fuel from accumulating in the exhaust and backfiring and to control emissions.

In an MFI system, fuel is injected near the intake ports and the manifold passages contain air from the air cleaner and throttle body. Because the intake manifold passages contain mostly air, intake manifold or intake air heating are not required. MFI systems usually have an IAC motor to control idle speed.

SEQUENTIAL FUEL INJECTION (SFI)

In other injection systems, the injectors are grounded individually by the PCM. This type of system is called **sequential fuel injection (SFI)** system **(Figure 38-9)**. In some SFI systems with a DI system, the number 1 blade that rotates past the distributor pickup is narrower than the other blades. This narrow blade provides a different signal to the PCM. When this signal is received, the PCM begins opening the injectors in the engine firing order. On systems without a distributor, a camshaft and crankshaft sensor are used to tell the exact position of the pistons and the camshaft to the PCM. The PCM opens each injector a significant number of crankshaft degrees just before the intake valve actually opens. This action fills the intake port with fuel vapor and ensures the proper supply of air-fuel mixture to the cylinder when the intake valve opens. Timing the spray of fuel to happen milliseconds before the intake valve opens allows the finely atomized fuel to be immediately pulled into the cylinder and makes for more efficient burn, whereas having the fuel sitting behind the valve allows it to possibly pool and reduces efficiency making for higher emissions and less power. Deposit buildup on intake valves was a common problem of systems that had fuel waiting behind the valves.

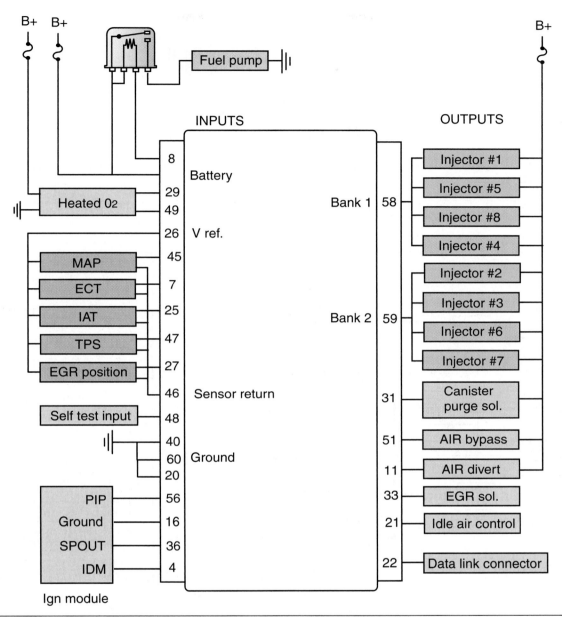

Figure 38-8 A multiport fuel injection (MFI) system.

Some SFI systems have an ASD relay and a fuel pump relay. The PCM operates both relays and the ASD relay supplies voltage to the injectors. Battery voltage is supplied to the PCM and ignition switch voltage is connected to another terminal at the PCM. The PCM must be supplied with the correct voltage and must be connected to a good chassis ground with a minimum amount of resistance. Excessive resistance in the voltage supply or ground wires may affect PCM operation and cause drivability problems.

As PCMs have become more complex and capable of handling more tasks with greater efficiency, more inputs are monitored by the PCM to maintain and adjust idle speed. Inputs that were normally not actively monitored are now under full input status and control. Items such as the park/neutral switch, power steering pump, and the A/C switch are sampled several times a second to determine if a change has been made in the system that could warrant an adjustment in fuel to maintain drivability or emission standards **(Figure 38-10)**.

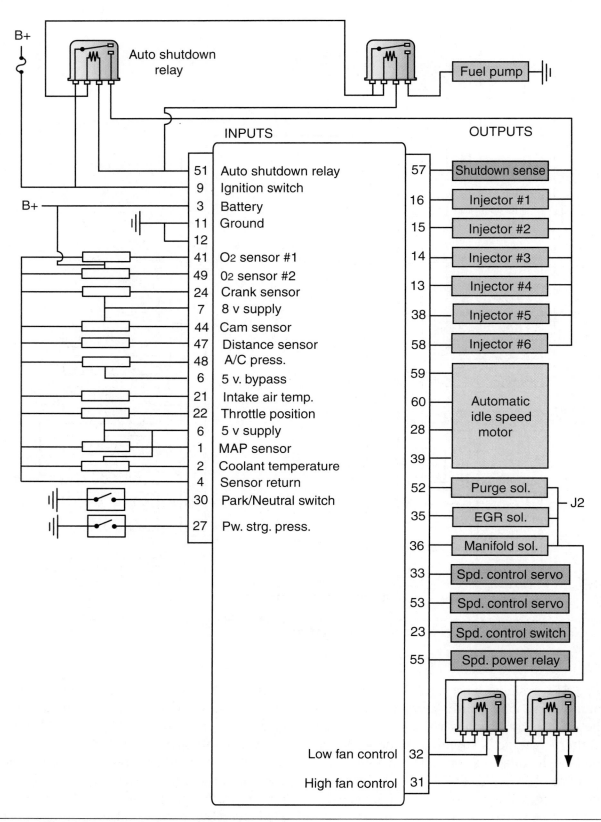

Figure 38-9 A sequential fuel injection (SFI) system.

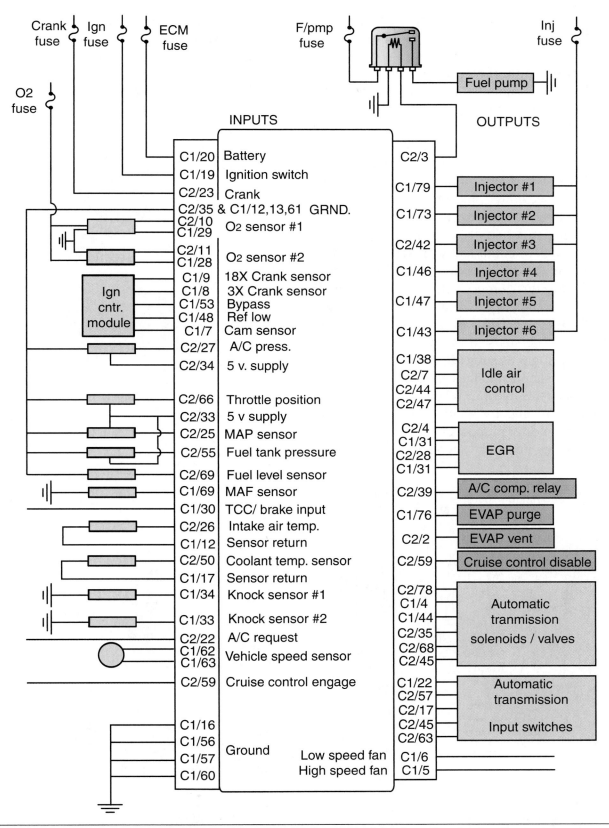

Figure 38-10 All systems are monitored by the PCM to provide data for fuel adjustment.

CENTRAL PORT INJECTION (CPI) SYSTEMS

In a **central port injection (CPI)** system, one central injector is mounted in the intake manifold. The upper half of the intake manifold is bolted to the lower half. This upper half may be removed to access the CPI components **(Figure 38-11)**. The pressure regulator is mounted with the central injector **(Figure 38-12)**. The fuel supply and return lines are connected to the central injector and the pressure regulator. These fuel lines are sealed with rubber grommets where they enter the intake manifold to prevent vacuum leaks. A port in the pressure regulator allows intake manifold vacuum to be supplied to this component. A poppet nozzle is located in each intake port and a nylon tube is connected from the central port injector to each poppet injector. The seats in the poppet nozzles are closed by spring pressure **(Figure 38-13)**.

Voltage is supplied to the central injector and the PCM grounds this injector. When the PCM grounds the central port injector winding, the coil magnetism lifts the injector plunger on the lower end of the plunger. Lifting this valve supplies fuel pressure to all the poppet nozzles simultaneously. When the fuel pressure at each poppet nozzle reaches 39 psi (269 kPa), the fuel pressure lifts the poppet nozzle seats and fuel is discharged from the poppet nozzles into the intake ports **(Figure 38-14)**. The PCM supplies the proper central injector pulse width to provide the correct air-fuel ratio.

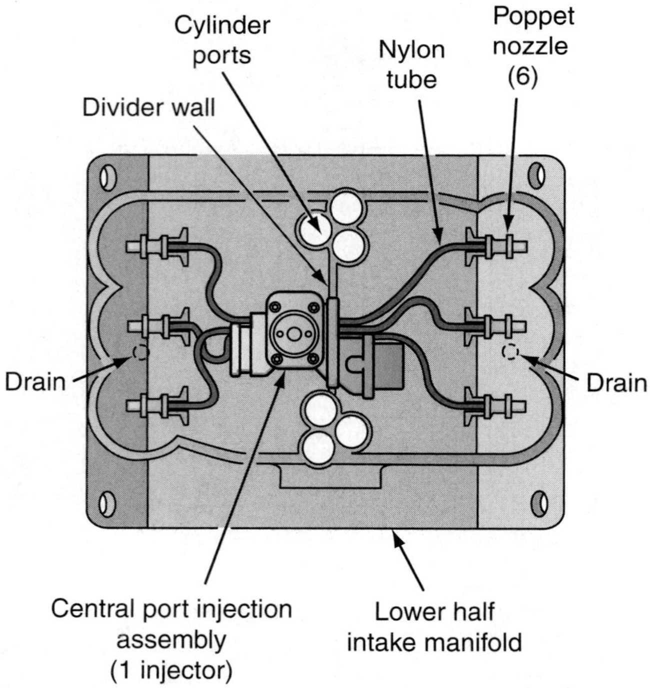

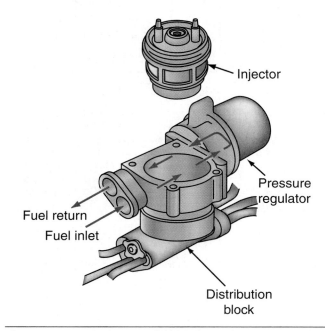

Figure 38-12 A central injector in a CPI system.

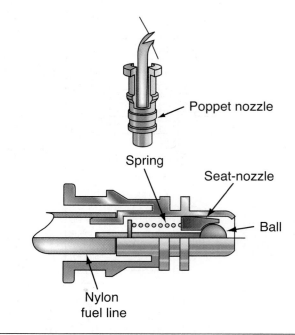

Figure 38-13 A poppet nozzle in a CPI system.

Figure 38-11 A central port injection (CPI) system.

CENTRAL MULTIPORT FUEL INJECTOR

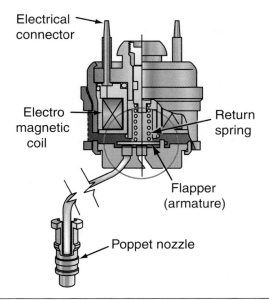

Figure 38-14 Central port injector operation.

On-Board Diagnostic I (OBD I) and On-Board Diagnostic II (OBD II) Systems

In 1988, the CARB adopted OBD I regulations for cars sold in California. These regulations required the engine computer system to monitor fuel control, EGR, emission components, and the PCM. OBD I systems also had to monitor all sensors used for fuel and emission control for opens and shorts. The PCM also had to monitor fuel trim, EGR, and HO2S. These systems were required to illuminate the MIL if a defect occurred and a DTC had to be set in the PCM memory. In the late 1980s, the CARB began investigating the possibility of using OBD system diagnosis to supplement inspection maintenance (I/M) emission programs. Simple OBD diagnostic tests could be performed with much less expensive equipment compared to tailpipe emission tests. In 1990, CARB implemented the first OBD requirements combined with emission testing. However, this combination experienced some problems. One of the difficulties in using OBD diagnosis for emission testing was that the OBD systems did not have much standardization. Vehicle manufacturers were each using their own DLC. Different scan tool software was required for each manufacturer. At that time, OBD systems did not have the capability to monitor critical emission-related components such as catalytic converters, evaporative emission systems, and secondary air injection systems. The OBD systems did not even have the capability to perform a complete test of the oxygen sensor(s). CARB and the EPA realized that a number of OBD system improvements had to be implemented if OBD systems were to be useful in I/M emission test programs. The result of these discoveries was the installation of OBD II systems on all light-duty vehicles on 1996 and later models. In 1994 and 1995, some vehicles had partial OBD II systems. Federal requirements demand that OBD II systems have many standard features including a standard DLC in the same location on all vehicles.

The EPA passed legislation requiring the implementation of OBD II diagnosis with emission I/M programs. A significant number of states have implemented OBD II diagnosis with an I/M emission program.

OBD II Monitoring

According to EPA rules, an OBD II system must detect emission-related defects and alert the driver. Keep in mind that the PCM will under most circumstances attempt to correct the problem by adjusting the fuel system output in an attempt to bring emission faults back in to specifications and turn the MIL on after corrections have failed. Because newer late model cars are completely computer controlled nearly all systems will affect other systems to some degree. When using the scan tool to diagnose a fuel system concern keep in mind that the PCM has already attempted corrections and it is very common to see certain parameters at their maximums or minimums from the PCM corrections. The following are the areas that must be monitored and all will have some effect upon the fuel system if the PCM detects a fault and tries to make adjustments to bring the engine back into specifications:

1. The engine misfire monitor must detect any engine misfire condition that causes exhaust emissions of one and one-half times the standard for non-methane hydrocarbons (NMHC), CO, or NOx. The PCM monitors the crankshaft

position sensor signal in the misfire monitor. Each time a spark plug fires, the crankshaft should speed up momentarily. If a cylinder misfires, the crankshaft does not speed up, and the PCM detects this problem from the crankshaft position sensor signal.

2. The HO2S monitor must detect any HO2S defect that results in exhaust emissions of one and one half times the standard for NMHC, CO, or NOx. The HO2S monitor must also detect any other system component that makes the HO2S sensor incapable of performing its intended function in the system.

3. The catalytic converter monitor must detect any catalytic converter problem that results in exhaust emissions of one and one-half times the standard for NMHC using an average 4,000-mile aged converter. The PCM monitors the upstream and downstream HO2S to monitor the converter operation.

4. The evaporative system monitor must be able to detect if there is an absence of evaporative purge air flow from the evaporative system.

5. The exhaust gas recirculation monitor, secondary air injection monitor, and the fuel control monitor must be able to detect any malfunction or deterioration in the powertrain system or any component directly responsible for emission control that causes emissions of NMHC, CO, or NOx to be one and one-half times the standard for these emissions.

6. The comprehensive monitor must be capable of detecting any defect or deterioration in the emission-related powertrain control system. This includes any defects in any PCM input sensor or output control device. Defects in these components are defined as malfunctions that cause the component to not meet certain continuity, rationality, or functionality checks performed at specific intervals by the PCM.

7. The secondary air injection monitor checks the operation of the air pump. The PCM turns on the electric drive air pump and checks for a leaner HO2S signal in a specific time limit. If the HO2S signal does not change, the PCM considers the AIR pump to be defective.

8. The fuel system monitor checks the long-term fuel trim to determine if this parameter is at the high or low limit. The fuel system monitor also checks the short term fuel trim. If the PCM detects a long-term fuel trim at the high or low limit during the fuel system monitor, the PCM considers the fuel system to have failed the monitor test.

9. The thermostat monitor checks the operation of the engine thermostat. A timer circuit monitor in the PCM checks the length of time required for the engine coolant temperature to increase a specific number of degrees during engine warm up.

OBD II Definitions

> **Tech Tip** *The number of monitors on a vehicle may vary depending on the vehicle model year. The main difference between an OBD I and an OBD II system is in the PCM software. An OBD II PCM contains a number of monitors that test various systems that affect emissions. The most significant additional hardware on an OBD II system is the extra HO2S mounted downstream from the catalytic converter.*

A **drive cycle** may be defined as an engine startup and vehicle operation that allows the PCM to enter closed loop and allows all the monitors to complete their function. Specific driving conditions for certain lengths of time are required for all the monitors to complete their function **(Figure 38-15)**. An engine warm-up cycle may be defined as an engine startup and engine operation until the coolant temperature increases at least 40°F (22°C) from startup and reaches a minimum temperature of 160°F (70°C).

A trip may be defined as an engine startup and vehicle operation that allows all the monitors to complete their function except the catalytic converter monitor **(Figure 38-16)**.

OBD II Operation

Many OBD II system defects do not cause illumination of the MIL light until the PCM senses the

DIAGNOSTIC TIME SCHEDULE FOR I/M READINESS
(Total time 12 minutes)

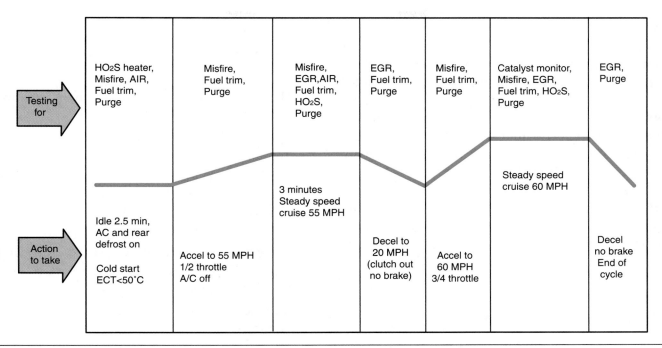

Figure 38-15 An OBD II drive cycle.

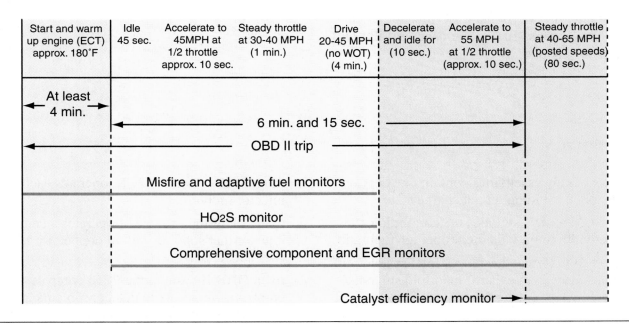

Figure 38-16 An OBD II trip.

defect on two consecutive drive cycles. However, many defects cause a DTC to be set in the PCM memory the first time the defect is sensed. When a defect occurs and a DTC is set in the PCM memory, the PCM also stores information in the freeze frame. This information may be retrieved with a scan tool during system diagnosis. If the engine switches to any default mode that is abnormal for present driving conditions, the MIL light must be turned on. If engine misfiring occurs that may cause catalytic converter damage, the MIL light flashes once per second.

When a defect detected by the fuel control monitor or engine misfire monitor does not recur in three consecutive trips under similar driving conditions, the PCM turns the MIL light off. History DTCs are cleared from the PCM memory after forty consecutive warm-up cycles have occurred without a malfunction. A scan tool connected to the DLC displays system status information indicating whether each monitor has completed its monitoring function. The vehicle must be driven through the specific drive cycle conditions for all the monitors to complete their function. This system status information is useful in verifying repairs. If all the monitors have completed their function and no DTCs are set in the PCM memory, there are no defects in the OBD II system.

OBD II Standardization

Vehicle manufacturers must comply with the following OBD II standardization requirements:

1. A standard 16-terminal DLC must be mounted on the driver's side of the dash and the DLC must be visible.

2. Standard DTCs are dictated by regulations, but the OEMs can insert some other DTCs.

3. The universal use of scan tools on all vehicle makes and test modes.

4. A standard communication protocol (data links).

5. Standard diagnostic test modes.

6. The PCM must transmit vehicle identification data to the scan tool.

7. The PCM must have the ability to erase DTCs from the PCM memory.

8. The PCM must have the ability to record and store data in snapshot form regarding conditions when a defect occurred.

9. The OBD II system must alert the driver if a defect occurs that increases emissions above a specific limit. The system must also store a DTC representing any such defect.

10. OEM terminology must be standard regarding electronic terms, acronyms, and definitions.

Summary

- Throttle body assemblies contain one or two injectors depending on the engine size.

- The PCM always supplies the correct injector pulse width to provide the proper air-fuel ratio.

- On a TBI system, a heated air inlet system pulls hot air through an exhaust manifold stove into the air cleaner during engine warm-up.

- An ISC motor stem moves the throttle linkage to control idle speed.

- An IAC motor controls the amount of air bypassing the throttle to control idle speed.

- In an MFI system, the PCM operates the injectors in groups of two to four.

- In an SFI system, the PCM operates each injector individually.

- In a CPI system, the PCM operates a central injector that supplies fuel to poppet nozzles in each intake port.

- In an OBD II system, the PCM controls a number of monitors to detect problems in various systems and components.

- Many OBD II features are regulated by SAE standards.

Review Questions

1. Technician A says in a TBI system, the PCM increases the injector pulse width as engine speed increases. Technician B says in a TBI system, the PCM maintains a 14.7:1 air-fuel ratio at part throttle. Who is correct?

   A. Technician A

   B. Technician B

   C. Both Technician A and Technician B

   D. Neither Technician A nor Technician B

2. Technician A says some throttle body systems require a heated air inlet system. Technician B says a heated air inlet system maintains air cleaner temperature at 160°F (71°C). Who is correct?

   A. Technician A

   B. Technician B

   C. Both Technician A and Technician B

   D. Neither Technician A nor Technician B

3. Technician A says if a signal is received from the power steering pressure switch, the PCM shuts off the AC compressor. Technician B says the power steering pressure switch signal informs the PCM to increase the spark advance. Who is correct?

   A. Technician A

   B. Technician B

   C. Both Technician A and Technician B

   D. Neither Technician A nor Technician B

4. Technician A says in a CPI system, the poppet injectors are operated by the PCM. Technician B says in a CPI system, a vacuum hose is connected from the intake manifold to the pressure regulator. Who is correct?

   A. Technician A

   B. Technician B

   C. Both Technician A and Technician B

   D. Neither Technician A nor Technician B

5. In a TBI system, the PCM supplies the proper air-fuel ratio by controlling the:

   A. fuel pump pressure.

   B. IAC motor.

   C. AIR pump.

   D. injector pulse width.

6. While discussing TBI systems with dual injectors, Technician A says the PCM alternately grounds each injector winding. Technician B says when the ignition switch is turned on, 12 volts are supplied to both injectors. Who is correct?

   A. Technician A

   B. Technician B

   C. Both Technician A and Technician B

   D. Neither Technician A nor Technician B

7. All of these statements about an IAC motor are true *except*:

   A. The IAC motor regulates the amount of air that is bypassing the throttles.

   B. If the air conditioning is turned on, the PCM moves the IAC plunger toward the open position.

   C. When the transmission is shifted from DRIVE to PARK, the PCM moves the IAC plunger toward the closed position.

   D. When the engine coolant is cold, the PCM moves the IAC plunger toward the closed position.

8. All of these statements about MFI and SFI systems are true *except*:

   A. In an MFI system, the PCM grounds two or more injectors simultaneously.

   B. In an SFI system, the PCM grounds each injector individually.

C. In an SFI system, the PCM opens an injector when the intake valve is opening in the same cylinder.

D. Some SFI systems have an automatic shutdown relay that supplies voltage to the injectors when the ignition switch is turned on.

9. In a CPI system:

A. The PCM grounds each injector individually.

B. Voltage is supplied directly from the PCM to each injector.

C. A V8 engine has eight outlets on the central port injector.

D. The pressure regulator is mounted separately from the central port injector.

10. In a CPI system, the pressure required to open the poppet nozzles is:

A. 10 psi.

B. 26 psi.

C. 32 psi.

D. 39 psi.

11. An IAC motor controls idle speed by regulating the amount of _____ that is bypassing the throttle.

12. Some ISC motors have an _____ _____ in the motor stem.

13. The PCM uses the park/neutral switch to control _____.

14. Define a drive cycle in an OBD II system.

39 Fuel Injection System Service and Repair

Learning Objectives

After you have read, studied, and practiced the contents of this unit, you should be able to:

- Perform injector and fuel pump electrical tests.

- Perform fuel pump pressure and injector leak tests.

- Remove and replace fuel pumps and fuel injectors.

Key Terms

Current DTCs
Flash DTCs
History DTCs

INTRODUCTION

The control of the fuel injection system affects engine performance and economy more than any other system. The complexity, coupled with the number of different fuel delivery systems, often prevents the average car owner from changing any of their filters let alone performing any repair or diagnostic work. Therefore, the demand for technicians who can quickly and accurately diagnose and repair these systems will continue to rise because future systems are likely to be more complex, not less.

FUEL SYSTEM MAINTENANCE

During routine under-hood and under-body service and repair, the fuel system should be inspected for fuel leaks, loose electrical connections, worn wiring insulation, and vacuum leaks. Fuel leaks reduce fuel mileage and create the possibility of a fire. Loose electrical connections or worn insulation on fuel injector wiring may cause injector misfiring that reduces engine performance and economy. Vacuum leaks may cause erratic idle operation. A vacuum leak in the fuel pressure regulator hose increases fuel pressure, which causes more fuel to be injected **(Figure 39-1)**. This situation results in reduced fuel economy and an engine that constantly runs rich, often resulting in a DTC that indicates a rich condition.

In any fuel injection system, the fuel filter should be replaced at the vehicle manufacturer's recommended service interval. The fuel filter may be located under the vehicle or in the engine compartment. The fuel pressure should be relieved before loosening any of the fuel lines. On MFI and SFI systems, fuel pressure may be relieved by connecting a fuel pressure gauge to the fuel line Schrader valve and bleeding the pressure off into a

Figure 39-2 A fuel pressure gauge connected to the Schrader valve on the fuel rail.

container **(Figure 39-2)**. On TBI systems, disconnect the injector connector and use two jumper wires to momentarily supply power and a ground to the injector terminals.

Tech Tip *Energizing an injector winding for more than 5 seconds may burn out the winding because these windings are normally energized by the PCM for a few milliseconds at a time.*

After the fuel system pressure is relieved, use a clean shop towel to wipe any dust and debris from the fuel line connections to prevent this material from entering the fuel lines. Loosen and remove the line fittings from the filter and remove the filter. Install the new filter in the proper direction and tighten the fittings to the specified torque. After the filter installation, always start the engine and inspect the filter and line fittings for leaks.

Tech Tip *The inlet and outlet fittings on most filters are identified, or the filter housing has an arrow to indicate the proper direction of fuel flow. Some fuel lines have quick disconnect fittings. On some fuel lines, a special tool is required to release the connectors.*

Figure 39-1 Disconnecting the vacuum line raises line pressure.

FUEL SYSTEM DIAGNOSIS

When diagnosing fuel injection systems, one of the most important tests is fuel pressure. Improper fuel pressure causes many drivability problems. For example, if the fuel pressure is higher than specified, the injectors deliver more fuel, resulting in reduced fuel economy. When the fuel pressure is less than specified, the injectors supply less fuel and this may cause drivability problems such as acceleration stumbles and loss of engine power or cutting out at high speed. The PCM diagnostics do not include fuel pressure and the PCM operates under the assumption that fuel pressure is correct. The PCM assumes the specified fuel pressure and volume are available at the injectors and will attempt to make corrections based upon this.

> **Tech Tip** *Excessive fuel pressure may be caused by a restricted fuel return line.*

Fuel Pressure and Volume Tests

Be sure there is an adequate supply of fuel in the fuel tank. On MFI, SFI, and CPI systems, connect the fuel pressure gauge to the Schrader valve on the fuel rail **(Figure 39-3)**. On TBI systems, the fuel pressure gauge must be connected in series in the fuel supply line at the TBI assembly. Cycle the ignition switch on and off several times and read the fuel pressure on the gauge. If the fuel pressure is less than specified, the fuel filter may be severely restricted, the voltage supply or ground on the fuel pump may have high resistance, or the fuel pump may be defective. If the pressure is at specifications, test the volume of fuel available. A partially plugged filter or marginal pump may allow the system to pass a pressure test but when the engine is being used may not be able to supply enough fuel to maintain pressure and keep up with demand. With the engine running use the pressure bypass port on the test gauge to drain off fuel into a measured container. As a general rule the engine should not die and the pump should be able to flow an additional half-quart in about 15 seconds **(Figure 39-4)**. It is a good idea to perform this test on a known good vehicle with the gauge setup you will be using, as different gauge setups will sometimes flow different amounts of fuel. If it fails to flow enough fuel, then further electrical diagnostics are needed to determine if it is a pump or filter fault. If the pump has to be replaced it is a good practice to also replace the filter to ensure the faulty pump did not pass on materials that may ruin the injectors. Keep in mind that with today's

Figure 39-3 Using a fuel pressure gauge to bleed off fuel pressure.

Figure 39-4 Volume testing a fuel pump.

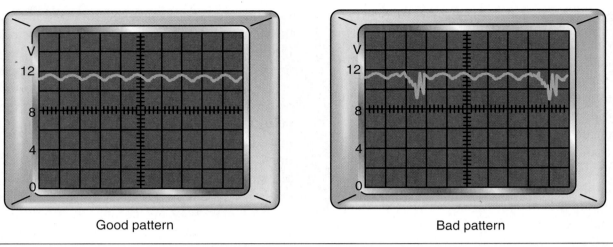

Good pattern   Bad pattern

Figure 39-5 Using a scope to test a fuel pump wave pattern.

fuels it is not normal for a filter to plug unless it is a high mileage car that has had no maintenance. If the filter is plugged, a sampling of the fuel should be taken to determine if there are large amounts of particles in the tank. This will help avoid a customer return for the same fault.

> **Tech Tip** *A leaking pressure regulator diaphragm on an MFI or SFI system causes fuel to be pulled through this diaphragm and the vacuum hose into the intake manifold. This results in excessive fuel consumption and erratic idle operation. A leaking nylon fuel line on a CPI system causes engine misfiring at idle because it reduces the fuel pressure so this pressure no longer opens the poppet nozzle.*

Electrical Fuel Pump Testing

With the engine running, use a digital voltmeter to test the voltage at the fuel pump in the tank. With the engine running, the voltage at this location should be 11.5 volts or more. If this voltage is lower than specified, the wire from the fuel pump relay to the fuel pump has excessive resistance. To measure the voltage drop across the fuel pump ground, connect the voltmeter from the fuel pump ground terminal to a good chassis ground. The voltage drop across this circuit should not exceed 0.2 volt. In many shops it is common practice to hook an oscilloscope to the fuel pump electrical system and watch the pattern displayed from

the motor. The scope will show problems in the motor long before it fails and once you become really familiar with the electrical patterns, you can even tell the exact type of failure in the pump **(Figure 39-5)**. The key to using a scope when testing components is to look for patterns that do not match the other like components. If the fuel filter and the fuel pump voltage supply and ground are satisfactory, replace the fuel pump.

Injector Testing

A rough idle problem may be caused by faulty injectors. Some fuel injector problems, such as an open winding or severely restricted discharge orifices, cause cylinder misfiring at idle. When the engine speed is increased, the misfiring is less noticeable. If the orifices in an injector are partially restricted, a rough idle problem results. With the air cleaner removed and the engine idling on some TBI systems, the fuel spray from the injectors may be observed visually. If an injector is not spraying fuel, a stethoscope pickup may be placed against the injector body with the engine idling **(Figure 39-6)**. If a clicking noise is heard, the PCM is operating the injector and the injector winding is probably satisfactory. When the injector is not clicking with the engine idling, test the injector voltage supply and ground wire to the PCM and test the injector winding with an ohmmeter. TBI or port injectors may be tested with an ohmmeter. A simple injector wiring tester, often

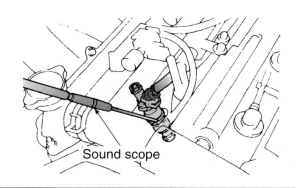

Figure 39-6 Injector testing with a stethoscope.

Figure 39-7 An injector "NOID" light.

called a "NOID" light, can be used to determine if the PCM is cycling the injector **(Figure 39-7)**.

To ohm test an injector, turn the ignition switch off and remove the injector connector. Connect the ohmmeter leads to the injector terminals **(Figure 39-8)**. The injector must have the specified resistance. If the resistance is less than specified, the injector winding is shorted. An infinite ohmmeter

reading indicates an open injector winding. If the injector does not have the specified resistance, replacement is necessary.

> **Tech Tip** *A shorted injector winding causes high current through the injector and the PCM driver. This high current flow may damage the injector driver in the PCM. In some PCMs, the injector drivers sense high current flow and, rather than allowing injector driver damage, they shut the circuit off. After a cool down period, the injector driver may allow injector operation.*

An injector balance test may be performed to test the injectors. The injector balance tester energizes each injector for a specific short time period. Each injector terminal must be disconnected and the lead from the balance tester connected to the injector terminals. The power leads on the injector balance tester must be connected to the battery terminals with the correct polarity **(Figure 39-9)**. During the injector balance test, the fuel pressure gauge must be connected to the Schrader valve on the fuel rail. Cycle the ignition switch several times until the specified pressure is indicated on the fuel gauge. Press the button on the balance tester to energize the injector for a specific time period. Record the fuel pressure displayed on the fuel gauge. Repeat this

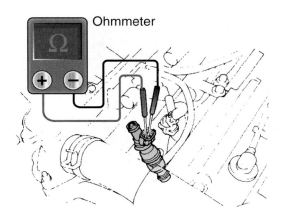

Figure 39-8 Testing an injector with an ohmmeter.

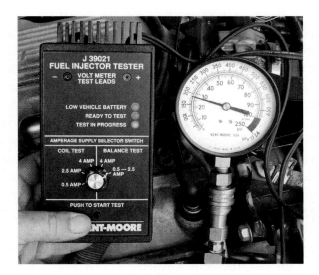

Figure 39-9 An injector balance test.

CYLINDER	1	2	3	4	5	6
HIGH READING	225	225	225	225	225	225
LOW READING	100	100	100	90	100	115
AMOUNT OF DROP	125	125	125	135	125	110
RESULTS	OK	OK	OK	Faulty, rich (too much fuel drop)	OK	Faulty, lean (too little fuel drop)

Figure 39-10 Interpreting injector balance test results.

procedure on each injector. Obtain the maximum difference in pressure drop on each injector from the vehicle manufacturer's specifications. If an injector has considerably less pressure drop compared to the other injectors, this injector has restricted orifices. When an injector has considerably more pressure drop compared to the other injectors, the injector plunger is sticking open **(Figure 39-10)**.

Injectors with restricted orifices may have leaking seats that drip fuel from the injector tips into the intake manifold during idle operation and after the engine is shut down. This condition causes rough idle operation and hard starting after a hot engine has been shut down for a few minutes. When testing fuel pump pressure or injector balance, cycle the ignition switch several times until the specified fuel pressure appears on the fuel gauge. After 15 minutes, observe the fuel pressure. If the fuel pressure slowly decreases, the injectors may be leaking or the pressure regulator valve may have a leak. This decrease in fuel pressure may also be caused by a leaking one-way check valve in the fuel pump. To eliminate these leak sources and locate the problem, repeat the leak down test and use a pair of straight-jawed vise grips to close each line one at a time. If the fuel pressure still leaks down with the fuel return line and the fuel supply line blocked, the injectors are dripping fuel. If the leak down problem is eliminated when the fuel return line is closed, the pressure regulator valve is leaking. When the leak down problem is eliminated with the fuel supply line closed, the one-way check valve in the fuel pump is leaking.

> **Tech Tip** *It's always a good idea to perform tests on vehicles in good working order. This helps speed up later diagnostic procedures and allows a good understanding of what things should look like and how they should test.*

Fuel injectors may be tested with a lab scope. Connect the scope leads to the injector terminals. A satisfactory waveform from a conventional injector is shown in **Figure 39-11**. A satisfactory waveform from a peak-and-hold injector is illustrated in **Figure 39-12**, and a waveform from a pulse-modulated injector is pictured in **Figure 39-13**. If you are unable to determine which type of injector, then you should test all injectors and compare the wave patterns. If an injector has an abnormal waveform compared to the others, that injector has a problem.

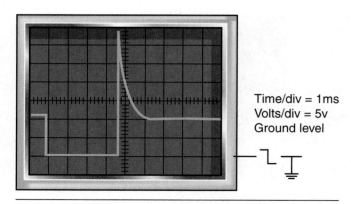

Time/div = 1ms
Volts/div = 5v
Ground level

Figure 39-11 Waveform from a conventional fuel injector.

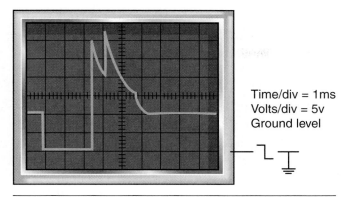

Figure 39-12 Waveform from a peak-and-hold fuel injector.

Time/div = 1ms
Volts/div = 5v
Ground level

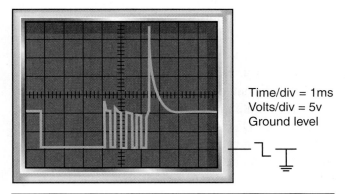

Figure 39-13 Waveform from a pulse-modulated fuel injector.

Time/div = 1ms
Volts/div = 5v
Ground level

Tech Tip *The PCM supplies normal current flow to a peak-and-hold injector to open the injector and then reduces the current flow to hold the injector open for a very short time period. The PCM initially supplies normal current flow for a pulse-modulated injector and then pulses the injector on and off very quickly.*

Injector Cleaning

Injectors may be cleaned with a special injector cleaner placed in a pressurized container **(Figure 39-14)**. The injector cleaning solution is a mixture of cleaning solution and unleaded gasoline. After the cleaning solution is placed in the injector cleaning tool, the tester is pressurized with shop air pressure until the pressure

Figure 39-14 Injector cleaning equipment.

gauge on the tester is slightly below the specified fuel pressure. At this pressure, the pressure regulator valve does not open during the injector cleaning process.

During the injector cleaning process, the fuel pump must be inoperative. Disconnect the wires from the fuel pump relay to prevent fuel pump operation. If the fuel pump circuit has an oil pressure switch, disconnect the wires from this switch. Locate a section of rubber hose in the fuel return line and use a pair of straight-jawed vise grips to close this line. This action prevents the possibility of the injector cleaning solution flowing through the fuel return line into the fuel tank. Connect the hose on the injector cleaner to the Schrader valve on the fuel rail and open the cleaner valve. Start and run the engine on the injector cleaning solution. The engine usually runs for 15 to 20 minutes on the cleaning solution. After the injectors have been cleaned, the engine may still experience a rough idle problem because the PCM has to re-learn a new fuel strategy with the cleaned injectors. With the engine at normal operating temperature, drive the vehicle for 5 minutes to allow the PCM to re-learn regarding the cleaned injectors.

> **Tech Tip** *Some vehicle manufacturers do not recommend cleaning certain types of injectors in their vehicles. One example of this type of injector is the Multec injectors in some General Motors vehicles. These injectors have a sharp-edged orifice plate at the injector tip that provides a self-cleaning action for the injectors.*

Engine Control System Scan Tool Diagnosis

When diagnosing fuel injection systems, observe the MIL in the instrument panel. If this light is illuminated with the engine running, the PCM has detected a problem in the system and a DTC is set in the PCM memory. In the 1980s and early 1990s, each vehicle manufacturer used a different DLC. On some of these systems, a jumper wire could be used to jump across two terminals in the DLC and obtain flash DTCs from the MIL. For example, if the MIL flashed quickly three times followed by a pause and three more quick flashes, DTC 33 was indicated. However, as more DTCs were programmed into the systems, it became very difficult to read flash DTCs. For example, some three-digit DTCs were as high as 999, which is difficult to read via flash DTCs. **Flash DTCs** are displayed by the flashes of the MIL. As the ability to read live data to and from the PCM became widespread, the need to interpret this data along with the DTCs made flash codes obsolete.

A scan tool is connected to the DLC under the dash to read the DTCs and other input and output data. When performing a scan tool diagnosis, the engine should be at normal operating temperature. With the ignition switch off, connect the scan tool to the DLC. On some vehicles, you have to enter the vehicle make, model year, and engine code into the scan tool. On other vehicles, the PCM supplies this information to the scan tool. When READ CODES is selected on the scan tool, the DTCs are displayed. Some scan tools supply an interpretation for each DTC. When using other scan tools, it is necessary to obtain the DTC interpretation from the vehicle manufacturer's information. DTCs may be identified as current or history. **Current DTCs** are present at the time of testing. **History DTCs**

represent intermittent faults that occurred sometime in the past, but the fault is no longer present.

A DTC indicates a problem in a certain area. For example, a DTC representing an MAF sensor may indicate a defective sensor, defective wires from the sensor to the PCM, or a PCM that is not able to receive this signal. You must perform specific tests, such as voltmeter or ohmmeter tests, to locate the exact cause of the problem as outlined in the proper repair manual. OBD I systems used two- or three-digit numbered DTCs. OBD II DTCs are formatted according to SAE standard J2012. This standard requires five-digit alphanumeric DTC identification. The prefix in each DTC indicates the DTC function: P—powertrain, B—body, C—chassis, U—network communication. The first number in the DTC represents the group responsible for the code. If the first number is 0, the code is designated by SAE. When the first number is 1, the vehicle manufacturer is responsible for the DTC. Many DTCs are dictated by SAE but the vehicle manufacturer may insert some codes. The third digit in the DTC indicates the subgroup to which the code belongs as follows:

1—air-fuel control

2—air-fuel control, injectors

3—ignition system, misfire

4—auxiliary emission controls

5—idle speed control

6—PCM input and output

7—transmission/transaxle

8—transmission/transaxle

The fourth and fifth digits in the DTC indicate the area in which the defect is located. For example, in DTC P0155, P indicates it is a powertrain code; 0 indicates it is designated by SAE; 1 indicates the code is in the air-fuel control subgroup; and 55 indicates the code represents a malfunction in the HO2S heater, bank 2, sensor 1. Bank 2 indicates the bank of a V6 or V8 engine that is opposite to the bank where number 1 cylinder is located and sensor 1 is the upstream HO2S sensor located ahead of the catalytic converter. A bank 1 HO2S is located on the same side of the engine as number 1 cylinder and sensor 2 is mounted downstream from the catalytic converter.

Using a scan tool often allows you to perform in-depth testing of the fuel circuit, depending on the scan tool and the types of tests the manufacturer has programmed into the system. For example, injector balance tests are common. Some provide for pulsing individual injectors, turning the pump on and off, and any number of tests. The best way to use a scan tool to diagnose a system is to become familiar with the particular tool you will be using on several different makes of cars, keeping in mind that the tests available will be determined by make, model, year, and the scan tool itself. Often scan tools from the manufacturer provide more testing ability than aftermarket tools.

Engine Control System Service

Always be sure the exact cause of the customer's complaint is diagnosed before performing any engine control system service. For example, before replacing a fuel pump, be sure no other causes of improper fuel pump operation are present. Therefore, be sure the fuel filter, fuel lines, and the fuel pump voltage supply and ground wires are satisfactory before replacing the fuel pump. Before replacing any electrical component, connect a 12-volt power supply to the cigarette lighter socket and then disconnect the negative battery terminal. Disconnecting the battery prevents accidental shorts to ground from damaging components or starting a fire when servicing vehicle components or systems. The power supply connected to the cigarette lighter socket keeps the PCM memory alive during the service procedure. This action prevents drivability problems after the vehicle is restarted and prevents customer complaints if you forget to reset all preprogrammed settings.

Fuel Pump Replacement

When replacing most electric in-tank fuel pumps, the fuel tank must be removed. On some vehicles, a fuel pump access cover is located in the trunk directly above the fuel pump in the tank. After this access door is removed, the fuel pump may be taken out of the fuel tank. If the fuel tank must be removed to gain access to the fuel pump, use an electric or hand pump to pump the gasoline out of the tank

Figure 39-15 An approved fuel recovery system.

into an approved gasoline container **(Figure 39-15)**. Be careful not to spill any fuel. If fuel is accidentally spilled, use the proper absorption material and clean up the spill immediately. Raise the vehicle on a lift or use a floor jack to raise the vehicle onto safety stands. Disconnect the fuel lines and electrical connector near the fuel pump and remove the fuel tank straps. Slowly lower the fuel tank onto the floor and remove it from under the vehicle. Use an air gun to blow debris from the top of the fuel tank and the fuel pump assembly. Remove the fuel pump retaining ring and lift the fuel pump assembly from the tank **(Figure 39-16)**. In many vehicles the fuel pump, fuel intake filter, and fuel tank sensing unit are replaced

Figure 39-16 A typical in-tank fuel pump.

as an assembly. In some applications, the fuel pump may be replaced separately. Always use a new seal between the fuel pump assembly and the top of the fuel tank. Install the new fuel pump assembly and be sure the fuel pickup is very close to or touching the bottom of the tank.

SAFETY TIP *Do not drag a fuel tank across the cement floor; this could cause a spark and a serious fire. If it must be moved have someone help carry it.*

Install the fuel pump assembly retaining ring in the top of the fuel tank. Install the fuel tank in its proper location under the vehicle and install the holding straps. Tighten the holding strap fasteners to the specified torque. Reconnect the fuel lines and electrical connector to the fuel pump assembly. Tighten the fuel lines to the specified torque. Lower the vehicle onto the shop floor. Reconnect the negative battery cable and remove the power supply connector from the cigarette lighter. Start the engine and test the fuel pump to be sure it has the specified pressure with no fuel leaks.

Injector Replacement

Always relieve the fuel system pressure before replacing the injectors. Connect a 12-volt power supply to the cigarette lighter and disconnect the negative battery cable. Release the lock ring and remove each injector terminal.

Disconnect the fuel inlet and outlet lines on the fuel rail and the vacuum hose on the pressure regulator. Remove the fuel rail retaining bolts. Remove any other components that interfere with the fuel rail removal. Remove the fuel rail and injectors **(Figure 39-17)**.

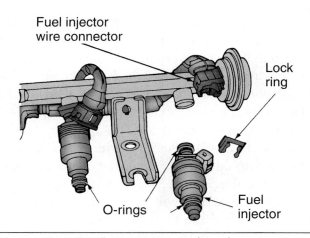

Figure 39-17 Injector removal and replacement.

Always install new injector O-rings when removing and replacing injectors. Place a small amount of engine oil on each upper injector O-ring and install the injectors into the fuel rail. Use engine oil to lightly coat the lower injector O-rings and install the fuel rail and injectors in the intake manifold. Be sure the injectors are fully seated in the fuel rail and the intake manifold. Reconnect the injector wiring terminals. Install the fuel rail fasteners, and tighten these fasteners to the specified torque. Reconnect the fuel inlet and outlet lines and tighten the fittings to the specified torque. Install the vacuum hose on the pressure regulator. Reconnect the battery negative cable and disconnect the power supply from the cigarette lighter. Retest the system to ensure the fault has been repaired.

SAFETY TIP *Always be sure to install the vehicle manufacturer's recommended injectors. Installing other injectors may cause drivability problems.*

Summary

- The fuel filter should be changed at the vehicle manufacturer's specified intervals.

- The fuel system pressure should be relieved before disconnecting any fuel system component.

- The PCM assumes the specified fuel pressure is available when calculating the proper amount of fuel to be injected.

- The PCM diagnostics do not provide any indication of improper fuel pressure.

- Before replacing a fuel pump, always be sure the voltage supply and ground wires to the pump are satisfactory.

- Fuel injectors may be tested by measuring the resistance in the injector windings.

- An injector balance test measures the fuel pressure drop when each injector is opened for a specific length of time.

- If the fuel system pressure gradually decreases after the engine is shut off, the one-way check valve in the fuel pump may be leaking, the pressure regulator valve may be leaking, or the injectors may be dripping fuel.

- Some injectors may be cleaned by connecting a pressurized container filled with injector cleaner to the Schrader valve on the fuel rail.

- A DTC indicates a fault in a certain area of the engine control system.

- In an OBD II DTC, the first digit indicates the area to which the DTC belongs. For example, P indicates powertrain.

- The second digit in an OBD II DTC indicates whether the DTC is established by SAE standards or introduced by the vehicle manufacturer.

- The third digit in an OBD II DTC indicates the subgroup to which the DTC belongs.

- The fourth and fifth digits in an OBD II DTC represent the fault in the system.

- On most vehicles, the fuel tank must be removed to access the fuel pump.

- When removing and replacing any electrical/electronic component, always connect a 12-volt power supply to the cigarette lighter socket and disconnect the negative battery cable.

Review Questions

1. Technician A says in a P0155 DTC, the 0 indicates the DTC is developed by the vehicle manufacturer. Technician B says the 1 indicates the subgroup to which the DTC belongs. Who is correct?

   A. Technician A

   B. Technician B

   C. Both Technician A and Technician B

   D. Neither Technician A nor Technician B

2. Technician A says higher-than-specified fuel pump pressure causes a rich air-fuel ratio. Technician B says higher than-specified fuel pressure may be caused by a leaking one-way check valve in the fuel pump. Who is correct?

   A. Technician A

   B. Technician B

   C. Both Technician A and Technician B

   D. Neither Technician A nor Technician B

3. Technician A says during an injector balance test, if the pressure drop on one injector is higher than on the other injectors, this injector has restricted orifices. Technician B says during an injector balance test, each injector should have nearly the same pressure drop. Who is correct?

   A. Technician A

   B. Technician B

   C. Both Technician A and Technician B

   D. Neither Technician A nor Technician B

4. In an MFI system, lower-than-specified fuel pressure may cause all of these problems *except*:

   A. a lean air-fuel ratio.

   B. engine surging and cutting out at high speed.

   C. hesitation during engine acceleration.

   D. engine backfiring during deceleration.

5. While discussing TBI system diagnosis, Technician A says the PCM has the capability to diagnose fuel pump pressure. Technician B says when testing fuel pressure with a gauge, connect the gauge hose to the Schrader valve. Who is correct?

   A. Technician A

   B. Technician B

   C. Both Technician A and Technician B

   D. Neither Technician A nor Technician B

6. All of these problems may cause low fuel pump pressure *except*:

   A. low voltage at the fuel pump.

   B. a restricted fuel filter.

   C. a restricted fuel return line.

   D. a restricted fuel supply line.

7. When diagnosing an SFI system with the fuel pressure gauge connected to the system, the ignition switch is cycled several times until the specified fuel pressure appears on a fuel pressure gauge. The fuel supply line and the return fuel line are then completely blocked. After 15 minutes, the fuel pressure has dropped from 45 psi to 10 psi and there are no indications of external fuel leaks. The most likely cause of this pressure reading is:

   A. dripping fuel injectors.

   B. a leaking pressure regulator valve.

   C. a leaking one-way check valve in the fuel pump.

   D. a leaking fuel filter.

8. When performing an injector cleaning procedure:

   A. The fuel return line must be open.

   B. The fuel pump must be in operation.

   C. The injector cleaning solution container is connected to the Schrader valve.

   D. The engine should be running for 1 hour.

9. While diagnosing an OBD II fuel system, a current P0155 DTC is obtained on a scan tool. All of these statements about this DTC are true *except*:

   A. This DTC represents a problem in the powertrain.

   B. This DTC is designated by the vehicle manufacturer.

   C. This DTC is in the air-fuel control subgroup.

   D. This DTC represents a defect that is present during the diagnosis.

10. During the injector cleaning process, the fuel pump must be _____.

11. If an OBD II DTC begins with a B, the DTC is in the _____ category.

12. If the third digit in an OBD II DTC is 3, the DTC is related to _____.

13. Explain the reason for connecting a 12-volt power supply to the cigarette lighter socket before disconnecting a vehicle battery.

14. Describe the proper fuel pressure gauge connection to test fuel pump pressure on a TBI system.

15. When the fuel pump pressure is less than specified, describe the diagnostic procedure to locate the exact cause of the problem.

16. Describe the possible results of a shorted injector on the PCM.

CHAPTER

40

Vehicle Emissions and Emission Standards Fundamentals

Learning Objectives

After you have read, studied, and practiced the contents of this unit, you should be able to:

- List the three main automotive pollutants and explain how each is formed.

- Describe how photochemical smog is formed.

- Explain the origin of crankcase and evaporative emissions.

- Explain the difference between basic and enhanced emission test programs.

- List the items to be inspected during a visual emission pre-inspection.

- Explain the items to be checked in an emissions component inspection.

- Explain an idle test emission procedure.

- Describe an IM240 emission test procedure.

- Describe a BAR-31 emission test procedure.

- Describe an ASM emission test procedure.

Key Terms

Hydrocarbons (HC)

Inspection/maintenance (I/M)

Oxides of Nitrogen (NOx)

Positive Crankcase Ventilation (PCV)

Photochemical smog

Quench area

INTRODUCTION

Over the last 40 years automotive emissions standards have become increasingly stringent, greatly reducing the amount of emissions produced by automobiles. This reduction in automotive emissions provides cleaner air to breathe, reducing health hazards such as smog that can lead to respiratory diseases. It is very important for technicians to understand the types of vehicle pollutants and the allowable standards for these pollutants. You should understand how each automotive emission control device operates to reduce harmful emissions and keep engine performance at its peak. Many new technicians are required each year to diagnose and repair vehicles that have failed compulsory emission tests. Your understanding of emissions and vehicle emission systems is absolutely essential to perform this job quickly and accurately.

AIR POLLUTION AND VEHICLE EMISSIONS

Vehicle emissions have been regulated since the early 1960s, since auto emissions were determined to be the major cause of smog in Los Angeles. **Photochemical smog** is formed by the reaction of sunlight with **hydrocarbons (HC)** and **oxides of nitrogen (NOx)** in the atmosphere. Smog appears as a light-brown haze in the air. Ozone is the primary component in photochemical smog. Ozone contains three oxygen atoms (O_3). Ozone occurs naturally in the upper atmosphere and serves to prevent harmful ultraviolet rays given off by the sun from reaching the earth. Hot air near the ground usually rises and is then cooled by cooler air at a higher altitude. This action usually clears smog from the air at ground level. However, in some locations such as in a valley or over a large metropolitan area, the warm air near the ground becomes trapped by an inversion layer that tends to hang over the area. When this action occurs, the smog is trapped near the ground and the smog becomes more concentrated because it does not escape. Smog is known to aggravate lung conditions such as asthma.

Vehicles manufactured today emit less than 5 percent of the pollution emitted by vehicles manufactured in the 1960s because of emission control devices and improved engine design. There are many sources of air pollution other than vehicles. These sources include industry, homes, and airplanes.

The main regulated automotive pollutants are unburned HC, CO, and NOx. Other exhaust emissions include microscopic soot and dust particles that are small enough to remain suspended in the air. Particulates are a primary concern with diesel engines. Other tailpipe emissions that are often measured during vehicle diagnosis include CO_2

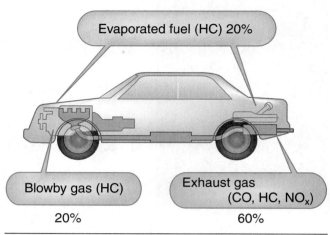

Figure 40-1 Automobile emission sources.

Composition Type of Gas	CO	HC	NOx
Exhaust gas	100%	55%	100%
Blow-by gas	—	25%	—
Evaporated fuel	—	20%	—

Figure 40-2 The percentage of emissions from automobile sources.

and O_2. These emissions are not pollutants but are useful in the diagnostic process. There are three sources of emissions on a vehicle: exhaust emissions from the tailpipe, evaporative emissions from the fuel tank, and crankcase emissions.

All three pollutants are present in exhaust emissions, whereas evaporative or crankcase emissions contain largely HC **(Figure 40-1)**. On a typical vehicle, 60 percent of the HC emissions come from the exhaust, 20 percent come from the crankcase, and 20 percent come from evaporative sources **(Figure 40-2)**.

Hydrocarbons (HC)

HC emissions are caused by unburned air-fuel mixture in the combustion chamber. At the end of each combustion event, some unburned air-fuel mixture is left on the cooler combustion chamber surfaces when the flame front is quenched near this surface **(Figure 40-3)**. This action results in HC emissions. A larger combustion chamber surface provides more **quench area**.

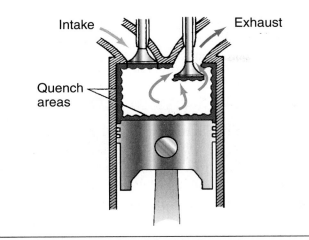

Figure 40-3 Unburned fuel left on the combustion chamber surface.

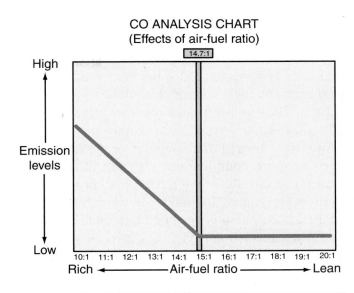

Figure 40-4 CO emissions in relation to air-fuel ratio.

On some engines, the air-fuel ratio becomes richer during deceleration and this increases HC emissions. If a cylinder misfires, the whole air-fuel charge in the combustion chamber is forced out through the exhaust system and HC emissions increase significantly. Thus, HC emissions are an indicator of how much fuel is unburned at the end of the combustion process. In the United States, approximately 27 percent of the manmade HC emissions are emitted from vehicles.

Carbon Monoxide (CO)

CO emissions are a by-product of combustion. Carbon monoxide is a deadly gas that is odorless and colorless. CO is an unstable pollutant that readily combines with oxygen to become harmless CO_2. When breathed in, CO will combine with the oxygen in a person's bloodstream and become CO_2, which results in oxygen deprivation. The symptoms are often dizziness, sleepiness, and headaches.

CO emissions occur when the air-fuel mixture is not completely burned. CO emissions are lowest at the ideal or stoichiometric air-fuel ratio of 14.7:1. As the air-fuel ratio becomes richer than 14.7:1, CO emissions increase in proportion to the richness of the air-fuel ratio **(Figure 40-4)**. In the United States, it is estimated that 80 percent of the CO emissions come from vehicles. In urban areas, this figure may be as high as 90 percent due to reduced natural airflow.

SAFETY TIP *Carbon monoxide is a poisonous gas. In lower concentrations it results in headaches and nausea. In high concentrations, carbon monoxide may be fatal for human beings.*

Oxides of Nitrogen (NOx)

NOx emissions are caused by oxygen and nitrogen in the air-fuel mixture uniting at combustion chamber temperatures above 2,500°F (1,371°C). NOx is an air pollutant that is primarily manmade and has no natural sources of readily measurable size. High combustion chamber temperatures, such as an overly lean condition, are the main cause. NOx emissions are reduced when combustion chamber temperatures are lowered. NOx combines with rainwater to produce acid rain that pollutes rivers and lakes, killing fish and vegetation. In the United States, 32 percent of the NOx emissions come from vehicles.

EVAPORATIVE AND CRANKCASE EMISSIONS

The source of evaporative emissions is mainly the fuel tank. Since gasoline is a fuel that will readily vaporize, the more open the fuel system, the more fuel vapors can escape. The ambient air temperature also plays a part in this; the hotter it is, the faster fuel will vaporize. For many years, vehicles have been equipped with evaporative emission systems to reduce the escape of fuel vapors from the fuel tank and the engine to the atmosphere. The emission laws regarding evaporative emissions have become increasingly stringent. In response to these laws, evaporative systems have become more complex and efficient in reducing these emissions. Refueling is another source of evaporative emissions. In many newer vehicles, the evaporative system is designed to reduce evaporative emissions while refueling. Gasoline refueling pumps have also been designed to reduce refueling evaporative emissions **(Figure 40-5)**. On older vehicles, the carburetor was a major source of evaporative emissions; for example, on hot days it was not uncommon to smell the gasoline vapors coming from the carburetor after the vehicle was shut off, and the heat from the engine would help this process by heating the fuel that was stored in the carburetor.

Crankcase emissions are caused by combustion blowby past the piston rings into the crankcase. All internal combustion engines have a small amount of blowby. This condition causes HC emissions to enter and accumulate in the crankcase. The **positive crankcase ventilation (PCV)** system moves crankcase gases back into the intake manifold to greatly reduce crankcase emissions. However, excessive blowby because of worn piston rings and/or cylinders causes too much crankcase pressure, which may result in high crankcase emissions. A defective PCV system may also result in high crankcase emissions.

EMISSION STANDARDS

In the United States, emission standards began in 1961 when all new cars sold in California were required to have a PCV system to reduce crankcase emissions. In 1966, new cars sold in California were required to have exhaust emission systems. Some of these systems involved the use of a belt-driven air pump to force air into the exhaust manifolds **(Figure 40-6)**. The oxygen in this air allowed the air-fuel mixture left in the exhaust to continue burning in the exhaust manifolds. This action reduced HC emissions. In 1970, the United States Congress passed the Clean Air Act that established maximum emission levels for HC, CO, and NOx. The Clean Air Act has been amended several times since to increasingly tighten emission standards. The implementation of emission standards is usually administered by federal or state organizations such as the EPA or the CARB. The EPA establishes and implements emission standards across the United States, but some states such as California have set their own more stringent emission standards.

Figure 40-6 A belt driven air pump.

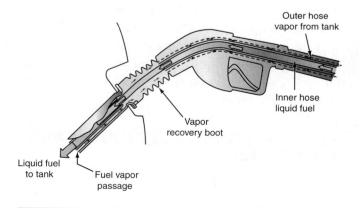

Figure 40-5 Modern refueling station equipped with a vapor recovery system.

Outer hose vapor from tank

Inner hose liquid fuel

Vapor recovery boot

Liquid fuel to tank

Fuel vapor passage

	HC	CO	NOx
1990 - U.S.	0.41	3.4	1.0
1994 - U.S.	0.25**	3.4	0.4
1993 - Calif.	0.25	3.4	0.4
1994 - TLEV	0.125	3.4	0.4
1997 - LEV	0.075	3.4	0.2
2000 - ULEV	0.040	1.7	0.2
** non-methane HC			

Figure 40-7 Emission standards.

Often states will set more stringent standards in response to local pollution and air quality problems. Emission standards have been established for Transitional Low Emission Vehicles (TLEVs), Low Emission Vehicles (LEVs), and Ultra Low Emission Vehicles (ULEVs) **(Figure 40-7)**. In recent years California has required that no fewer than 2 percent of all cars sold in that state be zero emission vehicles, and raising the percent of ZLEVs is currently being considered. At this time only electric vehicles meet the requirements to be called zero emission vehicles.

Emission Testing

In the early 1970s, **Inspection/Maintenance (I/M)** emission programs were identified as an effective method for reducing emissions on in-use vehicles. By having vehicles inspected and repaired if necessary, emissions could be reduced to the level required for that model year. In 1977, the Clean Air Act was amended to require mandatory implementation of I/M programs in those areas of the United States that were not in compliance with National Ambient Air Quality Standards (NAAQS). Areas of the country that do not meet NAAQS standards are called non-attainment areas. If the states containing non-attainment areas did not implement I/M programs, they could be subjected to federal sanctions including the loss of funds for highway construction and improvement and the ability to grant permits for new industrial development. Early I/M tests were usually basic two-speed idle tests because these tests were easy to perform and required a minimum amount of test equipment.

> ***Did You Know?*** *Centralized I/M programs are operated by governments or contractors working for governments. De-centralized I/M programs are operated by privately owned repair shops or franchised dealerships licensed by the state authorities.*

The EPA monitored these I/M tests and discovered that the test and repair procedures did not reduce emissions as much as expected. The EPA decided that more sophisticated test procedures were necessary for the newer computer-controlled vehicle systems. The 1990 Clean Air Act amendments addressed these concerns by requiring enhanced I/M programs in all non-attainment areas identified as serious, severe, and extreme. In 1992, the EPA established the requirements for enhanced I/M programs. The purpose of enhanced I/M programs is to reduce HC, CO, and NOx emissions from in-use vehicles. Enhanced I/M programs include these features: road simulation tests run on a dynamometer for tailpipe emissions, evaporative emission system tests, more stringent program enforcement, and a $450.00 waiver for repair requirements.

In 1995, the United States Congress passed an amendment to the National Highway Systems Designation Act that allowed the state authorities considerable flexibility in designing enhanced I/M programs. As a result of this legislation, a variety of enhanced I/M programs have been developed and introduced.

Types of Inspection/Maintenance (I/M) Programs

Most enhanced I/M programs include a visual pre-inspection safety check **(Figure 40-8)**. This safety check includes the following items:

1. Tires—The vehicle is rejected if any tire is excessively worn or the space-saver spare tire is mounted on one of the wheels. The tires must be safe to operate on the dynamometer during the enhanced emission test.

I/M INSPECTION SHEET

ELECTRICAL SYSTEM CHECKS

☐ Visual inspection
battery wiring

☐ Battery condition
_____ Clean
_____ Corroded
_____ Damaged
_____ State of charge

LIGHTS

☐ _____ Park   _____ Emergency
_____ Signal   _____ Dash
_____ Brake   _____ Back-up

Headlights

_____ High left
_____ Low left
_____ High right
_____ Low right
_____ Horn operation

COOLING SYSTEM CHECKS

☐ _____ Level
Anti-freeze (protection to _____°)

Condition of hoses
_____ Radiator
_____ Heater
_____ Thermostat bypass

_____ Pressure test cap
_____ Pressure test radiator
_____ Water pump and belt

BRAKE SYSTEM INSPECTION

☐ _____ Pedal travel
_____ Emergency brake
_____ Brake hoses and lines
_____ Master / wheel cylinder leakage
_____ Fluid level

FUEL SYSTEM CHECKS

☐ _____ Condition of hoses
_____ Gas cap condition
_____ Air cleaner
_____ Crankcase vent filter
_____ Fuel filter (miles until change)

ON-GROUND
STEERING / SUSPENSION

☐ _____ Steering wheel free play
_____ Power steering fluid level
_____ Shock absorber bounce test

	Good	Bad
Front	_____	_____
Rear	_____	_____

UNDER CAR
STEERING / SUSPENSION

☐ _____ Inspect steering linkage
_____ Inspect shock absorbers
_____ Inspect suspension bushings
_____ Inspect ball joints
_____ Wear
_____ Seals
_____ Ride height

TIRES

☐ Inflate to _____ lb.

Tire condition

	Good	Fair	Unsafe
RF	_____	_____	_____
LF	_____	_____	_____
RR	_____	_____	_____
LR	_____	_____	_____

_____ Check spare

EXHAUST

☐ _____ Mufflers and pipes
_____ Pipe hangers
_____ Exhaust leaks
_____ Heat riser

DRIVE LINE CHECKS

☐ _____ Check universal or CV joints

Inspect gear cases
_____ Transmission
_____ Transfer (4X4)
_____ Differential

_____ Inspect motor mounts
_____ Check ATF level

Figure 40-8 Typical I/M vehicle inspection sheet.

2. Brakes—If defective brake conditions prevent the vehicle from maintaining the drive trace during an enhanced emission test, the vehicle is rejected.

3. Exhaust system—Vehicles with leaking exhaust systems are rejected because this condition affects tailpipe emission readings and the test operators may be subjected to excessive CO.

4. Steering and suspension—The vehicle should be rejected if any suspension or steering component is worn or damaged so its normal function is adversely affected.

5. Fuel system—Vehicles with fuel system leaks are rejected because they are a fire hazard.

6. Instrument readings—Vehicles that are overheating or leaking coolant should be rejected.

An emissions component inspection is also performed. This inspection includes tampering. Included in the emissions component inspection are the fuel inlet restrictor in the fuel tank, the catalytic converter, and emission control items in the under-hood area. The emissions component inspection also includes a gas tank filler cap inspection and test on applicable vehicles. The inspector tests the filler cap with a special pressurized tester.

Basic Idle Test

The basic idle test is performed with the engine at normal operating temperature and an exhaust gas analyzer pickup inserted in the tailpipe. This analyzer must be properly calibrated. This type of analyzer uses infrared rays to measure the HC, CO, and NOx emission levels in the exhaust. The transmission selector must be in PARK for an automatic transmission or in NEUTRAL for a manual transmission. The emission levels are recorded at idle. The next step is to record the emission levels after engine operation at 2,500 rpm for a specified time period. The engine is then returned to idle and the emissions recorded again. In some jurisdictions, the pass requirement demanded that the tailpipe emissions remain below applicable emission standards at all three readings. In other areas, the pass requirement was based on the lowest of the two emission readings taken at idle. Jurisdictions establish

emission test standards based on the standards for the vehicle year being tested.

IM240 Test

The IM240 test is an enhanced emission test that requires the use of a dynamometer and a sophisticated constant volume sampling (CVS) exhaust emissions analyzer **(Figure 40-9)**. All the exhaust is routed through the CVS analyzer. During the IM240 test, the vehicle is safety-locked onto the dynamometer and the inspector drives the vehicle to maintain a bright dot on the IM240 drive trace displayed on a computer monitor **(Figure 40-10)**. The inspector must drive the vehicle to maintain the bright dot on the drive trace. If the bright dot goes off the drive trace to any extent, the computer aborts the test and the test procedure must be repeated.

Figure 40-9 A typical IM240 emission test station.

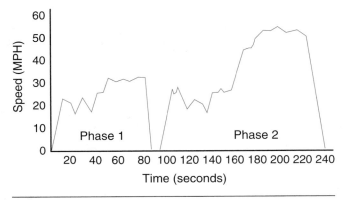

Figure 40-10 An IM240 emission test drive trace.

Completing the drive trace requires 240 seconds and includes a wide variation in operating conditions, including acceleration and deceleration and speeds up to 56 mph (90 kmh).

> ***Did You Know?*** *The IM240 test procedure is taken from various parts of the Federal Test Procedures (FTP). Compared to the IM240 test, the FTP is a much longer test procedure run on a dynamometer and uses more sophisticated exhaust emission measurement equipment. The EPA uses the FTP to certify new vehicles sold in the United States.*

After the drive cycle is completed, the emissions analyzer prints out a trace for HC, CO, and NOx. All readings are in grams per mile (GPM). These traces show the cut point for each pollutant and the actual emission level throughout the test **(Figure 40-11)**. When diagnosing the cause of high emissions, these emission traces can be very helpful because they indicate the driving conditions that produced the high emission level above the cut point. The pass emission requirements are set by each jurisdiction and are based on EPA guidelines. During the IM240 test, the emissions tester measures the purge flow in the evaporative system.

BAR-31 Test

The BAR-31 test requires the same equipment as the IM240 test and the exhaust emissions are measured in the same way. The BAR-31 test requires only 31 seconds because it has a simpler drive trace. In most jurisdictions an emissions pass/fail decision is based on three BAR-31 tests.

Acceleration Simulation Mode (ASM) Test

The acceleration simulation mode (ASM) test is a steady-state test completed with a constant throttle opening. During the test, the vehicle is driven on a dynamometer at a constant speed and load setting. Two different ASM tests are used in emission testing. During the ASM 5015 test, the vehicle is driven at 15 mph (24 kmh) and at 50 percent of the maximum acceleration load during the federal test procedure cycle. The ASM 2525 test measures emissions at 25 mph (40 kmh) and at 25 percent of the maximum acceleration load during the federal test procedure cycle.

The tester used to record the emissions during an ASM test uses the same technology as the tester for the idle tests. The tester measures HC and NOx in parts per million and CO in percentage.

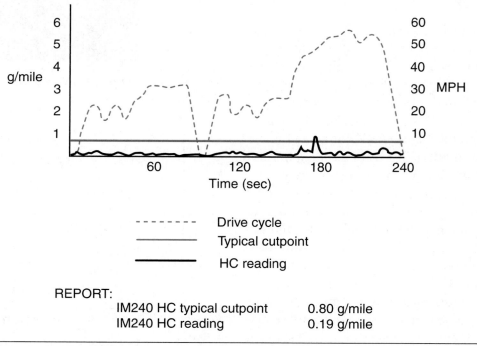

REPORT:
IM240 HC typical cutpoint	0.80 g/mile
IM240 HC reading	0.19 g/mile

Figure 40-11 A hydrocarbon trace from an IM240 test.

A significant number of states use the ASM test procedure for emission testing.

NOTE: The ASM tester may be called a *BAR-97 analyzer*. Exhaust gas analyzers are usually built to specifications developed by the California Bureau of Automotive Repair (BAR). BAR specifications are labeled by the year they became effective. For example, specifications for the BAR-97 became effective in 1997.

Many jurisdictions demand an OBD II check on applicable vehicles during an emissions test. This OBD II test usually involves checking for proper operation of the MIL with the ignition switch on and with the engine running. The OBD II test also involves connecting a scan tool to the DLC and determining that all the monitors have completed their function and there are no DTCs in the PCM memory. Since OBD II systems accurately monitor emissions levels at all times, several states have considered using the OBD II system to do emission system testing.

Did You Know? *Due to the accuracy of current OBDII systems for monitoring emissions, proposals in several variations for OBDIII systems have the car doing its own testing. If a failure is detected, the vehicle sends a report to a governing body to monitor that repairs are made.*

Summary

- Photochemical smog is formed by sunlight reacting on HC and NOx in the atmosphere.
- The main component in photochemical smog is ozone.
- The main regulated emissions are HC, CO, and NOx.
- Vehicle emissions may come from the tailpipe, crankcase, or fuel tank.
- HC emissions are a result of unburned fuel after the combustion process.
- CO emissions are caused by incomplete burning of the air-fuel mixture.
- NOx emissions are caused by oxygen and nitrogen combining at combustion chamber temperatures above 2,500°F (1,371°C).

- Emission standards are implemented and administered by the EPA, but states may introduce their own, more stringent emission standards.
- A vehicle safety inspection and emissions component inspection are performed before an emission test.
- A basic idle emission test includes emissions recorded at idle speed and at 2,500 rpm.
- Enhanced emission tests are run with the vehicle on a dynamometer.
- Enhanced emission tests include the IM240 test, BAR-31 test, and ASM test.

Review Questions

1. Technician A says that HC emissions increase as the air-fuel ratio becomes richer. Technician B says cylinder misfiring causes high HC emissions. Who is correct?

   A. Technician A
   B. Technician B
   C. Both Technician A and Technician B
   D. Neither Technician A nor Technician B

2. Technician A says NOx emissions are formed at low combustion chamber temperatures. Technician B says NOx is one of the pollutants that form smog. Who is correct?

A. Technician A

B. Technician B

C. Both Technician A and Technician B

D. Neither Technician A nor Technician B

3. Technician A says the PCV system reduces crankcase emissions. Technician B says crankcase emissions contain mainly HC. Who is correct?

A. Technician A

B. Technician B

C. Both Technician A and Technician B

D. Neither Technician A nor Technician B

4. Technician A says most evaporative emissions come from the fuel tank. Technician B says evaporative emissions contain NOx and CO. Who is correct?

A. Technician A

B. Technician B

C. Both Technician A and Technician B

D. Neither Technician A nor Technician B

5. Photochemical smog is formed by the reaction of sunlight with:

A. CO and CO_2.

B. NOx and HC.

C. HC and CO.

D. NOx and CO_2.

6. All of these statements about HC emissions are true *except*:

A. HC tailpipe emissions are caused by unburned air-fuel mixture in the combustion chamber.

B. When the air-fuel ratio becomes richer, the HC emissions increase.

C. In a typical vehicle, 90 percent of the HC emissions come from the tailpipe.

D. Approximately 27 percent of the total HC emissions come from vehicles.

7. When discussing CO emissions, Technician A says CO emissions increase when one cylinder is misfiring. Technician B says CO emissions increase when the air-fuel ratio becomes richer than 14.7:1. Who is correct?

A. Technician A

B. Technician B

C. Both Technician A and Technician B

D. Neither Technician A nor Technician B

8. NOx emissions increase considerably when combustion chamber temperatures are above:

A. 800°F.

B. 1,200°F.

C. 1,650°F.

D. 2,500°F.

9. The most likely cause of excessive HC emissions from the crankcase is:

A. worn piston rings.

B. worn valve stem seals.

C. a PCV valve that is stuck open.

D. a lean air-fuel ratio.

10. While discussing emission test procedures, Technician A says during a BAR-31 emission test the vehicle is driven at a constant speed on a dynamometer. Technician B says during an IM240 emission test the vehicle is operated on a dynamometer at varying speed and load conditions. Who is correct?

A. Technician A

B. Technician B

C. Both Technician A and Technician B

D. Neither Technician A nor Technician B

11. Emission I/M facilities operated by a government are called _____ facilities.

12. An idle emission test is performed at idle speed and _____ rpm.

13. During an IM240 emission test, CO, HC, and NOx are recorded in _____ _____.

14. During a BAR-31 emission test, the vehicle is run on a _____.

15. Describe the IM240 emission test procedure.

16. Describe the ASM emission test procedure.

17. List the items that are inspected during an emissions component inspection.

18. Describe a BAR-31 emission test procedure.

41 Emission Systems

Learning Objectives

After you have read, studied, and practiced the contents of this unit, you should be able to:

- Explain the operation of the PCV system.

- Describe the operation of a typical EGR valve.

- Describe the operation of positive backpressure and negative backpressure EGR valves.

- Explain the operation of a pulsed air injection system.

- Describe the operation of belt-driven and electric-drive air pumps.

- Explain the operation of conventional and enhanced EVAP systems.

- Describe the operation of oxidation, dual-bed, and three-way catalytic converters.

Key Terms

Catalyst

Evaporative (EVAP) system

Exhaust gas recirculation (EGR) system

Monolith-type catalytic converter

Pellet-type catalytic converter

Pressure feedback electronic (PFE) sensor

Pulse width modulated (PWM) signal

INTRODUCTION

Improper emission system operation may result in high emissions, drivability complaints, and reduced fuel economy. When diagnosing emission system problems, the first requirement is to understand the purpose of these systems and how they operate. Not understanding the operation of emission control systems and their diagnosis will likely result in time-consuming and inaccurate results.

POSITIVE CRANKCASE VENTILATION (PCV) SYSTEM DESIGN AND OPERATION

The Positive Crankcase Ventilation (PCV) system removes crankcase gases containing HCs from the engine and directs them into the intake manifold. This action prevents crankcase gases containing HCs from escaping into the atmosphere. The tapered, spring-loaded PCV valve is mounted so the intake manifold vacuum pulls this valve toward the closed position and the spring tension pushes the valve open. The PCV system pulls clean air from the air cleaner into the engine.

Tech Tip *Blowby gas flows past the piston rings into the crankcase. There is always a small amount of blowby even in a new engine.*

The PCV valve usually fits snugly into a rubber grommet in the rocker arm cover. A hose is connected from the PCV valve to the intake manifold. A clean air inlet hose is connected from the rocker arm cover to the air cleaner. In a V6 or V8 engine, the clean air hose is in the opposite rocker arm cover from the PCV valve **(Figure 41-1)**. A filter is positioned on the air cleaner end of the clean air hose and a steel mesh is mounted in the fitting where the clean air hose enters the rocker arm cover.

When the engine is idling, the high manifold vacuum pulls the PCV valve toward the closed position. At idle speed, the engine produces less blowby gas and a reduced PCV valve opening is adequate under this condition. If the engine is operating at part throttle, the manifold vacuum decreases slightly and the spring pushes the PCV valve open to allow additional blowby gases to be pulled from the crankcase into the intake manifold **(Figure 41-2)**. When the engine is operating at wide-open throttle, the manifold vacuum decreases and the spring forces the PCV valve further toward the open position **(Figure 41-3)**. This action allows the additional

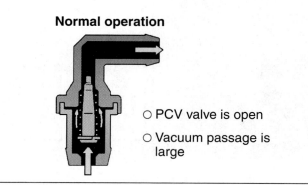

Normal operation

○ PCV valve is open

○ Vacuum passage is large

Figure 41-2 PCV valve position at part throttle.

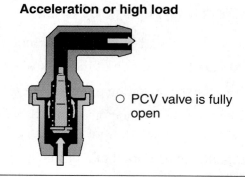

Acceleration or high load

○ PCV valve is fully open

Figure 41-3 PCV valve position at wide open throttle.

blow-by gas produced under this condition to be pulled through the PCV valve into the intake manifold and burned in the combustion chamber.

If the engine backfires into the intake manifold, the filter and steel mesh screen in the clean air hose prevent any flame front from entering the engine through the clean air hose. An engine backfire into the intake manifold seats the PCV valve against the valve housing and prevents any flame front from entering the engine through the PCV valve **(Figure 41-4)**.

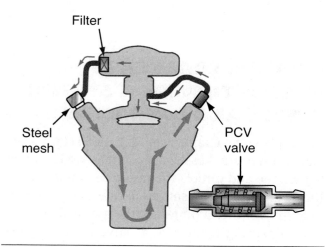

Figure 41-1 Airflow through a PCV system.

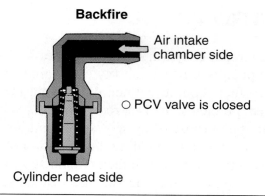

Backfire

Air intake chamber side

○ PCV valve is closed

Cylinder head side

Figure 41-4 PCV valve position during engine backfire.

EXHAUST GAS RECIRCULATION (EGR) SYSTEM DESIGN AND OPERATION

The **Exhaust Gas Recirculation (EGR) system** is designed to allow a small amount of relatively inert exhaust gas to be returned to the intake manifold and put into the cylinder with the incoming air-fuel mixture. Adding a small amount of exhaust gas effectively cools the combustion chamber and performs two functions: prevents NOx formation and reduces detonation. In this chapter, we will deal exclusively with the EGR's reduction of NOx.

Figure 41-6 An EGR valve solenoid.

NOTE: The EGR solenoid may be called an *electronic vacuum regulator* (*EVR*).

A tapered valve is mounted in the lower end of the EGR valve that is linked to a diaphragm mounted in a sealed chamber on top of the valve **(Figure 41-5)**. Manifold or port vacuum is usually supplied through a PCM-operated solenoid and vacuum hose to open the EGR valve diaphragm **(Figure 41-6)**. Earlier systems often relied upon ported vacuum to operate with no PCM control.

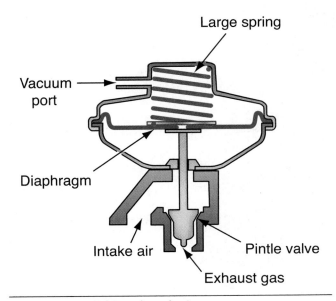

Figure 41-5 EGR valve design.

These systems used vacuum passages that had little to no vacuum at idle and would supply full vacuum only at RPMs above a preset level to avoid idle running problems.

For all late-model EGR systems, when exhaust gas recirculation is required to lower NOx emissions, the PCM energizes the EVR solenoid. Under this condition, the solenoid opens and supplies vacuum to the EGR valve. When vacuum is supplied to the EGR valve diaphragm, the EGR valve opens and allows a specific amount of exhaust gas to recirculate into the intake manifold. The PCM opens the EGR valve under certain engine operating conditions. The PCM does not open the EGR valve when the engine coolant is cold because NOx emissions are not a problem on a cold engine. The EGR valve is not opened at idle because this results in rough idle and engine stalling. The PCM usually opens the EGR valve when the engine is operating at normal temperature and speeds above approximately 35 mph. The PCM does not open the EGR valve at wide-open throttle because the flow of exhaust gas through the EGR valve into the cylinders tends to reduce engine power because a rich fuel mixture burns cooler and it is not necessary.

On some EGR systems, the PCM pulses the EVR solenoid on and off to supply the precise exhaust flow required to lower NOx emissions without creating drivability problems. Other types of EGR systems have an orifice located in the exhaust

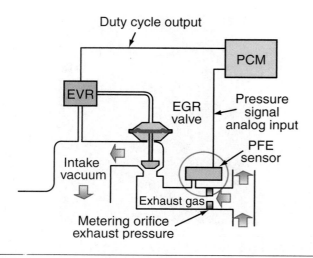

Figure 41-7 An EGR valve system with metering orifice and PFE sensor.

stream near the EGR valve **(Figure 41-7)**. This orifice creates a pressure drop and a small exhaust pipe is connected from the area between the orifice and the EGR valve to the **pressure feedback electronic (PFE) sensor**. The PFE sensor sends a voltage signal to the PCM in relation to the amount of exhaust flow through the EGR system.

If the exhaust flow through the EGR system does not match the EGR flow demanded by the PCM input signals, the PCM makes a correction in EGR flow.

Positive Pressure EGR Valves

Some EGR systems have a positive backpressure-type EGR valve. These EGR systems have a PCM-operated solenoid that supplies vacuum to the EGR valve, but this type of valve has an internal, normally open control valve **(Figure 41-8)**. Exhaust gas is supplied from the lower end of the valve through a hollow valve stem to the control valve. With the engine idling, exhaust pressure is not high enough to close the control valve and this valve remains open. When the control valve is open, any vacuum in the diaphragm chamber is bled off through the valve. When the vehicle speed reaches approximately 35 mph (56 kmh), the exhaust pressure forces the control valve closed. Under this condition, vacuum supplied through the EGR valve solenoid lifts the diaphragm and opens the EGR valve.

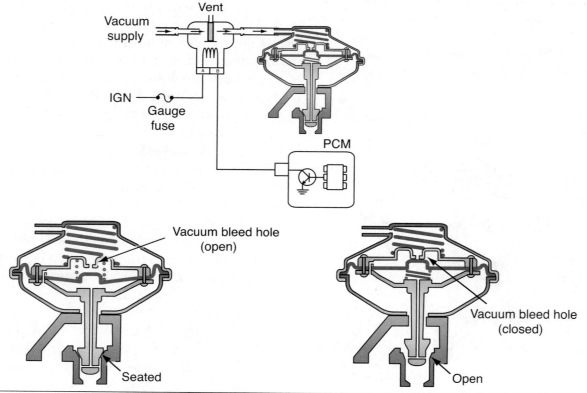

Figure 41-8 A positive backpressure EGR valve.

Negative Pressure EGR Valves

NOTE: High-pressure exhaust pulses may be called *positive pulses* and low-pressure exhaust pulses may be referred to as *negative pulses*.

Other EGR valves have a negative backpressure-type EGR valve with an internal, normally closed control valve **(Figure 41-9)**. Exhaust pressure is supplied through a hollow valve stem to the control valve. Each time a cylinder fires there is a high-pressure pulse in the exhaust, but between these high-pressure pulses there is a low pressure. At idle speed, there is a longer time between the high-pressure pulses in the exhaust and the low-pressure pulses are more predominant. These low-pressure pulses pull the control valve open in the EGR valve. Any vacuum supplied to the diaphragm is bled off through the control valve. When the vehicle speed is approximately 35 mph (56 kmh), the high-pressure pulses in the exhaust are closer together and the negative exhaust pulses are less dominant. Under this condition, the EGR control valve closes and the vacuum supplied through the solenoid to the EGR valve diaphragm opens the EGR valve. Positive or negative type EGR valves are identified by an "N" or a "P" stamped on the top of the EGR valve housing **(Figure 41-10)**.

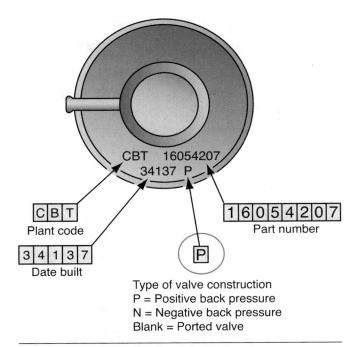

Figure 41-10 EGR valve identification.

Electronic EGR Valves

Many vehicles are now equipped with electronic EGR valves **(Figure 41-11)**. The PCM pulses the EGR valve winding on and off to provide the precise EGR flow required by the engine under all operating conditions.

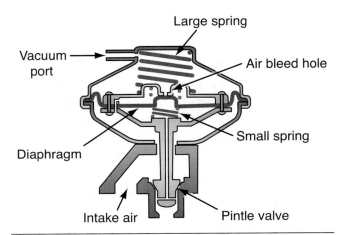

Figure 41-9 A negative backpressure EGR valve.

Figure 41-11 An electronic-type EGR valve.

AIR INJECTION SYSTEM DESIGN AND OPERATION

Some older vehicles have a pulsed air injection system. In this system, a pipe with a one-way check valve is connected into each exhaust port. These basic systems' main function is to reduce HC at the exhaust by adding additional air into the exhaust to help finish the combustion process. All the one-way check valves are attached to a common reservoir and a clean air hose is connected from this reservoir to the air cleaner **(Figure 41-12)**. At low engine speeds, the low-pressure pulses in the exhaust manifold open the one-way check valves and pull air into the exhaust manifold. The oxygen in the air mixes with the HC emissions in the exhaust manifold. This mixture ignites and burns in the exhaust manifold to reduce HC emissions from the tailpipe. When engine speed reaches a specific rpm, the pulsed air injection system becomes ineffective because the low-pressure pulses in the exhaust are greatly reduced.

Some vehicles have a belt-driven air pump. In some of these pumps, air enters the pump through a centrifugal filter behind the pump pulley **(Figure 41-13)**. Air is delivered from the pump to a bypass valve and a diverter valve **(Figure 41-14)**. These valves are opened and closed by vacuum supplied through vacuum/electric solenoids operated by the PCM. When the engine coolant is cold, the PCM does not energize either of the vacuum/electric solenoids. Under this condition, air from the pump is

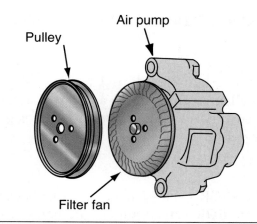

Figure 41-13 A belt-driven air pump.

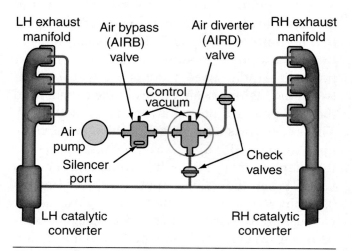

Figure 41-14 A belt-driven air pump system.

bypassed to the atmosphere at the bypass valve. When the engine coolant reaches a specific temperature, the PCM energizes the bypass valve solenoid, which supplies vacuum to the bypass valve. This action opens the bypass valve and allows air from the pump to flow through the bypass valve to the diverter valve. The PCM also energizes the diverter valve solenoid, which supplies vacuum to the diverter valve. Under this condition, the diverter valve is positioned so it delivers air from the pump to the exhaust ports. This airflow into the exhaust ports provides oxygen to mix with unburned hydrocarbons in the exhaust manifold. When this mixing occurs, the hydrocarbons are hot enough to ignite and burn. This action reduces hydrocarbon emissions during engine warm-up. One-way check valves in the hoses connected to the exhaust manifold pipes prevent exhaust gas from blowing back into the air pump system when the air pump is not blowing

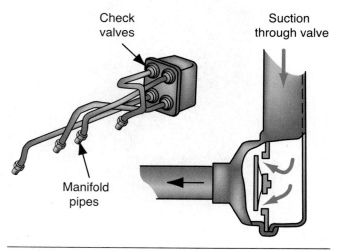

Figure 41-12 A pulsed air injection system.

air into the exhaust manifolds. If exhaust gas entered the air pump system, the pipes and hoses would be severely burned and damaged. When the engine reaches normal operating temperature, the PCM de-energizes the diverter valve solenoid and shuts off vacuum to the diverter valve. Under this condition, the diverter valve is positioned so the airflow from the pump is directed to the catalytic converter. This airflow into the catalytic converter is essential for efficient converter operation.

Some newer vehicles have an electric-drive air pump **(Figure 41-15)**. Air is delivered from the pump to an air bypass valve and a diverter valve. These valves are controlled by vacuum supplied through a solenoid operated by the PCM. The air bypass and diverter valves operate in basically the same way as in the belt-driven air pump system. During engine warm-up, the PCM energizes the diverter valve solenoid, which supplies vacuum to the diverter valve. Under this condition, the diverter valve is positioned to deliver air pump air to the exhaust ports. When the engine coolant reaches normal operating temperature, the PCM shuts off the solenoid and vacuum supply to the diverter valve. Under this condition, the diverter valve is positioned so it delivers air from the pump to the catalytic converter **(Figure 41-16)**. The operation of the air pump may vary depending on the engine application. Keep in mind that not all vehicles have an air pump; most newer vehicles do not because newer designed catalytic converters often do not need them in order to work efficiently.

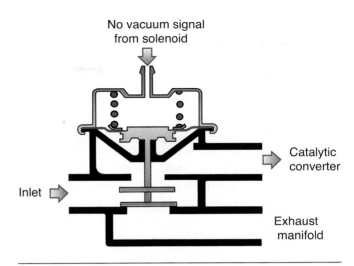

Figure 41-16 A diverter valve for an electric-drive air pump.

EVAPORATIVE (EVAP) EMISSION SYSTEM DESIGN AND OPERATION

The **evaporative (EVAP) system** reduces the volume of fuel vapors containing HC that escape to the atmosphere. In a typical EVAP system, a hose is connected from the top of the fuel tank to a charcoal canister that is usually located in the engine compartment **(Figure 41-17)**. The fuel tank has a domed top to allow fuel vapors to collect in the top of the tank. The tank pressure control valve (TPCV) is located in the hose between the canister

Figure 41-15 An electric-drive air pump.

Figure 41-17 A charcoal vapor recovery canister.

and the fuel tank. With the engine running, intake manifold vacuum is supplied to the TPCV. Under this condition, the valve opens and allows fuel vapors to flow from the fuel tank into the canister. With the engine stopped and no vacuum supplied to the TPCV, this valve is closed. If the tank pressure exceeds a specific value, as may happen on a hot summer day, the pressure pushes the TPCV open and allows vapors to flow into the canister. The fuel tank is equipped with a pressure/vacuum valve **(Figure 41-18)**. This valve is similar to a radiator cap, but it operates at a much lower pressure or vacuum.

A normally closed purge solenoid is mounted in the hose between the canister and the intake

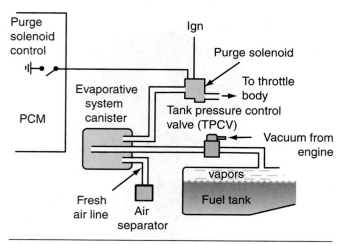

Figure 41-18 A fuel tank filler cap with pressure relief and vacuum valves. Courtesy of Steffy Ford, Columbus NE.

Figure 41-19 Atypical evaporative (EVAP) system.

manifold **(Figure 41-19)**. When the purge solenoid is closed, fuel vapors from the tank flow into the canister. The charcoal in the canister holds the fuel vapors. The PCM energizes and opens the purge solenoid when the engine is at or near normal operating temperature and the vehicle speed and engine speed and are above specific values. When the purge solenoid is open, manifold vacuum is supplied to the canister. This vacuum purges fuel vapors out of the canister into the intake manifold to be added to the incoming air/fuel mixture. As the canister is purged, air flows through the air separator into the canister.

Newer vehicles must meet more stringent evaporative standards and these vehicles have enhanced EVAP systems. Some enhanced EVAP systems have the canister mounted near the fuel tank and a vent valve positioned in the air inlet to the canister **(Figure 41-20)**.

An enhanced EVAP system has the capability to check for leaks in the system. These systems are used on OBD II systems.

A fuel tank pressure sensor is mounted in the top of the fuel tank and is connected electrically to the PCM. This sensor sends a voltage signal to the PCM in relation to fuel tank pressure. The fuel tank vent valve replaces the previous vent valve in the filler cap. A conventional purge solenoid is mounted in the engine compartment. The PCM operates the purge solenoid with a **pulse width modulated (PWM) signal** to control the vapor flow from the canister to the intake manifold. The PCM

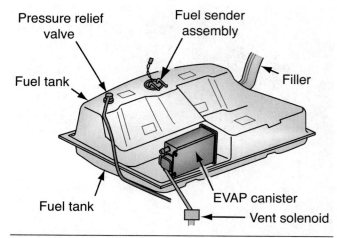

Figure 41-20 An enhanced evaporative (EVAP) system.

operates the purge solenoid and the vent solenoid to maintain the specified fuel tank pressure. A PWM signal is a digital signal with a variable on time. Periodically, the PCM tests the evaporative system for leaks. To test the system, the PCM de-energizes and closes the purge solenoid and then energizes the vent solenoid to seal the system. The PCM then scans the fuel tank pressure sensor signal for any decrease in fuel tank pressure, which indicates a leak in the evaporative system. If a leak is detected on two occasions, the PCM sets a DTC in memory. Depending on the year of the vehicle, the PCM must be able to detect a leak equivalent to a 0.040-inch (1.016-millimeter) or 0.020-inch (0.508-millimeter) diameter opening in the EVAP system. Some EVAP systems on OBD II vehicles have an on-board refueling vapor recovery (ORVR) system. The ORVR system reduces the escape of fuel vapors during refueling. The fuel tank filler neck has a smaller diameter neck that provides a liquid seal in the filler neck to prevent fuel vapors from escaping while refueling. A one-way check valve in the end of the filler pipe allows fuel to flow into the tank, but this valve prevents fuel from spitting back out of the filler neck. While refueling, fuel vapors flow through the fill limiter vent valve (FLVV) on top of the fuel tank into the canister **(Figure 41-21)**. A larger canister is used on a vehicle with an ORVR system. The FLVV closes and prevents fuel from escaping in a vehicle rollover situation.

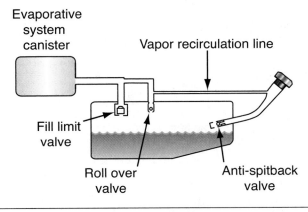

Figure 41-21 An on-board refueling vapor recovery (ORVR) system.

CATALYTIC CONVERTER DESIGN AND OPERATION

HC, CO, and NOx emissions can be converted to harmless gases at temperatures above 1,832°F (1,000°C). However, it is impossible to maintain this exhaust gas temperature. A catalyst in the exhaust system containing platinum (Pt), palladium (Pd), and/or rhodium (Rh) allows conversion of HC, CO, and NOx to harmless gases at temperatures of 572° to 1,652°F (300° to 900°C). This temperature range can be achieved in the exhaust system. A **catalyst** is a material that accelerates a chemical reaction without being changed itself.

All types of catalytic converters must reach a temperature of approximately 600°F (315°C) to light off and start the chemical reaction that oxidizes and reduces pollutants. Catalytic converters may be pellet type or monolith type. A **pellet-type catalytic converter** contains somewhere between 100,000 and 200,000 small pellets. The design of a **monolith-type catalytic converter** is similar to a honeycomb. All newer model catalytic converters are of the monolith-type of design.

A monolith catalytic converter has a surface area about the size of ten football fields or 500,000 square feet. Metal monoliths are manufactured from an iron, chrome, and aluminum alloy. A material called corierite is used in ceramic monoliths. An aluminum oxide called alumina is sprayed on the ceramic or metal monoliths. The alumina is also sprayed on the pellets in a pellet-type converter. The alumina is highly porous and contains many small openings called micropores. Base metals such as cerium (Ce) or iron (Fe) are dispersed in the alumina to help prevent shrinkage. The monolith or pellets are coated with the noble metals platinum, palladium, and/or rhodium. These noble metals are dispersed into the alumina. The base metals in the alumina can chemically absorb O_2 and CO.

Oxidation Converter Operation

The first catalytic converters installed on vehicles in the 1970s were oxidation converters. This type of converter oxidizes HC and CO into CO_2

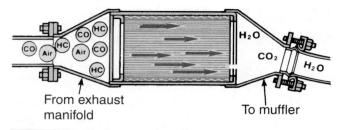

Figure 41-22 An oxidization-type catalytic converter.

and H_2O **(Figure 41-22)**. The oxidation converter does not reduce NOx emissions, so these engines are equipped with an EGR valve to reduce this pollutant. On engines with an oxidation catalytic converter, the fuel system is calibrated to deliver an air-fuel ratio that is leaner than the 14.7:1 stoichiometric air-fuel ratio. This lean air-fuel ratio provided additional oxygen for the converter.

Dual-Bed Converter Operation

Beginning in the mid-1980s, most engines have a dual-bed converter. This type of converter has a dual bed containing two monoliths or two pellet beds. The first bed in a dual-bed converter contains rhodium and palladium. This bed reduces NOx to N_2 and CO_2 **(Figure 41-23)**. The second bed in a dual-bed converter contains platinum and palladium. This bed oxidizes HC and CO to H_2O and CO_2. Most engines with a dual-bed converter have an air pump that delivers air to the converter between the two beds. This airflow provides additional oxygen to help oxidize the HC and CO in the rear bed. On an engine with a dual-bed converter, the fuel system is calibrated to deliver an air-fuel ratio of 14.6:1, which is slightly richer than stoichiometric. A typical dual-bed converter reduces HC, CO, and NOx emissions by 80 percent.

Three-Way Converter Operation

Beginning in the 1990s, many vehicles were equipped with a three-way catalytic converter. These converters contain platinum and/or palladium and rhodium in a single bed. The three-way converter operates in a similar manner to the dual-bed converter. On engines with a three-way converter, the fuel system must deliver an average air-fuel ratio

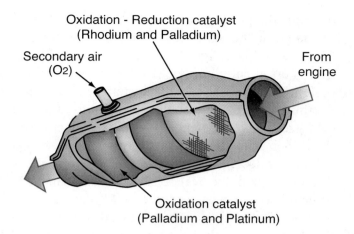

From engine		Between chamber	Exhaust
CO	0.1 - 1.0%	0 - 0.5%	0
HC	50 - 200 ppm	0 - 50	0
CO2	13.6 - 14.3%	14.7 -15.5%	9.6 - 12%
O2	0.3 - 0.7%	0 - 0.5%	2.5 - 5.5%
NOx	1200 ppm	120	120
			Water vapor
			Nitrogen

Figure 41-23 A dual-bed catalytic converter.

that is at or very close to the stoichiometric air-fuel ratio of 14.7:1. When this air-fuel ratio is available, the converter operates at peak efficiency in oxidizing HC and CO and reducing NOx **(Figure 41-24)**.

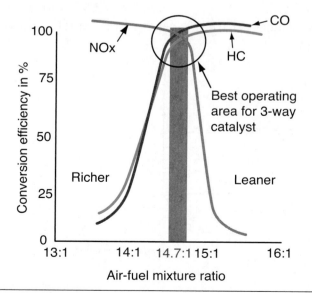

Figure 41-24 A three-way catalytic converter requires a average 14.7:1 air-fuel ratio for efficient converter operation.

If the air-fuel ratio is leaner than stoichiometric, combustion temperature increases and the converter does not reduce all the NOx. When the air-fuel ratio is richer than stoichiometric, the converter does not oxidize all the HC and CO. If the engine is operating at a 14.7:1 air-fuel ratio, a three-way converter provides a 90 percent reduction in HC, CO, and NOx emissions.

The air/fuel ratio is cycled between slightly rich to slightly lean in order to provide the optimum operating environment for the catalysts to work. When the air-fuel ratio cycles slightly to the lean side of 14.7:1, there is more oxygen in the exhaust stream. The base metals in the converter temporarily absorb this oxygen. When the air-fuel ratio cycles slightly to the rich side of 14.7:1, there is less oxygen in the exhaust stream and the base metals in the converter release the stored oxygen. The oxygen storage capacity of the converter must be matched to the engine cubic inch displacement (CID) and the fuel system calibration. Going slightly rich allows the catalytic converter to reduce the NOx emissions. When it goes to the slightly rich ratio the catalytic converter will reduce CO and HC emissions. The ratio is cycled back and forth because the conditions to reduce CO and HC are the opposite that are required to reduce NOx. Cycling the fuel ratio back and forth allows one catalytic converter to do both jobs.

Summary

- The PCV system prevents crankcase gases from escaping to the atmosphere.
- The EGR system recirculates exhaust gas into the intake manifold to reduce NOx emissions.
- The EGR valve is open in the cruising speed range with the engine at normal operating temperature.
- EGR valves may be conventional type, positive backpressure type, negative backpressure type, or electronically operated.
- A pulsed air injection system is operated by low pressure pulses in the exhaust system at low engine speeds.
- A belt-driven or electric-drive air pump delivers air to the exhaust ports during engine warm up to reduce HC emissions.
- A belt-driven or electric-drive air pump typically delivers air to the catalytic converter with the engine at normal operating temperature.

- EVAP systems reduce the amount of fuel vapors that escape from the fuel tank to the atmosphere.
- In an enhanced EVAP system, the PCM has the capability to test for leaks in the system.
- Catalytic converters may be oxidation type, dual-bed, or three-way type.
- An oxidation converter oxidizes HC and CO into H_2O and CO_2.
- A dual-bed or three-way catalytic converter performs the same function as an oxidation converter, but these converters also reduce NOx to N_2 and CO_2.
- A three-way catalytic converter requires an average air-fuel ratio of 14.7:1.

Review Questions

1. Technician A says the PCV valve reduces NOx emissions. Technician B says the PCV valve opening increases as the throttle opening becomes wider. Who is correct?

    A. Technician A
    B. Technician B
    C. Both Technician A and Technician B
    D. Neither Technician A nor Technician B

2. Technician A says the EGR valve should be closed at idle. Technician B says a solenoid operated by the PCM supplies vacuum to the EGR valve. Who is correct?

   A. Technician A

   B. Technician B

   C. Both Technician A and Technician B

   D. Neither Technician A nor Technician B

3. Technician A says in a positive backpressure-type EGR valve, the normally-open, internal control valve is closed by exhaust pressure at a specific throttle opening. Technician B says recirculating exhaust through the EGR valve into the intake manifold reduces HC emissions. Who is correct?

   A. Technician A

   B. Technician B

   C. Both Technician A and Technician B

   D. Neither Technician A nor Technician B

4. Technician A says on some vehicles, a belt-driven air pump system delivers air to the catalytic converter during engine warm up. Technician B says a belt-driven air pump reduces HC emissions during engine warm up. Who is correct?

   A. Technician A

   B. Technician B

   C. Both Technician A and Technician B

   D. Neither Technician A nor Technician B

5. All of these statements about PCV systems are true *except*:

   A. The PCV system reduces HC crankcase emissions.

   B. Manifold vacuum pulls the PCV valve toward the closed position.

   C. When throttle opening increases, PCV valve opening decreases.

   D. The PCV valve spring moves the valve toward the open position.

6. All of these statements about EGR valves are true *except*:

   A. Vacuum is supplied to some EGR valves through a PCM-operated solenoid.

   B. The EGR valve is opened during cold engine operation.

   C. When the EGR valve is open, exhaust gas is recirculated into the intake manifold.

   D. The EGR valve reduces NOx emissions by lowering combustion temperature.

7. The PFE sensor in an EGR system:

   A. sends a voltage signal to the PCM in relation to exhaust temperature.

   B. sends a voltage signal to the PCM in relation to EGR flow.

   C. provides a higher voltage signal when the EGR valve opening is reduced.

   D. is supplied with exhaust pressure directly from the exhaust manifold.

8. When discussing various types of EGR valves, Technician A says a negative pressure EGR valve has a normally-closed, internal valve that is opened by low-pressure pulses in the exhaust system at lower engine speeds. Technician B says a positive pressure EGR valve has a normally open, internal valve that is closed by positive pressure in the exhaust system above a certain vehicle speed. Who is correct?

   A. Technician A

   B. Technician B

   C. Both Technician A and Technician B

   D. Neither Technician A nor Technician B

9. On an OBD II vehicle, the enhanced EVAP system has all of these features *except*:

   A. a pressure sensor mounted in the fuel tank.

   B. a purge solenoid connected between the canister and the intake manifold.

   C. a vent solenoid connected on the canister air intake connection.

   D. the capability to detect a 0.005-inch diameter leak in system.

10. With the engine at normal operating temperature, a fuel injection system and a three-way catalytic converter maintain the air-fuel ratio at:

    A. 14.0:1.

    B. 14.4:1.

    C. 14.7:1.

    D. 15.2:1.

11. A typical EVAP system purges vapors from the charcoal canister when the vehicle speed is above _____ mph.

12. An enhanced EVAP system has a(n) _____ sensor mounted in the fuel tank.

13. An enhanced EVAP system has the capability to detect a leak in the system equivalent to a(n) _____-inch or_____-inch opening, depending on the vehicle model year.

14. An oxidation catalytic converter oxidizes _____ and _____ into _____ and _____.

15. Explain the operation of a dual-bed catalytic converter.

16. Describe the air-fuel ratio requirements for a three-way catalytic converter.

17. Describe the operation of a negative backpressure type EGR valve.

18. Explain the operation of a typical electric-drive air pump.

CHAPTER

42

Emission System Maintenance, Diagnosis, and Service

Learning Objectives

After you have read, studied, and practiced the contents of this unit, you should be able to:

- Maintain, diagnose, and service PCV systems.
- Maintain, diagnose, and service EGR systems.
- Maintain, diagnose, and service air injection systems.
- Maintain, diagnose, and service EVAP systems.
- Maintain, diagnose, and service catalytic converters.

Key Terms

EVAP emission leak detector (EELD)

Ultrasonic leak detector

INTRODUCTION

A restricted PCV system causes pressure buildup in the engine, which results in oil vapor being forced through the clean air hose into the air cleaner. This excessive pressure in the engine may also cause oil leaks at engine gaskets and seals. If the PCV valve is sticking in the open position, excessive air flows through this valve, resulting in a lean air-fuel ratio and erratic idle operation. Proper maintenance, diagnostics, and service of the PCV system and all other emission control systems is important to maintain satisfactory engine drivability and provide proper life of system components.

PCV SYSTEM MAINTENANCE, DIAGNOSIS, AND SERVICE

PCV System Maintenance

When basic under-hood service is performed, the PCV system should be inspected. Check the clean air hose and inspect the PCV valve hose for cracks, splits, deterioration, and internal restriction.

Inspect the air cleaner and the PCV inlet air filter for oil contamination. If there is oil contamination in these components, the PCV valve or hose is restricted or the engine has excessive blowby. Replace the PCV inlet air filter if it is dirty. Be sure the PCV valve fits snugly into the rocker arm cover grommet. Remove the PCV valve and shake it beside your ear. The plunger in the valve should rattle during this action. Inspect the PCV valve for sludge deposits and replace the valve if necessary.

PCV System Diagnosis

With the engine idling, remove the oil filler cap and watch for oil vapors escaping from the oil filler opening. Accelerate the engine and continue watching for oil vapors at this location. If an excessive volume of oil vapors is escaping from the oil filler opening, the PCV system is restricted or the engine has excessive blowby. With the engine idling, place your hand over the oil filler opening and wait for 2 minutes. When you remove your hand, you should feel a vacuum in the engine. If there was no vacuum buildup in the engine, the PCV system is restricted or some of the engine gaskets and/or seals are leaking air into the engine.

> **Tech Tip** If engine gaskets allow the PCV system to pull air into the engine, some dirt particles will be suspended in the air depending on the location of the vehicle. If dirt particles enter the engine from this source, engine wear is accelerated. When engine gaskets allow air to enter the engine, they will likely allow oil to leak from the engine, especially after the engine is shut off.

Check the seals around the oil filler cap and dipstick tube because these are taken on and off on a regular basis for maintenance and can become damaged and provide an air leak. With the PCV valve disconnected from the rocker arm cover grommet, start the engine and allow the engine to idle. A hissing noise from the PCV valve indicates it is not plugged. Vacuum should be felt when your finger is placed over the end of the PCV valve. If there is very little or no vacuum at the PCV valve, check the valve, hose, and vacuum inlet at the manifold for a plugged condition.

PCV SYSTEM SERVICE

The PCV system does not require any adjustments. PCV system service involves replacing the PCV valve and the inlet air filter at the vehicle manufacturer's specified intervals. Be sure the PCV hose is in good condition. This hose must be fit tightly on the PCV valve and the intake manifold fitting. A large clip usually retains the inlet air filter to the air cleaner housing. After the inlet air filter is replaced, be sure this clamp is in place and check the fit of the clean air hose on the filter and the fitting in the rocker arm opening. If there is a steel mesh in the rocker arm clean air fitting, be sure this mesh is not contaminated. If necessary, wash this mesh in an approved solvent. When the PCV valve needs to be replaced, the properly calibrated replacement must be used. Different PCV systems have different requirements for the proper amount of spring tension and air hole size; installing the wrong one can lead to idle problems and sludge buildup in the crankcase.

EGR SYSTEM MAINTENANCE, DIAGNOSIS, AND SERVICE
EGR System Maintenance

If the EGR valve is open with the engine idling, engine idle is rough and the engine may stall. If the EGR valve is stuck open the engine may be difficult to start or may not start at all. When the EGR valve is not opening under any condition, NOx emissions are high and engine detonation may occur because the combustion chamber temperature is increased. Engine detonation will likely result in reduced fuel economy since the PCM will retard the timing to compensate. EGR system maintenance involves inspecting all the vacuum hoses in the EGR system for cracks, deterioration, oil contamination, and looseness. Be sure the vacuum hoses are properly connected. Most vehicles have an under-hood vacuum hose routing diagram to illustrate the proper vacuum hose connections. Inspect the wires on the EGR valve solenoid, PFE sensor, or electronic EGR valve for loose connections and worn insulation. If the engine has a PFE sensor, inspect the small exhaust pipe from the exhaust manifold to this sensor for leaks and kinks. PFE sensors often have a small vacuum hose that connects them to the intake manifold; carefully inspect this hose for deterioration. If replacement is needed a special type of silicon vacuum line is necessary to withstand the heat from the exhaust; regular vacuum hose will quickly melt after a short drive.

EGR System Diagnosis

To test a conventional EGR valve, disconnect the vacuum hose from the EGR valve. Connect a vacuum hand pump to the EGR valve and start the engine. With the engine idling, operate the hand pump to supply 20 in. Hg to the EGR valve. Under this condition, the EGR valve should open and the engine speed should decrease at least 100 rpm or the engine should stall **(Figure 42-1)**. If the engine slows down over 100 rpm or stalls, the EGR valve is operating. When the engine does not slow down more than 100 rpm or does not stall, the EGR valve is defective or the EGR exhaust passages are restricted.

To test a negative backpressure EGR valve, disconnect the EGR valve vacuum hose and connect a hand vacuum pump to the EGR valve. Supply 18 in. Hg to the EGR valve with the hand vacuum pump. Under this condition, the EGR valve should open and hold the vacuum for 20 seconds. With 20 in. Hg supplied to the EGR valve, start the engine. The vacuum should quickly drop to zero and the valve should close. Replace the EGR valve if it does not operate properly.

Figure 42-2 Hand-operated vacuum pump.

To test a positive backpressure-type EGR valve, start the engine and be sure the engine is at normal operating temperature. Disconnect the vacuum hose from the EGR valve and connect a hand vacuum pump to the EGR valve. When the hand pump is operated, the vacuum should be bled off through the EGR valve. Accelerate the engine to 2,000 rpm and apply 18 in. Hg to the EGR valve with the vacuum hand pump **(Figure 42-2)**. Under this condition, the EGR valve should open and the vacuum reading should be maintained on the hand pump gauge. When the engine is returned to idle speed, the vacuum should be bled off and the EGR valve should close.

Scan Tool Diagnosis of EGR Systems

Be sure the ignition switch is off and the engine at normal operating temperature. Connect a scan tool to the DLC. Some engines have a digital EGR valve with two or three EGR solenoids that operate EGR valve plungers of different sizes. The scan tool may be used to command the PCM to energize each EGR valve plunger. Select EGR control on the scan tool followed by solenoid number 1. When the PCM energizes and opens this solenoid plunger, the engine should slow down. The engine should slow down as each EGR solenoid is selected on the scan tool. If the engine does not slow down

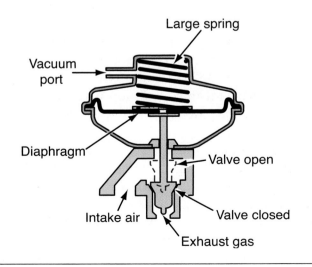

Figure 42-1 Conventional EGR valve operation.

when any EGR solenoid is energized, the solenoid winding may be open, the plunger is sticking, or the EGR passages are restricted. If the engine has a linear EGR valve, the scan tool may be used to command the PCM to provide a specific EGR valve opening. The scan tool also reads the actual percentage of EGR valve opening. If the actual EGR valve opening is not within 10 percent of the commanded opening, there is a defect in the system. On engines with digital or linear EGR valves, check for DTCs on the scan tool. If there are DTCs displayed on the scan tool representing EGR problems, the exact cause of the DTC must be diagnosed. A DTC indicates a problem in a certain area, not in a specific component. For example, if a DTC is present representing solenoid number 2 in a digital EGR valve, the solenoid winding may be defective or there may be a defect in the wires from the solenoid to the PCM. Always be sure there are 12 volts supplied to the EGR valve. The solenoid windings may be tested for opens, shorts, and grounds with an ohmmeter **(Figure 42-3)**. If the solenoid windings are satisfactory, test the wires between the EGR valve and the PCM.

> **Tech Tip** *When a scan tool indicates a solenoid is on or off, it indicates the PCM command only to the solenoid. The PCM may provide the proper command, but an electrical or mechanical defect in the solenoid may prevent the solenoid from carrying out the command. In this situation a DTC should be set, but may not always, in the PCM memory.*

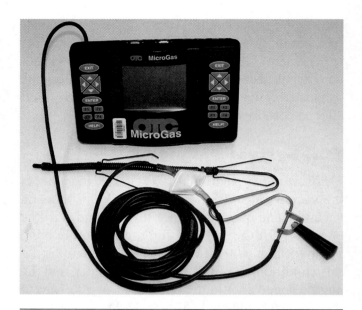

Figure 42-4 A five-gas exhaust analyzer.

The EGR system operation may be tested with a five gas analyzer **(Figure 42-4)**. If the EGR system is not operating properly, NOx emissions are high. For example, the EGR valve may be operating normally, but restricted EGR exhaust passages may be reducing EGR flow. Under this condition, NOx emissions as indicated on the five-gas analyzer are high. Under most operating conditions, NOx emissions should be below 1,000 parts per million (ppm).

EGR System Service

EGR systems do not require any periodic adjustments. EGR system service involves replacing system components after they have been diagnosed as being defective. EGR system service also includes removing carbon from the EGR valve exhaust passages if this condition is present. Some technicians may attempt to clean the tapered EGR valve, but this may not be practical because of high labor costs. When replacing an EGR valve, always install a new gasket under the valve. If the EGR passages are being cleaned, extra care should be taken to remove all of the loose deposits. Leftover deposits can cause the valve to stick open resulting in possibly worse running condition than before the service.

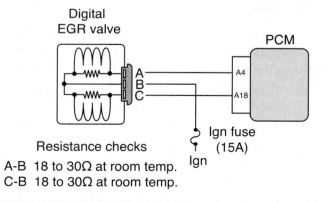

Figure 42-3 Testing digital EGR valve windings.

AIR INJECTION SYSTEM MAINTENANCE, DIAGNOSIS, AND SERVICE

Air Injection System Maintenance

On a pulsed air injection system, inspect the pipes for signs of burning, which indicate that the one-way check valves are leaking. Inspect all the pipes and hoses for leaks. If the air injection system has a belt-driven or electric-drive air pump, inspect all the air hoses and pipes for signs of burning, leaks, loose clamps, or contact with rotating components. Be sure to inspect the hose connected from the diverter valve to the catalytic converter on belt-driven air pump systems. Replace or repair the pipes and hoses as necessary. Inspect the air pump drive belt for wear, cracks, oil contamination, and proper tension. Inspect all the vacuum hoses and electric wires on the bypass and diverter valves.

Air Injection System Diagnosis

The diverter valve may be tested with the engine running at normal operating temperature and idle speed. Connect a vacuum hand pump to the diverter valve vacuum connection. With 20 in. Hg supplied to the diverter valve, the airflow from the pump should be delivered from the diverter valve outlet to the exhaust ports. When there is no vacuum supplied to the diverter valve, the airflow from the pump should be delivered from the diverter valve outlet to the catalytic converter **(Figure 42-5)**. If the air injection system has a bypass valve, it may be tested in the same way as the diverter valve. With no

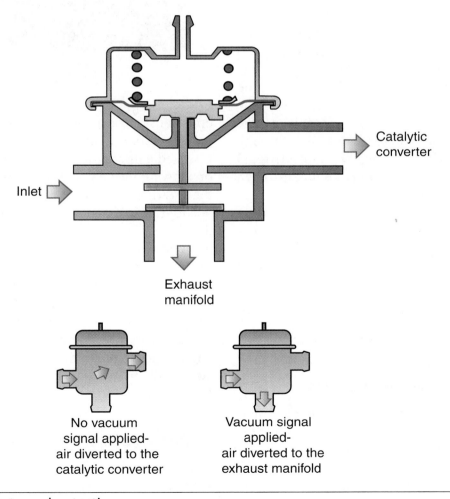

Figure 42-5 Diverter valve testing.

vacuum supplied to the bypass valve, airflow from the pump should be bypassed from the bypass valve to the atmosphere. When vacuum is supplied to the bypass valve, airflow from the pump should be delivered from the bypass valve outlet connected to the diverter valve. A scan tool may be used to diagnose belt-driven or electric-drive air pump systems. If there is an electrical defect in the bypass or diverter solenoids or in the electric drive air pump, a DTC is set in the PCM memory. The scan tool indicates the status of the bypass and diverter valve solenoids. When the vehicle is driven on a road test with the scan tool connected, the bypass and diverter valve solenoids should be energized and de-energized under the appropriate vehicle operating conditions.

Air Injection System Service

The air injection system does not require any periodic adjustments except for the belt tension on a belt-driven air pump. If a belt-driven or electric-drive air pump has internal defects, replace the pump. The PCM grounds an electric drive air pump relay winding. These relay contacts supply voltage to the electric-drive air pump. Before replacing the pump, always be sure voltage is available through the relay to the pump. The pulley and centrifugal filter may be replaced separately on a belt-driven air pump. If the bypass or diverter valve solenoids have electric defects, replace the solenoid. When a bypass or diverter valve fails to pass the tests mentioned previously, replace the valve.

EVAPORATIVE (EVAP) SYSTEM MAINTENANCE, DIAGNOSIS, AND SERVICE

EVAP System Maintenance

If the EVAP canister has a replaceable filter, this filter should be replaced at the vehicle manufacturer's specified intervals. Replace the canister if it is cracked or damaged **(Figure 42-6)**. Inspect all the EVAP system hoses for cracks, kinks, looseness, and oil contamination. Replace these hoses as required. Leaking purge hoses cause HC

Figure 42-6 An EVAP system canister.

emissions to escape to the atmosphere and cause a gasoline odor. Leaking vacuum hoses may cause rough idle operation, improper idle speed, and inoperative EVAP components. Inspect the wiring in the EVAP system for worn insulation, loose connections, breaks, and contact with rotating components. Repair the wiring as necessary. Vehicle owners who regularly overfill their gas tank run the risk of forcing fuel into the EVAP system. If no leaks are found but a strong fuel odor is present, check the charcoal canister to see if it is full of fuel. If it is, it will have to be replaced. This problem can also give performance problems when the engine tries to purge the canister of vapors and instead pulls in raw fuel, resulting in an overly rich fuel mixture.

EVAP System Diagnosis

A scan tool may be connected to the DLC to diagnose the EVAP system. If there is an electrical defect in the purge solenoid, vent solenoid, or fuel tank pressure sensor, a DTC will likely be set in

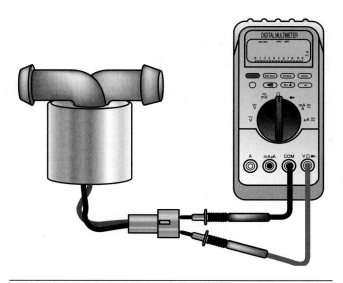

Figure 42-7 Testing the purse solenoid windings with an ohmmeter.

Figure 42-8 An EVAP system service port.

the PCM memory. If a DTC is displayed representing a fault in a solenoid, the solenoid windings may be tested with an ohmmeter to verify the solenoid condition **(Figure 42-7)**. The vehicle may be road tested with the scan tool connected to the DLC. The purge and/or vent solenoids should be energized under the appropriate vehicle operating conditions.

On vehicles with OBD II systems and enhanced EVAP systems, the PCM has the capability to test the EVAP system of a 0.040-inch (0.016 millimeter) or 0.020-inch (0.508 millimeter) diameter leak depending on the vehicle model year. If the PCM detects such a leak, a DTC is set in the PCM memory. The PCM performs this leak test as it monitors the EVAP system. To perform a large leak test, the PCM closes the vent valve and opens the purge solenoid to build vacuum in the system. When a specific vacuum is present in the EVAP system as indicated by the fuel tank pressure (FTP) sensor, the PCM closes the purge solenoid to close the system. The PCM then checks the rate of vacuum decrease in the EVAP system over a specific time period. If the vacuum decrease in the EVAP system is excessive, the PCM repeats the test. When the vacuum decrease is excessive on two tests, the PCM sets a DTC in memory. The PCM also performs a small leak test using the same procedure for a longer time period. A scan tool may be used to command the PCM to perform the large and small EVAP leak tests. Many OBD II

systems allow for individual control of the solenoids for test purposes.

If a leak is present in the EVAP system, some equipment manufacturers supply leak detection equipment that pressurizes the EVAP system with dry nitrogen. The leak detection tester is connected to an EVAP system service port in the engine compartment **(Figure 42-8)**. This service port is usually green in color. An **ultrasonic leak detector** is then positioned near any suspected EVAP system leaks. This tester produces a beep when a leak is present. When performing an EVAP system leak test, the fuel tank must be between 20 percent and 80 percent full.

> **Tech Tip** *A vacuum/pressure gauge that reads inches of water column (WC) may be connected to the EVAP system service port to check the EVAP system pressure.*

Other equipment manufacturers supply an **EVAP emission leak detector (EELD)** that produces and forces smoke into the EVAP system at the EVAP service port to locate system leaks. Smoke may be seen escaping at the EVAP system leak location. Some EELD testers have a portable light that produces a bright white beam to make smoke detection easier.

Tech Tip *When using an ultrasonic leak detector or an EELD, always allow enough time during the test for the EVAP system to become filled with nitrogen or smoke.*

EVAP System Service

If a DTC and a leak detection test indicate a leak in an EVAP system hose or component, replace the hose or component as required. When the FTP sensor voltage is not within specifications and a DTC indicates a defect in this area, always check the wires from the PCM to this sensor before replacing the sensor. In most applications, the fuel tank must be removed to access the FTP. When performing EVAP system service, always check the fuel filler cap for proper operation of the pressure relief and vacuum valves. An improperly installed fuel tank cap will often set a DTC as an EVAP leak.

CATALYTIC CONVERTER MAINTENANCE, DIAGNOSIS, AND SERVICE
Catalytic Converter Maintenance

A catalytic converter requires a minimum amount of maintenance. Inspect the outer shell of the converter for burning or damage and check the converter for exhaust leaks. If the converter has an exhaust leak or the exterior shell is severely burned or damaged, converter replacement is necessary. Be sure there are no leaks in the exhaust pipe or other exhaust system components. Use a rubber mallet to lightly tap the outer surface of the converter and listen for loose internal converter components that cause a rattling noise. If the converter has loose

Tech Tip *A continual rich air-fuel ratio causes severe overheating of the catalytic converter. This condition may cause the internal components in the converter to actually melt, which makes the converter incapable of reducing emissions. The exterior of the converter may also appear overheated and burned.*

internal components, it should be replaced because these loose components will likely restrict the exhaust flow through the converter.

Catalytic Converter Diagnosis

When testing the catalytic converter on pre-OBD II vehicles, the converter inlet and outlet temperatures may be measured to determine the converter condition. The easiest way to test a catalytic converter is by comparing temperatures at different points with an infrared temperature sensor **(Figure 42-9)**; usually just the inlet pipe and the outlet pipe are sufficient.

SAFETY TIP *The catalytic converter is extremely hot. Wear protective gloves and use caution during this test.*

Follow these steps to measure the converter inlet and outlet temperatures:

1. Disable the air injection system. Most are easily disabled by unplugging the vacuum control system. Or it can be done by squeezing the air pump outlet hose with a pair of straight-jawed vice grips until this hose is completely flat.

Figure 42-9 An infrared temperature sensor.

Figure 42-10 Testing a catalytic converter with an infrared temperature sensor.

2. Run the engine until it is at normal operating temperature and then operate the engine at 2,500 rpm for 2 minutes to heat the converter and exhaust system. Monitor the temperature as the catalytic converter warms up. There will be a sudden jump in temperature as the converter 'lights off' **(Figure 42-10)**.

3. Shut the engine off and disconnect and ground one spark plug wire in each cylinder bank on V6 or V8 engines. On a four-cylinder engine, disconnect and ground one spark plug wire.

4. Run the engine at 1,000 rpm. Either use a scan tool to set the idle or disconnect the IAC motor electrical connector.

5. Use an infrared temperature scanner to check temperatures on the inlet and outlet side on the converter.

Shut the engine off, and compare the converter inlet and outlet temperatures. The outlet temperature should be higher than the inlet temperature. When this temperature difference is less than 50°F (28°C), replace the converter. Reconnect the spark plug wire and the IAC motor connector. Restore air pump operation. Use a scan tool to clear any DTCs set during the test.

When diagnosing the catalytic converter on OBD II vehicles, observe the MIL in the instrument panel. If this light is flashing, the PCM has detected a serious problem such as a very rich air-fuel ratio that may result in rapid and permanent

catalytic converter damage. Connect a scan tool to the DLC under the dash and check for any DTCs indicating catalytic converter defects. Check for DTCs indicating defects in the HO_2S heaters. Defects in these heaters prevent the sensors from reaching operating temperature in the required time and this affects the sensor voltages. Observe the voltage waveforms on the upstream and downstream HO_2S to determine the converter condition. These voltage waveforms may be observed on a lab scope, a graphing voltmeter, or a scan tool with graphing capabilities. If the catalytic converter operation is satisfactory, the downstream HO_2S voltage is considerably lower and cycling slower from low voltage to high voltage compared to the upstream HO_2S. When there is very little difference between the upstream and downstream HO_2S voltage signals, the catalytic converter is not reducing HC, CO, and NOx emissions properly and the converter should be replaced **(Figure 42-11)**. Keep in

GOOD CATALYST

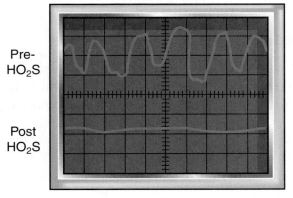

Pre-HO_2S

Post HO_2S

BAD CATALYST

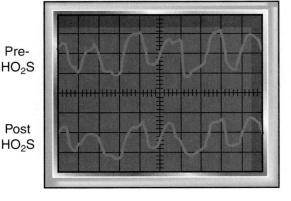

Pre-HO_2S

Post HO_2S

Figure 42-11 Catalytic converter diagnosis using HO_2S waveforms.

mind that this test uses the assumption that the HO_2S's are good; faulty sensors will give faulty readings. If the engine has HO_2S in pairs, such as one on each manifold, they should be replaced as a pair. Older HO_2S will cycle between high and low slower than new ones and will set DTCs because the PCM will compare the rate at which they cycle.

A plugged or partially plugged converter will give power loss problems. Plugged converters are rarely a problem at or near idle; symptoms tend to show up at cruising speeds and/or when under load. Often these can be detected by tapping on the converter shell to check for loose and broken internal components. An exhaust backpressure test will confirm the amount of exhaust flow blockage. Each manufacturer states an acceptable amount of backpressure. Too little causes EGR flow problems and too much causes performance issues. Typically it is around 1 lb of pressure. To quickly test for suspected backpressure problems remove the upstream HO_2 sensors. If the performance problem clears up immediately, suspect a plugged converter.

Catalytic Converter Service

When a catalytic converter is replaced, always diagnose the cause of the converter failure. If the cause of the converter failure is not diagnosed and corrected, the replacement converter will fail in a short time. For example, if the inside of the converter is melted, the air-fuel ratio may be excessively rich **(Figure 42-12)**. This problem can be caused by excessive fuel pressure or a defective input sensor such as the ECT sensor or lazy HO_2S. Cylinder misfiring also causes excessive HC in the exhaust that overheats the converter. The cause of the rich air-fuel ratio must be diagnosed and corrected or the new converter will fail in a short time. If the inside of the converter is oil contaminated, diagnose and correct the cause of the engine oil consumption. After the converter is replaced,

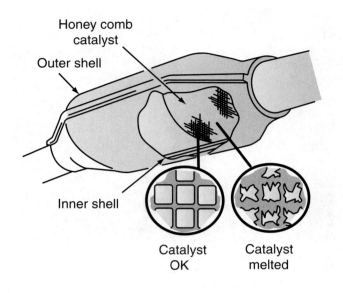

Figure 42-12 A melted section of monolithic type catalytic converter.

always be sure the voltage waveforms of the upstream and downstream HO_2S are satisfactory. If there is a rich air-fuel ratio or ignition misfiring, the upstream HO_2S will not be satisfactory.

The Clean Air Act allows the EPA to regulate the sale, use, and installation of aftermarket converters to ensure the replacement converters provide the same amount of emission reduction as the original converter on the vehicle.

When installing an aftermarket converter, EPA regulations must be followed. According to these regulations, the replacement converter must be the same type as the original converter and it must be specified by the converter manufacturer for the vehicle being serviced. The replacement converter must be in the same location in the exhaust system. Copies of the converter invoice must be retained for 6 months and the replaced converter must be kept for 15 days. It is a generally good idea to replace the HO_2S with the converter, particularly if they have high mileage.

Summary

- The PCV system helps to prevent crankcase vapors from escaping to the atmosphere.

- When a conventional EGR valve is opened at idle, the engine should slow down at least 100 rpm or stall.

- If the EGR valve is open during idle operation, rough idle will result.

- A scan tool may be used to command digital and linear EGR valves to open.

- A scan tool indicates the status of solenoids, such as the bypass and diverter valve solenoids, in an air injection system.

- A scan tool may be used to command the PCM to perform large and small EVAP leak tests.

- During an EVAP system leak test, the PCM builds vacuum in the system and seals the system. The PCM then monitors the change in EVAP system vacuum over a period of time.

- You may check for EVAP system leaks with equipment that pressurizes the EVAP system with dry nitrogen or forces smoke into the EVAP system.

- If the catalytic converter is operating properly, the outlet temperature should be 50°F (28°C) higher than the inlet temperature.

- If the catalytic converter is operating properly in an OBD II system, the voltage on the downstream HO_2S should be cycling much slower from low voltage to high voltage compared to the upstream HO_2S.

Review Questions

1. Technician A says oil contamination in the air cleaner may indicate a restricted PCV system. Technician B says oil contamination in the air cleaner may indicate the engine has excessive blowby. Who is correct?

   A. Technician A

   B. Technician B

   C. Both Technician A and Technician B

   D. Neither Technician A nor Technician B

2. On an OBD II vehicle, the upstream and downstream HO_2S have similar voltage waveforms. Technician A says the downstream HO_2S is defective. Technician B says the catalytic converter is defective. Who is correct?

   A. Technician A

   B. Technician B

   C. Both Technician A and Technician B

   D. Neither Technician A nor Technician B

3. On a pre-OBD II vehicle, the catalytic converter outlet temperature is 100°F (55°C) hotter than the converter inlet temperature.

   Technician A says the converter is defective. Technician B says the air-fuel ratio is excessively rich. Who is correct?

   A. Technician A

   B. Technician B

   C. Both Technician A and Technician B

   D. Neither Technician A nor Technician B

4. An engine has a rough idle problem at normal operating temperature. Technician A says the EGR valve may be stuck open. Technician B says the EVAP purge solenoid may be open. Who is correct?

   A. Technician A

   B. Technician B

   C. Both Technician A and Technician B

   D. Neither Technician A nor Technician B

5. An EGR valve diaphragm is forced upward manually to open the EGR valve with the engine idling and there is no change in engine speed. The most likely cause of this problem is:

A. a vacuum leak in the EGR valve diaphragm.

B. blocked EGR exhaust passages.

C. a vacuum leak in the EGR valve vacuum hose.

D. a defective EGR solenoid.

6. When 20 in. Hg is supplied to a conventional EGR valve from a vacuum hand pump with the engine idling, the engine stalls. This action indicates:

A. the EGR valve diaphragm has a vacuum leak.

B. the EGR valve is operating normally.

C. the EGR exhaust passage is restricted.

D. the EGR valve is not opening fully.

7. All of these statements about scan tool diagnosis of digital and linear EGR valves are true *except*:

A. A scan tool may be used to command any of the solenoids open in a digital EGR valve.

B. A digital EGR valve has a vacuum solenoid connected between the PCM and the EGR valve.

C. High NOx emissions may indicate the EGR valve is not opening properly.

D. On some linear EGR valves, the scan tool indicates the commanded and the actual EGR valve opening.

8. When diagnosing a positive backpressure-type EGR valve, a vacuum hand pump should be used to open the EGR valve when the engine is running at:

A. idle speed.

B. 1,000 rpm.

C. 1,200 rpm.

D. 2,000 rpm.

9. The hoses connected to a belt-driven AIR pump are burned. The most likely cause of this problem is:

A. a leaking one-way check valve in the AIR system.

B. a slipping AIR pump drive belt.

C. a defective AIR pump.

D. a plugged AIR pump inlet filter.

10. When diagnosing an enhanced EVAP system, a DTC indicates a leak in the EVAP system. Technician A says to pressurize the EVAP system with dry nitrogen gas and use an ultrasonic leak detector to locate the leak. Technician B says to use a smoke producing machine to force smoke into the EVAP system and locate the leak source. Who is correct?

A. Technician A

B. Technician B

C. Both Technician A and Technician B

D. Neither Technician A nor Technician B

11. When a vacuum hand pump is used to supply 20 in. Hg to a positive backpressure-type EGR valve with the engine idling, the valve should

_____ _____.

12. When a scan tool is used to command one solenoid to open in a digital EGR valve, the engine should _____

_____.

13. When 20 in. Hg is supplied from a vacuum hand pump to a diverter valve, the valve should be positioned to deliver air from the pump to the _____.

14. When performing an EVAP system leak test, the fuel tank should be between _____ and _____ percent full.

15. Describe the operation of a PCV valve in relation to engine speed.

16. Explain the test procedure for a negative backpressure-type EGR valve.

17. Describe two methods of leak testing an EVAP system.

18. Explain the cause of a flashing MIL light on an OBD II vehicle.

43 Engine Control Computers and Output Controls

Learning Objectives

After you have read, studied, and practiced the contents of this unit, you should be able to:

- Describe an analog voltage signal.
- Describe a digital voltage signal.
- Describe the purpose of a microprocessor chip in a computer.
- Explain how a computer controls the system outputs.
- Describe adaptive strategy.
- Describe the operation of the fuel pump circuit.

Key Terms

Adaptive strategy

Analog/digital (A/D) converter

Analog voltage signal

Bit

Byte

Digital voltage signal

Inertia switch

Integrated circuit (IC)

Port fuel injection (PFI)

Rheostat

INTRODUCTION

It is extremely important for technicians to understand the function of engine computers and output controls. When the understanding of these components is inadequate, fast and accurate diagnosis of component malfunctions becomes more difficult. If you understand the function of these components, it is much easier to recognize and diagnose improper component operation.

ANALOG VOLTAGE SIGNALS

When a **rheostat** is used to control a 5-volt bulb and the rheostat voltage is 1 volt, a small amount of current flows through the bulb to produce a dim light from the lamp **(Figure 43-1)**. If the rheostat voltage is 5 volts, higher current flow produces increased lamp brilliance. As the rheostat voltage is decreased, the light becomes dimmer. The rheostat voltage may be anywhere between 0 volts and 5 volts. This is an

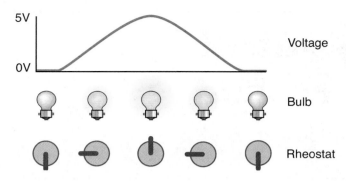

Figure 43-1 An analog voltage signal.

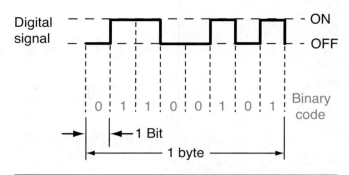

Figure 43-3 Binary coding.

example of an **analog voltage signal**. Many of the input sensors in an engine computer system produce analog voltages. Analog voltage is continuously variable within a specific range.

DIGITAL VOLTAGE SIGNALS

If a conventional on/off switch is connected to a 5-volt bulb and the switch is off, 0 volts are available at the bulb. When the switch is turned on, a 5-volt signal is sent from the switch to the bulb and the bulb is illuminated to full brilliance **(Figure 43-2)**. If the switch is turned off, the voltage at the bulb returns to 0 volts and the bulb goes out. The voltage signal from the switch is either 0 volts or 5 volts, or we could say the voltage signal is either high or low. This type of signal is called a **digital voltage signal**. If the switch is turned on and off rapidly, a square wave digital voltage signal is applied from the switch. Many computers have the capability to vary the on-time of digital signals. A digital voltage signal is either on or off.

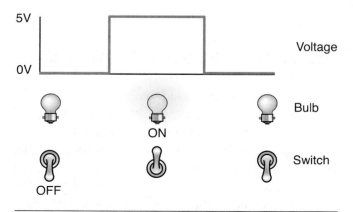

Figure 43-2 A digital voltage signal.

BINARY CODE

A numeric value may be assigned to digital signals. For example, an off or low digital signal may be given a value of 0, and an on or high digital signal may be given a value of 1 **(Figure 43-3)**. This assignment of numeric values to digital signals is called binary coding. The word binary means two values and in the binary coding system, the two values are 0 and 1.

In an automotive computer, information is transmitted in binary codes. Conditions, numbers, and letters can be represented by a series of zeroes and ones. Each individual on or off condition is called a **bit**. In Figure 43-3 the computer transmitted the value of 01100101. A group of bits is called a **byte**. The computer can process bytes of information according to a program.

Many computer input sensors operate in the 0 to 5 volts range. The TPS may produce the following voltages: at closed throttle 0.6 volt, at part open throttle 2.5 volts, and at wide open throttle 4.8 volts. A numeric value may be assigned to each of these voltages by the engine computer. The computer may assign a value of 0 for closed throttle and a value of 255 at wide-open throttle.

NOTE: In computer language, each zero and each one represent a bit of information. Eight bits make a byte, which may be called a *word*. Electronic information is exchanged in bytes.

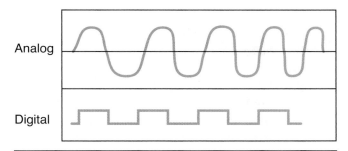

Figure 43-4 An analog signal must be converted to a digital signal for the computer to process.

INPUT SIGNAL CONDITIONING

Because many input sensors produce analog voltage signals and the microprocessor in the computer operates on digital signals, something must change these analog signals to digital signals **(Figure 43-4)**. This job is performed by the input amplification and signal conversion chip in the computer **(Figure 43-5)**. The input amplification and signal conversion chip may be called an **analog/digital (A/D) converter**. This chip continually scans the input sensor signals, assigns numeric values to the signal voltages, and then translates the numeric values to a binary code.

Some input sensors such as the O_2 sensor produce a very low voltage signal with a low current flow. This type of signal must be amplified or increased before it is sent to the microprocessor. The input amplification and signal-conditioning chip also provides the necessary signal amplification.

MICROPROCESSORS

The microprocessor is the calculating and decision-making chip in a computer. Millions of miniature transistors and diodes are contained in the microprocessor. These transistors act as electronic switches that are either on or off. The components in the microprocessor are etched on an **integrated circuit (IC)** that is small enough to fit on a fingertip. The silicon chip containing the IC is mounted in a flat, rectangular protective box. Metal connecting pins extend from each side of the microprocessor container **(Figure 43-6)**.

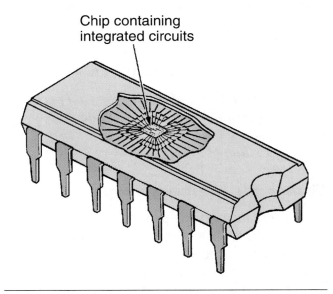

Figure 43-6 A chip containing an integrated circuit.

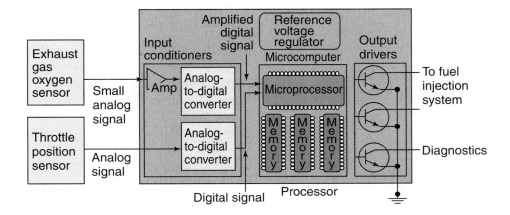

Figure 43-5 A computer with an input conditioning chip, a microprocessor, and memory chips.

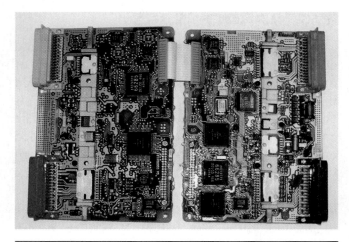

Figure 43-7 An internal view of an automotive computer using a variety of integrated circuit chips.

These pins connect the microprocessor to the circuit board in the computer **(Figure 43-7)**.

> **Did You Know?** *Some of the first personal computer chips in 1979 had 30,000 transistors. Today microprocessors use many millions of transistors.*

Computer Memory Chips

Various memory chips that store information and help the microprocessor in making decisions support the microprocessor. These memory chips are similar in appearance to the microprocessor chip. Computer memory chips are called by various names, including random access memory (RAM), read only memory (ROM), programmable read only memory (PROM), electronically erasable programmable read only memory (EEPROM), and keep alive memory (KAM) **(Figure 43-8)**.

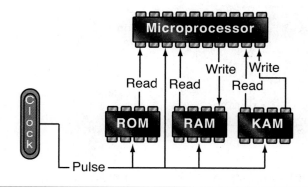

Figure 43-8 A computer block diagram showing ROM, RAM, and KAM.

NOTE: Many vehicle manufacturers now supply new computers that are not programmed. These computers must be programmed when they are installed or the engine will not start. This computer programming is done with the appropriate scan tool, personal computer (PC), or the vehicle manufacturer's recommended test equipment such as Daimler Chrysler's Mopar Diagnostic System (MDS). This computer programming may be called *flash programming*. This programming also applies to other new computers, such as ABS computers.

The memory chips store information regarding the ideal operating conditions of various components and systems and the input sensors inform the computer about the engine and vehicle operating conditions. The microprocessor reads the ideal operating conditions in the memory chip programs and compares this information with the sensor inputs. After this comparison, the microprocessor makes the necessary decisions and operates the various components to meet the ideal operating conditions in the computer program.

Computer Output Drivers

The computer operates many different output controls, such as relays and solenoids. The computer contains a number of drivers, or single transistors, that switch the output controls on and off. The microprocessor commands the drivers to operate the output controls.

For example, if the ECT sensor indicates the engine temperature is high enough to require cooling fan operation, the microprocessor commands the appropriate driver to ground the cooling fan relay winding. This action closes the relay contacts that supply voltage to the cooling fan motor **(Figure 43-9)**.

NOTE: The driver may sometimes be called a *quad driver*. A quad driver has four driver circuits on one chip and may be used to operate up to four outputs.

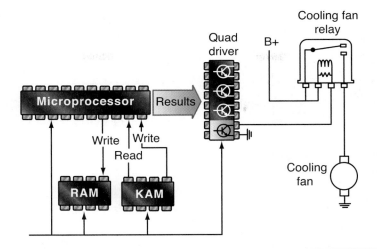

Figure 43-9 Computer output controls.

Adaptive Strategy

Many automotive computers have an **adaptive strategy** that allows the computer to compensate for various changes in some of the inputs or outputs. For example, if the injector orifices become partially restricted, the injectors deliver less fuel and the air-fuel mixture becomes lean. Under this condition, the O2 sensor voltage signal is continually low. The computer senses this condition and automatically provides a slight increase in injector pulse width to allow the injector fuel delivery to return to normal air-fuel ratios.

Injector pulse width is the length of time in milliseconds that the computer keeps the injectors open. If these injectors are replaced, the computer still supplies the increased injector pulse width and the air-fuel mixture tends to be rich. Under this condition the engine may experience rough idle operation. After about 5 minutes of driving with the engine at normal operating temperature, the computer learns about the injector replacement and returns the injector pulse width to normal. The adaptive strategy in a computer is erased if the battery voltage is disconnected from the computer. If this condition occurs, the engine operation may be erratic until the computer re-learns the system and restores the adaptive strategy.

Fuel Pump Electric Circuit

Fuel-injected vehicles have an electric fuel pump in the fuel tank. The PCM operates a relay that supplies voltage to the fuel pump. When the ignition switch is turned on, the PCM supplies voltage to the fuel pump relay winding **(Figure 43-10)**. Under this condition, current flows from the PCM through the fuel pump relay winding to ground and the relay

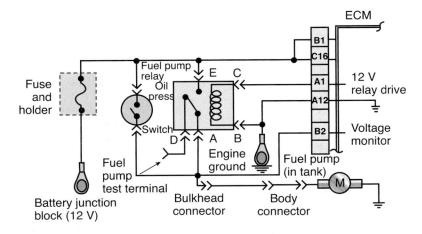

Figure 43-10 A fuel pump electrical circuit.

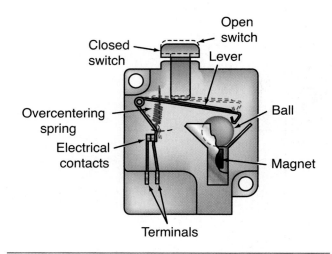

Figure 43-11 An inertia fuel pump switch; sometimes called an IFS switch (inertia fuel shutoff switch).

contacts close. Voltage is supplied through these relay contacts to the in-tank fuel pump.

If the engine is not cranked or started within 2 seconds, the PCM shuts off the current flow to the fuel pump relay winding and the fuel pump stops operating. While the engine is cranking or running, the PCM maintains the voltage supply to the fuel pump relay winding and the fuel pump relay contacts remain closed to maintain fuel pump operation. The 2-second shutoff feature on the fuel pump turns the pump off if the vehicle is in a collision and the engine stalls while the ignition switch is left in the on position or if the computer fails to see an RPM signal.

> **Tech Tip** *In some fuel pump systems, the PCM switches the ground side of the fuel pump relay on and off.*

Many fuel pump circuits on Ford Motor Company vehicles have an **inertia switch** in the circuit between the relay and the fuel pump **(Figure 43-11)**. If the vehicle is in even a moderate collision, the inertia switch opens the fuel pump circuit and turns the pump off. A reset button on top of the inertia switch must be pressed to restore the switch operation. On cars the inertia switch is in the trunk area and on light-duty trucks the inertia switch is under the dash.

NOTE: Ford Motor Company calls the inertia switch an *IFS (inertia fuel switch).*

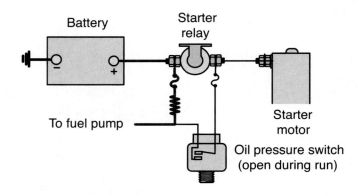

Figure 43-12 A fuel pump oil pressure switch.

On some General Motors vehicles, the fuel pump circuit has an oil pressure switch connected in the fuel pump circuit **(Figure 43-12)**. If the fuel pump relay fails, current is supplied through the oil pressure switch to maintain fuel pump operation.

> **Tech Tip** *If the fuel pump relay fails in this circuit, then extended cranking time could result as the oil pressure would need to build up to the point where the oil pressure switch closes and supplies current to the fuel pump.*

Fuel Pump Pressure Regulation

On **port fuel injection (PFI)** systems, the fuel line is connected from the filter to the fuel rail **(Figure 43-13)**. Fuel pump pressure is supplied to the injectors and to the pressure regulator in the TBI assembly or fuel rail. At the specified fuel pressure, the valve opens in the pressure regulator and some fuel is allowed to return through the

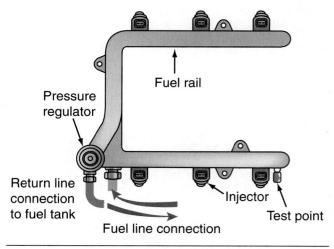

Figure 43-13 A pressure regulator on a fuel rail.

return line to the fuel tank (Figure 43-13). When the fuel pressure drops slightly, the pressure regulator valve closes and fuel pressure increases to open this valve again. The average fuel pressure on PFI systems is 35 to 65 psi (241 to 448 kPa) depending on the vehicle application.

> **Tech Tip** *Fuel pressure is extremely important to maintain proper fuel system operation. If the fuel pressure is less than specified, the injectors deliver a reduced amount of fuel and the air-fuel mixture is lean. Excessive fuel pressure causes the injectors to deliver too much fuel and the air-fuel mixture is rich.*

On most PFI systems, a vacuum hose is connected from the intake manifold to the pressure regulator. When the engine is running, this hose supplies manifold vacuum to the upper side of the regulator diaphragm. If the engine is idling or operating at moderate throttle opening, manifold vacuum is high and this vacuum above the regulator diaphragm allows the diaphragm to move upward at a lower pressure and open the regulator valve. When the engine is operating at wide-open throttle, manifold vacuum is low and a higher fuel pressure is required to move the regulator diaphragm upward and open the regulator valve. Therefore, fuel pressure indicated on the gauge is higher when operating at wide-open throttle. The pressure difference between the injector inlet and the injector outlet remains the same. The fuel pressure changes to match pressure changes in the intake manifold.

Some vehicles have a return-less fuel system. In these systems, the pressure regulator is mounted on top of the fuel tank or inside the top of the fuel tank **(Figure 43-14)**. Fuel is returned from the pressure regulator directly into the fuel tank. On these systems, there is no return line. Since the fuel is not returned from the engine compartment, the fuel will not be warmed, which will help to reduce fuel tank vapor pressure.

Some Ford vehicles have a return-less fuel system using a computer-controlled fuel pump driver module **(Figure 43-15)**. The module receives signals from the computer that increases duty cycle to the pump under various conditions. The computer receives pressure signals from a fuel pressure sensor that takes the place of the fuel pressure regulator mounted on the fuel rail to assist the PCM in deciding when to crank up voltage to the pump.

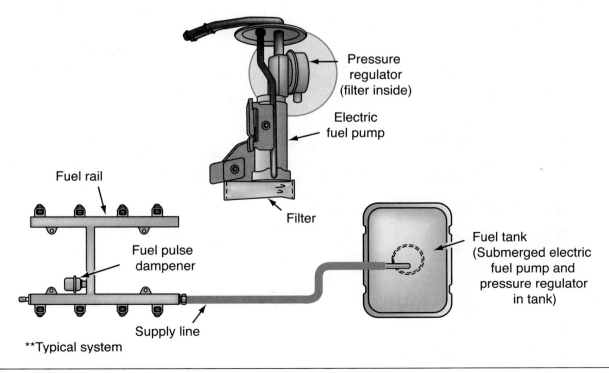

Figure 43-14 A returnless fuel system.

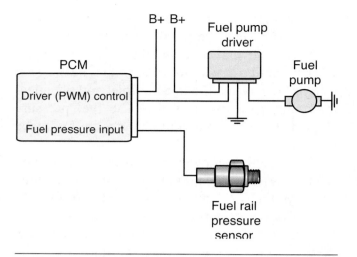

Figure 43-15 The computer can control fuel pressure using a fuel pump driver module.

Tech Tip *Some vehicles have a fuel pressure sensor in the fuel rail. These vehicles do not have a pressure regulator. In response to the fuel pressure sensor signal, the PCM pulses the fuel pump on and off to maintain the specified fuel pressure.*

Powertrain Control Module (PCM) Outputs

The PCM controls a variety of solenoids, relays, and motors to operate the output devices. In most cases, the PCM provides a ground to operate these output controls. Many computers operate the output controls by opening and closing the ground circuit on the relay or solenoid. Computer outputs include the fuel injectors, EGR solenoid, canister purge solenoid in the EVAP emission system, ignition system, and others **(Figure 43-16)**.

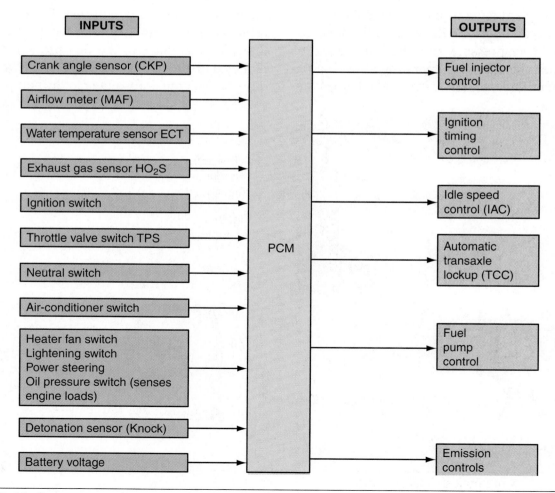

Figure 43-16 Common computer inputs and outputs.

Summary

- Analog voltage signals are variable within a specific range.

- Digital voltage signals are either on or off.

- Binary coding is the assignment of numbers to digital signals.

- The microprocessor is the decision making and calculating chip in a computer.

- Computers usually operate outputs by providing a ground.

- Adaptive strategy allows a computer to compensate for various changes in some of the inputs or outputs.

- The PCM operates a relay that supplies voltage to the electric in-tank fuel pump.

- If the ignition switch is on and the engine is not cranked or started, the PCM shuts off the fuel pump relay and fuel pump in 2 seconds.

- The fuel pressure regulator regulates fuel pressure.

- The fuel pressure can be controlled by the computer.

Review Questions

1. Technician A says that analog signals do not need to be converted to digital signals for the computer to process. Technician B says throttle position sensors produce digital signals. Who is correct?

   A. Technician A

   B. Technician B

   C. Both Technician A and Technician B

   D. Neither Technician A nor Technician B

2. Technician A says analog signals are either on or off. Technician B says digital signals are continuously variable. Who is correct?

   A. Technician A

   B. Technician B

   C. Both Technician A and Technician B

   D. Neither Technician A nor Technician B

3. Technician A says the fuel pump in a PFI system will be running if the ignition switch is on for 5 minutes and the engine is not cranked or started. Technician B says many fuel injection systems return excess fuel from the pressure regulator to the fuel tank. Who is correct?

   A. Technician A

   B. Technician B

   C. Both Technician A and Technician B

   D. Neither Technician A nor Technician B

4. Technician A says an inertia switch opens the fuel pump circuit if this circuit overheats. Technician B says an oil pressure switch in the fuel pump circuit supplies voltage to the fuel pump if the relay fails. Who is correct?

   A. Technician A

   B. Technician B

   C. Both Technician A and Technician B

   D. Neither Technician A nor Technician B

5. All of these statements about digital signals are true *except*:

   A. A digital signal varies continuously within a specific voltage range.

   B. A digital signal is either on or off.

   C. A digital signal may be called a square-wave signal.

   D. A conventional on/off light switch produces a digital voltage signal.

6. The input signal conditioning chip in a computer may perform all of these functions *except*:

   A. Change analog signals to digital signals.

   B. Amplify input voltage signals.

   C. Assign numeric values to input voltage signals.

   D. Reduce input voltage signals.

7. Technician A says the adaptive strategy is erased if the battery voltage is disconnected from the PCM. Technician B says after 5 minutes of driving the vehicle, the adaptive strategy relearns about component replacements in the PCM system. Who is correct?

   A. Technician A

   B. Technician B

   C. Both Technician A and Technician B

   D. Neither Technician A nor Technician B

8. The process of flash programming an engine computer:

   A. installs new information only in the RAM.

   B. may be necessary when a computer is replaced.

   C. may be done with a lab scope and a digital voltmeter.

   D. is required before replacing engine computer system components.

9. Technician A says the adaptive strategy in an engine computer is erased if the computer is disconnected from the battery. Technician B says the adaptive strategy in an engine computer allows the computer to adapt to minor system defects. Who is correct?

   A. Technician A

   B. Technician B

   C. Both Technician A and Technician B

   D. Neither Technician A nor Technician B

10. All of these statements about a typical electric fuel pump circuit in a port fuel injection system are true *except*:

    A. The fuel pump operates for 2 seconds if the ignition switch is turned on and the engine is not cranked.

    B. Voltage is supplied through the fuel pump relay contacts to the fuel pump.

    C. The inertia switch opens the fuel pump circuit if the vehicle is in a collision.

    D. The fuel pump is usually mounted on the frame rail under the vehicle.

11. An analog voltage signal is continuously _____ in a specific voltage range.

12. A digital voltage signal may be considered _____ or _____.

13. The input amplification and conditioning chip in a computer changes _____ signals to _____ signals.

14. The amount of fuel discharged from an injector is determined by the _____ _____ _____.

15. Explain adaptive strategy in a computer.

16. Describe the operation of the fuel pump circuit if the ignition is left in the on position and the engine is not started.

17. Explain how the computer operates many of its outputs.

18. Explain the difference between a bit and a byte.

CHAPTER

44 Input Sensors for Engine Control Systems

Learning Objectives

After you have read, studied, and practiced the contents of this unit, you should be able to:

- Describe the operation of an oxygen sensor.
- Explain open loop and closed loop computer operation.
- Define a speed density fuel injection system.
- Define a mass flow fuel injection system.
- Describe the operation of various input sensors.

Key Terms

Closed loop

Engine coolant temperature (ECT) sensor

Fuel trim

Heated oxygen sensors (HO$_2$S)

Intake air temperature (IAT) sensor

Knock sensor

Manifold absolute pressure (MAP) sensor

Mass air flow (MAF) sensor

Open loop

Speed density system

Thermistor

Throttle position (TP) sensor

Titania O$_2$ sensor

Zirconia O$_2$ sensor

INTRODUCTION

A thorough understanding of input sensors provides the basis for fast, accurate diagnosis of these sensors. If you do not understand input sensor operation and typical input sensor voltage signals, diagnostic procedures for these sensors are often time consuming.

OXYGEN SENSORS

The oxygen sensor is threaded into the exhaust manifold or exhaust pipe. OBD II systems have oxygen sensors upstream and downstream from the catalytic converter. The upstream oxygen sensor is primarily used for fuel trim while the downstream oxygen sensor is used for monitoring the efficiency of the catalytic converter. **Fuel trim** occurs when the computer increases or decreases injector pulse width based on information from the upstream oxygen sensor. Fuel trim is necessary to maintain the proper air-fuel ratio during engine closed-loop operation.

An oxygen-sensing element is positioned in the center of the O_2 sensor **(Figure 44-1)**. The oxygen-sensing element is coated with zirconia or titania and this element is mounted in an insulator and contained in a metal case. Zirconia and titania-type O_2 sensors have a similar design but different operating principles.

The sensor case has external threads that fit matching threads in the exhaust system. The lower end of the sensor is mounted in the exhaust stream. A protective cap on the lower end of the sensor contains slots that allow exhaust to be applied to the outside of the sensing element.

On **zirconia O_2 sensors**, internal air slots in the protective boot and an air port in the side of the sensor allow ambient air to enter the inside of the sensing element. **Titania O_2 sensors** may have a protective rubber boot over the top of the sensor or the air is sealed inside the sensing element.

The sensing element is connected to the terminal on top of the O_2 sensor and a wire is connected from the sensor terminal to the PCM. Some O_2 sensors have a ground wire connected from the sensor to the PCM. On newer vehicles, the O_2 sensors have four wires. These sensors have an internal electric heater that has a voltage supply wire and a ground wire. These sensors are called **heated oxygen sensors (HO$_2$S)**. Voltage may be supplied to the HO$_2$S heater directly from the ignition switch, through a relay, or from the PCM. If the HO$_2$S heaters are powered from the ignition switch or through a relay, voltage is supplied to these heaters while the ignition switch is on. When the HO$_2$S heaters are powered by the PCM, the PCM supplies voltage to these heaters only when necessary, such as during engine warm-up and at idle and low engine speeds. If the engine is at normal operating temperature and running at higher speeds, the exhaust flow maintains HO$_2$S sensor temperature and the PCM shuts off the sensor heaters.

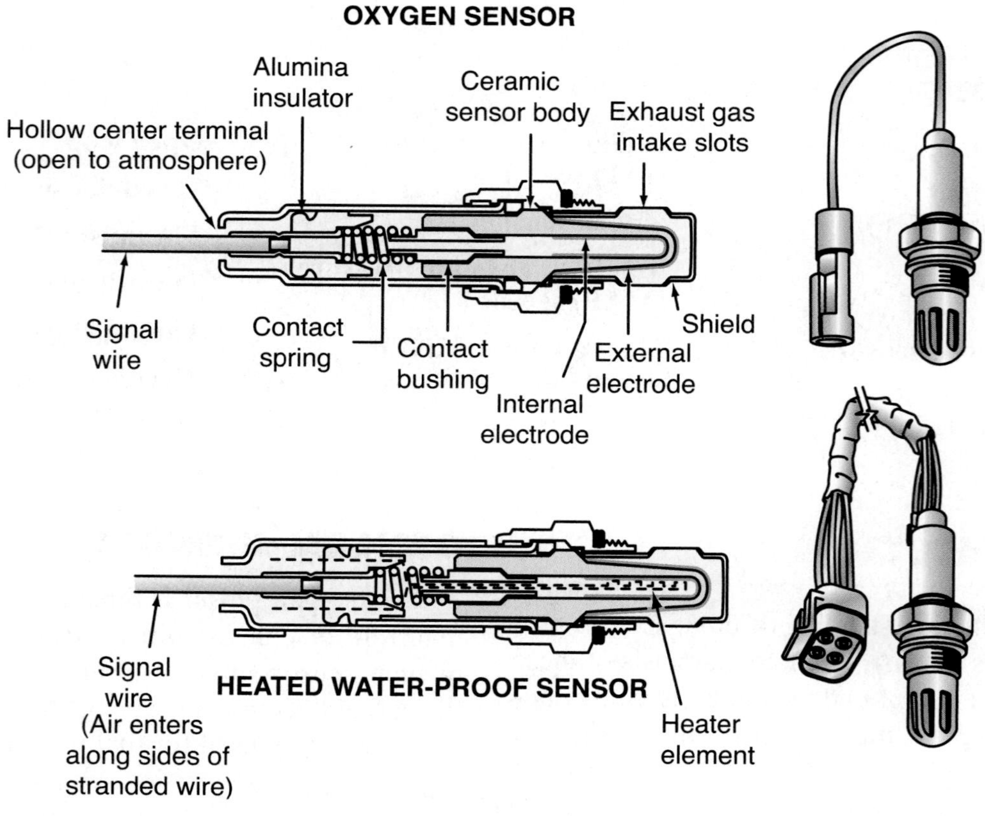

OXYGEN SENSOR

HEATED WATER-PROOF SENSOR

Figure 44-1 An oxygen sensor design.

> **Tech Tip** *Always use the proper oxygen sensor for the vehicle as specified by the manufacturer.*

Zirconia Oxygen Sensor Operation

When the engine exhaust temperature is cold, a zirconia-type HO_2S does not produce a satisfactory voltage signal. When the engine exhaust approaches normal operating temperature, this sensor begins to produce a satisfactory signal. During engine's warm-up (engine temperature below 140°F), the PCM operates in **open loop (Figure 44-2)**. In this mode, the PCM ignores the HO_2S signal and a program in the PCM controls the air-fuel ratio. In

open loop, the PCM supplies a preprogrammed amount of fuel depending on engine temperature and throttle opening. At a specific engine temperature, the PCM enters the **closed loop** mode. In closed loop, the PCM uses the HO_2S signal and other inputs to control the air-fuel ratio.

If the PCM is operating in closed loop and the air-fuel ratio is lean, nearly all of the fuel injected has been combined with air and burned in the combustion chambers and excess oxygen is left over. Under this condition, the exhaust stream flowing past the HO_2S has high oxygen content. This causes oxygen to be supplied to both sides of the oxygen-sensing element in the HO_2S, resulting in a very low voltage signal (100 mV) from this sensor to the PCM. When this signal is received, the PCM increases the injector pulse width and provides a richer air-fuel ratio.

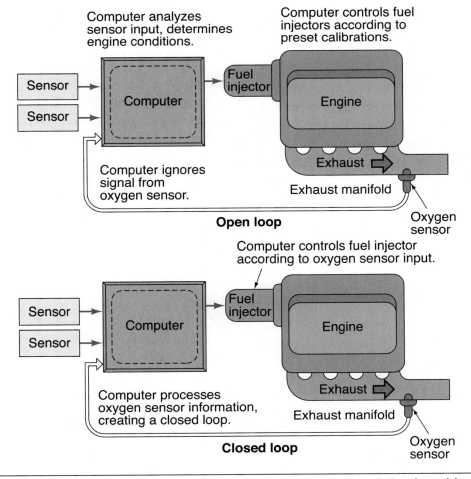

Figure 44-2 The O_2 sensor is ignored in open loop. When the system is in closed loop the computer is adjusting A/F based on O_2 sensor readings.

Tech Tip *Excess oxygen levels in the exhaust usually indicate a lean air-fuel ratio. Low oxygen levels usually indicate a rich air-fuel ratio.*

A rich air-fuel ratio causes all the oxygen in the intake air to be mixed with fuel and excessive fuel is left over. Under this condition, the exhaust stream has very low oxygen content. This condition causes low oxygen content on the outside of the HO_2S element, and high oxygen content is present inside this element. When this condition occurs, the HO_2S produces a higher voltage (900 mV) **(Figure 44-3)**. In response to this HO_2S signal, the PCM reduces the injector pulse width and provides a leaner air-fuel ratio. The HO_2S signal cycles very quickly from low voltage to high voltage and the PCM uses this signal to control the air-fuel. This rate of signal cycling from low to high voltage is called cross counts. Cross counts are the number of times that the HO_2S cycles from rich to lean in a given time period, usually 1 second.

Tech Tip *A defective spark plug, sparkplug wire, or other condition may cause a misfire in which unburned air and fuel to be exhausted from the cylinder will cause the oxygen sensor to detect high levels of unburned oxygen. This usually causes the computer to richen the air-fuel ratio as the computer thinks the exhaust is lean.*

Tech Tip *An oxygen sensor can sense only oxygen; it is not a fuel sensor.*

A zirconia oxygen sensor will not perform if the sensor unit is covered in mud, dirt, grease, oil, or other debris blocking the ability of the sensor to compare outside air with the exhaust gases.

Tech Tip *Exhaust leaks can also affect oxygen sensor operation by introducing outside air near the sensor.*

Titania Oxygen Sensor Operation

Titania sensors are made from aluminum titanate (also known as titanium dioxide, TiO_2). A titania-type HO_2S modifies a reference voltage whereas a zirconia-type HO_2S generates voltage. The PCM supplies voltage to the zirconia-type HO_2S and this voltage is lowered by a resistor in the circuit.

The resistance of the titania sensor varies as the air-fuel ratio cycles from lean to rich. If the air-fuel ratio is lean, the titania resistance is high (20,000 ohms) and the sensor voltage signal is low. When the air-fuel ratio is rich, the titania resistance is low (1,000 ohms) and the sensor voltage signal is high **(Figure 44-4)**. The titania-type HO_2S produces a satisfactory signal almost immediately after a cold engine is started. This action provides improved air-fuel ratio control during

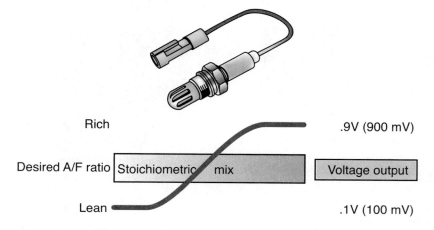

Figure 44-3 A zirconia-type oxygen sensor voltage signal.

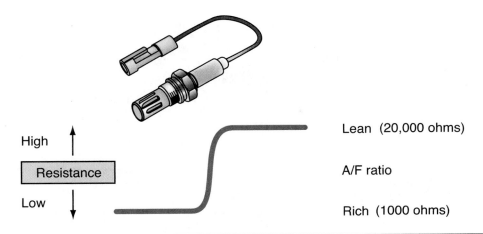

Figure 44-4 A titania-type oxygen sensor voltage signal.

cold engine operation. The titania HO_2S can also cycle faster from low voltage to high voltage.

Titania oxygen sensors do not depend upon outside air for an oxygen reference when doing its job. They may be used in vehicles that are regularly used in harsh environments and off-road circumstances. Care must be taken that the other oxygen sensor, zirconia oxygen sensor, is not accidentally installed in the vehicle. Obviously, these two oxygen sensors are not interchangeable.

ENGINE COOLANT TEMPERATURE AND INTAKE AIR TEMPERATURE SENSORS

Engine coolant temperature (ECT) sensors and **Intake Air Temperature (IAT) sensors** both contain a thermistor. When a **thermistor** is cold it has very high resistance. This resistance decreases if the thermistor is heated. Some ECT sensors have 35,000 ohms of resistance when the engine is very cold and the same sensor has less than 1,000 ohms of resistance if the sensor is at normal operating engine temperature. Two wires are usually connected from the PCM to the ECT or IAT sensors. One of these wires is a signal wire and the other wire is a ground wire.

The PCM supplies a 5-volt reference through the signal wire to each sensor and the PCM senses the voltage drop across the sensor. When the engine is cold and sensor resistance is high, the voltage drop across the sensor may be 4.5 volts. If the engine is at normal operating temperature, the voltage drop across the ECT or IAT sensor is very low **(Figure 44-5** and **Figure 44-6)**. The relationship between temperature and resistance can be seen in **Figure 44-7**.

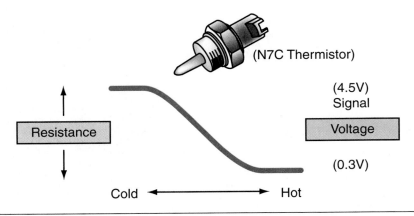

Figure 44-5 An ECT sensor voltage signal.

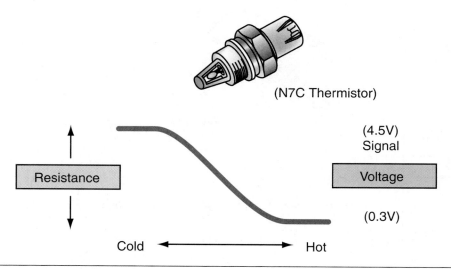

(N7C Thermistor)

Figure 44-6 An IAT sensor voltage signal.

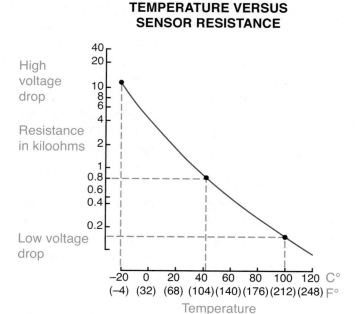

TEMPERATURE VERSUS SENSOR RESISTANCE

High voltage drop

Resistance in kiloohms

Low voltage drop

Figure 44-7 This chart shows that as temperature increases, the sensor's resistance increases. IAT and ECT are sensors that typically behave in this manner.

The PCM uses the ECT and IAT sensor signals to control many of its outputs. For example, the PCM supplies a richer air-fuel ratio when the ECT sensor indicates that the engine coolant is cold. The PCM uses the ECT sensor and HO_2S signals to determine when to enter closed loop.

MANIFOLD ABSOLUTE PRESSURE SENSORS

The **Manifold Absolute Pressure (MAP) sensor** is usually mounted in the engine compartment. Some MAP sensors are threaded directly into the intake manifold. Other MAP sensors have a vacuum hose connected from the intake manifold to the sensor.

MAP sensors have three wires connected from the sensor to the PCM **(Figure 44-8)**. These wires are a 5-volt reference wire, a signal wire, and a ground wire connected to the PCM. The PCM supplies a constant 5 volts through the reference wire to the MAP sensor.

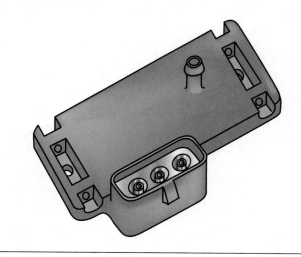

Figure 44-8 A MAP sensor.

In some MAP sensors, the manifold vacuum is supplied to a silicon diaphragm in the sensor. When the manifold vacuum increases, it stretches this diaphragm and this action changes the voltage signal to the PCM. A typical MAP sensor provides a 1-volt signal to the PCM with the engine idling and high vacuum supplied to this sensor. At wide-open throttle, the vacuum decreases and the voltage signal from the MAP sensor is approximately 4.5 volts. A fuel injection system with a MAP sensor may be called a speed density system.

In a **speed density system,** the PCM determines the amount of air entering the engine from the engine RPM signal, and the MAP sensor signal, which indicates the density of the air in the intake manifold. The PCM must know the amount of air entering the engine to calculate the amount of fuel to be injected.

Some MAP sensors produce a digital voltage with a continually varying frequency. This frequency increases as throttle opening and engine load increase. The signal from this type of MAP sensor is measured in cycles per second or hertz. The MAP sensor signal may vary from 109 hertz at idle to 153 hertz at wide-open throttle **(Figure 44-9)**.

Did You Know? *Some MAP sensors contain a barometric pressure sensor that sends a signal to the PCM in relation to barometric pressure each time the ignition switch is turned on. Barometric pressure varies depending on atmospheric conditions and altitude. When the PCM receives this signal, it corrects the air-fuel ratio in relation to altitude and air pressure.*

MASS AIR FLOW SENSORS

The **mass air flow (MAF) sensor** is a device that measures the mass of the air entering the engine. The MAF sensor is mounted in the air intake hose between the air cleaner and the throttle body. A MAF sensor housing may have an arrow that indicates the direction in which air must flow through the sensor. Most sensors are constructed so that they may only be installed in the proper direction.

Did You Know? *Mass is the quantity of matter and is often measured in grams or kilograms.*

Many engines have a hot wire-type MAF sensor. In this type of sensor, a hot wire is positioned in the air stream through the sensor **(Figure 44-10)**. An ambient temperature-sensing wire is mounted beside the hot wire. When the ignition switch is turned on, the module in the MAF sends enough current through the hot wire to maintain the temperature of this wire at 392°F (200°C) above the

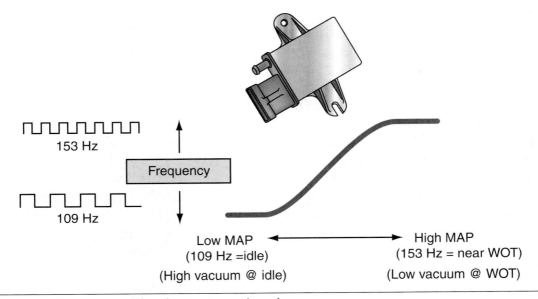

Figure 44-9 A MAP sensor with a hertz-type signal.

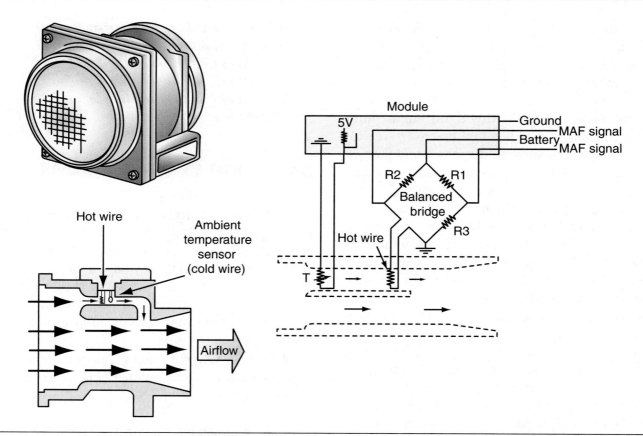

Figure 44-10 A hot wire-type MAF sensor.

temperature of the cold wire. When the engine is suddenly accelerated, the rush of air through the MAF tries to cool the hot wire. When the module senses the cooling of the hot wire, it immediately increases the current through this wire to maintain the wire temperature at 392°F (200°C). The module sends this increasing current signal to the PCM. In response to this signal, the PCM supplies the correct amount of fuel to go with the additional air entering the engine.

The MAF signal informs the PCM regarding the amount of air entering the engine and the PCM provides the correct amount of fuel to mix with the air and maintain the proper air-fuel ratio. MAF systems are much more flexible, as compared to speed density, in their ability to compensate for engine changes since they actually measure airflow instead of computing it based on preprogrammed assumptions. This system is called mass flow.

Some MAF sensors contain a vane-type sensor. These may also be known as Vane Air Flow (VAF)

meter. In this type of sensor, a moveable, spring-loaded vane is located in the air intake hose between the air cleaner and the throttle body **(Figure 44-11)**. This normally closed vane is connected to a variable

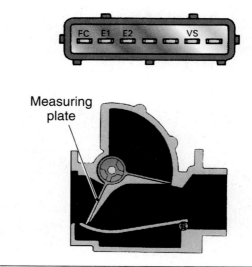

Figure 44-11 A vane-type MAF sensor.

resistor in the sensor housing. The PCM supplies 5 volts to this resistor and the voltage signal from the sensor to the PCM varies as the vane moves a contact on the variable resistor.

When the engine is started, the airflow through the air intake opens the vane slightly. If the engine is accelerated, the increased airflow through the air intake opens the vane farther and moves the contact on the variable resistor. This action changes the voltage signal sent from the MAF to the PCM.

> **Tech Tip** *Some MAF circuits have a burn-off circuit containing a burn-off relay. When the ignition switch is turned off, the burn-off relay closes. This relay activates a burnoff circuit that provides a high current flow through the hot wire for a short time. This action burns contaminants off the wire. Even though the MAF sensor is mounted after the air cleaner, the air still contains very small particles that the air filter does not remove.*

THROTTLE POSITION SENSORS

The **throttle position (TP) sensor** has three wires connected from the sensor to the PCM. These wires are similar to the wires used on a MAP sensor. A 5-volt reference is supplied to both the MAP and TP sensors. The TP sensor also has a signal wire and a ground wire **(Figure 44-12)**. The TP sensor contains a variable resistor that is connected to the throttle shaft. As the throttle is opened, a sliding contact moves on the variable resistor.

A typical TP voltage is 0.5 volt to 1 volt at idle and 4.5 volts at wide-open throttle **(Figure 44-13)**. As the throttle is opened, the TP voltage must increase smoothly. A defective variable resistor may cause erratic TP voltage and a hesitation during engine acceleration.

KNOCK SENSORS

The **knock sensor** is often threaded into the engine block or cylinder head. The knock sensor

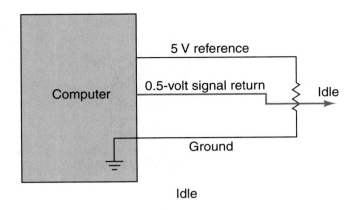

Idle

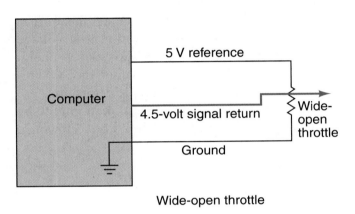

Wide-open throttle

Figure 44-12 The TPS is supplied a 5 V reference and a ground. The TPS supplies a signal return to the computer.

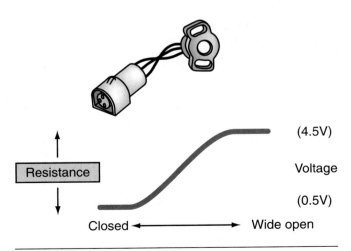

Figure 44-13 ATPS voltage signal.

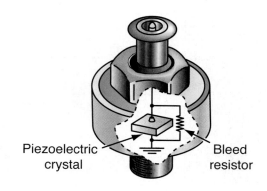

Figure 44-14 A knock sensor.

contains a piezoelectric crystal **(Figure 44-14)**. When the engine detonates or knocks, a vibration is present in the block and cylinder head. The knock sensor changes this vibration to a voltage signal. When the PCM receives a knock sensor signal indicating the engine is detonating, the PCM reduces the spark advance to stop the detonation.

More power can be produced in an engine by increasing spark advance, up to a limit. However, too much advance causes engine knock. Once an engine starts to knock, performance decreases and serious engine damage could result. Knock gets its name because it causes the air fuel to burn too rapidly in the cylinder. Knock causes a "pinging" noise that sounds like small ball bearings bouncing on metal. Mild knock reduces power as the piston cannot move along with the burning air/fuel. Knocking also increases emissions. Engine knock can destroy internal engine parts including the pistons, connecting rods, valves, head gaskets, and spark plugs.

> **Tech Tip** *Some knock sensor signals are sent to an electric spark control (ESC) module and then to the PCM. The ESC module modifies the knock sensor signal and changes it from analog to digital.*

Summary

- When the air-fuel ratio is rich, the HO_2S provides a high voltage signal and a lean air-fuel ratio causes a low voltage signal from this sensor.

- A typical HO_2S voltage signal varies from 0.2 volt to 1 volt.

- An ECT sensor has high resistance when cold and low resistance when hot.

- The PCM senses the voltage drop across the ECT and IAT sensors to determine temperature.

- Some MAP sensors contain a silicon diaphragm that stretches as vacuum is applied to the diaphragm.

- A typical MAP sensor voltage signal is 1 volt at idle and 4.5 volts at wide-open throttle.

- Some MAP sensors provide a hertz-type voltage signal.

- Some MAP sensors act as barometric pressure sensors when the ignition switch is first turned on.

- Some MAF sensors have a hot wire and an ambient temperature-sensing wire.

- In a hot wire-type MAF sensor, the sensor module keeps the hot-wire 392°F (200°C) hotter than the ambient temperature-sensing wire.

- Some MAF sensors have a moveable vane in the air intake hose. This vane is connected to a variable resistor in the sensor.

- The TP sensor contains a variable resistor and a moveable contact connected to the throttle shaft slides on this resistor.

- A typical TP sensor voltage signal is 0.5 volt to 1 volt at idle and 4.5 volts at wide-open throttle.

- The knock sensor changes a vibration to a voltage signal. The PCM then uses this information to adjust ignition timing to eliminate knock.

Review Questions

1. Technician A says a rich air-fuel ratio provides an exhaust stream that has an excess amount of oxygen. Technician A says with a rich air-fuel ratio, the HO_2S voltage signal is low. Who is correct?

   A. Technician A

   B. Technician B

   C. Both Technician A and Technician B

   D. Neither Technician A nor Technician B

2. Technician A says if oxygen is present on both sides of the zirconia HO_2S sensing element, this sensor produces a low voltage. Technician B says a lean air-fuel ratio results in high oxygen content on the exhaust side of the HO_2S sensing element. Who is correct?

   A. Technician A

   B. Technician B

   C. Both Technician A and Technician B

   D. Neither Technician A nor Technician B

3. Technician A says when the engine coolant is cold, the ECT sensor resistance is high. Technician B says when the engine coolant is cold, the ECT sensor voltage drop is high. Who is correct?

   A. Technician A

   B. Technician B

   C. Both Technician A and Technician B

   D. Neither Technician A nor Technician B

4. Technician A says with the engine idling, a satisfactory MAP sensor voltage signal is 4.5 volts. Technician B says the MAP sensor sends a barometric pressure signal to the PCM when the ignition switch is first turned on. Who is correct?

   A. Technician A

   B. Technician B

   C. Both Technician A and Technician B

   D. Neither Technician A nor Technician B

5. All of these statements about the HO_2S are true *except*:

   A. OBD II systems have the HO_2S upstream and downstream from the catalytic converter.

   B. A rich air-fuel ratio causes a zirconia HO_2S to produce a low voltage signal.

   C. A typical HO_2S has four wires connected to the sensor.

   D. When the engine is cold, a zirconia-type HO_2S does not produce a satisfactory voltage signal.

6. A typical ECT sensor:

   A. contains a titania element.

   B. has a higher resistance when the engine coolant is cold.

   C. has three wires connected to the sensor.

   D. provides a digital voltage signal to the PCM.

7. When discussing typical MAP sensors, Technician A says the MAP sensor voltage signal increases when the engine speed and/or load increases. Technician B says a MAP sensor may contain a barometric pressure sensor. Who is correct?

   A. Technician A

   B. Technician B

   C. Both Technician A and Technician B

   D. Neither Technician A nor Technician B

8. All of these statements about mass air flow sensors are true *except*:

   A. An arrow on the MAF housing indicates the proper direction of airflow.

   B. The MAF sensor may contain a hot wire and a cold wire.

   C. The MAF module maintains a constant hot wire temperature.

   D. Some MAF sensors have a burn-off circuit that cleans the air passage through the sensor.

9. A typical TP sensor voltage signal with the throttle in the wide open position is:

   A. 2.1 volts.

   B. 2.8 volts.

   C. 3.1 volts.

   D. 4.5 volts.

10. When the PCM receives a knock sensor signal, the PCM:

   A. provides a richer air-fuel ratio.

   B. reduces the spark advance.

   C. reduces the injector pulse width.

   D. ignores the HO_2S signal.

11. A titania-type HO_2S changes _____ as the air-fuel ratio cycles from lean to rich.

12. The PCM uses the voltage signals from the _____ and _____ sensors to determine when to enter closed loop.

13. The PCM ignores the _____ sensor signal when operating in open loop.

14. In a speed density fuel injection system, the PCM determines the amount of air entering the engine from the _____ and _____ signals.

15. Explain the operation of a zirconia-type HO_2S in relation to exhaust gas oxygen.

16. Explain the relationship of fuel trim in relation to a zirconia-type HO_2S voltage.

17. Explain the design and wiring connections of a TP sensor.

18. Describe the operation of the TP sensor voltage signal in relation to throttle opening.

19. Describe the design and operation of a knock sensor.

CHAPTER

45

Input Sensor Maintenance, Diagnosis, and Service

Learning Objectives

After you have read, studied, and practiced the contents of this unit, you should be able to:

- Perform O_2 sensor tests.
- Describe the proper operation of an O_2 sensor.
- Perform fuel pump pressure testing.
- Perform TP sensor tests.
- Perform MAF sensor tests.
- Perform knock sensor tests.

Key Terms

Cross counts

Transition time

INTRODUCTION

An understanding of input sensor operation is absolutely essential to provide fast, accurate sensor diagnosis. This understanding includes satisfactory and unsatisfactory sensor readings indicated by waveforms and ohmmeter or voltmeter readings. You must understand how other components can affect sensor readings or your diagnosis will often be inaccurate. For example, you must understand satisfactory and unsatisfactory HO_2S waveforms and you must also understand the effect that ignition or fuel system defects have on the HO_2S waveforms.

INPUT SENSOR MAINTENANCE

When minor under-hood service is performed, all the sensor wiring should be inspected for loose connections, worn insulation, and contact with hot components such as exhaust manifolds. Repair or replace loose sensor connections and worn insulation as required. Sensor wires must be secured away from hot components. When diagnosing input sensors, the fuel pressure, ignition system, and the data from all the other input sensors should be tested because problems in these areas and components affect some of the input sensor signals.

Check the MAP sensor vacuum hose or manifold grommet for leaks and kinks. If necessary, replace this hose and check all the other vacuum hoses for leaks. A vacuum leak in any vacuum hose decreases the manifold vacuum supplied to the MAP sensor. Inspect the engine for oil leaks that may soak input sensor wires and contribute to a shorted condition between the wires. Inspect the battery area for indications of battery acid leaks, which contaminate input sensor and other wiring harnesses.

> **Tech Tip** *A loose or missing dipstick, or other vacuum leak in the crankcase, can affect the vacuum balance in an engine and thus affect the MAP sensor signal to the computer.*

If the engine has an MAF sensor, check the hose between this sensor and the throttle body for leaks and loose clamps. Leaks in this hose will allow air to enter the air intake without flowing through the MAF sensor. This causes an incorrect MAF sensor signal that indicates there is less air entering the engine. This signal results in a lean air-fuel ratio and engine performance problems such as hesitation during acceleration.

> **Tech Tip** *It is important to check all of the ground circuits for the vehicle. A loose or defective ground can cause many problems in the computer system.*

OXYGEN SENSOR DIAGNOSIS

When diagnosing any input sensor problem, always check the MIL in the instrument panel. If the MIL is illuminated with the engine running, the PCM has detected a defect in the system and a DTC is probably set in the PCM memory. With the ignition switch off, connect a scan tool to the DLC under the dash and retrieve any DTCs in the PCM memory. These DTCs indicate the general area where the defect is located. For example, a P0131 DTC indicates a low voltage in HO_2S bank 1 sensor 1.

> **Tech Tip** *In a P0131 DTC, the P indicates a powertrain code and the 0 indicates this code is mandated by the SAE standards. If the second digit is a 1, the code is a vehicle manufacturer's code. When the third digit is a 1, the code belongs to the fuel and air metering subgroup and 31 is the code representing the low HO_2S voltage. The vehicle manufacturer's service information provides DTC interpretations.*

> **Tech Tip** *A bank 1 HO_2S is located on the same side of the engine as number 1 cylinder and sensor 1 is positioned upstream from the catalytic converter. A bank 2 sensor is on the opposite side of the engine from number 1 cylinder and sensor 2 is mounted downstream from the catalytic converter.*

The next step in sensor diagnosis is to inspect the sensor wiring for worn insulation and loose connections because these wiring defects may be the cause of the problem. If the sensor wiring and vacuum hose are satisfactory, further sensor diagnosis is required. With the engine operating at normal temperature, the HO_2S sensor may be tested with a scan tool or a digital multimeter on the DC voltage function.

Before testing this sensor, the engine speed should be maintained at 2,000 rpm for 2 minutes to be sure the sensor is hot. Then allow the engine to idle and observe the upstream HO_2S voltage displayed on the scan tool. The HO_2S should provide a voltage signal that switches quickly from low voltage to high voltage. A typical HO_2S voltage cycles between 0.3 volt and 0.8 volt. If the sensor voltage is not cycling quickly in the proper range, the air-fuel ratio is improper, the ignition system has a defect causing misfiring, or the HO_2S sensor is defective.

> **Tech Tip** *Testing an HO_2S with an analog voltmeter will ruin the sensor.*

If the engine thermostat is stuck open and the engine temperature is always low, the air-fuel ratio is rich and again the HO_2S voltage signal is always high. Therefore, when diagnosing engine computer

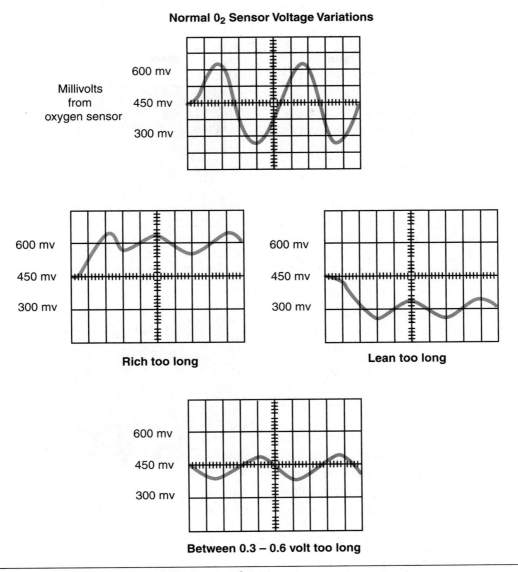

Figure 45-1 Normal and abnormal HO$_2$S waveform.

systems and input sensors, one of the first tests is to prove that the fuel system has the specified fuel pressure. On most vehicles there is nothing in the computer diagnostics to indicate fuel pressure. Many PCMs assume that the fuel pressure is within the specified range.

An accurate HO$_2$S test may be performed with a graphing voltmeter or a lab scope. Some scan tools display voltage graphs. A satisfactory HO$_2$S should provide an acceptable waveform. A normal and some abnormal O$_2$ sensor waveforms are shown in **Figure 45-1**. The HO$_2$S must have the proper number of cross counts, and display a fast transition time from lean to rich and rich to lean **(Figure 45-2)**. **Cross counts** are the number of times that the HO$_2$S

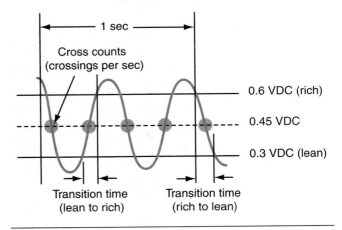

Figure 45-2 An HO$_2$S waveform must have the proper cross counts and the correct lean-to-rich transition time.

cycles from rich to lean in a given time period, usually 1 second. The HO$_2$S **transition time** is the time required for this sensor to switch from lean to rich and rich to lean. Good transition times are less than 100 milliseconds.

> **Tech Tip** *Cross counts can be displayed on a scan tool. Cross counts should be two to three at 2,000 RPM with no load.*

A defective spark plug that causes cylinder misfiring provides an HO$_2$S voltage waveform as indicated in **Figure 45-3**. A leaking injector that results in a rich air-fuel ratio causes an HO$_2$S voltage waveform as displayed in **Figure 45-4**.

The HO$_2$S heater may be tested. Identify the heater terminals on the O$_2$ sensor **(Figure 45-5)**. Connect an ohmmeter across the heater voltage supply and ground terminals **(Figure 45-6)**. The heater should have the specified resistance, which is usually about 10 ohms. If the ohmmeter reading is infinite, the heater has an open circuit and the sensor must be replaced.

An inoperative HO$_2$S heater causes slow sensor warm-up and the PCM remains in open loop for a longer time. The PCM is also more likely to go back into open loop with the engine idling. This results in a richer air-fuel ratio and reduced fuel economy.

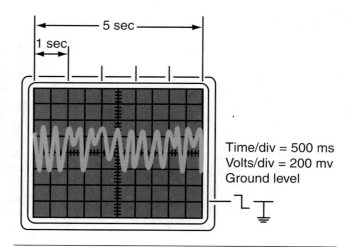

Figure 45-3 An unsatisfactory HO$_2$S waveform caused by a defective spark plug.

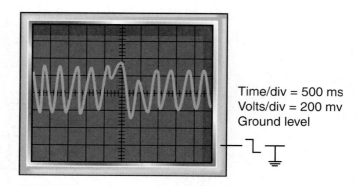

Figure 45-4 An unsatisfactory HO$_2$S waveform caused by a leaking injector.

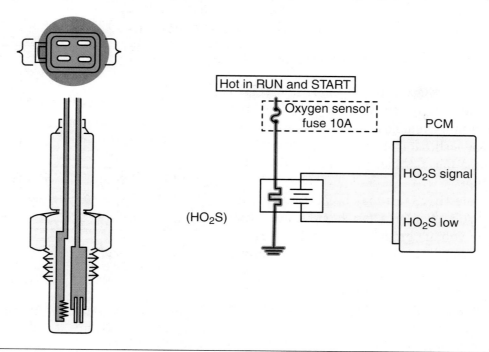

Figure 45-5 HO$_2$S heater terminals.

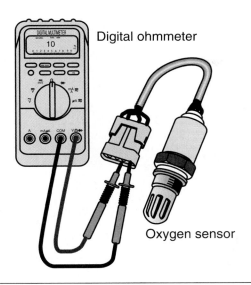

Figure 45-6 Testing the HO₂S heater for proper resistance.

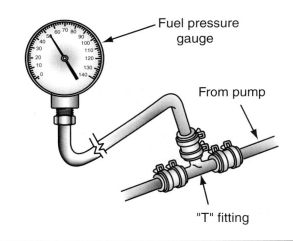

Figure 45-8 On fuel systems that do not have a Schrader valve, it will be necessary to connect a tee fitting in the fuel pressure line.

> **Tech Tip** *Downstream HO₂Ss have a lower voltage signal and cycle slower from low voltage to high voltage compared to upstream HO₂Ss. If the upstream and downstream HO₂Ss have the same voltage waveform, the catalytic converter is defective.*

FUEL PUMP PRESSURE TESTING

Connect the appropriate fuel pressure gauge to the Schrader valve on the fuel rail or a tee fitting at the fuel inlet line at the fuel rail **(Figure 45-7 and**

Figure 45-8). Cycle the ignition switch on and off several times and read the fuel pressure. The manufacturer may also provide specifications for fuel pressure at idle and at WOT. The fuel pressure must equal the vehicle manufacturer's specifications. A defective fuel pump, a restricted fuel filter, an air leak in a fuel line, low voltage at the fuel pump, or a high resistance in the fuel pump ground circuit may cause low fuel pressure. Before replacing a fuel pump, always be sure the other causes of low fuel pressure do not exist. A fuel pump flow test should also be performed to indicate the amount of fuel the pump supplies in a specific length of time.

ENGINE COOLANT TEMPERATURE (ECT) SENSOR AND INTAKE AIR TEMPERATURE (IAT) SENSOR DIAGNOSIS

The diagnoses of the ECT sensor and IAT sensor are similar because both of these sensors contain thermistors. With the ignition switch turned off and the sensor connector disconnected, an ohmmeter may be connected to the sensor terminals to test the ECT or IAT sensor. The sensor to be tested may be removed and placed with a thermometer in a

Figure 45-7 Connect the fuel pressure gauge to the Schrader fitting on the fuel rail.

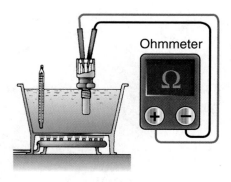

Figure 45-9 Testing an ECT sensor with an ohmmeter.

Cold-10 K-ohm resistor		Hot-909-ohm resistor	
-20 degree F	4.7v	110 degree F	4.2 v
0 degree F	4.4v	130 degree F	3.7v
20 degree F	4.1v	150 degree F	3.4v
40 degree F	3.6v	170 degree F	3.0v
60 degree F	3.0v	180 degree F	2.8v
80 degree F	2.4v	200 degree F	2.4v
100 degree F	1.8v	220 degree F	2.0v
120 degree F	1.25	240 degree F	1.62v

Figure 45-11 ECT sensor voltage vs. temperature specifications.

container filled with water **(Figure 45-9)**. When the water is heated, the sensor must have specified resistance at various temperatures **(Figure 45-10)**.

> **Tech Tip** *In some ECT circuits, the PCM changes the resistance in the circuit by switching an internal resistor into the circuit at 120°F (48.8°C). This resistance change causes a significant change in the voltage drop across the sensor.*

If either sensor does not have the specified resistance, replace the sensor. The ECT or IAT sensor may also be tested with the engine running by measuring the voltage across the sensor terminals. These sensors must have the specified voltage drop in relation to the sensor temperature as specified by the manufacturer **(Figure 45-11)**.

> **Tech Tip** *When diagnosing input sensors, you must understand the effect of improper sensor signals on engine performance and other sensor signals. If the ECT sensor resistance is higher than specified, the voltage signal is also high. Under this condition, the air-fuel ratio is too rich and the PCM does not enter closed loop properly. High ECT sensor resistance may be caused by an engine thermostat that is continually open.*

MANIFOLD ABSOLUTE PRESSURE SENSOR DIAGNOSIS

When a MAP sensor problem is encountered, be sure the vacuum hose to this sensor is not kinked or leaking. Before testing the MAP sensor, test the reference voltage to the sensor. With the ignition switch on, connect a voltmeter from the 5-volt reference wire to a good engine ground. The voltage at this terminal should be 4.8 volts to 5.2 volts. If

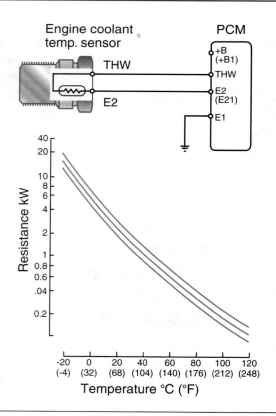

Figure 45-10 Resistance specifications for an ECT sensor at various temperatures.

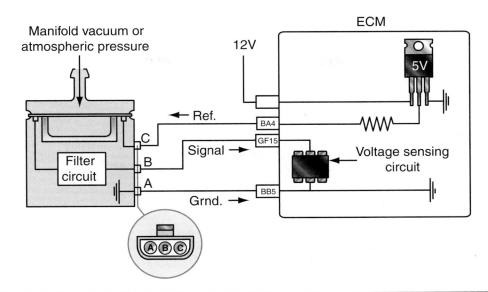

Figure 45-12 MAP sensor terminals.

this voltage is not present, check the battery voltage and the 5-volt reference wire from the sensor to the PCM.

With the ignition switch on, connect a voltmeter from the MAP sensor ground wire to ground. The voltage drop across this wire should be less than 0.2 volt. Connect a voltmeter to the MAP sensor signal wire and ground and connect a vacuum hand pump to the sensor vacuum port **(Figure 45-12)**. With the ignition switch on, the sensor must have the specified voltage in relation to the vacuum supplied to the sensor. If the MAP sensor does not have the specified voltage signal, replace the sensor.

Vacuum Applied	Output Frequency
0 in. Hg	152–155 Hz
5 in. Hg	138–140 Hz
10 in. Hg	124–127 Hz
15 in. Hg	111–114 Hz
20 in. Hg	93–98 Hz

Figure 45-13 MAP sensor hertz specifications in relation to vacuum.

When testing a MAP sensor that provides a hertz-type signal, connect a digital multimeter from the MAP sensor signal wire to ground. Switch the multimeter to the hertz scale and turn the ignition switch on. Connect a vacuum hand pump to the MAP sensor vacuum port and supply the specified vacuum to the sensor. The MAP sensor must have a specified hertz signal in relation to the vacuum supplied to the sensor **(Figure 45-13)**.

THROTTLE POSITION SENSOR DIAGNOSIS (TP)

When diagnosing a TP sensor, test the 5-volt reference and ground wires to the sensor as explained

Tech Tip *In some EFI systems, a defective MAP sensor may cause a DTC indicating a problem in the EGR system. In some systems when the PCM opens the EGR valve, the PCM checks the change in the MAP sensor signal caused by the change in manifold vacuum when the EGR valve opened. If the MAP sensor is defective and not producing a proper signal, the MAP signal does not change when the EGR valve opens. The PCM interprets this as an EGR fault because the PCM thinks the EGR valve did not open.*

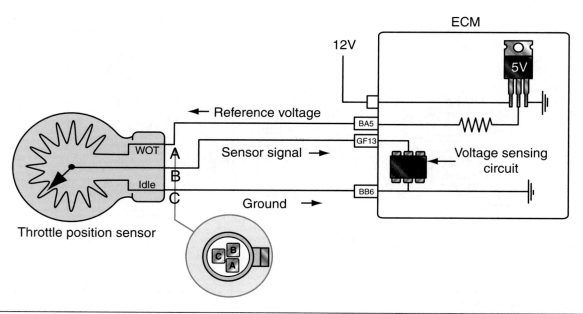

Figure 45-14 TPS terminals.

in the MAP sensor diagnosis **(Figure 45-14)**. Turn the ignition switch on and observe the TP sensor voltage on the scan tool as the throttle plate is opened as slowly and smoothly as possible. The TP sensor voltage signal should increase from 0.5 volt to 1 volt at idle to 4.5 volts at wide-open throttle. Always use the specifications for the vehicle being diagnosed. If the TP sensor voltage does not increase smoothly and gradually, replace the sensor.

NOTE: Many times technicians may refer to the throttle position sensor as a *TP*, *TPS*, or *TP sensor*.

Tech Tip *A lab scope will provide better results than a scan tool when checking the TP sensor. The lab scope is much faster and makes a direct measurement of the TP sensor voltage.*

A TP sensor may develop a worn spot on the variable resistor, especially on vehicles driven for long periods of time at a constant speed. This worn spot on the variable resistor may cause a very momentary glitch in the TP sensor voltage signal that is not noticeable on a digital voltmeter. A lab scope can

provide a voltage sweep across the screen that is in seconds to milliseconds. This instrument will pick out momentary glitches in TP sensor voltage signals.

Connect the lab scope to the TP sensor signal wire and slowly increase the throttle opening with the ignition switch on. Observe the scope for a satisfactory TP voltage waveform **(Figure 45-15)**. If an unsatisfactory TP waveform with a momentary glitch is observed, then the TP sensor will need to be

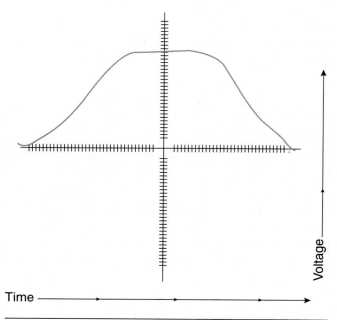

Figure 45-15 A satisfactory TPS waveform.

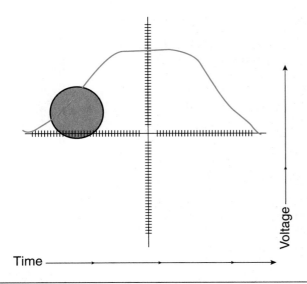

Figure 45-16 An unsatisfactory TPS waveform.

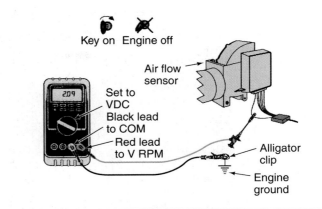

Figure 45-17 A voltmeter connected to a MAF.

replaced **(Figure 45-16)**. A momentary glitch in a TP voltage signal may cause acceleration stumbles.

MASS AIRFLOW SENSOR DIAGNOSIS

To test a hot-wire-type MAF sensor, observe the MAF on the scan tool. This information may be displayed as grams of air per second or in pounds per minute. Hold the accelerator pedal steady at several different engine speeds and observe the grams per second. At a constant throttle opening, the grams-per-second reading should remain the same.

With the ignition switch on, connect a voltmeter from the MAF sensor ground wire to ground. The voltage drop across this wire should be less than 0.2 volt **(Figure 45-17)**. The readings may also be observed as the engine is running. These readings can be graphed at different RPMs. The readings should be in a nearly straight line **(Figure 45-18)**. If the sensor does not have the specified voltage signal, replace the sensor.

When testing a sensor that provides a hertz-type signal, connect a digital multimeter from the sensor signal wire to ground. Switch the multimeter to the hertz scale and turn the ignition switch on. The MAF sensor must have a specified hertz signal **(Figure 45-19)**.

The sensor may also be lightly tapped with a screwdriver while observing the readings. If this

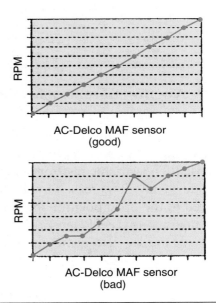

Figure 45-18 Acceptable and unacceptable MAF readings.

reading fluctuates up or down, the sensor is defective. The result of this problem is usually rough idle and hesitation during acceleration.

The sensor may be cleaned with a spray-type throttle body cleaner, but care should be taken not to spray other components that may be sensitive to the cleaner. Some MAF sensors may be disassembled and the hot wire may be cleaned with a Q-tip dampened in throttle body cleaner. Consult the vehicle manufacturer's recommendations regarding this service procedure. If the hot wire cannot be cleaned, sensor replacement is required.

A vane-type MAF sensor may be tested with a digital multimeter connected to the MAF voltage

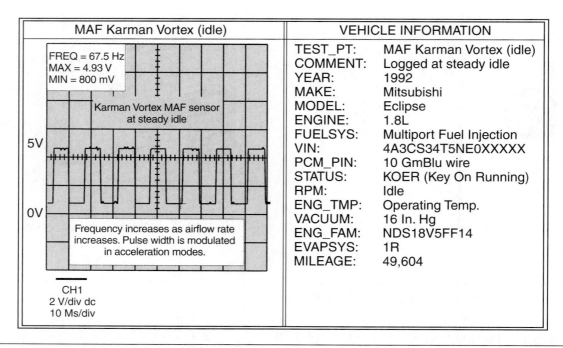

Figure 45-19 A normal trace for a hertz type MAF sensor.

signal wire. Switch the multimeter to DC volts. With the ignition switch on, slowly push the vane open and closed. The variable resistor in the sensor should provide a smooth increase in voltage as the vane is moved to the open position and functions in a similar manner as the TP sensor. If the MAF does not provide the specified, smooth voltage signal, replace the sensor.

NOTE: A vane-type MAF sensor may be referred to as a *VAF (vane-airflow) sensor.*

KNOCK SENSOR DIAGNOSIS

The knock sensor may be tested with the engine running and the engine knock data displayed on the scan tool. Some scan tools display YES or NO for the engine knock data. During engine operation at idle or a constant throttle opening, the knock data display should indicate NO. During hard engine acceleration at normal engine operating temperature, the knock data may indicate YES momentarily. The engine block may also be tapped with a hammer and the scan tool display observed.

The knock sensor may also be tested with a lab scope. The sensor should be tapped while the lab scope is attached to the output leads of the sensor. The sensor should produce a voltage that is observed on the scope when the sensor is tapped.

INPUT SENSOR SERVICE

When connecting electronic test equipment to input sensors, do not probe through insulation on wires to make a meter connection. This action may cause corrosion in the wire and eventually a resistance that affects the sensor signal. A back-probe kit may be used. A T-pin or other small probe may be used to back probe terminals on input sensors to complete a meter connection **(Figure 45-20)**. Always be careful not to damage the weather-pack seal on the wiring connector. A T-pin is like a common pin used in sewing, but the T-pin is T-shaped. T-pins are available in the sewing departments of some large retailers.

Tech Tip *Do not pierce the wiring insulation with a pin or probe. The opening left behind by the pin will allow corrosion to form inside the wire.*

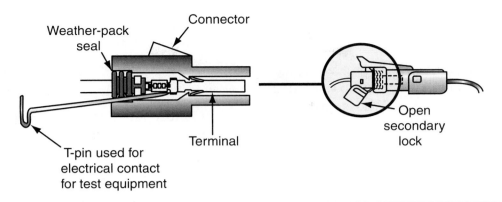

Figure 45-20 A T-pin may be used to back-probe an electrical connector.

When installing an HO$_2$S, always place a small amount of anti-seize compound on the sensor threads. This action keeps the sensor from seizing into the exhaust system, which makes it very difficult to remove the sensor the next time it requires replacement.

Summary

- A rich air-fuel ratio results in a low exhaust gas oxygen content which causes high HO$_2$S voltage and a lean air-fuel ratio causes a high exhaust gas oxygen content which causes low HO$_2$S voltage.

- An HO$_2$S may be tested with a digital voltmeter, a graphing voltmeter, or a lab scope.

- When diagnosing EFI systems, one of the first tests must be fuel pressure.

- ECT and IAT sensors may be tested with an ohmmeter, or a voltmeter may be used to measure the voltage drop across these sensors.

- Measuring the MAP voltage signal in relation to the vacuum supplied to the sensor may test a MAP sensor.

- A MAP sensor that produces a hertz-type signal may be tested using a multimeter with a hertz scale.

- The TP may be tested by measuring the TP voltage signal in relation to throttle opening.

- A hot-wire MAF sensor may be tested by measuring the grams-per-second airflow through the sensor with a scan tool.

- Measuring the voltage signal from the sensor in relation to vane opening may test a vane-type MAF sensor.

- A knock sensor may be tested with an lab scope or a scan tool.

Review Questions

1. Technician A says if the HO$_2$S voltage signal is continually low, the fuel pressure may be excessive. Technician B says if the HO$_2$S voltage signal switched rapidly, the fuel injectors may be restricted. Who is correct?

   A. Technician A

   B. Technician B

   C. Both Technician A and Technician B

   D. Neither Technician A nor Technician B

2. Technician A says the module in a hot-wire MAF sensor maintains the hot wire temperature at the same temperature as the cold wire. Technician B says at a constant throttle opening, the grams-per-second airflow reading on the MAF sensor should be fluctuating. Who is correct?

   A. Technician A

   B. Technician B

   C. Both Technician A and Technician B

   D. Neither Technician A nor Technician B

3. Technician A says a bank 1 sensor 2 HO_2S is located upstream from the catalytic converter. Technician B says a bank 2 sensor 1 HO_2S is located on the opposite side of the engine from number 1 cylinder. Who is correct?

   A. Technician A

   B. Technician B

   C. Both Technician A and Technician B

   D. Neither Technician A nor Technician B

4. Technician A says the resistance of the ECT and IAT sensors should decrease as the sensor temperature increases. Technician B says if the engine thermostat is stuck open, the air-fuel ratio will be rich. Who is correct?

   A. Technician A

   B. Technician B

   C. Both Technician A and Technician B

   D. Neither Technician A nor Technician B

5. An air leak is present in the air hose between the MAF sensor and the throttle body. The most likely result of this air leak is:

   A. excessive spark advance.

   B. a lean air-fuel ratio.

   C. low fuel pump pressure.

   D. a high voltage signal from the ECT sensor.

6. Technician A says an HO_2S with a bank 2, sensor 2 designation is located upstream from the catalytic converter. Technician B says an HO_2S with a bank 2, sensor 2 designation is located on the same side of the engine as number 1 cylinder. Who is correct?

   A. Technician A

   B. Technician B

   C. Both Technician A and Technician B

   D. Neither Technician A nor Technician B

7. At normal operating engine temperature, a typical HO_2S voltage signal should cycle between:

   A. 0.3 volt and 0.8 volt.

   B. 0.1 volt and 0.3 volt.

   C. 0.2 volt and 0.5 volt.

   D. 0.5 volt and 0.9 volt.

8. The most likely result of excessive fuel pump pressure in a port-injected fuel system is:

   A. reduced spark advance.

   B. engine detonation.

   C. an inoperative EGR valve.

   D. a rich air-fuel mixture.

9. An engine thermostat that is stuck in the open position may cause:

   A. a defective EGR valve.

   B. a rich air-fuel ratio.

   C. a knock sensor voltage signal.

   D. a lower-than-normal ECT sensor resistance.

10. A TP sensor has a worn spot near the center of the variable resistor. The most likely result of this problem is:

    A. hesitation during acceleration.

    B. engine surging and misfiring at high speed.

    C. high idle speed with the engine warmed up.

    D. erratic engine idle operation.

11. On a conventional MAP sensor, the voltage signal should _____ as the vacuum supplied to the sensor increases.

12. The TP voltage signal should be approximately _____ volts at wide-open throttle.

13. Before testing an HO_2S, the sensor should be _____ by running the engine at _____ rpm or _____ minutes.

14. An erratic grams-per-second airflow reading on a hotwire MAF sensor may indicate the _____ _____ is contaminated.

15. Explain the operation of a zirconia-type HO_2S in relation to exhaust gas oxygen.

16. Describe the operation of a hot-wire MAF sensor in relation to engine speed.

17. Explain the cleaning procedure for the hot wire in a MAF sensor.

18. Explain four causes of low fuel pump pressure.

46 Engine Control Systems

Key Terms

California Air Resources Board (CARB)

Central port injection (CPI) system

Drive cycle

Engine warm-up cycle

Injector pulse width

Intake manifold tuning valve (IMTV)

On-Board Diagnostics (OBD)

Port fuel injection (PFI) system

Sequential fuel injection (SFI) system

Stoichiometric air-fuel ratio

Throttle body injection (TBI) system

Trip

INTRODUCTION

Technicians must understand the various types of fuel injection systems to successfully diagnose these systems. For example, you must understand how the injectors are connected to the PCM in various injection systems and how the PCM maintains the proper air-fuel ratio. You must also understand the various monitors in OBD II systems and the requirements to illuminate the MIL in these systems. It is very important to understand OBD terminology, such as trip and drive cycle.

THROTTLE BODY INJECTION (TBI) SYSTEMS

In a **throttle body injection (TBI) system,** the throttle body assembly may contain one or two injectors depending on the engine displacement. Four-cylinder engines usually have a single injector in the throttle body assembly **(Figure 46-1)** and V6 or V8 engines often have dual injectors. The injector body is sealed to the throttle body with O-ring seals. When the ignition switch is turned on, 12 volts are supplied through a fuse to the injector(s)

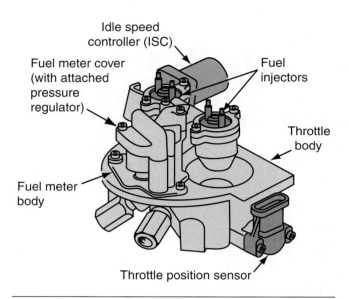

Figure 46-1 A single throttle body assembly.

(Figure 46-2). The other terminal on each injector is connected to the PCM. The PCM supplies a ground for the injector windings to open each injector. The PCM grounds an injector each time the ignition system fires a spark plug. When the next spark plug fires in a dual throttle body, the PCM grounds the opposite injector. Grounding an injector winding allows current to flow through the

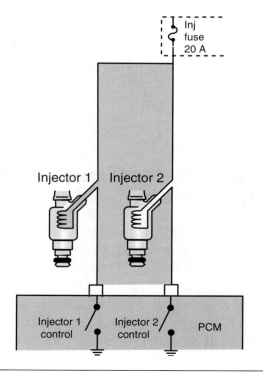

Figure 46-2 Dual throttle body injector wiring connections.

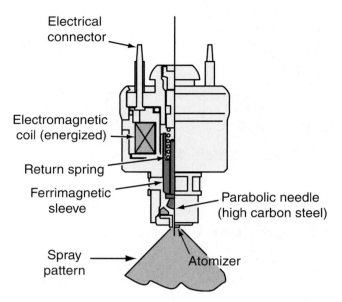

Figure 46-3 A throttle body injector.

winding and the resulting coil magnetism lifts the injector plunger. When the plunger tip is lifted off its seat, fuel is discharged from the injector orifice **(Figure 46-3)**.

The amount of fuel discharged by the injector is determined by the injector pulse width. **Injector pulse width** is the length of time in milliseconds that the PCM keeps the injectors open. At idle speed, the PCM may supply an injector pulse width of 2 milliseconds, whereas at wide-open throttle the injector pulse width may be 7 to 12 milliseconds. **Figure 46-4** shows a scope pattern of a fuel injector that is operating at a pulse width of 5 milliseconds.

The PCM always provides the correct injector pulse width to maintain the stoichiometric air-fuel ratio. The **stoichiometric air-fuel ratio** is the ideal air-fuel ratio at which the engine provides the best performance and economy. In a gasoline fuel system, this ratio is about 14.7 pounds of air to every 1 pound of fuel. This ratio is expressed as 14.7:1.

Some TBI systems have an idle speed control (ISC) motor mounted on the throttle body assembly. The ISC motor is a reversible motor that is controlled by the PCM. The stem of the ISC motor pushes against the throttle linkage to vary or adjust the throttle plate opening. The PCM operates the ISC motor to supply the proper idle speed at all engine temperatures. When the engine coolant is cold, the PCM operates the ISC motor to supply a faster

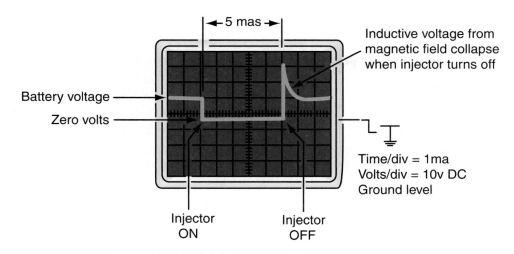

← 5 mas →

Inductive voltage from
magnetic field collapse
when injector turns off

Battery voltage

Zero volts

Time/div = 1ma
Volts/div = 10v DC
Ground level

Injector
ON

Injector
OFF

Figure 46-4 A fuel injector operating at a pulse width of 5 milliseconds.

idle speed. Some ISC motors have an idle switch in the ISC motor stem. When the throttle linkage contacts the ISC motor stem, the idle switch closes and sends a signal to the PCM. The PCM controls the ISC motor only when the throttle linkage is contacting the motor stem.

Some throttle body assemblies have an idle air control (IAC) motor mounted on the throttle body **(Figure 46-5)**. The IAC motor is operated by the PCM. This motor contains a plunger that opens and closes an air passage. The airflow in the IAC motor passage bypasses the throttle. The IAC motor plunger controls idle speed by regulating the amount of air bypassing the throttle. When the engine is cold, the PCM moves the IAC motor plunger to provide more airflow through the IAC

motor passage around the throttle. This action increases engine rpm. As the engine warms up, the PCM gradually moves the IAC motor plunger toward the closed position to reduce engine rpm.

PORT FUEL INJECTION (PFI) SYSTEM

In a **port fuel injection (PFI) system**, the injectors are positioned in the intake manifold and the injector tips are located near the intake ports **(Figure 46-6)**. An O-ring on the lower end of the injector seals the injector to the intake manifold and a second O-ring on the upper end of the injector seals the injector to the fuel rail. The injector is

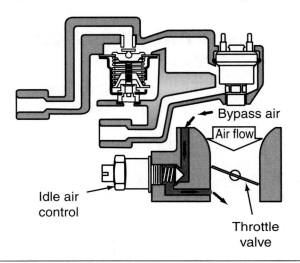

Bypass air

Air flow

Idle air
control

Throttle
valve

Figure 46-5 An idle air control (IAC) motor.

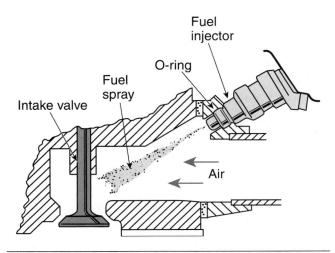

Fuel
injector

O-ring

Fuel
spray

Intake valve

Air

Figure 46-6 A port injector located in the intake port.

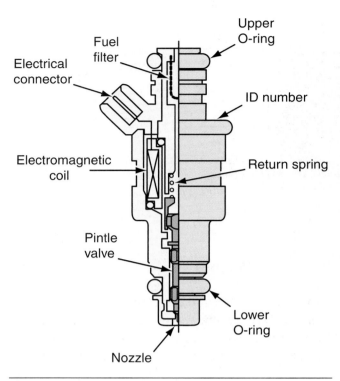

Figure 46-7 A typical fuel injector.

attached to the PCM through the electrical connector. The PCM will provide a ground for the injector windings to allow current to flow through the winding. The coil magnetism lifts the injector plunger and fuel is discharged from the injector orifice **(Figure 46-7)**.

In some PFI systems, the injectors are connected in pairs or groups to the PCM and the PCM operates two to four injectors simultaneously. In other injection systems, the injectors are grounded individually by the PCM **(Figure 46-8)**. This type of system is sometimes called **sequential fuel injection (SFI) system**. The PCM opens each injector a significant number of crankshaft degrees before the intake valve actually opens. This action fills the intake port with fuel vapor and ensures the proper supply of air-fuel mixture to the cylinder when the intake valve opens.

The PCM always supplies the correct injector pulse width to maintain 14.7:1 air-fuel ratio under part throttle engine operation. During cold engine operation, hard acceleration, or wide-open throttle conditions, the PCM supplies a richer air-fuel ratio to maintain engine performance. When the engine is decelerating, the PCM supplies a leaner air-fuel ratio to improve fuel mileage and emissions.

PFI systems usually have an IAC motor to control idle speed **(Figure 46-9)**. The IAC motor is operated by the PCM. Some of the idle air controllers are controlled by a signal that varies the pulse width **(Figure 46-10)**. In this system the idle air controller is a solenoid that provides the air for idle. As the pulse width increases, so does the airflow. If the PCM needs to reduce idle it then reduces the pulse width to the idle air controller.

Many PCMs have an input from the park/neutral switch. The signal from this switch is used by the PCM to help control idle rpm. Some PCMs have an input from the power steering pressure switch. When this signal is received during a turn, the PCM may increase the idle speed of the engine. This action helps to prevent engine stalling at low speeds from the compressor and power steering loads on the engine.

Some PCMs contain the cruise control module, so the inputs from the cruise control switches are sent to the PCM. Many PCMs operate the low fan control and high fan control relays to operate the cooling fan at low and high speed. When the PCM receives a signal from the ECT sensor indicating that cooling fan operation is required, the PCM energizes the appropriate cooling fan relay. When the relay contacts close, voltage is supplied through these contacts to the cooling fan motor.

Tech Tip *The PCM will also turn on the cooling fans when the AC system is turned on.*

CENTRAL PORT INJECTION (CPI) SYSTEMS

In a **central port injection (CPI) system**, one central injector module is mounted in the intake manifold. The upper half of the intake manifold is bolted to the lower half. This upper half may be removed to access the CPI components **(Figure 46-11)**. The pressure regulator is mounted with the central injector. The fuel supply and return lines are connected to the central injector module and the pressure regulator. These fuel lines are sealed with rubber grommets where they enter the intake manifold to prevent vacuum leaks. A port in the pressure regulator allows the intake manifold

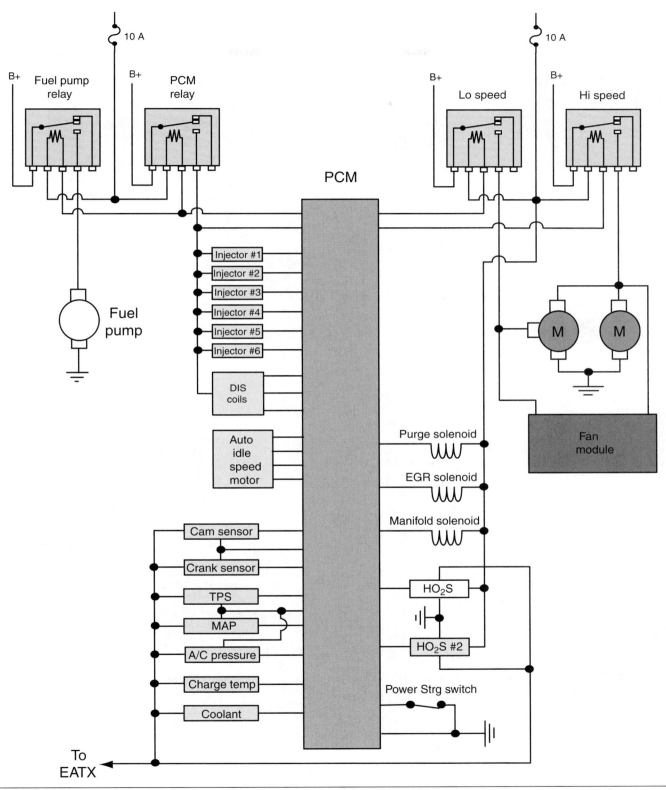

Figure 46-8 A port fuel injection (PFI) system.

Figure 46-9 Typical idle air controllers used in port fuel injection.

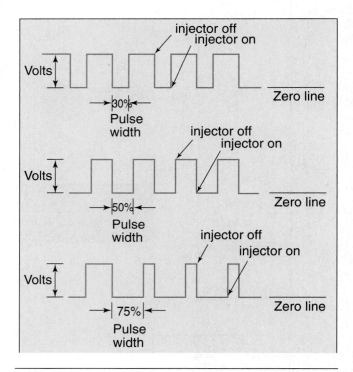

Figure 46-10 The PCM can vary pulse width to control an idle air solenoid.

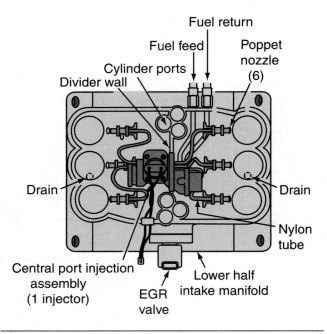

Figure 46-11 A central port injection (CPI) system.

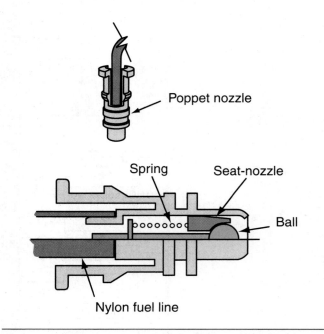

Figure 46-12 A poppet nozzle in a CPI system.

vacuum to be supplied to this component. A poppet nozzle is located in each intake port and a tube is connected from the central port injector to each poppet injector. The seats in the poppet nozzles are closed by spring pressure **(Figure 46-12)**.

Voltage is supplied to the central injector module and the PCM grounds this injector. When the PCM grounds the central port injector winding, the coil magnetism lifts the injector plunger and the flapper valve on the lower end of the plunger. Lifting this valve supplies fuel pressure to all the poppet nozzles

simultaneously. When the fuel pressure at each poppet nozzle reaches 39 psi (269 kPa), the fuel pressure lifts the poppet nozzle seats and fuel is discharged from the poppet nozzles into the intake ports. The PCM supplies the proper central injector module pulse width to provide the correct air-fuel ratio.

VARIABLE INTAKE MANIFOLD TUNING

Intake manifolds generally give better engine performance if the runners are longer for lower rpm and give better performance if the runners are shorter for higher rpm. When the intake valve is open on the engine, air is pushed into the engine by atmospheric pressure. Essentially the air in the intake runner is moving rapidly toward the cylinder. When the intake valve closes, this moving air suddenly stops and forms a high-pressure wave. This high-pressure wave then reflects back down the intake manifold runner. When it reaches the end of the intake runner it reflects back down the intake runner toward the cylinder. If the intake manifold runner is just the right length, the pressure wave will arrive at the intake valve just as it opens for the next intake stroke. This pressure wave can help to better fill the cylinder. A tuned intake manifold can benefit engine performance but can only do so at a narrow rpm range.

An intake manifold that can provide the best for low and high rpm was developed. A number of these variable intake manifold tuning designs are in production. A relatively simple design uses both long and short intake manifold runners **(Figure 46-13)**. A valve is used to separate the long and short runners. This valve is located between the two plenums in the intake manifold and is designed to block the flow of air between the two plenums.

A vacuum diaphragm or servo motor operates the **intake manifold tuning valve (IMTV)**. Vacuum is supplied to this diaphragm through a PCM-operated vacuum solenoid. At engine speeds below 3,600 rpm, there is no vacuum supplied to the IMTV and this valve allows normal airflow through the intake manifold. At speeds above 3,600 rpm, the PCM operates the IMTV solenoid and vacuum is supplied through this solenoid to the MTV diaphragm. This action opens the IMTV and this valve position provides shorter intake manifold air passages to improve engine performance.

ON-BOARD DIAGNOSTIC (OBD) AND ON-BOARD DIAGNOSTIC II (OBD II) SYSTEMS

In 1988, the **California Air Resources Board (CARB)** adopted **On-Board Diagnostics (OBD)** regulations for cars sold in California. These regulations required the engine computer system to monitor fuel control, EGR, emission components, and the PCM. OBD systems also had to monitor all sensors used for fuel and emission control for opens and shorts. The PCM also had to monitor fuel trim, EGR, and HO_2S. These systems were required to illuminate the MIL if a defect occurred and a DTC had to be set in the PCM memory.

> **Tech Tip** *The number of monitors on a vehicle may vary depending on the vehicle model year. The main difference between an OBD I and an OBD II system is in the PCM software. An OBD II PCM contains a number of monitors that test various systems that affect emissions. The most significant additional hardware on an OBD II system is the extra HO_2S mounted downstream from the catalytic converter.*

In the late 1980s, the CARB began investigating the possibility of using OBD system diagnosis to supplement inspection maintenance (I/M) emission programs. Simple OBD diagnostic tests could be

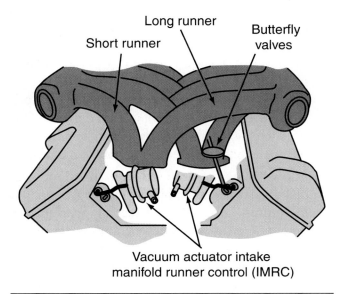

Long runner
Short runner
Butterfly valves

Vacuum actuator intake manifold runner control (IMRC)

Figure 46-13 An intake manifold tuning valve (IMTV).

performed with much less expensive equipment compared to tailpipe emission tests.

In 1990, CARB implemented the first OBD requirements combined with emission testing. However, this combination experienced some problems. One of the difficulties in using OBD diagnosis for emission testing was that the OBD systems did not have much standardization. Vehicle manufacturers were each using their own DLC. Different scan tool software was required for each manufacturer. At that time, OBD systems did not have the capability to monitor critical emission-related components such as catalytic converters, evaporative emission systems, and secondary air injection systems. The OBD systems did not even have the capability to perform a complete test of the oxygen sensor(s).

CARB and the EPA realized that a number of OBD system improvements had to be implemented if OBD systems were to be useful in I/M emission test programs. The result of these discoveries was the installation of OBD II systems on all light-duty vehicles on 1996 and later models. In 1994 and 1995, some vehicles had partial OBD II systems. Federal requirements demand that OBD II systems have many standard features including a standard DLC in the same location on all vehicles.

The EPA passed legislation requiring the implementation of OBD II diagnosis with emission I/M programs. A significant number of states have implemented OBD II diagnosis with an I/M emission program.

OBD II Monitoring

According to EPA rules, an OBD II system must detect the following emission-related defects and alert the driver:

1. The engine misfire monitor must detect any engine misfire condition that causes exhaust emissions of one and one-half times the standard for non-methane hydrocarbons (NMHC), CO, or NOx. The PCM monitors the crankshaft position sensor signal in the misfire monitor. Each time a spark plug fires, the crankshaft should speed up momentarily. If a cylinder misfires, the crankshaft does not speed up, and the PCM detects this problem from the crankshaft position sensor signal. A specific number of misfires must occur in 1,000 cylinder firings before the PCM takes the necessary action such as setting a DTC.

2. The HO_2S monitor must detect any HO_2S defect that results in exhaust emissions of one and one-half times the standard for NMHC, CO, or NOx. The HO_2S monitor must also detect any other system component that makes the HO_2S sensor incapable of performing its intended function in the system. The PCM also monitors each HO_2S heater for proper current flow. Low current flow indicates a burned-out heater or open heater wire and the PCM interprets this situation as a failed HO_2S heater monitor.

3. The catalytic converter monitor must detect any catalytic converter problem that results in exhaust emissions of one and one-half times the standard for NMHC using an average 4,000-mile aged converter. The PCM monitors the upstream and downstream HO_2S to monitor the converter operation. The voltage cycles on the downstream HO_2S should slow switching times compared to the upstream HO_2S. If the upstream and downstream HO_2S signals are similar then the catalytic converter is not functioning properly. The PCM will then take the necessary action, such as setting a DTC.

4. The evaporative system monitor must be able to detect if there is an absence of evaporative purge airflow from the evaporative system. This monitor must also be able to detect any leak in the evaporative system equal to or greater than a leak caused by a 0.040-inch (1.016 millimeters) or a 0.020-inch (0.508 millimeter) diameter orifice, depending on the vehicle model year. This leak detection capability excludes the tubing between the purge valve and the intake manifold.

5. The exhaust gas recirculation monitor, secondary air injection monitor, and the fuel control monitor must be able to detect any malfunction or deterioration in the powertrain system or any component directly responsible for emission control that causes emissions of NMHC, CO, or NOx to be one and one-half times the standard for these emissions.

6. The comprehensive monitor must be capable of detecting any defect or deterioration in the emission-related powertrain control system. This includes any defects in any PCM input sensor or output control device. Defects in these components are defined as malfunctions that cause the component to not meet certain continuity, rationality, or functionality checks performed at specific intervals by the PCM.

7. The secondary air injection monitor checks the operation of the air pump. The PCM turns on the electric drive air pump and checks for a leaner HO_2S signal in a specific time limit. If the HO_2S signal does not change, the PCM considers the AIR pump to be defective.

8. The fuel system monitor checks the long-term fuel trim to determine if this parameter is at the high or low limit. The fuel system monitor also checks the short-term fuel trim. If the PCM detects a long-term fuel trim at the high or low limit during the fuel system monitor, the PCM

considers the fuel system to have failed the monitor test.

9. The thermostat monitor checks the operation of the engine thermostat. A timer circuit monitor in the PCM checks the length of time required for the engine coolant temperature to increase a specific number of degrees during engine warm-up.

OBD II Definitions

A **drive cycle** may be defined as an engine startup and vehicle operation that allows the PCM to enter closed loop and allows all the monitors to complete their function. Specific driving conditions for certain lengths of time are required for all the monitors to complete their function (**Figure 46-14**). An **engine warm-up cycle** may be defined as an engine startup and engine operation until the coolant temperature increases at least 40°F (22°C) from startup and reaches a minimum temperature of 160°F (70°C).

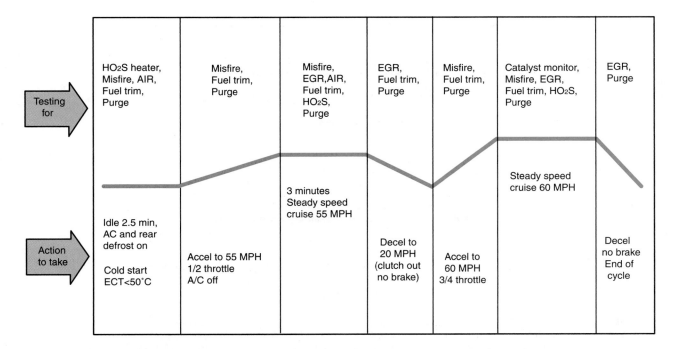

DIAGNOSTIC TIME SCHEDULE FOR I/M READINESS
(Total time 12 minutes)

Figure 46-14 An OBD II drive cycle.

A **trip** may be defined as an engine startup and vehicle operation that allows all the monitors to complete their function except the catalytic converter monitor **(Figure 46-15)**.

OBD II Operation

Many OBD II system defects do not cause illumination of the MIL light until the PCM senses the defect on two consecutive drive cycles. However, many defects cause a DTC to be set in the PCM memory the first time the defect is sensed. When a defect occurs and a DTC is set in the PCM memory, the PCM also stores information in the freeze frame. This information may be retrieved with a scan tool during system diagnosis. If the engine switches to any default mode that is abnormal for present driving conditions, the MIL light must be turned on. If engine misfiring occurs that may cause catalytic converter damage, the MIL light flashes once per second. Current DTCs represent defects that are continually present. Intermittent or history DTCs represent faults that were present for a short time. Pending DTCs represent faults that set a DTC in the PCM memory, but the fault did not occur the second time and turn on the MIL light.

When a defect detected by the fuel control monitor or engine misfire monitor does not recur in three consecutive trips under similar driving conditions, the PCM turns the MIL light off. Similar

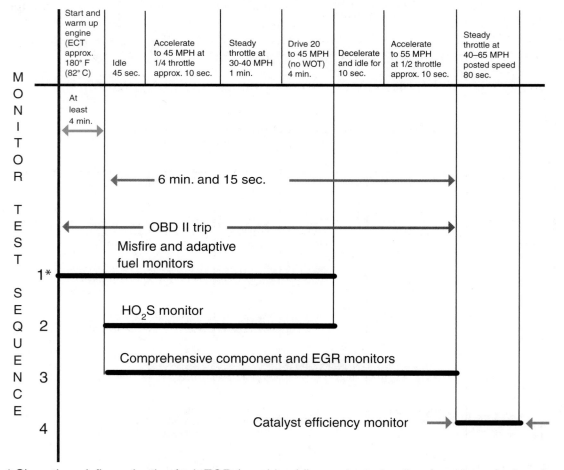

* Since the misfire, adaptive fuel, EGR (requiring idles and accelerations), and comprehensive monitors are continuously checked by the OBD-II system, the test sequence may vary on each vehicle due to outside ambient temperature, engine/vehicle performance temperature, and driving conditions.

Figure 46-15 An OBD II trip.

driving conditions are: engine speed within 375 rpm of the original defect occurrence, engine load within 20 percent, and engine warm-up conditions are the same and no new defects have been sensed. For defects other than those detected by the fuel control monitor and the misfire monitor, the MIL light is turned off if the defect is not sensed and no new problems are detected in three consecutive trips. History DTCs are cleared from the PCM memory after forty consecutive warm-up cycles have occurred without a malfunction.

A scan tool connected to the DLC displays system status information indicating whether each monitor has completed its monitoring function. The vehicle must be driven through the specific drive cycle conditions for all the monitors to complete their function. This system status information is useful in verifying repairs. If all the monitors have completed their function and no DTCs are set in the PCM memory, there are no defects in the OBD II system. When OBD II diagnosis is combined with I/M inspections, the system status information will be essential.

> **Tech Tip** *Many scan tools are available that can retrieve generic OBDII codes from the vehicle's computer.*

OBD II Standardization

Vehicle manufacturers must comply with these OBD II standardization requirements:

1. A standard 16-terminal DLC must be mounted on the driver's side of the dash and the DLC must be visible (SAE standard J1962). Twelve of the 16 terminals in the DLC are dictated by regulations and the OEMs may determine the connections to the other four terminals. Standard DTCs are dictated by regulations, but the OEMs can insert some other DTCs (SAE standard J2012).

2. SAE standard J1979 requires the universal use of scan tools on all vehicle makes and test modes.

3. SAE standard J1850 requires a standard communication protocol (data links).

4. SAE standard J2190 requires standard diagnostic test modes.

5. The PCM must transmit vehicle identification data to the scan tool.

6. The PCM must have the ability to erase DTCs from the PCM memory.

7. The PCM must have the ability to record and store data for later review regarding conditions when a defect occurred.

8. The OBD II system must alert the driver if a defect occurs that increases emissions above limits. The system must also store a DTC representing any such defect.

9. OEM terminology must conform to SAE standard J1930 regarding electronic terms, acronyms, and definitions.

Summary

- Throttle body assemblies contain one or two injectors depending on the engine size.

- The PCM always supplies the correct injector pulse width to provide the proper air-fuel ratio during closed loop operation.

- An idle air control motor controls the amount of air bypassing the throttle to control idle speed.

- In a PFI system, the PCM operates the injectors in groups of two to four or may operate each injector individually.

- In a CPI system, the PCM operates a central injector module that supplies fuel to poppet nozzles in each intake port.

- In an SFI system, the PCM operates an injector for each cylinder that supplies fuel in each intake port.

- In an OBD II system, the PCM performs a number of monitors to detect problems in various systems and components.

- In an OBD II system, the PCM informs the driver about any problem in a system or component that causes emissions of NMHC, CO, or NOx to be one and one-half times higher than the emission standards for that

model year. The defect usually has to occur during two consecutive drive cycles.

- Many OBD II features are regulated by SAE standards.

Review Questions

1. Technician A says in a TBI system, the PCM increases the injector pulse width as engine speed increases. Technician B says in a TBI system, the PCM maintains a 14.7:1 air-fuel ratio during closed loop operation. Who is correct?

   A. Technician A

   B. Technician B

   C. Both Technician A and Technician B

   D. Neither Technician A nor Technician B

2. Technician A says some fuel injection systems are throttle body. Technician B says a port fuel injection system has an injector for each cylinder. Who is correct?

   A. Technician A

   B. Technician B

   C. Both Technician A and Technician B

   D. Neither Technician A nor Technician B

3. Technician A says if a signal is received from the power steering pressure switch, the PCM shuts off the AC compressor. Technician B says the power steering pressure switch signal informs the PCM to increase the spark advance. Who is correct?

   A. Technician A

   B. Technician B

   C. Both Technician A and Technician B

   D. Neither Technician A nor Technician B

4. Technician A says in a CPI system, the poppet injectors are operated by electrical signals from the PCM. Technician B says in a CPI system, a vacuum hose is connected from the intake manifold to the poppet injectors. Who is correct?

   A. Technician A

   B. Technician B

   C. Both Technician A and Technician B

   D. Neither Technician A nor Technician B

5. In a TBI system, the PCM supplies the proper air-fuel ratio by controlling the:

   A. fuel pump pressure.

   B. IAC motor.

   C. AIR pump.

   D. injector pulse width.

6. While discussing TBI systems with dual injectors, Technician A says the PCM alternately grounds each injector winding. Technician B says when the ignition switch is turned on, 12 volts are supplied to both injectors. Who is correct?

   A. Technician A

   B. Technician B

   C. Both Technician A and Technician B

   D. Neither Technician A nor Technician B

7. All of these statements about an IAC motor are true *except*:

   A. The IAC motor regulates the amount of air that is bypassing the throttles.

   B. If the air conditioning is turned on, the PCM moves the IAC plunger toward the open position.

   C. When the transmission is shifted from DRIVE to PARK, the PCM moves the IAC plunger toward the closed position.

   D. When the engine coolant is cold, the PCM moves the IAC plunger toward the closed position.

8. All of these statements about PFI systems are true *except*:

   A. In a PFI system, the PCM may ground two or more injectors simultaneously.

   B. In a PFI system, the PCM may ground each injector individually.

   C. In a PFI system, the PCM opens an injector when the intake valve is opening in the same cylinder.

   D. PFI systems have a PCM that controls the fuel injectors.

9. In a CPI system:

   A. The PCM grounds each injector individually.

   B. Voltage is supplied directly from the PCM to each injector.

   C. A V6 engine has six outlets on the central port injector.

   D. The pressure regulator is mounted separately from the central port injector.

10. In a CPI system, the pressure required to open the poppet nozzles is:

    A. 10 psi.

    B. 26 psi.

    C. 32 psi.

    D. 39 psi.

11. An IAC motor controls idle speed by regulating the amount of _____ that is bypassing the throttle.

12. Some idle air controllers use solenoids and are controlled by _____ _____ signal from the PCM.

13. The PCM uses the park/neutral switch to control _____ _____ .

14. In an OBD II system, the misfire monitor senses cylinder misfiring from the _____ _____ sensor signal.

15. Explain how the oxygen sensors are used to sense converter condition in an OBD II system.

16. In an OBD II system, explain how the PCM monitors the air pump operation.

17. Define a drive cycle in an OBD II system.

18. Describe an engine warm-up cycle in an OBD II system.

47 Engine Control System Maintenance, Diagnosis, and Service

Learning Objectives

After you have read, studied, and practiced the contents of this unit, you should be able to:

- Perform fuel pump pressure and injector tests and interpret the results.
- Obtain DTCs on OBD II systems.

- Remove and replace fuel pumps.
- Remove and replace fuel injectors.

Key Terms

Current DTC

History DTC

Injector cleaner

INTRODUCTION

The engine control system affects engine performance and economy more than any other system. When an engine experiences reduced performance and/or fuel economy, the cause of the problem is usually located in the engine control system. Therefore, it is extremely important for technicians to be able to perform quick and accurate diagnoses on this system.

ENGINE CONTROL SYSTEM MAINTENANCE

During minor under-hood service, the fuel system should be inspected for fuel leaks, loose electrical connections, worn wiring insulation, and vacuum leaks. Fuel leaks reduce fuel mileage and create the possibility of a fire. Loose electrical connections or worn insulation on fuel injector wiring may cause injector misfiring that increases emissions, and reduces engine performance and economy. Vacuum leaks may cause erratic idle operation. A vacuum leak in the fuel pressure regulator hose increases fuel pressure, which causes more fuel to be injected. This situation results in reduced fuel economy and can affect emissions.

In any fuel injection system, the fuel filter should be replaced at the vehicle manufacturer's recommended service interval. The fuel filter may be located under the vehicle or in the engine compartment. The fuel pressure should be relieved before loosening any of the fuel lines.

> **Tech Tip** *Follow the manufacturer's recommended procedure for relieving fuel pressure from the fuel system before beginning work.*

After the fuel system pressure is relieved, use a clean shop towel to wipe any dust and debris from the fuel-line connections to prevent this material from entering the fuel lines. Loosen and remove the line fittings from the filter and remove the filter. Some vehicles may use a hand quick-disconnect fuel line **(Figure 47-1)**. These lines are removed either by rotating the line or by squeezing a set of retaining clips to remove the lines **(Figure 47-2)**. Other fuel filters may require a tool to remove the quick-disconnect fuel line **(Figure 47-3)**. This style typically uses a garter spring on the one fitting to retain the fuel fitting on the other side **(Figure 47-4)**.

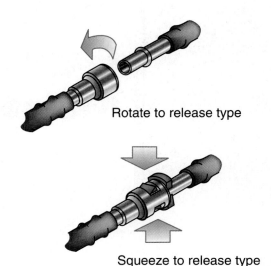

Rotate to release type

Squeeze to release type

Figure 47-2 Hand quick-disconnect fuel line removal.

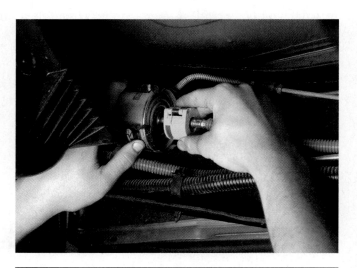

Figure 47-3 Quick-disconnect fuel line removal using a tool.

> **Tech Tip** *The inlet and outlet fittings on most filters are identified, or the filter housing has an arrow to indicate the proper direction of fuel flow. Some fuel lines have quick-disconnect fittings. On some fuel lines, a special tool is required to release the connectors.*

Fuel Filter

Fuel Tank

Figure 47-1 A fuel filter using hand quick-disconnect fuel lines.

Install the new filter in the proper direction and tighten the fittings to the specified torque. After the filter installation, always start the engine and inspect the filter and line fittings for leaks.

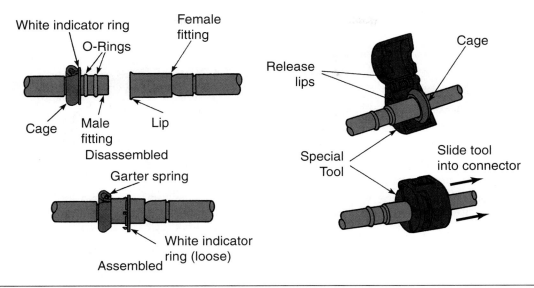

Figure 47-4 Quick-disconnect fuel line details.

ENGINE CONTROL SYSTEM DIAGNOSIS

When diagnosing fuel injection systems, one of the most important tests is fuel pressure. Improper fuel pressure causes many drivability and emission-related problems. For example, if the fuel pressure is higher than specified, the injectors deliver more fuel, resulting in reduced fuel economy.

Tech Tip *Excessive fuel pressure may be caused by a restricted fuel return line.*

Tech Tip *A leaking pressure regulator diaphragm on an MFI or SFI system causes fuel to be pulled through this diaphragm and the vacuum hose into the intake manifold. This results in excessive fuel consumption and erratic idle operation.*

When the fuel pressure is less than specified, the injectors supply less fuel and this may cause drivability problems such as acceleration stumbles and loss of engine power or cutting out at high speed. The PCM diagnostics do not include fuel pressure. The PCM assumes the specified fuel pressure is available at the injectors. Before testing fuel pressure, always relieve the fuel pressure in the system. Be sure there is an adequate supply of fuel in the fuel tank.

Tech Tip *Follow the manufacturer's recommended procedure for relieving fuel pressure from the fuel system before beginning work.*

On PFI, CFI, and TBI systems, connect the fuel pressure gauge to the Schrader valve on the fuel rail **(Figure 47-5)**. On TBI and some other systems, the fuel pressure gauge must be connected in series in the fuel supply line. Cycle the ignition switch on and off several times and read the fuel pressure on the gauge. If the fuel pressure is less than specified, the fuel filter may be severely restricted, the voltage supply or ground on the fuel pump may have high resistance, or the fuel pump may be defective.

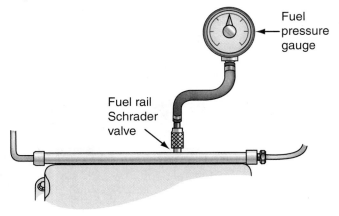

Figure 47-5 A fuel pressure gauge connected to the Schrader valve on the fuel rail.

The manufacturer may also provide specifications for fuel pressure at idle and at WOT.

With the engine running, use a digital voltmeter to test the voltage at the fuel pump in the tank. With the engine running, the voltage at this location should be 11.5 volts or more. If this voltage is lower than specified, the wire from the fuel pump relay to the fuel pump has excessive resistance. To measure the voltage drop across the fuel pump ground, connect the voltmeter from the fuel pump ground terminal to a good chassis ground. The voltage drop across this circuit should not exceed 0.2 volt. If the fuel filter and the fuel pump voltage supply and ground are satisfactory, replace the fuel pump.

Injector Testing

Faulty injectors may cause a rough idle problem. Some fuel injector problems, such as an open winding or severely restricted discharge orifices, cause cylinder misfiring at idle. When the engine speed is increased, the misfiring is less noticeable. If the orifices in an injector are partially restricted, a rough idle problem may result.

With the air cleaner removed and the engine idling on some TBI systems, the fuel spray from the injectors may be observed visually. If an injector is not spraying fuel, a stethoscope pickup may be placed against the injector body with the engine idling **(Figure 47-6)**. If a clicking noise is heard, the PCM is operating the injector and the injector winding is likely satisfactory. The injector may be unplugged and a noid light may be plugged into the injector harness **(Figure 47-7)**. Crank or start the engine and observe the light flashes. If the noid

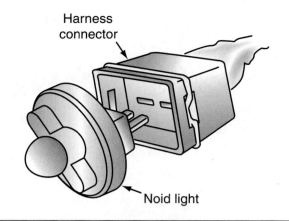

Figure 47-7 A noid light is used to check the injector harness for voltage pulses.

light does not flash then voltage signals are not reaching the injector harness.

> **Tech Tip** *A lab scope may also be used to test fuel injectors, as discussed later in this chapter.*

When the injector is not clicking with the engine idling, test the injector voltage supply and ground wire to the PCM and test the injector winding with an ohmmeter. Injectors may be tested with an ohmmeter. Turn the ignition switch off and remove the injector connector. Connect the ohmmeter leads to the injector terminals **(Figure 47-8)**. The injector

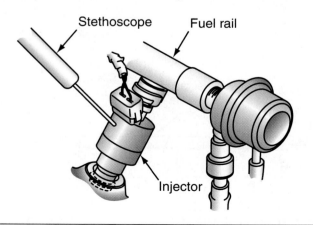

Figure 47-6 Injector testing with a stethoscope.

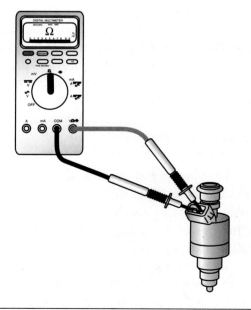

Figure 47-8 Testing an injector with an ohmmeter.

Figure 47-9 An injector tester.

must have the specified resistance. If the resistance is less than specified, the injector winding is shorted. An infinite ohmmeter reading indicates an open injector winding. It is very important to carefully measure the resistance of the injector. If the injector does not have the specified resistance, replacement is necessary.

An injector balance test may be performed to test the injectors. The injector balance tester energizes each injector for a specific short time period (**Figure 47-9**). Each injector terminal must be disconnected and the lead from the balance tester connected to the injector terminals. The power leads on the injector balance tester must be connected to the battery terminals with the correct polarity (**Figure 47-10**).

During the injector balance test, the fuel pressure gauge must be connected to the Schrader valve on the fuel rail. Cycle the ignition switch several times until the specified pressure is indicated on the fuel gauge. Press the button on the balance tester to energize the injector for a specific time period.

Tech Tip *Energizing an injector winding for more than 5 seconds may burn out the winding because these windings are normally energized by the PCM for a few milliseconds at a time.*

Record the fuel pressure displayed on the fuel gauge. Repeat this procedure on each injector. Obtain the maximum difference in pressure drop on each injector from the vehicle manufacturer's specifications. If an injector has less pressure drop compared to the other injectors, this injector has restricted orifices. When an injector has more

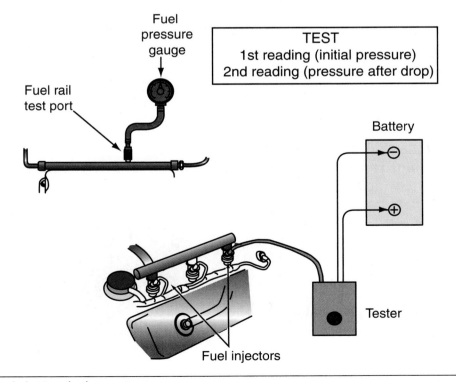

Figure 47-10 An injector balance test.

CYLINDER	1	2	3	4	5	6
HIGH READING	45PSI	45PSI	45PSI	45PSI	45PSI	45PSI
LOW READING	20PSI	20PSI	20PSI	15PSI	20PSI	20PSI
AMOUNT OF DROP	25PSI	25PSI	25PSI	30PSI	25PSI	25PSI
RESULTS	OK	OK	OK	Faulty, rich (too much fuel drop)	OK	Faulty, lean (too much fuel drop)

Figure 47-11 Interpreting injector balance test results.

pressure drop compared to the other injectors, the injector plunger is sticking open **(Figure 47-11)**.

Injectors with restricted orifices may have leaking seats that drip fuel from the injector tips into the intake manifold during idle operation and after the engine is shut down. This condition causes rough idle operation and hard starting after a hot engine has been shut down for a few minutes. When testing fuel pump pressure or injector balance, cycle the ignition switch several times until the specified fuel pressure appears on the fuel gauge. After 15 minutes, observe the fuel pressure. If the fuel pressure slowly decreases, the injectors may be leaking or the pressure regulator valve may have a leak. A leaking one-way check valve in the fuel pump may also cause this decrease in fuel pressure.

To eliminate these leak sources and locate the problem, repeat the leak down test and use a pair of straight-jawed vise grips to close each line one at a time. Do not pinch nylon fuel lines, as this will ruin the fuel line.

> **Tech Tip** *A nylon line cannot be repaired and must be completely replaced.*

If the fuel pressure still leaks down with the fuel return line and the fuel supply line blocked, the injectors are dripping fuel. If the leak down problem is eliminated when the fuel return line is closed, the pressure regulator valve is leaking. When the leak down problem is eliminated with the fuel supply line closed, the one-way check valve in the fuel pump is leaking.

Fuel injectors may be tested with a lab scope. Connect the scope leads to the injector terminals.

A satisfactory waveform from a conventional injector should be observed **(Figure 47-12)**. Some manufacturers use a peak-and-hold technique and the correct pattern for this type of injector should be observed **(Figure 47-13)**.

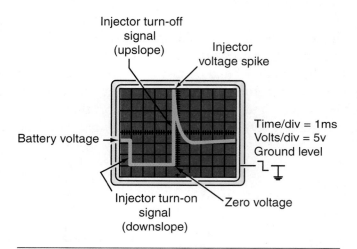

Figure 47-12 Waveform from a conventional fuel injector.

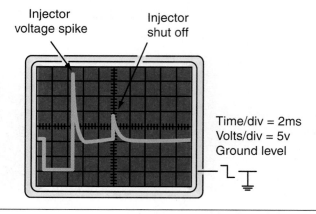

Figure 47-13 Waveform from a peak-and-hold fuel injector.

Tech Tip *The PCM supplies normal current flow to a peak-and-hold injector to open the injector and then reduces the current flow to hold the injector open for a very short time period. The PCM initially supplies normal current flow for a pulse-modulated injector and then pulses the injector on and off very quickly.*

Some manufacturers may use a pulse-modulated technique and the correct pattern for this type of injector should be observed **(Figure 47-14)**. If an injector has an abnormal waveform, the injector has a problem.

Tech Tip *A shorted injector winding causes high current through the injector and the PCM driver. This high current flow may damage the injector driver in the PCM. In some PCMs, the injector drivers sense high current flow and, rather than allowing injector driver damage, they shut the circuit off. After a cool-down period, the injector driver may allow injector operation.*

Injector Cleaning

Injectors may be cleaned with a special **injector cleaner** placed in a pressurized container

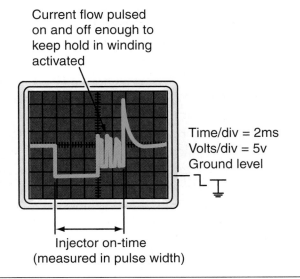

Current flow pulsed
on and off enough to
keep hold in winding
activated

Time/div = 2ms
Volts/div = 5v
Ground level

Injector on-time
(measured in pulse width)

Figure 47-14 Waveform from a pulse-modulated fuel injector.

Figure 47-15 Portable injector cleaning equipment.

(Figure 47-15). Larger cleaning equipment may be available for high volume work and is usually mounted in a cart **(Figure 47-16)**. The injector cleaning solution is a mixture of cleaning solution and unleaded gasoline. After the cleaning solution is

Figure 47-16 Injector cleaning cart.

placed in the injector-cleaning tool, the tester is pressurized with shop air pressure until the pressure gauge on the tester is slightly below the specified fuel pressure. At this pressure, the pressure regulator valve does not open during the injector cleaning process.

During the injector cleaning process, the fuel pump must be inoperative. Disconnect the wires from the fuel pump relay to prevent fuel pump operation. If the fuel pump circuit has an oil pressure switch, disconnect the wires from this switch. Locate a section of rubber hose in the fuel return line and use a pair of straight-jawed vise grips to close this line. This action prevents the possibility of the injector cleaning solution flowing through the fuel return line into the fuel tank. Connect the hose on the injector cleaner to the Schrader valve on the fuel rail and open the cleaner valve.

Start and run the engine on the injector cleaning solution. The engine usually runs for 15 to 20 minutes on the cleaning solution. After the injectors have been cleaned, the engine may still experience a rough idle problem because the PCM has to re-learn about the cleaned injectors. With the engine at normal operating temperature, drive the vehicle for 5 minutes to allow the PCM to re-learn regarding the cleaned injectors.

> **Tech Tip** *A hand pump on some injector cleaners is used to pressurize the container. Injector cleaner is also supplied in a pressurized container so the technician does not have to pressurize the container.*

> **Tech Tip** *Some vehicle manufacturers do not recommend cleaning certain types of injectors in their vehicles.*

ENGINE CONTROL SYSTEM SCAN TOOL DIAGNOSIS AND DIAGNOSTIC TROUBLE CODES

When diagnosing fuel injection systems, observe the MIL in the instrument panel. If this light is illuminated with the engine running, the PCM has detected a problem in the system and a DTC is set in the PCM memory. A scan tool is connected to the DLC under the dash to read the DTCs and other input and output data.

When performing a scan tool diagnosis, the engine should be at normal operating temperature. With the ignition switch off, connect the scan tool to the DLC. On some vehicles, you have to enter the vehicle make, model year, and engine code into the scan tool. On other vehicles, the PCM supplies this information to the scan tool. When READ CODES is selected on the scan tool, the DTCs are displayed. Some scan tools supply an interpretation for each DTC.

When using other scan tools, it is necessary to obtain the DTC interpretation from the vehicle manufacturer's information. DTCs may be identified as current or history. **Current DTCs** are present at the time of testing. **History DTCs** represent intermittent faults that occurred sometime in the past, but the fault is no longer present.

> **Tech Tip** *A history DTC can be cleared if the fault does not reappear during the next 40 warm-ups, the battery is disconnected, or the scan tool is used to clear the code.*

A DTC indicates a problem in a certain area. For example, a DTC representing a MAF sensor may indicate a defective sensor, defective wires from the sensor to the PCM, or a PCM that is not able to receive this signal. You must perform specific tests, such as voltmeter or ohmmeter tests, to locate the exact cause of the problem.

OBD I systems used two- or three-digit numbered DTCs. OBD II DTCs are formatted according to SAE standard J2012. This standard requires five-digit alphanumeric DTC identification. The first digit (or prefix) in each DTC indicates the DTC function: P—powertrain, B—body, C—chassis, U—network communication. The second digit in the DTC represents the group responsible for the code. If the number is 0, the code is designated by SAE. When the second digit is 1, the vehicle manufacturer is responsible for the DTC. Many DTCs are dictated by SAE but the vehicle manufacturer may insert some codes. The third digit in the DTC indicates the subgroup to which the code belongs as follows:

1—air-fuel control

2—air-fuel control, injectors

3—ignition system, misfire

4—auxiliary emission controls

5—idle speed control

6—PCM input and output

7—transmission/transaxle

8—transmission/transaxle

The fourth and fifth digits in the DTC indicate the area in which the defect is located.

For example, in DTC P0155, P indicates it is a powertrain code, 0 indicates it is designated by SAE, 1 indicates the code is in the air-fuel control subgroup, and 55 indicates the code represents a malfunction in the HO_2S heater, bank 2, sensor 1. The location of bank 2 sensor 1 would be on the side of the engine opposite of cylinder number 1 and upstream of the catalytic converter.

> **Tech Tip** *Bank 1 indicates the bank of a V6 or V8 engine that is on the same side where number 1 cylinder is located and bank 2 is on the opposite side of the cylinder number 1. Sensor 1 is the upstream HO_2S sensor located ahead of the catalytic converter and sensor 2 is mounted downstream from the catalytic converter.*

ENGINE CONTROL SYSTEM SERVICE

Always be sure the exact cause of the customer's complaint is diagnosed before performing any engine control system service. For example, before replacing a fuel pump, be sure no other causes of improper fuel pump operation are present. Therefore, be sure the fuel filter, fuel lines, and the fuel pump voltage supply and ground wires are satisfactory before replacing the fuel pump. Before replacing any electrical/electronic component, connect a 12-volt power supply or memory saver to the cigarette lighter socket and then disconnect the negative battery terminal. Disconnecting the battery prevents accidental shorts to ground from damaging components or starting a fire when servicing vehicle components or systems. The power supply connected to the cigarette lighter socket keeps the PCM memory alive during

the service procedure. This action prevents drivability problems after the vehicle is restarted.

FUEL PUMP REPLACEMENT

When replacing most electric in-tank fuel pumps, the fuel tank must be removed **(Figure 47-17)**. On some vehicles, a fuel pump access cover is located in the trunk directly above the fuel pump in the tank **(Figure 47-18)**. After this access door is removed,

Figure 47-17 Some fuel tanks must be removed from the vehicle to replace the fuel pump.

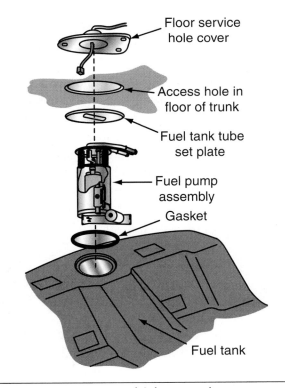

Floor service hole cover

Access hole in floor of trunk

Fuel tank tube set plate

Fuel pump assembly

Gasket

Fuel tank

Figure 47-18 Some vehicles may have an access cover that may be removed to gain access to the fuel pump.

the fuel pump may be taken out of the fuel tank. If the fuel tank must be removed to gain access to the fuel pump, use an electric or hand pump to pump the gasoline out of the tank into an approved gasoline container. Be careful not to spill any fuel. If fuel is accidentally spilled, use the proper absorption material and clean up the spill immediately.

Raise the vehicle on a lift or use a floor jack to raise and lower the vehicle onto safety stands. Disconnect the fuel lines and electrical connector near the fuel pump and remove the fuel tank straps. Slowly lower the fuel tank onto the floor and remove it from under the vehicle. Use an air gun to blow debris from the top of the fuel tank and the fuel pump assembly. Remove the fuel pump retaining ring and lift the fuel pump assembly from the tank. In many vehicles the fuel pump, fuel intake filter, and fuel tank sensing unit are replaced as an assembly. In some applications, the fuel pump may be replaced separately.

Always use a new seal between the fuel pump assembly and the top of the fuel tank. Install the new fuel pump assembly and be sure the fuel pickup is very close to or touching the bottom of the tank. Install the fuel pump assembly-retaining ring in the top of the fuel tank. Install the fuel tank in its proper location under the vehicle and install the holding straps. Tighten the holding strap fasteners to the specified torque. Reconnect the fuel lines and electrical connector to the fuel pump assembly. Tighten the fuel lines to the specified torque. Lower the vehicle onto the shop floor. Reconnect the negative battery cable and remove the power supply connector from the cigarette lighter. Start the engine and be sure the fuel pump has the specified pressure with no fuel leaks.

SAFETY TIP *Do not drag the fuel tank across the cement floor. This action may cause a spark and a serious fire.*

INJECTOR REPLACEMENT

Always relieve the fuel system pressure before replacing the injectors. Connect a 12-volt power supply to the cigarette lighter and disconnect the negative battery cable. It may be necessary to remove

part of the intake manifold on some vehicles to gain access to the injectors. Release the lock ring and remove each injector terminal.

Disconnect the fuel inlet and outlet lines on the fuel rail and the vacuum hose on the pressure regulator **(Figure 47-19)**. Remove the fuel rail retaining bolts. Remove any other components that interfere with the fuel rail removal. Remove the fuel rail and injectors **(Figure 47-20)**.

SAFETY TIP *Always be sure to install the vehicle manufacturer's recommended injectors. Installing improper injectors may cause drivability problems.*

Figure 47-19 Injector removal and replacement.

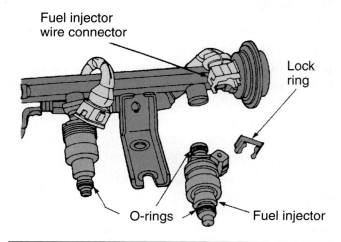

Figure 47-20 To remove the injectors from the fuel rail, remove the lock ring and remove the injector.

Always install new injector O-rings when removing and replacing injectors **(Figure 47-21)**. Place a small amount of engine oil on each upper injector O-ring and install the injectors into the fuel rail. Use engine oil to lightly coat the lower injector O-rings and install the fuel rail and injectors in the intake manifold. Be sure the injectors are fully seated in the fuel rail and the intake manifold.

Reconnect the injector wiring terminals. Install the fuel rail fasteners, and tighten these fasteners to the specified torque. Reconnect the fuel inlet and outlet lines and tighten the fittings to the specified torque. Install the vacuum hose on the pressure regulator. Reconnect the battery negative cable and disconnect the power supply from the cigarette lighter.

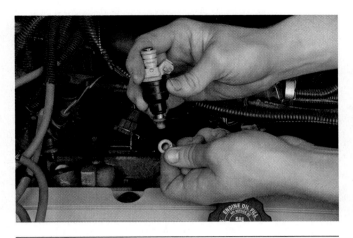

Figure 47-21 Always use new O-rings when replacing injectors.

Summary

- The fuel filter should be changed at the vehicle manufacturer's specified intervals.

- The fuel system pressure should be relieved before disconnecting any fuel system component.

- The PCM assumes the specified fuel pressure is available when calculating the proper amount of fuel to be injected.

- Some PCM diagnostics provide indication of improper fuel pressure.

- Before replacing a fuel pump, always be sure the voltage supply and ground wires to the pump are satisfactory.

- Measuring the resistance in the injector windings may test fuel injectors.

- An injector balance test measures the fuel pressure drop when each injector is opened for a specific length of time.

- If the fuel system pressure gradually decreases after the engine is shut off, the one-way check valve in the fuel pump may be leaking, the pressure regulator valve may be leaking, or the injectors may be dripping fuel.

- Some injectors may be cleaned by connecting a pressurized container filled with injector cleaner to the Schrader valve on the fuel rail.

- A DTC indicates a fault in a certain area of the engine control system.

- In an OBD II DTC, the first digit indicates the area to which the DTC belongs. For example, P indicates powertrain.

- The second digit in an OBD II DTC indicates whether the DTC is established by SAE standards or introduced by the vehicle manufacturer.

- The third digit in an OBD II DTC indicates the subgroup to which the DTC belongs. For example, 5 indicates the DTC is in the idle speed control subgroup.

- The fourth and fifth digits in an OBD II DTC represent the fault in the system.

- On many vehicles, the fuel tank must be removed to access the fuel pump.

- When removing and replacing electrical/electronic component, always connect a 12-volt power supply to the cigarette lighter socket and disconnect the negative battery cable.

Review Questions

1. Technician A says in a P0155 DTC, the 0 indicates the vehicle manufacturer developed this DTC. Technician B says the 1 indicates the subgroup to which the DTC belongs. Who is correct?

   A. Technician A

   B. Technician B

   C. Both Technician A and Technician B

   D. Neither Technician A nor Technician B

2. Technician A says a bank 2, sensor 2 HO_2S is located upstream from the catalytic converter. Technician B says this HO_2S is located on the same side of the engine as number 1 cylinder. Who is correct?

   A. Technician A

   B. Technician B

   C. Both Technician A and Technician B

   D. Neither Technician A nor Technician B

3. Technician A says higher-than-specified fuel pump pressure causes a rich air-fuel ratio. Technician B says higher than a leaking one-way check valve in the fuel pump may cause specified fuel pressure. Who is correct?

   A. Technician A

   B. Technician B

   C. Both Technician A and Technician B

   D. Neither Technician A nor Technician B

4. Technician A says during an injector balance test, if the pressure drop on one injector is greater than on the other injectors, this injector has restricted orifices. Technician B says during an injector balance test, each injector should have nearly the same pressure drop. Who is correct?

   A. Technician A

   B. Technician B

   C. Both Technician A and Technician B

   D. Neither Technician A nor Technician B

5. In an MFI system, lower-than-specified fuel pressure may cause all of these problems *except*:

   A. a lean air-fuel ratio.

   B. engine surging and cutting out at high speed.

   C. hesitation during engine acceleration.

   D. engine backfiring during deceleration.

6. While discussing TBI system diagnosis, Technician A says the PCM has the capability to diagnose fuel pump pressure. Technician B says when testing fuel pressure with a gauge, connect the gauge hose to the Schrader valve. Who is correct?

   A. Technician A

   B. Technician B

   C. Both Technician A and Technician B

   D. Neither Technician A nor Technician B

7. All of these problems may cause low fuel pump pressure *except*:

   A. low voltage at the fuel pump.

   B. a restricted fuel filter.

   C. a restricted fuel return line.

   D. a restricted fuel supply line.

8. When diagnosing an SFI system with the fuel pressure gauge connected to the system, the ignition switch is cycled several times until the specified fuel pressure appears on a fuel pressure gauge. The fuel supply line and the return fuel line are then completely blocked. After 15 minutes, the fuel pressure has dropped from 45 psi to 10 psi and there are no indications of external fuel leaks. The most likely cause of this pressure reading is:

   A. dripping fuel injectors.

   B. a leaking pressure regulator valve.

   C. a leaking one-way check valve in the fuel pump.

   D. a leaking fuel filter.

9. When performing an injector cleaning procedure:

A. the fuel return line must be open.

B. the fuel pump must be in operation.

C. the injector cleaning solution container is connected to the Schrader valve.

D. the engine should be running for 1 hour.

10. While diagnosing an OBD II fuel system, a current P0155 DTC is obtained on a scan tool. All of these statements about this DTC are true *except*:

A. This DTC represents a problem in the powertrain.

B. This DTC is designated by the vehicle manufacturer.

C. This DTC is in the air-fuel control subgroup.

D. This DTC represents a defect that is present during the diagnosis.

11. During the injector cleaning process, the fuel pump must be _____ .

12. While reading PCM flash DTCs, the MIL flashes quickly four times followed by a pause and one flash. This indicates DTC

_____ .

13. If an OBD II DTC begins with a B, the DTC is in the _____ category.

14. If the third digit in an OBD II DTC is 3, the DTC is related to _____

_____ .

15. Explain the reason for connecting a 12-volt power supply to the cigarette lighter socket before disconnecting a vehicle battery.

16. Describe the proper fuel pressure gauge connection to test fuel pump pressure on a TBI system.

17. When the fuel pump pressure is less than specified, describe the diagnostic procedure to locate the exact cause of the problem.

18. Describe the possible ways a history DTC may be cleared.

CHAPTER

48 Manual Transmission/ Transaxle

Learning Objectives

After you have read, studied, and practiced the contents of this unit, you should be able to:

- Describe two different types of transaxle gears.

- Explain gear reduction and overdrive ratios and describe how gear ratios are calculated.

- Describe synchronizer design and operation.

- Explain manual transmission operation in first, second, third, fourth, and fifth speed and reverse.

- Describe the lubricants required in manual transmissions and transaxles.

- Perform manual transmission/transaxle maintenance, including a visual inspection, leak diagnosis, and shift linkage adjustments.

- Diagnose manual transmission/transaxle problems. Remove and replace manual transmissions/transaxles.

Key Terms

Backlash

Engine support fixture

Gear ratio

Helical gear

Interlock

Splines

Spur gear

Synchronizers

Transaxle vent

INTRODUCTION

Manual transmissions and transaxles transfer engine torque to the differential and provide gear reductions for smooth vehicle acceleration. In third or fourth gear, depending on the transaxle, engine torque is transmitted directly without providing a reduction or an overdrive. In fifth or sixth gear, depending on the transaxle or transmission, an overdrive gear ratio can be provided to improve fuel economy at cruising speed. The transmission gears and components must be able to withstand high engine torque during hard acceleration. Proper transmission/transaxle maintenance is essential to provide normal unit life. Accurate transmission/transaxle diagnosis is very important to determine the cause of operational problems. Professional transmission/transaxle service is critical to repair operational problems quickly and accurately.

GEARS

Transaxle or transmission gears may be **helical** or **spur**. Helical gears always have more than one tooth in mesh at once to provide additional gear strength **(Figure 48-1)**. Helical gear teeth create a wiping action as they engage or disengage with other gear teeth. This action provides quieter operation than spur gears. Helical gear teeth create axial thrust on the gear. Helical gears have teeth that are cut at an angle to the centerline of the gear. Straight-cut gear teeth are noisier than helical gears, but they do not cause axial thrust **(Figure 48-2)**.

Figure 48-1 Helical-cut gears have teeth cut at an angle to the centerline and are stronger and quieter.

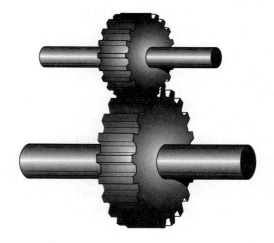

Figure 48-2 Straight-cut gears have teeth cut parallel to the centerline.

Tech Tip *Helical gears are cut at an angle to their rotational axis. Straight-cut gears have teeth cut perpendicular to their rotational axis.*

Tech Tip *Spur gears are sometimes used for reverse. Most gears used in the transmission are helical.*

Gear teeth have a drive and a coast side. While the engine is supplying torque to the transmission, the drive side of the gear teeth transmitting torque is in contact with the teeth of another gear. During engine deceleration, the drive wheels are transmitting torque to the engine and the coast side of the gear teeth that are transmitting torque is in contact with the teeth of another gear. A typical five-speed transmission with helical and straight-cut gears is shown in **Figure 48-3**. The teeth on many gears remain in constant mesh with each other. Synchronizers are used to engage or disengage constant mesh gears.

Backlash is the amount of movement between the meshed teeth on two gears. Some backlash is required to allow proper lubrication of the gear teeth and compensate for tooth expansion when the gear temperature increases **(Figure 48-4)**. Excessive backlash may indicate gear wear and this condition may cause gear tooth damage and noisy operation.

Tech Tip *Needle bearings are positioned between the counter shaft and the counter shaft gear to reduce friction.*

GEAR RATIOS

A gear reduction occurs when a smaller diameter gear drives a larger diameter gear. A gear reduction provides an increase in torque and a decrease in output shaft speed. When a large-diameter gear drives a smaller-diameter gear, an overdrive condition is present. An overdrive gear ratio reduces engine torque and increases output shaft speed. **Gear ratio** is the size comparison between the drive and

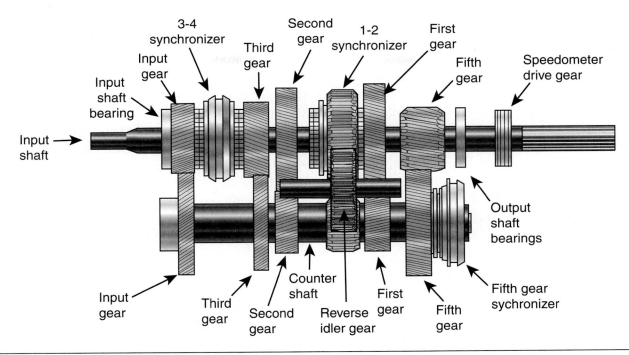

Figure 48-3 A typical five-speed transmission with helical- and straight-cut gears.

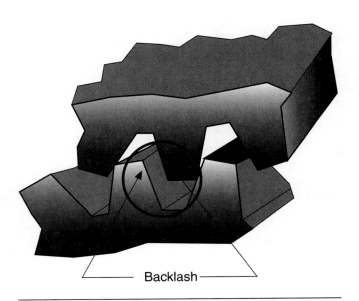

Figure 48-4 Backlash is the clearance between meshing gear teeth.

driven gears. Gear ratio is calculated by dividing the number of drive gear teeth into the number of driven gear teeth. When the drive gear has ten teeth and the driven gear has thirty teeth, the gear ratio is 3:1. If the drive gear has ten teeth and the driven gear has eight teeth, the gear ratio is $8 \div 10 = 0.8{:}1$. In Figure 48-3, the fifth gear on the counter shaft is

larger than the fifth driven gear on the input shaft, creating an overdrive ratio between these gears.

If torque is transmitted through two gear ratios, the total gear ratio is determined by multiplying the two ratios. For example, if the first speed gear ratio is 3.40:1 and the differential gear ratio is 3.72:1, the total gear ratio is 12.648:1.

SYNCHRONIZERS

Synchronizers prevent gear engagement until the two gears rotating at different speeds are rotating at the same speed, which provides smooth shifting. In many current transaxles and transmissions, all the forward gears are synchronized. Some older transmissions, especially on trucks, do not have synchronizers on some of the gears. Some transmissions and transaxles have synchronizers on all forward gears and reverse, but on other transmissions and transaxles reverse gear does not have a synchronizer.

The blocker-type synchronizer is presently the most common type of synchronizer. A blocker-type synchronizer has a hub with internal and external **splines**. The internal splines are mounted on matching splines on the transmission output shaft **(Figure 48-5)**. A synchronizer sleeve with internal splines is mounted on the external hub splines **(Figure 48-6)**. Therefore, the sleeve can slide forward or rearward on the hub splines.

When the transmission is assembled, a shifter fork is positioned in a wide groove in the outer diameter of the sleeve. Three inserts are mounted in hub slots. A narrow raised area in each insert is mounted in an internal groove in the sleeve. A circular spring on each side of the hub holds the inserts outward against the sleeve. A brass-blocking ring is mounted on each side of the synchronizer.

The inserts fit into wide grooves in the blocking rings. The inside diameter of the blocking ring is a female cone-shaped area with sharp grooves on the surface. This cone-shaped area matches a male cone-shaped area on the gear next to each side of the synchronizer. A series of beveled teeth are positioned around the outer diameter of each blocking ring. The teeth on the blocking ring match beveled teeth on the gear beside the synchronizer. (See Figure 48-6.)

When the transmission is in neutral and the engine is idling with the clutch engaged, the input shaft and counter shaft are rotating. Because the

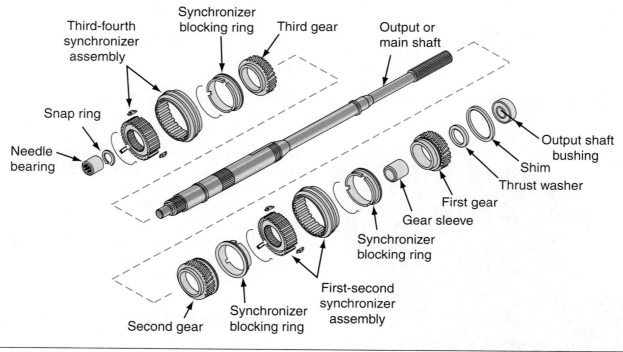

Figure 48-5 An output shaft with related gears and synchronizers.

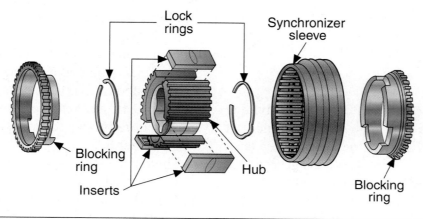

Figure 48-6 Synchronizer components.

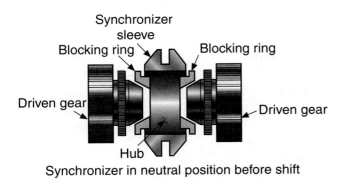

Synchronizer in neutral position before shift

Figure 48-7 Synchronizer in the neutral position.

gears on the counter shaft are meshed with gears on the output shaft, the gears on the output shaft also rotate. However, the synchronizers are in the neutral position and no torque is transferred from the counter shaft gears to the gears on the output shaft **(Figure 48-7)**.

When a shift occurs, the shifter shift linkage moves the fork, which moves the synchronizer sleeve toward the selected gear. The inserts move with the sleeve and these inserts push the blocking ring toward the selected gear. The grooved surface on the blocking ring cone cuts through the film of lubricant on the selected gear cone **(Figure 48-8)**.

When the blocking ring grooves cut through the lubricant, friction occurs between the blocking ring grooves and the cone on the selected gear. The friction developed by the contact between these two components causes the blocker ring to rotate

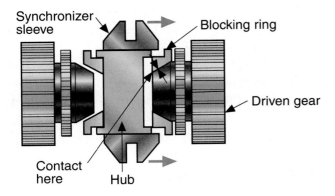

During synchronization—blocking ring and gear shoulder contacting

Figure 48-8 Synchronizer position during a shift.

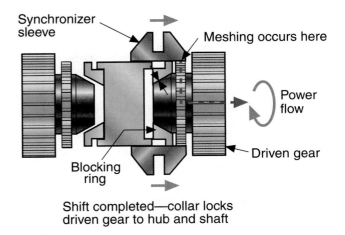

Shift completed—collar locks driven gear to hub and shaft

Figure 48-9 Synchronizer position when a shift is completed.

slightly thus blocking the movement of the synchronizer sleeve until the synchronizer hub and the selected gear are at the same speed. With the two components rotating at the same speed, the sleeve slides over the dog teeth on the blocking ring and the dog teeth on the selected gear lock the gear to the synchronizer ring, hub, and output shaft together **(Figure 48-9)**. Under this condition, engine torque is transmitted from the counter shaft gear to the output shaft gear, synchronizer sleeve, synchronizer hub, and output shaft.

> **Tech Tip** *Some transmissions may use a very low first-gear ratio for improved acceleration.*

TRANSMISSION TYPES

In current car production, the five-speed transmission is most commonly used. However, transmissions may be three speed, four speed, five speed, or six speed. In a four-speed transmission, the fourth gear usually has a 1:1 ratio.

A five-speed transmission is similar to a four-speed transmission, but a fifth overdrive speed with a synchronizer is added. In a five-speed transmission, the four-speed gear ratio is usually 1:1 and the fifth-speed gear ratio is usually between 0.70:1 and 0.90:1.

Some performance cars have a six-speed transmission. The fifth and sixth speeds have an overdrive

Gear	Ratio	Overall ratio
First	2.66:1	9.10:1
Second	1.78:1	6.10:1
Third	1.30:1	4.45:1
Fourth	1.00:1	3.42:1
Fifth	0.74:1	2.53:1
Sixth	0.50:1	1.71:1

Figure 48-10 Gear ratios in a typical six-speed transmission.

ratio **(Figure 48-10)**. A sixth-speed gear ratio of 0.50:1 allows the engine to run at a lower rpm during highway cruising. In this type of transmission, all forward gears and reverse are synchronized.

TRANSMISSION OPERATION

Shifting mechanisms and shifter forks are used to change gears. Most shift forks have two legs fit into a groove in the synchronizer sleeve **(Figure 48-11)**. Each shift fork is secured to a shift rail with a tapered pin. Each shift rail has notches near the end of the rail and a spring-loaded ball or bullet rides on these notches. If the shift lever is mounted on top of the transmission, this lever is connected to a socket with the bottom end fitted into slots in the shift rails **(Figure 48-12)**.

Figure 48-11 Shift forks.

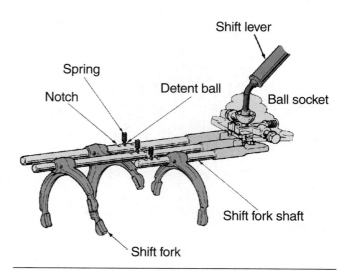

Figure 48-12 Shift forks and shift rails.

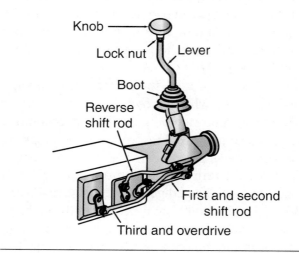

Figure 48-13 Shift lever and linkages.

If the gear shift lever is mounted on the side of the transmission, linkages are connected from this lever to the shift rails **(Figure 48-13)**. When the gear shift lever is moved to perform a shift, this lever moves a shift rail.

When a shift is completed, the spring-loaded ball fits into a shift rail detent to provide a firm shift feel and maintain the shift fork position. An **interlock** mechanism on the shift rails prevents the engagement of two gears at the same time **(Figure 48-14)**.

First-Gear Operation

In first gear, engine torque is transmitted from the input shaft and gear to the counter shaft gear in mesh with the input shaft gear. The 1–2 synchronizer sleeve is moved rearward by the shift fork.

Right interlock plate is
moved by the 1-2 shift
rail into the 3-4 shift rail slot

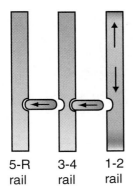

5-R   3-4   1-2
rail rail rail

Left interlock plate is moved
by lower tab of the right interlock
plate into the 5-R shift rail slot

The 3-4 shift rail pushes both
the interlock plate outward
into the slots of the 5-R and
1-2 shift rails

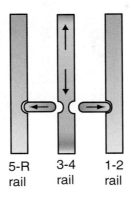

5-R   3-4   1-2
rail rail rail

3-4 rail

Right interlock plate is moved
by lower tab of the left interlock
plate into the 1-2 shift rail

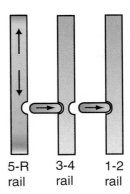

5-R   3-4   1-2
rail rail rail

Left interlock plate is moved
by the 5-R shift rail into the
3-4 shift rail slot

Figure 48-14 Interlock mechanism in a transmission shift assembly.

This synchronizer hub is now connected to the first-speed gear. Engine torque is now transmitted from the counter shaft gear, first-speed gear, synchronizer sleeve, and synchronizer hub to the output shaft **(Figure 48-15)**. All other gears on the output shaft are not locked to the shaft and rotate freely.

Second-Gear Operation

In second gear, the shift fork moves the 1–2 synchronizer sleeve forward. This sleeve connects the second speed gear to the synchronizer hub **(Figure 48-16)**. Engine torque is transferred from

the input shaft to the counter shaft gear in mesh with the input shaft gear. Engine torque is transferred to the second-speed gear from the counter shaft gear in mesh with the second-speed gear. From the second-speed gear, engine torque is transmitted to the sleeve and hub on the 1–2 synchronizer and the output shaft.

Third-Gear Operation

In third gear, the 1–2 synchronizer is moved to the neutral position and the 3–4 synchronizer sleeve

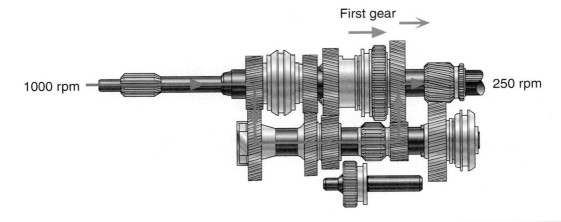

First gear

1000 rpm   250 rpm

Figure 48-15 Power flow in first gear.

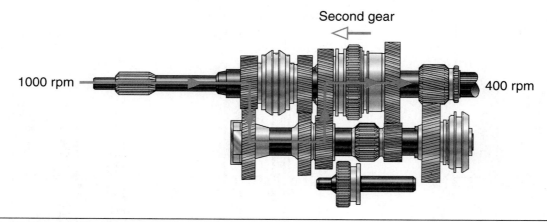

Figure 48-16 Power flow in second gear.

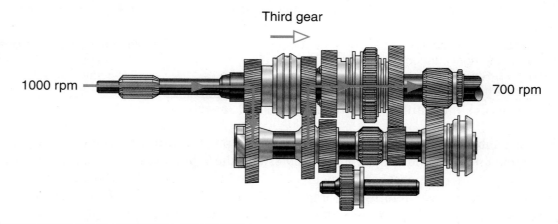

Figure 48-17 Power flow in third gear.

is moved rearward by the shift fork **(Figure 48-17)**. Under this condition, engine torque is transmitted through the input shaft gear, counter gear third-speed gear, and synchronizer sleeve and hub to the output shaft.

Fourth-Gear Operation

In fourth gear, the shift fork moves the 3–4 synchronizer sleeve forward. Engine torque is now transmitted from the input shaft gear, to the synchronizer sleeve and hub to the output shaft **(Figure 48-18)**.

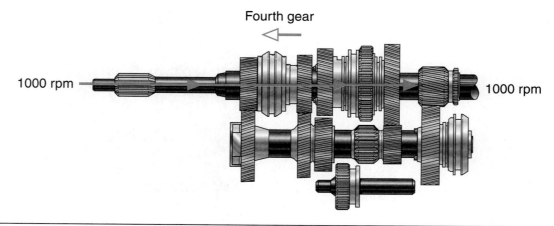

Figure 48-18 Power flow in fourth gear.

Fifth-Gear Operation

In fifth gear, the shift fork moves the fourth-speed synchronizer sleeve rearward to the neutral position and the fifth-speed synchronizer toward the fifth-speed gear. Under this condition, the fifth-speed gear is connected to the counter shaft. Engine torque is now transmitted from the input shaft gear, counter shaft, fifth-speed gear, and matching gear on the output shaft to output shaft (**Figure 48-19**).

Reverse Gear Operation

In reverse gear, all the synchronizers are shifted to the neutral position and a shift fork moves the reverse idler gear so it is in mesh with the matching gear on the counter shaft and the gear on the outer diameter of the 1–2 synchronizer sleeve (**Figure 48-20**). This movement allows engine torque to be transmitted from the input shaft gear to the matching gear on the counter shaft from the counter shaft

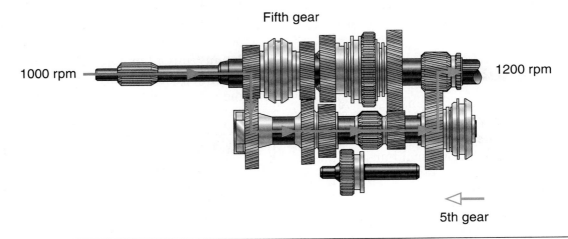

Fifth gear

1000 rpm

1200 rpm

5th gear

Figure 48-19 Power flow in fifth gear.

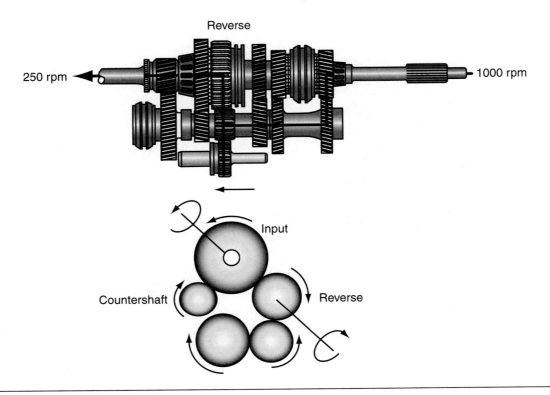

Reverse

250 rpm

1000 rpm

Input

Countershaft

Reverse

Figure 48-20 Power flow in reverse.

to the reverse idler gear. The input shaft rotates clockwise and the counter gear turns counterclockwise. Therefore, the reverse idler rotates clockwise.

The reverse idler gear is also in mesh with the reverse gear on the outer diameter of the 1–2 synchronizer sleeve. Clockwise reverse idler rotation turns the 1–2 synchronizer sleeve, hub, and output shaft counterclockwise to drive the vehicle in reverse. For clarity, Figure 48-19 shows the reverse idler gear out of its engaged position. The reverse idler gear would be in mesh with the reverse gear on the counter shaft and the reverse gear on the output shaft.

Transaxle Design and Operation

A transaxle is a combined transmission and differential that provides a compact assembly that is suitable for front-wheel drive vehicles and some rear-wheel drive vehicles. Manual transaxles have the same type of gears and synchronizers as manual transmissions. The shifting mechanism may have a selector rod connected from the gear shift lever to the transaxle shift forks **(Figure 48-21)**.

A stabilizer bar may be mounted between the transaxle case and the shifter housing. The shift forks in a manual transaxle are similar to those used in a manual transmission **(Figure 48-22)**. Some manual transaxles have two cables between the gear shift lever and the transaxle shift forks to provide proper shifting **(Figure 48-23)**. Gear shifting in various gears is similar in manual transmissions and transaxles.

Manual Transmission and Transaxle Lubrication

Lubrication in a manual transmission or transaxle is extremely important to protect all the moving metal parts. Bearing races, rollers, and other moving components must be covered with a thin film of lubricant to prevent bearing wear and overheating. A thin film of lubricant must also protect gear teeth, even when high engine torque is applying extreme pressure to the gears.

> **Tech Tip** *Some manual transaxles have a dipstick on top of the transaxle case. The fluid level should be at the specified mark on the dipstick.*

Vehicle manufacturers may specify 30W engine oil, automatic transmission fluid (ATF), or gear oil for their manual transmissions or transaxles. Always use the lubricant specified by the vehicle manufacturer.

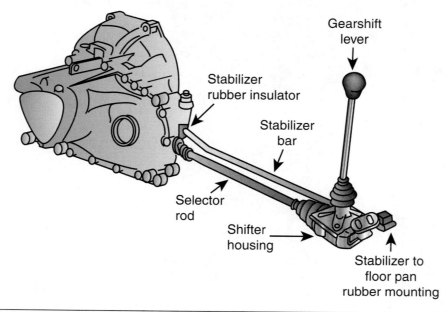

Figure 48-21 Transaxle gear shift linkage.

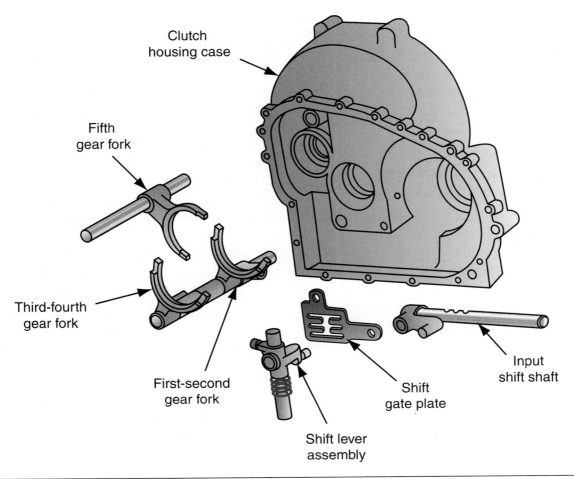

Figure 48-22 Transaxle shift forks.

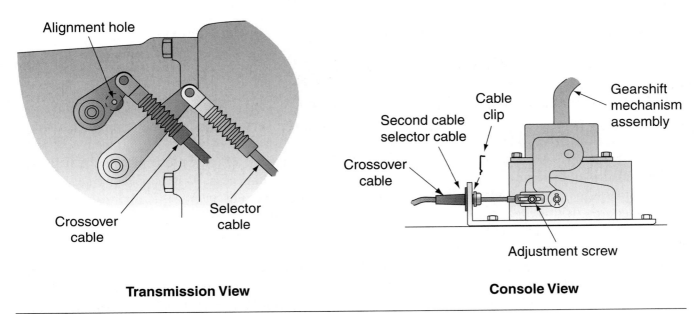

Transmission View   **Console View**

Figure 48-23 Transaxle shift mechanism with dual cables.

Most vehicle manufacturers do not recommend changing the manual transmission or transaxle lubricant at specified intervals. Some vehicle manufacturers recommend changing the manual transmission or transaxle lubricant at specified intervals if the vehicle is operated under severe conditions, such as trailer towing.

Manual Transmission and Transaxle Maintenance

Manual transmissions or transaxles require a minimum amount of maintenance. When a chassis lubrication is performed, the fluid level in the transmission or transaxle should be checked. When the filler plug is removed, the lubricant should be level with the bottom of the filler plug opening. If necessary, add the manufacturer's specified lubricant to obtain the proper fluid level. Always use the lubricant recommended by the vehicle manufacturer.

Some vehicle manufacturers recommend changing the transaxle lubricant at specified intervals if the vehicle is operated under severe driving conditions such as trailer towing, continual driving in city traffic during extremely hot conditions, or continual driving in mountainous terrain. During under-car service, inspect the transaxle for leaks at the drive axle seals. Before replacing a drive axle seal, always check the transaxle vent to be sure it is not plugged. A **transaxle vent** is located in the top of the case. When the transaxle components become hot during normal service, the vent allows air to escape from the transaxle to prevent pressure buildup in the transaxle case. A plugged vent causes pressure buildup in the transaxle that contributes to seal leaks.

To replace the drive axle seals, the drive axles must be removed. Always use the proper seal puller and driver to remove and install the seal. When replacing any seal, always inspect the seal lip contact area. If this area is scored, the component must be replaced. Inspect the lower part of the flywheel housing for leaks. A leak in this area may be caused by a leak at the transaxle input shaft seal or at the rear main bearing seal in the engine. If the transaxle fluid is low and the engine has not been losing any oil, the leak at the flywheel housing is likely at the transaxle input shaft seal.

If the vehicle has a transmission, inspect the output shaft seal for fluid leaks. If a leak is present, always be sure the transmission vent is not restricted before replacing the seal. To replace the output shaft seal, the drive shaft must be removed. Use the proper seal puller and driver to remove and install the seal.

With the engine idling, depress the clutch pedal and check for noise and proper clutch pedal free-play. Some clutch problems may be confused with transaxle problems. For example, excessive clutch pedal free-play, worn synchronizers, broken engine/transaxle mounts, or improper shift cable/linkage adjustments, may cause hard transaxle shifting.

When diagnosing this problem, the clutch pedal free-play, engine/transaxle mounts, and cable/linkage adjustments can be inspected and measured to determine if any of these causes are the source of the problem. If these three causes are eliminated, worn synchronizers are the likely cause of hard shifting.

The adjustment procedure for the shift linkages or cables varies depending on whether the transaxle has shift cables or linkages. This procedure also varies depending on the vehicle make and model year.

> **Tech Tip** *Road test the vehicle to verify proper shifting and that no unusual noise or vibration are present.*

MANUAL TRANSMISSION AND TRANSAXLE DIAGNOSIS

When diagnosing transmission/transaxle problems, it is very important to determine whether the problem is caused by an external problem or a defect in the transaxle. If these problems are not diagnosed accurately, unnecessary and expensive service may be performed. For example, if a hard shifting problem is caused by excessive clutch pedal free-play and the transaxle is overhauled to correct this problem, a great deal of unnecessary, expensive service work has been performed. Before diagnosing transmission/transaxle problems, always check the transmission/transaxle fluid level and condition. Low fluid level or contaminated fluid may cause bearing or gear noises in a transmission/transaxle.

> **Tech Tip** *There are many similarities between transmission and transaxle diagnosis. In this discussion we refer to diagnosing transaxle problems, but most of the diagnosis also applies to transmissions.*

> **Tech Tip** *Some transmissions and transaxles have a certain amount of noise with the clutch pedal released and the engine idling in neutral. This noise is aggravated by improper engine idle speed. Always be sure the engine is idling at the specified speed.*

Transaxle Does Not Shift into a Certain Gear

If the transaxle does not shift into one gear, the problem could be in the shift linkage. Be sure the shift linkage is properly adjusted and there is no interference between the shift linkage and the floor shift mechanism and the console or other components. The shifter rails in the transaxle could also cause this problem. When the transaxle does not shift into one gear, the synchronizer and/or related gear may be severely worn or damaged.

Transaxle Jumps out of Gear

If a transaxle jumps out of gear, the shift linkage may be the cause of the problem. An improperly adjusted or worn shift linkage may not be allowing the synchronizer and shift rail detent to shift completely into position. Check the shift cables or linkage for proper adjustment and wear. Be sure there is no interference between any of the shift mechanisms and other components. Worn engine/transaxle mounts may cause improper transaxle position that prevents complete synchronizer movement into a certain gear. A detent, shift rail, or shift fork defect may cause the same problem. Severely worn dog teeth on a gear and a worn synchronizer may cause a transaxle to jump out of gear. Excessive end play on a gear may cause the transaxle to jump out of gear.

The Transaxle Locks in One Gear

If a transaxle locks in one gear, the shift linkage may be binding on some other component. Internal problems, such as worn shift rails and detents, may cause the transaxle to lock in one gear. Severely worn synchronizer components, such as the sleeve and hub splines or spacers, may cause locking in one gear. A blocking ring seizing onto the gear cone may also cause this problem. Low fluid level in the transaxle or contaminated fluid may cause the blocking rings to seize onto the gear cones.

Transaxle Noise

Always road test the vehicle to be sure the noise is actually in the transaxle. If the transaxle is noisy in one gear only, the problem is likely wear or broken teeth on the gears that are meshed when the noise occurs. If a rattling noise occurs only with the clutch pedal depressed and the engine running with the transaxle in neutral, the pilot bearing or bushing is worn.

When a growling noise occurs as the clutch pedal is depressed, the clutch release bearing is defective. If the vehicle has a constant-running release bearing, a defective release bearing also causes a growling noise with the clutch pedal released and the transaxle in neutral. If the vehicle does not have a constant-running release bearing and a growling noise occurs only with the clutch pedal released and the engine running in neutral, the input shaft bearing is defective. If a rattling noise is heard under these same conditions, the input shaft gear and matching gear on the counter shaft may be worn or damaged.

Excessive Vibration

Although a vibration may seem to be coming from the transaxle, the vibration is likely caused by something external to the transaxle, such as worn inner CV joints or imbalanced tires and wheels. On a rear-wheel drive vehicle with a transmission, vibration may be caused by worn universal joints or improper drive shaft angles. On either front-wheel drive or rear-wheel drive vehicles, if the pressure plate is to be reused, it is recommended that the pressure plate be punch-marked and reinstalled on the flywheel in the original position. Vibration from internal transaxle problems is usually accompanied by noise problems, which are explained in the preceding discussion.

MANUAL TRANSMISSION AND TRANSAXLE SERVICE

After a transmission/transaxle problem is accurately diagnosed, the necessary service must be performed to correct the problem. Because of high labor rates, it is a common practice in the automotive service industry to replace rather than rebuild manual transaxles.

Transmission/Transaxle Removal

Before attempting to remove a transmission or transaxle, connect a memory saver or a 12-volt supply to the cigarette lighter socket and disconnect the negative battery cable. Place a drain pan under the transaxle and remove the drain plug to drain the fluid. Reinstall the drain plug. Support the engine so it does not drop downward when the transaxle or transmission is removed. An **engine support fixture** is usually the preferred method of supporting the engine during the removal and installation process **(Figure 48-24)**. On front-wheel drive vehicles, a support fixture is placed under the hood and attached to the engine **(Figure 48-25)**.

> **Tech Tip** *On some front-wheel drive vehicles, it is easier to remove the engine and transaxle as an assembly rather than removing the transaxle separately. Consult the service manual for specific details on transaxle removal.*

Raise the vehicle on a lift. Disconnect all wiring connectors and the speedometer cable from the transaxle. Remove all the linkages from the transaxle. If you are removing the transaxle from a front-wheel

Figure 48-25 An engine support fixture is used to support the engine during transmission/transaxle removal.

drive vehicle, remove the front drive axles from the transaxle. On some front-wheel drive vehicles, half of the engine cradle must be removed prior to transaxle removal.

Use a transmission jack to support the transmission or transaxle **(Figure 48-26)**. Be sure the unit is positioned securely on the transmission jack and remove the transmission or transaxle-to-engine

Figure 48-26 A transmission/transaxle jack.

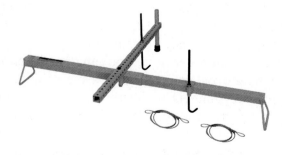

Figure 48-24 An engine support fixture.

Figure 48-27 A transmission/transaxle jack is used to remove/install the transmission/transaxle from the vehicle.

retaining bolts **(Figure 48-27)**. Slide the jack and transaxle to remove the input shaft from the clutch and pressure plate. Slowly lower the jack to remove the transaxle from the vehicle. Place the transaxle in a transaxle stand or on the workbench. Be sure the transaxle is securely bolted to the stand.

> **Tech Tip** *If the transmission/transaxle has been removed, it is highly recommended that the clutch assembly be carefully inspected. The added cost of servicing the clutch assembly is relatively low compared to replacing it at a later time and repeating removal of the transmission/transaxle.*

Transmission/Transaxle Installation

Before installing the transaxle, always shift the transaxle through all the gears and be sure the input shaft rotates freely in all gears. Install a new clutch release bearing before installing the transaxle.

Raise the transaxle with a transmission jack and push the transaxle toward the engine so the input shaft enters the pressure plate and clutch disc. Be sure the transaxle housing is fully seated against the engine block. Do not force the transaxle into the clutch disc because this action may damage the clutch disc or transaxle input shaft or housing. If the transaxle does not slide easily into place, remove it and recheck the input shaft splines. Be sure the clutch disc is aligned with the pilot bearing opening. After the transaxle is completely seated against the engine, install the flywheel housing-to-engine bolts and tighten these bolts to the specified torque. Remove the transmission jack. Connect all electrical wires and the speedometer cable to the transaxle. Connect the shift linkages to the transaxle. Install the front drive axles into the transaxle and be sure all fasteners are tightened to the specified torque. If part of the engine cradle was removed, install this cradle and tighten all fasteners to the specified torque.

Fill the transaxle with the specified lubricant to the bottom of the filler plug opening or to the full mark on the dipstick. Install the filler plug or dipstick. Remove the engine support fixture and reconnect the negative battery cable. Disconnect the memory saver or the 12-volt power supply from the cigarette lighter socket.

You should road test the vehicle to verify the correction of the original customer complaint.

Summary

- Transmissions and transaxles provide gear reductions to allow smooth vehicle acceleration.
- In fifth or sixth gear, a transmission provides overdrive gear ratios, which supply improved fuel economy at cruising speed.
- A gear ratio is calculated by dividing the number of gear teeth on the drive gear into the number of teeth on the driven gear.
- If torque is transmitted through two gear ratios, the total gear ratio is determined by multiplying the two ratios.

- During a shift, the synchronizer blocking ring grooves contact the cone on the selected gear to bring the synchronizer and the selected gear to the same speed.

- Vehicle manufacturers may recommend engine oil, automatic transmission fluid, or gear oil in manual transmissions and transaxles.

- Worn synchronizers, excessive clutch pedal free-play, may cause hard shifting, or improper shift linkage adjustment.

- Jumping out of gear may be caused by an improperly adjusted shift linkage, worn engine or transaxle mounts, worn dog teeth on a gear, worn synchronizer, or excessive end play on a gear.

- If a growling noise occurs as the clutch pedal is depressed, the clutch release bearing may be defective.

- When a rattling noise is evident with the clutch pedal depressed, the pilot bearing or bushing may be worn.

- If a growling noise is present only with the clutch pedal released, the input shaft bearing may be defective.

- Before the transaxle or transmission is removed, the engine must be supported so it does not drop downward.

- The transaxle or transmission should be supported on a transmission jack during the removal procedure.

Review Questions

1. Technician A says the fifth speed in a transmission has an overdrive ratio. Technician B says a gear reduction is provided when a smaller diameter gear is driving a larger diameter gear. Who is correct?

   A. Technician A

   B. Technician B

   C. Both Technician A and Technician B

   D. Neither Technician A nor Technician B

2. Technician A says a spur gear does not create thrust on the gear. Technician B says backlash is the amount of movement between the teeth on two gears. Who is correct?

   A. Technician A

   B. Technician B

   C. Both Technician A and Technician B

   D. Neither Technician A nor Technician B

3. A second-speed drive gear has fourteen teeth and the second-speed driven gear has thirty-six teeth. The second-speed gear ratio is:

   A. 2.1:1.

   B. 2.5:1.

   C. 2.8:1.

   D. 3.2:1.

4. Technician A says a synchronizer sleeve is mounted on splines on the synchronizer hub. Technician B says a synchronizer hub is mounted on splines on the transmission output shaft. Who is correct?

   A. Technician A

   B. Technician B

   C. Both Technician A and Technician B

   D. Neither Technician A nor Technician B

5. A gear set has nine teeth on the drive gear and thirty-seven teeth on the driven gear. The gear ratio is:

   A. 3.6:1.

   B. 3.9:1.

   C. 4.1:1.

   D. 4.3:1.

6. All of these statements about gears and gear sets are true *except*:

   A. Helical gear teeth are cut at an angle in relation to the gear centerline.

   B. In an overdrive gear set, the driven gear is larger than the drive gear.

   C. Backlash is the amount of movement between the teeth on two gears.

   D. Helical gear teeth create axial thrust on the gear.

7. When shifting gears that have a synchronizer, Technician A says during the shift, the grooved blocking ring cone cuts through the lubricant on the selected gear cone to allow metal-to-metal contact. Technician B says the synchronizer action brings the selected gear to the same speed as the synchronizer hub. Who is correct?

   A. Technician A

   B. Technician B

   C. Both Technician A and Technician B

   D. Neither Technician A nor Technician B

8. All of these problems may cause a manual transaxle to lock in one gear *except*:

   A. a blocking ring seized onto a gear cone.

   B. worn shift rail detents.

   C. severely worn synchronizer sleeve hub splines.

   D. excessive clutch pedal free-play.

9. A vehicle with a manual transaxle has a growling noise only when the clutch pedal is released and the transmission is in neutral. The most likely cause of this problem is:

   A. a defective transaxle input shaft bearing.

   B. a rough clutch release bearing.

   C. a worn pilot bushing.

   D. worn splines on the input shaft and clutch hub.

10. A four-speed manual transmission is very hard to shift into second gear. This problem may be caused by all of these defects *except*:

    A. an improperly adjusted shift linkage.

    B. severely worn 1–2 synchronizer and second speed gear.

    C. insufficient clutch pedal free-play.

    D. worn or damaged shifter rails.

11. An overdrive gear ratio _____ torque and _____ output shaft speed.

12. If torque is transmitted through two gear ratios of 3.5:1 and 4.08:1, the total gear ratio is _____.

13. In a three-speed transmission, the third-speed gear ratio of the transmission is _____.

14. List three possible lubricants for manual transmissions/transaxles.

    1. _____

    2. _____

    3. _____

15. Explain the purpose of a synchronizer.

16. Describe the operation of a synchronizer.

17. Explain the causes of hard transaxle shifting.

18. Describe the causes of jumping out of gear.

Automatic Transmission/Transaxle

Learning Objectives

After you have read, studied, and practiced the contents of this unit, you should be able to:

- Describe the purpose of a torque converter.

- Explain the torque converter operation.

- Explain planetary gear set design.

- Explain how a planetary gear set provides reverse.

- Describe how a planetary gear set provides a 1:1 gear ratio.

- Describe how a planetary gear set provides a gear reduction.

- Describe how a planetary gear set provides a gear overdrive.

- Describe the vehicle operation encountered with a slipping stator one-way clutch.

- Describe the vehicle operation encountered with a seized stator one-way clutch.

- Explain the inputs used by the PCM to provide torque converter clutch lockup.

- Explain the operation of a multiple disc clutch.

- Describe the operation of a band.

- Explain the operation of an overrunning clutch.

- List the different types of transaxle pumps and explain the operation of the pressure regulator valve.

- Explain the operation and purpose of the governor.

- Describe the operation of the throttle cable and throttle valve.

- Explain the operation of a shift valve.

- Explain the operation of an accumulator.

- Describe basic transaxle electronic controls including shift control, pressure control, inputs, and outputs.

- Change automatic transmission or transaxle fluid and filter.

- Check automatic transmission fluid level and condition.

- Diagnose automatic transmission and transaxle problems.

- Perform automatic transmission and transaxle service procedures.

Key Terms

Accumulator	Planetary gearset	Torque converter clutch
Coupling point	Rotary flow	Viscous coupling
Flexplate	Servo piston	Vortex flow
Line pressure	Shift valve	

INTRODUCTION

Transmissions and transaxles have become complex mechanical devices controlled by hydraulics and electronics. It is essential for technicians to be familiar with the mechanical principles of multiple disc clutches, bands, and planetary gear sets to gain an understanding of automatic transmissions. You must also be familiar with torque converters. You need to be familiar with the electronic controls in today's transaxles and transmissions. When these mechanical, hydraulic, and electronic principles are clearly understood, diagnosing and servicing transaxle and transmission problems becomes much easier.

NOTE: The transmission and transaxle contain many of the same components and operate on the same principles. The terms *transmission* and *transaxle* may be used interchangeably when discussing the main operating characteristics and components.

TORQUE CONVERTER PURPOSE AND DESIGN

A torque converter contains three main components: the impeller pump, the turbine, and the stator. The blades on the impeller and the turbine are curved in opposite directions. The converter cover is welded to the impeller pump to seal these members in a housing. The converter cover is bolted to the engine **flexplate**, which is bolted to the crankshaft **(Figure 49-1)**. The impeller section of the torque converter is driven by the engine and the turbine section is driven by the transmission fluid and transfers power from the engine to the transmission **(Figure 49-2)**.

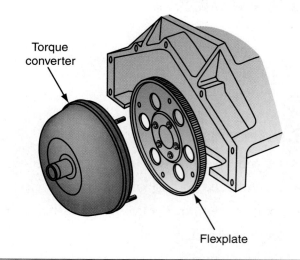

Figure 49-1 The torque converter is bolted to the flexplate.

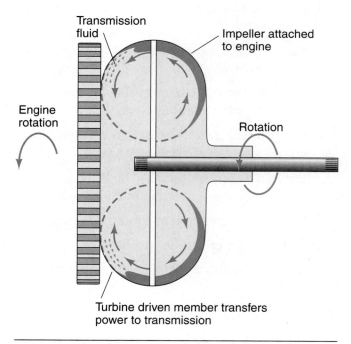

Figure 49-2 The impeller drives the turbine, which in turn transfers power to the transmission.

When the engine is running, the converter cover and impeller pump must rotate with the crankshaft. A hub on the rear of the impeller has two notches or flats that fit into matching grooves in the inner member of the transmission oil pump. Therefore, the converter drives the transmission pump. The pump seal lips contact the smooth machined surface on the outer diameter of the converter hub. The impeller pump hub is supported on a bushing in the oil pump of the transmission.

The turbine, inside the torque converter, is splined to the transaxle input shaft and the stator is connected to the reaction shaft through an overrunning clutch. The reaction shaft extends forward from the transmission pump through the converter hub into the stator. The reaction shaft is stationary and cannot rotate **(Figure 49-3)**.

The stator hub is splined to the reaction shaft. The stator overrunning clutch contains a series of spring-loaded rollers in tapered grooves. This overrunning clutch allows the stator to turn freely in one direction and lock up in the opposite direction **(Figure 49-4)**.

In a typical transaxle, the front end of the input shaft is splined to the turbine and the rear end of this shaft is splined to a drive sprocket. The engine torque is transmitted from the input shaft and drive sprocket through a chain and driven sprocket into the transaxle **(Figure 49-5)**.

In recent years, most torque converters are lockup type and are referred to as torque converter clutch (TCC). Many lockup converters contain a lockup disc between the turbine and the cover. The

lockup disc is splined to a short extension on the front of the turbine and a friction material about 1 inch (25.4 millimeters) wide is molded onto the

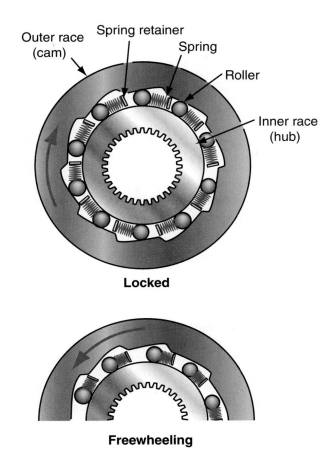

Figure 49-4 The overrunning clutch allows the stator to turn freely in one direction and lock up in the opposite direction.

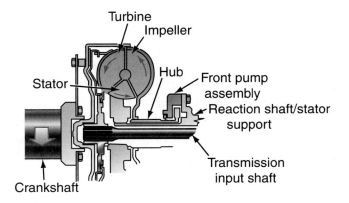

Figure 49-3 Torque converter components.

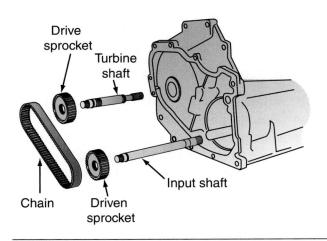

Figure 49-5 In a typical transaxle engine, torque is transmitted through the input shaft drive sprocket, chain, and driven sprocket.

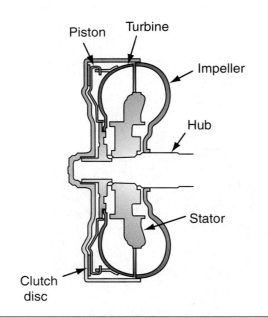

Figure 49-6 A lockup torque converter.

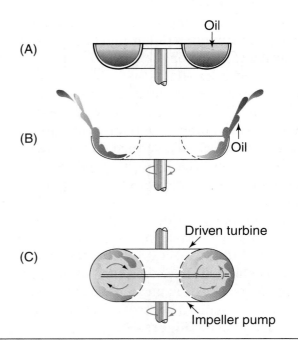

Figure 49-7 (A) Fluid in the torque converter at rest, (B) Fluid discharged from the impeller pump vanes, (C) Fluid flowing from the turbine back to the impeller pump.

front of the lockup disc near the outer diameter of the disc **(Figure 49-6)**.

> **Tech Tip** *The impeller pump and the turbine rarely turn at exactly the same speed because there is some slippage in a torque converter.*

The lockup disc is free to move a short distance on the turbine splines. Torsional springs are mounted in the lockup disc between the outer part of the disc and the hub. Some lockup discs have a silicone clutch between the outer part of the disc and the hub. The silicone clutch provides smoother TCC engagement compared to a TCC with torsional springs.

Torque Converter Operation

The torque converter contains automatic transmission fluid (ATF). When the engine is running, the impeller picks up ATF at its center and centrifugal force causes this fluid to be discharged on to the blades at the impeller rim **(Figure 49-7)**. As the oil strikes the turbine blades, the energy of the impeller oil striking the turbine causes it to rotate. If the engine is idling at the specified speed, the force of the fluid striking the turbine blades is not high enough to rotate the turbine. When the engine

speed is increased slightly, more rotational force is exerted on the turbine blades and the turbine begins to rotate in the same direction as the impeller pump. The impeller pump and the turbine rotate in the same direction. Engine torque is now transmitted through the turbine and input shaft to the transaxle gear train, drive axles, and drive wheels.

At lower engine speeds, the fluid movement from the impeller pump to the turbine is redirected by the stator blades so this fluid helps to rotate the turbine **(Figure 49-8)**. Under this condition, the overrunning clutch prevents stator rotation. If the stator did not perform this function, this fluid movement would work against the turbine rotation and reduce the effect of engine power and torque. The fluid follows the contour of the turbine blades so it leaves the turbine in the opposite direction to turbine rotation. Because the direction of this fluid motion is opposite the engine and impeller pump rotation, the fluid leaving the turbine and re-entering the impeller pump would work against the engine and reduce effective engine power and torque.

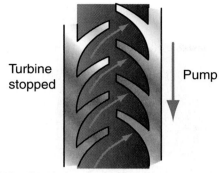

(A)   Oil is thrown against pump vanes.

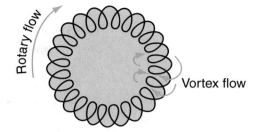

Oil flows inside the torque converter from the pump, to the turbine, to the stator, and back to the pump.

Figure 49-9 The spiral pattern of fluid flow is called vortex.

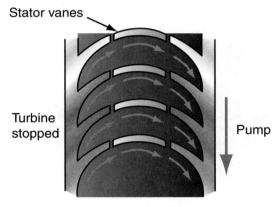

(B)   Oil path is changed by stator.

Figure 49-8 The stator redirects the fluid flowing from the turbine back into the impeller pump.

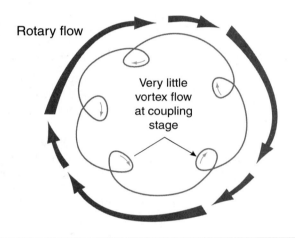

Figure 49-10 The coupling phase has very little vortex flow.

The term **vortex flow** is used to describe this fluid movement in the converter when the stator is redirecting the fluid movement from the turbine to the impeller pump. This action of redirecting the fluid flow enhances the power flow and provides torque multiplication, hence the name torque converter **(Figure 49-9)**.

When the engine speed increases, the speed of the turbine approaches the speed of the impeller pump. This condition is called the **coupling point**. When the impeller pump and turbine reach the coupling point, the fluid returning from the turbine into the impeller pump begins striking the back of the stator blades. Under this condition, the overrunning clutch in the stator allows the stator to rotate freely, in the same direction, with the impeller pump and turbine **(Figure 49-10)**.

The impeller pump, turbine, and stator are now rotating as a unit and the converter is no longer multiplying torque. When the torque converter reaches the coupling point, the fluid movement in the converter is changed from vortex flow to **rotary flow (Figure 49-11)**.

The torque converter clutch lockup system is electronically operated by the PCM. The main purpose of the **torque converter clutch** is to lock up the torque converter to improve fuel economy. When the vehicle is driven at lower speeds, the torque converter clutch is not locked up. Under this condition, fluid is directed into the converter so it flows through the hollow transmission input shaft and out in front of the lockup disc. The fluid forces the lockup disc away from the front of the converter so the friction material on the lockup disc does not contact the front of the converter.

Turbine   **Stator**   **Impeller**

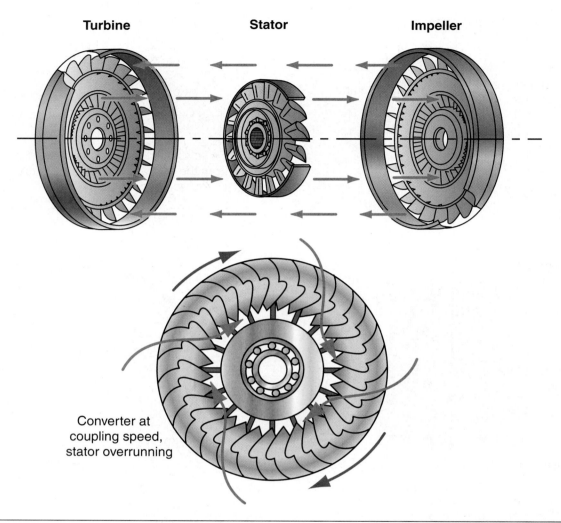

Converter at
coupling speed,
stator overrunning

Figure 49-11 Converter fluid flow in the coupling phase with the stator overrunning.

The torque converter clutch is always unlocked when the engine coolant is cold.

When the engine reaches the proper coolant temperature and the correct vehicle speed, the PCM operates the lockup valve in the transaxle valve body. Movement of the lockup valve and switch valve now direct fluid into the rear of the torque converter hub **(Figure 49-12)**. Fluid is forced over the outer diameter of the turbine and in behind the lockup plate. This fluid movement forces the lockup plate against the front of the converter. This action connects the flex plate and the front of the converter directly through the lockup disc to the turbine hub and input shaft. Because the impeller pump is also bolted to the flex plate, the impeller pump and the turbine must rotate at the same speed and the converter is locked up.

PLANETARY GEARSET DESIGN AND OPERATION

Most automatic transmissions use planetary gearsets to provide forward gear reductions and reverse. Each planetary gearset contains three elements: an outer ring gear with internal teeth, a planet carrier containing planetary pinions, and a sun gear in the center **(Figure 49-13)**. Planetary gearsets are compact, strong, and quiet running.

In a **planetary gearset**, the planet carrier is the largest gear as the planet gears it carries has the most number of teeth. The ring gear is the second largest, and the sun gear is the smallest gear. If a planetary gearset has an input but no member of the gear set is held stationary, the gear set is in neutral. Under this condition, no torque is transmitted from the

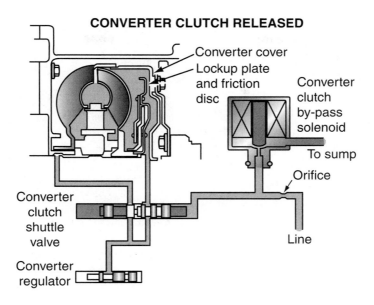

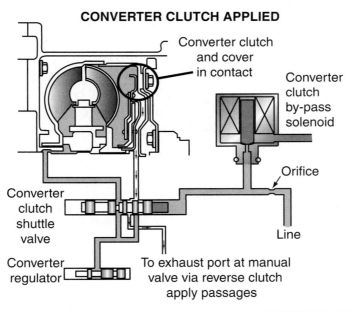

Figure 49-12 Hydraulic system for PCM controlled torque converter clutch lockup.

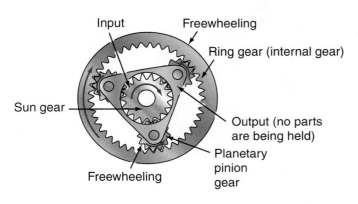

Figure 49-13 Planetary gear set in neutral.

input shaft. To provide gear reductions or overdrive, one member or the planetary gearset must be held or locked and the input must be supplied to one of the other members. Various planetary gearset members are held by one-way clutches, groups of clutch plates, or bands.

NOTE: In a planetary gearset, the ring gear may be called an *internal gear* or an *annulus gear*.

When the sun gear is held and the ring gear is the input and the planet carrier is the output, a forward gear reduction is provided because the smaller ring gear is driving the larger planet carrier **(Figure 49-14)**.

When the sun gear is the input, the ring gear is held, and the planet carrier is the output, a forward gear reduction is provided because the smaller sun gear is driving the larger planet carrier **(Figure 49-15)**. Under this condition, a typical gear reduction is 3.3:1, which may be suitable for low gear.

If the planet carrier is the input, the sun gear is held, and the ring gear is the output, an overdrive

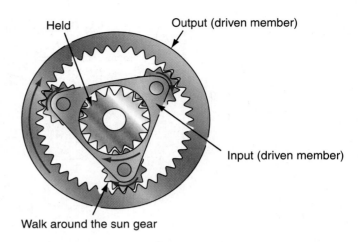

Walk around the sun gear

Figure 49-16 A forward overdrive gear ratio is provided if the carrier is the input, the sun gear is held, and the ring gear is the output.

gear ratio is provided because the larger planet carrier is driving the smaller ring gear **(Figure 49-16)**.

In a planetary gearset, a direct drive or a 1:1 gear ratio is provided by locking two members together. If both the ring gear and sun gears are inputs, these two members are effectively locked together and the planet carrier is the output **(Figure 49-17)**. The complete planetary gear set must turn together.

When the sun gear is the input, planet carrier is held, and the ring gear is the output, a reverse gear reduction is provided because the smaller sun gear

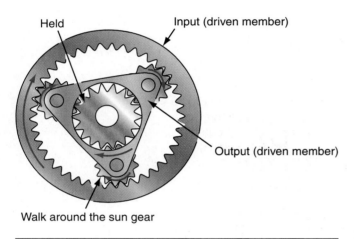

Walk around the sun gear

Figure 49-14 A forward gear reduction is provided when the ring gear is the input, the sun gear is held, and the carrier is the output.

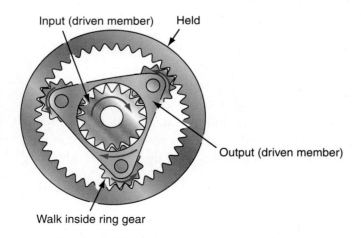

Walk inside ring gear

Figure 49-15 A forward gear reduction is provided when the sun gear is the input, the ring gear is held, and the carrier is the output.

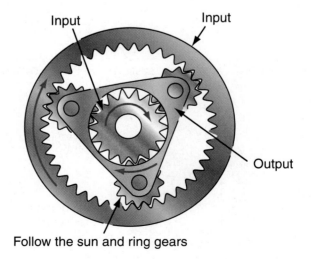

Follow the sun and ring gears

Figure 49-17 Direct drive is provided when two planetary members are locked together or two members are inputs.

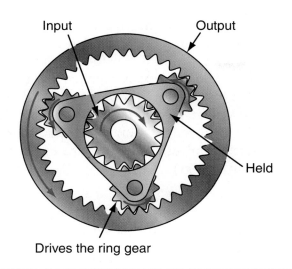

Figure 49-18 A reverse gear reduction is provided when the sun gear is the input, the carrier is held, and the ring gear is the output.

is driving the larger ring gear **(Figure 49-18)**. Because the planet carrier is held, the ring gear must turn in the opposite direction to the sun gear rotation.

Planetary gears provide quiet, smooth operation in automatic transmissions and transaxles. Planetary gears are also strong and durable as they are contained within the internal gear. Defective or worn planetary gears may cause excessive noise, improper shifting, and locking of the transmission. The diagnosing and servicing of planetary gears is an integral part of complete transmission and transaxle diagnosis and service.

NOTE: Planetary gears are helical-cut gears.

MULTIPLE-DISC CLUTCHES

Automatic transaxles contain several multiple-disc clutches that apply and release the members in a planetary gear set. A multiple-disc clutch assembly contains a hub, driving discs, driven plates, apply piston, seals, pressure plate, release springs, and clutch assembly container. These components are held in a clutch drum by a snap ring.

A friction material is usually bonded to the driving discs, which have internal serrations mounted on the hub splines. The hub that fits inside the driving discs is from another member in the transaxle. The sides of the driven discs do not usually have any friction material and these plates have external tangs mounted in the clutch drum slots or transmission case **(Figure 49-19)**.

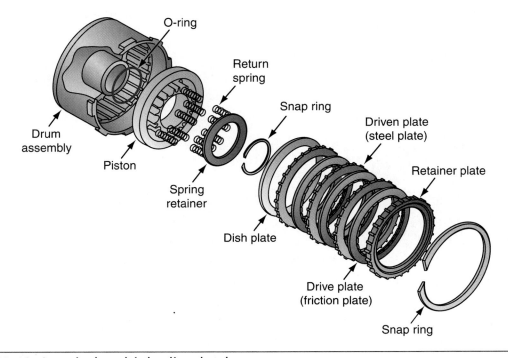

Figure 49-19 A typical multiple-disc clutch.

The driving and driven discs are placed alternately in the clutch drum. The spring-released apply piston has internal and external seals between the piston and the clutch drum hub. When fluid pressure is not supplied to the apply piston, the drive and driven plates are free to turn within each other. If fluid pressure is applied behind the apply piston, the piston is moved against the spring pressure. This piston movement forces the clutch discs together and the hub is connected through the driving and driven clutch discs to the clutch drum.

BANDS AND ONE-WAY CLUTCHES

Bands are used in automatic transmissions and transaxles to hold one member in a planetary gear set. The bands are made from flexible steel with friction material on the inner surface. The band is usually mounted over the outer surface of a clutch drum. One end of the band is anchored in the transaxle case and the other end is connected to a servo piston stem or a strut and apply lever **(Figure 49-20)**.

A **servo piston** is in contact with the servo piston stem. The servo piston is sealed in a machined bore and is retained in the bore by a retainer and snap ring. A spring is positioned between the servo

piston and the end of the bore next to the band. When fluid pressure is not supplied to the servo piston, the clutch drum is free to rotate inside the band. When fluid pressure is supplied to the servo piston, this piston moves toward the band. This piston movement pushes the servo piston stem against the band and this action tightens the band on the clutch drum, effectively locking the drum in place **(Figure 49-21)**.

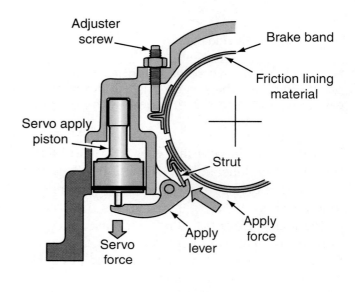

Figure 49-21 A band, strut, and servo assembly.

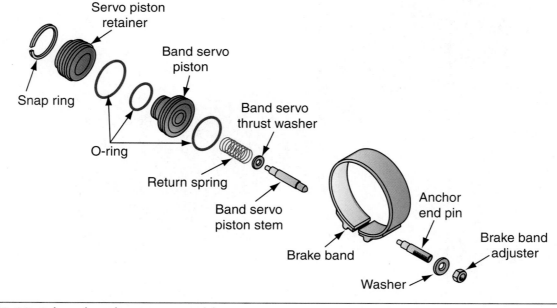

Figure 49-20 A band and servo assembly.

Some members in an automatic transaxle are mounted on a one-way clutch. The one-way clutch acts as a locking device to prevent rotation in one direction and allow rotation in the opposite direction. The one-way clutch contains a hub and an outer race. The hub is splined to one of the transmission members. A series of tapered grooves are cut in the inner surface of the outer race. Spring-loaded steel rollers are mounted in these tapered grooves.

***

NOTE: A one-way clutch may be called an *overrunning clutch*.

***

If the hub is rotated counterclockwise, the rollers move to the narrow part of the tapered grooves **(Figure 49-22)**. Under this condition, the rollers jamb between the hub and the outer race. This action locks these two units and prevents hub rotation in a counterclockwise direction. If the hub is turned in a clockwise direction, the rollers move against the spring tension into the wider part of the tapered grooves. Under this condition, the hub can rotate freely.

OIL PUMP AND PRESSURE REGULATOR VALVE

Fluid pressure to operate the various clutch packs and bands in an automatic transmission or transaxle is supplied from the oil pump. On many transmissions and transaxles, the oil pump is driven by the torque converter hub. Two grooves on the converter hub fit into matching notches in the oil pump inner rotor **(Figure 49-23)**. When the engine starts, the oil pump builds up pressure immediately.

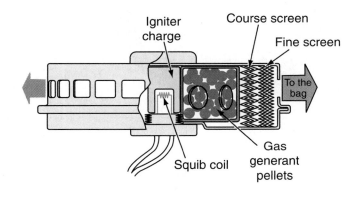

Figure 49-23 A rotor-type oil pump.

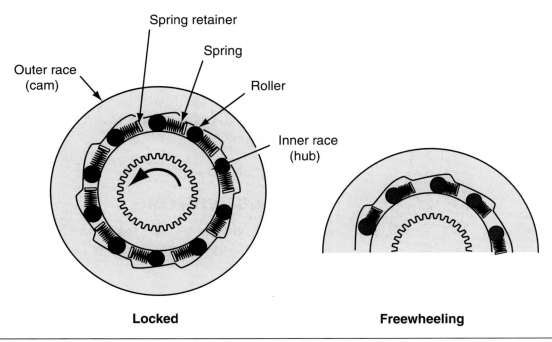

Figure 49-22 An overrunning clutch.

Many transaxles have a gear-type pump **(Figure 49-24)**, but some transaxles have a vane-type pump **(Figure 49-25)**. Some transaxles have a drive shaft extending from the front of the converter through a hollow transaxle input shaft to the oil pump hub **(Figure 49-26)**.

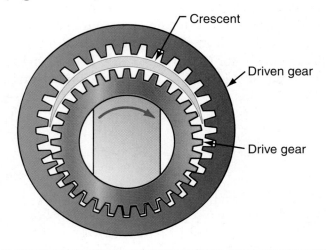

Figure 49-24 A gear-type pump.

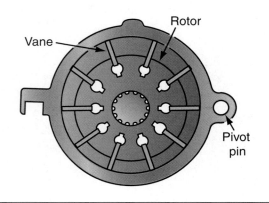

Figure 49-25 A vane-type pump.

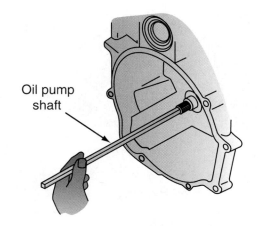

Figure 49-26 A transmission oil pump drive shaft.

Tech Tip *On many transaxles, the pump is driven by a drive shaft extending from the converter hub rather than driving the pump directly from the back of the converter hub.*

The oil pump intakes oil from the transaxle reservoir through the filter and delivers oil to the transaxle components **(Figure 49-27)**. The filter is usually positioned near the bottom of the reservoir **(Figure 49-28)**.

The pressure regulator valve regulates fluid pressure by returning some fluid to the reservoir. When the pump pressure increases, the fluid pressure forces the pressure regulator valve against the spring tension. This valve movement allows more fluid to return to the reservoir. This pressure regulator valve movement controls line pressure. The fluid pressure delivered from the pressure regulator valve is called **line pressure** or mainline pressure.

Line pressure is delivered from the pressure regulator valve to the manual valve. The manual valve is operated by the gear shift linkage. In the various gear selector positions, the manual valve delivers line pressure to the other transaxle components. Fluid pressure must be high enough to apply clutches and bands without slipping. However, the fluid pressure must be limited to prevent harsh shifting and damage to clutch pack seals.

Fluid is also delivered to the torque converter. The return fluid from the torque converter flows through the oil cooler and then to the lubrication system. The lubrication system provides lubrication to bushings and bearings and other transaxle components.

GOVERNORS

Governors are used in transaxles and transmissions that are not computer controlled.

NOTE: The governor is a speed-sensitive valve that assists in the shifting of the transmission.

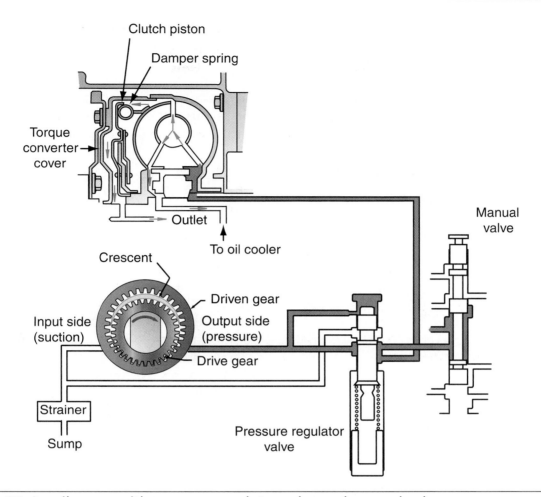

Figure 49-27 An oil pump with pressure regulator valve and manual valve.

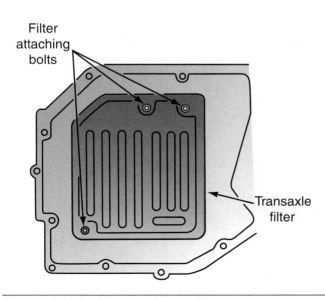

Figure 49-28 A transmission filter.

As the governor increases in speed it will cause the transmission to up-shift. The governor is gear driven, usually from the output shaft. The governor has pivoted primary and secondary weights. The primary weight flies outward at low speeds and the lighter secondary weights move outward at higher speeds. As the governor weights move outward, they move a spool valve in the governor bore. Line pressure is supplied to the governor and governor pressure is delivered from the governor to one end of the shift valves.

At low vehicle speeds, the governor valve is positioned so some of the fluid is exhausted at the governor exhaust port and governor pressure remains low **(Figure 49-29)**. When the vehicle speed increases, the governor weights move outward and gradually close the exhaust port allowing governor pressure to increase. Governor pressure can reach a point where it is equal to line pressure, but it cannot exceed the line pressure.

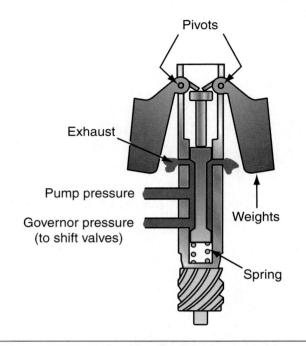

Figure 49-29 Governor operation.

THROTTLE LINKAGES AND THROTTLE VALVES

Throttle valves are mainly used in transaxles and transmissions that are not computer controlled. They are used to tailor shift points and clutch pressures to match engine load.

Tech Tip *The throttle valve is sensitive to changes in throttle opening, which accurately indicates engine load to the transmission.*

As the throttle valve is opened it tends to delay transmission up-shift. These transmissions have a throttle cable connected from the throttle linkage at the throttle body or carburetor to the throttle valve in the transaxle or transmission. The throttle cable moves the TV plunger against the spring tension in relation to throttle opening. This increased spring tension is applied to the throttle valve **(Figure 49-30)**. Line pressure is supplied from the pressure regulator and manual valves to the throttle valve.

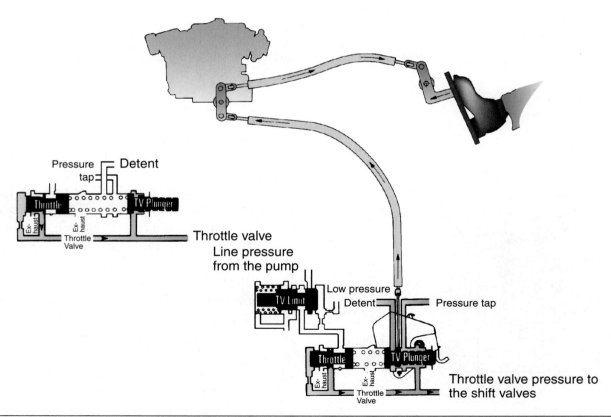

Figure 49-30 A throttle valve and cable.

As the throttle is opened, the increasing pressure on the throttle valve moves this valve so the line pressure gradually flows past the valve lands into the throttle valve pressure passage to increase throttle pressure. This increasing throttle pressure is supplied to the opposite end of the shift valves to which the governor pressure is supplied.

MODULATORS

Some transmissions and transaxles have a vacuum modulator containing a diaphragm in a sealed chamber.

> **Tech Tip** *The modulator is a valve that senses relative engine load. It often acts in conjunction with the pressure regulator to increase mainline pressure in proportion to engine load.*

The modulator is mounted in the side of the transaxle or transmission case. A stem on the transaxle side of the diaphragm contacts a modulator valve in the transaxle **(Figure 49-31)**.

When the engine is running, manifold vacuum is supplied through a hose to the outer side of the modulator diaphragm and the inner side of the diaphragm is vented to the atmosphere. The diaphragm is connected to the modulator valve. At wide throttle openings, the vacuum decreases and the movement of the modulator diaphragm and

valve directs modulator pressure to the spring ends of the shift valves to delay the shifts until the engine reaches a higher rpm.

ACCUMULATORS

Many transmissions and transaxles contain accumulators. An **accumulator** is a spring-loaded piston mounted in a machined bore. A seal or O-ring is positioned in a groove on the accumulator piston **(Figure 49-32)**. Accumulators act as shock absorbers to cushion the application of multiple-disc clutches and bands. An accumulator cushions sudden increases in fluid pressure by allowing the pressure to flow into the accumulator bore against the spring-loaded piston. As the accumulator piston moves and the fluid fills the piston bore, the

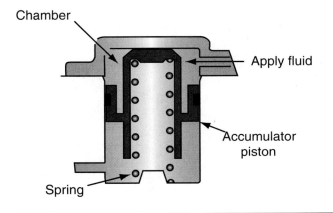

Figure 49-32 An accumulator assembly.

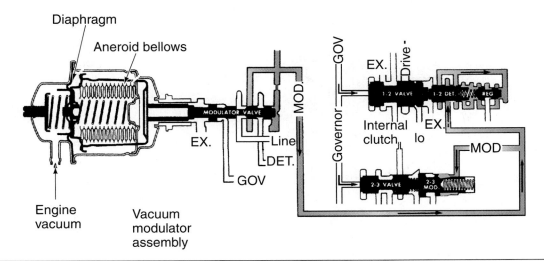

Figure 49-31 A vacuum modulator.

fluid pressure builds up gradually. This action delays the application of the clutch discs or band to change the feel of the shift such as preventing harsh shifting.

VALVE BODY AND SHIFT VALVES

The valve body is usually mounted on the bottom of the transmission or transaxle case in the reservoir. The valve body contains spool valves and check balls **(Figure 49-33)**. The **shift valves** in the valve body control the shifting.

In non-computer-controlled transmissions, governor pressure is supplied to one end of each shift valve with throttle and modulator pressure is supplied to the opposite end of the shift valves. When vehicle speed increases with a steady throttle opening, the governor pressure reaches a point where it overcomes throttle pressure on the opposite end of the 2–3 shift valve.

When governor pressure overcomes throttle pressure on the 2–3 shift valve, the valve moves and allows line pressure to flow past the shift valve lands to the appropriate multiple clutch set or band **(Figure 49-34)**. When the clutch or band is applied, one of the planetary gear members is locked while the other two members are the input and output to supply the proper gear.

ELECTRONIC TRANSMISSION AND TRANSAXLE CONTROLS

Modern vehicles usually have computer-controlled automatic transmissions or transaxles. The computer that controls the transaxle functions is combined in the PCM, or a separate transmission controller may be used.

Many computer-controlled transaxles or transmissions have a turbine speed sensor and an output speed sensor mounted in the transaxle case. A park, reverse, neutral, drive, low (PRNDL) switch

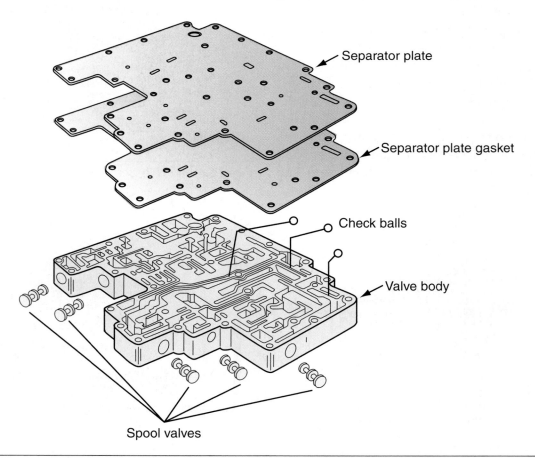

Figure 49-33 Valve body with spool valves, check balls, separator and gasket.

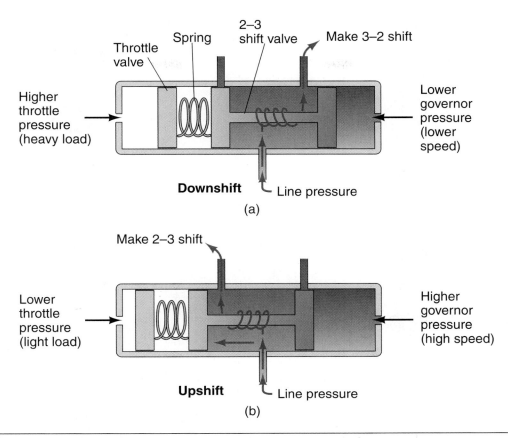

Figure 49-34 Shift valve operation during upshifts and downshifts.

is mounted on the transaxle and operated by the gear shift linkage. This switch informs the transmission computer regarding the gear selection.

Some transaxles have pressure switches that send feedback information to the transmission computer when a shift occurs. The transmission computer shares some input sensors with the engine computer. These inputs may include ECT sensor, TP sensor, MAP sensor, VSS, and brake switch **(Figure 49-35)**. These input signals are usually sent to the PCM and relayed on the interconnecting data links to the transmission controller. The computer controls the solenoid operation for the various gears of the transmission **(Figure 49-36)**.

> **Tech Tip** *See the chapters on computer engine controls for more details on computer inputs.*

A governor is not required in a computer-controlled transaxle, because the computer operates two or more solenoids to supply pressure to the shift valves to control shifting based on sensor inputs such as throttle position, vehicle speed, rpm, and other inputs **(Figure 49-37** and **Figure 49-38)**. The computer also controls an electronic pressure control valve to control transaxle pressure. The electronic pressure control valve replaces the pressure regulator valve. The computer operates a TCC solenoid that supplies pressure to the TCC control valve. This valve controls the fluid supplied to the converter.

FINAL DRIVES AND DIFFERENTIALS

In rear-wheel drive vehicles, engine torque is transmitted through the transmission and drive shaft to the pinion shaft and ring gear in the differential. The pinion gear is meshed with the differential ring gear. A small shaft retains two small differential gears inside the differential case and drive the side gears in turn. The two side gears are splined to the axles. These side gears are meshed with the small

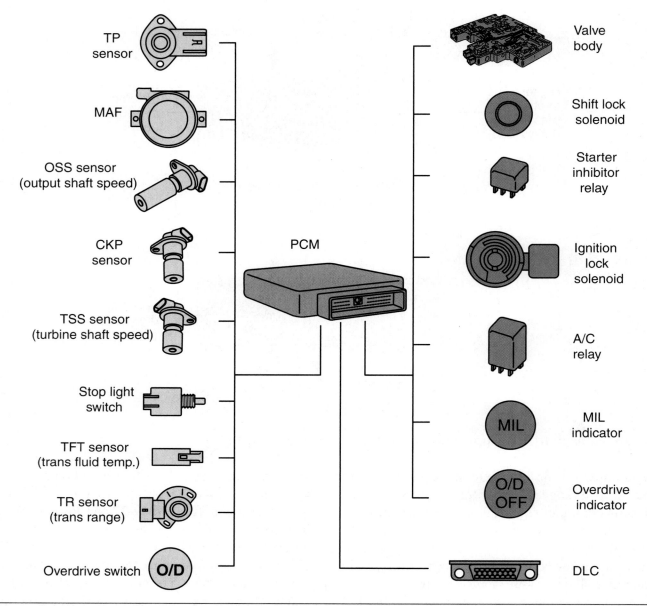

TP sensor

MAF

OSS sensor (output shaft speed)

CKP sensor

TSS sensor (turbine shaft speed)

Stop light switch

TFT sensor (trans fluid temp.)

TR sensor (trans range)

Overdrive switch **O/D**

PCM

Valve body

Shift lock solenoid

Starter inhibitor relay

Ignition lock solenoid

A/C relay

MIL

MIL indicator

O/D OFF

Overdrive indicator

DLC

Figure 49-35 Typical computer-controlled transmission/transaxle inputs and outputs.

pinion gears **(Figure 49-39)**. A steel pin retains the small pinion shaft in the differential case.

When the vehicle is driven straight ahead, the pinion gear drives the ring gear and differential case. Because the pinion gear is much smaller than the ring gear, the differential provides a gear reduction. The differential case with the small pinion gears, side gears, and axle shafts rotates as a unit. Engine torque is transmitted from the pinion gear to the ring gear, differential case, side and spider gears, and drive axles to the drive wheels. When the vehicle turns a corner, the side gears and small pinion gears rotate and allow the outside

drive wheel to turn faster than the inside drive wheel.

Tech Tip *Spare tires that are smaller than the regular vehicle tires can cause excess wear to the differential if the tire is installed on the driving axle.*

In some front-wheel drive vehicles with an automatic transaxle, a gear on the transmission output shaft is meshed with a gear on the transfer shaft. A gear on the opposite end of the transfer shaft is meshed with the differential ring gear

SHIFT SOLENOID OPERATION CHART					
Transaxle range selector lever position	Powertrain Control Module Gear commanded	Eng braking	AX4N solenoids		
			SS 1	SS 2	SS 3
P / N [a]	P / N	NO	OFF [b]	[b]	OFF
R (Reverse)	R	YES	OFF		OFF
Overdrive	1	NO	OFF		OFF
	2	NO	OFF	OFF	OFF
	3	NO		OFF	
	4	YES			
D (Drive)	1	NO	OFF		OFF
	2	NO	OFF	OFF	OFF
	3	YES		OFF	OFF
Manual 1	2 [c]	YES	OFF		OFF
	3 [c]	YES	OFF	OFF	OFF
		YES		OFF	OFF

a When transmission fluid temperature is below 50° then SS1=OFF, SS2=ON, SS3=ON to prevent cold creep.

b Not contributing to powerflow

c When a manual pull-in occurs above calibrated speed the transaxle will downshift from the higher gear until the vehicle speed drops below this calibrated speed.

Figure 49-36 Typical transmission solenoid shift operation.

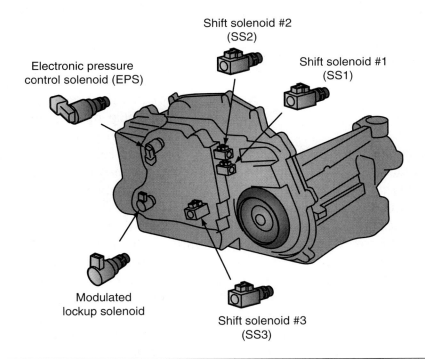

Figure 49-37 Typical transmission solenoid locations.

SOLENOID LOCATIONS

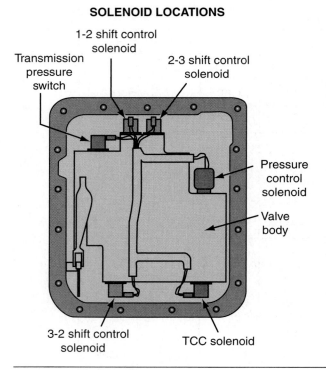

Figure 49-38 Typical transaxle solenoid locations.

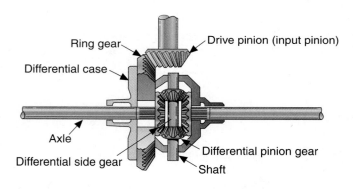

Figure 49-39 Basic differential.

(Figure 49-40). Engine torque is transmitted from the transmission output shaft through the transfer shaft to the differential ring gear.

Some front-wheel drive vehicles with automatic transaxles have a planetary differential **(Figure 49-41)**. In this type of differential, the ring gear is splined to the transaxle case and the transaxle output shaft drives the differential sun gear. The differential sun gear drives the carrier which is connected to the drive axles. A gear reduction is provided when the smaller sun gear drives the larger planet carrier.

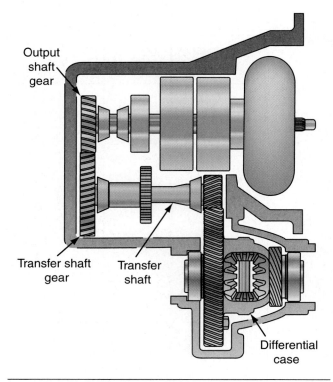

Figure 49-40 Final drive assembly in a transaxle.

FOUR-WHEEL DRIVE (4WD)

A typical four-wheel drive (4WD) vehicle has a front, longitudinally mounted engine, a manual or automatic transmission, front and rear drive shafts, front and rear drive axle assemblies, and a transfer case. The transfer case is usually mounted behind or beside the transmission. The transfer case may be bolted to the rear of the transmission **(Figure 49-42)**.

Engine torque is transmitted from the transmission output shaft through a short drive shaft or a gear or chain drive in the transfer case. A shift lever or electric switch inside the vehicle allows the driver to select two-wheel drive or four-wheel drive. In two-wheel drive, engine torque is transmitted through the transfer case only to the drive shaft connected to the rear wheels. If the driver selects four-wheel drive, engine torque is transmitted through the transfer case to both the front and rear drive shafts **(Figure 49-43)**.

The driver may also shift the transfer case to four-wheel low range. If this transfer case shift is completed, the transfer case provides a gear reduction, which increases torque when driving in adverse conditions such as off-road or rough terrain.

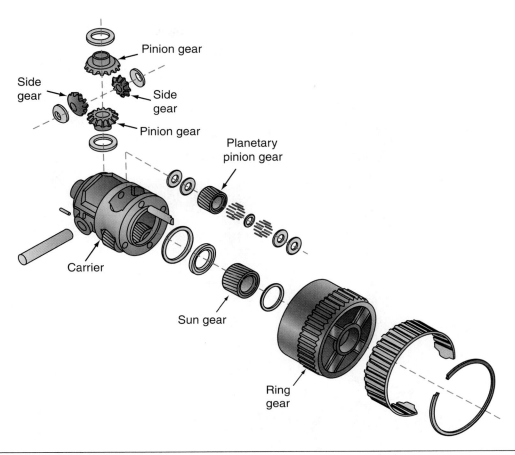

Figure 49-41 Planetary differential in a transaxle.

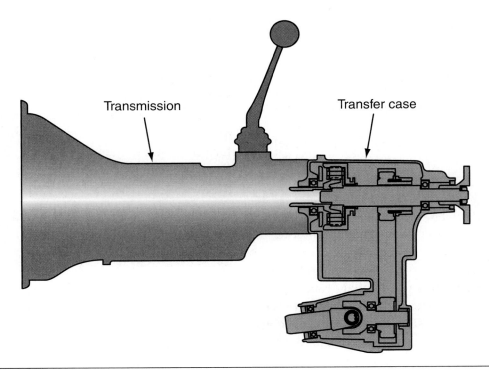

Figure 49-42 A typical transfer case.

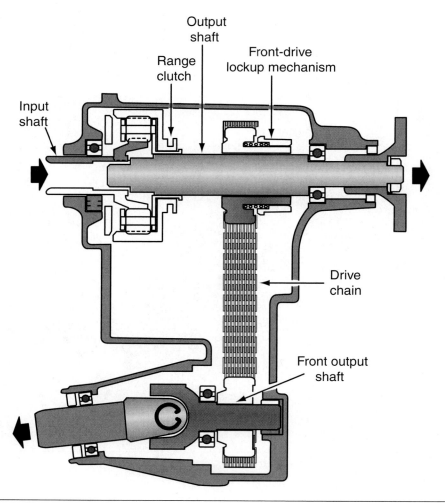

Figure 49-43 Power flow through a typical transfer case.

ALL-WHEEL DRIVE

A growing number of vehicles are equipped with all-wheel drive. These systems are in four-wheel drive at all times and the driver cannot select between four-wheel drive and two-wheel drive. All-wheel drive systems improve traction and vehicle handling when driving on snow-covered or icy road surfaces. All-wheel drive vehicles are not intended for adverse conditions encountered in off-road operation.

Many all-wheel drive systems are a front-wheel drive vehicle with a center differential designed into the transaxle to drive the rear wheels **(Figure 49-44)**. Some all-wheel drive systems have a viscous coupling that allows some variation in front and rear axle speeds. If one wheel begins to spin on an icy road surface, the viscous coupling immediately transfers torque to the other wheels that are not spinning. A **viscous coupling** is a sealed chamber containing a thick honey-like fluid. It also contains a number of discs much like a clutch pack. As the input member turns, the output member is turned through the friction of the fluid inside the coupler.

TORQUE CONVERTER MAINTENANCE AND DIAGNOSIS

Torque converters are a sealed assembly. When the fluid is drained from the transaxle or transmission, some of the fluid is also drained from the torque converter. On modern cars and light-duty trucks, the torque converter cannot be drained separately from the transmission or transaxle.

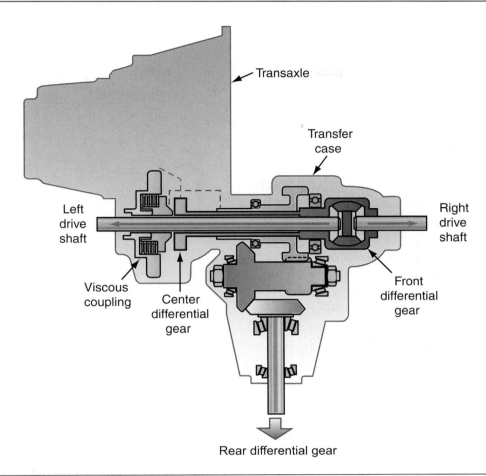

Figure 49-44 An all-wheel drive transaxle with rear differential drive gear and viscous coupling.

On some vehicles, the access cover or plug may be removed from the bottom of the flywheel housing to inspect the torque converter. If ATF is leaking from the flywheel housing, the pump seal in the transmission is the most likely source of the leak. The transaxle or transmission must be removed from the vehicle to remove the torque converter to replace the pump seal.

On transmissions or transaxles with torque converter clutch lockup and/or computer-controlled shifting, visually inspect all the wires and wiring connectors entering the transmission.

One-Way Stator Clutch Diagnosis

If the transaxle passes the visual inspection, road test the vehicle. When the vehicle lacks power during acceleration, the exhaust system may be restricted or the one-way stator clutch in the torque converter may be slipping.

To test for a restricted exhaust system, connect a vacuum gauge to the intake manifold. Operate the engine at 2,500 rpm for 3 minutes. If the vacuum reading on the gauge slowly decreases, the exhaust is restricted. This decrease in vacuum is even more pronounced during a road test. If the exhaust system is not restricted, the one-way stator clutch may be slipping. If the engine speed flares up during acceleration in DRIVE and the vehicle does not have normal acceleration, the clutches or bands in the transaxle or transmission are slipping. When the engine speed does not flare up but the vehicle accelerates sluggishly, the one-way stator clutch is likely slipping.

A one-way stator clutch that is seized causes reduced vehicle acceleration vehicle acceleration at low speed and also at high speed. A seized one-way stator clutch may cause transmission/transaxle and engine overheating.

Torque Converter Clutch Diagnosis

During the road test, test the operation of the TCC lockup. Typically the PCM uses inputs from the ECT sensor, VSS, TPS, and brake switch to operate the TCC lockup. The engine must be warmed up and operating at a certain speed before the PCM provides TCC lockup.

Many transaxles/transmissions allow lockup only in third or fourth gear. If the brake pedal is depressed when the TCC is locked up, the brake switch sends a voltage signal to the PCM. When the PCM receives this signal, it immediately unlocks the TCC. When TCC lockup occurs, a slight "bump" in engine operation is felt. The TCC lockup should be smooth without any shudder or vibration. If a shudder occurs during the TCC lockup, the friction material on the TCC lockup disc may be worn or the fluid pressure in the transaxle may be low.

Broken damper springs in the TCC lockup disc also cause a shudder during lockup. When the shudder is only noticeable after the TCC lockup has occurred, the problem is likely in the engine or other systems in the transaxle. For example, intermittent ignition misfiring may cause a shudder after TCC lockup occurs.

Scan Tool Diagnosis of Converter Clutch Lockup

A scan tool may be connected to the DLC under the dash to check the TCC lockup system and other transaxle electronics **(Figure 49-45)**. Be sure the scan tool contains the proper module for the vehicle being tested and be sure the ignition switch is off when connecting the scan tool. Some scan tools allow the technician to test each gear in the transmission and compare the results to the shift solenoid chart, as shown earlier in this chapter.

Tech Tip *If the scan tool is connected or disconnected with the ignition switch on, the scan tool or on-board electronic components may be damaged.*

With the engine and transaxle at normal operating temperature, check for DTCs on the scan tool

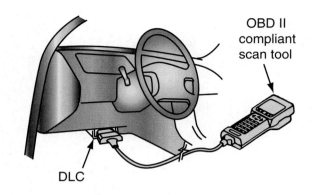

Figure 49-45 A scan tool connected to the data link connector (DLC).

display. If there are any defects in the TCC electrical system, such as an open wire or an open or shorted TCC solenoid winding, a DTC representing these faults is displayed on the scan tool.

Check the scan tool display for other DTCs related to TCC operation. For example, if a defective ECT sensor always indicates cold engine temperature, the TCC does not lock up. A defective VSS sensor that always indicates low vehicle speed prevents TCC lockup. If a faulty VSS indicates higher-than-actual vehicle speed, the TCC locks up sooner than specified.

Tech Tip *When reading transmission pressures during a road test on vehicles with computer-controlled transmissions or transaxles, specific vehicle speed and sensor outputs may be required. For example, a specific voltage from the TPS may be required for the computer to operate properly.*

Diagnosing Improper Torque Converter Clutch Release Problems

Fluid normally flows through the torque converter and then through the transmission cooler before flowing back into the transmission. If the fluid passages in the cooler become restricted, the fluid cannot flow out of the torque converter. This condition may cause the TCC lockup to remain applied

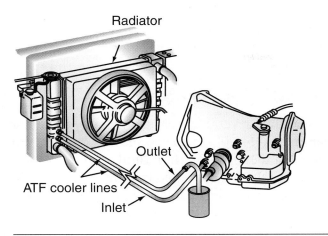

Figure 49-46 Testing fluid flow through the transmission cooler.

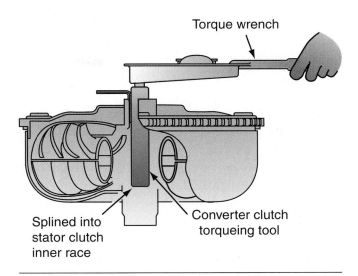

Figure 49-47 Testing the stator one-way clutch.

during deceleration and idle. This condition may cause engine stalling during deceleration and idle or when the transmission selector is moved from PARK to DRIVE or REVERSE.

To test for restricted cooler fluid passages, disconnect the cooler return line at the transmission. Place a length of hose from the disconnected line to an empty drain pan **(Figure 49-46)**. Start and run the engine at idle speed for 20 seconds. Shut the engine off and measure the fluid in the drain pan. One quart of fluid should have been discharged from the cooler return line in 20 seconds. If there is less than one quart of fluid in the drain pan, the cooler passages are restricted. To blow out the cooler passages, disconnect the cooler inlet line and alternately supply 50 psi air pressure to the inlet and return lines.

TORQUE CONVERTER SERVICE

After the transmission or transaxle is removed, grasp the converter and pull it from the front of the transaxle. Inspect the converter welds after it is removed. Inspect the converter hub for scoring in the pump seal contact area. If scoring is present, the converter must be replaced. Inspect the lugs on the converter hub that drive the transaxle oil pump. If these lugs are damaged or worn, replace the converter.

If the converter has an oil pump drive shaft, check the shaft support bushing in the converter for wear and inspect the drive shaft splines.

Diagnosis of a Stator One-Way Clutch

Insert a special tool for checking the one-way stator clutch into the converter hub and connect a torque wrench to the tool **(Figure 49-47)**. This tool holds the inner race of the one-way clutch and allows the technician to apply torque to the outer race. While holding the inner race, rotate the torque wrench in both directions. The torque wrench should rotate easily and smoothly in one direction and lock up immediately in the opposite direction. If the one-way clutch allows rotation in both directions or if the rotation is erratic, replace the converter.

When the special tool for checking the one-way stator clutch is not available, a pair of snap ring pliers with long, thin jaws may be inserted into the inner race of the one-way clutch. This method of checking the one-way clutch is not as accurate because it does not apply heavy torque to this clutch.

AUTOMATIC TRANSMISSION AND TRANSAXLE MAINTENANCE

Automatic transmissions and transaxles require a minimum amount of maintenance. Always follow the vehicle manufacturer's recommended change intervals for the transmission fluid and filter.

Some vehicle manufacturers recommend different fluid and filter change intervals depending on the vehicle operating conditions.

One vehicle manufacturer recommends changing transmission fluid and filter at 50,000 miles (80,000 kilometers) if the vehicle is over 8,600 gross vehicle weight rating (GVWR) or operated under one or more of the following conditions: frequent trailer towing; taxi, police, or delivery service; driving continuously in hilly or mountainous terrain; and driving continually in city traffic when the atmospheric temperature is regularly above 90°F (32°C).

If the vehicle has a GVWR below 8,600 and is not operated under any of the above conditions, the manufacturer recommends changing the transmission fluid and filter at 100,000 miles (161,000 kilometers).

AUTOMATIC TRANSMISSION AND TRANSAXLE FLUIDS

Always use the automatic transmission fluid specified by the vehicle manufacturer. There are many different specifications for ATF. For example, Daimler Chrysler recommends Mopar® ATF+4 type 9602 in many current automatic transaxles. Dexron III/Mercon ATF is recommended for many General Motors and Ford automatic transaxles and transmissions.

> **Tech Tip** *With the rapid changes in the ATF used by various manufacturers, it is very important to use only the automatic transmission fluid specified by the vehicle manufacturer.*

Some vehicle manufacturers provide warnings in their service manuals indicating that the use of an ATF other than the specified ATF may cause torque converter shudder or clutch slipping. Many vehicle manufacturers also have warnings in their service manuals indicating that ATF additives should never be used. The ATF contains many additives such as friction modifiers, oxidation and corrosion inhibitors, extreme pressure lubricants, anti-foaming agents, detergents and dispersants, and pour-point depressants.

> **Tech Tip** *Additional additives are neither recommended nor needed.*

Checking Fluid Level

The fluid level in an automatic transmission or transaxle should be between the add and the full marks on the dipstick. High fluid level may cause fluid to be thrown out of the filler tube. A low fluid level may cause erratic transmission operation and damage to the clutches and bands.

> **Tech Tip** *Low fluid level can cause the pump to cavitate and suck in air causing the fluid to become aereated. This can seriously affect the operation of the transmission as it is designed to operate on a steady flow of fluid.*

Before checking the fluid level, ensure that the vehicle is parked on a level surface, the engine and transmission are at normal operating temperature, and the parking brake is applied and the transmission is in PARK.

> **Tech Tip** *Special instructions for checking ATF levels are often stamped on the dipstick.*

Checking Fluid Condition

After the dipstick is removed, always check the condition of the fluid on the dipstick. The fluid should be red, clear, and it should wipe off the dipstick easily. Abnormal fluid conditions are the following:

1. Milky fluid on the dipstick indicates that coolant has leaked into the transmission, usually through a leaking cooler.

2. Burned fluid is dark and has a burned smell. This condition is caused by overheating and other components in the transmission such as clutches are likely damaged. Transaxle overhaul is usually required when this condition is present.

3. If there is varnish on the dipstick, the ATF cannot be wiped easily from the dipstick. Varnish

on the dipstick indicates oxidized ATF which is caused by overheating and/or lack of ATF changes. This condition usually requires a transmission overhaul.

VISUAL INSPECTION

The automatic transmission or transaxle should be inspected for fluid leaks at the seals. If ATF is leaking from the bottom of the flywheel housing, the pump seal is likely leaking. Because the ATF is red, it should not be confused with an engine oil leak at the rear main bearing. Inspect the transmission cooler lines for leaks and contact with other components that may wear a hole in a line.

On a transaxle, inspect the drive axle seals for leaks. The rear extension housing seal in a transmission should be checked for leaks. Before replacing any seals, be sure the vent in the case is not restricted. Check for a leak at transmission or transaxle components such as the speedometer drive, vehicle speed sensor, turbine speed sensor, output speed sensor, and dipstick tube. Inspect the area around the shift linkage for a fluid leak. Check for fluid leaks around the oil pan and also around the area where the electrical connection enters the case.

If the transaxle has a vacuum modulator, be sure the vacuum hose to the modulator is not leaking and is firmly attached to the modulator. Be sure the full intake manifold vacuum is available at the modulator with the engine idling.

AUTOMATIC TRANSMISSION AND TRANSAXLE DIAGNOSIS

The first step in transmission diagnosis is to check the fluid level and condition. Be sure the fluid level is correct and visually inspect the transmission. When the fluid level is correct and the visual inspection does not indicate any problems, road test the vehicle to verify the customer's complaint.

Listen for abnormal transmission noises during the road test. A defective pump may cause a humming or buzzing noise that increases with engine speed. This noise occurs in any gear selector position.

A grinding noise that increases with vehicle speed and load may be caused by a damaged planetary gear set, worn bushing, or a rough needle bearing. Refer to **Figure 49-48** for transmission and transaxle noise and vibration diagnosis.

The flexplate normally has a flex movement of approximately 0.080 to 0.100 inch (2.03 to 2.54 millimeters). A cracked flexplate causes a knocking noise that changes with engine speed and load.

> **Tech Tip** *The manufacturer will provide diagnosis and application charts for clutches and bands to assist in the testing of the transmission. Refer to the manufacturer's service manuals for these charts.*

During the road test, check the vehicle speed at which the shift points occur. If the upshifts and downshifts do not occur at the specified speed, there may be a problem with the governor, throttle cable, or modulator. When the shift points do not occur at the proper vehicle speed or some shifts are missing, perform a pressure test on the transmission.

Automatic Transmission and Transaxle Pressure Testing

Prior to performing a pressure test, the transaxle must be at normal operating temperature and have the correct fluid level. Some transaxles and transmissions have several plugs in the case that may be removed to test pressure in certain gears. For example, there may be plugs in the case that can be removed to check line pressure, forward clutch pressure, and other pressures **(Figure 49-49)**.

Raise the vehicle on a lift and remove the desired pressure test plug from the transmission. Connect the appropriate pressure gauge to the opening from which the plug was removed. Operate the vehicle on the lift or road test the vehicle and observe the pressure in each gear **(Figure 49-50)**.

If the pressure is low in all gears including reverse, the pump or pressure regulator valve may be defective. When the pressure is low in only one gear, the seals on the components applied in that gear are leaking. The leaking seals are typically clutch drum seals or band servo seals.

Problem	Probable Cause(s)
Ratcheting noise	The return spring for the parking pawl is damaged, weak, or misassembled
Engine speed-sensitive whine	Torque converter is faulty A faulty pump
Popping noise	Pump cavitation—bubbles in the ATF Damaged fluid filter or filter seal
Buzz or high-frequency rattle Whine or growl	Cooling system problem Stretched drive chain Broken teeth on drive and/or driven sprockets Nicked or scored drive and/or driven sprocket bearing surfaces Pitted or damaged bearing surfaces
Final drive hum	Worn final drive gear assembly Worn or pitted differential gears Damaged or worn differential gear thrust washers
Noise in forward gears	Worn or damaged final drive gears
Noise in specific gears	Worn or damaged components pertaining to that gear
Vibration	Torque converter is out of balance Torque converter is faulty Misaligned transmission or engine The output shaft bushing is worn or damaged Input shaft is out of balance The input shaft bushing is worn or damaged

Figure 49-48 Transmission and transaxle noise and vibration diagnosis.

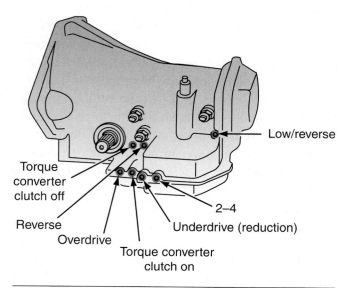

Figure 49-49 Pressure taps on a typical transaxle.

Diagnosis of Computer-Controlled Transmissions and Transaxles

When diagnosing a computer-controlled transmission or transaxle, a scan tool may be connected to the DLC under the dash **(Figure 49-51)**. If faults are present in the transmission electronic system, DTCs are stored in the transmission computer memory. These DTCs may be retrieved with the scan tool.

DTCs may indicate defects in certain circuits such as a shift solenoid circuit, torque converter clutch solenoid circuit, or a pressure control valve circuit. A DTC indicates a problem in a certain circuit, but further diagnosis with a voltmeter and ohmmeter may be necessary to determine if the problem is in the solenoid winding or in the wires from the computer to the solenoid.

Transmission Pressure with TPS at 1.5 Volts and Vehicle Speed above 8 Km/h (5 mph)					
Gear	EPC Tap	Line Pressure Tap	Forward Clutch Tap	Intermediate Clutch Tap	Direct Clutch Tap
1	276–345 kPa (40–50 Psi)	689–814 kPa (100–118 Psi)	620–745 kPa (90–108 Psi)	641–779 kPa (93–113 Psi)	0–34 kPa (0–5 Psi)
2	310–345 kPa (45–50 Psi)	731–869 kPa (106–126 Psi)	662–800 kPa (96–116 Psi)	689–827 kPa (100–120 Psi)	655–800 kPa (95–116 Psi)
3	341–310 kPa (35–45 Psi)	620–758 kPa (90–110 Psi)	0–34 kPa (0–5 Psi)	586–724 kPa (85–105 Psi)	551–689 kPa (80–100 Psi)

Figure 49-50 Pressure test specifications for a typical computer-controlled transmission.

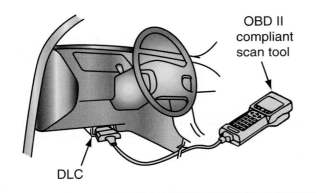

Figure 49-51 A data link connector under the dash.

All 1996 and newer vehicles under 8,500 GVWR are equipped with OBD II electronic systems. In these systems, the DTCs have a standard format. For example, in a PI711 DTC, the P indicates a powertrain code and the 1 indicates this code is supplied by the manufacturer. If the second digit in the code is a 0, the code is a standard code defined by the SAE and is used by all vehicle manufacturers.

When the third digit in the code is a 7, the code belongs to the transmission subgroup. The last two digits in the code indicate the exact circuit where the problem is located. In the previous example, the last two digits, 11 in the P1711 code, indicate a defect in the transmission oil temperature (TOT) sensor **(Figure 49-52)**. Data may be displayed on the scan tool from all the sensors in the transmission including the turbine speed sensor, output speed sensor, and transmission oil temperature sensor.

AUTOMATIC TRANSMISSION AND TRANSAXLE SERVICE

Most oil pans do not have a drain plug. To change the transmission fluid, raise the vehicle on a lift and place a large diameter drain pan under the transmission pan. Place the drain pan on a stand so it is closer to the transmission oil pan to reduce oil splashing out of the pan **(Figure 49-53)**.

Loosen the oil pan retaining bolts more on one side of the pan than on the opposite side and allow the pan to drop downward on one side. Some of the fluid will begin to run into the drain pan. Remove the oil pan bolts on the lower side of the oil pan and allow the pan to drop further downward to drain more fluid from the pan. Carefully hold the oil pan and remove the remaining bolts from the pan. Lower the pan and drain the remaining fluid into the drain pan.

Check the oil pan for clutch and band material and aluminum or copper deposits. An excessive amount of clutch and band material in the oil pan indicates that the clutches and/or bands are worn and a complete overhaul is required now or in the near future. Excessive aluminum deposits in the

The SAE J2012 standards specify that all DTCs will have a five-digit alphanumeric numbering and lettering system. The following prefixes indicate the general area to which the DTC belongs:

1. P — powertrain
2. B — body
3. C — chassis

The first number in the DTC indicates who is responsible for the DTC definition.

1. 0 — SAE
2. 1 — manufacturer

The third digit in the DTC indicates the subgroup to which the DTC belongs. The possible subgroups are:

0 — Total system
1 — Fuel-air control
2 — Fuel-air control
3 — Ignition system misfire
4 — Auxiliary emission controls
5 — Idle speed control
6 — PCM and I/O
7 — Transmission
8 — Non-EEC power train

The fourth and fifth digits indicate the specific area where the trouble exists. Code P1711 has this interpretation:

P — Powertrain DTC
1 — Manufacturer-defined code
7 — Transmission subgroup
11 — Transmission oil temperature (TOT) sensor and related circuit

Figure 49-52 OBD II diagnostic trouble code format.

Figure 49-53 Transmission drain tray on an oil drain unit.

oil pan indicate damage to the aluminum castings, case, or torque converter. Copper cuttings in the oil pan indicate damaged bushings. Wash the oil pan in an approved solvent and use a scraper to remove the old gasket from the oil pan. Use a plastic scraper to remove old gasket material from the oil pan mating surface on the transmission case.

Tech Tip *Using a metal scraper to remove old gasket material from an aluminum casting may scratch the casting so it is difficult or impossible for the gasket to prevent oil leaks between the two mating surfaces.*

Remove the filter retaining clips and remove the filter from the transaxle **(Figure 49-54)**. Some

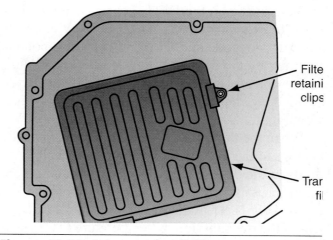

Filter retaini clips

Tran fi

Figure 49-54 A transaxle oil filter.

transmissions may use an external filter. Lubricate the filter tube O-ring and install it on the new filter. Install the new filter and the retaining clips. Install the oil pan with a new gasket and tighten the oil pan bolts to the specified torque. Install the specified type of fluid in the transaxle until the proper fluid level is indicated on the dipstick with the engine and transaxle at normal operating temperature.

Band Adjustments

On some transmissions or transaxles, the band adjustments are on the outside of the case, whereas on other vehicles the oil pan must be removed to access these adjustments.

When the band adjustments are external, locate the band adjusting nut and remove any dirt around the nut. Loosen the band adjustment locknut and back it off approximately five turns while preventing the adjusting screw from turning **(Figure 49-55)**. Tighten the band adjustment to the specified torque with an inch-pound torque wrench **(Figure 49-56)**. Back out the adjustment screw the specified number of turns. Hold the band adjusting screw in this position and tighten the locknut to the specified torque.

> **Tech Tip** *It is important to hold the adjusting screw in position while tightening the locknut. Failure to do so may cause the band to be improperly adjusted and transmission damage may result.*

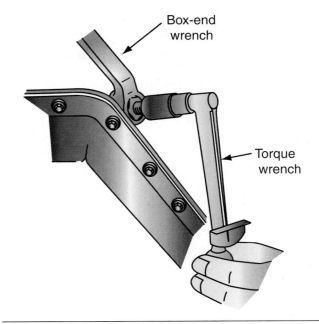

Figure 49-56 Adjusting a transmission or transaxle band.

Seal Replacement

The converter must be removed to access the pump seal. Use the proper puller to remove the pump seal. Clean the seal mounting area in the pump and be sure there are no metal burrs in this area. Some pump seals have sealant applied to the outside diameter of the seal case. If this sealant is not present, apply sealant to the outer edge of the seal case. Use the proper driver to install the pump seal and be sure the seal is started squarely into the pump **(Figure 49-57)**.

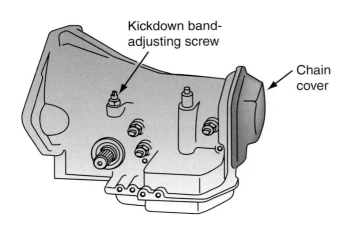

Figure 49-55 A typical band adjustment screw and locknut.

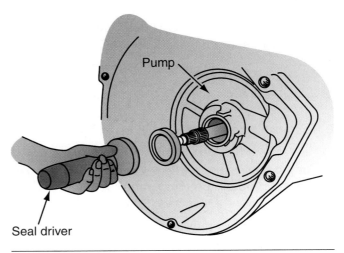

Figure 49-57 Installing a transmission or transaxle pump seal.

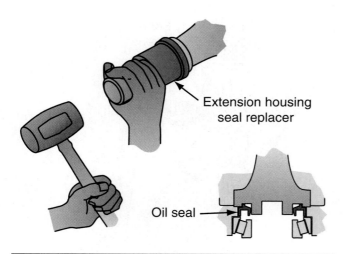

Figure 49-58 Extension housing seal installation.

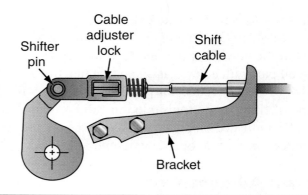

Figure 49-59 Shift cable adjuster lock assembly.

The transmission extension housing seal must be pulled from the housing with a special puller. Clean the seal mounting area and be sure there are no metal burrs in this area. Place a thin layer of sealant on the outer diameter of the seal case and use the proper driver to install the seal **(Figure 49-58)**.

The rear extension housing also contains a bushing. If the seal requires frequent replacement then the bushing should be checked for excess wear and replaced if necessary. Always follow the manufacturer recommended procedure for bushing inspection and replacement.

Linkage Adjustments

Misadjusted transmission linkage will not allow proper manual valve position, resulting in improper fluid supply to some of the clutch packs or bands. Under this condition, the clutches or bands may slip and experience excessive wear. To adjust a cable-type transaxle linkage, follow this procedure:

1. Place the shift lever in PARK.
2. Loosen the clamp bolt on the shift cable bracket.
3. Pull the shift lever to the front detent (PARK) position and then tighten the clamp bolt on the shift cable bracket.
4. If the cable has an adjuster, move the shift lever into the PARK position and rotate the adjuster to adjust the cable **(Figure 49-59)**. The adjuster provides a click when the lock is fully adjusted.

After making the transmission linkage adjustment, it may be necessary to adjust the neutral safety switch. Try to start the vehicle with the gear selector in the PARK, NEUTRAL, OD, and REVERSE positions. The starter should operate only with the gear selector in PARK or NEUTRAL. If the starter operates in other gear selector positions, the neutral safety switch requires adjusting.

Summary

- The torque converter is a device that transfers engine torque to the transmission.

- Torque converter lockup occurs with the engine at normal operating temperature and the vehicle operating at or above a certain speed.

- Various forward reduction and overdrive gear ratios are provided by holding one member of a planetary gearset and providing input to another member with the third member as the output.

- When two members of a planetary set are locked or two members are inputs at the same speed, the gear set provides a 1:1 ratio (direct).

- A slipping one-way stator clutch causes a loss vehicle acceleration at low speed and a seized one-way stator clutch causes a loss of vehicle acceleration at low and high speeds.

- A multiple-disc clutch assembly contains a hub, driving discs, driven plates, apply piston, seals, pressure plate, release springs, and clutch assembly container.

- Bands are used to hold a clutch drum or other component.

- An overrunning clutch allows one race to rotate freely in one direction but lock up in the opposite direction.

- The pressure regulator valve limits pump pressure.

- An accumulator is used to cushion a shift and prevent harsh shifting.

- In an electronically controlled transmission/transaxle, the pressure is electronically controlled.

- In an electronically controlled transmission/transaxle, the computer controls three or more solenoids to control the shifting and fluid pressure. The computer also controls another solenoid to operate the TCC lockup.

- In a four-wheel drive vehicle, the driver may select two-wheel drive or four-wheel drive operation. The driver may also select four-wheel low range operation.

- In an all-wheel drive system, if one wheel spins on a slippery road surface, the viscous coupling immediately transfers torque to the wheels that are not spinning.

- When checking the automatic transmission fluid level, the engine and transmission should be at normal operating temperature.

- Milky fluid on the dipstick indicates the transmission fluid has been contaminated with coolant, whereas dark fluid with a burned smell is usually caused by overheating and burned clutch discs.

- Transmission/transaxle fluid leaking from the bottom of the flywheel housing indicates a leaking pump seal.

- A cracked flexplate causes a knocking noise that changes with engine speed and load.

- When transmission/transaxle pressure is low in all gears, including reverse, the pump or pressure regulator valve may be defective.

- When transmission/transaxle pressure is low in only one gear, the seals on the components applied in that gear are leaking.

- A scan tool may be connected to the DLC under the dash to diagnose computer-controlled transmission/transaxle.

- Proper transmission/transaxle linkage adjustment provides the correct manual valve position.

Review Questions

1. Technician A says in a multiple-disc clutch, the driving discs usually have friction material on their surfaces. Technician B says in a multiple-disc clutch, the driving discs never have internal splines. Who is correct?

   A. Technician A

   B. Technician B

   C. Both Technician A and Technician B

   D. Neither Technician A nor Technician B

2. Technician A says the bands in an automatic transaxle may be applied by accumulators. Technician B says a band is used to hold one member in a planetary gear set. Who is correct?

   A. Technician A

   B. Technician B

   C. Both Technician A and Technician B

   D. Neither Technician A nor Technician B

3. Technician A says the oil pump in an automatic transaxle supplies fluid pressure to the pressure regulator valve. Technician B says the pressure regulator valve limits fluid pressure by reducing the oil pump speed. Who is correct?

   A. Technician A

   B. Technician B

   C. Both Technician A and Technician B

   D. Neither Technician A nor Technician B

4. Technician A says the manual valve is controlled by fluid pressure from the oil pump. Technician B says the gear shift linkage controls the modulator valve. Who is correct?

   A. Technician A

   B. Technician B

   C. Both Technician A and Technician B

   D. Neither Technician A nor Technician B

5. When discussing torque converter operation, Technician A says at lower vehicle speeds the stator redirects fluid leaving the turbine so this fluid flow helps to rotate the impeller pump. Technician B says rotary fluid flow occurs in a torque converter at lower vehicle speeds. Who is correct?

   A. Technician A

   B. Technician B

   C. Both Technician A and Technician B

   D. Neither Technician A nor Technician B

6. The coupling point occurs in a torque converter when the:

   A. stator is rotating faster than the turbine.

   B. stator is being held by the overrunning clutch.

   C. stator is rotating faster than the impeller pump.

   D. the speed of the turbine approaches the speed of the impeller pump.

7. In a planetary gear set, the planet carrier is the input, the ring gear is the output, and the sun gear is held. This combination provides:

   A. a forward gear overdrive.

   B. a reverse gear overdrive.

   C. a reverse gear reduction.

   D. a forward gear reduction.

8. ATF is leaking from the bottom of the flywheel housing. The most likely cause of this problem is:

   A. a leaking torque converter weld.

   B. a leaking rear main bearing seal.

   C. a leaking transmission pump seal.

   D. a misaligned transmission.

9. When performing a pressure test on an automatic transmission, the pressure is low in all gears including reverse. Technician A says the pressure regulator valve may be sticking or defective. Technician B says the seal on the reverse clutch piston may be leaking. Who is correct?

   A. Technician A

   B. Technician B

   C. Both Technician A and Technician B

   D. Neither Technician A nor Technician B

10. When diagnosing an automatic transmission in an OBD II vehicle, the scan tool displays DTC P1711. All of these statements about DTC 1711 are true *except*:

    A. This DTC belongs to the powertrain area.

    B. This DTC is a standard code defined by SAE.

    C. This DTC belongs to the transmission subgroup.

    D. This DTC indicates a defect in the transmission oil temperature sensor or circuit.

11. List some of the possible causes for noises in the automatic transmission.

12. Explain how to test the stator one-way clutch.

13. To provide better fuel economy a
    _____ _____
    _____ is used.

14. An accumulator is used to
    _____ the shifts in a transmission or transaxle.

15. Explain how torque is amplified in the torque converter.

16. Describe the general shifting operation of the computer-controlled automatic transmission.

17. List and explain the inputs for the computer-controlled automatic transmission.

18. Explain the general steps for band adjustment.

CHAPTER

50 Clutch Maintenance, Diagnosis, and Service

Learning Objectives

After you have read, studied, and practiced the contents of this unit, you should be able to:

- Describe clutch disc, flywheel, and pressure plate design.

- Explain clutch release bearing and fork design.

- Explain clutch cables and linkages.

- Describe hydraulic clutch system design.

- Explain clutch operation during engagement and disengagement.

- Describe the clutch pedal free-play adjustment.

- Explain the results of improper clutch pedal free-play adjustment.

- Describe the indications and causes of clutch slipping.

- Describe the indications and causes of clutch chatter.

- Explain the causes of various clutch noises.

- Describe the indications and causes of clutch vibration.

- Describe the indications and causes of clutch dragging.

- Explain the causes of a pulsating clutch pedal.

- Remove and replace clutches and pressure plates.

Key Terms

Belleville spring

Clutch cable

Clutch chatter

Clutch pedal free-play

Clutch slipping

Constant-running release bearing

Flexible clutch disc

Hydraulic clutch

Mechanical linkage

Rigid clutch disc

INTRODUCTION

On vehicles with manual transmissions, the clutch is mounted between the engine and transmission. The purpose of the clutch is to connect and disconnect engine torque from the transmission. The clutch must perform this function smoothly without grabbing, slipping, or chattering. During clutch engagement, the clutch must withstand a very high torque load from the engine and it must do this repeatedly without damaging clutch components. Technicians must understand clutch purpose and normal operation to be able to diagnose clutch problems quickly and accurately.

CLUTCH DISC DESIGN

The clutch plate or disc contains a splined hub that is mounted on matching splines on the transmission input shaft. A steel plate is attached to the clutch hub and frictional materials are riveted or bonded to both sides of this steel plate. Asbestos was the most common material used in clutch facings until the health hazards related to breathing asbestos dust became known.

> **Tech Tip** *Because of repeated engagement and disengagement, clutch facings gradually wear and dust particles from the clutch facings are distributed on clutch components and around the flywheel housing.*

Clutch facings are now made from paper-based or ceramic materials mixed with cotton, brass, and wire particles. Grooves are cut diagonally across the clutch facings to increase clutch cooling and provide a place for facing dust to accumulate **(Figure 50-1)**. The grooves in the clutch facings also provide smoother clutch engagement. The facings are attached to wave springs that gradually flatten out during clutch engagement. This action causes the contact pressure on the clutch facings to increase gradually. This action provides smoother clutch engagement.

Engines may have a **flexible clutch disc** or a **rigid clutch disc**. A flexible clutch disc has torsional springs and a friction ring between the hub

Figure 50-1 A clutch disc.

and the facings. When the clutch is first engaged, engine torque is transmitted from the clutch facings through the torsional springs and friction ring to the clutch hub and the transmission input shaft. The torsional springs allow some movement between the facings and the clutch hub to cushion the clutch application **(Figure 50-2)**. Stop pins limit the movement between the facings and the hub. The torsional springs also prevent engine power pulses from being transferred to the transmission input shaft. The number of torsional springs and the amount of tension they have depends on the engine torque and vehicle weight. A flexible clutch disc allows some movement between the facings and the hub. A rigid clutch disc does not allow any movement between the facings and the hub.

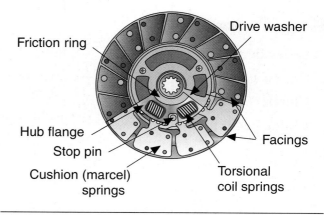

Figure 50-2 Clutch disc components.

FLYWHEEL DESIGN

The flywheel is a heavy, circular steel component that is bolted to the rear crankshaft flange **(Figure 50-3)**. The front clutch disc facings fit against a smooth, machined surface on the flywheel. A pilot bearing or bushing in the center of the rear crankshaft flange supports a machined extension on the transmission input shaft **(Figure 50-4)**. A ring gear is pressed onto the outside diameter of the flywheel and the teeth on this ring gear mesh with the starter drive teeth when the starting motor is engaged.

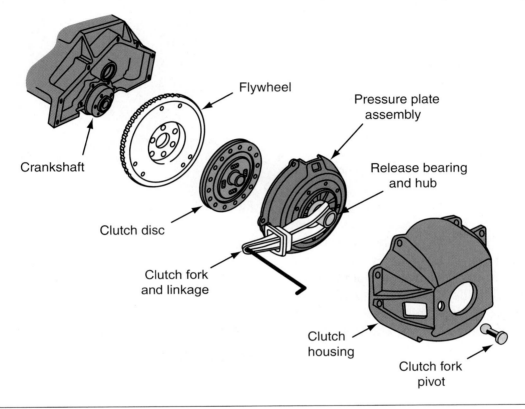

Figure 50-3 Flywheel and clutch components.

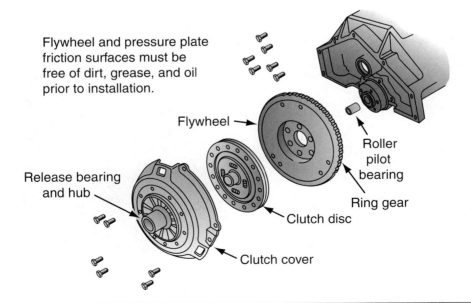

Figure 50-4 Pilot bearing and related clutch components.

When the clutch is engaged, the clutch disc facings are pressed against the flywheel surface and engine torque is transmitted from the flywheel to the clutch facings and hub. The flywheel acts as a balancer for the engine and also dampens engine vibrations caused by cylinder firings. The flywheel provides inertia for the crankshaft between cylinder firings.

Engines with automatic transmissions have a flexplate in place of a flywheel. Because a clutch is not used with an automatic transmission, the smooth, machined surface on the flywheel is not required **(Figure 50-5)**. The torque converter in an automatic transmission is bolted to the flexplate and the weight of the torque converter and the fluid inside it dampens engine pulses and provides inertia for the crankshaft.

Some light-duty trucks with diesel engines and some luxury or sport cars have dual-mass flywheels **(Figure 50-6)**. This type of flywheel reduces crankshaft oscillations before they reach the transmission. This action provides smoother transmission shifting and reduces gear noise. A dual-mass flywheel has two rotating plates connected by a spring and damper mechanism. The front rotating plate is bolted to the crankshaft flange and the pressure plate is mounted on the rear rotating plate. Engine torque is transferred from the front flywheel plate through the damper mechanism to the rear flywheel plate. The damper mechanism absorbs

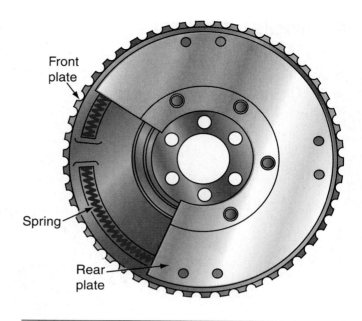

Figure 50-6 A dual-mass flywheel.

engine torque spikes during hard acceleration to prevent these spikes from being transferred to the transmission.

PRESSURE PLATE ASSEMBLY DESIGN

The pressure plate has a smooth machined surface facing toward the clutch plate **(Figure 50-7)**. When the clutch is engaged, this machined surface squeezes the clutch plate facings against the machined flywheel surface. Under this condition,

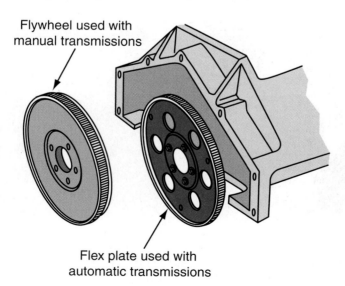

Figure 50-5 A flywheel and flexplate.

Figure 50-7 A pressure plate.

engine torque is transferred from the flywheel and pressure plate to the clutch disc and transmission input shaft. To disengage the clutch, the pressure plate is moved away from the clutch facings so these clutch facings are free to rotate. When this action takes place, engine torque is no longer transferred from the flywheel and pressure plate to the clutch disc.

The pressure plate assembly typically contains a pressure plate, clutch cover, a diaphragm spring, spacer bolts, release levers, and rivets **(Figure 50-8)**. Rivets attach the diaphragm spring to the cover. The center part of the diaphragm spring is slotted to form fingers that act as release levers. The clutch release bearing is mounted near the rear surface of these release levers.

A **Belleville spring** is a diaphragm spring made from thin sheet metal that is formed into a cone shape. Some pressure plate assemblies have coil springs instead of a diaphragm spring **(Figure 50-9)**.

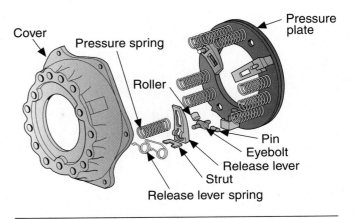

Figure 50-9 A pressure plate with coil springs.

This type of pressure plate has three pivoted release levers that pull the pressure plate away from the clutch facings to release the clutch. When the clutch is engaged, the coil springs force the pressure plate against the clutch facing **(Figure 50-10)**. Some coil

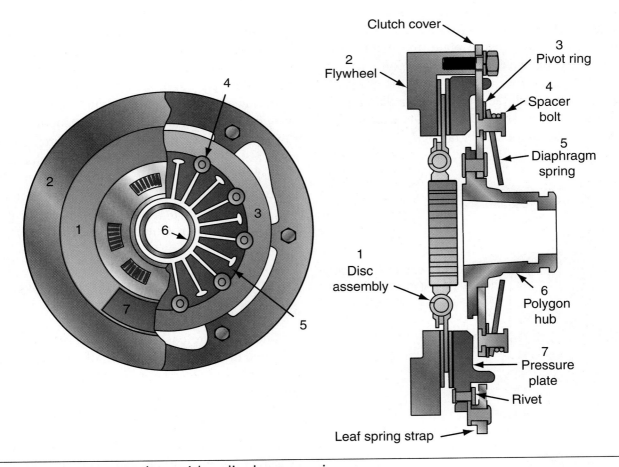

Figure 50-8 A pressure plate with a diaphragm spring.

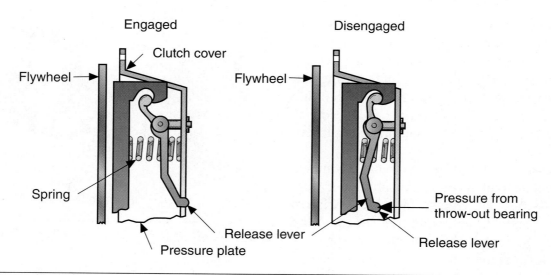

Figure 50-10 Operation of a coil spring-type pressure plate.

spring pressure plates have a semi-centrifugal design. This type of pressure plate has pivoted weights that move outward from centrifugal force as the engine speed increases. When these weights fly outward, they increase the clamping force on the pressure plate and clutch disc **(Figure 50-11)**. Compared to the spring-type pressure plate, the diaphragm-type pressure plate is lighter, more compact, requires less pedal effort, and has fewer moving parts to wear. As the clutch facings wear on a diaphragm-type pressure plate, the pressure plate exerts more squeezing force on the clutch facings.

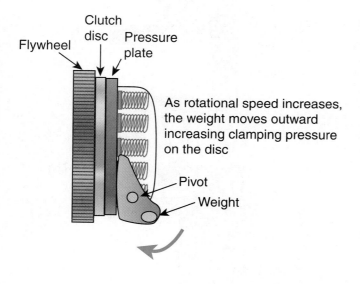

Figure 50-11 Semi-centrifugal pressure plate operation.

NOTE: A Belleville spring may be called a *clutch diaphragm spring.*

When the clutch release bearing is forced against the release levers, these release levers are forced toward the flywheel. This release lever movement causes the diaphragm spring to pivot over the fulcrum ring and the outer rim on the diaphragm moves away from the flywheel. This movement on the outer rim of the diaphragm spring pulls the pressure plate away from the flywheel to release the clutch. When the release bearing moves away from the release levers, the diaphragm spring pivots over the fulcrum ring and forces the pressure plate against the clutch facing to engage the clutch **(Figure 50-12)**.

CLUTCH RELEASE BEARING AND LEVER

The clutch release bearing has a ball bearing mounted on a hub and the front of the release bearing contacts the release levers on the pressure plate **(Figure 50-13)**. The clutch release bearing is a pre-lubricated, sealed bearing that does not require lubrication in service. A clutch release fork fits into a machined groove in the rear of the clutch release bearing. The clutch release-bearing hub slides forward and rearward on a machined extension on

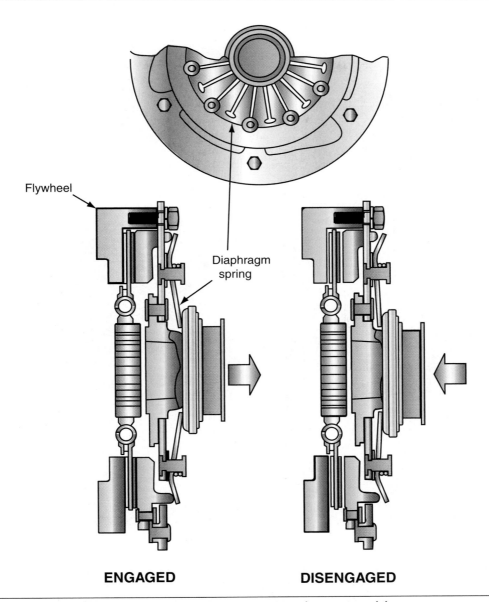

Flywheel

Diaphragm
spring

ENGAGED   **DISENGAGED**

Figure 50-12 Operation of a diaphragm spring pressure plate assembly.

the front bearing retainer in the transmission or transaxle **(Figure 50-14)**.

The flywheel housing is bolted to the rear of the engine block. This housing encloses the clutch assembly. The transmission is bolted to the rear of the flywheel housing. Some clutch release bearings must be adjusted so they are close to the pressure plate release levers, but they do not contact these levers when the clutch is engaged. This type of clutch has an adjustable clutch linkage. Many clutches have a **constant-running release bearing** that is designed to maintain light contact with the release levers when the clutch is engaged **(Figure 50-15)**. This type of

clutch has a self-adjusting clutch cable or a hydraulically operated clutch.

NOTE: The flywheel housing can also be called the *clutch housing* or the *bell housing.*

CLUTCH LINKAGES

Some clutches have a **mechanical linkage** that is connected from the release fork to the clutch pedal. A linkage from the clutch pedal is connected to a

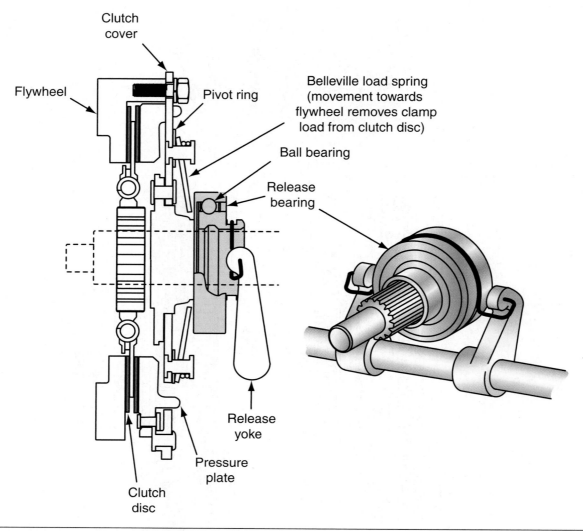

Figure 50-13 A clutch release bearing.

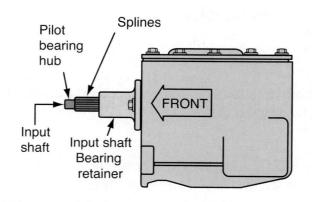

Figure 50-14 Transmission front bearing retainer with machined extension that supports the clutch release bearing.

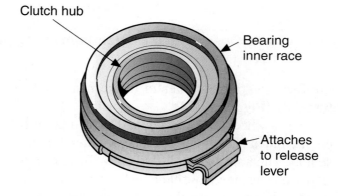

Figure 50-15 A constant-running clutch release bearing.

pivoted equalizer lever and a second linkage is connected from this lever to the release fork **(Figure 50-16)**. When the pedal is depressed, the linkage movement forces the release lever rearward, which in turn forces the pivoted release fork forward against the release bearing to release the clutch. When this type of clutch is engaged, a retracting spring pulls the release lever so the release bearing does not

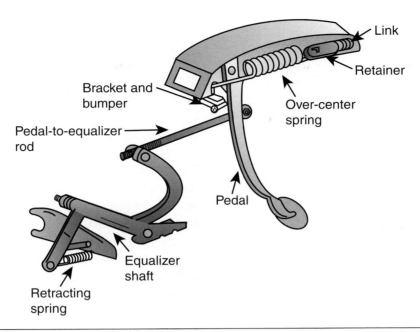

Figure 50-16 A mechanical lever-type clutch linkage.

contact the pressure plate release levers. This type of clutch linkage requires proper adjustment to position the release bearing so it does not contact the release levers when the clutch is engaged.

Some clutches have a **clutch cable** in place of a linkage connected between the clutch pedal and the release fork **(Figure 50-17)**. Some clutch cables are self-adjusting and these clutches have a

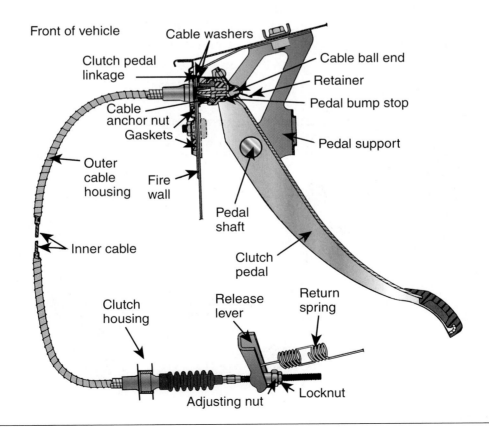

Figure 50-17 A clutch cable system.

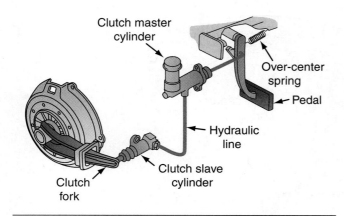

Figure 50-18 A hydraulic clutch system.

constant-running release bearing. Other clutch cables require adjusting and these clutches require proper adjustment to keep the release bearing away from the pressure plate release levers when the clutch is engaged.

Many late-model vehicles are equipped with a **hydraulic clutch**. In these systems, a clutch master cylinder is bolted to the firewall and a slave cylinder is mounted near the outer end of the clutch release fork **(Figure 50-18)**. A pushrod is connected from the clutch pedal to the master cylinder and another pushrod is connected from the slave cylinder to the release fork. A hydraulic line is connected from the master cylinder to the slave cylinder.

When the clutch pedal is depressed, the pushrod is forced against the piston in the master cylinder **(Figure 50-19)**. This action creates hydraulic pressure in the master cylinder. This pressure is supplied through the line to the slave cylinder. Hydraulic pressure in the slave cylinder moves the slave cylinder piston, pushrod, and clutch release fork against the release bearing to disengage the clutch. Some vehicles may be equipped with a slave cylinder mounted behind the clutch release bearing **(Figure 50-20)**. Some hydraulic clutches have a constant-running release bearing and do not require adjustment.

Some hydraulic clutches have a small damper cylinder in the line between the master cylinder and the slave cylinder. This damper absorbs vibrations during clutch engagement and disengagement **(Figure 50-21)**.

CLUTCH OPERATION

When the driver depresses the clutch pedal, the linkage, cable, or hydraulic pressure forces the release fork rearward. Because the release fork is pivoted near the center, the inner end of this fork is moved forward. This forward movement slides

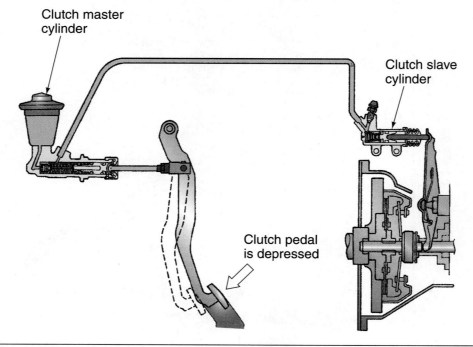

Figure 50-19 Operation of a hydraulic clutch system.

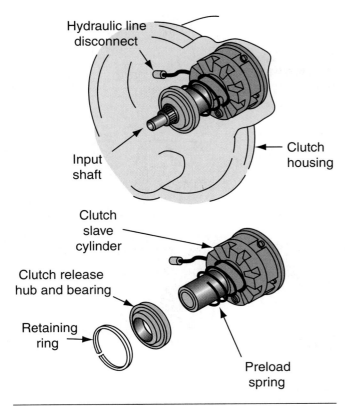

Figure 50-20 A hydraulic clutch slave cylinder mounted behind the clutch release bearing.

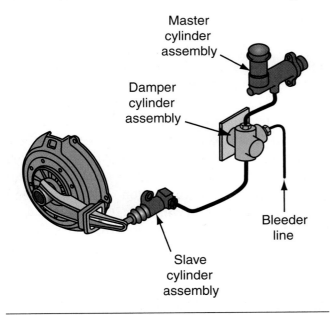

Figure 50-21 A hydraulic clutch system with a damper cylinder assembly.

the release-bearing hub forward. When the release bearing hub slides forward on the front transmission-bearing retainer, the release bearing is forced against the pressure plate release levers to release the clutch

disc. When the clutch disc is released, the pressure plate and flywheel continue to rotate while the engine is running. Because there is no firm contact between the clutch disc facing and the pressure plate, no torque is transferred to the clutch disc and transmission input shaft. To engage the clutch, the driver slowly allows the clutch pedal to move upward off the vehicle floor. This action allows the clutch release fork to move forward and the inner end of this release fork moves rearward. This release fork movement slowly allows the pressure plate spring(s) to squeeze the clutch disc facings between the pressure plate and the flywheel. Under this condition, engine torque is gradually transferred from the flywheel and pressure plate through the clutch disc to the transmission input shaft. On front-wheel drive vehicles, engine torque is transferred through the transaxles and drive axles to the drive wheels. As the driver allows further upward clutch pedal movement, the pressure plate exerts more pressure on the clutch disc facings. The clutch is fully engaged when the driver's foot is released from the clutch pedal.

Tech Tip *If the clutch facings are contaminated with oil, the source of the oil leak into the flywheel housing must be corrected before the clutch is replaced. Otherwise, the new clutch will quickly become oil contaminated and the slipping problem will return.*

CLUTCH MAINTENANCE

A clutch requires a minimum amount of maintenance. If the clutch has an adjustable linkage or cable, the **clutch pedal free-play** should be checked at the vehicle manufacturer's recommended intervals. The free-play is the pedal movement before from the fully released position to the point where some resistance is felt when the release bearing contacts the pressure plate release levers **(Figure 50-22)**. If the clutch pedal free-play is less than 1/2 inch (13 millimeters), a free-play adjustment should be performed. Excessive clutch pedal free-play may not allow the clutch to release when the pedal is fully depressed. This condition may cause hard transmission shifting and/or gear clashing

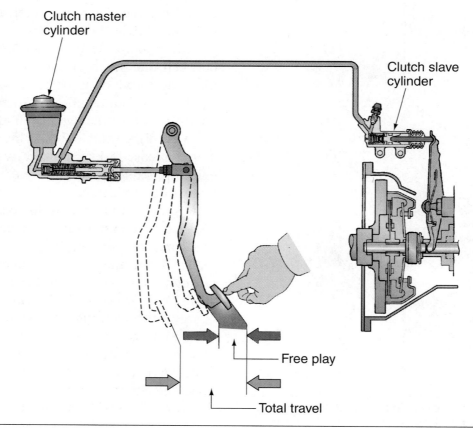

Figure 50-22 Clutch pedal free-play.

when shifting. Insufficient clutch pedal free-play may not allow the pressure plate to exert enough pressure against the clutch disc facings. This condition may cause clutch slipping, especially during hard acceleration.

Before performing a clutch pedal free-play adjustment, always inspect the clutch pedal linkage or cable for worn, bent, or loose linkages, and worn pivots. Observe the linkage while depressing and releasing the clutch pedal. Replace any worn, bent, or damaged linkage components before performing the free-play adjustment. To perform the clutch free-play measurement, place a ruler beside the clutch pedal and slowly depress the clutch pedal. Clutch pedal free-play is the amount of pedal movement before the release bearing contacts the pressure plate release levers. On many vehicles, this measurement should be between ½ and 1½ inches (13 to 44 millimeters). If the free-play is not correct, loosen the locknut and rotate the adjustment on the linkage to obtain the proper

free-play **(Figure 50-23)**. Tighten the locknut after the freeplay adjustment is complete.

When performing the free-play adjustment on a clutch with an adjustable cable, loosen the locknut and rotate the adjusting nut on the lower end of the cable **(Figure 50-24)**. When performing the clutch pedal free-play adjustment, always check the total pedal travel and be sure the stop for the upper end of the pedal is in good condition. On clutches with a self-adjusting cable or some hydraulic clutches, a free-play adjustment is not required **(Figure 50-25)**. On these applications if the clutch does not engage or disengage properly, inspect the self-adjusting cable mechanism or check the hydraulic clutch system for low fluid level or air in the system. If the fluid level is low, inspect the hydraulic clutch system for leaks. If fluid is leaking from the master cylinder or slave cylinder, overhaul or replace these components. When fluid is leaking from the line between the master cylinder and slave cylinder, repair or replace the line.

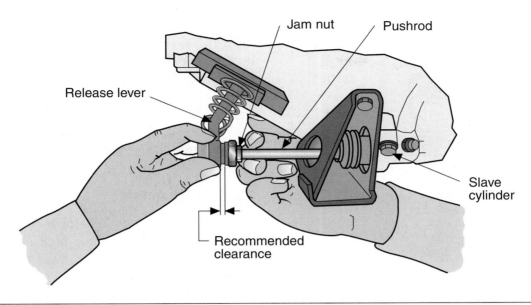

Figure 50-23 Free-play adjustment procedure.

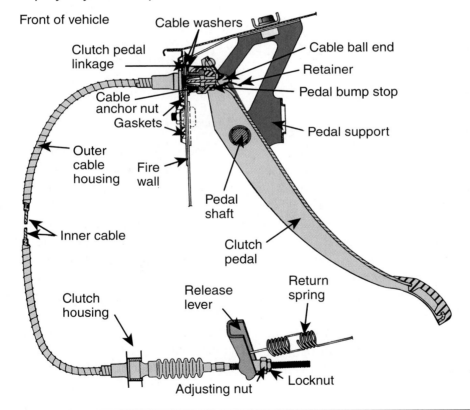

Figure 50-24 Free-play adjustment procedure on a clutch cable.

CLUTCH DIAGNOSIS

When diagnosing clutch problems, always obtain as much information as possible from the customer. Always be sure you are aware of the exact tomer. Always be sure you are aware of the exact

customer complaint related to the operation of the clutch. If possible, find out the past service history of the vehicle. This information may help you to diagnose the clutch problem. A road test may be necessary to identify and verify the customer's

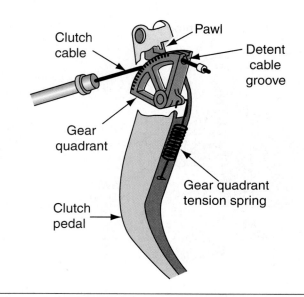

Figure 50-25 Self-adjusting clutch cable.

complaint. Some of the most common clutch problems are listed in **Figure 50-26**.

Clutch Slipping

Clutch slipping occurs when the engine speed increases without the proper corresponding increase in vehicle speed. The engine rpm increases too much for the speed at which the vehicle is moving. If clutch slipping is present, check the clutch pedal free-play. If the free-play is less than specified, clutch slipping may occur. Be sure the clutch pedal is coming all the way upward. A binding linkage or some other component restricting linkage movement may prevent the pedal from coming fully upward to engage the clutch. On a hydraulic clutch, be sure the return port in the master cylinder is not restricted, which prevents the clutch from fully engaging. Raise the vehicle on a lift and look for any sign of oil leaking from the bottom of the flywheel housing. If an oil leak at the rear main bearing or at the front of the transmission is contaminating the clutch facings, slipping will occur. Some vehicles have a small, removable pan on the bottom of the flywheel housing that can be removed to check for oil leaks in this area. Look for clutch facing material in the flywheel pan. If there is facing material in this area, the clutch facings are worn out. Clutch

slipping may be the result of worn clutch disc facings. If this condition is present, the clutch must be replaced. The clutch must also be replaced if the facings are contaminated with oil.

If the clutch slips completely and does not transfer any engine torque to the transmission with the pedal fully released, the facings may be completely worn off the clutch disc.

Clutch Chatter

Clutch chatter is a shuddering action as the clutch engages. Once the clutch is engaged, the shuddering action stops. Road test the vehicle and check for clutch chatter each time the clutch is engaged. Broken engine or transaxle mounts, worn and/or scored pressure plate and flywheel surfaces, and oil on the clutch facings and/or pressure plate and flywheel surfaces may cause clutch chatter. Clutch chatter may also be caused by misalignment of the flywheel housing in relation to the engine and flywheel.

Clutch Noises

One of the most common clutch noises is a growling noise when the clutch pedal is depressed. If the vehicle has a constant-running release bearing, the growling noise is also heard with the engine running, the transmission in neutral, and the clutch pedal released. If the vehicle has an adjustable clutch linkage or cable, the growling noise is present only when depressing or releasing the clutch pedal. A defective release bearing causes this noise. When a growling noise is present only with the engine running, the transmission in neutral, and the clutch pedal released, the bearing on the transaxle or transmission input shaft may be defective. A worn pilot bearing in the back of the crankshaft may cause a growling, rattling noise with the clutch pedal depressed and the clutch disengaged. If a severe scraping, rattling noise is heard when the clutch is engaged and no torque is transferred from the engine to the transmission, the hub may be broken out of the clutch disc.

A heavy knocking or thumping noise just above idle speed may be caused by loose flywheel retaining bolts.

CLUTCH TROUBLE DIAGNOSIS

PROBLEM	DIAGNOSIS	SERVICE
Fails to release (pedal pressed to floor)	• Improper linkage adjustment • Improper pedal travel • Loose linkage or worn cable • Faulty pilot bearing • Faulty drive disc • Fork off ball stud • Clutch hub binding on splne • Clutch disc warped • Pivot rings loose, or damaged	• Adjust linkage • Trim bumper and adjust • Replace as necessary • Replace bearing • Replace disc • Install properly and lube • Repair or replace gear or disc • Replace disc • Replace cover and pressure plate
Slipping	• Improper adjustment • Oil soaked driven disc • Worn or damaged facing • Warped flywheel or pressure plate • Weak diaphragm spring • Drive plate not seated in • Driven plate overheated	• Adjust linkage • Install new disc and correct leak • Replace disc • Replace flywheel or pressure plate • Replace pressure plate • Make 30 to 40 normal starts • Allow to cool
Grabbing (chattering)	• Oil on facing • Worn spine on clutch • Loose engine mounts • Warped flywheel or pressure plate • Resin on flywheel or pressure plate	• Install new disc and correct leak • Replace clutch • Tighten or replace mounts • Replace flywheel or pressure plate • Sand oil or replace component
Ratling or transmission click	• Weak retracting springs • Release fork loose • Oil in drive plate damper • Driven plate damper spring bad	• Replace pressure plate • Check ball stud and retainer • Replace driven disc • Replace driven disc
Release bearing noise (clutch fully engaged)	• Improper adjustment • Bearing binding on retainer • Insufficient fork spring tension • Fork improperly installed • Weak return spring	• Adjust linkage • Clean and lubricate • Replace fork • Install properly • Replace spring
Noisy	• Worn release bearing • Fork off ball stud • Pilot bearing loose	• Replace bearing • Install properly and lube • See bearing lits section
Pedal stays on floor when disengaged	• Bind in cable • Weak pressure plate springs • Springs over traveled	• Replace cable • Replace pressure plate • Adjust linkage
Hard pedal effort	• Bind in linkage • Worn driven plate • Friction in cable	• Lube and free linkage • Replace driven plate • Replace cable

Figure 50-26 Use this chart to help determine the cause of clutch problems.

Clutch Vibration

Clutch vibration may be felt in any clutch pedal position. The vibration may be felt on the clutch pedal and in the passenger compartment. If clutch vibration is present, raise the vehicle on a lift and inspect the engine and transaxle mounts. Watch for broken mounts that allow the engine or transaxle to contact the chassis. When all the mounts are in satisfactory condition, the vibration may be caused by excessive flywheel runout and improper flywheel or pressure plate balance.

Dragging Clutch

A dragging clutch occurs when the clutch pedal is fully depressed but the clutch disc does not completely release. This condition may cause hard transmission shifting and/or gear clashing while shifting. To check for a dragging clutch, allow the engine to idle and fully depress the clutch pedal. Shift the transmission into first gear and keep the pedal depressed. Shift the transmission into neutral, wait 10 seconds, and then shift into reverse. If gear clashing is heard, the clutch is dragging. When this problem is present, be sure the clutch pedal free-play adjustment is within specifications. Excessive clutch pedal free-play causes a dragging clutch.

Raise the vehicle on a lift and inspect the clutch linkage for looseness, wear, damage, or a binding condition. Repair the linkage as required. If the vehicle has a hydraulic clutch, check the fluid level in the master cylinder and bleed the air from the hydraulic system. When the linkage or hydraulic system is satisfactory, a warped clutch disc or pressure plate or a defective release lever may cause clutch drag.

Binding Clutch

If a clutch has a binding condition, it may cause erratic clutch engagement or grabbing. When this condition is present, inspect the clutch linkage or cable for wear and binding. Be sure the release fork is properly positioned on its pivot. Inspect the sleeve on which the release bearing slides. Ridges and wear on this sleeve may cause a binding clutch.

Pulsating Clutch Pedal

Clutch pedal pulsation is a rapid upward and downward pedal movement. To check for this problem, slowly depress the clutch pedal. If noticeable pulsations are present, there is a problem in the clutch. Pedal pulsations are usually caused by uneven pressure plate release levers, a misaligned flywheel housing, a warped pressure plate, or excessive flywheel runout. The clutch must be disassembled, inspected, and various measurements performed to test for these conditions.

CLUTCH SERVICE

Accurate clutch service is absolutely essential to provide a clutch that operates smoothly without chattering and unwanted noises.

Clutch and Pressure Plate Removal

To remove the clutch and pressure plate, it is necessary to remove the transaxle or transmission. Before removing the transaxle or transmission, the engine must be supported so it does not drop downward and damage other components. Follow the vehicle manufacturer's recommended procedure in the service manual to support the engine prior to transaxle or transmission removal. On some front-wheel drive vehicles it is easier to remove the engine and transaxle as an assembly and then remove the transaxle from the engine. Always follow the vehicle manufacturer's recommended procedure in the service manual regarding engine and/or transaxle removal.

Tech Tip *Before removing a pressure plate from the flywheel, always place index marks on these components so the pressure plate can be reinstalled in the same position to maintain proper flywheel and pressure plate balance if the pressure plate is to be reused.*

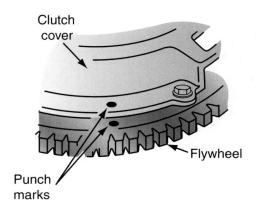

Figure 50-27 Placing index marks on the pressure plate and flywheel as this will help maintain flywheel and pressure plate balance if the pressure plate is to be reused.

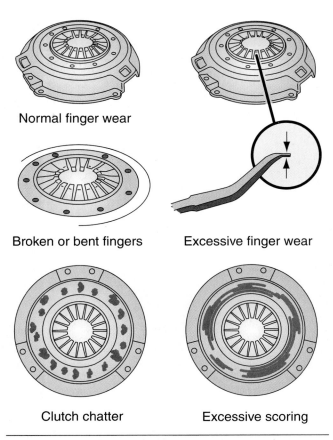

Normal finger wear

Broken or bent fingers   Excessive finger wear

Clutch chatter   Excessive scoring

Figure 50-28 Some common pressure plate problems.

If the engine is removed from the vehicle, bolt the engine securely onto an engine stand and then remove the transaxle from the engine. If the pressure plate is to be reinstalled, then always place index marks on the flywheel and pressure plate before removing the pressure plate from the flywheel **(Figure 50-27)**.

After the pressure plate and clutch plate are removed from the flywheel, inspect these components and other related parts such as the release bearing, fork, and pilot bearing.

The flywheel should be carefully inspected for any defects. The flywheel may be removed and resurfaced to correct minor defects in the surface finish of the flywheel. Always place index marks on the flywheel and the crankshaft to make reinstallation easier. Some flywheels are balanced and will only mount to the crankshaft in one position.

Clutch and Pressure Plate Installation

Install a new pilot bushing or bearing as necessary. Always use the proper tool to install the pilot bushing or bearing **(Figure 50-28)**.

Replace all worn or damaged components. Some common pressure plate problems are shown in **Figure 50-29**. Most clutch plates and pressure plates are replaced as a matched pair and are purchased as a set.

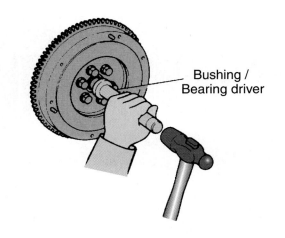

Bushing / Bearing driver

Figure 50-29 Use a driver to install the pilot bushing or bearing.

Inspect and lubricate all the pivot or sliding points in the clutch system **(Figure 50-30)**. Failure to do so may cause a binding clutch linkage or may cause improper functioning of the release bearing.

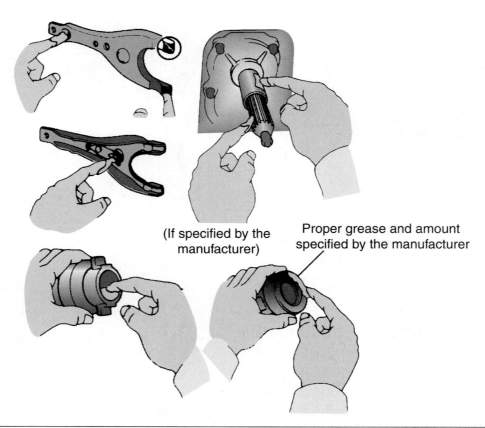

(If specified by the manufacturer)

Proper grease and amount specified by the manufacturer

Figure 50-30 Lubricate all contact points in the release mechanism using the lubricant specified by the manufacturer.

Figure 50-31 A universal clutch alignment tool.

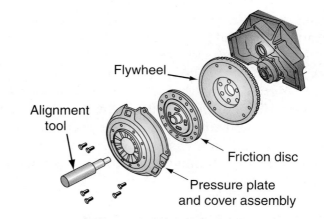

Figure 50-32 Use a clutch alignment tool when installing the clutch and pressure plate.

When installing the clutch plate and pressure plate, insert a clutch plate alignment tool through the clutch plate into the pilot bearing to ensure proper clutch plate alignment with the pilot bearing **(Figure 50-31** and **Figure 50-32)**.

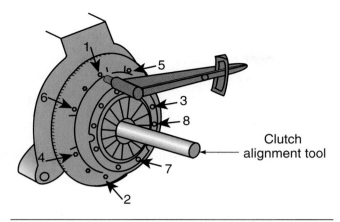

Figure 50-33 Torque the pressure plate bolts in the proper order and to the proper torque.

Always tighten the pressure plate bolts to the manufacturer specified torque and in the proper order **(Figure 50-33)**. Follow the vehicle manufacturer's recommended procedure in the service manual during transmission reinstallation. If the engine was removed from the vehicle, then reinstall the transaxle onto the engine and reinstall the assembly into the vehicle.

Verify that the clutch is properly adjusted and the vehicle shifts properly before returning the vehicle to the customer.

Summary

- When the clutch is engaged, the pressure plate squeezes the clutch facings between the pressure plate and the flywheel so engine torque is transferred from the flywheel and pressure plate to the clutch disc and transmission input shaft.

- If the clutch is released, the pressure plate is pulled away from the clutch disc and the flywheel and pressure plate rotate without transferring engine torque to the clutch disc.

- The torsional springs in the clutch disc prevent engine pulsations from being transferred from the engine to the transmission input shaft.

- The pressure plate may contain a diaphragm spring or coil springs.

- To disengage the clutch, the release bearing is moved forward against the pressure plate fingers.

- The pivoted release fork transfers clutch linkage or cable movement to the release bearing.

- A constant-running release bearing lightly contacts the pressure plate fingers with the clutch engaged.

- The clutch may have a mechanical linkage, a cable, or a hydraulic clutch system between the pedal and the release fork.

- Some clutch cables are self-adjusting.

- Hydraulic clutch systems have a master cylinder operated by the clutch pedal and a slave cylinder that operates the release lever.

- Clutches with mechanical linkages or adjustable cables require a free-play adjustment.

- Hydraulic clutches or clutches with self-adjusting cables do not require a free-play adjustment.

- Clutch problems include slipping, chattering, noise, vibration, dragging, binding, or pulsating.

- Flywheel runout may be measured with a dial indicator.

- Flywheel housing face and bore alignment may be measured with a dial indicator.

- When installing the clutch disc and pressure plate, a clutch disc alignment tool is used to position the clutch disc properly.

Review Questions

1. Technician A says the flywheel provides inertia for the crankshaft between cylinder firings. Technician B says on a vehicle with an automatic transmission, the torque converter dampens engine pulses. Who is correct?

   A. Technician A

   B. Technician B

   C. Both Technician A and Technician B

   D. Neither Technician A nor Technician B

2. Technician A says a pressure plate may have a diaphragm spring or coil springs. Technician B says when the clutch is engaged, the pressure plate is held away from the clutch disc facings. Who is correct?

   A. Technician A

   B. Technician B

   C. Both Technician A and Technician B

   D. Neither Technician A nor Technician B

3. Technician A says the clutch release bearing requires periodic lubrication. Technician B says a constant-running release bearing is used with a clutch that has an adjustable cable. Who is correct?

   A. Technician A

   B. Technician B

   C. Both Technician A and Technician B

   D. Neither Technician A nor Technician B

4. Technician A says in a hydraulic clutch system, the clutch pedal operates the slave cylinder pushrod. Technician B says air in a hydraulic clutch system may cause improper clutch disengagement. Who is correct?

   A. Technician A

   B. Technician B

   C. Both Technician A and Technician B

   D. Neither Technician A nor Technician B

5. All these statements about clutch plate design are true *except*:

   A. Diagonal grooves cut in the clutch facings improve clutch cooling.

   B. Diagonal grooves cut in the clutch facings provide smoother clutch engagement.

   C. The clutch facings are attached to wave springs that gradually flatten out during clutch engagement.

   D. A rigid-type clutch disc has torsional springs and a friction ring between the facings and the hub.

6. A dual-mass flywheel has all of these design features *except*:

   A. inner and outer rotating plates connected by a damper mechanism.

   B. an inner plate that is bolted to the crankshaft flange.

   C. an outer plate on which the clutch plate facing makes contact.

   D. a pressure plate that is bolted to the inner and outer flywheel plates.

7. When discussing clutch operation, Technician A says when the clutch is engaged; the clutch plate facings are jammed between the flywheel and the pressure plate. Technician B says the clutch release-bearing hub slides forward and rearward on a machined extension on the transmission front bearing retainer. Who is correct?

   A. Technician A

   B. Technician B

   C. Both Technician A and Technician B

   D. Neither Technician A nor Technician B

8. In a hydraulic clutch system:

   A. The clutch pedal pushrod is in contact with the master cylinder piston.

   B. Fluid pressure from the slave cylinder operates the master cylinder.

   C. There is a specified clearance between the clutch release bearing and the pressure plate diaphragm spring.

   D. A pushrod is connected between the slave cylinder and the clutch release bearing.

9. The clutch in a vehicle with a mechanical clutch linkage does not release properly and hard gear shifting is experienced. The most likely cause of this problem is:

   A. a worn pilot bearing.

   B. a worn transmission input shaft bearing.

   C. a rough clutch release bearing.

   D. excessive clutch pedal free-play.

10. A vehicle with a hydraulic clutch has a slipping clutch problem that allows the engine rpm to increase without the proper increase in vehicle speed. The clutch master cylinder has the proper level and type of fluid. The cause of this problem could be:

    A. air in the hydraulic clutch system.

    B. oil contamination on the clutch facings.

    C. improper clutch free-play adjustment.

    D. fluid leaking past the slave cylinder piston.

11. To disengage a clutch, the release fork pushes the _____ _____ toward the pressure plate.

12. Excessive clutch pedal free-play may cause improper clutch _____.

13. Clutch slipping may be caused by _____ clutch pedal free-play.

14. Broken engine mounts may cause clutch _____.

15. On a clutch with an adjustable linkage, explain the cause of a grinding noise only when the clutch pedal is depressed.

16. Describe the causes of clutch chatter.

17. Describe the causes of clutch slippage.

18. Explain the causes of clutch vibration.

CHAPTER

51 Driveshaft and Driveline

Learning Objectives

After you have read, studied, and practiced the contents of this unit, you should be able to:

- Explain drive shaft and universal joint design.
- Describe the purpose of drive shaft.
- Describe the changes in drive shaft speed during shaft rotation.
- Describe the noises that may be produced by worn universal joints.
- Remove and replace drive shafts and universal joints.
- Explain the motion requirements for inner and outer CV joints.
- Explain how equal length drive axles are possible on some front-wheel drive vehicles.

- Describe front drive axle purpose.
- Explain the purpose and importance of CV joint boots.
- Explain how worn engine or transaxle mounts may affect CV joints.
- Describe the diagnostic procedure for inner and outer CV joints.
- Describe the necessary repair action when the grease in a CV joint is contaminated with dirt or moisture.

Key Terms

Canceling angles

Cardan joints

Constant velocity (CV) joints

Critical speed

Double Cardan universal joint

Ellipse

Electronic vibration analyzer (EVA)

Fixed CV joint

Hotchkiss drive

In-phase

Plunging CV joint

Rzeppa joint

Torque steer

Transversely mounted

INTRODUCTION

Technicians must understand drive shaft and universal joint design and operation to diagnose these components accurately and quickly. For ex-

ample, if you do not understand drive shaft angles and the proper procedures for diagnosing drive shaft vibrations, it will likely be difficult and time-consuming to diagnose this problem. You must also understand the design, operation, and proper diagnostic and service procedures for front drive axles on FWD vehicles. When you have this knowledge and the related skills, you will be able to diagnose front-wheel drive axle problems accurately and quickly. With these skills, you will be able to repair drive axle problems correctly the first time the repair is performed.

DRIVE SHAFT DESIGN

On rear-wheel drive vehicles with the engine in the front, the drive shaft connects the transmission output shaft to the differential. Engine torque is transferred from the transmission output shaft through the drive shaft to the pinion shaft in the differential **(Figure 51-1)**. A typical drive shaft has a slip yoke with internal splines that fit over the splines on the transmission output shaft. The slip yoke has a smooth, machined outer surface that rides in a bushing pressed into the extension housing. A seal in the transmission extension housing rides on this surface to prevent oil leaks at this location. A universal joint is connected between the front slip yoke and the drive shaft and another universal joint is connected between the drive shaft and the differential **(Figure 51-2)**.

Figure 51-1 The drive shaft connects the transmission output shaft to the differential.

NOTE: The piece in the center of a universal joint that supports the bearing cups is called a *cross* or *spider.*

NOTE: The conventional universal joint we have explained may be called a *Cardan joint.*

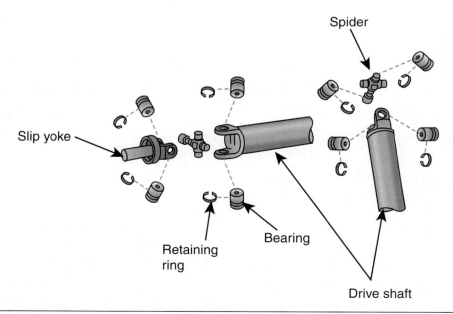

Figure 51-2 Typical drive shaft and universal joints.

Figure 51-3 Drive shaft center bearing.

Some drive shafts on trucks are a two-piece unit with a center support bearing in the middle between the two shafts **(Figure 51-3)**. This type of drive shaft has three universal joints and a slip yoke on the rear shaft fits over splines on the rear of the front shaft.

Four-wheel drive vehicles have two drive shafts. One shaft drives the rear wheels and the other shaft drives the front wheels **(Figure 51-4)**.

Most drive shafts are made from seamless steel tubing with universal joint yokes welded on each end of this tubing. Some drive shafts are made from aluminum or composite fiberglass. A composite drive shaft is lighter, stronger, and has improved balance characteristics compared to a steel drive shaft.

DRIVE SHAFT PURPOSE

The purpose of the drive shaft is to transfer engine torque from the engine and transmission to the differential **(Figure 51-5)**. The front of the drive shaft is connected to the transmission and the transmission and engine are supported on mounts attached to the chassis. These mounts allow very limited engine and transmission movement. The differential is mounted on springs that are supported on the chassis. As the rear wheels strike road irregularities, the rear wheels and differential move upward and downward. This differential action changes the angle of the drive shaft between the differential and the transmission. It also changes the distance between these two components. The universal joints on each end of the drive shaft permit the drive shaft angle to change and the slip

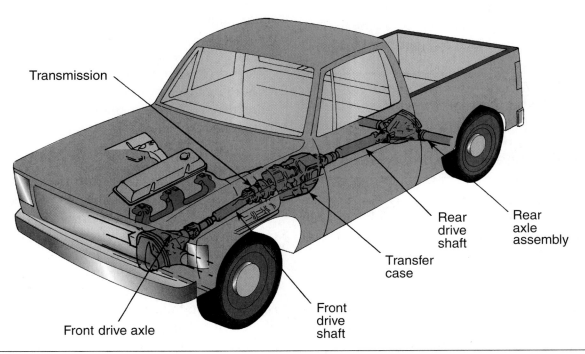

Figure 51-4 Front and rear drive shafts on a four-wheel drive vehicle.

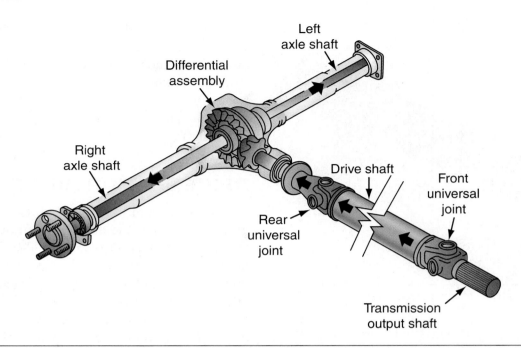

Figure 51-5 The drive shaft transmits engine torque from the transmission to the differential.

yoke in the front of the drive shaft can slip forward or rearward on the transmission output shaft to change the length of the drive shaft.

The drive shaft rotates about three to four times faster than the vehicle drive wheels because of the differential gear ratio. Because of this high rotational speed, the drive shaft tends to vibrate at its critical speed. **Critical speed** is the rotational speed at which a component begins to vibrate. This vibration is usually caused by centrifugal force.

> **Tech Tip** *The differential gear ratio describes the ratio between the drive shaft and the wheels. For example, if the differential ratio is 3.23:1, then the drive shaft will turn 3.23 times for every 1 turn of the wheels.*

Drive shaft diameters are as large as possible and the drive shaft is as short as possible to keep the critical speed frequency above the normal rotational speed of the drive shaft. To minimize this problem the drive shaft is balanced. The most common method of drive shaft balancing is to weld balance weights to the shaft during the manufacturing process. Care must be taken not to damage or otherwise remove these weights during driveshaft service.

TYPES OF DRIVE SHAFTS

The most common type of drive shaft is the **Hotchkiss drive (Figure 51-6),** which has an open drive shaft and two **Cardan joints**, otherwise known as universal joints **(Figure 51-7).**

On some trucks, the Hotchkiss drive has two drive shafts with a center support bearing. The center support bearing supports the drive shaft on the chassis **(Figure 51-8)**. A single drive shaft would be too long and vibrate more easily than shorter drive shafts. The use of two shorter shafts helps to reduce vibration. The two drive shafts require the use of the center support bearing for these longer vehicles.

UNIVERSAL JOINTS

A universal joint is held by two Y-shaped yokes. One of these yokes is on the drive shaft and on a front universal joint; the other yoke is on the slip joint. On the rear universal joint, the other yoke is on the differential companion flange **(Figure 51-9)**. Each yoke contains a machined opening. A cross is mounted in the center of the yokes at each end of the drive shaft. Bearing cups containing needle bearings are mounted in each yoke opening. The needle bearings in each bearing cup rides on the

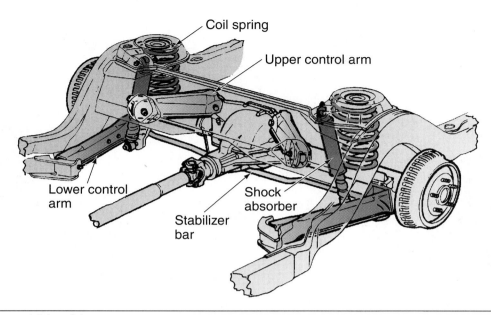

Figure 51-6 Hotchkiss drive.

machined surface on the cross. A retaining ring fits into a groove in each bearing cup to retain this cup in the yoke **(Figure 51-10)**. Therefore, the cross and its rollers are supported in the center of the yokes and the cross and needle bearings allow the slip yoke to move sideways or vertically in relation to the drive shaft. This universal joint action allows the drive shaft to rotate smoothly and also permits the drive shaft angle to change in relation to the slip yoke and transmission output shaft.

A universal joint transmits torque through an angle. The operating angle of a universal joint is determined by the angular difference between the transmission output shaft and the drive shaft **(Figure 51-11)**. The input shaft speed is constant, but the speed of the output shaft accelerates and decelerates twice during each revolution. The input yoke rotates at a constant speed during each revolution. However, the output yoke quadrants alternate between longer and shorter distances of travel compared to the input yoke quadrants that remain at constant distances. The rotational path of the output yoke appears like an ellipse. An **ellipse** is a compressed form of a circle. When one point of the output yoke travels the same distance in a shorter time, it must move at a slower rate. When one point in the output yoke moves the same distance in a longer time, it must travel faster. The face of a clock may be used to illustrate the ellipse action of a drive shaft **(Figure 51-12)**. If the hand on the clock moves the longer distance from top to bottom on the outer circle, it must move faster compared to movement from top to bottom on the ellipse.

This effect can be easily observed by the use of two extensions with a universal joint in between **(Figure 51-13)**. As the angle is increased on the universal joint the severity of the drive shaft speed variation becomes more pronounced. Turn one extension with the universal joint in an angle and observe the rotational variation in the other extension.

Figure 51-7 Typical Cardan joint.

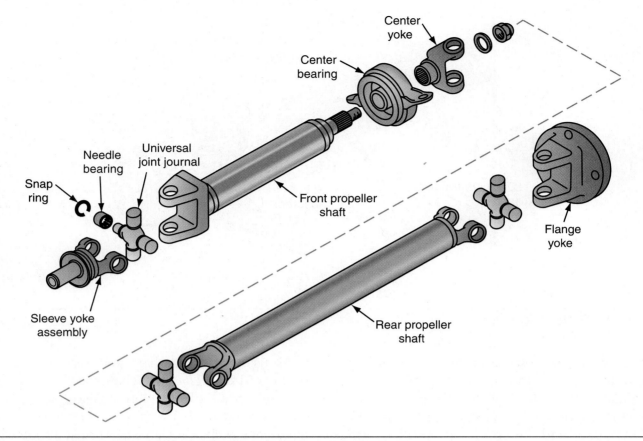

Figure 51-8 Two-piece drive shaft with a center support bearing.

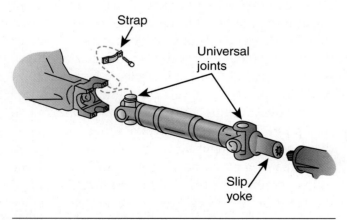

Figure 51-9 Drive shaft connections to the transmission and differential.

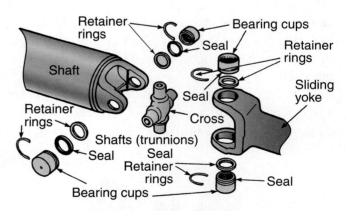

Figure 51-10 Typical Cardan universal joint assembly.

The output yoke in a universal joint is continually falling behind and catching up with the input yoke **(Figure 51-14)**. The resulting deceleration and acceleration causes fluctuating torque and torsional vibrations. A steeper drive shaft angle causes increased torsional vibrations. Using canceling angles may reduce drive shaft vibration. The front and rear universal joint angle should measure the

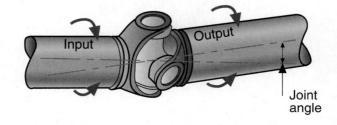

Figure 51-11 Drive shaft angles.

Figure 51-12 The face of a clock may be used to illustrate the elliptical movement of the drive shaft.

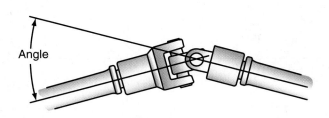

Figure 51-13 The variations in drive shaft speed can be observed with two extensions and a universal joint.

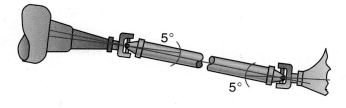

Figure 51-15 Equal front and rear drive shaft angles cancel drive shaft vibrations.

the front universal joint. **Canceling angles** are present when the vibrations from one universal joint are canceled by an equal and opposite vibration from another universal joint.

Universal joints on a drive shaft should be on the same plane **(Figure 51-16)**. When this condition is present, the universal joints are said to be in-phase with each other. If the drive shaft has a yoke that is splined to the drive shaft and the universal joints are disassembled, the splined yoke should be marked in relation to the drive shaft to maintain the in-phase condition. If this action is not taken, drive shaft vibration may occur. Universal joints on a drive shaft are **in-phase** when they are on the same plane.

Some rear-wheel drive vehicles have a **double Cardan universal joint** containing two individual Cardan universal joints mounted close together connected by a centering socket yoke. A round socket on the rear of the drive shaft fits into a ball socket on the rear yoke **(Figure 51-17)**. A short

same and offset one another **(Figure 51-15)**. If this condition is present, when the front universal joint accelerates and produces a vibration, the rear universal joint decelerates and produces an equal and opposite vibration that cancels the vibration from

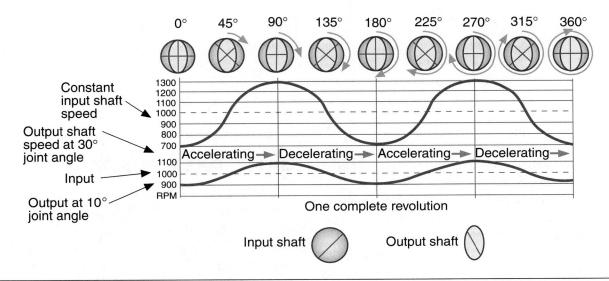

Figure 51-14 Variations in drive shaft speed in one revolution.

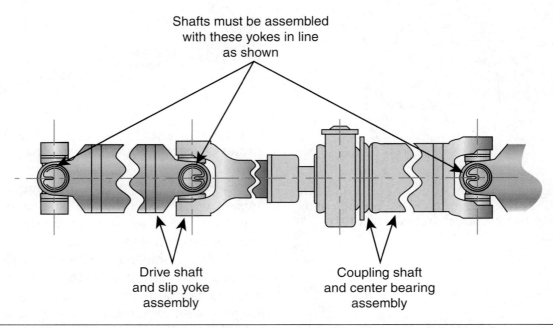

Shafts must be assembled
with these yokes in line
as shown

Drive shaft
and slip yoke
assembly

Coupling shaft
and center bearing
assembly

Figure 51-16 Universal joints in the same plane are in phase with each other.

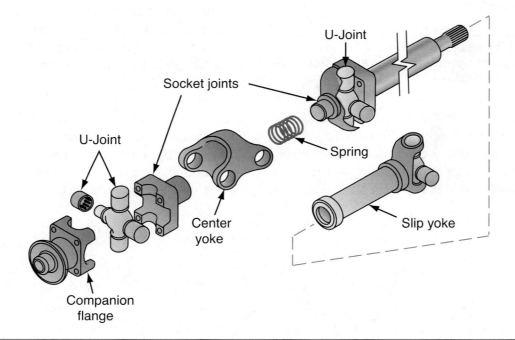

U-Joint

Socket joints

Spring

U-Joint

Center
yoke

Slip yoke

Companion
flange

Figure 51-17 Double Cardan universal joint.

center yoke connects the two universal joints. The ball and socket of the double Cardan universal joint splits the angle of the two drive shafts equally between the two joints **(Figure 51-18)**. When these two universal joints operate at the same angle, the normal fluctuations from the acceleration and deceleration of one joint are cancelled out by the equal and opposite fluctuations at the other joint.

DRIVE AXLE PURPOSE

Each drive axle must transfer engine torque from the differential in the transaxle to the drive wheels. These drive axles and CV joints must transfer uniform torque at a constant speed and allow the angles of the drive axles to change considerably as the front wheels are steering and move up

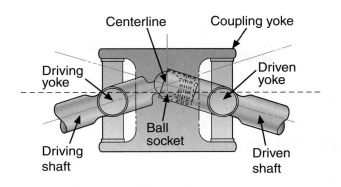

Figure 51-18 The ball and socket in a double Cardan universal joint splits the angle of the two shafts.

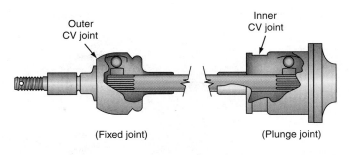

Figure 51-20 Front drive axle with inboard and outboard CV joints.

and down. The drive axles and CV joints must allow a drive axle angle up to 20° for vertical suspension movement and up to 40° for steering **(Figure 51-19)**. The drive axles and CV joints must also allow the drive axle length to change as the front wheels move vertically.

DRIVE AXLE DESIGN

A typical front-wheel drive vehicle has two drive axles. Each drive axle has three main components. Inner and outer **constant velocity (CV) joints** are attached to the axle shaft. The axle shafts are made from solid steel or steel tubes. Each CV joint connects the axle shaft to the CV joint housing. The inner CV joint must allow axle rotation while allowing the axle to move vertically and horizontally as the front wheel moves up and down and the axle length change while cornering **(Figure 51-20)**.

NOTE: Drive axles may be called *half-shafts* by some vehicle manufacturers.

Constant Velocity (CV) Joints

The outer CV joint must allow axle rotation and also allow vertical axle movement as the front wheel strikes road irregularities. The outer CV joint must also allow the front wheel to turn transfer a uniform torque and a constant speed while operating at a wide variety of angles.

The housing on the outer joint has a stub axle that is splined into the wheel hub. The inner joint housing has a stub axle that is splined into the differential side gear in the transaxle **(Figure 51-21)**. Some inner axle joint housings have a flange that is bolted to a matching flange on the transaxle. Flexible boots are clamped to the drive axle and to each joint housing to keep contaminants out of the joint and keep grease in the joint.

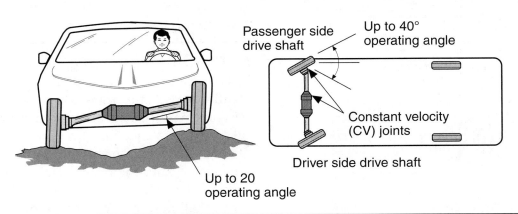

Figure 51-19 CV Joints allow the wheels to move up and down and turn.

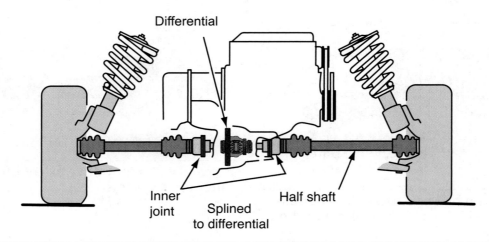

Differential

Inner joint   Splined to differential   Half shaft

Figure 51-21 The inner axle joint housing is splined into the differential side gear.

CV joints are used on the front wheels of many four-wheel drive vehicles **(Figure 51-22)**. Many rear-wheel drive vehicles with independent rear suspension also use CV joints.

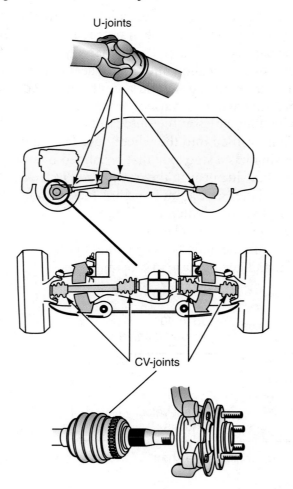

U-joints

CV-joints

Figure 51-22 Four-wheel drive vehicles may use CV joints.

In a typical front-wheel drive vehicle, the engine is transversely mounted and the transaxle is bolted to the rear of the engine. This configuration makes it impossible to center the transaxle in the under-hood area. A **transversely mounted** engine is positioned crosswise. Because the transaxle is not centered in the vehicle, the front drive axles are unequal in length **(Figure 51-23)**. The CV joints on the longer drive axle usually operate at a smaller angle compared to the angle on the shorter drive axle. When engine torque is supplied from the differential to the drive axle during hard acceleration, the longer drive axle has less resistance to turning and more engine torque is supplied to this axle and drive wheel. When more engine torque is supplied to one front drive wheel, the steering tends to pull to one side. This condition is called torque steer. **Torque steer** is a condition in which the vehicle steering pulls to one side during hard acceleration.

> **Tech Tip** *Torque steer causes the vehicle to pull to the side with the longer axle.*

On some front-wheel drive vehicles, the long drive axle is replaced with an intermediate drive shaft between the transaxle and the drive axle **(Figure 51-24)**. This allows the drive axles to be designed with equal lengths to reduce torque steer. The drive axles in front-wheel drive vehicles are connected after the differential. Therefore, these drive axles rotate much slower compared to the drive shaft in rear-wheel drive vehicles. Because of the slower rotation, the balance is not critical on

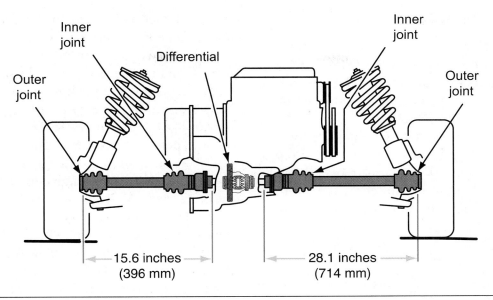

Figure 51-23 Unequal length drive axle shafts.

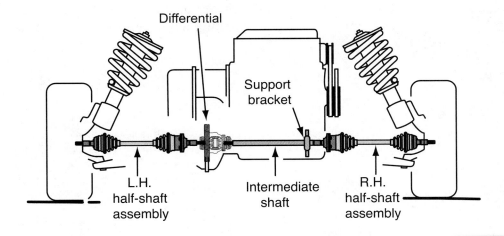

Figure 51-24 Equal length front drive axles with intermediate shaft.

drive axles. Some front-wheel drive vehicles with unequal length drive axles have a solid axle shaft on the shorter axle and a tubular portion in the longer drive axle **(Figure 51-25)**. This design allows the drive axles to twist equal amounts during hard acceleration and reduce torque steer.

Some front-wheel drive vehicles with unequal length drive axles may use a torsional damper on the longer drive axle to dampen torsional vibrations **(Figure 51-26)**.

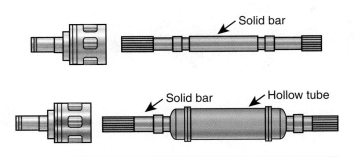

Figure 51-25 Solid and tubular drive axles.

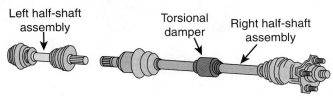

Figure 51-26 A front drive axle with torsional damper.

Types OF CV Joints

CV joints come in several types and may be seen in a variety of combinations **(Figure 51-27)**. These different combinations give the automotive engineer the flexibility and ability to design drive axles that best suit the vehicle.

One of the most common CV joints is the ball type. This type of joint may be called a **Rzeppa joint** after the original designer, Alfred. H. Rzeppa. The ball-type CV joint is commonly used for the outboard drive axle joint. In this type of CV joint,

the inner race is splined to the drive axle and a circlip retains this race on the axle **(Figure 51-28)**. A series of equally spaced machined grooves is located on the outer surface of the inner race. Steel balls are mounted in the inner race grooves. These balls are held in place by a ball cage. The number of balls in a CV joint may vary from three to six, usually six, depending on the torque applied to the CV joint. The outer surface of the balls fits into grooves in the outer race. When engine torque is supplied to the drive shaft, the inner race and balls are forced to rotate. The balls transfer the torque to the outer race that is splined to the front wheel hub. The grooves in the inner and outer races allow the angle between the drive axle and the outer race to change as the front wheels move up and down or turn in either direction. When viewed from the side, the balls in the CV joint always bisect the angle formed by the shafts on either side of the joint regardless of the operating angle **(Figure 51-29)**. This action reduces

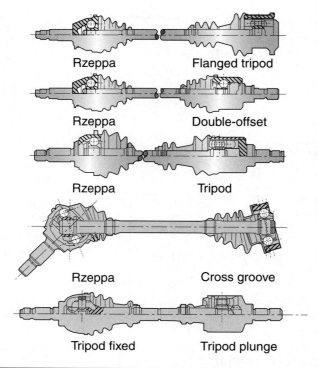

Figure 51-27 FWD drive axle angles.

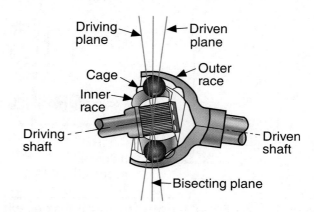

Figure 51-29 In a Rzeppa (ball-type) CV joint, the balls bisect the angle of the joint.

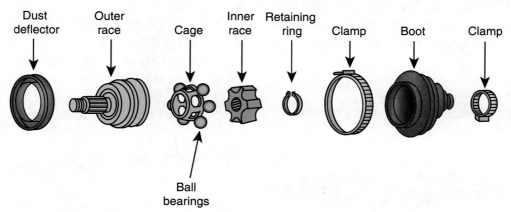

Figure 51-28 A Rzeppa (ball-type) CV joint.

the effective angle of the joint by 50 percent and eliminates vibration problems. The ball-type joint is a fixed CV joint. A **fixed CV joint** does not allow any inward and outward drive axle movement to compensate for changes in axle length.

A **plunging CV joint** allows inward and outward drive axle movement to compensate for changes in axle length as the front wheel moves upward or downward. The inner drive axle joint is commonly a tripod type. This type of CV joint has a central hub or cross with three trunnions and spherical rollers. Needle bearings are mounted between these rollers and the trunnions. A snap ring retains the cross on the drive axle shaft **(Figure 51-30)**. The outer surfaces of the spherical rollers are mounted in grooves in the outer race or joint housing. These grooves are long enough to allow inward and outward drive axle movement to compensate for changes in axle length as the front wheel moves upward and downward. Tripod joints may be of an open or of a closed design **(Figure 51-31)**.

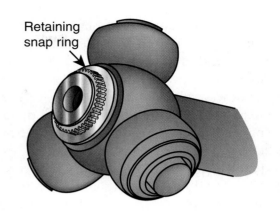

Figure 51-30 A tripod-type CV joint.

The outer housing or race on the inner CV joint is splined to the differential side gear. Engine torque is supplied from the differential side gear to the outer CV joint housing. From this housing, torque is transferred through the spherical rollers to the needle bearings and trunnions. Torque is then transferred from the trunnions to the drive axle. On vehicles with ABS, an exciter ring or tone

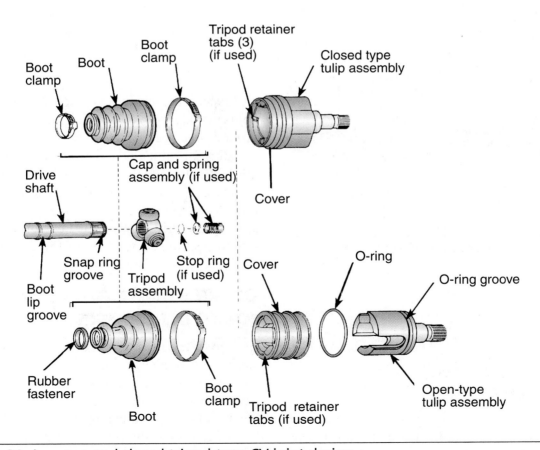

Figure 51-31 An open and closed tripod-type CV joint design.

wheel is pressed onto one of the CV joint outer housings **(Figure 51-32)**. These rings rotate past a sensor that sends a voltage signal to the ABS computer in relation to wheel speed.

A bellows-type neoprene boot covers each drive axle joint. The boot is clamped to the drive axle shaft and to the joint housing **(Figure 51-33)**. The boots are extremely important to keep grease in and moisture and dirt out of the CV joints. If a cracked boot allows moisture and dirt into a CV joint, the joint life is very short. If the drive axle boots are located close to the exhaust system, the boots are usually made from silicone or thermoplastic materials to withstand the additional heat. The appearance of the boot indicates the boot manufacturer. Always use the type of boot specified by the vehicle manufacturer. Various types of boot clamps are used to retain the boots. Always use the clamps specified by the vehicle manufacturer and follow the manufacturer's recommended procedure to install and tighten these clamps.

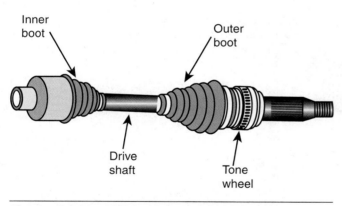

Figure 51-32 ABS exciter ring on CV joint housing.

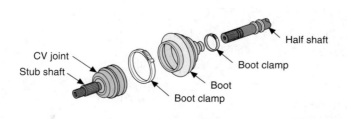

Figure 51-33 Clamps seal the boot against the drive axle and the CV joint stub shaft.

DRIVE SHAFT AND UNIVERSAL JOINT MAINTENANCE

When performing chassis lubrication, always check each universal joint to see if it has a grease fitting in the joint cross. If a grease fitting is present, use the grease gun to inject a small amount of grease into the joint. Applying excessive amounts of grease to a universal joint may damage the seal in each bearing cup. Some universal joints have a plug that may be removed to insert a grease fitting. These plugs should be removed and a grease fitting installed to allow greasing the universal joint at the vehicle manufacturer's recommended service interval. When performing an under vehicle inspection, check the universal joints for looseness. Grasp the drive shaft and determine if there is any vertical movement in the universal joint. Try and turn the drive shaft with your hand and watch for movement between the drive shaft and the yoke. If there is any vertical or rotary movement in a universal joint, it must be replaced.

NOTE: A grease fitting may be called a *zerk fitting*.

Inspect the drive shaft for damage, such as dents or missing balance weights. Inspect the area around the rear of the transmission extension housing for oil leaks. An oil leak in this area usually indicates a leaking extension housing seal. If this seal is leaking, the drive shaft must be removed and the seal replaced. When the vehicle has a two-piece drive shaft with a center support bearing, inspect the center bearing for looseness.

DRIVE SHAFT AND UNIVERSAL JOINT DIAGNOSIS

One of the most common complaints on drive shafts and universal joints is a squeaking noise whose frequency increases with vehicle speed. The frequency of the noise is the differential ratio times the wheel speed. For example, if the differential ratio is 3.23, then the noise would be occurring at 3.23 times the wheel speed. This noise is usually caused by a dry, worn universal joint and it is most noticeable at low vehicle speeds. Another common problem caused by a worn universal joint is a clanging noise when the transmission is shifted from park to drive or reverse. The clanging noise may also occur during initial acceleration. A severely worn universal joint may cause a clunking noise at low vehicle speeds. A universal joint that causes any of these noise complaints must be replaced. A defective center support bearing causes a steady growling noise that is more pronounced at low speed. A vibration that increases and decreases with vehicle speed may be caused by an unbalanced drive shaft. This condition may be caused by a dented drive shaft, missing balance weights, or undercoating and other material adhered to the drive shaft. Improper drive shaft angles may cause a drive shaft vibration that is most noticeable between 30 to 35 miles per hour (48 to 56 kilometers per hour). Worn engine or transmission mounts may cause improper drive shaft angles. Sagged springs lower the vehicle chassis and cause improper drive shaft angles **(Figure 51-34)**.

> **Tech Tip** *Do not check front CV joints with the vehicle raised on a lift and the front wheels dropped downward. This wheel position may cause a loose CV joint to feel tight.*

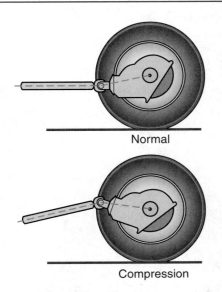

Normal

Compression

Figure 51-34 Sagged rear springs lower the chassis and cause improper drive shaft and universal joint angles.

> **Tech Tip** *A CV joint cannot be checked accurately for looseness by grasping the axle and/or outer joint housing and checking for movement. CV joints may have some looseness because of their design. The most accurate way to check a CV joint is to road test the vehicle and listen for unacceptable noises and observe the CV joint operation.*

DRIVE SHAFT AND UNIVERSAL JOINT SERVICE

Proper drive shaft and universal joint service is very important to prevent vibration and provide driver and passenger comfort. Vibration in the vehicle interior can greatly increase driver and passenger fatigue. Proper drive shaft and universal joint service also eliminates the annoying noises described previously.

Drive Shaft Removal and Installation

Before removing a drive shaft, always mark the drive shaft and yoke with a chalk or paint stick so these components can be reassembled in the same position in relation to each other **(Figure 51-35)**. This is done to maintain any match mounting that the factory may have used during vehicle assembly. If the vehicle has a two-piece drive shaft with a center support bearing, mark each yoke in relation to the drive shaft. If this action is not taken, drive shaft vibration may occur.

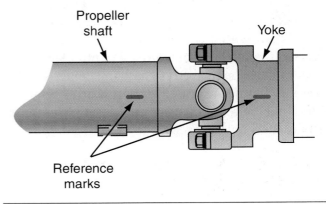

Propeller
shaft

Yoke

Reference
marks

Figure 51-35 Marking the drive shaft and yokes before disassembly.

SAFETY TIP *When removing a two-piece drive shaft, if the front half of the drive shaft falls downward, it may damage the shaft or cause personal injury.*

To remove the drive shaft, raise the vehicle on a lift. Place an oil drain pan under the rear of the transmission extension housing to catch any oil that leaks from this area when the drive shaft is removed. Use the proper wrench to remove the bolts that retain the rear universal joint to the differential flange. It may be necessary to bend the metal locking tabs away from these retaining bolts. Slide the drive shaft forward and lower the rear end of the shaft so it is below the differential flange. Pull the drive shaft rearward so the front slip yoke comes off the transmission output shaft. Place the drive shaft on the workbench and install a plug on the transmission output shaft to keep oil from continuing to leak from this source. If the vehicle has a two-piece drive shaft, remove the center support bearing retaining bolts first and have an assistant hold the center and the front half of the drive shaft as the complete drive shaft is removed.

When installing a drive shaft, slide the front slip yoke over the transmission output shaft and align the rear universal joint with the differential flange. Rotate the drive shaft so the index marks on the drive shaft and rear yoke are aligned. Position the bearing cups properly into the differential flange. Install the rear universal joint retaining bolts and be sure to install the bolt locks if applicable. Tighten these retaining bolts to the specified torque and bend the lock tabs up against the bolt heads. Use a grease gun to lubricate all the universal joints if they have grease fittings. If the vehicle has a two-piece drive shaft with a center support bearing, have an assistant guide the front slip yoke into place while you hold up the rear drive shaft and center support bearing. Install the center support bearing retaining bolts and the rear universal joint fasteners. Tighten all fasteners to the specified torque.

Universal Joint Disassembly and Reassembly

Defective universal joints must be replaced. Do not attempt to change parts from one universal joint to another. Regardless of the type of universal

joint, the disassembly and reassembly procedures are similar. Follow this procedure to disassemble a universal joint:

1. Clamp the front slip yoke in a soft-jawed vise across the bearing cup area and support the other end of the drive shaft.
2. Remove the lock rings that retain the bearing cups in the drive shaft. Mark the yoke in relation to the drive shaft **(Figure 51-36)**.
3. Select a socket that has an internal opening large enough for the bearing cup to fit into and select another socket that is small enough to fit into the bearing cup opening in the drive shaft.
4. Position the large socket over one bearing cup opening on the drive shaft and locate the smaller socket against the opposite drive shaft bearing cup **(Figure 51-37)**.

Figure 51-36 Disassembling a single U-joint (Cardan joint).

Figure 51-37 Disassembling a single U-joint (Cardan joint).

5. Tighten the vise so the smaller socket is forced against the bearing cup and pushes the opposite bearing cup out of the drive shaft and into the large socket.

6. Turn the drive shaft over and position the large socket over the remaining bearing cup opening on the drive shaft and locate the smaller socket against the opposite cross of the U-joint **(Figure 51-38)**.

7. Tighten the vise so the smaller socket is forced against the cross of the U-joint and pushes the opposite bearing cup out of the drive shaft and into the large socket.

8. Turn the drive shaft over and use a brass drift and a hammer to drive the cross and the remaining bearing cup downward out of the drive shaft opening.

9. Remove the slip yoke and cross and support the yoke in a vise on the sides of the yoke **(Figure 51-39)**.

10. Use a brass drift and a hammer to drive the bearing cups and cross out of the yoke if necessary **(Figure 51-40)**.

> **Tech Tip** *Never allow the vise jaws to contact the machined surface on the front slip yoke. This action ruins the slip yoke.*

These steps may be followed to assemble a universal joint:

1. Clean the bearing cup areas, including the snap ring grooves in the front slip yoke **(Figure 51-41)**.

Figure 51-38 Disassembling a single U-joint (Cardan joint).

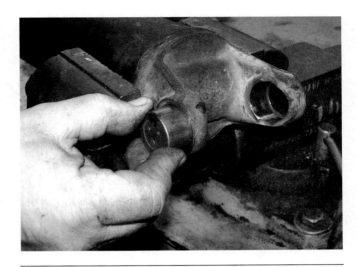

Figure 51-39 Disassembling a single U-joint (Cardan joint).

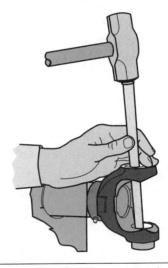

Figure 51-40 Disassembling a single U-joint (Cardan joint).

Figure 51-41 Assembling a single U-joint (Cardan joint).

Figure 51-42 Assembling a single U-joint (Cardan joint).

2. Remove the bearing cups from the new universal joint **(Figure 51-42)**.

3. Install the new cross in the front slip yoke and move it to one side as far as possible **(Figure 51-43)**. Note the direction of the grease zerk hole if provided; it should be toward the center of the driveshaft.

4. Install a bearing cup over the end of the cross and carefully place the assembly in the vise so the jaws are tightened against the bearing cup areas **(Figure 51-44)**.

5. Tighten the vise jaws so the bearing cup is partially pressed into the yoke opening **(Figure 51-45)**.

6. Remove the yoke from the vise and move the cross toward the opposite side of the yoke **(Figure 51-46)**. Install a second bearing cup over the end of the cross **(Figure 51-47)**.

Figure 51-44 Assembling a single U-joint (Cardan joint).

Figure 51-45 Assembling a single U-joint (Cardan joint).

Figure 51-43 Assembling a single U-joint (Cardan joint).

Figure 51-46 Assembling a single U-joint (Cardan joint).

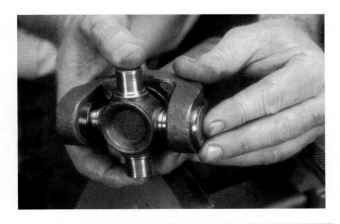

Figure 51-47 Assembling a single U-joint (Cardan joint).

Figure 51-48 Assembling a single U-joint (Cardan joint).

7. Place the slip yoke in the vise so the jaws contact the bearing cups and tighten the vise jaws to push the bearing cups into the yoke **(Figure 51-48)**.

8. Use the small socket selected to remove the bearing cups and use it to press the bearing caps to the proper depth. Be sure the bearing cups are fully installed into the slip yoke openings and install the bearing cup retaining rings.

9. Install the slip yoke and cross into the drive shaft with the index marks aligned and install the other two bearing cups and snap rings. Be sure all the snap rings are fully seated.

When disassembling a double Cardan universal joint, place index marks on all components so they can be reassembled in the same position. Inspect the ball and socket for wear, cracks, and scoring. On some drive shafts, the ball and socket can be replaced **(Figure 51-49)**. Always be sure the ball and socket are properly lubricated with the grease supplied with the replacement ball and socket.

Measuring Drive Shaft Runout

A bent drive shaft causes vibration and noise. To measure drive shaft runout, be sure the drive shaft is clean in the center area. Mount a dial indicator

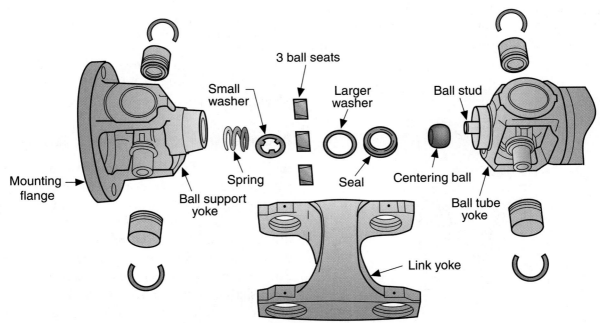

Figure 51-49 Double Cardan universal joint with replaceable ball and ball nuts.

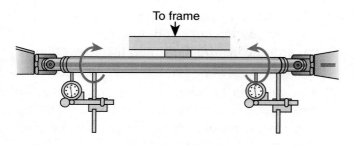

Figure 51-50 Measuring drive shaft runout.

to the chassis so the indicator stem is contacting the center of the drive shaft (**Figure 51-50**). Zero the dial indicator and slowly rotate the drive shaft by hand. The drive shaft runout indicated on the dial indicator must not exceed manufacturer's specifications. Perform drive shaft runout readings near the front and rear of the drive shaft. Replace the drive shaft if the runout is excessive.

Testing Drive Shaft Balance

Drive shaft imbalance causes undesirable vibration. An **electronic vibration analyzer** (**EVA**) may be use to detect driveshaft vibration (**Figure 51-51**). The EVA can display the frequency and strength or amplitude of the vibration (**Figure 51-52**). You would place a transducer at the end of the drive shaft for which vibration is to be checked (**Figure 51-53**). A strobe light may be used in conjunction with the EVA to check and correct drive shaft balance. Additional training is necessary to perform

Figure 51-51 An electronic vibration analyzer (EVA).

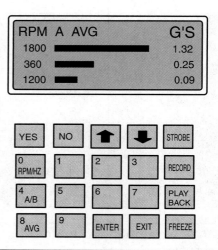

Figure 51-52 The EVA can display vibration frequency and amplitude.

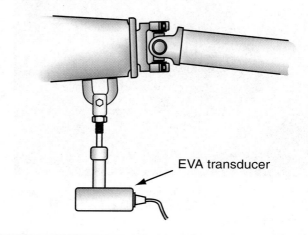

Figure 51-53 A transducer is used with the EVA to check drive shaft vibrations.

this procedure and to correctly diagnose frequency and amplitude readings that EVA is capable of displaying.

Measuring Drive Shaft Angles

The angle of the drive shaft is the difference between the angle of the drive shaft and the angles of the transmission and differential (**Figure 51-54**). Worn or sagged engine or transmission mounts or improper rear axle position may cause improper drive shaft angles. Improper drive shaft angles cause vibration. An inclinometer is used to measure drive shaft angles (**Figure 51-55**). Electronic inclinometers or electronic levels may also be used to measure drive shaft angles. The use of an electronic level is very similar to that of the inclinometer.

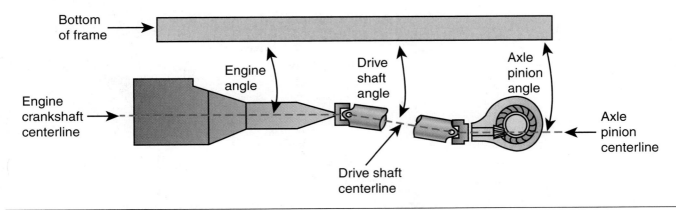

Figure 51-54 The installation angle of a drive shaft.

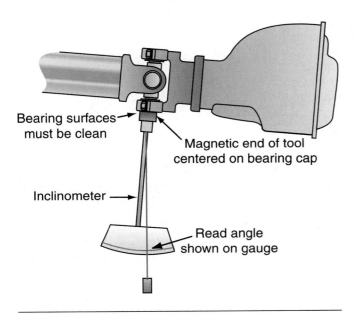

Figure 51-55 Measuring the operating angle of the rear universal joint.

DRIVE AXLE AND CV JOINT MAINTENANCE

The CV joint boots should be inspected regularly. These boots keep lubricant in the axle joints and also prevent moisture and dirt from entering the joint. Raise the vehicle on a lift and check for cracked or torn CV joint boots. Check these boots for loose or damaged clamps. The boots must be airtight. Inspect the CV joints and the areas around the joints for evidence of grease that has been thrown out of the boot. If grease has been thrown out of a CV joint, the boot is cracked or torn or the clamps are loose. If any of these boot or clamp conditions are present, the boots and clamps must be replaced. Drive axle boots may fail for any of the following reasons:

1. improper service on a suspension component
2. failure of the boot clamps or improper clamp installation
3. deterioration caused by fluid leaks or environmental damage
4. ice buildup on the chassis near the boot
5. improper vehicle connection to a tow truck
6. interference from damaged sheet metal
7. cuts or tears from objects on the road surface
8. improper drive axle removal procedure

Satisfactory drive axle boots do not ensure that the axle joints are in good condition. It is possible to have satisfactory boots and worn axle joints.

Inspect the front drive axles for damage or a bent condition. Rotate the wheels slowly and watch for a bent drive axle. Some runout is acceptable without causing vibration problems. Always refer to the vehicle manufacturer's specifications. Grasp the inner joint near the transaxle and check for movement. Excessive movement indicates worn transaxle output shaft bushings. This condition may be evidenced by oil leaks at the output shaft seals in the transaxle case. If there is excessive movement of the inner CV joint housing in the transaxle, the output bushings and seals must be replaced. If the inner CV joint housing is bolted to a flange on the transaxle, be sure the retaining bolts are tight in these components. Check for looseness of the outer CV joint in the front wheel

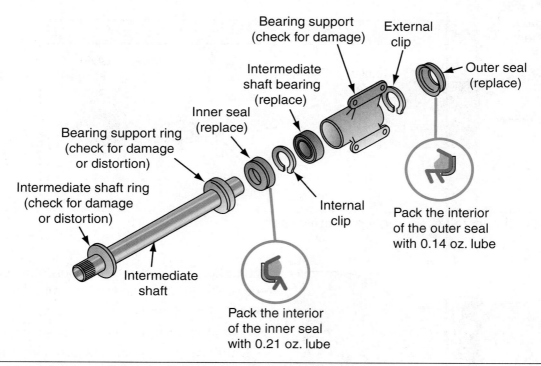

Figure 51-56 An intermediate shaft with support bearing.

hub. When the vehicle has an intermediate axle shaft, check the support bearing on this shaft for looseness **(Figure 51-56)**.

During a CV joint inspection, check all other components that could affect the drive axle and CV joint operation. Inspect the engine and transaxle mounts for wear, looseness, and deterioration. Worn mounts may cause improper engine and transaxle position that may cause abnormal CV joint angles. This condition may accelerate CV joint wear, noise, or vibration. Inspect the front suspension control arms, bushings, ball joints, sway bar and bushings, and struts for wear and looseness. Inspect the steering and brake components for looseness. Inspect the tires for proper inflation pressure and excessive or unusual wear patterns.

DRIVE AXLE AND CV JOINT DIAGNOSIS

The first step in drive axle and CV joint diagnosis is to talk to the customer and be sure the customer's complaint is identified. The next step is to road test the vehicle and verify the customer's complaint. Drive the vehicle in an area that is not congested with traffic. The road test area must also

allow the driver to perform turns at lower speeds. Drive the vehicle slowly through several sharp turns in both directions. A rhythmic clicking, popping, or clunking noise while turning a corner is usually caused by a worn outer CV joint. Listen to this noise when the vehicle is turned to the right and left. During the turn when the clicking noise is most noticeable, the worn CV joint is on the inside of the turn **(Figure 51-57)**.

If there are no unusual noises while cornering, accelerate and decelerate the vehicle. A clunking noise during acceleration or deceleration may indicate a worn inner CV joint. A worn inner CV joint, loose CV joint flange bolts, or a worn drive axle damper may also cause a shudder or shaking during engine torque changes. Accelerate the vehicle to 50 to 55 miles per hour (80 to 88 kilometers per hour). If a vibration is evident at 45 miles per hour (72 kilometers per hour) and increases with vehicle speed, it is likely caused by worn output shaft bushings in the transaxle. When the vibration pulses, the problem is likely a worn inner tripod CV joint or an out-of-balance wheel. The pulsating action is caused by worn roller tracks in the tripod joint that cause pulsating, erratic drive axle rotation. Decelerate the vehicle and coast through a sharp turn. During this driving condition, a clicking

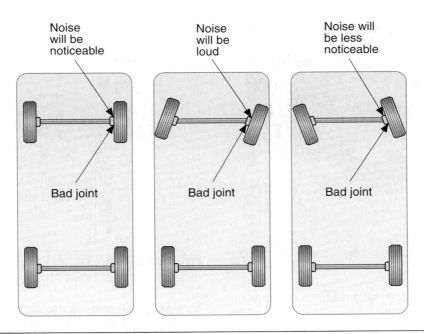

Noise will be noticeable

Noise will be loud

Noise will be less noticeable

Bad joint

Bad joint

Bad joint

Figure 51-57 A defective outer CV joint on the inside of a turn makes a noticeable noise.

noise indicates a worn outer CV joint or a defective wheel bearing. If a shudder during the turn is accompanied by a clunking noise, a worn inner CV joint is indicated. If a shudder or vibration is intermittent while driving straight ahead, a defective drive axle damper or bent drive axle may be the cause of the problem.

DRIVE AXLE AND CV JOINT SERVICE

Drive axle and CV joint service is very important to provide normal CV joint life and provide quiet, vibration-free drive axle operation. Drive axle and CV joint service includes drive axle removal and replacement, CV joint inspection on the bench, and CV joint replacement.

Removing and Installing a Drive Axle

Follow this procedure to remove and reinstall a drive axle:

1. With vehicle wheels on the shop floor, remove the axle hub nut and loosen the wheel nuts.

2. Raise the vehicle on a lift and remove the tire and wheel assembly.

3. Remove the brake caliper and use a length of wire to suspend the caliper from a chassis member. Do not allow the brake caliper to hang on the end of the brake hose.

4. Remove the brake rotor.

5. Disconnect the ends of the stabilizer bar from the lower control arms.

6. Remove the pinch bolt that retains the lower ball joint to the steering knuckle and remove the ball joint stud from the knuckle (**Figure 51-58**). If the lower ball joint is a taper type, then follow the manufacturer's recommended procedure for removal (**Figure 51-59**).

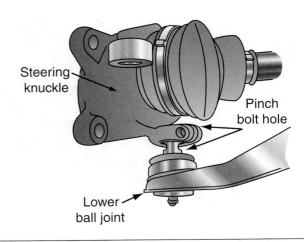

Steering knuckle

Pinch bolt hole

Lower ball joint

Figure 51-58 Removing the ball joint stud from the steering knuckle.

Figure 51-59 Removing the ball joint stud from the steering knuckle.

7. Use a slide hammer-type puller to remove the inner drive axle joint from the transaxle (**Figure 51-60**). On some vehicles it may be permissible to remove the inner CV joint by the use of a pry bar (**Figure 51-61**).

8. Remove the outer end of the drive axle from the front wheel hub. On some drive axles a special puller is required for this operation (**Figure 51-62**).

9. Remove the drive axle from the chassis and place the drive axle on the workbench.

10. Install the splines on the inner drive axle joint into the differential side gear. Push or lightly tap the joint with a soft hammer until it is fully seated in the side gear. Be sure not to damage the CV boot. On some transaxles, the inner CV

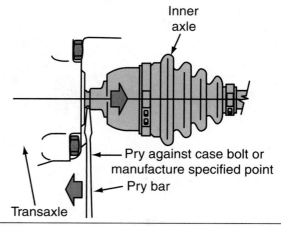

Figure 51-61 Using a pry bar to remove the inner CV joints from the transaxle.

Figure 51-62 Removing the outer end of the drive axle from the wheel hub.

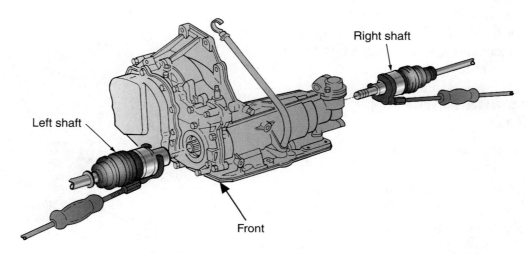

Figure 51-60 Using a slider hammer-type puller to remove the inner CV joints from the transaxle.

joint is bolted to a flange on the transaxle. If this type of flange is present, tighten the joint to flange retaining bolts to the specified torque.

11. Pull the spindle outward and slide the outer CV joint splines into the front wheel hub. Turn the drive axle slightly to align the splines on these two components. Pull the CV joint into the wheel hub as far as possible.

12. Install the ball joint stud into the steering knuckle.

13. Tighten the pinch bolt and connect the ends of the stabilizer bar to the lower control arms.

14. Install the hub nut washer and a new self-locking nut.

15. Tighten this nut as far as possible by hand.

16. Install the brake rotor and caliper. Be sure the caliper bolts are tightened to the specified torque.

17. Install the wheel and tire assembly and the wheel nuts.

18. Tighten the wheel nuts by hand.

19. Lower the vehicle onto the shop floor and tighten the hub nut to the specified torque. Some front drive axle nuts must be staked after they are tightened.

20. Tighten the wheel nuts to the specified torque.

Did You Know? In some early-model transaxles, the inner axle joints were held in the differential with circlips. The differential cover had to be removed and these circlips pulled out to allow drive axle removal.

CV Joint Inspection on the Bench

After the drive axle is removed, clamp the drive axle in a soft-jawed vice. Mark the inner end of the boot where it is positioned on the axle unless this end of the boot is mounted in a groove in the axle. Remove the boot clamps and remove the outer CV joint boot. Because the boot is always replaced, the boot may be cut from the axle shaft. Wipe the excess grease off the inner joint race and cage.

Inspect the grease for grit, which indicates dirt contamination. When the grease is contaminated with grit, the boot has been leaking at some time. If the grease has a milky appearance, it is contaminated with moisture. When the grease is contaminated with grit or moisture, the joint must be replaced. Inspect the balls, cage, inner race, and housing grooves for cracks, chips, pits, rust, and worn areas. Inspect the CV joint splines for wear, damage, and chips. If any of these conditions are present on these components, the joint must be replaced.

Remove the inner joint boot and check the grease in this joint for contamination. Inspect this tripod joint for cracks, wear, discoloration, or scoring on the spherical rollers and roller grooves in the outer housing. Check the CV joint splines for chips, damage, and wear. If any of these conditions are present, the joint must be replaced.

Removing and Replacing an Outer CV Joint and Boot

If a CV joint is in satisfactory condition, the boot may be replaced. Boot kits are available that include the boot, clamps, circlip, and lubricant (**Figure 51-63**). Whenever a CV joint is serviced, always install a new boot and clamp. Do not reuse a boot.

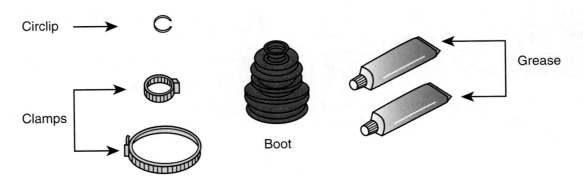

Figure 51-63 A boot kit for a CV joint.

The old boots may be cut from the drive axle kit **(Figure 51-64)**. There are two methods of retaining outer CV joints to the axle shaft. Some of these CV joints have an external ring that fits into a bearing race. This clip must be expanded to remove the CV joint from the axle. Other outer CV joints have an internal retaining clip that fits into an axle shaft groove and expands into the inner bearing race. This type of CV joint can be removed from the axle shaft by tapping it with a soft hammer. After the CV joint is removed from the axle, remove the old grease from the joint.

Clean the joint and install the new boot and all the grease supplied with the boot kit **(Figure 51-65)**. After the boot has been installed on the drive shaft and CV joint stub shaft, burp any excess air from the boot kit **(Figure 51-66)**. After the CV joint is installed on the axle, be sure the boot clamps are properly tightened **(Figure 51-67)**.

Figure 51-64 Remove the clamp and boot.

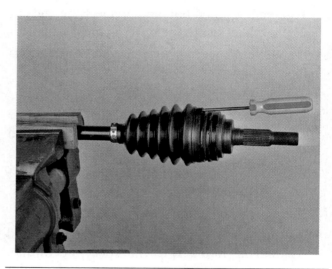

Figure 51-66 Remove any excess air from the boot.

Figure 51-65 Installing the proper amount of lubricant in the CV joint housings and boots.

Figure 51-67 Properly tighten the clamps.

Summary

- A drive shaft transmits engine torque from the transmission output shaft to the differential.

- Universal joints connect the front of the drive shaft to the front slip yoke and the rear of the drive shaft to the differential flange.

- As the drive shaft transmits torque through an angle, the drive shaft continually accelerates and decelerates.

- When universal joints in a drive shaft are on the same plane, they are in-phase.

- A squeaking noise that increases with vehicle speed may be caused by a dry, worn universal joint.

- A clanging noise when the transmission is shifted from park to drive or reverse may be caused by a worn universal joint.

- An unbalanced drive shaft causes a vibration that increases and decreases with vehicle speed.

- Before removing a drive shaft, always mark the drive shaft in relation to each yoke.

- Before disassembling a splined drive shaft, always be sure the yokes are marked in relation to the drive shaft.

- Excessive drive shaft runout, improper drive shaft angles, or drive shaft imbalance may cause vibration.

- The inner CV joint allows vertical and horizontal axle movement.

- The outer CV joint allows vertical wheel movement and rotation forward and rearward as the front wheels are steered.

- Many front-wheel drive vehicles have unequal length front drive axles.

- Some front-wheel drive vehicles have equal length drive axles and an intermediate shaft on one side.

- Torque steer is the tendency of the steering to pull to one side during hard acceleration.

- Cracked or torn drive axle boots allow moisture and dirt into the CV joints resulting in very short joint life.

- When a clicking noise is heard while cornering at low speed, one of the outer CV joints may be defective.

- If a pulsating vibration or shudder is experienced while accelerating or decelerating at medium vehicle speeds, one of the inner CV joints may be defective.

- If the grease in a CV joint is contaminated, the joint must be replaced.

- When replacing a CV boot, the grease supplied with the boot must be installed in the boot and joint housing.

Review Questions

1. Technician A says the drive shaft rotates three to four times faster than the rear wheels. Technician B says the drive shaft angle remains constant as the vehicle is driven. Who is correct?

   A. Technician A

   B. Technician B

   C. Both Technician A and Technician B

   D. Neither Technician A nor Technician B

2. Technician A says the distance between the differential and the transmission can change as the vehicle is driven. Technician B says sagged rear springs change the drive shaft angle. Who is correct?

   A. Technician A

   B. Technician B

   C. Both Technician A and Technician B

   D. Neither Technician A nor Technician B

3. Technician A says a universal joint transmits torque through an angle. Technician B says as a universal joint transmits torque through an angle the output drive shaft speed does not change as this shaft rotates. Who is correct?

A. Technician A

B. Technician B

C. Both Technician A and Technician B

D. Neither Technician A nor Technician B

4. Technician A says universal joints that are out of phase may cause a drive shaft vibration. Technician B says universal joints are in-phase when all the joints in a drive shaft are on the same plane. Who is correct?

A. Technician A

B. Technician B

C. Both Technician A and Technician B

D. Neither Technician A nor Technician B

5. All of these statements about drive shaft angles and vibration are true *except*:

A. Canceling angles between the front and rear universal joints may reduce drive shaft vibration.

B. A steeper drive shaft angle causes increased torsional vibrations.

C. When universal joints are not on the same plane, they are in-phase.

D. When a drive shaft is disassembled, the yoke should be marked in relation to the drive shaft.

6. A rear-wheel drive vehicle with a one-piece drive shaft provides a squeaking noise when accelerating from a stop. The noise frequency increases with vehicle speed. The most likely cause of this noise is:

A. a dry, worn universal joint.

B. a scored outer surface on the front slip yoke.

C. a dry, worn center support bearing.

D. a drive shaft that is out-of-phase.

7. A rear-wheel drive vehicle has a vibration between 50 and 60 mph, and this vibration changes with vehicle speed. Technician A says the drive shaft may have excessive runout. Technician B says the rear springs may be sagged resulting in an improper drive shaft angle. Who is correct?

A. Technician A

B. Technician B

C. Both Technician A and Technician B

D. Neither Technician A nor Technician B

8. All of these statements about front drive axles are true *except*:

A. The drive axles rotate at the same speed as drive shafts in rear-wheel drive vehicles.

B. Balance is critical on front drive axles.

C. Many vehicles have unequal length front drive axles.

D. Some vehicles have an intermediate shaft and equal length front drive axles.

9. On a front-wheel drive vehicle, torque steer is most noticeable:

A. during slow acceleration after the vehicle is stopped.

B. while cruising at a steady speed.

C. during hard acceleration at lower speed.

D. during deceleration from high speed.

10. Technician A says the inner drive axle joint allows inward and outward axle movement. Technician B says a tripod-type joint is used for the outer drive axle joint. Who is correct?

A. Technician A

B. Technician B

C. Both Technician A and Technician B

D. Neither Technician A nor Technician B

11. A double Cardan universal joint has a ball and socket between the front _____ _____ and the rear _____.

12. A dry, worn universal joint may cause a
_____ noise whose
_____ increases with vehicle
speed.

13. A worn universal joint may cause a
_____ noise when the
transmission is shifted from park to drive.

14. An unbalanced drive shaft may cause a vibration that _____ and
_____ with vehicle speed
changes.

15. Explain the causes of improper drive shaft angles.

16. Describe the drive shaft runout measurement procedure.

17. Describe the result of a worn outer CV joint in a front-wheel drive vehicle.

18. Describe the results of worn engine mounts on a front-wheel drive vehicle.

CHAPTER 52

Heating, Ventilation, and Air Conditioning Systems

Learning Objectives

After you have read, studied, and practiced the contents of this unit, you should be able to:

- Explain the effects of R-12 refrigerant on the earth's ozone layer.

- Describe the harmful effects of excessive ultraviolet radiation on humans.

- Describe the refrigerant oil used with R-134a refrigerant.

- Explain the SAE standards for the purity of R-12 and R-134a recycled refrigerant.

- Describe how the air temperature entering the passenger compartment is controlled in an HVAC system.

- Explain latent heat of vaporization.

- Explain latent heat of condensation.

- Describe the changes in the refrigerant state as it flows through the refrigeration system.

- Explain the basic operation of a compressor clutch.

- Explain how the temperature door is controlled in a manual air conditioning (A/C) system.

- Describe how the temperature door is controlled in a semi-automatic A/C system.

- Explain how the blower speed is controlled in an automatic A/C system.

- Describe the results of excessive compressor clutch clearance.

Key Terms

Air conditioning

Air Conditioning and Refrigeration Institute (ARI)

Cabin filter

Chlorofluorocarbon (CFC)

Cycling clutch orifice tube system (CCOT)

Heating, ventilation, and air conditioning (HVAC)

Hydro fluorocarbon (HFC)

Latent heat of condensation

Latent heat of vaporization

Orifice tube (OT)

Synthetic lubricant

Thermostatic expansion valve (TXV)

INTRODUCTION

The term **air conditioning** refers to the control of air temperature, movement of air, control of humidity, and the filtration of air. Many homes use a full air conditioning system that controls all of these operations. While some homes may not control humidity during the heating (winter) season, when the humidity level usually drops, they do indirectly control humidity during the cooling (summer) season. The airconditioning system will remove moisture from the air as a secondary function of cooling the air.

The temperature control system in a vehicle is called a **heating, ventilation, and air conditioning (HVAC)** system. During hot weather operation, heat in the vehicle interior comes from several sources, such as heat from the sun radiating on the vehicle glass and painted surfaces. Heat also enters the vehicle interior from the engine and exhaust system, especially the catalytic converter. Heat also comes from the outside air, and passengers add to the interior vehicle heat mostly from their breath. Therefore in hot weather, the HVAC system must cool and ventilate the vehicle interior to maintain passenger comfort.

In cold weather, the HVAC system must heat the vehicle interior for passenger comfort. In hot or cold weather, the HVAC system provides ventilation by forcing outside air into the vehicle interior. Air from the vehicle interior can be recirculated through the HVAC system back into the interior of the vehicle. Some vehicles also provide for the filtration of air entering the cabin or passenger compartment. These air filters are usually referred to as cabin filters.

HEATING SYSTEM DESIGN

In an HVAC system, the heating system contains a heater core mounted with the air conditioning (A/C) evaporator in the HVAC case. If the vehicle does not have A/C, the evaporator is omitted from this case. Hoses are connected from the heater core to the engine cooling system to provide coolant flow through the heater core.

Temperature controls on some systems use a control valve that regulates the coolant flow through the heater core **(Figure 52-1)**. The coolant control valve may be operated by vacuum or by a mechanical cable. Because the engine thermostat maintains coolant temperature, proper thermostat operation

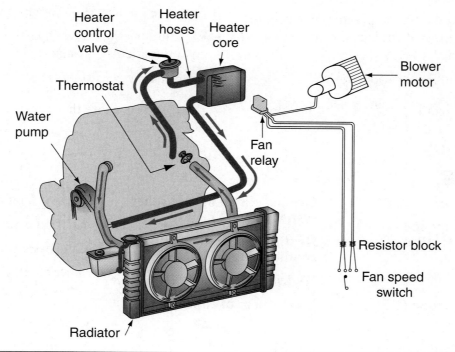

Figure 52-1 A heating system.

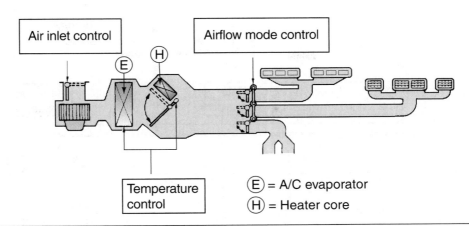

Figure 52-2 HVAC mode control doors.

is very important to provide adequate heating of the vehicle interior. The blower motor moves air from the outside into the HVAC case.

Other vehicles control the temperature through the use of a temperature door. The position of the temperature door determines the airflow through the evaporator and the heater core **(Figure 52-2)**. In a manual HVAC system, a cable connected to the temperature control lever in the HVAC control panel operates the temperature door.

NOTE: The temperature door may be called a *blend-air door*.

In most automatic HVAC systems, the temperature door is operated by an electric motor. In the maximum heat position, the temperature door is positioned so all the air flows through the evaporator and then through the heater core. Under this condition, maximum interior heating is provided. In the maximum A/C mode, the temperature door is positioned so it blocks airflow through the heater core but the air flows through the evaporator.

During normal HVAC operation, the air flows through the evaporator and the temperature door is positioned so a portion of the air flows through the heater core. Mode control doors in the HVAC case determine whether the air is directed from this case to the A/C outlets, floor outlets, or defrost outlets. Vacuum actuators or electric motors usually operate mode doors. On some manually controlled

systems, cables operate the mode doors. A recirculation door determines whether air flows from the outside into the HVAC case or from the vehicle interior into this case **(Figure 52-3)**.

CABIN FILTERS

Pollen and harmful substances inside a vehicle can be up to six times more concentrated than outside. Vehicle cabin filters can remove pollen and harmful substances from entering the vehicle interior and keep passengers protected.

A **cabin filter** works much like the filter used with your furnace and air conditioning system in your home. The filter is usually located in the air conditioning system of the vehicle and filters the air before it reaches the evaporator **(Figure 52-4)**. The cabin filter can remove most of the airborne contaminants before they enter the interior of your vehicle through the heating and air conditioning system. The filter performs "mechanical filtration." This means that fiber material with pores of a specific size will trap particles larger than the size of the pores.

The cabin filter is a type of air filter that has a large surface area and many times looks like an engine air filter. Some cabin filters may contain additional components such as a charcoal layer. This charcoal layer can filter out some odors and chemical compounds before they enter the passenger compartment. The charcoal is "activated," which means it is treated with chemicals and heat to give it specific odor-control properties.

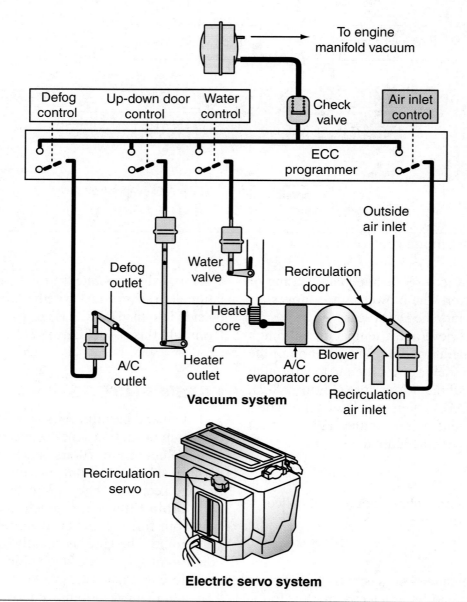

Figure 52-3 HVAC system recirculation door.

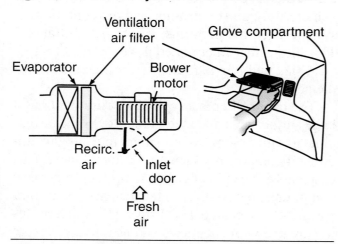

Figure 52-4 HVAC cabin filter.

HEAT ABSORPTION PRINCIPLES

Materials may be in a liquid, solid, or gaseous state. When materials change from one state to another, large amounts of heat may be transferred. If the temperature of water is reduced to 32°F (0°C), the water changes from a liquid to a solid (ice). Ice at 32°F (0°C) requires heat to change it into water, which will also be at 32°F (0°C). When the water temperature is increased to 212°F (100°C), it requires an additional 970 British thermal units (BTUs) of heat per pound to make the water boil. This additional heat does not register on a thermometer. This additional heat is called latent heat of vaporization.

Heat is defined as the rapid motion of atoms. As heat is added the atoms in a given material will move or vibrate faster. Cooling may be defined as taking away heat. As materials have their heat removed, the atoms move or vibrate slower.

During evaporation, heat is absorbed. A British thermal unit (BTU) is the amount of heat required to raise the temperature of 1 pound of water 1°F. **Latent heat of vaporization** is the amount of heat required to change a liquid to a gas after the liquid reaches the boiling point. When a gas condenses and becomes a liquid, it releases its latent heat. When steam condenses into water, 970 BTUs per pound are released. The heat released during condensation is called **latent heat of condensation**. Refrigerant rather than water is used in the air conditioning system, but the principles are the same. The latent heat of vaporization and the latent heat of condensation of refrigerant are principles used in any air conditioning system.

AIR CONDITIONING REFRIGERANTS AND OILS

Technicians need to understand the serious effects refrigerants have on the earth's ozone layer. You must also have knowledge of the regulations and standards regarding these refrigerants. Refrigerant system retrofitting procedures must also be familiar to service technicians as many of these vehicles are still on the road.

R-12 and the Ozone Layer

R-12 was introduced as a refrigerant during the 1930s. At that time it was considered the perfect refrigerant because it is non-toxic, easily produced, cheap to manufacture, and a very stable chemical.

However, scientific tests indicate that we may have a high penalty to pay for the comfort we have enjoyed as a result of automotive air conditioning. R-12 is a **chlorofluorocarbon (CFC)**-based chemical and scientific tests prove that CFCs are responsible for damaging the earth's protective ozone layer located in the stratosphere 10 to 30 miles

above the planet's surface. The ozone layer filters out most of the sun's harmful ultraviolet (UV) rays.

EPA sources indicate that 30 percent of all the CFCs released into the atmosphere come from mobile air conditioners. Some of these CFCs escape into the atmosphere from refrigerant system leaks, but most of the CFCs are released into the atmosphere during air conditioning service. When CFCs are released to the atmosphere, they travel high into the stratosphere where they can linger for 100 years or more. CFCs destroy the ozone layer through chemical reaction. When influenced by sunlight, a chlorine atom is released from a CFC molecule and reacts with an oxygen atom in the ozone layer to form chlorine monoxide and free oxygen, neither of which can filter out the sun's UV rays **(Figure 52-5)**.

For each one percent of ozone reduction, one and one-half to two percent more UV radiation reaches the earth's surface. The EPA suggests that the risks associated with exposure to excessive UV radiation include increased incidence of skin cancer, an increase in the number of eye cataracts, and damage to immune systems. EPA studies also indicate that excessive UV radiation damages vegetation, adversely affects organisms living in the oceans, and increases ground-level ozone, which is a contributor to smog. Research indicates that excessive UV radiation can be harmful to crops and plankton in the oceans. Plankton is floating or

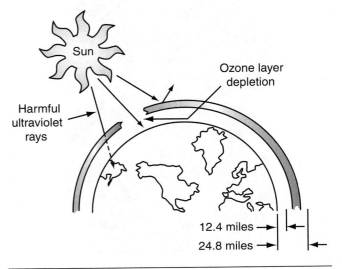

Figure 52-5 The ozone layer filters out harmful UV radiation.

weakly swimming animal and plant life, which some marine animals use for food. Loss of this source of food could have serious effects on marine life.

R-134a Refrigerant

Automotive manufacturers began to install R-134a, an ozone-friendly refrigerant, in new vehicles beginning in 1992. R-134a is classified as a **hydro fluorocarbon (HFC)**. Although R-134a does not contain CFCs, it is a greenhouse gas. R-134a is tetrafluoroethane and is a non-flammable, colorless gas.

By 1994, the automotive manufacturers completed the transition to R-134a refrigerant on all new vehicle production. The amendments to the Clean Air Act effective January 1, 1992, required certification for technicians servicing automotive air conditioners. Certification was provided through the ASE.

The Clean Air Act amendments also demanded use of approved refrigerant recycling equipment. Mandatory recovery of R-134a refrigerant was required after November 15, 1995 and venting of R-134a to the atmosphere became illegal. Recycling of R-134a using EPA-approved equipment became mandatory on January 29, 1998 **(Figure 52-6)**.

Figure 52-6 A typical EPA approved recovery/recycling unit.

Refrigerant Oils

Oil is used in the air conditioning system to lubricate the compressor. Mineral oils are used in R-12 refrigerant systems, but mineral oil is not compatible with R-134a refrigerant. Two new synthetic lubricants, poly-alkylene glycol (PAG) oil and polyol ester (ester) oil, have been developed for R-134a refrigerant. A **synthetic lubricant** is one that is developed in a laboratory.

When selecting a refrigerant oil for an R-134a refrigerant system, always follow the vehicle manufacturer's or compressor manufacturer's instructions. New vehicles usually have PAG oil in the R-134a refrigerant. Installing the proper refrigerant oil has a direct bearing on system performance and compressor life.

> **Tech Tip** *PAG and ester oils are hygroscopic, which means they easily absorb moisture from the atmosphere. Always keep oil containers tightly sealed and in a dry place.*

> **SAFETY TIP** *PAG and ester oils may damage painted surfaces and plastic parts.*

REFRIGERANT RECOVERY AND RECYCLING STANDARDS

The SAE has drafted a number of standards that apply to the recovery and recycling of refrigerant. SAE standard J2210 regulates the recovery of R-134a. These standards provide specifications for the equipment hardware such as hoses and fittings. Service hoses must have shut-off valves within 12 inches of the hose ends to prevent the unnecessary release of refrigerant.

The SAE standard for purity of R-134a requires a limit of 15 parts per million (ppm) moisture by weight, 500-ppm refrigerant oil by weight, and 150-ppm non-condensable gases (air) by weight. Recovery/ recycling equipment must be tested by an independent standards organization to be sure these standards are met.

SAE standards are also established for R-134a recovery-only machines. This type of machine cannot recycle refrigerant or recharge a system. The recovery equipment is used in operations such as salvage yards. Recovered refrigerant must be shipped to an off-site facility where the refrigerant is reclaimed not recycled. Reclaimed refrigerant must meet a like-new purity standard referred to as AR1700-93 and is established by the **Air Conditioning and Refrigeration Institute (ARI)**.

On recovery/recycling machines for R-134a, the low-side hose is solid blue with a black stripe, the high-side hose is solid red with a black stripe, and the utility hose is yellow with a black stripe. To reduce the possibility of mixing R-12 and R-134a refrigerant while servicing refrigerant systems, R-12 has 7/16 inch–20 fittings for connection to manifold gauges or recovery/recycling equipment, whereas R-134a systems must have 0.5 inch–16 ACME fittings at these locations.

REFRIGERATION SYSTEM OPERATION AND CONTROLS
Operating Principles

When the compressor clutch is engaged and the compressor is operating, it pressurizes gaseous refrigerant. A tube conducts the refrigerant to the condenser. As the refrigerant flows down through the condenser, the airflow through the condenser cools the refrigerant and this causes the refrigerant to condense into a liquid. The refrigerant transfers a large amount of heat to the air flowing through the condenser during the condensation process.

The liquid refrigerant flows through the tube connected from the condenser to the receiver/drier. The receiver/drier filters and removes moisture from the refrigerant on the high side. Some refrigerant systems have an accumulator on the low side between the evaporator and the compressor in place of a receiver/drier.

> **Tech Tip** *The high side of the system is between the compressor, through the condensor, and to the TXV or OT. The high side lines are smaller than the low side lines.*

The refrigerant flows through a tube connected between the receiver/drier and the evaporator.

Either a **thermostatic expansion valve (TXV)** or an **orifice tube (OT)** is located in the evaporator inlet tube. The expansion valve or the OT provides a restriction for the refrigerant flow. The orifice tube is used in conjunction with a pressure switch mounted on the accumulator and is used to cycle the compressor clutch on and off. This is referred to as the **cycling clutch orifice tube system (CCOT)**. The restriction of the TXV or the OT reduces the refrigerant pressure. The liquid refrigerant immediately boils into a vapor when the pressure is reduced. When this boiling action takes place, the refrigerant absorbs a large amount of heat from the air flowing through the evaporator into the vehicle interior and this provides cooling for the vehicle interior.

> **Tech Tip** *The low side is between the TXV or OT, through the evaporator, to the compressor. The low side lines are larger than the high side lines.*

A refrigeration system with a thermostatic expansion valve (TXV) is illustrated in **Figure 52-7**, and **Figure 52-8** shows a refrigeration system with an orifice tube. Notice that the orifice tube (OT) system has an accumulator, whereas the TXV system has a receiver/drier. The refrigeration principles are the same in nearly all air conditioning systems.

Compressors

A V-belt or a multi-ribbed belt drives the A/C compressor **(Figure 52-9)**. Drive belt condition and tension are very important to provide proper compressor operation.

There are many different types of air conditioning compressors **(Figure 52-10)**. The main categories of compressors are reciprocating and rotary. Reciprocating compressors use pistons and come in many variations. Rotary compressors can use vanes, scrolls, screws, or centrifugal.

In a conventional piston compressor, a crankshaft moves a piston up and down in the compressor cylinder. It is constructed in a similar fashion as an engine. Inlet and outlet reed valves are located in the cylinder head. When the piston

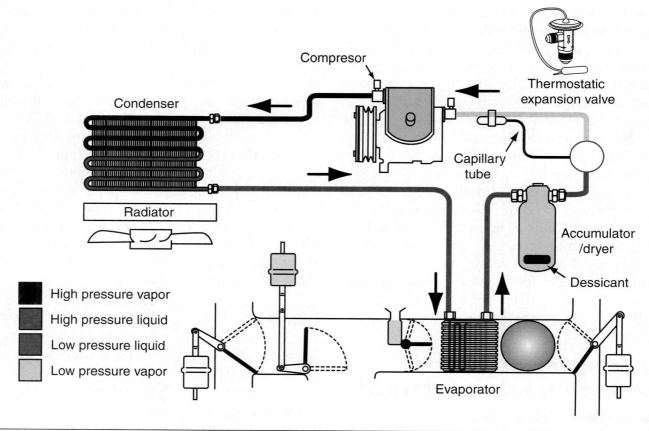

Figure 52-7 Refrigeration system with a thermostatic expansion valve.

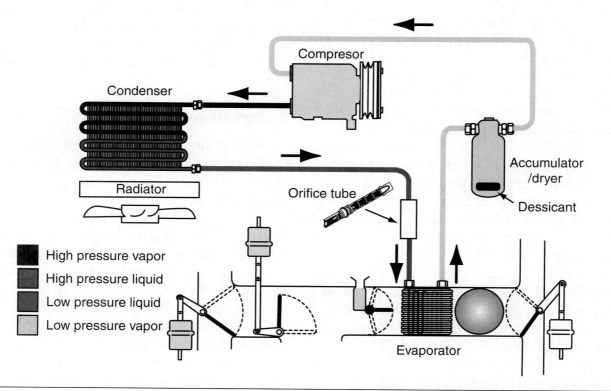

Figure 52-8 Refrigeration system with an orifice tube.

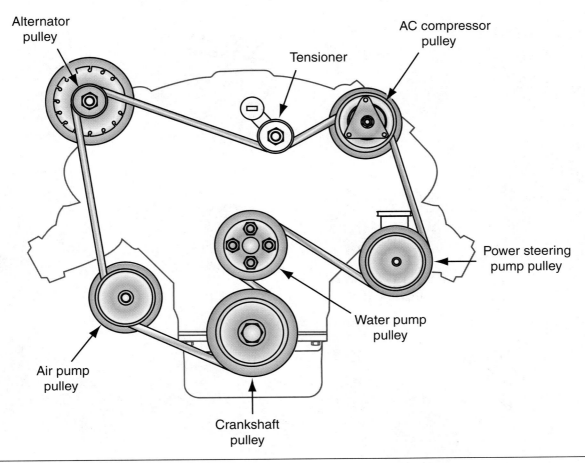

Figure 52-9 A/C compressor drive belt.

moves down, vacuum in the cylinder moves refrigerant through the inlet valve into the cylinder **(Figure 52-11)**. As the piston moves upward in the cylinder, the refrigerant is pressurized and forced out through the outlet valve.

The radial compressor is a different type of piston compressor. The pistons are mounted radially in the compressor body. These pistons have a slotted center area that fits over an eccentric plate attached to the compressor drive shaft. As the drive shaft rotates the eccentric plate, this plate moves the pistons back and forth in their radial cylinders **(Figure 52-12)**. Reed-type outlet valves are located in the ends of the cylinders and inlet valves are located in the pistons themselves.

Another variation is the variable displacement compressor. The variable displacement compressor can change displacement to better control air conditioner cooling capability and reduce the power needed to operate. This compressor changes its displacement by changing the length of the pistons stroke. It does this by changing the angle of the swash plate that moves the pistons back and forth. There is a control valve that controls the swash plate **(Figure 52-13)**.

Vane compressors work due to the effect of the vanes located inside the cylinders. The vanes rotate on a cam in the center of the compressor. The vanes are responsible for the movement of the gas and operate in a similar fashion as a vane type power steering pump **(Figure 52-14)**.

Scroll compressors use two scrolls. One scroll is stationary, and the other orbiting (but not rotating) around the first. Thanks to this motion, the gas contained between the two elements reaches a very

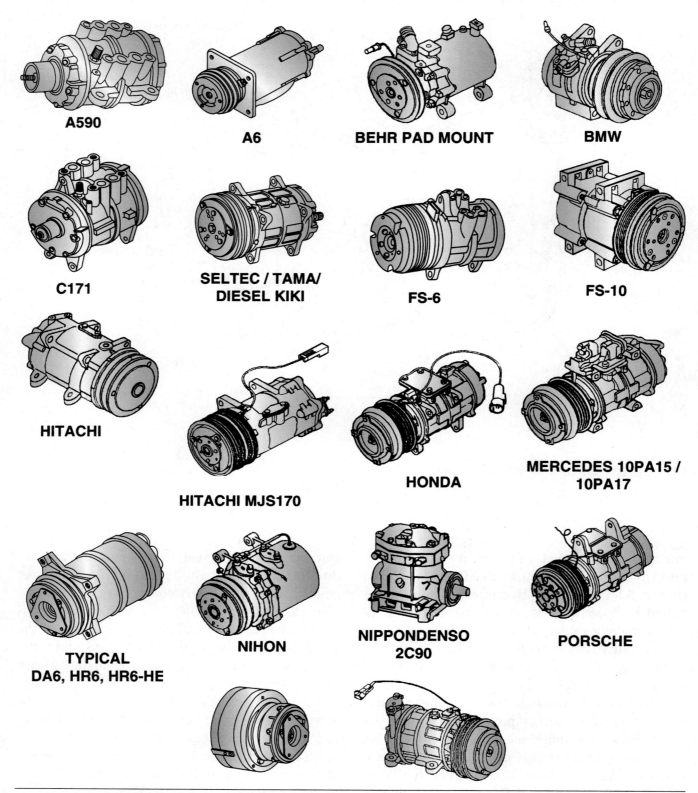

Figure 52-10 Some of the many compressors used in automobile air conditioning.

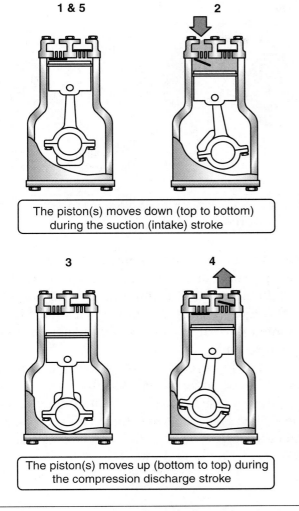

1 & 5

2

The piston(s) moves down (top to bottom) during the suction (intake) stroke

3

4

The piston(s) moves up (bottom to top) during the compression discharge stroke

Figure 52-11 A conventional piston compressor.

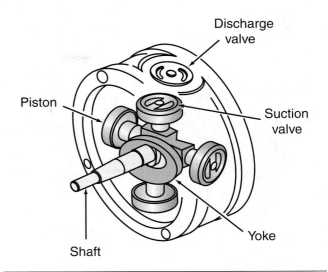

Discharge valve

Piston

Suction valve

Yoke

Shaft

Figure 52-12 A radial type compressor.

high pressure and is discharged through a hole in the center **(Figure 52-15)**.

Screw compressors operate much like a supercharger and a centrifugal compressor operates much like the compressor section of a turbocharger. These compressors are not used in automotive air conditioning but are usually used only in very large commercial and industrial air conditioning systems.

Compressor Clutches

A compressor clutch on the front of the compressor shaft is used to engage and disengage the

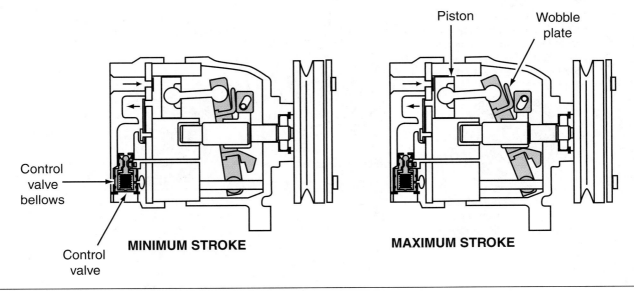

Piston

Wobble plate

Control valve bellows

Control valve

MINIMUM STROKE

MAXIMUM STROKE

Figure 52-13 A variable displacement piston compressor.

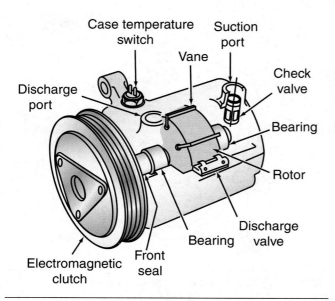

Figure 52-14 A vane-type compressor.

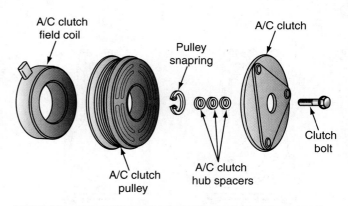

Figure 52-16 A/C compressor clutch components.

compressor. The clutch contains a large field coil that is pressed onto the front compressor housing and retained with a snap ring.

The compressor clutch pulley is mounted on a double-row bearing that is also pressed onto the front compressor housing. The pulley is also retained with a snap ring. A clutch hub and drive

plate is mounted on the compressor drive shaft and a key prevents this hub from turning on the shaft. A locknut retains the hub and drive plate to the shaft **(Figure 52-16)**. The drive plate on the hub is mounted on leaf springs so this plate can move toward the pulley.

If the clutch coil is not energized, then the clutch is not engaged and the drive belt rotates the pulley on the bearing. When voltage is supplied to the clutch coil the clutch coil is energized, and current flow through this coil creates a strong

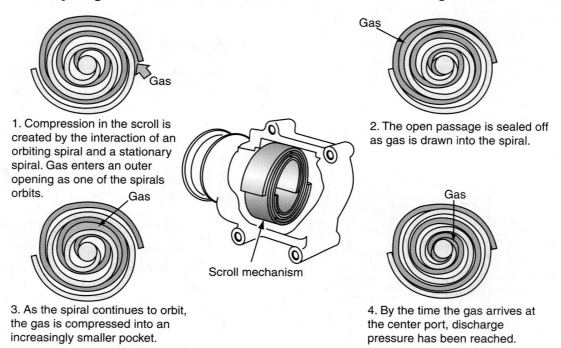

1. Compression in the scroll is created by the interaction of an orbiting spiral and a stationary spiral. Gas enters an outer opening as one of the spirals orbits.

2. The open passage is sealed off as gas is drawn into the spiral.

3. As the spiral continues to orbit, the gas is compressed into an increasingly smaller pocket.

4. By the time the gas arrives at the center port, discharge pressure has been reached.

Figure 52-15 A scroll-type compressor.

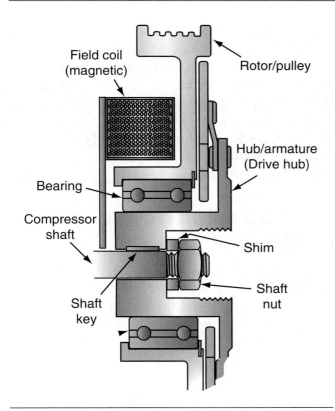

Figure 52-17 A cutaway view of an A/C compressor clutch.

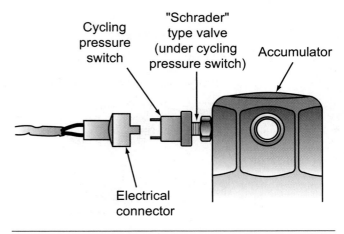

Figure 52-18 Accumulator and cycling switch.

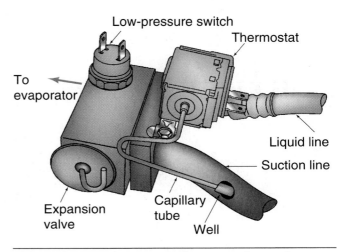

Figure 52-19 Thermostatic expansion valve with thermostatic switch.

magnetic field. This magnetic field attracts the drive plate toward the pulley surface and holds this plate firmly against the pulley. Under this condition, the drive plate, hub, and compressor drive shaft must rotate with the pulley **(Figure 52-17)**. The clearance between the drive plate and the pulley is adjustable by removing or installing shims between the inner end of the hub and the compressor shaft.

In an orifice tube system, the clutch is usually cycled on and off by a pressure switch in the accumulator **(Figure 52-18)**. The normally closed cycling switch cycles the compressor on and off to provide adequate passenger compartment cooling without excessive ice formation in the evaporator.

In a TXV system, the compressor clutch is cycled on and off by a thermostatic switch. A capillary tube is connected from the thermostatic switch so it contacts the suction line between the evaporator and the compressor **(Figure 52-19)**. The thermostatic switch cycles the compressor on and off in relation to evaporator outlet temperature. This switch performs the same function as the cycling

switch in the accumulator. A low-pressure switch in the TXV shuts the compressor off if the refrigerant pressure becomes excessively low because of a refrigerant leak. This protects the compressor from lack of lubrication as the refrigerant circulates the oil through the system.

MANUAL A/C SYSTEMS

In a manual A/C system, a cable is connected from the temperature lever on the A/C control panel to the temperature door. In some older systems, cables are connected from the other control panel levers to the mode control doors on the A/C heater case **(Figure 52-20)**. In other manual A/C systems, the push buttons in the A/C control panel

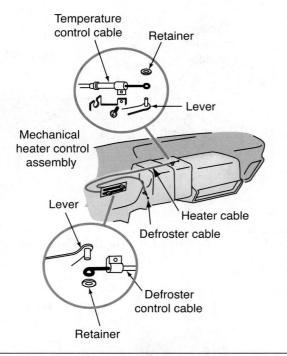

Figure 52-20 Manual A/C system control cables.

operate vacuum switches that shut the vacuum on and off to the mode door actuators.

The blower speed control switch switches different resistors into the blower circuit to provide the desired blower speed. When high blower speed is selected, the blower switch supplies voltage to the high blower relay winding. This action closes the relay contacts and 12 volts are supplied through the relay contacts to the blower motor **(Figure 52-21)**.

SEMI-AUTOMATIC A/C SYSTEMS

In a semi-automatic A/C system, the temperature control lever operates a variable resistor. A servomotor rotates the temperature door and an electronic module is located in the servomotor. An in-car sensor is mounted in the instrument panel and an out-

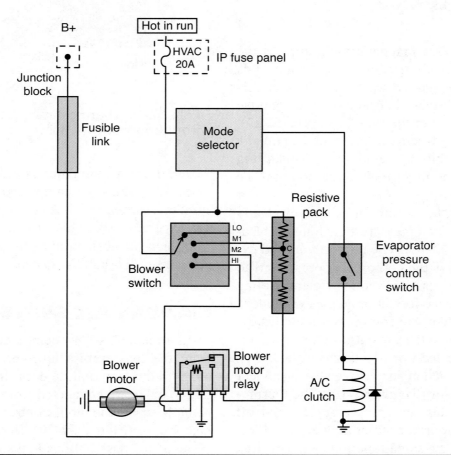

Figure 52-21 Blower motor circuit.

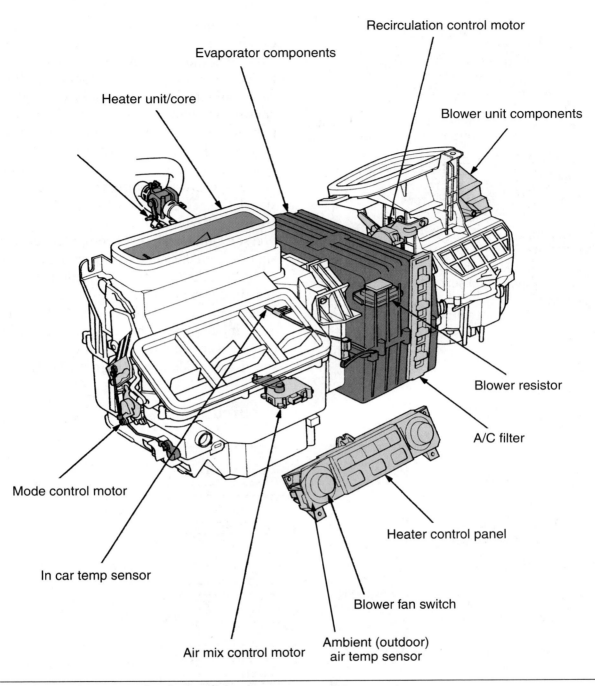

Figure 52-22 Semi-automatic air conditioning system components.

side air temperature sensor is located on the outside surface of the A/C heater case **(Figure 52-22)**.

If the temperature lever is moved to a higher temperature, the resistance of the variable resistor increases. This variable resistor and the sensors send voltage signals to the module in the temperature door servomotor. In response to these signals, the servomotor operates the temperature door to provide the temperature selected by the driver. The other control functions, such as blower motor, are similar to those on a manual system.

AUTOMATIC A/C SYSTEMS

In an automatic A/C system, the A/C computer operates all the functions. The temperature door is operated by an electric motor that is controlled by the A/C computer. The A/C computer may be

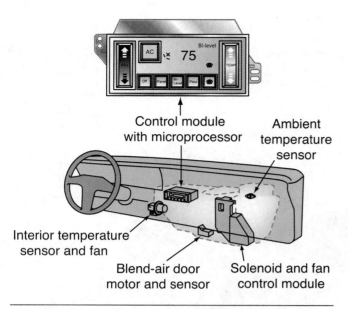

Figure 52-23 A/C control panel and computer.

mounted in the A/C control panel **(Figure 52-23)**, or this computer may be mounted separately.

In most automatic A/C systems, the mode doors are operated by electric motors controlled by the A/C computer. A blower speed controller usually controls the blower motor, but the computer commands the blower speed controller to provide the necessary blower speed. On many automatic A/C systems, the blower speed controller controls the blower speed with a pulse width modulated (PWM) signal.

The A/C computer also operates the A/C compressor clutch. The A/C computer receives input signals from the sunload sensor, in-car sensor temperature sensor, ambient (outdoor temperature) sensor, engine coolant temperature sensor, driver input, and A/C pressure switch. In response to these input signals, the A/C computer controls the compressor clutch, temperature door, blower speed, and mode doors **(Figure 52-24)**.

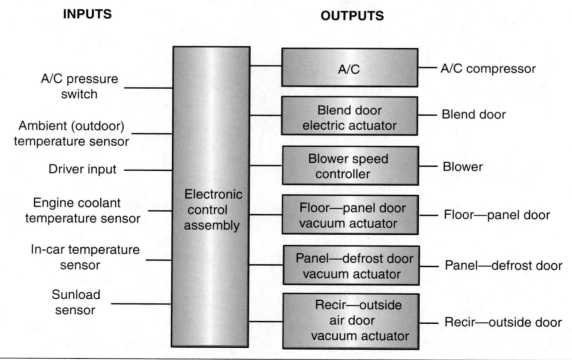

Figure 52-24 A/C computer inputs and outputs.

Summary

- The heater coolant control valve is usually operated by a cable or by a vacuum actuator.
- The position of the temperature door determines whether inlet air flows through the heater core or the evaporator.
- The mode control doors direct the inlet airflow through the floor outlets, A/C outlets, or defrost outlets.
- Vacuum actuators or electric motors usually operate the mode control doors.
- Latent heat of vaporization is the heat required to change a liquid to a vapor.
- Latent heat of condensation is the heat released when a vapor condenses into a liquid.
- R-12 refrigerant contains chlorofluorocarbons (CFCs) that are responsible for damaging the earth's ozone layer.
- Damage to the earth's ozone layer causes excessive UV radiation to reach the earth's surface.
- Excessive UV radiation contributes to skin cancer, eye cataracts, damage to crops, and to some forms of marine life.
- R-12 refrigerant has been replaced by R-134a on all new vehicles since 1994.
- Venting of any refrigerant to the atmosphere is illegal and refrigerant recovery and recycling is now mandatory.
- Refrigerant identifiers are available to inform the technician regarding the type of refrigerant in a system.

- The SAE has drafted standards for refrigerant recovery/recycling machines and the purity of recovered refrigerant.
- R-134a refrigerant systems have different service fittings compared to R-12 refrigerant systems.
- The compressor pressurizes the refrigerant and forces it through the system.
- The refrigerant changes from a vapor to a liquid in the condenser.
- An orifice tube or thermostatic expansion valve is mounted at the evaporator inlet.
- When refrigerant flows through the orifice tube or thermostatic expansion valve, the pressure of the refrigerant is reduced and the refrigerant immediately boils and changes from a liquid to a gas.
- The compressor clutch connects and disconnects the pulley from the compressor drive shaft.
- In a manual A/C system, a cable connected to the temperature lever operates the temperature door.
- In a semi-automatic A/C system, the temperature door is operated by an electric motor that is controlled by a module. This module receives inputs from the variable resistor connected to the temperature lever, the in-car sensor, and the outside air temperature sensor.
- In an automatic A/C system, the A/C computer operates the temperature door, mode doors, blower speed, and compressor clutch.

Review Questions

1. Technician A says it is legal to vent R-134a to the atmosphere. Technician B says R-134a refrigerant is a greenhouse gas. Who is correct?

    A. Technician A

    B. Technician B

    C. Both Technician A and Technician B

    D. Neither Technician A nor Technician B

2. Technician A says no oil is necessary for 134a refrigerant. Technician B says 134a refrigerant is illegal in many states. Who is correct?

    A. Technician A

    B. Technician B

    C. Both Technician A and Technician B

    D. Neither Technician A nor Technician B

3. Technician A says the temperature door controls the airflow through the heater core and evaporator. Technician B says a cable may operate the temperature door. Who is correct?

    A. Technician A

    B. Technician B

    C. Both Technician A and Technician B

    D. Neither Technician A nor Technician B

4. All of these statements about SAE standard J2210 regulating the recovery of R-134a are true *except*:

   A. a limit of 15 million ppm moisture by weight.

   B. a limit of 500-ppm refrigerant oil by weight.

   C. a limit of 150-ppm non-condensable gases (air) by weight.

   D. a limit of 1500-ppm refrigerant by weight.

5. When refrigerant flows through the condenser:

   A. the refrigerant is cooled and the refrigerant pressure increases.

   B. the refrigerant condenses from a gas to a liquid.

   C. the refrigerant pressure decreases and the refrigerant temperature increases.

   D. the refrigerant absorbs heat from the airflow through the condenser.

6. In a refrigeration system with a TXV, the receiver/drier is connected:

   A. between the condenser and the evaporator.

   B. between the compressor and the condenser.

   C. between the evaporator and the compressor.

   D. at the evaporator inlet.

7. The clearance between the compressor pulley drive plate and the pulley may be adjusted by:

   A. loosening or tightening the pulley retaining nut.

   B. removing or installing shims between the inner end of the pulley hub and the compressor shaft.

   C. installing a pulley with a different pulley hub length.

   D. installing shims between the pulley retaining nut and the pulley.

8. In a refrigeration system with an orifice tube, the compressor clutch is cycled on and off by:

   A. a high-pressure cut-off switch in the compressor.

   B. a low-pressure cut-off switch in the compressor.

   C. a pressure cycling switch in the accumulator.

   D. a dual contact switch in the evaporator inlet.

9. In a fully automatic A/C system, a blower speed controller controls the blower speed using a:

   A. pulse-width modulated voltage signal.

   B. a low-voltage DC signal.

   C. a high-voltage AC signal.

   D. a low-digital voltage signal.

10. Technician A says when the A/C compressor clutch is engaged, the drive plate is held against the pulley surface. Technician B says current flow through the compressor clutch coil engages the compressor clutch. Who is correct?

    A. Technician A

    B. Technician B

    C. Both Technician A and Technician B

    D. Neither Technician A nor Technician B

11. In an orifice tube refrigeration system, the compressor clutch is cycled on and off by the

    _____ _____.

12. In a TXV refrigeration system, a capillary tube is connected from the thermostatic switch to the _____

    _____ .

13. If the pressure becomes excessively low in a TXV refrigeration system, the compressor is shut off by _____

    _____ _____ .

14. In a semi-automatic A/C system, the temperature lever is connected to a

    _____

    _____.

15. Explain how the blower speed is controlled in an automatic A/C system.

16. Describe the temperature door position in the maximum A/C mode.

17. Explain the latent heat of vaporization.

18. Describe the changes that occur in the refrigerant as it flows through an orifice tube refrigerant system.

CHAPTER

53

Heating, Ventilation, and Air Conditioning System Service

Learning Objectives

After you have read, studied, and practiced the contents of this unit, you should be able to:

- Describe the basic maintenance to perform on the heating system.

- Describe the requirements to perform service work on an air conditioning (A/C) system.

- Describe the procedure to complete a performance test on an A/C system.

- Explain the causes of excessive high-side pressure in a refrigerant system.

- Explain the procedure for refrigerant recovery.

- Describe the procedure for evacuating a refrigeration system.

- Explain the procedure for charging a refrigeration system.

Key Terms

Charging station

Compound gauge

Desiccant

Schrader-type service valves

Static pressure

Vacuum pump

INTRODUCTION

Proper heating and air conditioning system maintenance, diagnosis, and service are extremely important to maintain passenger comfort. If the A/C system is properly maintained and serviced, it will provide the interior vehicle temperature desired by the driver and passengers.

When the A/C system is not properly maintained and serviced, it may suddenly quit working and this causes discomfort and frustration for the driver and passengers. Technicians must understand the proper heater and A/C maintenance, diagnosis, and service procedures to properly maintain these systems and provide the system dependability expected by the customer.

AIR CONDITIONING CERTIFICATION

Since January 1, 1993, any person repairing or servicing motor vehicle air conditioners shall certify to the EPA that such person has acquired, and is properly using, approved equipment, and that each individual is authorized to use the equipment. This requirement that the technician is properly trained and certified is mandated under Section 609 of the Clean Air Act. In addition, only Section 609 Certified Motor Vehicle A/C technicians can purchase refrigerants in any size container including containers of 20 pounds or less. Section 609 applies only to MVAC (mobile vehicle air conditioning). MVAC is defined as those air conditioning systems used to cool passengers in a vehicle.

The EPA has an online test that may be accessed at the following website: http://www.epatest.com/e_609cert.html. The technician must answer 21 of 25 questions correctly and submit the appropriate test fee. ASE also has an online test and may be accessed at http://www.asecert.org. The technician must answer 24 of 30 questions correctly and submit the appropriate fee.

Some of the main requirements of section 609 are as follows:

- Since August 13, 1992, persons repairing or servicing air conditioning systems on vehicles must be properly trained and certified by an EPA-approved 609 certification program.

- Since November 15, 1995, it is illegal to vent substitutes for CFC and HCFC refrigerants.

- Since January 29, 1998, it has been mandatory to recycle HFC-134a as well as any other automotive refrigerant.

- Any person who owns approved refrigerant recycling equipment certified for MVAC use must maintain records of the name and address of any facility to which refrigerant is sent.

- Any person who owns approved MVAC refrigerant recycling equipment must retain records demonstrating that all persons authorized to operate the equipment are 609 Certified.

- All records must be maintained for 3 years.

- Service shops must certify to the EPA that they own approved HFC-134a equipment.

HEATING SYSTEM MAINTENANCE, DIAGNOSIS, AND SERVICE

When performing minor under-hood service, the heater hoses should be inspected for unsatisfactory conditions such as cracks, soft spots, collapsed areas, loose clamps, and leaks. Loose heater hose clamps may be tightened if the hose is in satisfactory condition. If any of the other unsatisfactory hose conditions are present, replace the hoses.

If equipped, the cabin filter should be inspected, cleaned, and replaced if necessary. The cabin filters are usually accessed through the engine compartment side **(Figure 53-1)** or through the passenger compartment side **(Figure 53-2)**.

Inspect the area on the front passenger floor mat directly under the HVAC case for any evidence of coolant. If the heater core is leaking, the coolant usually drips out of the HVAC case onto the front floor mat. When the heater core is leaking, some of the coolant must be drained from the cooling system. In many vehicles, the HVAC case must be removed from under the dash to access the heater core. The heater core may be repaired in a radiator repair shop or it may be replaced.

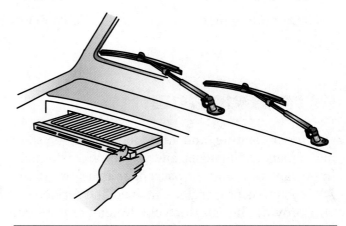

Figure 53-1 A cabin filter accessed through the engine compartment.

Glove compartment

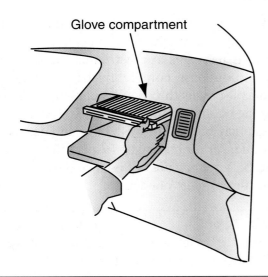

Figure 53-2 A cabin filter accessed through the passenger compartment.

If the customer complains about lack of heat in the vehicle interior, be sure the engine thermostat is operating properly. With the engine at normal operating temperature, use an infrared thermometer or a digital multimeter with a temperature probe to check the temperature of the upper radiator hose. This hose temperature should be near the temperature rating of the thermostat. If the engine coolant temperature is normal, touch each heater hose momentarily. Both of these hoses should be hot. If one heater hose is hot and the other heater hose is cool, the heater coolant control valve is closed or the heater core is partially plugged. The heater core may be back-flushed in the vehicle or removed and flushed out in a radiator repair shop.

SAFETY TIP *The heater hoses may be very hot. Use caution when touching them.*

Back-flushing may be used to dislodge any contaminants that may have collected in the inlet side of the heater core. Back-flushing uses water flowing in the opposite direction to help clean the heater core. To back-flush the heater core in the vehicle, drain some of the coolant from the cooling system and remove the heater hoses from the core. Connect a flushing gun to the heater core outlet and connect a length of hose from the inlet into a coolant drain pan. Use the flushing gun to back-flush the heater core. Continue back-flushing the heater core until the water flowing into the drain pan appears clear.

AIR CONDITIONING (A/C) SYSTEM MAINTENANCE

If the compressor drive belt is slipping, the compressor does not produce sufficient refrigerant pressure and interior vehicle cooling may be inadequate. Inspect the compressor drive belt for cracks, oil contamination, splits, missing chunks, and wear. If any of these conditions are present, replace the belt. Check the belt for proper tension and inspect the self-tensioner for proper operation **(Figure 53-3)**.

Tech Tip *An oil-contaminated drive belt will slip even if it has the specified tension.*

Tech Tip *Most ribbed V-belts have a spring-loaded belt tensioner. This type of belt does not usually have a tension adjustment as they are self adjusting.*

If the compressor clutch clearance is excessive, the clutch may slip and provide a scraping noise immediately after the clutch engages. Use a feeler gauge to measure the clearance between the clutch

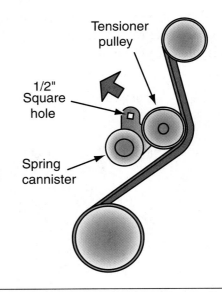

Figure 53-3 A spring-loaded belt tensioner on a ribbed V-belt.

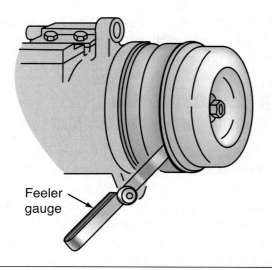

Figure 53-4 Measuring compressor clutch clearance.

drive plate and the pulley surface **(Figure 53-4)**. If the clearance is not within specifications, shims between the clutch hub and the compressor drive shaft may be removed or installed to adjust the clutch clearance **(Figure 53-5)**. If the clearance is excessive then the clutch surfaces should be inspected for excessive wear. An excessively worn clutch should be replaced.

Listen to the compressor with the engine running and the clutch engaged and disengaged. If a growling noise is heard with only the clutch engaged, there is a defective bearing in the compressor. When a growling noise occurs with the clutch disengaged or engaged, the pulley bearing is defective. Inspect the refrigeration system for damaged tubing and components.

Check the condenser for bugs or debris that restrict airflow through the condenser. If the condenser air passages are restricted, the refrigerant may not condense completely from a vapor to a liquid as it passes through the condenser. This reduces the cooling ability of the A/C system. A water hose may be used to wash bugs and debris from the condenser air passages. Inspect the condenser fins to be sure they are not bent so they restrict the airflow through the condenser.

> **Tech Tip** *The condenser is usually placed in front of the radiator so that it may be able to expend heat more efficiently.*

With the A/C system in operation, check for frosted components or lines. Frost on a refrigeration system line or component usually indicates an internal restriction. Inspect the refrigeration system for leaks indicated by an oil smudge in the leak area. If refrigerant has made its way through a leak, then so does the oil in the system. If the refrigerant has escaped through a system leak, a partial refrigerant charge may be installed to locate the leak.

An electronic leak detector can be used to check for leaks in the refrigeration system tubing and components **(Figure 53-6)**. They can be used to quickly find a leak, or to find the area in which the leak exists, in a sealed system when you don't even know where to start. An electronic leak detector gets you close to the leak. After you find the area in which the leak is detected, you can usually

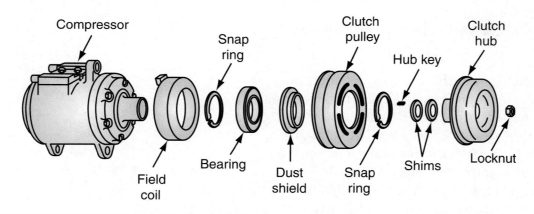

Figure 53-5 Shims between the clutch hub and the compressor drive shaft are used to adjust clutch drive plate clearance.

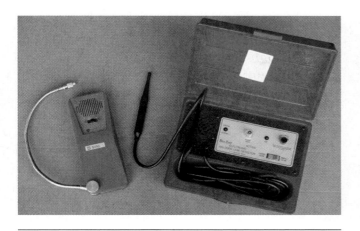

Figure 53-6 An electronic leak detector.

decrease the sensitivity of detectors to pin point the area of the leak.

> **Tech Tip** *Be sure the electronic leak detector is suitable for the refrigerant in the system. Some leak detectors work only on R-12 or R-134a.*

> **Tech Tip** *Refrigerant leaks leave an oil smudge in the leak area because some oil leaks out with the refrigerant.*

The leak area can then be coated with soap solution to verify the exact point of the leak **(Figure 53-7)**. There are soap solutions specially formulated for this purpose.

Figure 53-7 A leak can be detected with the use of a soap solution.

Figure 53-8 A UV leak detection kit.

Ultraviolet leak detection dyes are also available. These kits include an ultraviolet lamp, ultraviolet dye, and some method of injecting the dye into the system while it is under pressure **(Figure 53-8)**. These dye methods may be more time-consuming because of the time it takes to leak the dye and must be visible to the human eye. A leak in the evaporator is not readily observable to the technician.

> **Tech Tip** *Fluorescent dye may be added to a refrigeration system to help locate a leak source. After the dye is added to the system, a black light is used to check for leaks. The dye around a leak source appears as a luminous yellow-green color. Refrigerant containing dye may be purchased from some suppliers, or the dye may be purchased separately in a small pressurized container.*

AIR CONDITIONING (A/C) SYSTEM DIAGNOSIS

A manifold gauge set is connected to the service ports in the refrigeration system to read system pressures **(Figure 53-9)**. The pressure readings on the low side and high side of the refrigeration system are very useful when diagnosing system problems. R134a systems have **Schrader-type service valves (Figure 53-10)**. R-134a refrigeration systems have quick disconnect service valves **(Figure 53-11)**. Remove the protective caps from the service valves.

Figure 53-9 An R-134a manifold gauge set.

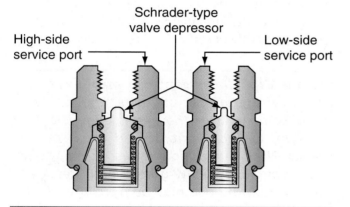

Figure 53-10 R-134a Schrader type service valve.

Figure 53-11 R-134a service port adapters.

The high-side gauge registers pressure up to 500 psi on most gauges. The high-side gauge is open to high-side pressure and the hand valve opens and closes the high-side hose to the center hose connected to the refrigerant supply when charging the system.

> **Tech Tip** *Manifold gauge sets are usually color-coded. The low side is color-coded blue and the high side is color-coded red.*

The low-side gauge is a **compound gauge** that reads both pressure and vacuum. The gauge is always open to the low (suction) side of the refrigeration system. The hand valve opens and closes the low-side hose from the hose in the center of the gauge.

Be sure the hand valves in the manifold gauge set are closed. The valves near the end of the gauge set hoses must also be closed. Connect the high-side hose to the high-side service valve port and connect the low-side hose to the low-side service port. These fittings may be found in various places but the low-side fitting will usually be near the accumulator or compressor inlet and the high-side fitting will be near the compressor or condenser.

> **Tech Tip** *To avoid contamination of refrigeration systems, do not use the same manifold gauge set for R-12 and R-134a systems. R-12 systems use 1/4 in. SAE fittings, while R-134a systems use quick disconnect fittings.*

PERFORMANCE TEST

A performance test may be completed to determine if the A/C system operation is satisfactory. Follow these steps to complete the performance test:

1. Start the engine and maintain engine speed at 2,000 rpm.

2. Set the temperature control in the full cold position, select MAX A/C on the A/C control panel, and move the blower switch to the high-speed position.

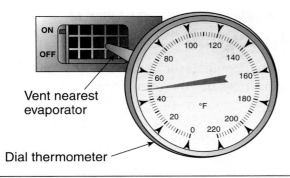

Figure 53-12 A thermometer placed in an A/C outlet.

3. Close the vehicle doors and windows and place a thermometer in the A/C outlet closest to the evaporator **(Figure 53-12)**.

4. Place an auxiliary fan in front of the condenser.

5. Continue running the engine for 5 to 10 minutes.

6. Check the temperature on the thermometer. This temperature should be 35°F to 45°F (1.6°C to 7.2°C) with an atmospheric temperature of 80°F (27°C).

7. If the temperature is too high, check the compressor cycling time and check the pressure indicated on the manifold gauge set. If the high-side pressure is excessive, over 300 psi, the system may be contaminated with air or moisture, the system may have an overcharge of refrigerant, or the condenser air passages may be restricted.

8. Check the clutch cycling time. If the clutch cycles on and off too rapidly, the system is likely low on refrigerant. Cycling times of less than 30 seconds may not be normal. A low refrigerant charge is indicated by nearly the same temperature on the compressor suction and discharge

Tech Tip *The compressor clutch cycling varies depending on atmospheric temperature and the load on the A/C system. The clutch cycling time is the total time required for the clutch-on and clutch-off cycle. A typical clutch-on time is 45 seconds with the clutch on and 15 seconds with the clutch off.*

hoses. If the system is operating properly, the compressor discharge hose should be hot and the compressor suction hose should be cool or cold.

REFRIGERATION SYSTEM PRESSURE DIAGNOSIS

When the engine is not running, refrigeration system pressure equalizes on the high side and low side of the system. After the engine is shut off, the refrigerant can be heard hissing through the orifice tube or TXV until the pressure equalizes in both sides of the system. The refrigeration system pressure with the engine shut off is called **static pressure**. R-134a system static pressures are related to temperature **(Figure 53-13)**.

With the engine running, the refrigeration system pressures vary depending on the atmospheric

Pressure Temperature Chart

Temperature		Pressure	Temperature		Pressure
F	C	HFC-134a	F	C	HFC-134a
−60	−51.1	21.8	55	12.8	51.1
−55	−48.3	20.4	60	15.6	57.3
−50	−48.6	18.7	65	18.3	63.9
−45	−42.8	16.9	70	21.1	70.9
−40	−40.0	14.8	75	23.9	78.4
−35	−37.2	12.5	80	26.7	86.4
−30	−34.4	9.8	85	29.4	94.9
−25	−31.7	6.9	90	32.2	103.9
−20	−28.9	3.7	95	35.0	113.5
−15	−26.1	0.0	100	37.8	123.6
−10	−23.3	1.9	105	40.6	134.3
−5	−20.6	4.1	110	43.3	145.6
0	−17.8	6.5	115	46.1	157.6
5	−15.0	9.1	120*	48.9	170.3
10	−12.2	12.0	125	51.7	183.6
15	−9.4	15.0	130	54.4	197.6
20	−6.7	18.4	135	57.2	212.4
25	−3.9	22.1	140	60.0	227.9
30	−1.1	26.1	145	62.8	244.3
35	1.7	30.4	150	65.6	261.4
40	4.4	35.0	155	68.3	279.5
45	7.2	40.0	160	71.1	298.4
50	10.0	45.3	165	73.9	318.3

Red figures — in. Hg Vacuum   Gray figures — PSIG
*Do not heat can above 120F

Figure 53-13 Static refrigeration system pressures.

Ambient Temp °F	High Side PSIG, R-134a	Low Side PSIG, R-134a
60	120 - 170	7 - 15
70	150 - 250	8 - 16
80	190 - 280	10 - 20
90	220 - 330	15 - 25
100	250 - 350	20 - 30
110	280 - 400	25 - 40

Figure 53-14 Refrigeration system pressure in relation to atmospheric temperature.

temperature. Normal pressures for R-134a systems in relation to temperature are provided in **Figure 53-14**.

SAFETY TIP *The compressor discharge hose may be very hot if the atmospheric temperature is high and the A/C system has been operating for several minutes. Use extreme caution when touching this hose.*

With the engine and the A/C system at normal operating temperature, these pressure guidelines will help you to diagnose the causes of improper refrigeration pressures:

1. When the high-side pressure is higher than normal, there may be air in the system, a refrigerant overcharge, a high-side restriction, or reduced airflow through the condenser.

2. When the high-side pressure is lower than normal, the refrigerant charge may be low or the compressor may be defective.

3. When the low-side pressure is higher than normal, there may be a refrigerant overcharge, a defective compressor, or a faulty metering device.

4. When the low-side pressure is lower than normal, the metering device may be faulty, there may be a low-side restriction, or a refrigerant undercharge.

COMPRESSOR CLUTCH DIAGNOSIS

If the compressor clutch does not engage, the system may be low on refrigerant and the low-pressure shut-off switch may have opened the clutch circuit. Shut the engine off and disconnect the low-pressure shut-off switch connector. Connect a jumper wire across the terminals in the low-pressure shut-off switch connector and start the engine. If the clutch engages, the refrigeration system is low on refrigerant or the low-pressure shut-off switch is defective. Do not allow the compressor to run for an extended period of time as damage may result. Operate the system only momentarily to diagnose the problem.

When the compressor does not run with the jumper wire connected across the low-pressure shut-off switch wiring connector terminals, shut the engine off and connect a digital voltmeter from the voltage input terminal on the compressor clutch to ground. With MAX A/C selected on the control panel and the temperature set below the atmospheric temperature, 12 volts should be supplied to the compressor clutch. If this voltage is not present, diagnose the compressor clutch circuit. The first step in this diagnosis is to check the compressor clutch fuse.

In many modern A/C systems, the computer energizes a compressor relay winding. When this winding is energized, the relay contacts close and supply 12 volts to the compressor clutch. If 12 volts are supplied to the compressor clutch, shut the engine off and disconnect the compressor clutch electrical connector.

Connect a pair of ohmmeter leads to the compressor clutch terminals. An infinite ohmmeter reading indicates an open clutch coil and a reading lower than specified indicates a shorted clutch coil. Connect the ohmmeter leads from one of the clutch terminals to ground. A low ohmmeter reading indicates a grounded clutch coil, whereas an infinite reading proves the coil is not grounded. Replace the clutch coil if it is grounded, shorted, or open. Connect the ohmmeter leads from the ground wire in the clutch wiring connector to a ground on the compressor. A low ohmmeter reading indicates a satisfactory ground wire, whereas an infinite reading proves the ground wire is open.

AIR CONDITIONING (A/C) SYSTEM SERVICE

The most common A/C service procedures are refrigerant recovery, evacuation of the refrigerant system, and system charging. Most shops use a recovery/recycling machine to perform these operations.

Refrigerant Recovery

Follow these steps to perform the refrigerant recovery operation:

1. Connect the high-side and low-side hoses on the manifold gauge set to the appropriate service fittings **(Figure 53-15)**. If using a manifold

Figure 53-16 Turning on the compressor switch on the recovery/recycling machine.

gauge set, then connect the yellow hose to the recovery machine.

2. Connect the recovery/recycling machine electrical cord to a 120-volt outlet and turn the main switch on the machine ON.

3. Turn the compressor switch on the recovery machine ON **(Figure 53-16)**.

4. Operate the compressor until a partial vacuum is indicated on the gauge on the recovery/recycling machine **(Figure 53-17)**. If the machine does not have an automatic shutoff feature,

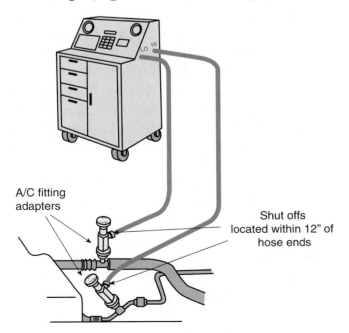

A/C fitting adapters

Shut offs located within 12" of hose ends

Figure 53-15 Connect the hoses of the recovery/recycling machine to the A/C system.

Figure 53-17 A vacuum must be indicated on the low-side gauge on the recovery/recycling machine.

turn the compressor off after achieving a system vacuum. If the machine has an automatic shutoff feature, then refrigerant recovery is completed.

5. Close all the service hose valves, and disconnect the manifold gauge set hoses and cap all fittings.

Refrigerant System Evacuation

When servicing a refrigerant system, it is very important to remove moisture from the system. Moisture reduces the cooling efficiency of the refrigeration system. Even a small amount of moisture may freeze in the orifice tube or TXV, preventing any cooling action from the system.

A **vacuum pump** or **charging station** is used to remove air from the refrigeration system **(Figure 53-18)**. A charging station will have a vacuum pump built into the unit. The vacuum pump also reduces the pressure in the system to the point where any moisture is boiled out of the system. Follow this procedure to evacuate a refrigeration system:

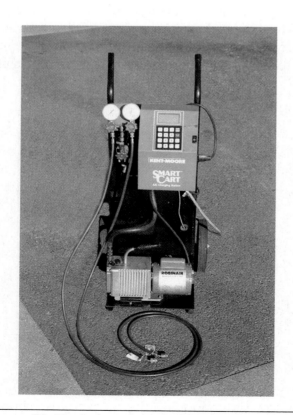

Figure 53-18 A charging station.

1. Be sure both hand valves on the manifold gauge set are closed and connect the manifold gauge set low-side and high-side hoses to the proper service fittings in the refrigeration system. If the refrigeration system has stem-type service valves, use a special wrench to rotate these valve stems so they are in the mid position.

2. Connect the charging station power cord to a 120-volt outlet and turn on the pump switch.

3. Open the low-side hand valve on the manifold gauge set and observe the reading on the low-side gauge. This gauge reading should indicate a slight vacuum immediately.

4. Observe both gauges after 5 minutes. The low-side gauge should indicate 20 in. Hg (33.8 kPa absolute) and the high-side gauge pointer should be slightly below zero unless the pointer movement is limited by a stop. If the high-side gauge does not read below zero, there is blockage in the refrigerant system. Discontinue the evacuation and repair the system blockage.

5. Operate the pump for 15 minutes and observe the gauges. The low-side gauge should indicate at least 26 in. Hg (13.5 kPa absolute). If this reading is not obtained, close the low-side hand valve and observe the low-side gauge reading. If the vacuum reading slowly moves toward zero on this gauge, the refrigeration system is leaking and must be repaired.

6. Be sure both hand valves on the manifold gauge set are open and operate the pump for 30 minutes. Close both hand valves, turn off the vacuum pump, and close all service hose shut-off valves. If the vacuum pump has a shut-off valve, close this valve. Disconnect all the service hoses and install the protective caps.

> **Tech Tip** *Some oil may be removed from the air conditioning system when the refrigerant is evacuated. Most machines are able to catch this oil and allow you to measure the amount that was removed. The same amount of new oil will need to be added back to the system when it is recharged.*

Refrigeration System Charging

After the refrigeration system is evacuated, it must be charged with refrigerant. This charging process is very important to provide proper A/C operation. A refrigerant overcharge or undercharge reduces the cooling efficiency of the refrigeration system. If using a charging station, be sure to follow all of the manufacturer's directions for the proper and safe operation of the unit. Follow this procedure to charge the refrigeration system:

1. Be sure all the service valves, hose shut-off valves, manifold gauge set hand valves, and the refrigerant source valve are closed. Connect the low-side and high-side hoses from the manifold gauge set to the proper service fittings.

2. Add the same amount of oil to the system that was removed during recovery if needed.

3. Set the charging station for the amount of refrigerant to add to the system.

4. Open the refrigerant cylinder hand valve. The refrigeration system is now under pressure from the refrigerant container to the service hose shutoff valve. A vacuum is present in the refrigeration system up to the low-side and high-side hose shutoff valves.

5. Open the service hose shutoff valve and open the low-side and high-side hose shutoff valves near the end of these hoses.

6. Note the gross weight of the cylinder and contents if charging manually.

7. Open the low-side manifold gauge set hand valve and allow the refrigerant to enter the system. *NOTE: Do not charge the refrigerant system through the high side.*

8. Allow the charging station to fill the system with the specified amount of refrigerant. If charging manually, closely observe the weight of the refrigerant container on the scale. When the decrease in the scale reading equals the specified amount of refrigerant for the system being charged, close the low-side manifold gauge set hand valve and the refrigerant source valve.

9. Perform a refrigerant system performance test as explained previously.

10. If the performance test results are satisfactory, turn all the A/C system controls off and shut the engine off.

11. Be sure all valves are closed including the service hose valve, manifold gauge set hand valves, and low-side and high-side hose valves.

12. Disconnect the manifold gauge set hoses and install all the protective caps.

> **Tech Tip** *In place of a scale that weighs the refrigerant, some shops use a clear cylinder with a graduated scale on the cylinder surface. This cylinder is connected between the refrigerant container and the center hose on the manifold gauge set. The refrigerant that is charged into the system is read on the graduated scale.*

> **Tech Tip** *R-12 and R-134a refrigerant are not compatible and must not be mixed. R-134a containers are light blue, whereas R-12 containers are white.*

Typical Refrigeration System Repairs

When it is necessary to change any refrigeration system component, the refrigerant must be recovered and the system should be evacuated and recharged. The receiver/drier or accumulator should be changed at the vehicle manufacturer's specified service intervals or when the system requires major service **(Figure 53-19** and **Figure 53-20)**. It may be necessary to drain the receiver/drier or accumulator and measure any remaining oil so that new oil equal to the amount removed may be added back to the system

The **desiccant** in a receiver/drier or accumulator removes moisture from the refrigeration system. If the desiccant in the receiver/drier or accumulator disintegrates, it moves through the system and usually plugs the orifice tube or TXV. If this happens, the refrigeration system requires flushing.

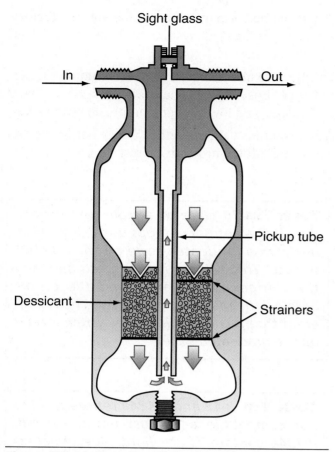

Figure 53-19 Receiver/drier internal design.

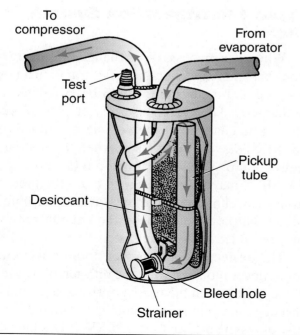

Figure 53-20 Accumulator internal design.

The following flush procedures are suggested for use with a pressured-type flush gun and flush solvent:

1. Fill flush gun with 6 oz. to 14 oz. of flush solvent.

2. Prepare A/C system for flushing by disconnecting hoses from condenser, evaporator, compressor, accumulator, or filter drier. Remove expansion valve or orifice tube from evaporator or liquid line. Do not flush these components. The accumulator or filter drier should be replaced.

3. Connect compressed air or nitrogen to flush gun container. *CAUTION: DO NOT EXCEED MANUFACTURER'S MAXIMUM ALLOWED SYSTEM PRESSURE.*

4. Flush evaporator and condenser. Care should be taken when flushing condensers and evaporators so that excessive flush is not left in the system. It is not advisable to flush the compressor as debris could be lodged in the compressor.

5. To flush, insert flush gun nozzle into inlet of component and flush thoroughly until clean flush solvent flows from the outlet.

6. Capture spent flush in an appropriate container.

7. Accumulators should be removed and drained. Measure the drained oil and replace with equivalent amount of new oil.

8. Pour approximately two ounces of the new oil into the new replacement filter drier if vehicle is so equipped.

9. Install the remainder of the new oil into the system as recommended by the vehicle manufacturer.

Tech Tip *Dispose of contaminated flush fluid in accordance with city, county, state, and federal laws.*

Tech Tip *The flushing procedure removes a substantial amount of contaminant particles, including the old oil and loose desiccant material.*

Figure 53-21 Different types of fitting seals.

If debris has entered the refrigerant system, an in-line filter may be installed between the condenser and the evaporator to help remove debris from the system.

Fittings in a refrigerant system are sealed with various types of gaskets and O-rings **(Figure 53-21)**. Many fittings in the refrigerant system use O-ring seals because this allows the connection to withstand vibration better than a rigid pipe joint. Always replace these seals if a fitting is disconnected. Most refrigerant system O-ring seals are now oval-shaped **(Figure 53-22)**. O-rings must be compatible with the refrigerant in the system. Most new O-rings are color-coded. O-rings should always be coated with the proper refrigerant oil before installation.

If an orifice tube is plugged, it will be necessary to replace this tube. After the evaporator inlet fitting is removed, a special tool is used to remove the orifice tube **(Figure 53-23)**. If the orifice tube breaks, a special extractor is available to remove the broken tube **(Figure 53-24)**. In some cases it

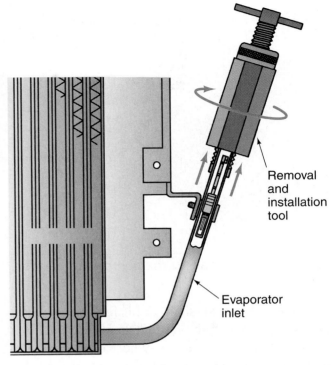

Figure 53-23 Removing an orifice tube.

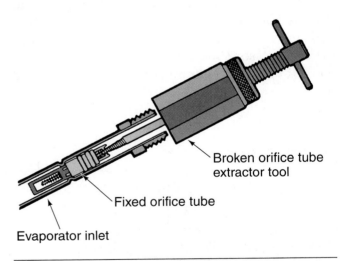

Figure 53-24 Removing a broken orifice tube.

may be necessary to replace the evaporator inlet fitting or the evaporator if the orifice tube cannot be removed.

> **Tech Tip** *Some in-line refrigerant filters contain an orifice tube. When installing this type of filter, remove the orifice tube at the evaporator inlet.*

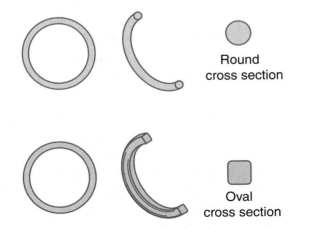

Figure 53-22 Refrigeration system O-rings.

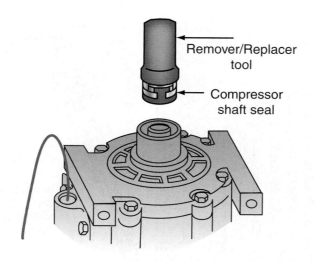

Figure 53-25 Replacing the compressor shaft seal.

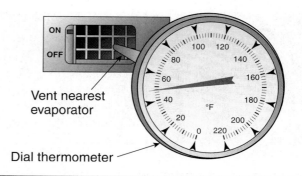

Figure 53-26 Checking outlet air temperature.

If the compressor shaft seal is leaking, the compressor pulley may be removed and this seal may be replaced. On some rear-wheel drive cars, this operation is possible without removing the compressor from the engine. Special tools are available to remove and replace the compressor shaft seal **(Figure 53-25)**.

If there is evidence of substantial oil loss, then the compressor should be checked for the proper level of oil in the compressor case. You will need

to follow the manufacturer's recommended procedure for checking the oil level.

If the compressor is to be replaced, then the system should be checked for any contamination from the old compressor. The system may need to be flushed and the receiver/drier or accumulator should be replaced.

Performance check the refrigerant system by connecting a manifold gauge set to the refrigerant system and installing a thermometer in the A/C outlet closest to the evaporator **(Figure 53-26)**. Run the engine for 15 minutes and be sure the system has the specified low-side and high-side pressure and the correct compressor clutch cycling time.

Summary

- A slipping A/C compressor drive belt reduces the cooling efficiency of the A/C system.

- Excessive compressor clutch clearance may cause clutch slipping and a scraping noise when the clutch engages.

- The compressor clutch clearance may be measured with a feeler gauge.

- Removing or installing shims between the clutch hub and the compressor shaft adjusts the compressor clutch clearance.

- Restricted airflow through the condenser reduces the cooling efficiency of the A/C system.

- During an A/C system performance test, the temperature of the air at the A/C outlet closest

to the evaporator should be 35°F to 45°F with an atmospheric temperature of 80°F.

- If the compressor clutch cycles too fast, the refrigeration system charge may be low.

- If the refrigeration system is low on refrigerant, the low-pressure shut-off switch opens the compressor clutch circuit.

- The compressor clutch coil may be tested with an ohmmeter.

- When removing the refrigerant from a refrigeration system, the refrigerant must be recovered with a recovery/recycling machine.

- The refrigeration system is evacuated with a vacuum pump to remove air and moisture.

The vacuum pump lowers the pressure so any moisture is boiled out of the system.

■ During the evacuation process, the refrigeration system is tested for leaks by observing if the air conditioning system can hold a vacuum.

■ During the charging process, the amount of refrigerant dispensed into the system may be measured by weight.

■ Fluorescent dye may be added to a refrigeration system to help locate leaks.

Review Questions

1. A vehicle has a lack of interior heat complaint. One heater hose feels hot while the other heater hose is cool. Technician A says the heater coolant control valve may be stuck closed. Technician B says the heater core may be plugged. Who is correct?

   A. Technician A

   B. Technician B

   C. Both Technician A and Technician B

   D. Neither Technician A nor Technician B

2. Technician A says a scraping noise when the compressor clutch engages may be caused by excessive clutch plate clearance. Technician B says the torque on the compressor pulley-retaining nut adjusts the clutch plate clearance. Who is correct?

   A. Technician A

   B. Technician B

   C. Both Technician A and Technician B

   D. Neither Technician A nor Technician B

3. Technician A says an oil smudge around a refrigeration system fitting indicates excessive oil in the refrigeration system. Technician B says if all of the refrigerant has escaped from a refrigeration system, a partial charge may be installed to locate the source of the leak. Who is correct?

   A. Technician A

   B. Technician B

   C. Both Technician A and Technician B

   D. Neither Technician A nor Technician B

4. Technician A says since November 15, 1995, it is illegal to vent substitutes for CFC and HCFC refrigerants. Technician B says since August 13, 1992, persons repairing or servicing air conditioning systems on vehicles must be properly trained and certified by an EPA-approved 609 certification program. Who is correct?

   A. Technician A

   B. Technician B

   C. Both Technician A and Technician B

   D. Neither Technician A nor Technician B

5. When diagnosing a lack of heat in the vehicle interior, one heater hose is hot and the other heater hose is cool with the engine at normal operating temperature. The most likely cause of this problem is:

   A. an engine thermostat that is stuck open.

   B. a coolant control valve that is stuck closed.

   C. partially restricted coolant passages in the radiator core.

   D. partially restricted air passages in the heater core.

6. All of these A/C system problems may cause insufficient vehicle interior cooling except:

   A. a low refrigerant charge.

   B. moisture in the refrigeration system.

   C. partially restricted air passages in the condenser.

   D. excessive compressor drive belt tension.

7. When an A/C system is operating, frost forms on the receiver/drier. Technician A says the refrigeration system is overcharged. Technician B says the receiver/drier is restricting the flow of refrigerant. Who is correct?

A. Technician A

B. Technician B

C. Both Technician A and Technician B

D. Neither Technician A nor Technician B

8. An air conditioning system cycles on and off every 15 seconds. The atmospheric temperature is 80°F. The most likely cause of this problem is:

A. a partially restricted condenser.

B. a defective compressor.

C. a refrigerant leak in the condenser.

D. a low refrigerant charge.

9. After a refrigeration system performance test is performed at an atmospheric temperature of 80°F, if the system is operating normally, the temperature on the thermometer should be:

A. 28°F to 32°F.

B. 35°F to 45°F.

C. 55°F to 65°F.

D. 70°F to 75°F.

10. After a refrigeration system is evacuated for 15 minutes, the low-side gauge indicates 27 in. Hg. When the vacuum pump is shut off and the low-side hand valve is closed, the low-side gauge reading decreases to 18 in. Hg in 5 minutes. Technician A says there is refrigeration oil left in the system and the air conditioning system must be flushed. Technician B says the refrigeration system has a leak and a partial charge should be installed to locate the leak. Who is correct?

A. Technician A

B. Technician B

C. Both Technician A and Technician B

D. Neither Technician A nor Technician B

11. The low-side gauge in a manifold gauge set is a compound gauge that reads _____ and _____.

12. Air or moisture in the refrigeration system causes high pressure in the _____ _____ of the refrigeration system.

13. A defective compressor may cause _____ high-side pressure and _____ low-side pressure.

14. Moisture is removed from a refrigeration system by creating a vacuum in the system, which causes the moisture to _____.

15. Explain how a refrigeration system leak is indicated while evacuating the system.

16. Describe the charging process.

17. Explain the necessary service procedures if debris has entered the refrigeration system.

18. Describe normal test results during an A/C system performance test.

54 Supplemental Restraint Systems

Learning Objectives

After you have read, studied, and practiced the contents of this unit, you should be able to:

- List the functions of conventional seat belts.
- List the components of the air bag system.
- Describe the operation of the seat belt pretensioner.
- Define the purpose of a clock spring electrical connector.

- Describe air bag sensor operation required to deploy an air bag.
- Describe the deployment of an air bag.
- Describe the normal operation of an air bag system warning light.
- Describe the clock spring centering procedure.

Key Terms

Active restraints

Air bag system diagnostic module (ASDM)

Passive restraints

INTRODUCTION

All vehicles manufactured and sold in the United States must have passive restraints. Passive restraints may be air bags or automatic seat belts. Most vehicles with air bags also have active restraints. **Passive restraints** operate automatically with no action required by the driver. **Active restraints** require action by the driver or passengers before the restraints provide any protection. Vehicle safety is one of the most important considerations of the average vehicle buyer today. In simple terms, safety sells vehicles! Therefore, vehicle manufacturers have spent a large amount of money

engineering improved safety systems. Passive restraints at the present time include the driver's air bag, passenger-side air bag, side-impact air bags, air bag curtains, and seat belt pretensioners. The seat belt and air bag systems are intended to work together to protect the driver and passengers. On an air bag-equipped vehicle, conventional seat belts perform these functions:

1. Hold the occupants in proper position when air bags inflate.
2. Reduce the risk of injury in a less severe collision in which the air bag(s) do not deploy.
3. Reduce the risk of occupant ejection from the vehicle and thus reduce the possibility of injury.

PASSIVE SEAT BELT RESTRAINTS

A passive seat belt system uses electric motors to automatically move the shoulder belts across the driver and front seat passenger. The upper ends of the belts are attached to a carrier mounted in a track just above the top of the doorframe. An inertia lock retractor that is mounted in the center console secures the other end of each shoulder belt. When a front door is opened, the outer end of the shoulder belts move forward in the door tracks to allow easy entry and exit from the vehicle **(Figure 54-1)**. When the door is closed and the ignition switch is turned on, the shoulder belts move rearward in the door tracks to secure the front seat occupants. The active lap belt must be buckled by the driver or passenger and must be worn with the shoulder belt.

AIR BAG SYSTEM COMPONENTS

Understanding air bag system components is essential to comprehending the complete system operation. A knowledge of air bag system operation is absolutely necessary to maintain, diagnose, and service air bag systems quickly and accurately.

SAFETY TIP *Many vehicles use a unique color, such as yellow, on all the wiring connectors in the air bag system so the system components are easily recognized. The wiring connectors on many air bag system components contain shorting bars that connect the wiring terminals together when the wiring connector is disconnected. These shorting bars help to prevent accidental air bag deployment from improper service procedures.*

Sensors

Some air bag system sensors contain a set of normally open, gold-plated contacts and a gold-plated ball that acts as a sensing mass **(Figure 54-2)**. This ball is mounted in a stainless steel-lined cylinder. A magnet holds the ball about 1/8 inch away from the contacts. When the vehicle is involved in a collision of sufficient force, the ball moves away from the magnet and closes the switch contacts. These contacts remain closed for 3 milliseconds before the magnet pulls the ball away from the contacts. The sensor is completely sealed in epoxy to prevent contaminants and moisture from entering the sensor. Sensors must be mounted with the forward marking on the sensor facing toward the front of the vehicle. To operate properly, sensors must be mounted in their original

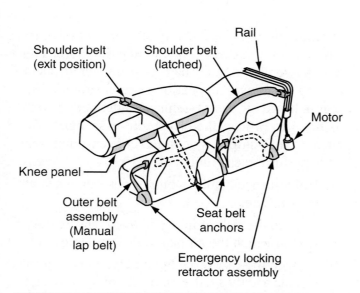

Figure 54-1 A passive seat belt restraint system.

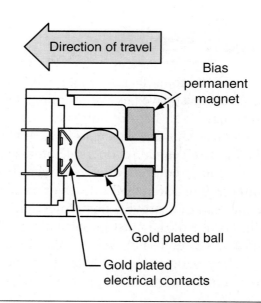

Figure 54-2 A mass-type air bag system sensor.

mounting position and sensor brackets must not be distorted. Some air bag sensors contain a roller on a ramp. This roller is held against a stop by small, retractable springs on each side of the roller. If the vehicle is involved in a collision of sufficient force, the roller moves up the ramp and strikes a spring contact completing the electrical circuit between the contact and the ramp **(Figure 54-3)**. Some air bag sensors contain an accelerometer that contains a piezoelectric element **(Figure 54-4)**. If the vehicle is involved in a collision, this element is distorted. The voltage signal from the sensor to the air bag system module depends on the force of the collision and the amount of element distortion.

SAFETY TIP *Loose sensor mounting bolts or distorted sensor brackets and mountings cause improper sensor operation and inaccurate air bag deployment.*

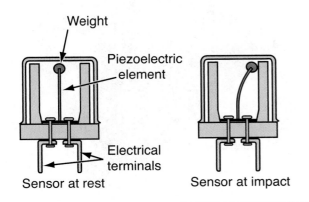

Figure 54-3 A roller-type air bag system sensor.

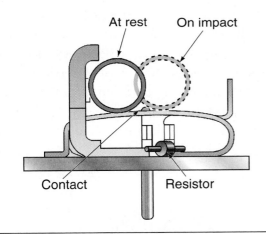

Figure 54-4 An accelerometer-type air bag system sensor.

Inflator Module

The inflator module contains the air bag, air bag container and base plate, inflator, and trim cover. The retainer and base plate are made from stainless steel and are riveted to the inflator module. The air bag is made from porous nylon and some air bags have a neoprene coating. The purpose of the inflator module is to inflate the air bag in a few milliseconds when the vehicle is involved in a collision. A typical driver's side air bag has a volume of 2.3 cubic feet. The driver's side air bag inflator module is usually retained with four bolts in the top of the steering wheel **(Figure 54-5)**. The passenger's side inflator module is mounted in the passenger's side of the instrument panel **(Figure 54-6)**. Because there is a

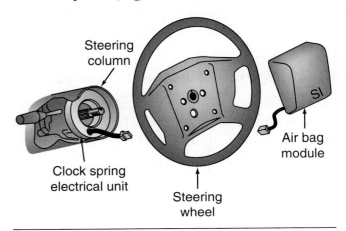

Figure 54-5 A driver's side air bag inflator module.

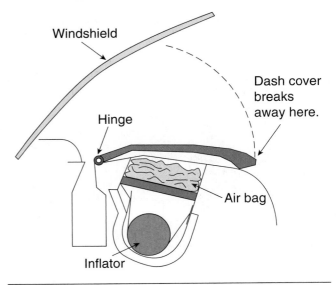

Figure 54-6 A passenger's side air bag inflator module.

greater distance between the passenger and the dash compared to the distance between the driver and the steering wheel, the passenger's side air bag's average volume is 7.4 cubic feet.

SAFETY TIP *If a child in a rearward facing child's seat is placed in the front passenger's seat and the air bag deploys, the deployed air bag may force the seat and the child against the vertical part of the front seat causing the child to suffocate.*

SAFETY TIP *During an air bag deployment, a small amount of sodium hydroxide is formed, which is an irritating caustic. You should always wear eye and hand protection when servicing deployed air bags.*

Clock Spring Electrical Connector

The clock spring electrical connector is mounted in the steering column directly below the steering wheel **(Figure 54-7)**. The clock spring electrical connector maintains electrical contact between the air bag inflator module and the air bag electrical system. The clock spring electrical connector contains a conductive ribbon that winds and unwinds as the steering wheel is turned. One end of the conductive ribbon is connected to the air bag electrical system and the opposite end of this ribbon is connected to the inflator module.

Air Bag System Diagnostic Module

The **air bag system diagnostic module (ASDM)** is often mounted under the instrument panel. Other mounting locations for the ASDM include the center console or under the front passenger's seat. Some ASDMs contain an air bag sensor(s). Many ASDMs contain a backup power supply that deploys the air bag(s) if battery voltage is disconnected from the ASDM during a collision **(Figure 54-8)**. Some air bag systems have a backup power supply that is external from the ASDM. A typical ASDM performs these functions:

1. Controls the air bag system warning light in the instrument panel.
2. Continuously monitors the complete air bag electrical system.
3. Controls air bag system diagnostic functions.
4. Provides a backup power supply to deploy the air bag(s) if battery voltage is disconnected from the air bag system during a collision.
5. On some air bag systems, the ASDM is responsible for deploying the air bag(s) when appropriate signals are received from the sensors.

If the ASDM detects an electrical defect in the air bag system, the ASDM illuminates the air bag system

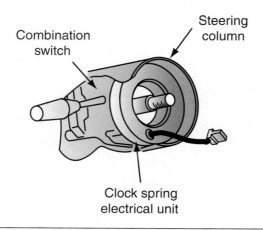

Figure 54-7 A clock spring electrical connector.

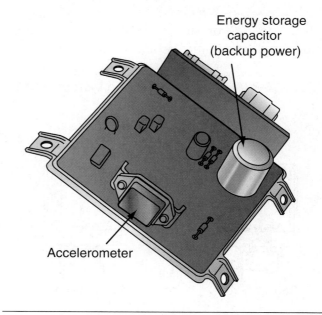

Figure 54-8 An air bag system diagnostic module (ASDM).

warning light in the instrument panel with the engine running. Some ASDMs have the capability to detect impact and record collision event information. The vehicle manufacturer uses special test equipment to download the information from the ASDM. This information indicates if the air bag system was operating properly at the time of the collision.

Air Bag System Warning Light

The air bag system warning light is mounted in the instrument panel. The air bag system warning light should come on when the ignition switch is turned on. If the ignition switch is left in the on position without starting the engine or when the engine is started, the air bag system warning light should flash a few times and then go out. This light action indicates the air bag system is satisfactory. If the air bag system warning light is on with the engine running, there is an electrical defect in the air bag system and the air bag system may not be operational in a collision.

AIR BAG SYSTEM OPERATION

Some air bag inflator modules use pyrotechnology (explosives) to inflate the air bag(s). The air bag inflator module contains a squib or igniter in the center of the module. Many air bag systems require the closing of two sensors to deploy the air bag(s). An arming sensor supplies voltage to the squib in the inflator module if this sensor closes during a collision. A passenger compartment discriminating sensor and a forward discriminating sensor are connected on the groundside of the squib **(Figure 54-9)**. One of these sensors must close to complete the circuit from the squib to ground. To close an arming sensor and a discriminating sensor and deploy the driver's side and passenger's side air bags requires the vehicle to be in a frontal collision where the maximum collision force is within 30° on either side of the vehicle centerline. When the arming sensor and one of the discriminating sensors close, approximately 1.75 amperes flow through the squib and the sensor circuit. This current flow heats the squib and ignites the igniter charge next to the squib. The burning igniter charge ignites the generant in the inflator module. This generant explodes

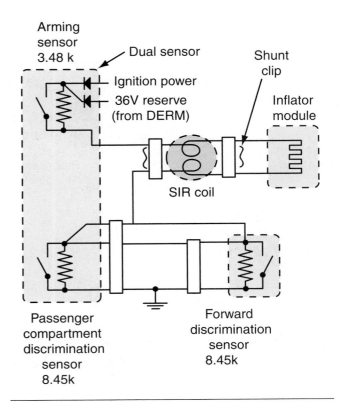

Figure 54-9 Typical squib and sensor wiring.

very quickly, producing large quantities of hot nitrogen gas which flows through the inflator module filter into the air bag in about 30 milliseconds **(Figure 54-10)**. The inflated air bag opens the tear seams in the inflator module cover and prevents the driver's or passenger's head and chest from contacting the windshield, steering wheel, or dash. Large openings under the air bag allow the bag to deflate in about 1 second so it does not block the driver's view if the vehicle is still in motion. This deflating action also

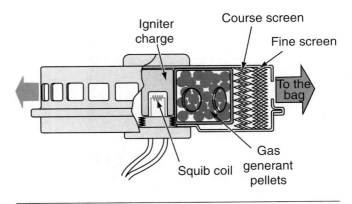

Figure 54-10 Inflator module using pyrotechnology.

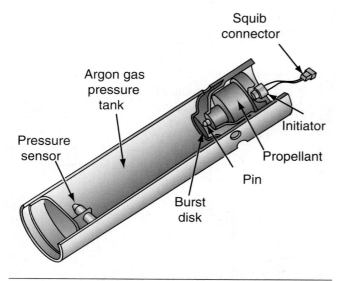

Figure 54-11 Inflator module containing pressurized argon gas.

prevents the driver or passenger from smothering in the air bag. The combustion temperature in the air bag may reach 2,500°F, but the air bag temperature is slightly above room temperature. Some hybrid inflator modules contain a squib (initiator) and a small amount of propellant. If the squib is energized during a collision, the exploding propellant pierces the larger argon gas container. The propellant heats the argon gas, which quickly escapes into the air bag **(Figure 54-11)**. A pressure sensor is mounted in the opposite end of the hybrid inflator module from the squib informs the ASDM if the pressure decreases in the gas chamber during normal vehicle operation. Some light-duty trucks have a switch in the instrument panel that allows the driver to insert the ignition key and turn the passenger's side air bag on or off should a child be placed on the passenger seat **(Figure 54-12)**.

Figure 54-12 Passenger's side air bag on/off switch.

Multi-Stage Air Bag Deployment

Many vehicles are now equipped with multi-stage air bags on both the driver's side and passenger's side. The inflator module in a multi-stage inflator module contains two squibs **(Figure 54-13)**. If the vehicle is involved in a collision at lower speeds that requires air bag deployment with reduced force, only one squib is ignited which inflates the air bag with less force. The second squib may be ignited 160 milliseconds later to use up the other igniter charge, but the air bag has already deployed, so this action does not affect air bag deployment. When the vehicle is involved in a severe collision at higher speeds, both squibs are fired close together so the air bag deploys faster with more force.

Smart Air Bag Systems

Some air bag systems have a switch in the passenger's side of the front seat. This switch informs the ASDM if anyone is sitting in the front passenger seat. If no one is sitting in this seat, the passenger's side air bag does not deploy if the vehicle is involved in a collision. Some smart air bag systems can detect this from the passenger's weight. If a child below a certain weight is occupying this seat, the air bag does not deploy. In some air bag systems, the weight of the person in the passenger's seat also determines the force with which the passenger's side air bag deploys. On some smart air bag systems, if a rearward facing child's seat is placed in the front passenger's seat, the passenger's side air bag does not deploy. Some vehicles have a small knee air bag that deploys out of the dash in front of the driver's knees. This air bag protects the driver from knee injury and also keeps the driver from sliding under the seat belt during an accident. This action maintains the driver in a better position to be protected by the driver's side air bag.

NOTE: The smart air bag system is sometimes called the *next generation air bag system.*

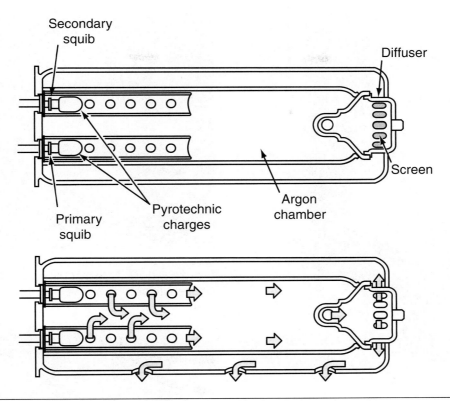

Figure 54-13 A multi-stage inflator module with dual squibs.

Side-Impact Air Bags

Some vehicles are now equipped with side-impact air bags that protect the driver or passengers during a side collision. The side-impact air bag systems are separate from the driver's side and passenger's side air bag systems. Side-impact air bag systems have their own sensors and ASDM. Separate ASDMs are usually installed for each side-impact air bag. The ASDMs for the side-impact air bags may be under the front seats or behind the B-pillar panels. Some vehicles have a side-impact air bag that deploys out of the door paneling. In other systems, the side air bags are mounted in the side of the seat back near the top. Some vehicles have a side air bag curtain that deploys out of the headliner just above the doors **(Figure 54-14)**. This type of air bag protects the front and rear seat occupants from head injury.

Figure 54-14 Side-impact air curtains.

> **Tech Tip** *Left and right side ASDMs for side-impact air bags are not interchangeable.*

SEAT BELT PRETENSIONERS

Some vehicles have seat belt pretensioners on the seat belts. The pretensioners contain materials similar to a single-stage air bag inflator module. The pretensioners may be mounted on the buckle

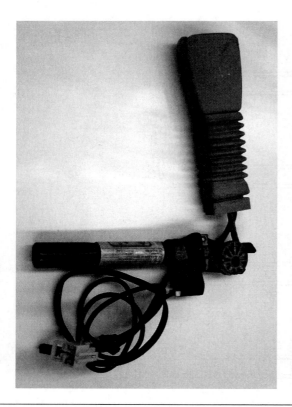

Figure 54-15 A buckle-mounted seat belt pretensioner.

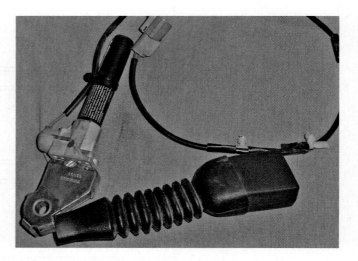

Figure 54-16 A retractor-mounted seat belt pretensioner.

Figure 54-17 Internal design of a retractor-mounted seat belt pretensioner.

side of the seat belt **(Figure 54-15)**. If the front air bags are deployed, the ASDM also fires the pretensioners. A thin cable is connected between the buckle and a small piston in the pretensioner. When a pretensioner is fired, the piston moves up the cylinder and the cable pulls the buckle tight. This action holds the occupant tightly against the seat and helps to prevent injury. Some pretensioners are on the retractor side of the seat belt **(Figure 54-16)**. When the pretensioner is fired, balls shoot out of the pretensioner against a fan wheel causing the retractor to rotate and tighten the seat belt **(Figure 54-17)**. The last ball to be shot out of the pretensioner is larger and sticks in the fan wheel to lock the seat belt. During any air bag system diagnostic check, the air bag warning light in the instrument panel should be observed for proper operation. As mentioned previously, this warning light should be illuminated when the ignition switch is turned on. It should flash a few times and go out when the engine starts. Some air bag warning lights remain on for a few seconds after the engine starts rather than flashing. If the air bag

warning light operates properly, the air bag system is operational. Inspect the seat belts for webbing damage. Fully extend each seat belt from the retractor and check the webbing for cuts, broken or pulled threads, cut loops at the belt edge, and bowed conditions **(Figure 54-18)**. If any of these conditions are present, replace the seat belt assembly. Check the seat belt buckles for damage and proper latching and unlatching. Check the seat belt retractors for proper operation and proper retention on the vehicle chassis. If the vehicle has passive seat belts, inspect the belts for free movement in the tracks. The electric motor for the passive seat belts is usually mounted behind the rear seat

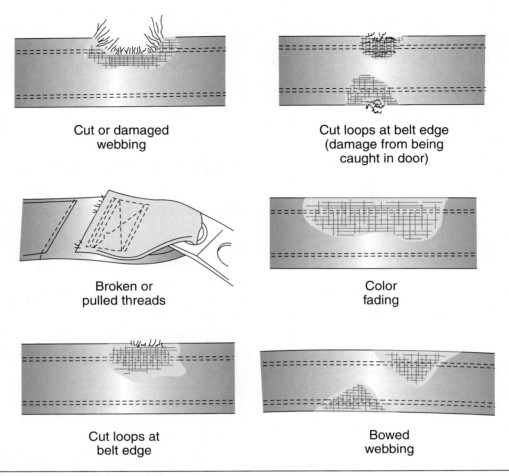

Cut or damaged
webbing

Cut loops at belt edge
(damage from being
caught in door)

Broken or
pulled threads

Color
fading

Cut loops at
belt edge

Bowed
webbing

Figure 54-18 Seat belt webbing defects.

trim panel. This motor operates the plastic tape connected to the seat belts. Be sure the seat belt moves through its complete travel without sticking when the door is opened and closed.

SUPPLEMENTAL RESTRAINT SYSTEM DIAGNOSIS

Most air bag systems are diagnosed with a scan tool. The scan tool and the scan tool module must be compatible with the vehicle and the air bag system. Follow this procedure for typical scan tool diagnosis of an air bag system:

1. Roll the driver's window down.
2. Connect the scan tool to the DLC under the dash.
3. Move the scan tool through the driver's window opening and stand outside the vehicle.

4. Reach into the vehicle and turn the ignition switch on. Be sure there is no one in the vehicle.
5. Select Air Bag System on the scan tool and read and record any air bag DTCs.

> **Tech Tip** *Always consult the service manual for the specific procedure and special equipment necessary for diagnosis of the air bag system.*

AIR BAG SYSTEM SERVICE

When servicing air bag system components, some vehicle manufacturers recommend disabling the air bag system to prevent accidental air bag deployment. A typical air bag system disabling procedure on a vehicle with driver's side, passenger's side, and side-impact air bags follows:

1. Turn the steering wheel until the front wheels are in the straight-ahead position.

2. Turn the ignition switch off and remove the ignition key from the switch.

3. Remove the air bag fuse from the fuse block. A connector position assurance (CPA) pin holds two wiring connectors together so they cannot be separated until the pin is removed.

4. Disconnect the following connectors:

   a) Driver's side air bag two-wire air bag system connector at the base of the steering column.

   b) Passenger's side air bag two-wire connector behind the passenger's side air bag inflator module.

   c) Driver's side-impact inflator module two-wire connector under the driver's seat.

   d) Passenger's side impact inflator module two-wire connector under the passenger's seat.

5. Perform the necessary service work on the air bag system.

6. Connect all the disconnected air bag system connectors and install the CPA pins in each connector.

7. Install the air bag system fuse.

8. Turn the ignition switch on and check the air bag system warning light for proper system operation.

SAFETY TIP *When servicing or diagnosing air bag systems, use only the vehicle manufacturer's recommended tools. Use of other tools, such as 12-volt test lights or self-powered test lights, may cause accidental air bag deployment.*

Tech Tip *Some air bag system wiring connectors are retained with a connector position assurance (CPA) pin. This pin must be removed before the connector can be disconnected.*

CLOCK SPRING CENTERING

Before installing a clock spring, it must be centered and installed with the front wheels straight ahead. To center the clock spring, turn the clock spring fully clockwise. Turn the clock spring fully counterclockwise counting the total number of turns. From the fully counterclockwise position, rotate the clock spring one-half the total number of turns clockwise. The clock spring is now in the centered position. Install the centered clock spring with the front wheels straight ahead.

Tech Tip *If a clock spring is not installed in the centered position with the front wheels straight ahead, the conductive ribbon in the clock spring will be broken when the steering wheel is rotated.*

Tech Tip *Some clock springs have a mark that indicates the centered position.*

Summary

- The law mandates the use of seat belts.

- Seat belts must be used with passive air bag systems.

- Many air bag system sensors complete an electrical circuit during a collision.

- Air bag system inflator modules use pyrotechnology or pressurized argon gas to inflate the air bag.

- The air bag system warning light indicates the readiness of the air bag system.

- Many vehicles now have multi-stage air bags.
- Side-impact air bags have separate sensors and inflator modules.
- Smart air bag systems can sense the presence and weight of a person in the passenger's seat.
- During a collision, seat belt pretensioners tighten the seat belts using pyrotechnology.

- A scan tool is used to diagnose the air bag system.
- Air bag systems must be disabled when servicing system components.
- The clock spring electrical connector must be installed in the centered position with the front wheels straight ahead.

Review Questions

1. Technician A says a mass-type air bag system sensor may be installed facing in either direction. Technician B says on a mass-type air bag system sensor, distortion of the sensor mounting area may affect sensor operation. Who is correct?

   A. Technician A

   B. Technician B

   C. Both Technician A and Technician B

   D. Neither Technician A nor Technician B

2. Technician A says the clock spring electrical connector contains spring-loaded copper contacts. Technician B says the clock spring electrical connector is mounted directly below the inflator module. Who is correct?

   A. Technician A

   B. Technician B

   C. Both Technician A and Technician B

   D. Neither Technician A nor Technician B

3. Technician A says the inflator module can deploy the air bag in about 30 seconds. Technician B says the passenger's side inflator module is mounted in the passenger's side of the instrument panel. Who is correct?

   A. Technician A

   B. Technician B

   C. Both Technician A and Technician B

   D. Neither Technician A nor Technician B

4. Technician A says multi-stage air bag inflator modules have dual squibs. Technician B says smart air bag systems can sense the presence of a person in the passenger's seat. Who is correct?

   A. Technician A

   B. Technician B

   C. Both Technician A and Technician B

   D. Neither Technician A nor Technician B

5. All these statements about passive seat belt systems are true *except*:

   A. When a door is opened, the outer end of the shoulder belts move forward in the door tracks.

   B. When a door is closed and the ignition switch is turned on, the shoulder belts move rearward in the door tracks.

   C. Electric motors move the shoulder belts across the driver and passenger.

   D. The lap belts are connected without any action from the driver or passenger.

6. All these statements about mass-type air bag system sensors are true *except*:

   A. The contacts and ball are gold-plated.

   B. A magnet holds the ball about 1/2 inch away from the contacts.

   C. The sensor is completely sealed in epoxy to eliminate contaminants.

   D. An arrow on the sensor housing must face toward the front of the vehicle.

7. In an air bag system:

   A. The passenger's side air bag inflator module is larger than the driver's side inflator module.

   B. The air bags are manufactured from porous plastic.

   C. The retainer and base plate are made from lead.

   D. The retainer and base plate are bolted to the inflator module.

8. When discussing air bag system operation, Technician A says the ASDM controls the air bag system warning light. Technician B says the ASDM continuously monitors the air bag electrical system. Who is correct?

   A. Technician A

   B. Technician B

   C. Both Technician A and Technician B

   D. Neither Technician A nor Technician B

9. In a vehicle with typical driver's side and passenger's side air bags, to deploy the air bags the vehicle must be involved in a frontal collision of sufficient force and the collision force must be within:

   A. 10° on either side of the vehicle centerline.

   B. 20° on either side of the vehicle centerline.

   C. 30° on either side of the vehicle centerline.

   D. 54° on either side of the vehicle centerline.

10. In a smart air bag system:

   A. The passenger's side air bag deploys when there is no one sitting in the front passenger's seat and the vehicle is in a severe frontal collision.

   B. The passenger's side air bag deploys regardless of the weight of the front passenger's seat occupant if the vehicle is in a severe frontal collision.

   C. On some smart air bag systems, the passenger's side air bag does not deploy if a rearward facing child's seat is in the front passenger's seat.

   D. The driver's side air bag does not deploy if the driver's weight is below 110 pounds and the vehicle is involved in a severe frontal collision.

11. Side-impact air bag systems have separate _____ and _____ _____.

12. If a rearward facing child's seat is placed in the front passenger's seat, a smart air bag system will not _____ the passenger's side air bag.

13. Seat belt pretensioners _____ the seat belts if the vehicle is involved in a collision of sufficient force.

14. Some air bag system disabling procedures require the removal of the air bag _____ before disconnecting any system connectors.

15. Explain the clock spring centering procedure.

16. Describe the sensor requirements to deploy an air bag.

17. Describe the purpose of the backup power supply in the inflator module.

18. Explain the operation of an inflator module containing pressurized argon gas during an air bag deployment.

Glossary
Glosario

···

Note: Terms are highlighted in color, followed by Spanish translation in bold.

Accumulator   A pressurized reservoir used to store fluid.
Acumulador   Depósito con sobrepresión interna que se usa para almacenar líquido.

ACFI   Air-conditioning and refrigeration institute.
ACFI   Instituto de Refrigeración y Aire Acondicionado.

Active restraints   A restraint system that does have active controls during an accident e.g. Air bags.
Atenuación real   Un sistema de atenuación que tiene controles activos durante un accidente, como por ejemplo las bolsas de aire.

Adaptive strategy   A computers ability to learn air-fuel corrections and apply them to a specific vehicle.
Estrategia de regulación   La habilidad de las computadoras para aprender correcciones de aire-combustible y aplicarlas a un vehículo específico.

Aiming pads   Small projections on the front of some headlights to which headlight aligning equipment may be attached.
Patines de alineación   Pequeñas proyecciones en la parte delantera de algunos faros a los cuales se puede conectar el equipo de alineación de faros.

Air bag diagnostic monitor (ASDM)   An automotive computer responsible for air bag system operation.
Monitor de bolsa de aire (ASD)   Una computadora automotiva que es responsable por la operación del sistema de la bolsa de aire.

Air-conditioning   The process of heating and cooling cabin air.
Aire acondicionado   El proceso de calentar y enfriar el aire de la cabina.

Air gap   Distance between two components.
Entrehierro   Distancia entre dos componentes.

Air purge   To remove trapped air.
Purga de aire   Para quitar el aire atrapado.

Alternating current   flows in one direction and then in the opposite direction.
Corriente alterna   fluye en una dirección y luego en la dirección opuesta.

Alternator output test   A test designed to test the full current output of the alternator.
Ensayo del alternador   Diseñado para probar la salida completa de corriente del alternador.

American Petroleum Institute (API) rating   A universal engine oil rating that classifies oils according to the type of service for which the oil is intended.
Evaluación del Instituto Americano de Petroleo (API)   Una evaluación universal del aceite automotivo que clasifica a los aceites según el tipo del servicio que se le requiere.

Amperes   A measurement for the amount of current flowing in an electric circuit.
Amperes   Una medida de la cantidad del corriente que fluye en un circuito eléctrico.

Analog/digital converter   A device that interprets analog data and transmits a converted signal in a digital format.
Convertidor D/A   Dispositivo que interpreta los datos análogos y transmite una señal convertida a un formato digital.

Analog meter   A meter with a pointer and a scale to indicate a specific reading.
Medidora análoga   Un medidor con una indicadora y una gama para indicar una lectura específica.

Angular load   A load applied at an angle somewhere between the horizontal and vertical positions.
Carga de soporte angular   Una carga aplicada en un ángulo que se encuentra entre las posiciones horizontales y verticales.

Aqueous   Water based or water soluble.
Acuoso   Con base de agua o soluble en agua.

Aspect ratio   The height of the tire is a proportion of the width.
Factor de forma   La altura de la llanta está en proporción con el ancho.

Atom   The smallest particle of an element.
Átomo   La partícula más pequeña de un elemento.

Atomize   To change from a liquid into a small series of very fine droplets.
Atomizar   Cambiar de un líquido a una serie pequeña de gotitas muy finas.

Automotive technician   A person trained and certified to repair automobiles.
Técnico automotriz   Persona entrenada y certificada para reparar automóviles

Automatic Transmission Rebuilders Association (ATRA) provides technical information to transmission shops and technicians.
Asociación de Reconstrutores de Transmisiones Automáticas (ATRA)   Provea la información técnica a los talleres de transmisiones y a los técnicos.

Automotive dealership   sells and services vehicles produced by one or more vehicle manufacturers.
Sucursal automotivo   vende y repara los vehículos producidos por un o varios fabricantes de automóviles.

Automotive Service Excellence (ASE)   An organization that certifies technicians
Excelencia de servicio automotriz (ESA)   organización que certifica a los técnicos

Automotive technician   A person trained and certified to repair automobiles.
Técnico automotriz   Persona entrenada y certificada para reparar automóviles.

Axial load   A load applied along the axis of rotation.
Carga axial   Una carga aplicada a lo largo del eje de rotación.

Axle tramp   The up and down movement of the axle or wheel assembly.
Bamboleo del eje   El movimiento hacia arriba y hacia abajo del eje o del montaje de la rueda.

Backlash   The free play between gear teeth.
Juego entre dientes   El juego que hay entre los dientes de un engrane.

Ball joint, load carrying   The ball joint that bears the vehicle weight through the spring.
Rótula esférica, transferencia de esfuerzo   La rótula esférica que lleva el peso del vehículo mediante el muelle.

Ball joint, non-load carrying   The ball joint that is opposite the load carrying joint or does not bear the vehicle weight.
Rótula esférica, sin transferencia de esfuerzo   La rótula esférica que está opuesta a la junta de transferencia de esfuerzo o que no lleva el peso del vehículo.

Ball socket   A joint that permits angular rotation about a central point in two directions.
Junta esférica   Unión que permite la rotación angular en un punto central en dos direcciones.

Battery plates   Lead and lead oxide plates that make up the conductive plates in a battery.
Placas de batería   Placas de acero y de óxido de acero que componen las placas conductoras en una batería.

Belleville spring   A diaphragm spring made from thin sheet metal that is formed into a cone shape.
Resorte Belleville   Un resorte de tipo diafragma hecho de una hoja delgada de metal que es en la forma de un cono.

Belt glazing   A hard slick coating on the friction surface of the belt.
Glaseado de la correa   Una capa gruesa y resbalosa en la superficie de fricción de una correa.

Bimetallic strip   contains two different metals fused together that expand at different rates and cause the strip to bend.
Tira bimetálica   tiene dos metales distintos fundidos que expanden en dos velocidades causando que la tira se dobla.

Bit   A single on-off signal.
Bit   Una señal sencilla de encendido y apagado.

Blowby   The amount of leakage between the piston rings and the cylinder walls.
Fuga   La cantidad de fuga entre los anillos de pistones y las paredes del cilíndro.

Bolt diameter   The measurement across the threaded area of the bolt.
Diámetro del perno   La medida a través del área enroscada del perno.

Bolt length   The distance from the bottom of the bolt head to the end of the bolt.
Longitud del perno   La distancia de la parte inferior de la cabeza del perno a la extremidad del perno.

Bore scope   A tool that allows one to see into a cylinder bore of an engine through the sparkplug hole.
Taladro de alcance   Herramienta que permite ver el calibre del cilindro de un motor por el orificio de la bujía.

Bounce test   A test designed to test the effectiveness of the shock absorber or strut.
Prueba de rebote   Evaluación diseñada para probar la eficacia de un amortiguador de choques o jabalcón.

Brake fade   Occurs when the brake pedal height gradually decreases during a prolonged brake application.
Pérdida de adherencia de los frenos   Ocurre cuando la altura del pedal de frenos baja gradualmente durante la aplicación prolongada del freno.

Brake pedal free-play   The amount of brake pedal movement before the booster pushrod contacts the master cylinder piston.
Juego del pedal del freno   La cantida del movimiento del pedal de freno antes de que la aumentadora de la varilla empujadora toca el piston del cilindro maestro.

Brake pressure modulator valve (BPMV)   An assembly containing the solenoid valves connected to each wheel in an antilock brake system (ABS).

Válvula modulador del presión del freno (BPMV) Una asamblea que contiene las válvulas del solenoide conectadas a cada rueda en un sistema de frenos antibloqueos (ABS).

Byte A group of bits.

Byte Un grupo de bits.

Cabin filter A filter used to clean the air as it enters the passenger compartment.

Filtro de la cabina Un filtro que se usa para limpiar el aire que va entrando en el compartimiento del pasajero.

Camber The outward tilt of the top of the tire from the body of the car as viewed from the front.

Inclinación de la rueda Es como se ve la inclinación externa de la parte superior de la llanta con relación al vehículo.

Canceling angles are present when the vibration from one universal joint is canceled by an equal and opposite vibration from another universal joint.

Ángulos de anulación se presentan cuando la vibración de una junta universal se anula por una vibración igual e opuersta de otra junta universal.

CARB California Air Resources Board

CARB Mesa Directiva de los Recursos de Aire de California

Caster The geometric line created by the pivot axis of the front tire as viewed from the side.

Avance del pivote de la rueda La línea geométrica que crea el eje del pivote de la rueda frontal como se ve de un lado.

Catalyst A material that accelerates a chemical reaction without being changed itself.

Catalizador Una materia que acelera una reación química sín ser cambiada sí misma.

Castellated nut A nut with grooves cut into the top of the nut to allow retention by a cotter pin.

Tuerca almenada Tuerca con ranuras cortadas en la parte superior para permitir detenerla con un pasador partido.

CCA Cold cranking amps rating ; the rating is the average amps delivered by a battery at 0 degree F for 30 seconds without dropping below 7.2 volts.

CCA Coeficiente de amperios del motor de arranque; el coeficiente es los amperios medios que conduce una batería a 0 grados F durante 30 segundos sin descender a menos de 7.2 voltios.

CCOT Cycling clutch orifice tube system.

Sistema CCOT (orificio calibrado) Sistema de refrigeración de embrague alternante-tubo de orificio.

CD Compact disk.

CD Disco compacto.

Cell group A battery cell group contains alternately spaced positive and negative plates kept apart by porous separators.

Grupo de células Un grupo de células de batería que contiene las placas positivas y negativas puestas alternativamente y separadas por separadores porosos.

Center link The link that is supported by the pitman arm and the idler arm.

Vástago central El vástago que se apoya en el brazo de mando de la dirección y en el brazo intermediario.

Center of gravity The point where weight is evenly distributed on all sides.

Centro de gravedad Punto en dónde el peso se distribuye equitativamente en todos los lados.

Central port injection (CPI) A fuel injection system with a central port injector that supplies fuel to a mechanical poppet injector in each intake port.

Inyección de puerta central (CPI) Un sistema de inyección de combustible con una puerta de inyector central que suministra1 el combustible a un inyector mecánico de contrapunto en cada puerta de entrada.

Charging station A machine designed to evacuate and recharge AC systems.

Estación de carga Máquina diseñada para evacuar o recargar los sistemas de AC.

Chemical radiator flush A chemical used to remove deposits from the inner cooling surfaces of a radiator and cooling system.

Lavado químico del radiador Químico que se usa para quitar los depósitos de las superficies refrigerantes interiores de un radiador y del sistema de enfriamiento.

Circuit breaker A mechanical device that opens and protects an electric circuit if excessive current flow is present.

Disyuntor Un dispositivo mecánico que abre para protejer un circuito eléctrico si se presenta un flujo de corriente excesivo.

CLC Chlorofluorocarbon.

CLC Clorofluorocarbonos.

Closed loop flushes the refrigeration system with the system intact, without allowing refrigerant to escape to the atmosphere.

Procedimiento de enjugado del bucle cerrado Enjuega el sistema de refrigeración con el sistma intacto, sin permitir fugar el refrigerante a la atmósfera.

Clutch cable A clutch operated by a cable mechanism.

Cable del embrague Embrague al que al que lo hace funcionar un mecanismo de cable.

Clutch chatter The undesirable noise or vibration caused by a clutch during engagement.

Vibración del embrague Ruido o vibración indeseable que causa un embrague durante el engranaje.

Clutch pedal free-play The amount of clutch pedal movement before the release bearing contacts the pressure plate release levers.

Juego del pedal del embrague La cantidad del movimiento del pedal del embrague antes de que el cojinete de desembrague toca las palancas de desembrague de la placa de presión.

Cohesion   The tendency of engine oil to remain on the friction surfaces of engine components.

Cohesión   La tendencia que tiene el aceite del motor de quedarse en las superficies de fricción de los componentes del motor.

Combination valve   A brake system valve that contains a metering valve, proportioning valve, and a switch to operate the brake system warning light.

Válvula de combinación   Una válvula del sistema del frenado que tiene una válvula medidora, una válvula dosificante, y un interruptor que opera la lámpara testigo del sistema del frenado.

Compound   A compound is a material with two or more types of atoms.

Compuesta   Una compuesta es una materia con dos tipos o más de átomos.

Compound gauge   has the ability to read two values.

Indicador compuesta   tiene la habilidad de leer dos valores.

Compression stroke   Movement of piston in the cylinder bore that compresses the fuel and air mixture.

Carrera de compresión   Movimiento del pistón en el orificio del cilindro que condensa el combustible y la mezcla de aire.

Conductor   A material that can carry electrical current e.g. copper.

Conductor   Material que puede llevar corriente eléctrica como el cobre.

Conicity   A manufacturing defect in a tire caused by improperly wound plies that causes the tire to be slightly cone-shaped.

Conicidad   Un defecto de fabricación del neumático debido a que los pliegues no sean envueltas correctamente causando que el neumático sea de una forma ligeramente cónica.

Constant ratio   A steering ratio in which turns of the steering wheel are proportional to the amount the wheels are steered.

Relación constante   Una relación direccional en la que las vueltas del volante son proporcionales a la cantidad que se mueven las ruedas.

Constant-running release bearing   makes light contact with the pressure plate release levers even when the clutch pedal is fully upward.

Cojinete de desembrague contínuo   mantiene un contacto ligero con las palancas de desembrague de la placa de presión aún cuando el pedal del desembrague esta completamente en alto.

Constant velocity (CV) joints   transfer a uniform torque and a constant speed while operating at a wide variety of angles.

Juntas de velocidad contínua   transfieren una torsión uniforme y una velocidad constante mientras que opera con una gran variedad de ángulos.

Coolant hydrometer   A tester that measures coolant specific gravity to determine the antifreeze content of the coolant.

Hidrómetro de refrigerante   Una probadora que mide la gravedad específica del refrigerante para determinar el contenido de anticogelante del refrigerante.

Coolant refractometer   Measures the refractive index of the coolant to determine its freeze protection temperature.

Refractómetro del enfriador   Mide el índice refringente del enfriador para determinar su temperatura de protección contra el congelamiento.

Coupling point   occurs in a torque converter when the impeller pump, turbine, and stator begin to rotate together.

Punto de acoplamiento   ocurre en un convertidor de torsión cuando la bomba impelador, la turbina, y el estátor comienzan a girar juntos.

Critical speed   The rotational speed at which a component begins to vibrate.

Velocidad crítica   La velocidad de rotación en la cual un componente comienza a vibrar.

Cross counts   The number of times the oxygen sensor signal switches from lean to rich in a given time period.

Cuenta de cruzadas   El número de veces en que cambia el señal del sensor de oxígeno de pobre a rico en un periodo específico.

Cross threaded   A defective thread condition caused by starting a fastener onto its threads when the fastener is tipped slightly to one side and the threads on the fastener are not properly aligned.

Rosca corrida   Una condición defectuosa de rosca causada al poner un fijador en su rosca mientras que el fijador esté un poco inclinado hacia un lado y las roscas del fijador no estan alineadas correctamente.

Current diagnostic trouble code (DTC)   represent faults that are present during the diagnosis.

Códigos diagnóstico de fallos corrientes (DTCs)   representan los fallos que se presentan durante el diagnósis.

CVRSS   Continuously variable road sensing suspension systemelectronically controlled shock absorbers or struts.

CVRSS   Sistema de suspensión con sensor de variables continuas del camino Amortiguadores de choques o jabalcones controlados electrónicamente.

Data link connector (DLC)   An electrical connector to which a scan tool may be connected to diagnose various electronic systems on a vehicle.

Conector de enlace de datos (DLC)   Un conector eléctrico al cual se puede conectar una herramienta exploradora para diagnosticar varios sistemas electrónicas de un vehículo.

Desiccant   A material that absorbs moisture.

Desecante   Material que absorbe la humedad.

Diagonally split brake system   A brake system in which the LF and RR are on one brake circuit and the RF and LR are on the other circuit.

Sistema de frenos de junta diagonal Un sistema de frenos en el que el frente izquierdo y la parte posterior derecha están sobre un circuito de frenos y el frente derecho y la parte posterior izquierda están en otro circuito.

Diagnostic procedure charts located in service manuals to provide the necessary diagnostic steps in the proper order to diagnose specific vehicle problems.

Gráfico de procedimientos diagnósticos ubicados en los manuales de servicio para proveer los pasos necesarios de diagnóstico en su orden correcto para diagnosticar prolemas específicos del vehículo.

Digital voltage signals A voltage signal that is either high or low.

Señal digital de voltaje Un señal de voltaje que es alto o bajo.

Diode trio A small device in some alternators that contains three diodes.

Triple diodo Un pequeño dispositivo en algunos alternadores que contiene tres diodos.

Direct pressure monitoring A system that measures tire pressure directly.

Monitoreo directo de la presión Sistema que mide directamente la presión de la llanta.

DLC Diagnostic link connector.

DLC o CED Conector de enlace de diagnóstico.

DOT Department of Transportation.

DOT Departamento de Transporte.

Double cardan universal joint A joint using two conventional u-joints to produce a constant velocity assembly.

Junta de doble cardán Una junta que usa dos conexiones u convencionales para producir un ensamblaje de velocidad constante.

Drive cycle Operation of a vehicle through a series of driving conditions to complete a predetermined series of tests.

Ciclo de conducción Operación de un vehículo mediante unas series de condiciones de manejo para completar una serie de pruebas predeterminadas.

Drop Center The depression in the center of the rim that allows the bead to be placed for tire installation or removal.

Canal profundo Depresión en el centro de la llanta que permite que se coloque el talón para instalar o quitar la llanta.

DTC Diagnostic trouble code.

CDF o DTC Código de diagnóstico de fallas.

Dual overhead cam (DOHC) An engine with the intake and exhaust valves mounted in the cylinder heads and two camshafts are mounted on each cylinder head.

Leva doble en cabeza (DOHC) Un motor que tiene las válvulas de entrada e escape montadas en las cabezas de los cilíndros y dos árboles de leva montados en cada cabeza del cilíndro.

Duty cycle The on time of a device.

Ciclo de servicio El tiempo de encendido de un dispositivo.

DVD digital versatile disk.

DVD disco video digital.

Dynamic wheel balance Spin balancing of the tire in two planes.

Equilibrio dinámico de la rueda Balanceo giratorio de la llanta en dos planos.

EGR Exhaust gas recirculation.

EGR o RGE Recirculación de los gases de escape.

Electrolysis Chemical and/or electrical corrosion.

Electrólisis Corrosión química y/o eléctrica.

Electrolyte A mixture of sulfuric acid and water in a lead acid battery.

Electrólito Una mezcla del ácido sulfúrico y el agua en una batería de plomo con ácido.

Electromagnetic induction The process of inducing a voltage in a conductor by moving the conductor through the magnetic field or vice versa.

Inducción electromagnético El proceso de inducir un voltaje en un conductor moviendo el conductor al través de un campo magnético o vice versa.

Electromagnetic pickup coil A pickup coil containing a permanent magnet surrounded by a coil of wire.

Devanado detector electromagnético Un devanado detector que contiene un imán permanente envuelto por un rollo de alambre.

Electronic control unit (ECU) A computer used to control vehicle functions.

Unidad de control electrónico (UCE) Computadora que se usa para controlar las funciones del vehículo.

Electrons Negatively charged particles located on the various rings of an atom.

Electrones Las partículas de carga negativa que se encuentran en los varios anillos de un átomo.

Element An element is a liquid, solid, or gas with only one type of atom.

Elemento Un elemento es un líquido, un sólido o un gas con un sólo tipo de átomo.

Ellipse A conical section.

Elipse Una sección cónica.

Engine coolant temperature (ECT) sensor A thermistor-type sensor that sends an analog voltage signal to the powertrain control module (PCM) in relation to coolant temperature.

Sensor de temperatura del refrigerante de motor Un sensor tipo termistor que manda un señal de voltaje analogo al módulo de control del trenmotor en relación a la temperatura del refrigerante.

Engine support fixture A device used to hold an engine usually when the transmission or transaxle is removed.

Dispositivo de fijación del motor Dispositivo que se utiliza para fijar el motor generalmente cuando se quitan la transmisión o el transaxle.

Environmental Protection Agency (EPA) A United States government agency in charge of all aspects of environmental protection.

Agencia de Protección del Medio Ambiente (EPA) Una agencia del gobierno de los Estados Unidos que se encarga de cada aspecto de la protección del medio ambiente.

EPS Electrical power steering.

EPS o DAE Dirección asistida eléctrica.

Equilibrium boiling point (ERBP) The temperature at which a liquid boils at a fixed pressure.

Punto de ebullición de equilibrio (PEE) Temperatura en la cual un líquido hierve a una presión fija.

EVA Electronic vibration analyzer.

EVA Analizador electrónico de vibración.

EVAP emission leak detector (EELD) produces a nontoxic smoke and blows this smoke into various components for leak-detection purposes.

Detector de fuga de emisión EVAP (EELP) produce un humo nontóxico y introduce el humo en varios componentes con el propósito de detectar las fugas.

Female-type quick disconnect coupling A coupling attached to the end of a shop air hose. An opening in the center of the coupling is inserted over the male part of the quick disconnect coupling. The female part of the coupling contains the locking and release mechanism that allows easy locking and release action with the male part of the coupling.

Acoplador de desconecta rápida tipo hembra Un acoplador conectado a la extremidad de una manguera de aire del taller. Una abertura en el centro del acoplador se enchufa con el parte macho del acoplador de desconecta rápida. La parte hembra del acoplador contiene el mecanismo de conexión y desconexión que permite una acción fácil de conexión y desconexión con el parte macho del acoplador.

Fixed (CV) joint does not allow any inward or outward drive axle movement to compensate for changes in axle length.

Junta de velocidad fija continua no permite que el movimiento hacia adentro o afuera compensa los cambios de longitud del eje.

Flash diagnostic trouble codes (DTCs) are displayed by the flashes of the malfunction indicator light (MIL) in the instrument panel.

Códigos diagnósticos instantáneos de fallos (DTCs) se exhiben por medio de los parpadeos de la lámpara indicadora de fallos (MIL)en el tablero de instrumentos.

Flash programming The process of downloading computer software from a scan tool or PC into an on-board computer.

Programación instantánea El proceso de instalar las programaciones de computadora de una herramienta exploradora o de una computadora portátil (PC) a una computadora a bordo.

Flash-to-pass feature allows a driver to move the signal light lever forward enough to operate the headlights on high beam to indicate he or she is going to pass the vehicle in front.

Opción de señales rápidos permite que un conductor mueva la palanca indicadora de señales hacia adelante lo bastante para operar los faros largos para indicar que el conductor va pasar el vehículo que va en frente.

Flex fan A cooling fan with flexible blades that straighten out as engine speed increases to reduce engine power required to turn the fan.

Ventilador flexible Un ventilador de enfriamiento que tiene las aletas flexibles que se enderezan con más velocidad al motor para reducir la cantidad de potencia del motor requirido para operar el ventilador.

Flexible clutch disc allows some movement between the clutch facings and the hub.

Embrague flexible permite algo de movimiento entre las caras del embrague y el cubo.

Flex plate The link between the engine crankshaft and the torque converter.

Placa flexible El enlace entre el cigüeñal y el convertidor de par.

Floating caliper A brake caliper that is designed so it slides sideways during a brake application.

Calibre flotante Un calibre de frenos diseñado para que se deslice horizontalmente durante la aplicación de los frenos.

Flushing gun Uses air and water to clean a cooling system.

Inyector de lavado Utiliza aire y agua para limpiar el sistema refrigerante.

FMVSS Federal Motor Vehicle Safety Standards

FMVSS Normas Federales de Seguridad de los Vehículos Automóviles.

Forward bias An electrical connection between a voltage source and a diode that results in current flow through the diode.

Polarización negativa frontal Una conexión entre la fuente del voltaje y un diodo que resulta en un flujo de corriente por medio del diodo.

Four-channel ABS has a pair of solenoids for each wheel to modulate the brake system pressure individually at each wheel.

Sistema de frenos antibloqueantes de cuatro canales (ABS) tiene un par de solenoides en cada rueda para modular la presión del sistema de frenos individualmente en cada rueda.

Four wheel alignment Alignment of all four to the thrust line of the vehicle.

Alineación de cuatro ruedas Aineación de las cuatro a la directriz de presiones del vehículo.

Frame contact lift   A lift that engages and lifts the vehicle from underneath.

Remontador por contacto   Un remontador que se une y levanta el vehículo por debajo.

Front/rear split brake system   A brake system in which the LR and RR are on one brake circuit and the RF and LF are on the other circuit.

Sistema de frenos en dos bloques frontales / traseros   Sistema de frenos en el que la parte posterior izquierda y derecha están sobre un circuito de frenos y el frente derecho e izquierdo están en otro circuito.

Fuel trim   The ability of the computer to make adjustments to the A/F ratio.

Regulación de combustible   La habilidad de una computadora de hacer ajustes a la relación aire/combustible.

Full fielding   The process of supplying full field current to an alternator to obtain maximum output from the alternator.

Campo completo   El proceso de proveer una corriente de campo completo a un alternador para obtener la salida máxima del alternador.

Fuse   A component that protects an electric circuit from excessive current flow.

Fusible   Un componente que proteja un circuito eléctrico de un flujo excesivo de corriente.

Gear ratio   The ratio between the drive and driven gears.

Relación de engranajes   La relación entre los engranajes de impulso y los de mando.

Glycol-based brake fluid   Brake fluids using glycols as the base chemical for manufacture, DOT 3, 4, and 5.1 are glycol based brake fluids.

Líquido de frenos con base de glicol   Líquido de frenos que se fabrican con una base química de glicol. El DOT 3, 4 y 5.1 son líquidos para frenos con base de glicol.

Grade   Tensile strength of a bolt.

Embalaje   Fuerza de resistencia de un perno.

Graphing voltmeter   A voltmeter that indicates voltage readings in graph form.

Volímetro gráfico   Un voltímetro que indica las lecturas del voltaje en una forma gráfica.

Hall effect switch   A switch that opens and closes in the presence or absence of a magnetic field.

Interruptor de efecto Hall   Un interruptor que se abre y se cierra en presencia o ausencia de un campo magnético.

Halogen   A term for a group of chemically treated nonmetallic elements including chlorine, fluorine, and iodine.

Halógeno   Un término para un grupo de elementos nometálicos tratados quimicamente que incluyen al cloruro, la fluorina y el iodo.

Hazardous waste   waste that poses substantial or potential threats to public health or the environment.

Desecho peligroso   desecho que presenta una amenaza sustancial o potencial a la salud pública o al medioambiente.

HC   Hydro carbon; fuel.

HC   Hidrocarburo; combustible.

HCU   Hydraulic control unit used in antilock brake systems.

HCU   Unidad de control hidráulico que se usa en los sistemas de frenos de antibloqueo.

Heated oxygen sensors   Oxygen sensors that have electric heaters to bring them to operating temperature quickly.

Sensores calientes de oxígeno   Sensores de oxígeno que tienen los calentadores eléctricos para que funcionen rápidamente a la temperatura de operación.

Heat range   The temperature at which the spark plug operates.

Intervalo de calor   La temperatura a la que opera una bujía.

Helical gear   have teeth that are cut at an angle to the center line of the gear.

Engranaje helicoidal   tiene las dientes cortadas en un ángulo a la línea central del engranaje.

HFC hydrofluorocarbon   A chemical compound with halogens links and is used a s refrigerant.

Fluorocarburo   Compuesto químico con conexiones de halógenos y se usa como refrigerante.

HID   High Intensity Discharge; used in some headlamps as a means of producing the light.

HID o DAI   Descarga de alta intensidad; la usan algunos faros para producir luz.

History diagnostic trouble codes (DTCs)   are caused by intermittent faults and these faults are not present during the diagnosis.

Códigos preexistentes de fallos diagnósticos (DTCs)   se causan por fallos intermitentes y éstos no están presentes durante un diagnósis.

Honesty   communicating and acting truthfully and with fairness

Honestidad   comunicarse y actuar verazmente y con franqueza

Horsepower   33,000 ft. lb/min.

Caballos de fuerza   33000 pies lb/min.

Hotchkiss drive   An open drive shaft with two universal joints.

Árbol Hotchkiss   Un árbol motor abierto que tiene dos juntas universales.

HVAC   Heating ventilation air conditioning.

HVAC o AAVC   Aire acondicionado ventilación y calefacción.

Hybrid organic additive technology (HOAT)   A special coolant additive package to help prevent cooling system corrosion.

Tecnología de aditivos orgánicos híbrido (HOAT)   Un paquete de aditivos refrigerantes especiales para ayudar en la prevención de la corrosión del sistema de enfriamiento.

Hydraulic clutch   A clutch operated by a hydraulic piston.

Embrague hidráulico   Embrague al que lo hace funcionar un pistón hidráulico.

Hydrometer   A tester that measures the specific gravity of a liquid.

Hidrómetro   Un probador que mide la gravedad específica de un líquido.

Hygroscopic   The ability to absorb water.

Higroscópico   Habilidad de absorber agua.

Idler arm   Arm that supports one end of the center link opposite the pitman arm.

Brazo Intermediario   Brazo que apoya una orilla del vástago central opuesto al brazo de mando de la dirección.

I/M   Inspection/ Maintainance.

I/M   Inspección/Mantenimiento.

IMTV   Intake manifold turning valve.

IMTV   Válvula calibradora del Múltiple de Admisión.

Incandescence   The process of changing electrical energy to heat energy in a light bulb to produce light.

Incandencia   El proceso de cambiar la energía eléctrica a la energía de calor en una bombilla eléctrica para producir la luz.

Included angle   The sum of the steering axis inclination and camber.

Angulo incluído   La suma de la inclinación del eje direccional y de la inclinación de la rueda.

Independent front suspension   An automobile suspemsion that allows the wheels to move independently of each other. One goes up and down without affecting the other.

Suspensión delantera independiente   Suspensión de automóvil que permite que las ruedas se muevan independientemente una de otra. Una va hacia arriba y hacia abajo sin afectar la otra.

Independent repair shop   An independent repair shop is privately owned and operated without being affiliated with a vehicle manufacturer, automotive parts manufacturer, or chain organization.

Taller de reparaciones independiente   Un taller de reparaciones independient es poseído e operado sín afilicación con el fabricante de un vehículo, el fabricante de partes automotivos o una organización de sucursal.

Indirect pressure monitoring   A monitoring system that depends on a difference in wheel speed to determine if a tire is low on air pressure.

Monitoreo indirecto de la presión   Sistema de control que depende de la diferencia en la velocidad de la rueda para determinar si la llanta está baja de presión de aire.

Inertia switch   A type of switch that reacts to acceleration or deceleration.

Conmutador de inercia   Tipo de conmutador o interruptor que reacciona a la aceleración y a la desaceleración.

Infinite ohmmeter reading   An open circuit.

Lectura infinita de ohmiómetro   Un circuito abierto.

Injector cleaner   A chemical used to clear deposits from fuel injectors.

Limpiador del inyector   Químico que se utiliza para limpiar los depósitos de los inyectores de combustible.

Injector driver   An electronic circuit used to pulse an injector.

Controlador del inyector   Un circuito electrónico que se usa para impulsar un inyector.

Injector pulse width   The length of time in milliseconds that an injector is open.

Apertura del pulso del inyector   La duración del tiempo en milisegundos que se queda abierto un inyector.

Inorganic additive technology (IAT)   An additive package when blended with ethylene glycol gives coolant a two year protection life protection.

Tecnología de aditivo inorgánico   Tecnología de aditivo inorgánico un paquete de aditivo que al mezclarse con etilenglicol le da al fluido refrigerador una protección de vida de dos años.

In-phase universal joints   are all on the same plane.

Juntas universales en fase   todos están en el mismo plano.

Insulator   An insulator has five or more valence electrons which do not move easily from atom to atom.

Aislador   Un aislador tiene cinco o más electrones de valencia que no se mueven fácilmente de un átomo a otro átomo.

Intake air temperature   A device that measures the air temperature entering the engine.

Temperatura del aire de entrada   Dispositivo que mide la temperatura del aire que entra en el motor.

Intake stroke   The movement of the piston in the cylinder bore that as the piston moves downward it draws in the fuel and air mixture from the intake manifold.

Tiempo de admisión   El movimiento del pistón en el calibre del cilindro que cuando el pistón baja, atrae el combustible y la mezcla de aire del colector de admisión.

Integrated circuit (IC)   A combined group of individual components within one component.

Circuito integrado CI o IC   Un grupo combinado de componentes individuales sin un componente.

Interlock   A mechanism used to prevent multiple gear selections in a manual transmission.

Enclavamiento   Mecanismo que se usa para prevenir selecciones múltiples de engranaje en una transmisión estándar.

Intermittent faults   A circuit fault.

Fallas intermitentes   Fallas de circuito.

International Automotive Technicians Network (iATN)   A large group of automotive technicians in many countries that share technical knowledge with other members through the Internet.

Red International de Técnico Automotivos (IATN) Un grupo grande de técnicos automotivos en muchos paises que comparten la experiencia técnica con otros miembros por medio del Internet.

ISC Idle speed control.

CVV Control de velocidad en vacío.

ISO International Organization for Standardization.

OIN Organización Internacional de Normalización.

Jounce travel Upward tire and wheel movement.

Sacudo Un movimiento hace arriba del pneumático y la rueda.

Kinetic energy Energy in motion.

Energía kinética La energía en movimiento.

Knock sensor A piezoelectric crystal that produces a voltage when it senses vibration in the engine.

Sensor de detonación Un cristal piezoeléctrico que produce un voltaje cuando siente vibración en el motor.

Lab scope A scope that displays various waveforms across the screen with a very fast trace.

Aparato óptico del laboratorio Un aparato que demuestra varias ondas en la pantalla con un registro muy rápido.

Latent heat of condensation The heat released during condensation from a gas to a liquid.

Calor latente del condensación El calor soltado durante la condensación de un gas a un líquido.

Latent heat of vaporization The amount of heat required to change a liquid to a gas after the liquid reaches the boiling point.

Calor latente de la vaporización La cantidad del calor requirido para cambiar un líquido a un gas después de que el líquido ha llegado a su punto de ebullición.

Lateral tire runout The amount of sideways wobble in a rotating tire.

Corrimiento lateral del pneumático La cantidad de bamboleo en un pneumático en rotación.

Left-hand thread A fastener with a left-hand thread must be rotated counterclockwise to tighten the fastener.

Rosca con paso a izquierdas Un fijador con un paso a izquierdas se tiene que girar en sentido contrario a la agujas del reloj para apretarlo.

Line pressure The pressure delivered from the pressure regulator valve in an automatic transmission.

Presión de la línea La presión que se entrega de la válvula reguladora de presión en una transmisión automática.

Lifting points Proper points to lift a vehicle safely.

Remontes Remontes apropiados para elevar un vehículo con seguridad.

Male-type quick disconnect coupling The part of a coupling that is threaded into an air-operated tool. This male coupling contains grooves and ridges to provide a locking action with the female part of the coupling.

Acoplador de desconecta rápida tipo macho La parte de un acoplador enroscada a una herramienta pneumática. El acoplador macho contiene las ranuras y los bordes para proveer una acción de conexión con la parte hembra del acoplador.

Malfunction indicator light (MIL) A warning light in the instrument panel that is illuminated by the PCM to indicate engine computer system defects.

Lámpara indicadora de fallos (MIL) Una lámpara de avisos en el tablero de instrumentos iluminada por el PCM para indicar los defectos del sistema de computadora del motor.

Manifold absolute pressure A device that measures the air pressure in the intake manifold of the engine.

Colector de presión absoluta Un dispositivo que mide la temperatura del aire en el colector de entrada del motor.

Mass air flow sensor A device that measures the air movement (mass) into the engine.

Sensor de la masa del flujo de aire Dispositivo que mide el movimiento del aire (masa) que entra al motor.

Match mount Placing the tire on the rim to achieve the least amount of necessary weight to dynamically balance the wheel.

Montaje de junta Colocar la llanta en la corona para lograr la mínima cantidad de peso necesario para equilibrar la rueda dinámicamente.

Material Safety Data Sheets provide all the necessary data about hazardous materials.

Hojas de Datos de Seguridad de los Materiales proveen todos los datos necesarios pertinente a las materiales peligrosas.

Mechanical linkage A clutch operated by a series of levers and rods.

Conexión mecánica Embrague al que lo hacen funcionar una serie de palancas y bielas.

Memory saver A small device used to provide battery power to retain volatile memory in vehicle systems.

¿Guardador de memoria? Un dispositivo pequeño que se usa para proporcionar potencia a la batería para retener memoria volátil en los sistemas automotrices.

Message center A digital display in the instrument panel where various warning messages are displayed.

Centro de avisos Una presentación digital en el tablero de instrumentos en donde aparecen varios indicadores de aviso.

Metering valve A valve designed to hold off front brake pressure in a disc/drum brake system.

Válvula reguladora Válvula diseñada para sujetar la presión del freno frontal en un sistema de frenos de disco o de tambor.

MFI Multi-port fuel injection.

MFI o IDL Inyector multipunto.

Mineral based brake fluid   Brake fluid based on ether compounds.
Líquido de frenos con base mineral   Líquido de frenos con base de compuestos de éter.

Molecule   A molecule is the smallest particle that a compound can be divided into and still retain its characteristics.
Moécula   Una molécula es la partícula más pequeña al que se puede dividir una compuesta y aún mantener sus características.

Monolith-type catalytic converter   contains a catalyst-coated monolith that is similar to a honeycomb.
Convertidor catalítico tipo monolítico   contiene un monolito parecido a una panal cubierto de una catalista.

Multiport fuel injection (MFI)   A fuel injection system that opens two or more injectors simultaneously.
Inyección de combustible de puertas múltiples   Un sistema de inyección de combustible que abre dos o más inyectores a la vez.

Neutral flame   A flame that has complete combustion of gasses.
Flama neutra   Flama que tiene una combustión completa de gases.

Neutrons   Particles with no electrical charge that are located with the protons in the nucleus of an atom.
Neutrones   Las partículas que no tienen una carga eléctrica que se encuentran con los protones en el núcleo de un átomo.

Non-directional finish (NDF)   A finish applied to a brake rotor to eliminate the machining grooves left by the carbide cutting tool.
Acabado adireccional (AA)   Un acabado que se aplica al rotor del freno para eliminar las ranuras de fabricación que deja la herramienta cortadora con plaquita de carburo.

Normally open   A switch that is open in its normal state e.g. horn switch.
Normalmente abierto   Un interruptor abierto en su estado normal, tal como el interruptor del cláxon.

Normally open contacts   are open when no pressure is supplied to the unit.
Contactos normalmente abiertos   están abiertos mientras que no se les aplica la presión.

NOx   Nitrous oxides; created under high temperature combustion conditions.
NOx   Óxidos de nitrógeno; creados bajo condiciones de combustión a alta temperatura.

Occupational Safety and Health Administration (OSHA)   regulates working conditions in the United States.
Acta de Seguridad y Salud en el Lugar de Trabajo (OSHA)   Regula las condiciones del trabajo en los Estados Unidos.

Octane   Resistance to detonation; also the relative burn rate of a fuel.
Octanaje   Resistencia a la detonación; también el índice relativo de combustión de un combustible.

OEM   Original equipment manufacturer
Equipo original del fabricante

ohmmeter   Electrical test equipment used to measure the resistance of an electrical circuit or device.
óhmmetro   Equipo de prueba eléctrica que se usa para medir la resistencia de un circuito o un dispositivo eléctricos.

Ohms   A measurement for resistance in an electric circuit.
Ohmios   Una medida de la resistencia en un circuito eléctrico.

On-board diagnostic II (OBD II)   A type of computer system installed in all cars and light-duty trucks manufactured since 1996.
Diagnóstico A bordoII (OBDII)   Un tipo de sistema de computadora que ha sido instalado en todos los coches y las camionetas de carga ligera fabricados des desde el 1996.

One channel ABS   A rear wheel only antilock brake unit.
ABS de un canal   Unidad de freno de antibloqueo de una sola rueda trasera.

Open circuit   An unwanted break in a electric circuit.
Circuito abierto   Un corte no deseado en un circuito eléctrico.

Open circuit voltage   A voltage test that is performed with no electrical load on the battery.
Prueba de circuito abierto   Una prueba de voltaje que se efectúa mientras que no tenga una carga eléctrica en la batería.

Open loop   A computer operating mode that occurs during engine warmup. In this mode, the computer ignores the oxygen sensor signal and the computer program and other parameters control the air-fuel ratio.
Bucle abierto   Un modo de operación de una computadora que ocurre durante el calentamiento del motor. En este modo, la computadora no hace caso a la señal del sensor de oxígeno y la programa de computadora y otros parámetros controlan la relación de combustible al aire.

Open pickup coil   A defective pickup coil with an open circuit in the coil winding; too high a coil resistance.
Bobina captadora abierta   Una bobina captadora defectuosa con un circuito abierto en el embobinado; una resistencia de bobina muy alta.

Organic additive technology (OAT)   An additive package when blended with ethylene glycol gives coolant a five year protection life protection.
Tecnología de aditivo orgánico   Tecnología de aditivo orgánico un paquete de aditivo que al mezclarse con etilenglicol le da al fluido refrigerador una protección de vida de cinco años.

Orifice tube   A device used in an AC system to restrict the low of liquid refrigerant into the evaporator thus allowing the refrigerant to evaporate.

Tubo de tocada   Dispositivo usado en un sistema AC para restringir la baja del líquido refrigerante dentro del evaporador, permitiendo así que se evapore el refrigerante.

Outside micrometer   A precision measuring instrument designed to measure the outside diameter of various components.

Micrómetro del exterior   Un instrumento de medidas precisas diseñado a medir el diámetro exterior de varios componentes.

Overhead cam (OHC)   An engine with the intake valves, exhaust valves, and the camshaft mounted in the cylinder heads.

Árbol de leva en cabeza (OHC)   Un motor con las válvulas de entrada, las válvulas de salida y el árbol de levas montado en la cabeza del cilindro.

Overhead valve (OHV)   An engine with the valves mounted in the cylinder heads and the camshaft and valve lifters located in the engine block.

Válvula en cabeza (OHV)   Un motor con las válvulas montadas en las cabezas del cilindro y las levantaválvulas ubicadas en el bloque motor.

Oxidation   A process that occurs when some engine oil combines with oxygen in the air to form an undesirable compound.

Oxidación   Un proceso que ocurre cuando un poco del aceite del motor combina con el oxígeno en el ambiente para formar una compuesta no deseada.

Parallelogram steering   A steering system that uses a center link, pitman arm and an idler arm to create a parallel movement of the center link to steer the front wheels.

Dirección en paralelogramo   Sistema de dirección que utiliza el vástago central, el brazo de mando de la dirección y el brazo intermediario para crear un movimiento paralelo del vástago central para dar vuelta a las ruedas frontales.

Passive restraints   A restraint system that does not have active controls during an accident e.g. seat belts.

Atenuación pasiva   Un sistema de atenuación que no tiene controles activos durante un accidente, como por ejemplo los cinturones de seguridad.

Pellet-type catalytic converter   contains 100,000 to 200,000 small pellets coated with a catalyst material.

Convertidor catalítico tipo pastilla   contiene 100,000 a 200,000 pastillitas cubiertas con una materia catalítica.

Periodic table   A listing of the known elements according to their number of protons and electrons.

Tabla periódica   Una lista de los elementos conocidos según su número de protones e electrones.

Photochemical smog   is formed by sunlight reacting with hydrocarbons (HC) and oxides of nitrogen (NO_x).

Contaminación fotoquímico   formado por uaa reacción del luz del sol con los hidrocarburos (HC) y los óxidos de nitrógeno

Pitch   Distance from one bolt thread peak to the next bolt thread peak.

Declive   Distancia entre un punto máximo de la rosca de un perno al siguiente punto máximo de la rosca de un perno.

Pitman arm   Arm that supports one end of the center link opposite the idler arm, is attached to the steering box.

Brazo de mando de la dirección   Brazo que apoya una orilla del vástago central opuesta al brazo intermediario, y está pegado al cárter de la dirección.

Planetary gear set   A gear set that uses an internal gear, planet gears , and a sun gear.

Engranaje planetario   Engranaje que utiliza una rueda interna, trenes epicicloidales y un eje planetario.

Plunging (CV) joint   allows inward and outward drive axle movement to compensate for changes in axle length.

Junta de velocidad descendida continua (CV)   Permite el movimiento hacia adentro y hacia afuera del eje para compensar la longitud del eje.

Pneumatic tools   Air operated tools.

Herramientas neumáticas   Herramientas que se operan con aire.

Port fuel injection (PFI)   A system where the fuel is injected just behind the intake valve in the intake port runner.

Puertos de admisión de inyección del combustible (PFI)   Sistema en donde el combustible se inyecta justo detrás de la válvula de admisión en el cursor del puerto de admisión.

Puertos de admisión de inyección de combustible (PFI)   Sistema en donde el combustible se inyecta justo detrás de la válvula de admisión en el cursor del puerto de admisión.

Positive crankcase ventilation (PCV) valve   A valve that controls the amount of crankcase vapors flowing through the valve into the intake manifold.

Válvula de ventilación positiva del cárter (PCV)   Una válvula que controla la cantidad de los vapores del cárter que fluyen por la válvula a la manívela de entrada.

Potentiometer   A variable resistor.

Potenciómetro   Un resistor variable.

Power stroke   The movement of the piston in the cylinder bore that occurs after the fuel mixture is ignited forces the piston to move downward.

Tiempo de combustión   El movimiento del pistón en el calibre del cilindro que sucede cuando la mezcla del combustible se enciende, fuerza al pistón a bajar.

Preignition The ignition of the air-fuel mixture in the combustion chamber by means other than the spark plug, such as a hot piece of carbon.

Pre encendido El encendido de la mezcla del aire-combustible en la cámara de combustion por otro metodo que por la bujía, tal como por un pedazo caliente del carbono.

Pressure differential valve A valve designed to illuminate a brake warning lamp if there is a pressure loss in one half of a split braking system.

Válvula diferencial de presión Válvula diseñada para iluminar una lámpara señal de freno en caso de que haya una pérdida de presión en la mitad de un sistema de frenos de junta diagonal.

Pressure tester Used to pressure test cooling systems and pressure caps.

Probador de la presión Se usa para probar la presión de los sistemas de refrigeración y las fundas de presión.

Printed circuit board A thin, insulating board used to mount and connect various electronic components such as resistors, diodes, switches, capacitors, and microchips in a pattern of conductive lines.

Placa de circuítos impresa Una placa aisladora delgada que se usa para montar y conectar a varios componentes electrónicos tal como los resistores, los diodos, los interruptores, los capacitores, y los microchips en un dibujo de líneas conductivas.

Professionalism act in a manner that builds positive public relations

Profesionalismo actuar de manera que se mantengan relaciones públicas positivas

Proportioning valve A valve designed to regulate rear braking pressure to prevent rear wheel lockup.

Válvula de dosificación Una válvula diseñada para regular la presión del freno trasero para prevenir el bloqueo de la rueda trasera.

Protons Positively charged particles located in the nucleus of an atom.

Protones Las partículas de carga positiva ubicados en el núcleo de un átomo.

Proving circuit A circuit used to prove the functionality of a circuit, usually

Circuito de comprobación Circuito que se usa para probar la funcionalidad de un circuito, generalmente

Pulse width modulation (PWM) An on/off voltage signal with a variable on time.

Modulación de separación de impulsos (PWR) Una señal de apagado/prendido con un tiempo prendido variado.

Quench area The area in the combustion chamber near the metal surfaces where the flame front is extinguished.

Área extinctor El área en la cámara de combustión cerca de las superficies de metal en dónde se extingue la llama delantera.

Quick disconnect coupling An air coupler that can be easily attached and unattached from various air tools.

Desconexión rápida Acoplador de aire que puede fácilmente pegarse y despegarse de varias herramientas de aire.

Quick-lube shop specializes in automotive lubrication work.

Taller de lubricación rápida especializa en el trabajo de lubricacián automotivo.

Rack-and-pinion steering A steering device using a pinion attached to a steering wheel that drives a rack that is in turn attached to the front wheels.

Dirección por cremallera Un dispositivo de dirección que usa un piñón unido a la rueda directriz que mueve la cremallera que a su vez está unida a las ruedas frontales.

Radial load Load perpendicular to the axis of rotation.

Carga radial Carga perpendicular al eje de rotación.

Radial runout The amount of diameter variation in a tire.

Corrimiento radial La cantidad de variación del diámetro de un pneumático.

Rebound travel Downward tire and wheel movement.

Viaje de rebote Un movimiento hacia abajo de la rueda y el pnuemático.

Rectify To change from AC to DC.

Rectificar Cambiar de CA a CD.

Repair order A document detailing the repairs needed for an automobile.

Orden de reparación Documento que detalla las reparaciones que se necesitan hacer en un automóvil

Reputation is the general opinion of the public toward a person or business

Reputación es la opinión general del público hacia una persona o negocio

Resource Conservation and Recovery Act (RCRA) States that hazardous material users are responsible for hazardous materials from the time they become a waste until the proper waste disposal is completed.

Acta de Conservación y Recobro de Recursos (RCRA) Declara que los consumidores de materiales peligrosas son responsables por las materiales peligrosas del tiempo de que se convierten en deshechos hasta que se haya deshechado correctamente y completamente de ellas.

Reverse bias An electrical connection between a voltage source and a diode that causes the diode to block voltage.

Polarización reversa Una conexión entre una fuente de voltaje y un diodo que causa que el diodo bloquea el voltaje.

Rheostat A type of variable resistor that is used to control current.

Reostato Tipo de resistor que se utiliza para controlar la corriente.

Ribbed serpentine belt Multi-groove belt.

Correa en serpentina con nervadura Correa de ranuras múltiples.

Ride height The height from the road to specified point on a vehicle.

Altura de la pista La altura del camino a un punto específico de un vehículo.

Right-hand thread A fastener with a right-hand thread must be rotated clockwise when tightening the fastener.

Rosca a la derecha Un fijador con una rosca a la derecha debe girarse en el sentido de las agujas del reloj para apretarse.

Rigid clutch disc Does not allow any movement between the clutch facings and the hub.

Embrague rígido No permite ningún movimiente entre las caras del embrague y el cubo.

Rim offset The mounting plane of the rim can be offset from the center of the wheel assembly. When this is the case the rim is said to be offset.

Separación de la llanta El plano de montaje de la llanta puede estar separado del centro de la rueda equipada. Cuando esto sucede se dice que la llanta está separada.

Road crown The curvature of the road surface to facilitate water drainage.

Parte más alta de la pista La curvatura de la superficie de la carretera para facilitar el drenaje del agua.

Rosin core solder Lead/tin solder with a hollow core filled with rosin as the flux medium.

Soldadura de colofonia Soldadura de acero / estaño con un núcleo vacío que se llena de colofonia como fluidificante.

Rotary flow occurs when the fluid in a torque converter moves in a circular direction with the impeller pump and turbine.

Flujo rotatorio ocurre cuando el fluido en un convertidor del par mueva en una dirección circular con la bomba impulsor y la turbina.

Rotary valve A hydraulic valve used in power steering to direct fluid flow to steer the wheels.

Válvula rotativa Válvula hidráulica que se usa en la dirección asistida para enviar el flujo del líquido para dirigir las ruedas.

Rzeppa joint Constant velocity joint.

Junta Rzeppa Junta homocinética.

SAE Viscosity rating the thickness of an oil.

viscosidad SAE Indice del grosor de un aceite.

SAI Steering axle inclination.

SAI Inclinación del eje radial.

Safe working habits work in a safe manner following consistent safety practices.

Hábitos de trabajo seguros trabajar en una manera segura siguiendo prácticas consistentes de seguridad.

Schrader A valve used to allow access to a pressurized system, used in AC systems and tire valves.

Válvula tipo Schrader Válvula que se utiliza para permitir el acceso a un sistema de sobrepresión interna, que utilizan los sistemas AC y las válvulas de los neumáticos.

Semiconductor A semiconductor has four valence electrons and these materials are used in the manufacture of diodes and transistors.

Semiconductor Un semiconductor tiene cuatro electrones de valencia y estas materiales se usan en la fabricación de los diodos y los transistores.

Sequential fuel injection (SFI) An injection system in which each injector is opened individually.

Inyección de combustible en secuencia (SFI) Un sistema de inyección en el cual cada inyector se abre individualmente.

Serpentine belt Multi-groove belt.

Correa en serpentina Correa de ranuras múltiples.

Servo action occurs when the operation of the primary brake shoe applies mechanical force to help apply the secondary shoe.

Acción Servo ocurre cuando la operación de la zapata principal del freno aplica la fuerza mecánica para asistir en la aplicación de la zapata segundaria.

Servo piston A hydraulic piston used to apply bands.

Pistón del servomotor Pistón hidráulico que se usa para instalar correas.

Shift valve A hydraulic valve used to apply or divert hydraulic pressure to bands and servos.

Válvula de desplace Válvula hidráulica que se utiliza para aplicar o desviar la presión hidráulica a las correas y a los servomotores.

Short circuit An unwanted copper-to-copper connection in an electric circuit.

Corto circuito Una conexión no deseada de cobre-a-cobre en un circuito eléctrico.

Shorted pickup coil A defective pickup coil with an short circuit in the coil winding; too low a coil resistance.

Bobina captadora recortada Una bobina captadora defectuosa con un corto circuito en el embobinado; una resistencia de bobina muy baja.

Short to ground Electrical fault the allows current to go directly to ground bypassing the load or the control circuit.

Corto a tierra Falla eléctrica que permite que la corriente vaya directamente a tierra desviando la carga o el circuito de control.

Silicone-based brake fluid Brake fluids using silicone as the base chemical for manufacture, DOT 5 is a silicone based brake fluids and is usually purple in color.

Líquido de frenos con base de silicona Líquido para frenos que se fabrica con silicona como base química. El DOT 5 es un líquido de frenos con base de silicona y generalmente es de color morado.

SI torque-to yield bolt   A bolt whose maximum clamping force comes from deforming the bolt slightly.

Perno de torsión de deformación SI   Perno el cual su fuerza máxima de fijación se deriva de deformar un poco el perno.

Society of Automotive Engineers (SAE) rating   A universal oil rating that classifies oil viscosity in relation to that atmospheric temperature in which the oil will be operating.

Grado de la Asociación de Ingenieros Automotivos (SAE)   Un grado universal que clasifica la viscosidad del aceite con relación a la temperatura del ambiente en la cual va operar el aceite.

Solenoid   An electro-mechanical device used to effect a push-pull mechanical operation using electric current.

Solenoide   Un dispositivo electro-mecánico que sirve para efectuar una operación mecánica de tire-jale usando un corriente eléctrico.

Alcanze de la bujía   La distancia desde la extremidad inferior del cuerpo de la bujía hasta el collarín del cuerpo.

Spark plug electrodes   The center and ground electrodes of a spark plug.

Electrodos   Electrodos centrales y de tierra de una bujía.

Specialty shop   specializes in one type of repair work.

Taller especializada   especializa en un tipo de reparación.

Specific gravity   is the weight of a liquid, such as battery electrolyte, in relation to an equal volume of water.

Gravedad específica   es el peso de un líquido, tal como la electrólito de la batería, con relación a un volumen identico del agua.

Speed density system   A fuel injection system in which the two main signals used for air-fuel ratio control are engine speed and manifold absolute pressure (MAP).

Sistema de densidad de velocidad   Un sistema de inyección de combustible en el cual las dos señales principales usados para el control de la relación del aire-combustible son la velocidad del motor y la presión absoluta del manívela (MAP).

Spontaneous combustion   chemical reaction that results in sufficient heat to cause a fire.

Combustión espontánea   reacción química que resulta en calor suficiente para causar un incendio.

Sprung body weight   The weight of the chassis supported by the springs.

Peso provisto de muelles   El peso de la carrocería apoyado por los muelles.

Spur gear   A gear with straight cut gears.

Rueda cilíndrica   Rueda con dentada recta.

Stabilizer bar   A long, spring steel bar connected from the crossmember to each lower control arm to reduce body sway.

Barra estabilizadora   Una barra larga de acero de resorte conectada desde el miembro transversal a cada brazo de control inferior para reducir la oscilación de la carrocería.

Static balance   Balancing of the tire in a single rotational plane.

Equilibrio estático   Balanceo de una llanta en un plano rotativo sencillo.

Starter drive   A mechanical device that connects and disconnects the starter armature shaft and the flywheel ring gear.

Acoplamiento del motor de arranque   Un dispositivo mecánico que conecta y disconecta el árbol de la armadura del arranque y el anillo dentado del volante.

Static pressure   The pressure in a system when the system is inoperative.

Presión estática   La presión en un sistema cuando el sistema no está en operación.

Steering damper   A shock absorber to dampen steering oscillations.

Amortiguador de dirección   Amortiguador que suprime los movimientos de dirección.

Stoichiometric air-fuel ratio   The ideal air-fuel ratio of 14.7:1 on a gasoline engine that provides the best engine performance, economy, and emissions.

Relación estequiométrico de aire-combustible   La relación ideal de aire-combustible de 14.7:1 en un motor de gasolina que provee lo mejor en ejecución, economía y emisiones de un motor.

Straight-cut gears   have teeth that are parallel to the gear centerline.

Engranaje de corte recta   tiene los dientes que son paralelos a la línea central del engranaje.

Strut chatter   A clicking noise that is heard when the front wheels are turned to the right or left.

Traqueteo de las manguetas   Un ruido de chasquido que se oye cuando las ruedas delanteras se giran a la derecha o a la izquierda.

Synthetic lubricant   is developed in a laboratory.

Lubricante sintético   se desarrolla en un laboratorio.

Tech talk   use of technical terms in a conversation which the average person might not understand.

Plática de técnicos   uso de términos técnicos en una conversación en la que una persona común no comprendería.

Thermistor   A special resistor that increases its resistance as the temperature decreases.

Termistor   Un resistor especial que incrementa su resistencia al caer la temperatura.

Thread depth   The height of the thread from its base to the top of its peak.

Profundidad del hilado   La altura del filete desde su fondo a lo más alto de su pico.

Three-channel ABS   has a pair of solenoids for each front wheel, but only one pair of solenoids for both rear wheels.

Sistema de frenos antibloqueantes de tres canales (ABS)   tiene un par de solenoides en cada rueda delantera, pero sólo un par de solenoides para ambas ruedas traseras.

Throttle body injection (TBI) A system where the fuel is injected at the throttle body.

Cuerpo de inyección del combustible de (TBI) Sistema en donde el combustible se inyecta en el cuerpo del acelerador.

Throttle position sensor (TPS) A device (potentiometer) that measures the angle or opening of the throttle plates.

Sensor de posición de aceleración Dispositivo (potenciómetro) que mide el ángulo a la abertura de las placas del acelerador.

Thrust angle The sum of the toe of the rear tires.

Angulo de empuje La suma del talón de las llantas traseras.

Thrust load A load applied in a horizontal direction

Carga límite de empuje Una carga aplicada en una dirección horizontal.

Tie rods Spherical joints that attach steering knuckles to various steering mechanisms.

Tirantes Juntas esféricas que sujetan los bujes de dirección a varios mecanismos direccinales.

Tire contact patch The area of the tire that is in contact with the road surface.

Parche de contacto de rodamiento Area de la llanta que está en contacto con las superficie del camino

Titanium 02 sensor An oxygen sensor that varies resistance based on the content of oxygen in the exhaust gas.

Sensor de óxido de titanio Sensor de oxígeno que varía la resistencia basada en el contenido de oxígeno en los gases de escape.

Toe Relationship of the distance at the front of the tires and the back of the tires on the same axle, usually slightly closer at the front.

Talón Relación entre la distancia del frente y de la parte trasera de las llantas en el mismo eje, generalmente un poco más cerca en el frente.

Toe out on turn As a vehicle makes a turn it is necessary for the inside to turn sharper than the outside tire.

Divergencia en vuelta Al dar la vuelta un vehículo es necesario que el interior se vuelva más agudo que la llanta exterior.

Torque A twisting force.

Torsión Una fuerza que tuerce.

Torque converter clutch A clutch built into a torque converter to minimize converter slippage during cruise.

Embrague del convertidor de par Embrague incorporado a un convertidor de par para minimizar el deslizamiento del convertidor durante el crucero.

Torque steer A condition in which the vehicle steering pulls to one side during hard acceleration.

Dirección con torsión Una condición en la cual la dirección del vehículo jala a un lado durante una aceleración fuerte.

Torque-to yield bolt A bolt whose maximum clamping force comes from deforming the bolt slightly.

Perno de torsión de deformación Perno el cual su fuerza máxima de fijación se deriva de deformar un poco el perno.

Total toe The sum of the left toe and the right toe.

Talón total La suma del talón izquierdo y el talón derecho.

TPC Tire performance criteria.

TPC Criterio de rendimiento del neumático.

Transaxle vent is located in the transaxle case to allow air to escape from the transaxle.

Toma de aire del transeje se encuentra en el cárter del transeje para permitir escapar el aire del transeje.

Transition time The time required for the oxygen sensor to switch from lean to rich and rich to lean.

Tiempo de transición El tiempo requirido para que el sensor de oxígeno cambia de pobre a rico y de rico a pobre.

Transversely mounted An engine that is mounted sideways in the vehicle.

Motor montado transversalmente Un motor que esta montado de lado en el vehículo.

TSB technical service bulletin

TSB boletín informativo de servicio técnico

Two wheel alignment Wheel alignment performed using only the geometric center of the vehicle.

Alineación de dos ruedas Alineación de ruedas hecha usando sólo el centro geométrico del vehículo.

TXV Thermostatic expansion valve A device used to control the amount of refrigerant flow into the evaporator based on evaporator temperature, pressure or both.

Válvula de expansión térmica TXV Dispositivo que se utiliza para controlar la cantidad del flujo refrigerante al evaporador basado en la temperatura del evaporador, la presión o ambas.

Ultrasonic leak detector An electronic device that emits a beeping noise when it senses hydrocarbons (HC).

Detector de fuga ultrasónico Un dispositivo electrónico que emite un ruido pipiante cuando percibe los hidrocarburos (HC).

Ultraviolet dye An additive added to a fluid and when the fluid leaks will cause a yellow stain when exposed to an ultraviolet light.

Tinte ultravioleta Aditivo que se añade a un líquido y cuando éste gotea dejará una mancha amarilla cuando se expone a una luz ultravioleta.

Unsprung weight The weight of the suspension that is not supported by the springs.

Muelles sin peso El peso de la suspensión que no se soporten por los muelles.

USCS Unified national series.

USCS Serie nacional unificada.

UTQG   Uniform tire quality grading.

UTQG   Sistema Unificado de Clasificación de la Calidad del Neumático.

Valence ring   The outer ring on an atom.

Anillo de valencia   El anillo exterior de un átomo.

Vaporize   To change from a liquid (droplet) state into a vapor.

Evaporar   Cambiar de un estado líquido (gotitas) a vapor.

Variable ratio   A steering ratio in which turns of the steering wheel are not proportional to the amount the wheels are steered.

Relación variable   Relación direccional en la que las vueltas del volante no son proporcionales a la cantidad que se mueven las ruedas.

Vehicle emission control information   The emission-control information shown directly on a label on each vehicle.

Información del control de salida del vehículo   Información del control de salida del vehículo información del control de salida que se muestra directamente en una etiqueta en cada vehículo.

Vehicle identification number (VIN)   A number located in the top left-hand side of the dash indicating various vehicle data such as model, year, body style, engine type, and serial number.

Número de Identificación del Vehículo (VIN)   Un número ubicado en la parte superior a mano izquierda del tablero indicando varios datos pertinentes al vehículo tal como el modelo, el año, el estilo de chasis, el tipo del motor y el número de serie.

Vacuum brake boosters   A device uses engine vacuum to magnify or boost brake pressure.

Compresores de freno a depresión   Dispositivo que utiliza el vacío del motor para aumentar o comprimir la presión del freno.

Vacuum pump   A device used to place the AC system into a vacuum.

Bomba de vacío   Dispositivo que se utiliza para colocar el sistema AC dentro de un vacío.

Variable ratio   A steering ratio in which turns of the steering wheel are not proportional to the amount the wheels are steered.

Relación variable   Relación direccional en la que las vueltas del volante no son proporcionales a la cantidad que se mueven las ruedas.

V-belt   A single groove belt.

V-belt   Correa en v correa de una sola ranura.

Vehicle emission control information   The emission-control information shown directly on a label on each vehicle.

Información del control de salida del vehículo   Información del control de salida que se muestra directamente en una etiqueta en cada vehículo.

VES   Variable effort steering.

VES   Dirección de esfuerzo variable.

Viscous coupling   A sealed chamber containing a thick, honey-like liquid.

Acoplamiento viscoso   Una cámara sellada que contiene un líquido espeso parecido al miel.

Viscosity   The thickness of oil.

Viscosidad   El espesor del aceite.

Voltage   A measurement for electrical pressure difference.

Voltaje   Una medida de la diferencia en presión eléctrica.

Voltage drop   The difference in voltage at two different locations in a circuit.

Caída del voltaje   La diferencia en el voltaje en dos lugares distinctos en un circuito.

Vortex flow   occurs in a torque converter when the stator is redirecting fluid from the turbine back into the impeller pump.

Flujo de torbellino   ocurre en un convertidor del par cuando el estator esta volviendo a dirigir el fluido de la turbina hacia la bomba impulsor.

Waste spark   An electronic ignition (EI) system that simultaneously fires a spark plug on the compression stroke and another spark plug on the exhaust stroke.

Sistema de chispa resíduo   Un sistema de encendido electrónico (EI) que dispara una chispa de la bujía en la carrera de compresión simultáneamente con otra chispa de la bujía en la carrera de escape.

Wheel shimmy   A rapid, repeated lateral wheel movement.

Abaniqueo de las ruedas   Una vibración lateral rápida de la rueda.

Wheel tramp   The rapid, repeated lifting of the tire and wheel off the road surface.

Salto de rueda   La elevación rápida y repetida del pneumático y la rueda de la superficie del camino.

Work   When a force is appled to an object and the object is moved a distance.

Trabajo   Cuando se aplica una fuerza a un objeto y este se mueve una distancia.

Zirconia 02 sensor   An oxygen sensor that generates a voltage based on the content of oxygen in the exhaust gas.

Sensor de óxido de circonio   Un sensor de oxígeno que genera voltaje basado en el contenido de oxígeno en el gas residual de escape.

Index

ABS systems. *See* Antilock brake systems (ABS)
AC (alternating current), 182–183
AC generators. *See* Alternators
A/C systems. *See* Air conditioning (A/C) systems
Acceleration (fuel injection systems), 632
Acceleration circuit (carburetors), 631
Acceleration simulation mode (ASM) test (I/M inspections), 668–669
Accelerometers (CVRSS systems), 519
Accumulators (automatic transmissions), 795–796
Acetylene gas, safety tips, 87
Active restraints, 901
A/D (analog/digital) converters, 701
Adaptive strategy (engine control computers), 703
Additives, in gasoline, 628
Advances, distributor, 307–310
Aiming pads (headlights), 211. *See also* Headlights and headlight assemblies
Air bag systems. *See also* Restraint systems
  air bag system diagnostic module (ASDM), 43, 904–905
  clock spring centering, 910
  clock spring electrical connector, 904
  features, 902
  inflator module, 903–904
  multi-stage deployment, 906
  operating principles, 905–906
  sensors, 902–903
  service procedures, 909–910
  side-impact air bag, 907
  smart systems, 906
  warning lights, 905
Air compressors, 516–517
Air conditioning (A/C) systems
  automatic, 881–882
  certification requirements, 886
  diagnostic procedures, 889–890
  maintenance procedures, 887–889
  manual, 879–880
  performance testing, 890–891
  refrigerant charging process, 895
  refrigerant evacuation and recovery, 893–894
  refrigerant system repairs, 895–898

  refrigerants and oils, 871–872
  semi-automatic, 880–881
Air Conditioning and Refrigeration Institute (ARI), 873
Air gaps
  antilock brake systems, 381
  high intensity discharge headlights, 193
  spark plugs, 299
Air injection emission systems
  design and operation, 678–679
  diagnostic procedures, 691–692
  maintenance procedures, 691
  service procedures, 692
Air pollutants (vehicle emissions)
  carbon monoxide, 663
  hydrocarbons (HCs), 662–663
  monitoring, 29–30
  oxides of nitrogen (NO_x), 663
Air purge tools, 619
Air spring valves, 516
Alcohol, drinking, 29
Aligning and aiming headlights, 211–213
Allen wrench, 56
All-wheel drive vehicles, automatic transmissions, 802
Alternating current (AC), 182–183
Alternator output test, 292–293
Alternators
  belt inspections, 289–291
  bias diodes, 280–282
  components, 279–282
  maintenance and service procedures, 295
  operating principles, 282–284
Aluminum core radiators, 598
American Petroleum Institute (API) rating
  engine oil, 575–576
  lubricants, 128
Ammeters, 227–228, 237
Amperes, 182
Analog meters, 69
Analog voltage signals (CVRSS systems), 518, 699–700
Analog/digital (A/D) converters, 701
Angles, drive shaft, measuring, 856–857
Angular load (wheels), 459